NANOWIRES - NEW INSIGHTS

Edited by **Khan Maaz**

Nanowires - New Insights
http://dx.doi.org/10.5772/65179
Edited by Khan Maaz

Contributors

Simas Rackauskas, Nadia Barbero, Guido Viscardi, Claudia Barolo, Hao Zhu, Marcos Puydinger Dos Santos, Fanny Béron, Kleber Roberto Pirota, José Alexandre Diniz, Stanislav Moshkalev, Ozgur Ergul, Soumen Dhara, Shengmao Zhang, Ningning Zeng, Shuguang Fan, Jingyi Ma, Yujuan Zhang, Pingyu Zhang, Zhijun Zhang, İzzet Paruğ Duru, Caner Değer, Vahap Eldem, Salah Obayya, Mohamed Hameed, Mohamed Hussein, Pavel Geydt, Mikhail Dunaevskiy, Erkki Lähderanta, Rosaria A. Puglisi, Maxim Shkunov, Grigorios Rigas, Marios Constantinou

Published by InTech

Janeza Trdine 9, 51000 Rijeka, Croatia

Publishing Process Manager Mirena Calmic
Technical Editor SPi Global
Cover InTech Design team

First published July, 2017
Printed in Croatia
Legal deposit, Croatia: National and University Library in Zagreb

Additional hard copies can be obtained from orders@intechopen.com

Nanowires - New Insights, Edited by Khan Maaz
p. cm.
Print ISBN 978-953-51-3283-7
Online ISBN 978-953-51-3284-4

For EU product safety concerns:
IN TECH d.o.o., Prolaz Marije Krucifikse Kozulić 3, 51000 Rijeka, Croatia,
info@intechopen.com or visit our website at intechopen.com.

Meet the editor

Dr. Maaz Khan is working as a senior researcher in the Pakistan Institute of Nuclear Science and Technology (PINSTECH), Pakistan. He has done his PhD from Quaid-i-Azam University, Islamabad, and postdoctorates from South Korea and China. His research interests include fabrication of nanomaterials and their structural, magnetic, optical, and electrical characterizations. He has authored more than 70 articles in peer-reviewed journals. He is the author and editor of many books. Dr. Maaz is working as the editor in chief of *Journal of Materials, Processing and Design* and executive editor of *International Journal of Nano Studies & Technology*. Besides, he is among the editorial board member and reviewer of many journals.

Contents

Preface

This book describes understanding about the nanowires, their fabrication, characterization, and applications in the recent technology. The book is organized into eleven chapters in total, which include fabrication of various metallic and oxide nanowires using different chemical and electrochemical techniques and their physical characterizations. Some very important applications of the nanowires such as for dye-sensitized and plasmonic nanowire solar cells, field-effect transistors, DNA, power transmission, and photovoltaic applications are discussed and presented in this book. The writing of each chapter is generally clear and precise, and numerous illustrations provide an easier understanding of the phenomena described in the book. This book addresses not only researchers in specialized field but also PhD students, postdoctorate fellows, and experienced technical professionals.

All the contributors of the book are active researchers in their fields, and therefore, this book provides up-to-date information about the nanowires and their applications. The content of the book provides the fundamental preparation needed for further study of advanced topics in nanowires. At the end of each chapter, proper references have been included that can lead the readers to the best sources in the literature and help them to go into more details about the nanowires.

I am grateful to all the authors who are experts in their fields for helping me complete this project and also to the entire InTech's publishing team for making this project possible. I am very thankful to the Publishing Process Manager Ms. Mirena Calmic for her cooperative attitude during the entire reviewing and publishing processes. I hope that this book will provide an opportunity to the readers to strengthen their knowledge and research capabilities in the field of materials science and nanotechnology with their specializations in one-dimensional nanostructures.

Maaz Khan
Physics Division, PINSTECH,
Islambad, Pakistan

Chapter 1

Synthesis and Application of Copper Nanowires and Silver Nanosheet-Coated Copper Nanowires as Nanofillers in Several Polymers

Ningning Zeng, Shuguang Fan, Jingyi Ma,
Yujuan Zhang, Shengmao Zhang, Pingyu Zhang and
Zhijun Zhang

Additional information is available at the end of the chapter

http://dx.doi.org/10.5772/67782

Abstract

A large amount of copper (Cu) nanowires was synthesized through the reduction of $Cu(OH)_2$ by hydrazine in an aqueous solution containing NaOH and ethylenediamine. Besides, Cu nanowires coated by silver nanosheet (denoted as Cu@Ag nanowires) were prepared with a facile transmetalation reaction method. In the meantime, the as-prepared Cu and Cu@Ag nanowires were used as the nanofillers of polyvinyl chloride (PVC), ultra-high molecular weight polyethylene (UHMWPE) and epoxy resin (EP), and their effects on the thermal properties and mechanical properties as well as friction and wear behavior of the polymer-matrix composites nanocomposites were examined. Results indicate that the as-prepared Cu@Ag nanowires consist of Cu nanowires core and Ag nanosheet shell. The Ag nanosheet shell can well inhibit the oxidation of the Cu nanowires core, thereby providing the as-prepared Cu@Ag nanowires with good thermal stability even at an elevated temperature of 230°C. As compared with Cu nanowires, Cu@Ag nanowires could effectively increase the thermal stability of the PVC matrix composites. Moreover, due to the special morphology and microstructure, the as-prepared Cu@Ag nanowires can effectively improve the mechanical properties and wear resistance of PVC, UHMWPE, and EP.

Keywords: Cu nanowires, Cu@Ag nanowires, nanocomposite, mechanical properties, tribological properties

1. Introduction

Copper nanowires could have promising applications in electronics [1–6], optoelectronics [7], solar cells [8–10], photonics, magnetics, genetic engineering, chemical sensors, and lubrication [11, 12], due to their unique characteristics such as high electrical conductivity and thermal conductivity as well as high aspect ratio and tribological properties. Particularly, copper nanowires often can effectively improve the mechanical properties, thermal properties, and wear resistance of polymers such as polystyrene [13, 14] and polyamide 6 [15, 16].

A variety of methods such as solution-phase method and complex-surfactant-assisted hydrothermal reduction approach are currently available for the synthesis of Cu nanowires [17–35]. Here the solution-phase method uses hydrazine (N_2H_4) to reduce Cu^{2+} ions in the highly basic aqueous solution of copper nitrate ($Cu(NO_3)_2$) at 25–100°C in the presence of ethylenediamine (EDA, $C_2H_8N_2$) as the surface-capping agent [32, 33]. The resultant Cu nanowires have a length of >40 μm and a diameter in the range of 60–160 nm, where a high concentration of NaOH (pH = 13–14) is required to prevent copper ions from forming copper hydroxide precipitates. The so-called complex-surfactant-assisted hydrothermal reduction approach can achieve facile synthesis of metal copper nanowires with an average diameter of 85 nm and a length of several tens of micrometers [34], where the copper nanowires are formed through the reduction of CuII–glycerol complex ($Cu(C_3H_6O_3)$) by phosphite (HPO_3^{2-}) at 120°C in the presence of surfactant sodium dodecyl benzenesulfonate (SDBS).

The synthesis and application of Cu nanowires, however, are still challenging because Cu nanowires are liable to spontaneous oxidation in air. Therefore, it is imperative to adopt carbon materials [36–39], polymers [40], and metals [10, 41–45] as the sources to introduce nonoxidizable shells on the surface of Cu nanowires, thereby preventing them from oxidation.

Among various surface-capped Cu nanowires, the metal-capped ones are of special interest, because they can be well prevented from oxidation while basic characteristics of metals are retained. We are particularly interested in Ag as a potential surface-capping shell of Cu nanowires. This is because, with thermal conductivity and electrical conductivity similar to that of Cu, Ag exhibits excellent oxidation resistance [46–51]. A couple of Ag-capped Cu nanowires have been reported elsewhere, mainly involving Cu nanowires-Ag nanowires synthesized in solution phase by the heterogeneous nucleation and growth of Ag nanocrystals on presynthesized Cu nanowires [50] and Cu@Ag nanowires with rough surface obtained by the transmetalation reaction at the room temperature [51]. The Cu nanowire cores of these Ag-capped Cu nanowires indeed can be well prevented from oxidation. However, the Ag shell of Cu@Ag nanowires consists of Ag nanoparticles, which is unfavorable for achieving full contact between the nanofillers and polymer matrix.

http://dx.doi.org/10.5772/67782

Therefore, in the present research, we prepare Ag nanosheets on Cu nanowires, hoping to achieve a large contact area between the nanofillers and polymer matrix, thereby acquiring improved mechanical properties and wear resistance. This article reports the preparation of Cu@Ag nanowires as well as the evaluation of their oxidation resistance. Besides, it also deals with the effect of Cu@Ag nanowires on the mechanical properties, thermal properties, and wear resistance of polyvinyl chloride (PVC), ultrahigh molecular weight polyethylene (UHMWPE), and epoxy resin (EP).

2. Preparation and characterization of Cu nanowires

Cu nanowires were prepared through the reduction of $Cu(OH)_2$ by hydrazine in an aqueous solution containing NaOH and ethylenediamine (EDA), with which the approach developed by Zeng et al. was modified to scale up the reaction by 3900 times (yield of Cu nanowires from 0.006 g to 23.5 g of) for potential large-scale production [30]. Briefly, NaOH (18 L, 7 M), $Cu(OH)_2$ (0.36 mol), EDA (135 mL), and hydrazine (18 mL, 35 wt.%) were added to a 20 L reactor and heated at 80°C for 30 min to achieve the reduction of Cu^{2+} to metallic copper at a rate of 100% (**Figure 1**). The as-prepared Cu nanowire cake is present on the top of the solution, possibly due to the entrapping of nitrogen bubbles among the nanowires ($2Cu^{2+} + N_2H_4 + 4OH^- \rightarrow 2Cu + N_2 + 4H_2O$). **Figure 1a** shows the scanning electron microscopic (SEM) images of the as-prepared Cu nanowires. They are straight and have a diameter of 150–200 nm and a length of about 10 μm. **Figure 1b** shows the X-ray diffraction (XRD) pattern of the Cu nanowires. The peaks at $2\theta = 43.3, 50.4, 74.1$, and 89.9° correspond to the (1 1 1), (2 0 0), (2 2 0), and (3 1 1) crystal planes of cubic Cu (PDF No. 89-2838), which indicates that the copper nanowires exhibit face-centered cubic (fcc) structure. No XRD signals of other species are detected, which indicate that the as-prepared Cu nanowires are highly pure. However, when the as-prepared Cu nanowires are placed in air for 24 h, three XRD peaks emerge at $2\theta = 36.4$, 42.3 and 52.5° (**Figure 1c**), and they correspond to the (1 1 1), (2 0 0), and (2 1 1) crystal planes of Cu_2O (PDF No. 78-2076). This means that the as-prepared Cu nanowires are easily oxidized in air environment.

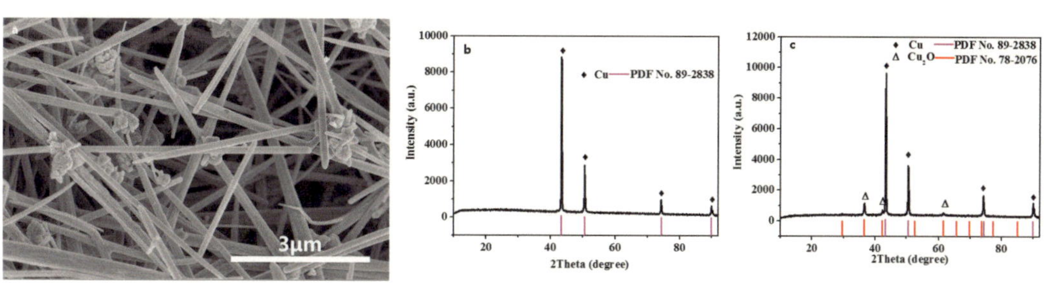

Figure 1. (a) SEM image of as-prepared Cu nanowires in mother liquor and XRD patterns, (b) as-prepared Cu nanowires, and (c) Cu nanowires after being stored in air for 24 h.

3. Preparation and characterization of Cu@Ag nanowires

The formation of Cu@Ag nanowires by the transmetalation reaction should be closely related to the different reduction potentials of Cu and Ag. First, 1.5 mmol $AgNO_3$ was dispersed in distilled water. Into the $AgNO_3$ solution was dropwise added the ammonium hydroxide (NH_4OH, 28–30%, BDH) aqueous solution under vigorous stirring until the yellow precipitate disappeared and the solution became clear. The as-obtained Ag-amine reagent solution was then directly added into 50 mL of Cu nanowires dispersion (with 0.5 g of Cu nanowires) under vigorous magnetic stirring at 40°C. After 2 h of reaction, the as-obtained product was washed with deionized water and collected by centrifugation, followed by redispersion in deionized water. During the synthesis progress, the Cu atoms on the surface of Cu nanowires quickly react with Ag^{+} ions to release Cu^{2+}. Simultaneously, Ag^{+} ions are reduced into Ag atoms and deposited on the surface of the Cu nanowires, during which the color of Cu nanowires dispersion turns grey along with the appearance of metallic Ag. Therefore, the morphology and microstructure of the as-prepared Cu@Ag nanowires should be highly dependent on the reaction conditions. Through comprehensive analysis, the optimal reaction condition is that temperature is 40°C; Cu/Ag molar ratio is 5:1; Cu dispersion concentration is 1% and silver ammonia reagent dropping speed is poured directly.

Figure 2 shows the SEM and TEM images as well as XRD pattern and EDS line map of the Cu@Ag nanowires prepared under the optimal reaction condition. It can be seen that the as-prepared Cu@Ag nanowires have a diameter of about 250 nm and a length of about 10 μm (**Figure 2a**). Besides, the major XRD peaks of the as-prepared Cu@Ag nanowires can be indexed to metallic Ag (PDF No. 87-0717) and Cu (PDF No. 89-2838) with fcc structure (**Figure 2b**). The

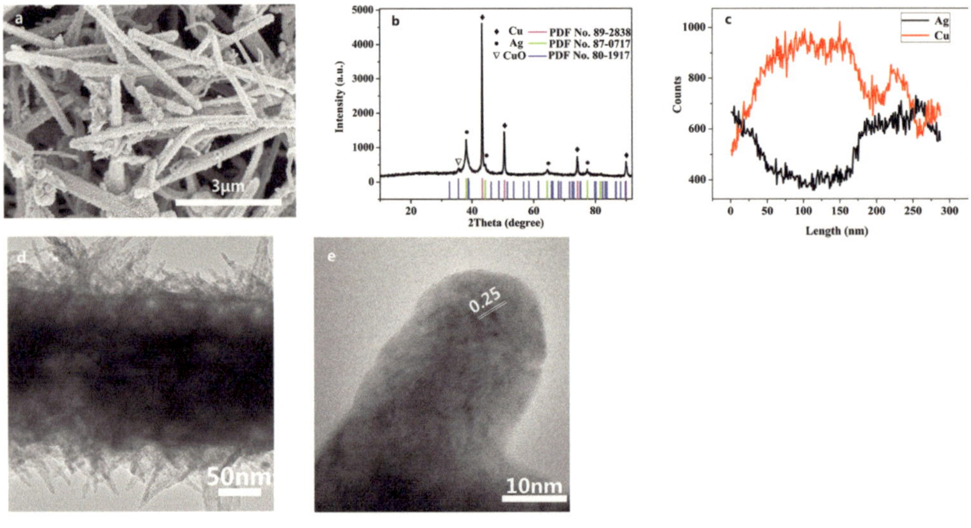

Figure 2. Characterization of Cu@Ag nanowires prepared under the optimal reaction condition: (a) SEM image, (b) XRD pattern, (c) EDS line map, (d) TEM image, and (e) HRTEM image.

small peak at $2\theta = 35.5°$ can be indexed to CuO (PDF No. 80-1917), which is possibly due to the oxidation of original Cu nanowires before coating with Ag nanosheets. In terms of the element line distribution, the copper element has a maximum value at the center part whereas the Ag shows a complementary profile with two maxima on both sides (**Figure 2c**). This indicates that the as-prepared Cu@Ag nanowires have a core-shell structure composed of Cu core and Ag shell. Such a core-shell structure of the as-prepared Cu@Ag nanowires is also supported by corresponding TEM image shown in **Figure 2d**. Namely, there is a layer of Ag nanosheets on Cu nanowires. Moreover, the high-resolution TEM (HRTEM) image of Cu@Ag nanowires indicates that the Ag crystal exhibits a fringe lattice spacing of 0.25 nm (**Figure 2e**), and it corresponds to the (200) plane of the Ag crystal.

The thermal stability of Cu@Ag nanowires with different treatments was measured by XRD. **Figure 3a** shows the XRD pattern of Cu@Ag nanowires prepared under the optimal condition after 2 month of storage in ambient condition. All the characteristic peaks remain unchanged after 2 month of storage in air, which indicates that the Ag shell layer can well inhibit the oxidation of the Cu nanowires core. **Figure 3b** shows the XRD patterns of the same product after being sintered at different temperatures. It can be seen that the XRD patterns have no change before 200°C contrasted with the XRD of as-prepared samples, while after being sintered at 300°C, a distinct peak of Cu_2O emerges and the major peaks of metallic Cu and Ag still remain, which indicates the Ag coating can protect Cu nanowires core from oxidation at high temperature. This means that the Cu@Ag nanowires show better thermal stability than Cu nanowires.

In order to further determine oxidation resistance of Cu@Ag nanowires, TG analysis was conducted in air flow. **Figure 4** shows the TG curves of the as-prepared Cu nanowires and Cu@Ag nanowires in air. The TG curves show Cu nanowires undergo a weight-gain around 100°C attributed to the oxidation thereat, while the Cu@Ag nanowires began to add the weight at around 230°C indicating that Cu@Ag nanowires have the higher oxidation resistance temperature, which also demonstrates that Cu@Ag nanowires exhibit much better thermal stability than Cu nanowires.

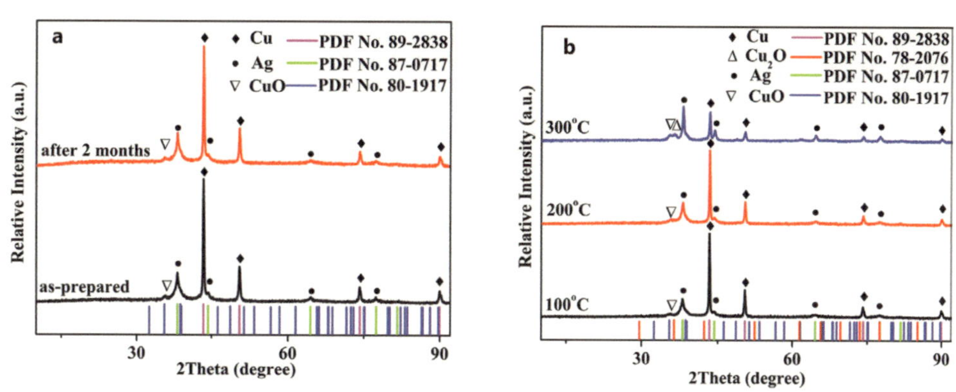

Figure 3. XRD paterns of Cu@Ag nanowires prepared under the optimal reaction condition: (a)after storage for different times in air, (b) after being sintered at different temperatures.

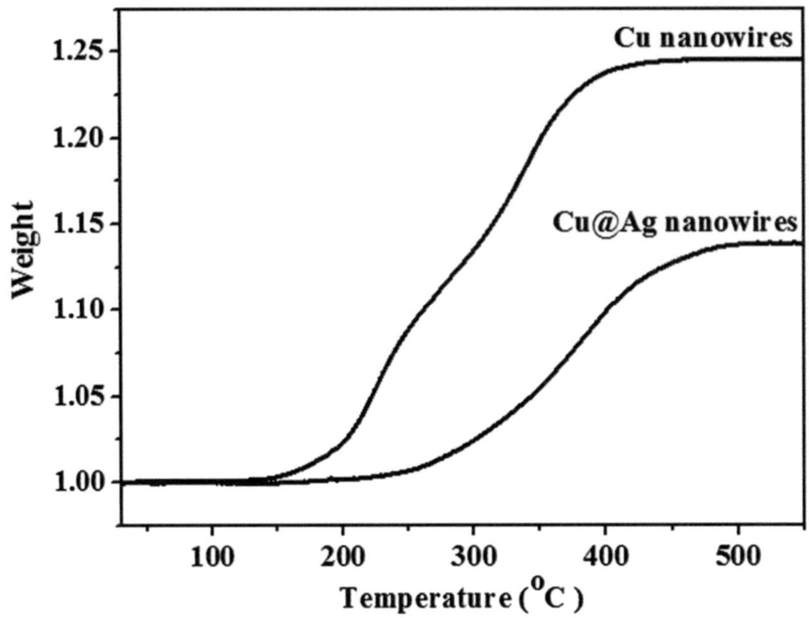

Figure 4. TG curves of Cu nanowires and Cu@Ag nanowires in air at a heating rate of 10°C/min.

4. Cu/Cu@Ag nanowires-PVC nanocomposites

The component and dosage for the preparation of Cu/Cu@Ag nanowires-PVC composites materials are listed in **Table 1**. The raw materials were mixed with mechanical stirring for 20 min, then melt blended by a laboratory two roll mill with a rotate speed of 50 rpm under 100°C for 10 min. The milled sheets were compressed by a press machine at 175°C and 20 MPa for 5 min. The samples for performance test were cut from the final sheets.

Nano-filler mass fraction (%)	Component (g)					
	PVC	Octadecanoic acid	Lead stearate	Calcium stearate	Dioctyl phthalate	Cu/Cu@Ag nanowires
0	160	1.28	0.64	0.64	32	0
0.1	160	1.28	0.64	0.64	32	0.195
0.5	160	1.28	0.64	0.64	32	0.978
1.0	160	1.28	0.64	0.64	32	1.97
3.0	160	1.28	0.64	0.64	32	6.02

Table 1. The component and dosage for the preparation of Cu nanowires- and Cu@Ag nanowires-PVC composites.

4.1. Thermal properties of Cu/Cu@Ag nanowires-PVC nanocomposites

Thermal properties of the composites were studied by differential scanning calorimetry (DSC). DSC curves were obtained by heating at 10°C/min rate under N_2 at a flow rate of 100 mL/min in the range of 20–120°C by using NETZSCH DSC 204HP instrument (Bavarian, Germany). The heated sample was cooled to 20°C and reheated for the second time using the same heating program.

An exemplary DSC heating curve for Cu nanowires-PVC nanocomposites is presented in **Figure 5**, with the indication of the procedure of glass transitions temperatures (Tg) determination as a central point value of the tangent line to the DSC curve, at the inflection point where the characteristic augment of the base line, related to the increase of the specific heat, was used for the determination of the glass transition. For all samples, the run of the DSC curve at the domain of the glass transition was similar.

The Tg of the Cu@Ag nanowires-PVC nanocomposites is presented in **Figure 6**. It can be seen that, as compared with Cu nanowires, the introduction of Cu@Ag nanowires results in a larger increase in the Tg value. Besides, the Tg of the Cu@Ag nanowires-PVC nanocomposites tends to rise with the increase of Cu@Ag nanowire content. This is because, with the increase of the Cu@Ag nanowire content, more Cu@Ag nanowires act as physical cross-linkers to be intertwined with macromolecular chains, thereby reducing the chain segmental mobility of PVC and increasing the Tg. However, when the content of Cu@Ag nanowires

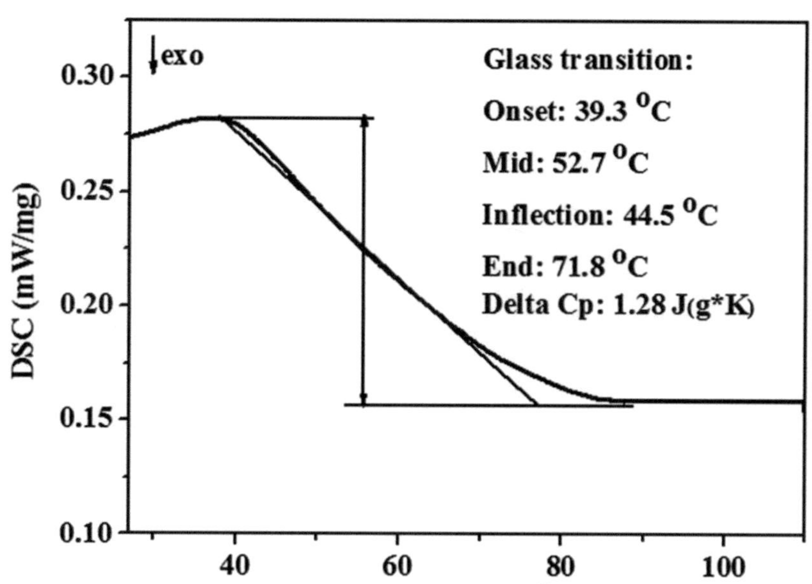

Figure 5. An exemplary DSC trace for Cu nanowires-PVC composite (Cu nanowires contain: 1 wt.%) at the domain of the glass transition.

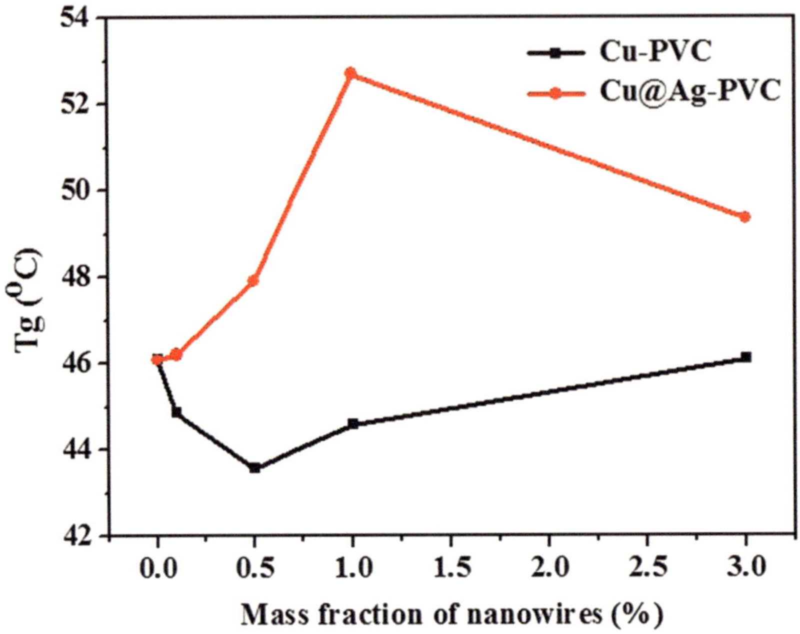

Figure 6. The glass transition temperature (Tg) of Cu/Cu@Ag nanowires-PVC nanocomposites with various mass fractions of the nanofillers.

is too high, the Tg values of the PVC-matrix nanocomposites declines to some extent. The reason lies in that a too high content of Cu@Ag nanowires would tend to agglomerate in the polymer matrix and weaken the interaction with the macromolecular chains, thereby limiting the movements of the macromolecular chains and reducing the Tg value of the PVC-matrix nanocomposites. Moreover, the presence of Cu@Ag nanowires in PVC matrix contributes to increasing the elasticity especially in the glass transition region, which corresponds to enhanced thermal stability of the PVC-matrix nanocomposites.

4.2. Mechanical properties of Cu/Cu@Ag nanowires-PVC nanocomposites

Mean values of five factors on mechanical properties were analyzed in this work. The tensile strength and elongation at break of these samples are shown in **Figure 7**. From **Figure 7**, it can be seen that the elongation at break of composite materials all increase to varying degrees by addition of the nanowires. And the effect of Cu@Ag nanowires is much larger than the Cu nanowires. Meanwhile, the tensile strength keeps no change on the whole.

4.3. Tribology properties of Cu/Cu@Ag nanowires-PVC nanocomposites

Figure 8 shows the friction coefficient and wear rate of Cu/Cu@Ag nanowires-PVC nanocomposites (load: 20 N; route: 5 mm; frequency: 2 Hz; time: 5 min; room temperature). The friction coefficient of Cu/Cu@Ag nanowires PVC nanocomposites has little change with

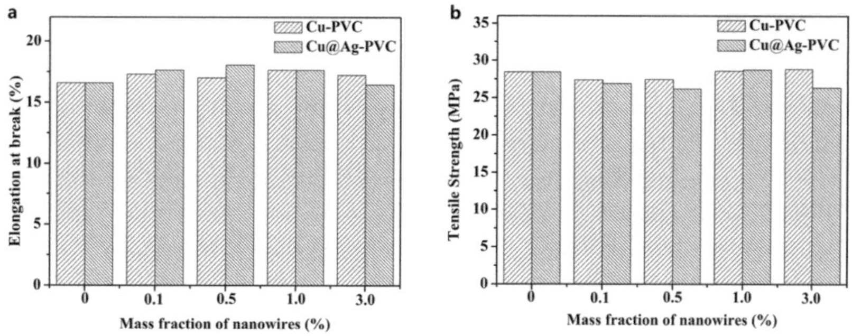

Figure 7. The elongation at break (a) and tensile strength and (b) Cu/Cu@Ag nanowires-PVC nanocomposites with different nanowires contents.

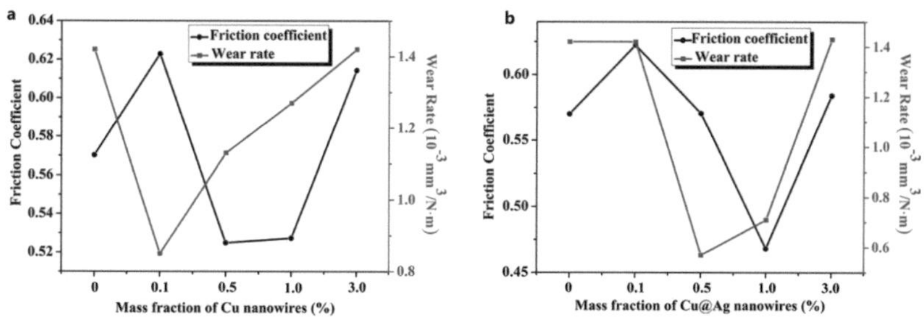

Figure 8. The friction and wear behavior (friction coefficient and wear rate) of PVC nanocomposites with different contents of (a) Cu nanowires and (b) Cu@Ag nanowires.

increasing amount of copper nanowire, while the wear rate changes greatly therewith. Namely, pure PVC (without Cu nanowires) has a larger wear rate as much as 1.42×10^{-3} mm^3/N m, while PVC composite containing 0.5 wt.% Cu@Ag nanowires has an obviously reduced wear rate of 0.57×10^{-3} mm^3/N m (reduced by nearly 60%). It can be clearly seen that lower or higher concentrations of fillers in the matrix are less effective in reducing the friction coefficient and wear rate. It can be inferred that, at a too low concentration, the nanofillers can hardly play its role on the friction-reducing and wear resistance in the friction process, thereby leading to relatively not very good friction property. When the additive concentration is too high, the nanofiller would tend to agglomerate and form a large number of weak interfaces in the matrix, which makes the nanofillers easily fall off from the matrix in the friction process causing a higher abrasive wear [52], thus resulting in relatively poor friction property. Therefore, it is suggested to keep the concentration of the nanofillers in polymer matrix as 0.5 wt.% Cu@Ag nanowires in order to effectively reduce the friction coefficient and wear rate. Moreover, the wear rate of Cu nanowires-PVC nanocomposites tends to decline with the increasing content of the Cu nanowires, and the minimum wear rate (0.85×10^{-3} mm^3/N m) is obtained at a Cu nanowires content of 0.1 wt.%.

5. Cu/Cu@Ag nanowires-UHMWPE nanocomposites

The component and dosage for the preparation of Cu@Ag nanowires-UHMWPE composites are listed in **Table 2**. The nanowires and UHMWPE raw material were mixed with ball milling for 1 h. The mixture was heated to 200°C and pressed under 20 MPa for 10–20 min, followed by cooling to room temperature in the mold to afford the Cu@Ag nanowires-UHMWPE nanocomposites. All the samples for performance test were cut from the final sheets.

5.1. Crystallization and melting behavior of Cu/Cu@Ag nanowires-UHMWPE nanocomposites

The crystallization and melting behavior of Cu/Cu@Ag nanowires-UHMWPE nanocomposites were studied by differential scanning calorimetry (DSC). DSC curves in the range of 20–200°C were measured with a NETZSCH DSC 204HP instrument (Bavarian, Germany) at a heating rate of 10°C/min under N_2 atmosphere (flow rate: 100 mL/min). **Figure 9** shows the crystallization and melting temperature of Cu/Cu@Ag nanowires-UHMWPE nanocomposites. It is seen from **Figure 9a** that Cu nanowires have a strong influence on the crystallization and melting temperature of UHMWPE matrix. When the content of Cu nanowires is 0.1 wt.%, the crystallization temperature of nanocomposite rises from 118.10 to 118.32°C, which could be because that the heterogeneous nucleation in the presence of copper nanowire served as nucleating agent contributes to improving the crystallinity of UHMWPE matrix. With the increasing of nanowires content, the crystallization temperature significantly decreases, which indicates that Cu nanowires hinder the orientation of PA6 crystals and restrict the movement of the molecular chains of UHMWPE. When the nanowires content is further increased, the effect of Cu nanowires on the crystallization and melting behaviors tends to decline, possibly due to the aggregation of the nanofillers in the polymer matrix. When the Cu nanowires content reaches 3 wt.%, however, the crystallization temperature decreases again, probably because that the interaction between excess nanowires and molecular chains of UHMWPE is enhanced to restrict the movement of the molecular chains. Correspondingly, a change in melting temperature occurs with the increasing of nanowires content.

Interestingly, the Cu@Ag nanowires have opposite effects on the crystallization and the melting temperature as compared with Cu nanowires, which could be because the Cu@Ag nanowires

Component	Mass fraction (%)						
	0	0.1	0.3	0.5	0.8	1.0	3.0
UHMWPE (g)	120	120	120	120	120	120	120
Cu/Cu@Ag nanowires (g)	0	0.120	0.361	0.603	0.968	1.21	3.71

Table 2. The component and dosage for the preparation of Cu/Cu@Ag nanowires-UHMWPE composites.

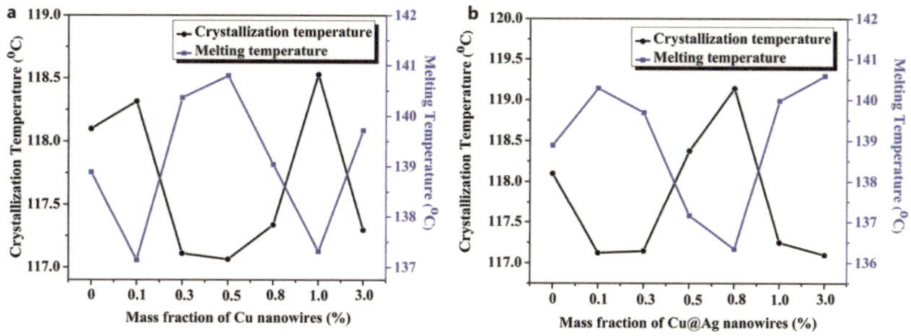

Figure 9. The crystallization and melting temperature of Cu/Cu@Ag nanowires-UHMWPE nanocomposites with different nanowires contents.

with unique morphology and microstructure exhibit significantly increased contact area with the UHMWPE matrix.

5.2. Mechanical properties of Cu/Cu@Ag nanowires-UHMWPE nanocomposites

The effect of Cu/Cu@Ag nanowires on the mechanical properties of UHMWPE is shown in **Figure 10**. We find that the elongation at break of composite materials gradually increases with increasing dosage of Cu/Cu@Ag nanowires; and in particular, the UHMWPE-based nanocomposites with 0.8 wt.% of Cu nanowires or with 0.3 wt.% of Cu@Ag nanowires possesses the maximum values of elongation at break (168 and 190%, respectively), larger than that of unfilled UHMWPE by 78.5 and 101.9% (**Figure 10a**). Besides, the elongation rate at break of the UHMWPE-based nanocomposites tends to decline with further increase of nanowires content, possibly due to the aggregation of the nanowires thereat. Moreover, the elongation rate at break of Cu@Ag nanowires-UHMWPE is higher than that of Cu nanowires-UHMWPE, which could be closely related to the special morphology and microstructure of the Cu@Ag nanowires. Furthermore, the UHMWPE-based nanocomposites change slightly with increasing dosage of Cu/Cu@Ag nanowires. The UHMWPE-based nanocomposites with 0.3 wt.% of Cu nanowires or with 0.5 wt.% of Cu@Ag nanowires possess the maximum values of tensile

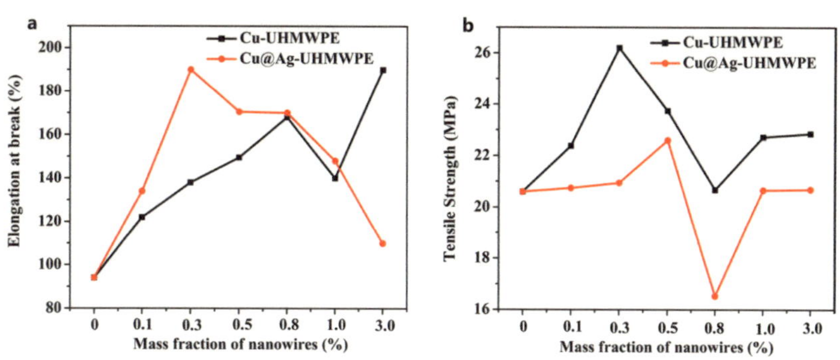

Figure 10. The mechanical properties of Cu/Cu@Ag nanowires-UHMWPE nanocomposites with different nanowires contents.

strength of 26.2 and 22.6 MPa, respectively, showing an increase by 27.1 and 9.7% in comparison with pure UHMWPE (20.6 MPa; **Figure 10b**).

5.3. Tribology properties of Cu/Cu@Ag nanowires-UHMWPE nanocomposites

Figure 11 shows the friction coefficient and wear rate of Cu/Cu@Ag nanowires-UHMWPE nanocomposites (load: 250 N; route: 5 mm; frequency: 2Hz; time: 5 min; room temperature). For Cu nanowires-UHMWPE nanocomposites (**Figure 11a**), the friction coefficient and wear rate of Cu nanowires-UHMWPE nanocomposites gradually decrease with increasing dosage of Cu nanowires; and the friction coefficient is rather large when the Cu nanowires content is below 0.3 wt.%. This may be because that when the content of Cu nanowires is too low to play its role on the friction-reducing and wear resistance in the friction process, thereby leading to relatively not very good friction property. Cu@Ag nanowires-UHMWPE nanocomposites show similar variation tendency as Cu nanowires-UHMWPE nanocomposites (**Figure 11b**) except the one with 3 wt.% of Cu@Ag nanowires is somewhat different.

6. Cu/Cu@Ag nanowires-EP nanocomposites

The Cu/Cu@Ag nanowires-EP (epoxy resin) nanocomposites were prepared by mixing the raw materials with high-speed mixers (1500 rpm, 3 min) in association with curing in the mold under ambient conditions. The content of each component is shown in **Table 3**.

6.1. The electrical conductivity Cu/Cu@Ag nanowires-EP nanocomposites

In general, the electrical conductivity of silver and copper is similar. The special structure of Cu@Ag nanowires makes it feasible to significantly improve the electrical conductivity of Cu nanowires by incorporating Ag shell because of their better oxidation resistance and more conductive network in the polymer matrix nanocomposites. We measured the electrical resistivity of EP-matrix composites filled with Cu and Cu@Ag nanowires using a Hall effect

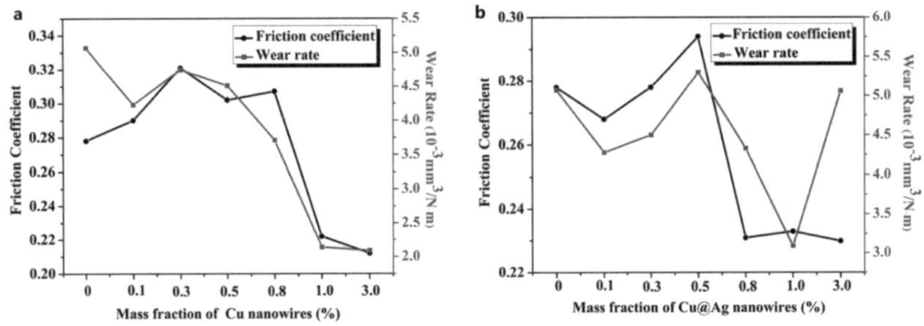

Figure 11. The friction and wear behavior (friction coefficient and wear rate) of UHMWPE nanocomposites with different contents of (a) Cu nanowires and (b) Cu@Ag nanowires.

Component	Mass fraction (%)						
	0	0.1	0.3	0.5	0.8	1.0	3.0
EP (g)	60	60	60	60	60	60	60
Curing agent (g)	15	15	15	15	15	15	15
Cu/Cu@Ag nanowires (g)	0	0.075	0.226	0.377	0.606	0.758	2.320

Table 3. The component and dosage for the preparation of Cu nanowires- and Cu@Ag nanowires-EP composites.

measurement system. The electrical resistivity of pure EP is about 10^{14} Ω cm, while the introduction of Cu/Cu@Ag nanowires results in greatly reduced electrical resistivity (**Figure 12**). Namely, the electrical resistivity decreases obviously with increasing content of Cu or Cu@Ag nanowires, and the EP-matrix nanocomposites filled with Cu@Ag nanowires exhibit lower electrical resistivity than the counterparts filled with Cu nanowires. For example, the resistivity of the EP composite containing 3 wt.% Cu nanowires is about 1.49×10^5 Ω cm, and it does not satisfy the conductive requirement for electrostatic conductive materials (10^4 Ω cm). The resistivity of Cu@Ag nanowires-EP composite (filler mass fraction: 3 wt.%) is reduced to 2.11×10^4 Ω cm, and it satisfies the conductive requirement for electrostatic conductive materials.

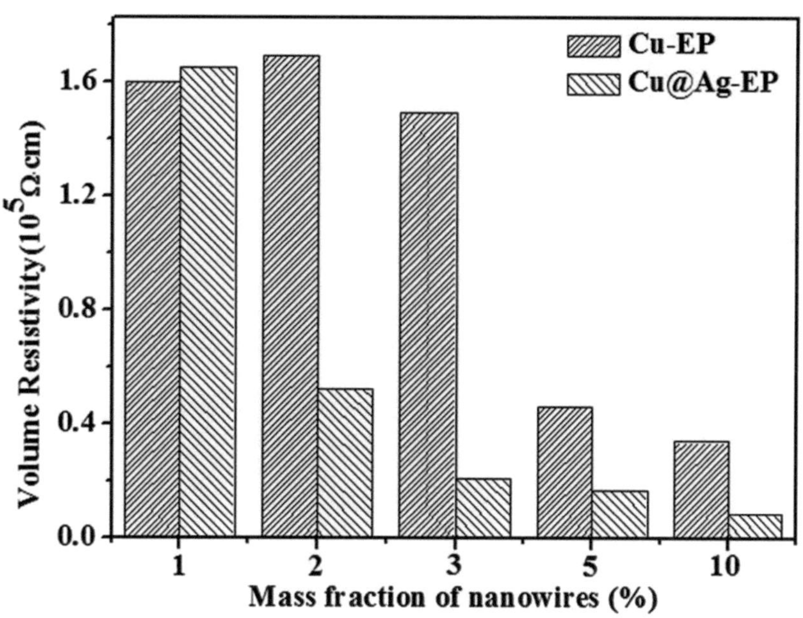

Figure 12. The electrical resistivity of Cu/Cu@Ag nanowires-EP nanocomposites with different nanowires contents.

In other words, encapsulating Cu nanowires with Ag nanosheets makes it feasible for the copper nanowires to be applied as potential electrostatic conductive filler, thereby broadening their application scope.

6.2. Mechanical properties of Cu/Cu@Ag nanowires-EP nanocomposites

The tensile strength and the elongation at break of pure EP and its nanocomposites filled with Cu nanowires or Cu@Ag nanowires are shown in **Figure 13**. Both the tensile strength and elongation at break increase after the introduction of Cu nanowires and Cu@Ag nanowires. Particularly, the EP-matrix nanocomposite with 0.8 wt.% of Cu nanowires exhibits the maximum tensile strength of 74.2 MPa, while the one with 1 wt.% of Cu@Ag nanowires exhibits the maximum tensile strength of 67.5 MPa. In addition, compared with the change of the elongation at break values, the tensile strength has little change with the increase of nanowires content. When the content of the nanowires is too high, the tensile strength and elongation at break are relatively low, which could be due to the aggregation of the nanowires in EP matrix at a high content.

6.3. Tribology properties of Cu/Cu@Ag nanowires-EP nanocomposites

Figure 14 shows the variation in the friction coefficient and wear rate of Cu/Cu@Ag nanowires-EP composites with the content of the nanofillers (load: 20 N; amplitude: 5 mm; frequency: 2Hz; time: 30 min; room temperature). Unfilled EP has a friction coefficient and wear rate of 0.646 and 4.82 × 10^{-3} mm^3/N m. Namely, the EP-matrix nanocomposites with less than 0.5 wt.% of Cu nanowires have reduced friction coefficient and wear rate than neat EP; and in particular, the EP-matrix composite with 0.1% of Cu nanowires exhibits the best friction coefficient and wear rate (0.555, 1.42 × 10^{-3} mm^3/N m; a reduction by 14.1 and 70.5% as compared with that of neat EP). However, the increase in the content of the Cu nanowires above 0.5 wt.% results in increased friction coefficient and wear rate, due to aggregation of the nanofillers thereat. Similar variation tendency is also observed for Cu@Ag nanowires-EP nanocomposites.

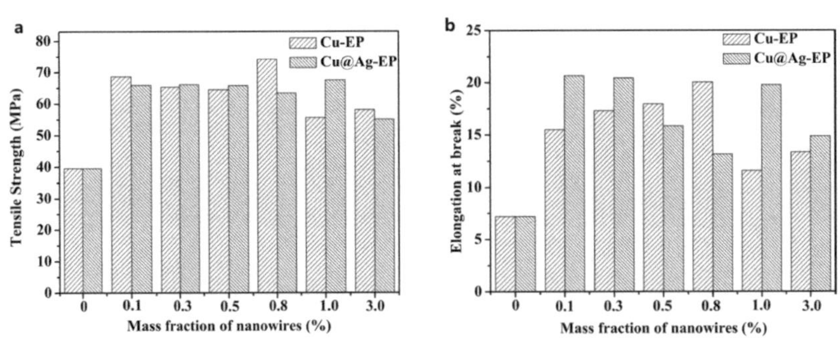

Figure 13. The mechanical properties of Cu/Cu@Ag nanowires-EP nanocomposites with different nanowires content.

http://dx.doi.org/10.5772/67782

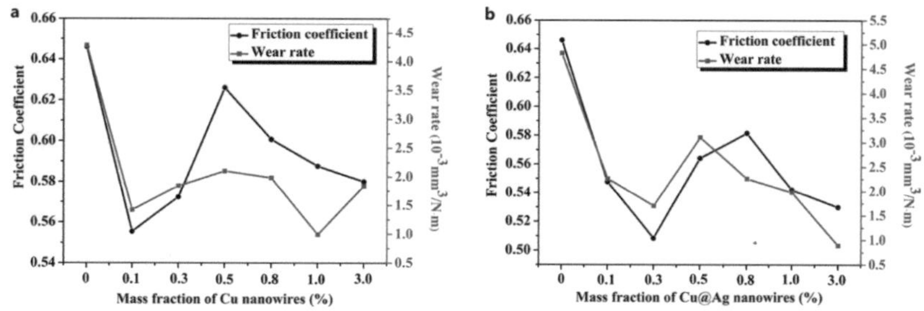

Figure 14. The friction and wear behavior (friction coefficient and wear rate) of EP-matrix nanocomposites with different contents of (a) Cu nanowires and (b) Cu@Ag nanowires.

7. Conclusion

A large amount of Cu nanowires is prepared by a solution-reduction approach, while Cu@Ag nanowires are prepared by a simple and efficient transmetalation reaction method. The morphology and microstructure of the as-prepared Cu@Ag nanowires can be well manipulated by properly adjusting the reaction condition. Resultant Cu@Ag nanowires consist of Cu nanocores and uniformly distributed Ag nanosheet shell, and they exhibit improved thermal stability and electrical conductivity as compared with Cu nanowires, due to the incorporation of the Ag shell. Besides, the as-prepared Cu@Ag nanowires can better improve the mechanical properties and wear resistance than the Cu nanowires.

Acknowledgements

We acknowledge the financial support provided by National Natural Science Foundation of China (Grant no. 51275154, 51405132, 21671053), the Plan for Young Scientific Innovation Talent of Henan Province (Grant no. 154100510018) and Natural Science Foundation of Henan Province (14A150006).

Author details

Ningning Zeng[1], Shuguang Fan[1], Jingyi Ma[1], Yujuan Zhang[1], Shengmao Zhang[1,2]*, Pingyu Zhang[1] and Zhijun Zhang[1]

*Address all correspondence to: zsm@henu.edu.cn

1 Engineering Research Center for Nanomaterials, Henan University, Kaifeng, P. R. China

2 Collaborative Innovation Center of Nano Functional Materials and Applications of Henan Province, Henan University, Kaifeng, P. R. China

References

[1] Hsu P, Wu H, Carney T J. Passivation coating on electrospun copper nanofibers for stable transparent electrodes. ACS Nano. 2012;**6**(6):5150-5156. DOI: 10.1021/nn300844g

[2] Kumar Ravi D V, Woo K, Moon J. Promising wet chemical strategies to synthesize Cu nanowires for emerging electronic applications. Nanoscale. 2015;**7**(41):17195-17210. DOI: 10.1039/c5nr05138j

[3] Yin Z, Lee C, Cho S. Facile synthesis of oxidation-resistant copper nanowires toward solution-processable, flexible, foldable, and free-standing electrodes. Small. 2014;**10**(24):5047-5052. DOI: 10.1002/smll.201401276

[4] He C, Liu G, Zhang W X. Tuning the structures and electron transport properties of ultrathin Cu nanowires by size and bending stress using DFT and DFTB methods. RSC Advances. 2015;**29**(5):22463-22470. DOI: 10.1039/C4RA15051A

[5] Nam V, Lee D. Copper nanowires and their applications for flexible, transparent conducting films: a review. Nanomaterials. 2016;**6**(3):47. DOI:10.3390/nano6030047

[6] Jung S M, Preston D J, Jung H Y. Porous Cu nanowire aerosponges from one-step assembly and their applications in heat dissipation. Advanced Materials. 2016;**28**(7):1413-1419. DOI: 10.1002/adma.201504774

[7] Im H G, Jung S H, Jin J. Flexible transparent conducting hybrid film using a surface-embedded copper nanowire network: a highly oxidation-resistant copper nanowire electrode for flexible optoelectronics. ACS Nano. 2014;**8**(10):10973-10979. DOI: 10.1021/nn504883m

[8] Zhao Y, Zhang Y, Li Y. Rapid and large-scale synthesis of Cu nanowires via a continuous flow solvothermal process and its application in dye-sensitized solar cells (DSSCs). RSC Advances. 2012;**2**(30):11544-11551. DOI: 10.1039/c2ra21224b

[9] Liang J, Bi H, Wan D. Novel Cu nanowires/graphene as the back contact for CdTe solar cells. Advanced Functional Materials. 2012;**22**(6):1267-1271. DOI: 10.1002/adfm.201102809

[10] Stewart I E, Rathmell A R, Yan L. Solution-processed copper–nickel nanowire anodes for organic solar cells. Nanoscale. 2014; **6**(11):5980-5988. DOI: 10.1039/c4nr01024h

[11] Stortini A M, Moretto L M, Mardegan A. Arrays of copper nanowire electrodes: preparation, characterization and application as nitrate sensor. Sensors and Actuators B: Chemical. 2015;**207, Part A**:186-192. DOI: 10.1016/j.snb.2014.09.109

[12] Huang J, Dong Z, Li Y. High performance non-enzymatic glucose biosensor based on copper nanowires–carbon nanotubes hybrid for intracellular glucose study. Sensors and Actuators B: Chemical. 2013;**182**:618-624. DOI: 10.1016/j.snb.2013.03.065

[13] Lin B, Gelves G A, Haber J A. Electrical, morphological and rheological study of melt-mixed polystyrene/copper nanowire nanocomposites. Macromolecular Materials and Engineering. 2008; **293**(7):631-640. DOI: 10.1002/mame.200800045

[14] Gelves G A, Sundararaj U, Haber J A. Electrostatically dissipative polystyrene nanocomposites containing copper nanowires. Macromolecular Rapid Communications. 2005;**26**(25): 1677-1681. DOI: 10.1002/marc.200500490

[15] Xu Q, Li X, Zhang S. Preparation and characterization of copper nanowire/polyamide 6 nanocomposites and its properties. Journal of Macromolecular Science, Part A. 2014;**51**(7):598-603. DOI: 10.1 080/10601325. 2014.916182

[16] Xu Q, Li X, Zhang S. Copper nanowire/PA6 composites prepared by in situ polymerization and its properties. Journal of Polymer Research. 2013;**20**(10):1-6. DOI: 10.1007/s10965-013-0257-7

[17] Cui F, Yu Y, Dou L. Synthesis of ultrathin copper nanowires using tris(trimethylsilyl) silane for high-performance and low-haze transparent conductors. Nano Letters. 2015;**15**(11): 7610-7615. DOI: 10.1021/acs.nanolett.5b03422

[18] Gelves G A, Murakami Z T M, Krantz M J. Multigram synthesis of copper nanowires using ac electrodeposition into porous aluminium oxide templates. Journal of Materials Chemistry. 2006;**16**(30):3075-3083. DOI: 10.1039/b603442j

[19] Inguanta R, Piazza S, Sunseri C. Influence of the electrical parameters on the fabrication of copper nanowires into anodic alumina templates. Applied Surface Science. 2009;**255**(21):8816-8823. DOI: 10.1016/j.apsusc.2009.06.062

[20] Gerein N J, Haber J A. Effect of ac electrodeposition conditions on the growth of high aspect ratio copper nanowires in porous aluminum oxide templates. The Journal of Physical Chemistry B. 2005;**109**(37):17372-17385. DOI: 10.1021/jp051320d

[21] Pate J, Zamora F, Watson S M D. Solution-based DNA-templating of sub-10 nm conductive copper nanowires. Journal of Materials Chemistry C. 2014;**2**(43):9265-9273. DOI: 10.1039/C4TC01632G

[22] Monson C F, Woolley A T. DNA-templated construction of copper nanowires. Nano Letters. 2003; **3**(3):359-363. DOI: 10.1021/nl034016+

[23] Zhang X, Cui Z. Synthesis of Cu nanowires via solventhermal reduction in reverse microemulsion system. Journal of Physics: Conference Series. 2009;**152**(1):012022. DOI: 10.1088/1742 6596/152/1/012022

[24] Shi Y, Li H, Chen L. Obtaining ultra-long copper nanowires via a hydrothermal process. Science and Technology of Advanced Materials. 2005;**6**(7):761-765. DOI: 10.1016/j.stam.2005.06.008

[25] Xu S, Sun X, Ye H. Selective synthesis of copper nanoplates and nanowires via a surfactant-assisted hydrothermal process. Materials Chemistry and Physics. 2010;**120**(1):1-5. DOI: 10.1016/j.matchemphys.2009.10.049

[26] Liu Y, Zhang M, Wang F. Facile microwave-assisted synthesis of uniform single-crystal copper nanowires with excellent electrical conductivity. RSC Advances. 2012;**2**(30):11235-11237. DOI: 10.1039/C2RA21578K

[27] Jiang Z, Tian Y, Ding S. Synthesis and characterization of ultra-long and pencil-like copper nanowires with a penta-twinned structure by hydrothermal method. Materials Letters. 2014;**136**:310-313. DOI: 10.1016/j.matlet.2014.08.033

[28] Ye E, Zhang S, Liu S. Disproportionation for growing copper nanowires and their controlled self-assembly facilitated by ligand exchange. Chemistry – A European Journal. 2011;**17**(11):3074-3077. DOI: 10.1002/chem.201002987

[29] Yang H, He S, Tuan H. Self-seeded growth of five-fold twinned copper nanowires: mechanistic study, characterization, and SERS applications. Langmuir. 2014;**30**(2):602-610. DOI: 10.1021/la4036198

[30] Wang W, Li G, Zhang Z. A facile templateless, surfactantless hydrothermal route to ultralong copper submicron wires. Journal of Crystal Growth. 2007;**299**(1):158-164. DOI: 10.1016/j.jcrysgro.2006.11.221

[31] Zhang X, Zhang D, Ni X. One-step preparation of copper nanorods with rectangular cross sections. Solid State Communications. 2006;**139**(8):412-414. DOI: 10.1016/j.ssc.2006.06.042

[32] Chang Y, Lye M L, Zeng H C. Large-scale synthesis of high-quality ultralong copper nanowires. Langmuir. 2005;**21**(9):3746-3748. DOI: 10.1021/la050220w

[33] Rathmell A R, Bergin S M, Hua Y L. The growth mechanism of copper nanowires and their properties in flexible, transparent conducting films. Advanced Materials. 2010;**22**(32):3558-3563. DOI: 10.1002/adma.201000775

[34] Liu Z, Yang Y, Liang J. Synthesis of copper nanowires via a complex-surfactant-assisted hydrothermal reduction process. The Journal of Physical Chemistry B. 2003;**107**(46):12658-12661. DOI: 10.1021/jp036023s

[35] Ye S, Stewart I E, Chen Z. How copper nanowires grow and how to control their properties. Accounts of Chemical Research. 2016;**49**(3):442-451. DOI: 10.1021/acs.accounts.5b00506

[36] Dou L, Cui F, Yu Y. Solution-processed copper/reduced-graphene-oxide core/shell nanowire transparent conductors. ACS Nano. 2016;**10**(2):2600-2606. DOI: 10.1021/acsnano.5b07651

[37] Mehta R, Chugh S, Chen Z. Enhanced electrical and thermal conduction in graphene-encapsulated copper nanowires. Nano Letters. 2015;**15**(3):2024-2030. DOI: 10.1021/nl504889t

[38] Zhao Y, Zhang Y, Li Y. A flexible chemical vapor deposition method to synthesize copper@carbon core–shell structured nanowires and the study of their structural electrical properties. New Journal of Chemistry. 2012;**36**(5):1161-1169. DOI: 10.1039/c2nj21026f

[39] Zhao Y, Wang J, Zhang Y. The investigation of a hydro-thermal method to fabricate Cu@C coaxial nanowires and their special electronic transport and heat conduction properties. New Journal of Chemistry. 2012;**36**(5):1255-1264. DOI: 10.1039/c2nj40036g

[40] Liu Y, Liu Z, Lu N. Facile synthesis of polypyrrole coated copper nanowires: a new concept to engineered core-shell structures. Chemical Communications. 2012;**48**(20):2621-2623. DOI: 10.1039/c2cc16961d

[41] Liu Z, Elbert D, Chien C. FIB/TEM characterization of the composition and structure of core/shell Cu–Ni nanowires. Nano Letters. 2008;**8**(8):2166-2170. DOI: 10.1021/nl080492u

[42] Alia S M, Yan Y. Palladium coated copper nanowires as a hydrogen oxidation electrocatalyst in base. Journal of the Electrochemical Society. 2015;**162**(8):F849-F853. DOI: 10.1149/2.0211508jes

[43] Alia S M, Jensen K, Contreras C. Platinum coated copper nanowires and platinum nanotubes as oxygen reduction electrocatalysts. ACS Catalysis. 2013;**3**(3):358-362. DOI: 10.1021/cs300664g

[44] Zhang S, Zeng H C. Solution-based epitaxial growth of magnetically responsive Cu@Ni nanowires. Chemistry of Materials. 2010;**22**(4):1282-1284. DOI: 10.1021/cm903105f

[45] Luo X, Gelves G A, Sundararaj U. Silver-coated copper nanowires with improved anti-oxidation property as conductive fillers in low-density polyethylene. The Canadian Journal of Chemical Engineering. 2013;**91**(4):630-637. DOI: 10.1002/cjce.21701

[46] Wei Y, Chen S, Lin Y. Cu–Ag core–shell nanowires for electronic skin with a petal molded microstructure. Journal of Materials Chemistry C. 2015;**3**(37):9594-9602. DOI: 10.1039/c5tc01723h

[47] Wang X, Wang R, Shi L. Synthesis of metal/bimetal nanowires and their applications as flexible transparent electrodes. Small. 2015;**11**(36):4737-4744. DOI: 10.1002/smll.201501314

[48] Stewart I E, Ye S, Chen Z. Synthesis of Cu–Ag, Cu–Au, and Cu–Pt Core–Shell nanowires and their use in transparent conducting films. Chemistry of Materials. 2015;**27**(22):7788-7794. DOI: 10.1021/acs.chemmater.5b03709

[49] Easow J S, Selvaraju T. Unzipped catalytic activity of copper in realizing bimetallic Ag@Cu nanowires as a better amperometric H_2O_2 sensor. Electrochimica Acta. 2013;**112**: 648-654. DOI: 10.1016/j.electacta.2013.09.033

[50] Han M, Liu S, Zhang L, et al. Synthesis of octopus-tentacle-like Cu nanowire-Ag nanocrystals heterostructures and their enhanced electrocatalytic performance for oxygen reduction reaction. ACS Applied Materials & Interfaces. 2012;**4**(12):6654-6660. DOI: 10.1021/am301814y

[51] Zhao J, Zhang D, Zhang X. Preparation and characterization of copper/silver bimetallic nanowires with core-shell structure. Surface and Interface Analysis. 2015;**47**(4):529-534. DOI: 10.1002/sia.5743

[52] Wetzel B, Haupert F, Zhang M. Epoxy nanocomposites with high mechanical and tribological performance. Composites Science and Technology. 2003;**63**(14):2055-2067. DOI: 10.1016/S0266-3538(03)00115-5

Second Harmonic Generation in ZnO Nanowires

Soumen Dhara and Stephen A. Lynch

Additional information is available at the end of the chapter

http://dx.doi.org/10.5772/intechopen.68150

Abstract

Second harmonic generation (SHG) is one of the most researched nonlinear material properties and finds applications in many fields ranging from laser projection to cancer detection to future optical switches for molecular devices. Studying SHG in ZnO nanostructures started few years ago and there is a long way to go to compete with the existing nonlinear crystals. Information gathered over the past few years in research on SHG of ZnO nanowires (NWs) is summarized in this chapter. Recent advancement in the growth techniques for various types of ZnO NWs used for SHG studies is also discussed. We present an extensive analysis and discussion on some key parameters that directly modify the efficiency of SHG in ZnO NWs. The key parameters considered for discussion are aspect ratio of NWs, doping, and external strain. At the end, current standing on the reported values of nonlinear coefficients and future outlook are presented.

Keywords: second harmonic generation, sum frequency generation, ZnO, symmetry deviation

1. Introduction

ZnO is a unique material with application prospects in areas ranging from medical to optoelectronic industry to astronomical radiation detection [1–5]. ZnO is a II–VI group wide-band gap semiconductor and is highly transparent in the visible region. In the hexagonal wurtzite phase of ZnO, each Zn ion is surrounded by a tetrahedron of four O ions and vice versa. This tetrahedral coordination results in a noncentrosymmetric crystal structure, which makes it a promising candidate to observe second harmonic generation (SHG) in ambient conditions. SHG is a second-order nonlinear coherent process in which two lower energy photons are up-converted to a single photon with frequency exactly twice the incident frequency. This interesting property of nonlinear materials has several important applications in various fields, e.g., probing the electronic and magnetic structure of crystals, cancer cell diagnostics, switches in molecular-scale memory devices, and many more [6–10]. It is a very sensitive all-optical and noninvasive technique that is compatible with bulk or surface detection under various circumstances.

Among the huge variety of ZnO nanostructures, nanorods and nanowires (NWs) have undoubtedly been the focus of most studies since their geometries allow the preparation of arrays of well-controlled uniformity and use as building blocks of many nanoscale devices. As well as for ZnO thinfilms, SHG from ZnO NWs grown by different methods has been widely investigated. Studying SHG in ZnO NWs started few years ago and lot of information/knowledge has been gathered over the years by the nanoscale research community. Previous studies reported efficient generation of SHG signal from various types of ZnO nanostructures, including NWs [11–20]. Therefore, with improvement in the fabrication technique for large area ZnO NWs, it can be widely used as one of the best SHG materials. It has been discovered that the magnitude of SHG is strongly influenced by crystal orientation, aspect ratio, crystal symmetry modification, and so on. Therefore, by controlling such parameters one can tune the SHG of ZnO NWs. However, effective implementation of the knowledge gathered over the past years to ZnO to get the best SHG performance is yet to be achieved. Strategies for raising the optical nonlinearities of materials are an active research theme with rich and broad implications/applications. Keeping this in mind, some previous studies attempted to improve the second-order nonlinearity of ZnO and developed some efficient methodologies. Very high values of susceptibility tensorial component of 22–30 pm/V were achieved by changing the crystallographic orientation or decorating the surface with metal nanoparticles followed by bicolor coherent treatment or doping-induced crystal symmetry deviation [20–23]. Recently, development of ZnO-based SHG microscopy has been successfully demonstrated [24, 25].

Presently there is no review article dedicated to SHG of ZnO NWs with extensive analysis of SHG parameters and up-to-date advancement in this field. In this chapter, we present an extensive review of recent advances in SHG from ZnO NWs including our own results. Following this brief introduction, crystal structure of ZnO and theoretical background of SHG from ZnO and nonvanishing second-order nonlinear tensorial components are presented in Section 2. Section 3 describes the methodology adopted to measure SHG and an extensive analysis of the associated susceptibility tensor components. Recent advancement in the growth techniques for various types of ZnO NWs used for SHG studies and their SHG characteristics is discussed in the next section. The effects of aspect ratio of NWs, doping, and external strain on SHG magnitude are extensively addressed in subsections. At the end, a summary of the current standing on the reported values of nonlinear coefficients and future outlook are presented.

2. Second harmonic generation

SHG is one of the most studied material properties since its discovery in the 1960s by Franken et al. [26]. It is also known as the sum frequency generation because of the frequency doubling effect. When a strong primary radiation beam (frequency ω) is fed into the NLO crystal, along with the transmitted primary beam an additional light beam (frequency 2ω) appears from the crystal with frequency twice that of the primary beam. In other words, wavelength of the SHG signal is exactly half of the wavelength of the incident primary beam. **Figure 1** depicts the schematic illustration of nonlinear optical process SHG. Most often, the polarization field is considered to be linearly related to the incident electric field

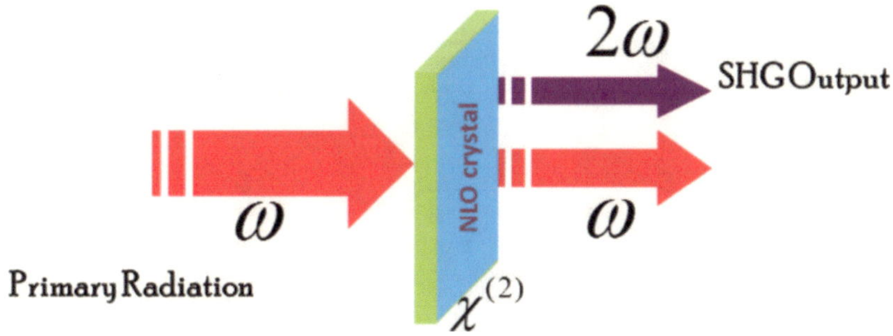

Figure 1. Schematic diagram of SHG in NLO crystal.

$$\vec{P} = \epsilon_0 \chi \vec{E} \tag{1}$$

where the electrical susceptibility, χ is a second-rank tensor. While this consideration tends to be sufficient when relating incident fields at low field strengths, it is a simplified approximation. In reality, the polarization field is more complicated than the linear relation given above. If the variation is small under strong electric field, comparable to interatomic electric fields, the polarization can be exactly expressed with the help of Taylor series

$$P_i = \epsilon_0(\chi_{ij}^{(1)}\, E_j + \chi_{ijk}^{(2)}\, E_j E_k + \chi_{ijkl}^{(3)}\, E_j E_k E_l + \ldots) \tag{2}$$

where terms are summed over repeated indices. The first coefficient $\chi_{ij}^{(1)}$ is the linear electric susceptibility component. The $\chi_{ijk}^{(2)}$ and $\chi_{ijkl}^{(3)}$ are second-order and third-order nonlinear susceptibilities responsible for SHG and third harmonic generation phenomenon. The second-order susceptibility tensor is also expressed in terms of nonlinear optical coefficients, $d_{ijk} = \frac{1}{2}\chi_{ijk}^{(2)}$. SHG arises only when the particular material has nonzero second-order susceptibility tensor, $\chi_{ijk}^{(2)}(-2\omega,\ \omega,\ \omega)$. The nonzero components exist only in the noncentrosymmetric crystal structure of the particular material. Furthermore, nanometer-sized centrosymmetric materials (nanostructures) also show weak SHG due to the breaking of space inversion symmetry at the boundary [27, 28].

The crystal structures shared by ZnO are wurtzite, zinc blende, and rocksalt; however, in ambient condition, only the wurtzite phase is thermodynamically stable. The wurtzite structure has a hexagonal unit cell with two lattice parameters, *a* and *c*, in a ratio of $c/a = 1.633$ and belongs to the space group of C6*v*4 or *P*63*mc*. The hexagonal lattice of ZnO is characterized by two interconnecting sub-lattices of Zn^{2+} and O^{2-}, such that each Zn ion is surrounded by a tetrahedron of O ions, and vice versa, which is shown in **Figure 2**. The tetrahedral coordination in ZnO results in noncentral symmetric structure. Another important characteristic of ZnO is the polar surfaces. The most common polar surface is the basal plane. The oppositely charged ions produce positively charged Zn-(0001) and negatively charged O-$(000\bar{1})$ polar surfaces, resulting in a normal dipole moment and spontaneous polarization along the *c*-axis. The (001) polar face along the *c*-axis is the primary growth orientation of ZnO.

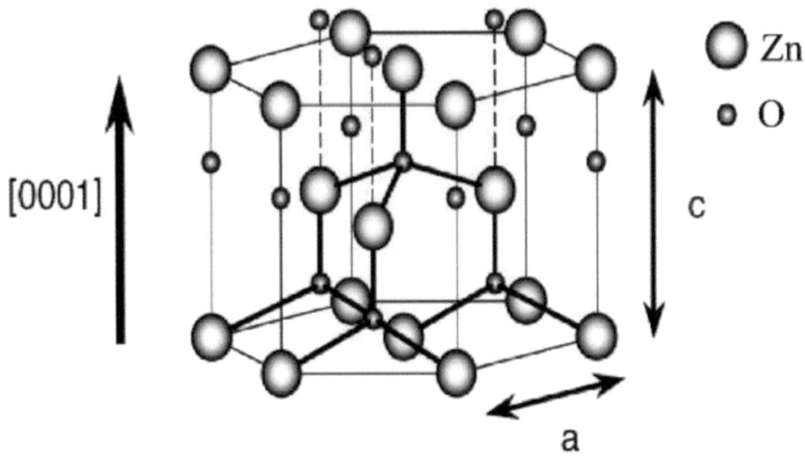

Figure 2. The wurtzite crystal structure model of ZnO having lattice constants *a* in the basal plane and *c* in the basal direction.

As a consequence of noncentrosymmetric crystal structure, ZnO possess nonvanishing second-order susceptibility tensor. The *c*-axis-grown vertically aligned ZnO NWs present five nonvanishing second-order susceptibility tensor elements, which are governed by point group symmetry of wurtzite crystal structure. The values of these second-order tensors or nonlinear optical coefficients are represented in pm/V. Five nonvanishing susceptibility tensorial components are $\chi^{(2)}_{15}$, $\chi^{(2)}_{24}$, $\chi^{(2)}_{31}$, $\chi^{(2)}_{32}$ and $\chi^{(2)}_{33}$ or equivalent in nonlinear optical coefficients d_{15}, d_{24}, d_{31}, d_{32}, and d_{33}. The point group symmetry *P63mc* of ZnO further reduces the nonvanishing terms to three components, d_{15} ($d_{15} = d_{24}$), d_{31} ($d_{31} = d_{32}$), and d_{33}. In ideal wurtzite structure, Kleinman's symmetry rule can be applied in a condition where the SHG wavelength is far from the absorption resonance. Within this approximation, $d_{15} = d_{31}$ is allowed and the number of independent d_{ij} terms further reduces to two.

3. Measurement technique and methodology

Since its introduction, several experimental techniques have been developed to calculate the magnitude of nonvanishing macroscopic nonlinear coefficients. Out of several techniques, the Maker fringes technique [29] is widely used to determine the magnitude of second-order susceptibility tensor elements. In brief, an SHG signal transmitted through the nonlinear crystal was measured as a function of angle of incidence of the fundamental beam with respect to the sample plane. This method is based on analysis of variation of SHG magnitude by incidence angle of the fundamental beam and crystal thickness. A schematic block diagram of the standard Maker fringes-based SHG measurement setup is shown in **Figure 3**. To generate SHG signal, we used an 800 nm, mode-locked femtosecond pulse light source from Ti:sapphire laser as a primary radiation. The polarization state of the fundamental beam and SHG signal are selected by using a polarizer (λ/2 plate) before the sample and an analyzer after the sample. The laser beam is tightly focused on the sample using suitable mirrors and lens assembly. The

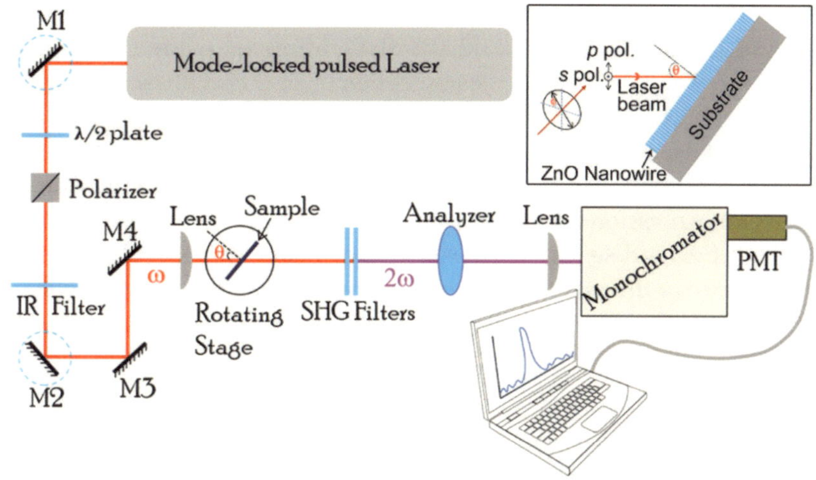

Figure 3. Schematic block diagram of SHG measurement setup according to the Maker fringes technique. Inset shows the ZnO nanowire film with the incident laser beam. θ and ϕ represent the incident angle and polarization angle of the laser beam, respectively.

variation of angle of incidence (θ) is achieved by placing the sample on a rotating stage. The generated SHG signal is analyzed using a monochromator and a highly sensitive detector. The angular dependence of SHG signal is measured either in P-in/P-out (*p-p*) or in S-in/P-out (*s-p*) polarization configuration.

The original Maker fringes methodology was based on the assumption of 100% transparency of the material in the SHG wavelength region. It is perfectly applicable only to a 100% transparent crystal. However, none of the real crystals are perfectly transparent up to that level. The original Maker fringes methodology failed to correctly estimate magnitude of the second-order d_{ij} values and large errors occur for absorbing materials. To take into account the absorption effect of the crystal in fundamental (ω) and SHG (2ω) wavelength regions, Herman and Hayden [30] presented a modified version of Maker fringes equation. According to Herman et al., the intensity of the second harmonic signal of Maker fringes, when the sample has significant absorption in the SHG wavelength region, is given by

$$\begin{aligned}
&P_{2\omega}^{x-p}(\theta,\lambda) = \frac{128\pi^3}{cA}\frac{[t_{af}^{\omega}]^4[t_{fs}^{2\omega}]^2[t_{sa}^{2\omega}]^2}{n_2^2c_2^2}\left(\frac{2\pi L}{\lambda}\right)^2 d_{eff}^2\,P_{\omega}^2 \times \exp[-2(\delta_1+\delta_2)]\frac{\sin^2\psi+\sinh^2\chi}{\psi^2+\chi^2}\\
&\psi = \left(\frac{2\pi L}{\lambda}\right)(n_1c_1 - n_2c_2)\\
&\delta_1 = \left(\frac{2\pi L}{\lambda}\right)\left(\frac{n_1\kappa_1}{c_1}\right), \delta_2 = \left(\frac{2\pi L}{\lambda}\right)\left(\frac{n_2\kappa_2}{c_2}\right), \chi = \delta_1 - \delta_2 \qquad (3)\\
&[t_{af}^{\omega}] = \begin{cases}\dfrac{2cos\theta}{c_1+n_1cos\theta}, & x=p\\[2mm] \dfrac{2cos\theta}{n_1c_1+cos\theta}, & x=s\end{cases}, [t_{fs}^{2\omega}] = \frac{2n_2c_2}{n_{2s}c_2+n_2c_{2s}}, [t_{sa}^{2\omega}] = \frac{2n_{2s}c_{2s}}{n_{2s}cos\theta + c_{2s}}
\end{aligned}$$

where $[t_{ij}^{k}]$ are the Fresnel transmission coefficients of fundamental (ω) and SHG (2ω) wavelength regions at different interfaces; *af*—air to film, *fs*—film to substrate and *sa*—

substrate to air. c and A are velocity of light and the area of the laser beam spot; P_ω and $P_{2\omega}$ are the power of the incident beam and the SHG output; L is the thickness of the film. Remaining terms in Eq. (3) arise from the absorption of ZnO and phase mismatch parameter (Ψ). The complex refractive index is $n_m = n_m(1+ ik_m)$ and $c_m = (1-(1/n_m)^2 \sin^2\theta)^{1/2}$. Subscripts 1, 2 are the fundamental beam and second harmonics, respectively. Finally, the term d_{eff} represents effective optical nonlinear coefficient, arising from the second-order nonlinear optical tensor. The general expression for d_{eff} is a bit complicated, being dependent on the polarization states of both first and second harmonic electric fields, respectively, and, of course, on the angle of incidence of the fundamental beam. Considering the ZnO group symmetry and Kleinman's symmetry approximation, the final expressions for d_{eff} for p and s polarization configurations are,

$$d_{eff} = \begin{cases} 2d_{31}c_1c_2 sin\theta/n_1 + d_{31}c_1^2 sin\theta/n_2 + d_{33}sin^3\theta/n_1^2 n_2, & x = p \\ \dfrac{d_{31}sin\theta}{n_2}, & x = s \end{cases} \tag{4}$$

Therefore, at a monochromated fundamental beam of fixed wavelength (λ), the intensity of the SHG in Eq. (3) can be simplified as a function of θ only. After fitting Eq. (3) into the measured SHG profile as a function of angle of incidence, one can estimate the associated d_{eff} value. In s-p configuration, d_{eff} depends only on d_{31}. By fitting the SHG signal data, one can obtain the value of the coefficient d_{31} which can be introduced into data fitting in the p-p configuration to calculate d_{33}. In another way, the d_{eff} value can be more accurately estimated by considering Eq. (3) as a function of wavelength (λ) only [14, 20]. This method minimizes error due to the Gaussian profile broadening of the fundamental beam with finite full width at half maximum (FWHM). At a fixed incident angle (θ), Eq. (3) can be adapted in the given form

$$\mathrm{P}_{2\omega}^{p-p}(\lambda) = [\mathrm{P}_{2\omega(\lambda)}^{p-p}]'\mathrm{P}_{\omega(\lambda)}^2 \left(\frac{2\pi \mathrm{L}}{\lambda}\right)^2 \mathrm{d}_{\mathrm{eff}}^2 \tag{5}$$

where $P_{\omega(\lambda)}$ is the relative power of the primary beam at a particular wavelength, λ.

Alternative to Maker fringes experimental configuration, the reflective second harmonic generation (RSHG) scheme involves the measurement of the SHG signal in reflection mode, at a fixed incidence angle and as a function of the azimuthal angle, which is the angle between the incidence plane and the x-axis on the sample surface. In brief, the sample is rotated along its surface normal, *i.e.*, along the z-axis, while the reflected SHG signal is systematically measured. The RSHG signals originating from the symmetry of ZnO crystal as a function of the azimuthal angle ϕ are given by

$$\mathrm{P}_{2\omega}^{x-p} = \begin{cases} d_{15}\mathrm{E}_0^2\hat{x} + \dfrac{1}{2}(d_{31} + d_{33})\mathrm{E}_0^2\hat{z}, & x = p \\ d_{31}\mathrm{E}_0^2\hat{z}, & x = s \end{cases} \tag{6}$$

The RSHG is independent of azimuthal angle, if the direction of ZnO (0002) is along the z-axis. However, a tilt of the ZnO (0002) direction may cause a variation of the resulting SHG on the azimuthal polar plot.

4. Second harmonic generation in ZnO nanowires

As a consequence of noncentrosymmetric structure, ZnO is expected to possess nonzero second-order optical nonlinearity parameters and hence room temperature SHG is expected. SHG from ZnO nanostructures including NWs has been experimentally demonstrated by several groups. Many efforts have been made to quantitatively estimate the second-order nonlinear coefficients of single ZnO NWs or ZnO NW arrays. A theoretical study on the estimation of SHG intensity from ZnO nanostructures was done by Attaccalite et al. [31]. The modified Maker fringes equation was employed by considering the dense NW arrays as a NW film. All the important experimental results on SHG of ZnO NWs are summarized in **Table 1**.

Chan et al. [13] show room temperature SHG from ZnO NWs with angular dependence exactly similar to thinfilms. Vertically aligned ZnO NWs were grown on the ZnO-seeded glass substrate by low temperature aqueous chemical solution method, which is shown in **Figure 4 (a)**. Following the reaction at 90°C, glass substrates were rinsed with de-ionized water to remove the residual salt on the surface, and then dried at 100°C. The average length of the ZnO NWs ranged from 50 to 700 nm. A Q-switched Nd:YAG laser was used as fundamental incident beam ($\lambda = 1064$ nm, 8 ns) at 10 Hz frequency. SHG signal from samples was detected by a photo-multiplier tube and then further processed with a signal-integrating oscilloscope. The variation of SHG intensity with the angle of incidence depicts periodic profile and obeys the Maker fringes methodology (**Figure 4(b)**). The second-order coefficients d_{31} and d_{33} were estimated after fitting to the experimental angular variation of SHG data by Maker fringes Eq. (3) and were found to be $d_{31} = 1.28$ and $d_{33} = -15.7$ pm/V.

The nonlinear second-order coefficients, d_{ij} can also be estimated precisely in another way by fitting the experimental SHG data with Maker fringes Eq. (5) [14, 20, 32]. The SHG signal was measured at fixed incidence angle, where SHG is maximum as a function of wavelength within the range of Gaussian beam profile. Broadband frequency-doubling properties of *c*-axis-oriented ZnO NW arrays grown by low-temperature chemical bath deposition method on glass substrate were investigated by Das et al. [14]. The measured spectral profile of SHG was found to be in good agreement with theoretical simulations. The generated SHG signals were analyzed using a spectrometer at pumping wavelength of 806 nm (Ti:sapphire laser oscillator at pulse durations ~13 fs and spectral FWHM ~100 nm) and pump intensity of 5.5×10^{10} W/cm^2. The maximum effective nonlinearity was found to be $d_{eff} = 15$ pm/V, which is about 7.5 times higher than that of a type-I beta-barium borate (BBO) crystal. Changes in the growth parameters in the chemical bath deposition also result in an improvement in the SHG intensity [41].

Our group used similar theoretical fitting method to retrieve the d_{eff} by measuring SHG signal as a function of wavelength. We prepared dense and *c*-axis-oriented ZnO NWs onto quartz substrate by modified aqueous chemical method as shown in **Figure 5(a)** [20]. Highly crystalline ZnO NWs with diameters of 70–100 nm and length of few microns were grown perpendicular to the substrate. The SHG signal was measured by passing a linearly polarized femtosecond pulse beam (at 800 nm) from a mode-locked Ti:sapphire laser through the sample and analyzing in a spectrometer with a spectral resolution of 0.2 nm. At an incident angle (θ) of 0° *i.e.*, when the incident beam is perpendicular to the substrate plane, the magnitude of

Growth technique	Substrate	Laser source	Aspect ratio	d_{ij} (pm/V)	Reference
Sonication and dispersion	Sapphire	fs Laser @800 nm	–	$d_{31} = 1.36$ $d_{33} = -14.3$	Wang et al. [33]
Chemical vapor deposition	Sapphire	fs Laser @800 nm	–	$\chi_{eff}^{(2)} = 5.5$	Johnson et al. [34]
Aqueous solution method	Glass	ns Laser @1064 nm	5.7 10.8	$d_{31} = 0.14$ $d_{33} = 7.8$ $d_{31} = 2.88$ $d_{33} = 18.0$	Chan et al. [13]
Low temperature chemical bath method	Glass	fs Laser @806 nm	23.2 14.7 10.7	$d_{eff} = 15.0$ $d_{eff} = 2.0$ $d_{eff} = 3.2$	Das et al. [14]
Aqueous solution method	–	fs Laser @810 nm	–	$d_{31} = -3.0$ $d_{33} = 0.56$ $d_{33} = 0.86$	Green et al. [35]
Chemical bath deposition	Glass	fs Laser @800 nm	8	$d_{31} = -31.4$ $d_{33} = 2.56$	Das et al. [36]
Hydrothermal method	ITO coated glass	fs Laser @1034 nm	6.0 13.2	$d_{31} = 0.29$ $d_{33} = -1.38$ $d_{31} = 0.99$ $d_{33} = -9.53$	Zhou et al. [37]
Chemical vapor deposition	Si	fs Laser @800 nm	–	$d_{15} = 10.2$	Liu et al. [38]
Modified aqueous chemical method, Europium doping	Quartz	fs Laser @800 nm	–	$d_{eff} = 12.57$ $d_{eff} = 19.09$ after 1% doping	Dhara et al. [20]
Aqueous solution method	Glass	fs Laser @1044 nm	10.0	$d_{33}/d_{31} = 1.11$	Liu et al. [39]
Hydrothermal synthesis, Co doping	Glass	fs Laser @1044 nm	15.5	$d_{31}/d_{31(\text{undoped})} = 1.17$ $d_{33}/d_{33(\text{undoped})} = 2.34$	Liu et al. [40]

Table 1. Summary of SHG results of ZnO NWs.

SHG is almost zero. Then SHG increases with the changes in the tilting axis of the substrate and shows a maximum at an incident angle of –55°, as expected. The polarization dependence of SHG shows a periodic variation (periodicity π for P-in/P-out and $\pi/2$ for S-in/P-out) with changes in the degree of polarization of the incident beam. The incident laser power dependence on the SHG intensity increases with the increase in incident laser power and shows very good linearity with power dependence factor of $\sim 1.91 \pm 0.02$, very close to the reported value (2.02 ± 0.10) for the bulk ZnO crystal [18]. Therefore, the SHG output closely follows the ideal square law dependence on the incident laser power. **Figure 5(b)** shows measured SHG spectral profile of the as prepared ZnO NWs in P-in/P-out polarization configuration at an angle of incidence (θ) of 50°. The estimated second-order nonlinear coefficient (d_{eff}) was found to be 12.57 ± 0.04 pm/V, which is larger than the reported results from ZnO thinfilms [12, 42, 43].

The microscopic SHG mapping of a single ZnO NW was measured for the first time by near-field scanning optical microscopy (NSOM) [34]. For NSOM studies, ZnO NWs were removed

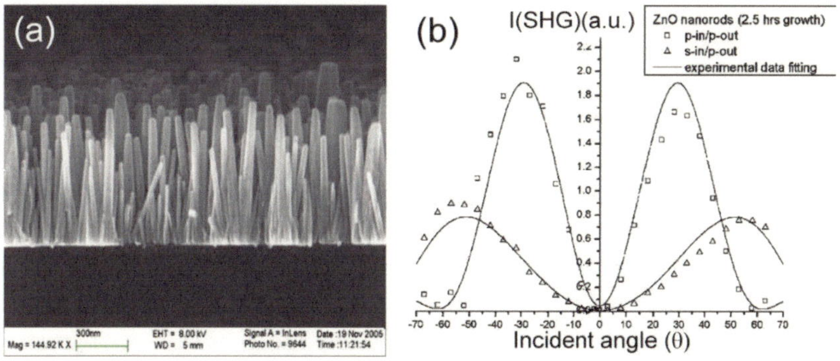

Figure 4. (a) Side-view SEM image of the ZnO nanorods/nanowires grown by aqueous chemical solution method. SHG vs. incident angle (θ) of ZnO nanorods in s-in/p-out and p-in/p-out configurations and their experimental data fittings according to the Maker fringes method. Reproduced with permission from Chan et al. [13] © Springer-Verlag 2006.

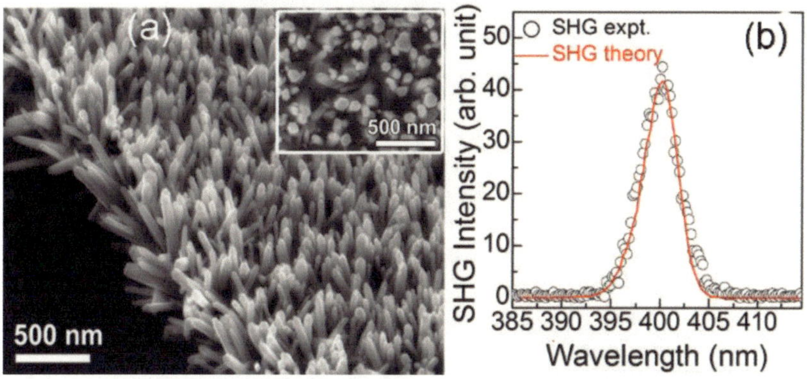

Figure 5. (a) FESEM image (tilted view) of the chemically grown vertically aligned undoped ZnO NWs. Inset shows the top view image of the same ZnO NWs. (b) Spectrum of the measured SHG signal (circle) of the ZnO NWs and theoretical fit (red line). Reproduced with permission from Dhara et al. [20] © IOP Publishing Ltd 2014.

from the substrate by sonication and dispersed onto a flat sapphire substrate for NSOM studies. Near-field SHG wave was collected using a scanning fiber probe at an oblique angle (θ, angle between the surface normal and the fundamental beam k-vector) of 55°. A large nonlinear SHG response with asymmetric variation across the diameter of the NW was presented (**Figure 6**). Furthermore, a strong polarization dependence was evidenced by the SHG images, which is ascribed to the asymmetry of the nonlinear susceptibility. The NW shows relatively efficient SHG with a larger $\chi^{(2)}_{eff}$ = 5.5 pm/V than a BBO crystal, $\chi^{(2)}_{eff} \approx$ 2.0 pm/V, a commonly used frequency doubling crystal. However, the estimated highest coefficient ($\chi^{(2)}_{33}$) is considerably lower than the reported bulk value (14.3 pm/V) [33]. One of the possible reasons for lower value is that the number of ZnO molecules probed for a single NW is less than those probed on a solid disk.

Studying SHG sometimes allows us to use a contactless surface as investigation tool to identify structural defects. Similar to thinfilms, if twin defects are present in the NWs, it can be experimentally investigated using SHG mapping to the individual rods [44]. The ZnO rods were grown on fused quartz by the aqueous solution method. The rods (length of several microns and diameter of 100–250 nm) were grown horizontally to the substrate where the polar axes of rods are parallel to the surface of fused quartz, as shown in **Figure 7(a)** and **(b)**. The SHG signal was generated using a mode-locked femtosecond pulse Ti:sapphire laser at approximately 810 nm. The transmitted SHG signals were measured under normal incidence with the polarization direction along the rod's axis. The far-field scattering patterns of the

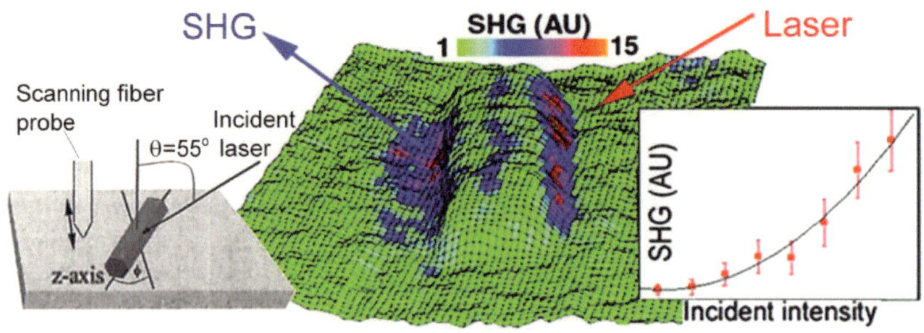

Figure 6. Near-field SHG image of single ZnO NW measured by NSOM. Insets show the illustration of the sample/beam geometry and incident laser power dependence of the SHG signal. Reproduced with permission from Johnson et al. [34] © American Chemical Society 2002.

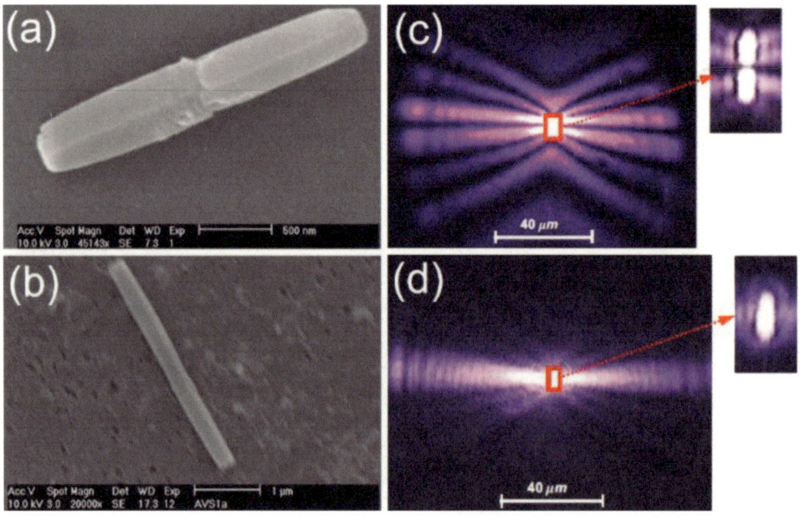

Figure 7. SEM images of two typical ZnO nanowires. (a) ZnO nanowire with twinning boundary structure and (b) ZnO nanowire without the twinning structure. SHG images and scattering patterns for single, (c) twinned ZnO nanowire of length 3.8 μm and (d) twin-free ZnO nanowire of length 1.5 μm. The insets show the enlarged view of above SHG images. Reproduced with permission from Liu et al. [44] ©The American Physical Society 2008.

transmitted SHG waves from single twinned rod were compared to the pattern arising from a twin-free ZnO rod to see if it is possible to differentiate between twinned and twin-free ZnO rods. **Figure 7(c)** and **(d)** shows typical SHG image for a single rod with twinned and twin-free structure, respectively. The images exhibit strong far-field scattering fringes resulting from the interferences of the SHG waves originating from different locations along the axis of the rod. Interestingly, a clear striking difference in the SHG fringes pattern was observed between SHG mapping images of twinned and twin-free rods. In particular, the zero-angle fringes highlight the different features of the two kinds of rods; a bright spot of SHG with a small dark gap for the twinned rod and very wide bright spot with no dark gap for the twin-free rods. A dark (bright) fringe at the 0° scattering angle was ascribed to destructive (constructive) interference of the SHG waves originating from each halve of the twinned (twin-free) rods. A small dark gap with low SHG efficiency (dark fringe) was observed only in the twinned rod, which indicates the existence of twin defects. Furthermore, use of polarization-dependent SHG microscopy to efficiently detect the lattice distortion in single-bent ZnO NWs has been demonstrated [25].

From SHG studies, it is found that SHG signal from bulk ZnO or even ZnO nanostructures is not so strong for application purpose. The reported value of an effective second-order tensorial component varies from 2 to 15 pm/V. However, it could be improved further by proper control of its crystal structure and disturbing crystal symmetry. Many studies have shown dependence of nonlinear parameters on various factors (internal or external to ZnO NWs) and also demonstrated several methods to improve the SHG further. Effects of some of the important parameters on the SHG are discussed in the following subsections. Recent advancement suggests that it is possible, in principle, for the researchers to identify the crystalline orientation, symmetry deviation, and polarities in more complicated ZnO nanostructures by the SHG patterns.

4.1. Wavelength dependence

When SHG was measured near the resonance region of ZnO, in most of the studies SHG signal was detected along with two-photon photoluminescence (2PL). Considering the different nonlinear mechanisms of both the process, final output intensities strongly depend on both pumping wavelength and light intensity. Competition between SHG and 2PL was observed and explained in several works [45–48]. Measuring emitted light as a function of fundamental beam wavelength is a useful way to distinguish the different contributions of SHG and 2PL to the emission spectrum. SHG wavelength changes according to the change in fundamental beam wavelength, while the wavelength of 2PL is fixed by the ZnO band gap energy. The competition between SHG and 2PL as a function of pump wavelength was described by Pedersen et al. [45] in randomly oriented ZnO NWs (**Figure 8(a)**). The monochromator was scanned over a broad region around SHG wavelength for a wide spectral range from 710 to 1000 nm. The contour plot that depicts the structure with highest intensity is the SHG signal at half the pump wavelength, while the weaker horizontal structure is multi-photon excited luminescence from the ZnO band gap. When the 2PL appears at shorter wavelengths than SHG (below the SHG line), it is presumably multiple-photon luminescence while the much stronger two-photon process is seen above the SHG line. The normalized $d_{eff,}$ retrieved from

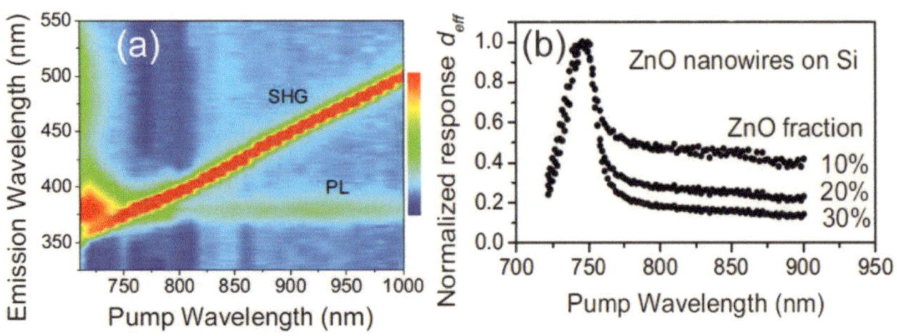

Figure 8. (a) Contour plot showing light emission intensity as a function of pump emission wavelength from ZnO nanowires. Structures due to SHG and multi-photon photoluminescence are marked. (b) Normalized effective nonlinear response of ZnO nanowires as a function of the pump wavelength. Reproduced with permission from Pedersen et al. [45] © WILEY-VCH Verlag GmbH & Co. KGaA 2008.

SHG profile in **Figure 8(a)** shows a strong wavelength-dependent behavior. At a pumping wavelength below two times of the band gap, d_{eff} gradually increases and reaches a maximum close to 760 nm and then varies slowly at higher pumping wavelength.

4.2. Dependence on aspect ratio

It is observed that the second-order nonlinear optical coefficients are strongly modified by dimensions of the NWs and aspect ratio. Chan et al. [13] and later Das et al. [14] demonstrated that changing of the aspect ratio of the ZnO nanorods could lead to a stronger SHG signal. ZnO NWs with different dimensions were grown by chemical method for different growth times, having diameter 10 to 62 nm and length 57 to 667 nm, respectively. Influence of aspect ratio on d_{31} and d_{33} is shown in **Figure 9** with aspect ratios ranging from 5.7 to 10.8. Interestingly, both the nonlinear second-order coefficients, d_{31} and d_{33}, were found to increase with increasing aspect ratio. The observed enhancement of d_{ij} is closely related to local field effects associated to the elongated nanorod structure. As we know, only d_{31} is excited when pumping beam is polarized along the rod's diameter. On the other hand, d_{33} being excited along the rod's longitudinal axis, the local field effect enhancement is predominantly connected to the rod's length. Similar effect on enhancement of d_{eff} was presented by Das et al. [14] for ZnO NWs with different length and diameters. Three different samples were grown with diameters 56, 59, and 95 nm and lengths 1300, 870, and 1020 nm, respectively. Calculated d_{eff} values are found to be 15.0, 2.0, and 3.2 pm/V, which is in correlation with the nanorod's length only. Later, a relationship between nanorod's diameter/length (aspect ratio) and corresponding $\chi^{(2)}$ values was established based on Lorentz local field, both theoretically and experimentally by Zhou et al. [37]. Theoretically, Lorentz local field-induced spectral red shift and the consequent dielectric constant modification were used to elucidate the size effect. The maximal value of $\chi_{31}^{(2)}$ of NW is the same as its bulk nonlinear susceptibility, while the maximum of $\chi_{33}^{(2)}$ of NW is significantly larger than its bulk counterpart. Therefore, the local field enhancement effect is more dominant along the longitudinal axis of NWs.

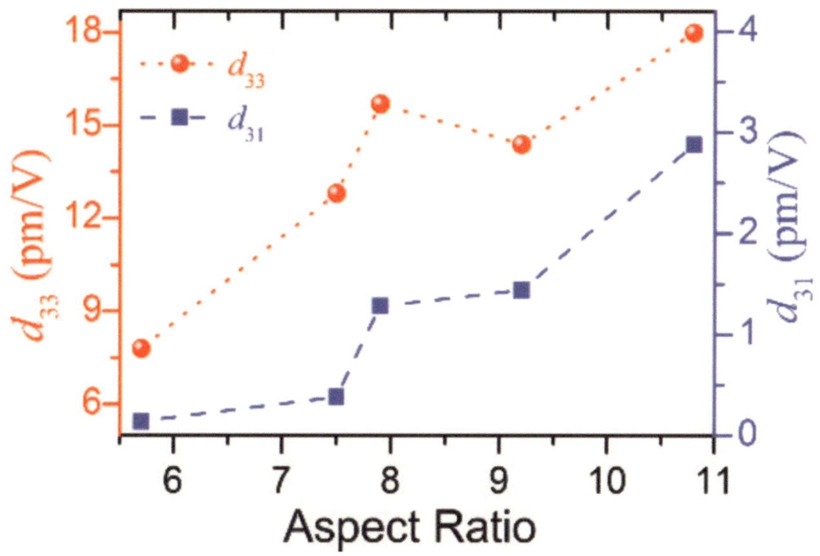

Figure 9. Dependences of d_{31} and d_{33} with the aspect ratio of ZnO nanowires.

4.3. The effect of doping

In order to improve the magnitude of the SHG in ZnO NWs, our group developed a technique to modify the crystal site symmetry of the ZnO crystal through rare earth element (Eu) doping [20]. We were able to improve the SHG about four times higher than the undoped ZnO NWs. That was the first ever report that used crystal symmetry modification technique to improve SHG. In the first step, Eu-doped ZnO (Eu:ZnO) NWs were grown vertically on the ZnO-seeded quartz substrate by a modified low temperature aqueous chemical method, using europium nitrate doping precursor. To facilitate the incorporation of larger Eu ions than the Zn ions into the ZnO lattice, a modification was done in the standard aqueous chemical method [49–52]. Eu doping was performed in a controlled way and a set of samples were prepared for different concentrations of Eu ranging from 0.5 to 5.0 at.%. Incorporation of Eu inside the ZnO lattice causes modification in the crystal site symmetry by disturbing the internal lattice arrangement. Extensive structural analysis using XRD data and high-resolution lattice images reveals expansion of lattice spacing and existence of several lattice distortions, as shown in the inset in **Figure 10**. Due to the larger ionic radius of Eu ions and the charge imbalance, incorporation of Eu completely disturbed the inside lattice arrangement. It is expected that the existence of a "self-purification process" [53] may further disturb the lattice arrangement. As a result, the formation of several lattice distortions in the Eu:ZnO NWs and hence degradation of structural quality and modification of site symmetry around the doped ions are expected.

The SHG intensity is found to increase after Eu doping. A nonmonotonic enhancement in the SHG is observed with increase in europium concentration. Maximum SHG was obtained from the 1 at.% Eu:ZnO NWs with an enhancement factor of 4.5. The effective second-order

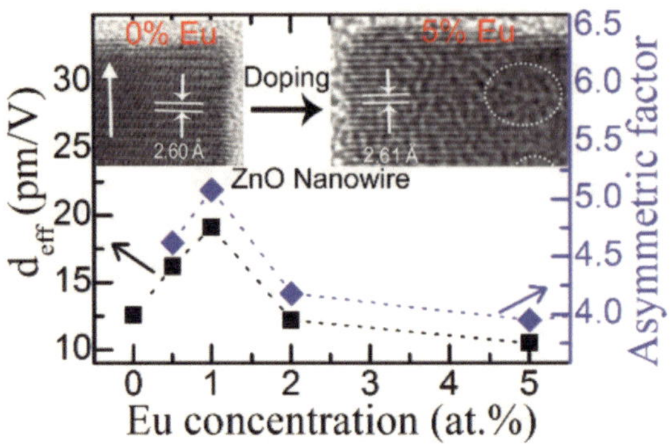

Figure 10. Plot of the d_{eff} and asymmetric factor (calculated from the PL intensity ratio) as a function of the Eu doping concentration for the Eu-doped ZnO nanowires. Inset shows high-resolution lattice image of the above undoped and 5 at.% Eu-doped ZnO NWs with growth orientation along <0001>. Formation of several lattice distortions in the Eu-doped ZnO nanowires is marked by dotted circles. Reproduced with permission from Dhara et al. [20] © IOP Publishing Ltd 2014.

nonlinear coefficient (d_{eff}) estimated from modified Maker fringes method is shown in **Figure 10** as a function of europium doping level. The highest d_{eff} (19.09 ± 0.11 pm/V) is obtained for the 1 at.% doped Eu:ZnO NWs, which is comparable to the value obtained by previous efforts to enhance the second-order susceptibility of ZnO thinfilms. Previous attempts to improve the second-order nonlinearity by changing the crystallographic orientation led to highest susceptibility ~30 pm/V [21] and decorating the surface with metal nanoparticles followed by bicolor coherent treatment led to the highest obtained susceptibility ~22.5 pm/V [54]. In comparison, our technique uses a simple process, whereas previous efforts were quite complicated from a technological point of view. Observed nonmonotonic enhancement was explained on the basis of Eu doping-induced symmetry deviation. The deviation in the crystal site symmetry (defined as the asymmetric factor) was estimated by employing photoluminescence (PL) spectroscopy and further supported by the microstructural characterizations. A strong correlation (linear dependence) is found between d_{eff} values and the magnitude of the asymmetric factor, which reveals that deviation in the local site symmetry around the Eu^{3+} ions in the Eu:ZnO crystal is the origin of the observed large second-order nonlinearity. Therefore, Eu doping on the ZnO NWs significantly influences the centrosymmetricity of the ZnO crystal, which results in an enhancement of the d_{eff} and hence SHG. This study opens up an avenue to improve the SHG of ZnO nanostructures through modification of local symmetry by incorporation of rare earth ions.

The effect of cobalt or thulium doping in ZnO NWs on the SHG characteristic was studied by other research group [40, 55]. Liu et al. [40] performed an SHG study after doping ZnO nanorods with cobalt. SHG was used to investigate bulk and surface structure quality of ZnO nanorods by measuring net dipole contribution for a different level of Co doping. Co-doped ZnO nanorods, $Zn_{1-x}Co_xO$ [$0 \leq x \leq 0.40$, x is the weight (wt.) % of Co in the growth solution], were fabricated by hydrothermal synthesis on seeded glass substrate. The SHG experiment was performed by varying incident angles in transmission mode (P-in/P-out configuration) with a femtosecond laser (1044 nm) light source. Ratios of d_{33} of Co-doped ZnO nanorods to the undoped ones are 2.34, 1.06, and 0.59 for Co concentration of 1, 15, and 35 wt.%, respectively. Similar analysis found d_{31} to be 1.17, 0.83, and 1.48 for Co concentration of 1, 15, and 35 wt.%, respectively. Dilute Co concentration (1 wt.%) leads to larger d_{33} coefficient with maximum enhancement that leads to enhancement in the bulk crystal quality of ZnO nanorods.

4.4. External strain dependence

As we know, second harmonic generation depends on the nonzero second-order susceptibility tensor, which indeed depends on the crystal symmetry of that material. Now the question arises, can we modify the SHG in ZnO by applying external strain? To verify this, our group investigated external strain (tensile and compressive) dependent SHG in ZnO NWs after bending the NWs by applying external force [56]. In the first step, ZnO NWs were transferred from substrate and dispersed into a PMMA solution. The solution mixture was spin coated on a thin steel substrate (thickness 0.5 mm) to prepare a thin layer (~200 nm) of PMMA with horizontally aligned ZnO NWs within it. The PMMA thinfilm was heated at ~60°C to remove residuals and get a continuous film. External force was applied at the center of the substrate (opposite side of the film) to bend it along the radius of curvature of the substrate. The bending of substrate led to bending of the attached ZnO NWs and it experienced a strain on the surface, as shown in the inset of **Figure 11**. The PMMA matrix was used to hold the ZnO NWs during bending of the substrate. Applied strain at the center of the substrate along the direction of the applied force was estimated according to the strain equation developed earlier [57, 58].

We measured RSHG at an angle of incidence of 45° using a p polarized fundamental femtosecond pulse light tightly focused at the center of the substrate. The RSHG signal in both the stretched side and compressed side of the ZnO NWs were measured with increasing strain and then with decreasing strain (**Figure 11**). Stretched side of the NWs experienced tensile strain whereas compressed side experienced compressive strain. With increasing tensile strain, the SHG intensity first increases up to a certain amount of tensile strain and then decreases to the initial value with further strain. The magnitude of SHG trace during release of tensile strain showed reproducible strain-dependent characteristic. Maximum enhancement was obtained

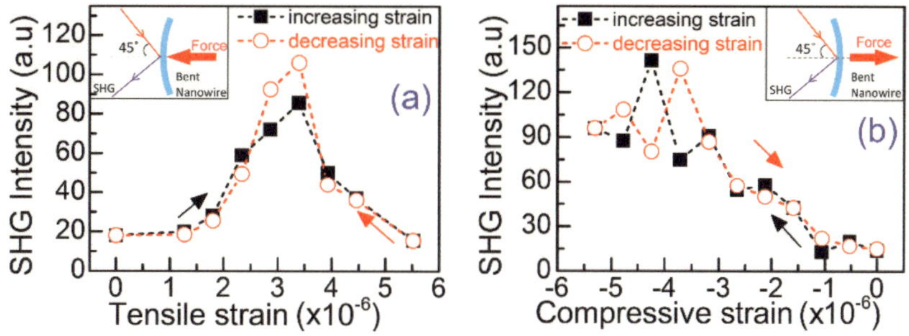

Figure 11. External strain-dependent RSHG intensity of bent ZnO NWs plotted as a function of (a) tensile strain and (b) compressive strain. SHG was measured by applying strain (black arrow, increasing strain) and then releasing strain (red arrow, decreasing strain). Insets show the measurement geometry of the bent ZnO NWs with external applied force (solid arrow).

about 400% at a tensile strain of 3.39×10^{-6}. On the other hand, SHG intensity gradually increases with increase in compressive strain. Reproducible SHG characteristic during release of compressive strain was also observed. At a compressive strain of 3.39×10^{-6}, obtained enhancement was about 430%. Therefore, SHG intensity in ZnO NWs could be enhanced and tuned through incorporation of external strain, either tensile or compressive.

5. Conclusion and future outlook

In this chapter, an up-to-date summary of important studies and results by several research groups worldwide on the SHG of ZnO NWs/nanorods is demonstrated. We present an extensive analysis and discussion on some key parameters that directly modify the efficiency of SHG in ZnO NWs. The key parameters considered for discussion are aspect ratio of NWs, doping, and external strain. Sample growth techniques, SHG measurement parameters, and extracted values of the nonlinear second-order coefficients from all the important studies are tabulated in **Table 1**. Most SHG studies are conducted on ZnO NWs with *c*-axis orientation (perpendicular to substrate) since their geometries allow to prepare *c*-axis–oriented vertically aligned structure, while only few studies deal with NWs with *c*-axis parallel to substrate. A significant advancement has been made in the improvement of nonlinearity of ZnO NWs, hence SHG intensity. Modified ZnO NWs now can compete with the existing standard nonlinear crystals. However, its needs further optimization and tailoring to get the maximum second-order nonlinearity. Fabrication of nonlinear optical devices using SHG characteristic of the ZnO NWs is certainly one of the future research directions.

Author details

Soumen Dhara[1]* and Stephen A. Lynch[2]

*Address all correspondence to: soumen5484@yahoo.co.in

1 Department of Condensed Matter Physics and Materials Sciences, S. N. Bose National Centre for Basic Sciences, Salt Lake, Kolkata, India

2 School of Physics and Astronomy, University of Cardiff, Cardiff, United Kingdom

References

[1] Kumar N, Dorfman A, Hahm J. Ultrasensitive DNA sequence detection using nanoscale ZnO sensor arrays. Nanotechnology. 2006;**17**:2875–2881

[2] Cui Y, Wei Q, Park H, Lieber CM. Nanowire nanosensors for highly sensitive and selective detection of biological and chemical species. Science. 2001;**293**:1289–1292

[3] Bano N, Zaman S, Zainelabdin A, Hussain S, Hussain I, Nur O, et al. ZnO-organic hybrid white light emitting diodes grown on flexible plastic using low temperature aqueous chemical method. Journal of Applied Physics. 2010;**108**:043103

[4] Fu X-W, Liao Z-M, Zhou Y-B, Wu H-C, Bie Y-Q, Xu J, et al. Graphene/ZnO nanowire/graphene vertical structure based fast-response ultraviolet photodetector. Applied Physics Letters. 2012;**100**:223114

[5] Johnson JC, Yan H, Schaller RD, Haber LH, Saykally RJ, Yang P. Single nanowire laser. The Journal of Physical Chemistry B. 2001;**105**:11387–11390

[6] Fiebig M, Pavlov VV, Pisarev RV. Second-harmonic generation as a tool for studying electronic and magnetic structures of crystals: Review. Journal of the Optical Society of America B. 2005;**22**:96−118

[7] Eisenthal KB. Second harmonic spectroscopy of aqueous nano- and microparticle interfaces. Chemical Review. 2006;**106**:1462–1477

[8] Campagnola P. Second harmonic generation imaging microscopy: Applications to diseases diagnostics. Analytical Chemistry. 2011;**83**:3224–3231

[9] Chen X, Nadiarynkh O, Plotnikov S, Campagnola PJ. Second harmonic generation microscopy for quantitative analysis of collagen fibrillar structure. Nature Protocols. 2012;**7**:654–669

[10] Castet F, Rodriguez V, Pozzo J-L, Ducasse L, Plaquet A, Champagne B. Design and characterization of molecular nonlinear optical switches. Accounts of Chemical Research. 2013;**46**:2656–2665

[11] Cao H, Wu JY, Ong HC, Dai JY, Chang RPH. Second harmonic generation in laser ablated zinc oxide thin films. Applied Physics Letters. 1998;**73**:572

[12] Griebner U, Kaindl RA, Elsaesser T, Seeber W. Frequency doubling and autocorrelation studies of 20-fs pulses using polycrystalline zinc oxide thin films. Applied Physics B. 1998;**67**:757–760

[13] Chan SW, Barille R, Nunzi JM, Tam KH, Leung YH, Chan WK, et al. Second harmonic generation in zinc oxide nanorods. Applied Physics B. 2006;**84**:351–355

[14] Das SK, Bock M, O'Neill C, Grunwald R, Lee KM, Lee HW, et al. Efficient second harmonic generation in ZnO nanorod arrays with broadband ultrashort pulses. Applied Physics Letters. 2008;**93**:181112

[15] Lo K-Y, Lo S-C, Yu C-F, Tite T, Huang J-Y, Huang Y-J, et al. Optical second harmonic generation from the twin boundary of ZnO thin films grown on silicon. Applied Physics Letters. 2008;**92**:091909

[16] Huang Y-J, Lo K-Y, Liu C-W, Liu C-C, Chu S-Y. Characterization of the quality of ZnO thin films using reflective second harmonic generation. Applied Physics Letters. 2009;**95**:091904

[17] Shi SL, Xu SJ, Xu Z-X, Roy VAL, Che C-M. Broadband second harmonic generation from ZnO nano-tetrapod. Chemical Physics Letters. 2011;**506**:226–229.

[18] Zheng CC, Xu SJ, Ning JQ, Zhang SF, Wang JY, Che CM, et al. Inner surface enhanced femtosecond second harmonic generation in thin ZnO crystal tubes. Journal of Applied Physics. 2011;**109**:013528

[19] Lu X, Zhou H, Salamo GJ, Tian ZR, Xiao M. Generation of exciton-polaritons in ZnO microcrystallines using second-harmonic generation. New Journal of Physics. 2012;**14**:073017

[20] Dhara S, Imakita K, Mizuhata M, Fuji M. Europium doping induced symmetry deviation and its impact on the second harmonic generation of doped ZnO nanowires. Nanotechnology. 2014;**25**:225202.

[21] Kityk IV, Ebothe J, Elchichou A, Addou M, Bougrine A, Sahraoui B. Linear electro-optics effect in ZnO–F film–glass interface. Physica Status Solidi B. 2002;**234**:553–562.

[22] Hyun JK, Kang T, Baek H, Oh H, Kim D-S, Yi G-C. Enhanced second harmonic generation by coupling to exciton ensembles in Ag-coated ZnO nanorods. ACS Photonics. 2015;**2**:1314−1319.

[23] Yang M, Shen S, Wang X, Yu B, Huang S, Xu D, et al. Plasmon-enhanced second-harmonic generation from hybrid ZnO-covered silver-bowl array. Journal of Physics: Condensed Matter. 2016;**28**:214003.

[24] Ren M-L, Agarwal R, Liu W, Agarwal R. Crystallographic characterization of II−VI semiconducting nanostructures via optical second harmonic generation. Nano Letters. 2015;**15**:7341−7346.

[25] Han X, Wang K, Long H, Hu H, Chen J, Wang B, et al. Highly sensitive detection of the lattice distortion in single bent ZnO nanowires by second-harmonic generation microscopy. ACS Photonics. 2016;**3**:1308−1314.

[26] Franken PA, Hill AE, Peters CW, Weinreich G. Generation of optical harmonics. Physical Review Letters. 1961;**7**:118.

[27] Rumpel A, Manschwetus B, Lilienkamp G, Schmidt H, Daum W. Polarity of space charge fields in second-harmonic generation spectra of Si(100)/SiO_2 interfaces. Physical Review B. 2006;**74**:081303.

[28] Jun B, White YV, Schrimpf RD, Fleetwood DM, Brunier F, Bresson N, et al. Characterization of multiple Si/SiO_2 interfaces in silicon-on-insulator materials via second-harmonic generation. Applied Physics Letters. 2004;**85**:3095–3097.

[29] Maker PD, Terhune RW, Nisenhoff M, Savage CM. Effects of dispersion and focusing on the production of optical harmonics. Physical Review Letters. 1962;**8**:21–22.

[30] Herman WN, Hayden LM. Maker fringes revisited: Second-harmonic generation from birefringent or absorbing materials. Journal of the Optical Society of America B. 1995;**12**:416–427.

[31] Attaccalite C, Nguer A, Cannuccia E, Gruning M. Strong second harmonic generation in SiC, ZnO, GaN two-dimensional hexagonal crystals from first-principles many-body calculations. Physical Chemistry Chemical Physics. 2015;**17**:9533–9540.

[32] Dhara S, Imakita K, Mizuhata M, Fujii M. Eu-doping induced improvement on the second harmonic generation of ZnO Nanowires. In: Jagadish C, editor. Materials Research Society Symposium Proceedings; 2014. pp. 95–100; DOI: 10.1557/opl.2014.367

[33] Wang G, Wong GL, Ketterson JB. Redetermination of second-order susceptibility of zinc oxide single crystals. Applied Optics. 2001;**40**:5436

[34] Johnson JC, Yan H, Schaller RD, Petersen PB, Yang P, Saykally RJ. Near-field imaging of nonlinear optical mixing in single ZnO nanowires. Nano Letters. 2002;**2**:279–283

[35] Geren K, Liu SW, Zhou HJ, Zhang Y, Tian R, Xiao M. Second-order susceptibilities of ZnO nanorods from forward second-harmonic scattering. Journal of Applied Physics. 2009;**105**:063531.

[36] Das M, Rana S, Sen P. Second harmonic generation in ZnO nanorods. The Journal of Nonlinear Optical Physics & Materials. 2010;**19**:445–458.

[37] Zhou G-Y, Hsu K-J, Su T-Y, Huang N-H, Chen Y-F, Chu S-W. Effect of Lorentz local field for optical second order nonlinear susceptibility in ZnO nanorod. Journal of Applied Physics. 2012;**111**:103112.

[38] Liu W, Wang K, Long H, Chu S, Wang B, Lu P. Near-resonant second-order nonlinear susceptibility in c-axis oriented ZnO nanorods. Applied Physics Letters. 2014;**105**:071906.

[39] Liu C-W, Chang S-J, Huang C-HHR-J, Lin Y-S, Su M-C, Wang P-H, et al. Probing surface structure quality of ZnO nanorods by second harmonic generation. IEEE Photonics Technology Letters. 2014;**26**:789–792.

[40] Liu C-W, Chang S-J, Brahma S, Hsiao C-H, Chang FM, Wang PH, et al. Enhancement in the structure quality of ZnO nanorods by diluted Co dopants: Analyses via optical second harmonic generation. Journal of Applied Physics. 2015;**117**:084315.

[41] Panda R, Bhattacharya S, Samal R, Singh A, Sahoo PK, Datta PK, et al. Second harmonic generation of femtosecond pulses using ZnO nanorods grown by chemical bath deposition with drop casted seed layer. Journal of Nonlinear Optical Physics & Materials. 2016;**25**:1650029.

[42] Kulyk B, Essaidi Z, Kapustianyk V, Turko B, Rudyk V, Partyka M, et al. Second and third order nonlinear optical properties of nanostructured ZnO thin films deposited on α-BBO and $LiNbO_3$. Optics Communications. 2008;**281**:6107–6111.

[43] Narayanan V, Thareja RK. Harmonic generation in ZnO nanocrystalline laser deposited thin films. Optics Communications. 2006;**260**:170–174.

[44] Liu SW, Zhou HJ, Ricca A, Tian R, Xiao M. Far-field second-harmonic fingerprint of twinning in single ZnO rods. Physical Review B. 2008;**77**:113311.

[45] Pedersen K, Fisker C, Pedersen TG. Second-harmonic generation from ZnO nanowires. Physica Status Solidi C. 2008;**5**:2671–2674.

[46] Prasanth R, Vugt LKV, Vanmaekelbergh DAM, Gerritsen HC. Resonance enhancement of optical second harmonic generation in a ZnO nanowire. Applied Physics Letters. 2006;**88**:181501.

[47] Zhang CF, Dong ZW, You GJ, Zhu RY, Qian SX, Deng H, et al. Femtosecond pulse excited two-photon photoluminescence and second harmonic generation in ZnO nanowires. Applied Physics Letters. 2006;**89**:042117.

[48] Das SK, Biswas M, Byrne D, Bock M, McGlynn E, Breusing M, et al. Multiphoton-absorption induced ultraviolet luminescence of ZnO nanorods using low-energy femtosecond pulses. Journal of Applied Physics. 2010;**108**:043107.

[49] Greene LE, Law M, Goldberger J, Kim F, Johnson JC, Zhang Y, et al. Low-temperature wafer-scale production of ZnO nanowire arrays. Angewandte Chemie International Edition. 2003;**42**:3031–3034.

[50] Xu S, Wang ZL. One-dimensional ZnO nanostructures: Solution growth and functional properties. Nano Research. 2011;**4**:1013–1098.

[51] Tak Y, Yong K. Controlled growth of well-aligned ZnO nanorod array using a novel solution method. The Journal of Physical Chemistry B. 2005;**109**:19263–19269.

[52] Dhara S, Giri PK. Ti nanoparticles decorated chemically grown ZnO nanowires: Photocurrent and photoluminescence properties. Journal of Experimental Nanoscience. 2013;**8**: 332–340.

[53] Gustavo MD, James RC. Self-purification in semiconductor nanocrystals. Physical Review Letters. 2006;**96**:226802.

[54] Ozga K, Kawaharamura T, Umar AA, Oyama M, Nouneh K, Slezak A, et al. Second order optical effects in Au nanoparticle-deposited ZnO nanocrystallite films. Nanotechnology. 2008;**19**:185709.

[55] Aloufy A, Ebothé J, Dumelié N, Chaki I, Abd-Lefdil M, AlZayed NS, et al. Optically stimulated optical effects in the ZnO:Tm nanorods. Physica E. 2015;**73**:96–99.

[56] Dhara S, Imakita K, Mizuhata M, Fuji M. External strain dependent anomalous second harmonic generation in ZnO nanowires film. (unpublish work).

[57] Liarokapis E, Papadimitriou D, Rumberg J, Richter W. Raman and RAS measurements of uniaxially strained thin semiconductor layer. Physica Status Solidi B. 1999;**211**:309–316.

[58] Zhao J-H, Chen Q-D, Chen Z-G, Jia G, Su W, Jiang Y, et al. Enhancement of second-harmonic generation from silicon stripes under external cylindrical strain. Optics Letters. 2009;**34**:3340–3342.

Chapter 3

Electrical Manipulation of a Single Nanowire by Dielectrophoresis

Marcos Vinicius Puydinger dos Santos, Fanny Béron,
Kleber Roberto Pirota, José Alexandre Diniz and
Stanislav Moshkalev

Additional information is available at the end of the chapter

http://dx.doi.org/10.5772/67386

Abstract

Nanowires (NWs), due to their unique highly anisotropic characteristics, hold a great promise to be used in wide technological fields, such as building blocks for data storage and memory, advanced scanning probes, and biotechnological applications. In addition, given the high sensitivity to their environment, NWs can be used as sensor for a number of applications. The fabrication and electrical characterization of NW-based devices can be achieved after proper placing of NWs between electrodes, which represents one of the major challenges in this field. The dielectrophoresis (DEP) method can be used to trap electrically neutral NWs by the application of an alternating electric field between a pair of electrodes. Here, we present a systematic study of DEP parameters as well as electrodes geometry for NW deposition. This method presents a suitable protocol for deposition in a useful and coherent fashion of post-growth electrodeposited NWs and further electrical characterization. This can be used for investigation of the fundamental transport properties of individual NWs and fabrication of NW-based devices, such as sensors and field-effect transistors.

Keywords: nanowire, dielectrophoresis, electrically-induced manipulation, electrical characterization

1. Introduction

Nanowires (NWs) represent adequate elements for electronic devices that require ultra-low power consumption, given their low current levels and high sensitivity they usually exhibit [1–5]. Their one-dimensional geometry, as well as unique possibilities for engineering of magnetic, electric and optic properties, make them to be promising nanostructures for a variety of applications, including chemical and biological sensors [6, 7], field-effect transistors [8], advanced scanning probes and magnetic sensors [9], light-emitting diodes [10], lasers [11] and photodetectors [12, 13]. Furthermore, NWs can be synthesized through a number of techniques, such as metal-organic epitaxy [14], focused-ion-beam (FIB) [8], focused-electron-beam-induced deposition (FEBID) [15], electron beam lithography [5] and electrodeposition [16, 17], yielding to unique attributes, such as particular crystallographic properties, geometry and axial/coaxial heterostructures [18].

However, despite significant advances that have been made in NWs synthesis and devices characterization, post-growth manipulation and placement of single NWs in a coherent and useful fashion remains one of the major challenges to fabricate and study the electrical transport properties of NW-based devices [19–21], especially the ones made of a single NW. So far, numerous techniques have been proposed to realize NW-based devices from already existing NWs assembly, such as atomic force microscopy nanomanipulation [22], Langmuir-Blodgett films [23], fluidic-directed assembly [24], dry-transfer printing [25], although with relatively low NW yield for precise positioning of functional system [18].

Alternatively, NWs suspended in a dielectric electrolytic solution can be manipulated through electric fields [1, 5, 26–31]. When an electrically neutral NW is subjected to a non-uniform electric field, the electric charges redistribution within the NW as well as in the solid-liquid interface builds up an electric dipole moment [32]. Thus, as the Coulomb forces on either sides of the dipole can differ in direction and intensity, a net force is exerted on the NW, which is known as the dielectrophoretic force [19, 27–32]. The force direction depends on the relative polarizabilities of the electrolyte and of the NW, inducing the latter move towards or against the region of higher electric field intensity. Such motion of nanoparticles is termed as dielectrophoresis (DEP) [19, 20, 26–32]. DEP has a tremendous advantage over the aforementioned techniques, as it relies on the possibility of working with NWs of a wide range of conductivities as well as arbitrary substrates including those that require low-temperature processing such as flexible substrates [18]. Since DEP directly depends on the dielectric properties of the particles and diluting medium, as well as the nanoparticles geometry, it allows high process selectivity [18, 32]. Moreover, large electrode arrays can be properly defined by lithography such that DEP can take place concomitantly in a large number of electrodes, leading to high throughput.

However, to fully realize these capabilities, DEP must be carefully controlled to ensure that wires are precisely placed at desired locations. Therefore, it is necessary to determine the relevant parameters that affect DEP experiment and how to control them. In this work, nickel nanowires (NiNWs) with length of around 4 μm and 35 nm of diameter, obtained by electrodeposition into pores of anodized alumina membrane and further dispersed in a dimethylformamide (DMF) electrolytic solution, were manipulated by DEP in order to make electrical contact between a pair of electrodes. Electrodes geometry and DEP electrical parameters were varied to evaluate the NiNW deposition efficiency by this technique. The materials for NW and diluting medium presented in this work were arbitrarily chosen to provide a proof-of-concept for the reader, though a selection of several NW materials and diluting medium is provided as well. In addition, COMSOL simulation supports experimental results on NWs deposition.

http://dx.doi.org/10.5772/67386

2. Dielectrophoresis

2.1. DEP theory

The electric force induced by non-uniform electric fields on polarizable and anisotropic nanoparticles can be used to properly align and trap individual liquid-suspended nanoparticles at predefined locations of a substrate. This force, termed as the DEP force, uses AC electric fields to selectively move neutral nanoparticles (e.g. NiNWs) dispersed in a dielectric diluting medium (e.g. DMF) (**Figure 1a**) [26, 32]. It relies on the polarizability differences between the NW and the DMF. The electrodes shape yields to a non-uniform electric field, which is proportional to the applied voltage. It creates a net force on the NW that exceeds the viscous force between the NW and the fluid, inducing a preferential NW movement towards the electrodes gap (**Figure 1b**). Although the discussion presented in this work considers only NW as example of particles, the analysis also applies to other anisotropic particles such as nanotubes, nanorods or sheets.

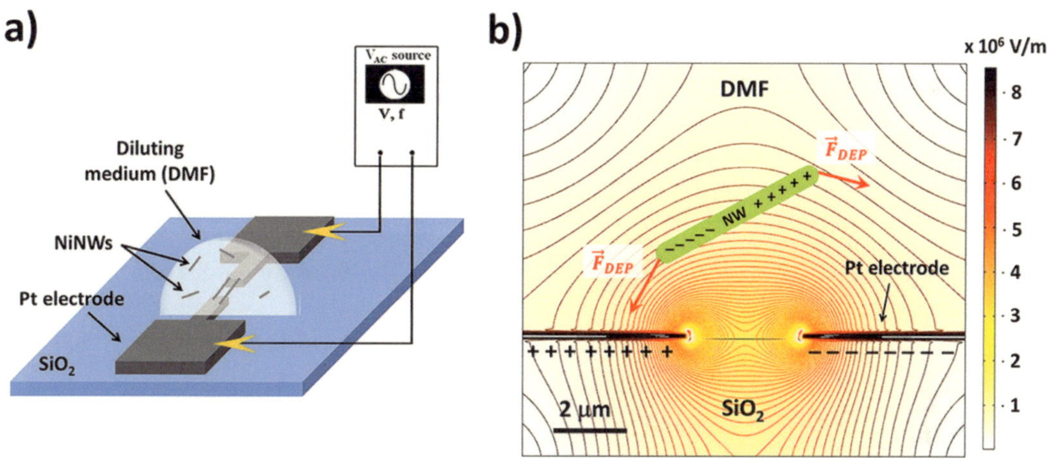

Figure 1. (a) Schematic of the DEP experiment. Adapted with permission from [33]. Copyright 2015 by American Vacuum Society. (b) DEP mechanism, in which the electric-field gradient induces attraction forces on the nanowire towards the gap between electrodes.

This force can be expressed by [5, 30, 33]

$$F_{\mathrm{DEP}} = \frac{1}{8}\pi\,\varphi^2 L\,\varepsilon_{\mathrm{DMF}}\,Re\{K\}\ \nabla\left|E\right|^2 \tag{1}$$

where φ and L represent, respectively, the NW diameter and length, *Re{}* is the real term and $K(\omega)$ the Clausius-Mossoti factor—or the so-called complex polarization factor—expressed as function of electrical permittivities of NiNW and DMF (respectively, $\varepsilon_{\mathrm{NiNW}}$ and $\varepsilon_{\mathrm{DMF}} = 36.7\,\varepsilon_0$, where ε_0 is the electrical permittivity of vacuum) [5]:

$$K(\omega) = \frac{\varepsilon^*_{\mathrm{NiNW}}(\omega) - \varepsilon^*_{\mathrm{DMF}}(\omega)}{\varepsilon^*_{\mathrm{NiNW}}(\omega) + 2\,\varepsilon^*_{\mathrm{DMF}}(\omega)}. \tag{2}$$

The imaginary component of the complex permittivity, ε^*, depends on the conductivity, σ, and the applied field angular frequency, ω:

$$\varepsilon^*(\omega) = \varepsilon - j\frac{\sigma}{\omega}. \quad (3)$$

Thus, the real-term frequency-dependent factor of the DEP force is given by

$$Re\{K\} = \frac{\omega^2(\varepsilon_{DMF}\,\varepsilon_{NiNW} - \varepsilon_{DMF}^2) + (\sigma_{DMF}\,\sigma_{NiNW} - \sigma_{DMF}^2)}{\varepsilon_{DMF}^2\,\omega^2 + \sigma_{DMF}^2}. \quad (4)$$

The complex polarization factor's dependence on the permittivity introduces a relationship between the dielectrophoretic force and the frequency of the applied field. Eqs. (5) and (6) present the high- and low-frequency limits, respectively:

$$\omega \to \infty : Re\{K\} = \frac{\varepsilon_{NiNW} - \varepsilon_{DMF}}{\varepsilon_{DMF}} \quad (5)$$

$$\omega \to 0 : Re\{K\} = \frac{\sigma_{NiNW} - \sigma_{DMF}}{\sigma_{DMF}}. \quad (6)$$

At the high-frequency limit, the DEP force is determined by the relative permittivity of the NW and the DMF, while at low frequencies the force will be a function of their relative conductivities. Within this frequency range, the transition between those two regimes is evidenced by plotting the DEP force against frequency. Inserting the appropriate NW and diluting liquid electrical conductivities (respectively, $\sigma_{NiNW} = 1.4 \times 10^7\ \Omega^{-1}\ m^{-1}$ and $\sigma_{DMF} = 2.5 \times 10^{-4}\ \Omega^{-1}\ m^{-1}$) in Eq. (4), one may calculate the frequency-dependence of F_{DEP}, where a reduction of DEP force is observed for frequencies higher than 100 kHz (**Figure 2a**).

At 500 kHz and 1 MHz, the force approximately decreases by one and two orders of magnitude, respectively. However, this model does not consider fluid dynamics effects, such as

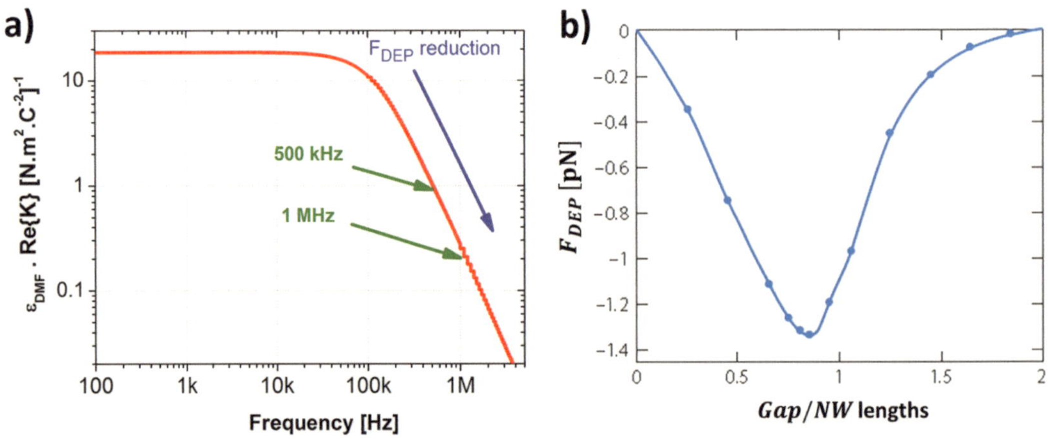

Figure 2. (a) Real part of the complex polarization factor (proportional to F_{DEP}), as function of frequency, for NiNW diluted in DMF medium, showing force reduction for frequencies above 100 kHz. Adapted with permission from [33]. Copyright 2015 by American Vacuum Society. (b) DEP force as function of the ratio between electrodes gap length and NW length, showing that the attraction force is maximized in the minimum of the force curve. Adapted with permission from [35]. Copyright 2012 by American Chemical Society.

http://dx.doi.org/10.5772/67386

AC electroosmosis [20]. For the electrostatics parameters of NiNW and DMF, this effect can reduce the force for frequencies above 1–10 kHz [19], as will be discussed in the next section. Other effects also act on the NW, such as viscous and frictional forces, fluid flow and NW-surface interactions [5, 34]. Therefore, the DEP force needs to overcome those effects in order to effectively perform deposition. Additionally, the DEP force will always depend on the ratio between the electrodes gap and NW lengths (**Figure 2b**), as this ratio influences the electric field lines density covering the NW and thus the applied torque [35]. The DEP force is maximized (attraction force, see in **Figure 2b**) for a ratio of around 0.8, since the electric field gradient and strength effects are the largest for this ratio [36]. For a smaller gap, the DEP force decreases because, despite that the electric field applied at the gap centre remains constant, it is reduced around the entire NW length. On the other hand, for a larger gap, the DEP force also decreases, simply because the electric field around the NW is less intense [5, 35, 36]. In this work, for NiNWs length of (4 ± 1) μm, we used a gap length of (2.5 ± 0.3) μm, yielding a ratio of (0.6 ± 0.2), which is near the maximum DEP force condition.

Furthermore, Eq. (1) exhibits a quadratic dependence of DEP force with the applied voltage, which increases the amount of deposited NWs in the gap region [5, 33]. In this work, we fixed the peak-to-peak voltage (V_{PP}) to 3 V, since it produces a reasonable DEP force without overheating and consequently damaging the NiNWs during DEP process.

Finally, although we have presented DEP analysis for NiNWs dissolved in a DMF solution, one might be interested in experiments involving NWs composed of different materials. Thus, we present **Table 1**, which provides an overview of reported parameters from the literature that optimizes DEP force.

Nanowire material	Medium	Potential (V)	Frequency (kHz)	Electrode gap (μm)	Electric field (V μm^{-1})
Au	Methanol	0.97	150	2	0.485
Au	Methanol	10	>100	2	5
Au-biotin	Methanol	0.18	1000	2	0.09
ZnO	Ethanol	5	1000	6–10	0.833–0.5
ZnO	Ethanol	5	1000	4	1.25
Ag	Ethanol	0.1	5	4	0.025
Ag or Au	H_2O or EtOH	0.2	100	30	0.00667
Rh rods	Acetone	10	Unknown	5–30	2–0.333
CNT[a]	Acetone	45	Unknown	5–30	9–1.5
p-Si	Benzyl alcohol	110	10	40	2.75
Si	IPA[b] and H_2O	0.35	0.5	2	0.175

[a]CNT, carbon nanotube.
[b]IPA, isopropanol.

Table 1. Overview of optimized parameters reported in literature for DEP alignment of different types of NW (adapted from [19]).

2.2. AC electroosmosis

Under the influence of an AC electric field, electrolytes on planar microelectrodes exhibit fluid flow tangential to the electrode surface. The Coulomb force generated from this electric field on the solution interfacial charges (**Figure 3a**) leads to fluid flow along the electrode surface away from the gap in a process termed AC electroosmosis [20, 37]. This effect is triggered by applying an alternating potential difference of the order of kHz at low voltages (around 1 V) to each pair of electrodes. It has been exploited in microfluidics to pump fluids in microcapillaries where surface forces preclude traditional pumping by pressure differentials.

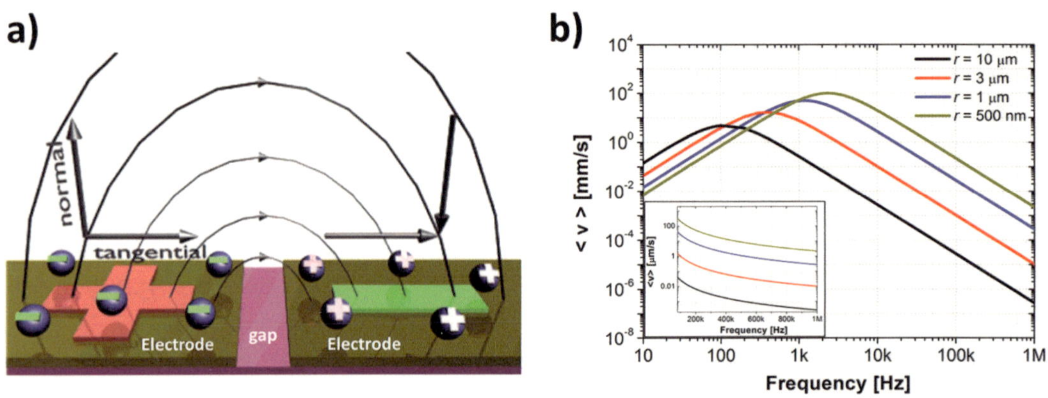

Figure 3. (a) Surface-bound interfacial charges (spheres), induced by electrodes, interact with the tangential component of electric field (curved lines), thus moving the fluid along the electrode surface away from the electrode gap. Field lines originating from the positively charged electrode and ending on the negatively charged one represent the force experienced by positive charge (negative charge experiences force in opposite direction). This force leads to fluid flow along the electrodes away from the gap. Adapted with permission from [20]. Copyright 2007 by American Chemical Society. (b) Average velocity of the fluid obtained for $V_0 = 3\ V_{PP}$ for several distances from the electrodes gap centre. *Inset*: Frequency working range utilized in this work, in which the average velocity decreases about two orders of magnitude between 100 kHz and 1 MHz for any distance, *d*.

The physical mechanism responsible for driving the flow relies on non-uniformities in the AC electric fields at the gap region. They produce an electric force on the surface-bound interfacial charges, yielding a non-zero time-averaged electroosmotic slip velocity at the electrodes surface [38]. The AC electroosmotic effect can produce fluid velocities from hundreds of μm/s up to a few mm/s in some cases [20, 37]. It should be noted that this fluid flow is distinct from the one originated from electro-thermal effects, which are found at higher frequencies and higher electrolyte conductivities [38].

For symmetrical, coplanar microelectrode gaps such as those reported in this work, AC electroosmosis produces fluid flow along the electrode surface directed away from the centre of the electrode gap. The average fluid velocity near the electrode surface is given by [20, 37, 38]

$$\langle v \rangle = \frac{1}{8} \frac{\varepsilon_{\mathrm{DMF}}}{\eta d} \frac{V_0^2 \Omega^2}{(1 + \Omega^2)^2}, \tag{7}$$

where d *is* the position along the electrodes surface with its origin at the gap centre, V_0 is the applied voltage between electrodes, η = 0.92 mPa s the electrolyte viscosity and the dimensionless frequency Ω is given by

$$\Omega = \omega d \frac{\varepsilon_{\mathrm{DMF}}}{\sigma_{\mathrm{DMF}}} \frac{\pi}{2} \kappa, \tag{8}$$

where $\kappa = (15\ \mathrm{nm})^{-1}$ is the reciprocal of electrolyte Debye length for induced double charge on the electrodes. Eq. (7) gives a bell-shaped profile for the frequency dependence of the velocity. The average velocity is small at both low and high frequencies. In the former case, most of the electric field relies in the interfacial charged layer between the electrode and the electrolyte, preventing the tangential field to extend very far into the solution. In the latter, the interfacial charged layer is thin because the charged species in the electrolyte are not fast enough to follow the rapidly changing polarities of the electrodes. On the contrary, at intermediate frequencies, the velocity can be considerably large [37]. **Figure 3b** shows average velocities of DMF for d = 500 nm, 1 μm, 3 μm and 10 μm from the electrodes gap centre, at 3 V. For these DEP conditions, the fluid velocity at the gap region will range from a few nanometres to tens of millimetres per second, depending on the frequency, which produces additional forces on the NWs and thus considerably influences NWs deposition.

3. Efficiency evaluation of DEP: a case study

In the next sections, DEP efficiency was evaluated for NiNW trapping on Pt electrodes, chosen due to the low-oxidation rate and relatively low-electrical resistivity that allows further two-wire electrical transport measurements on a single isolated NW. The DEP parameters, such as voltage and frequency, were varied in order to optimize the number of NWs that make contact between the electrodes, for which statistical evaluation was performed. The total electric field distribution over the gap area was simulated using COMSOL Multiphysics simulation tool to support experimental results.

3.1. Electrodes fabrication

Pt electrodes were defined on a SiO_2/Si structure. First, a 300 nm-thick SiO_2 layer was grown on an n^+-type Si (1 0 0) wafer (electrical resistivity of 1–10 Ω cm) by wet thermal oxidation in a conventional furnace, in order to act as a dielectric layer (**Figure 4a**). Then, photolithography was performed to define the electrodes region. An 80 nm-thick Pt layer was sputtered by a physical vapour deposition system, and lift-off process was carried out to define electrodes (**Figure 4b**). Three different electrode geometries were fabricated to evaluate the effect of electrode shape on DEP force, further denominated 1 (rectangular extremities), 2 (circular extremities) and 3 (narrow extremities) (**Figure 5**).

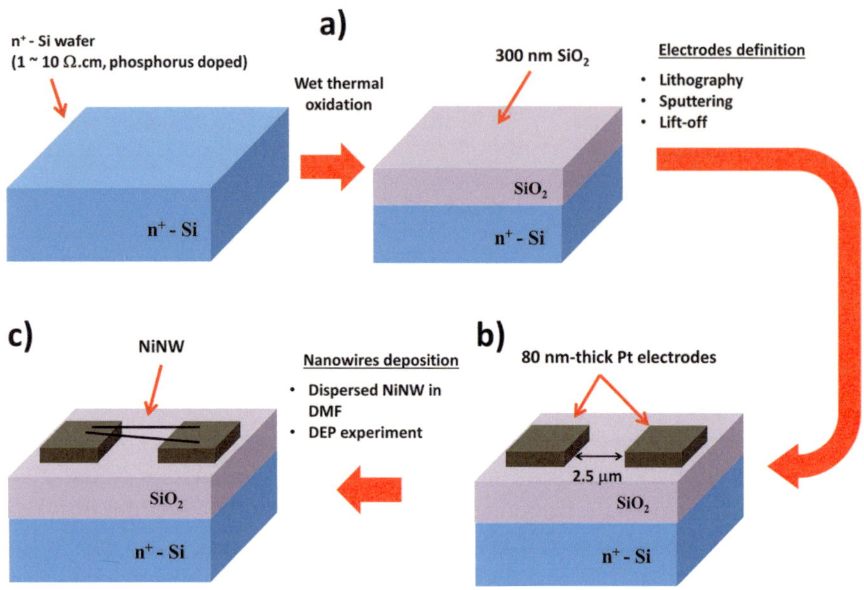

Figure 4. Schematics of experimental procedures: (a) dielectric layer formation on top of n⁺-Si wafer by thermal oxidation; (b) electrodes definition by photolithography and lift-off; (c) NiNW deposition on electrodes by DEP experiment. Adapted with permission from [33]. Copyright 2015 by American Vacuum Society.

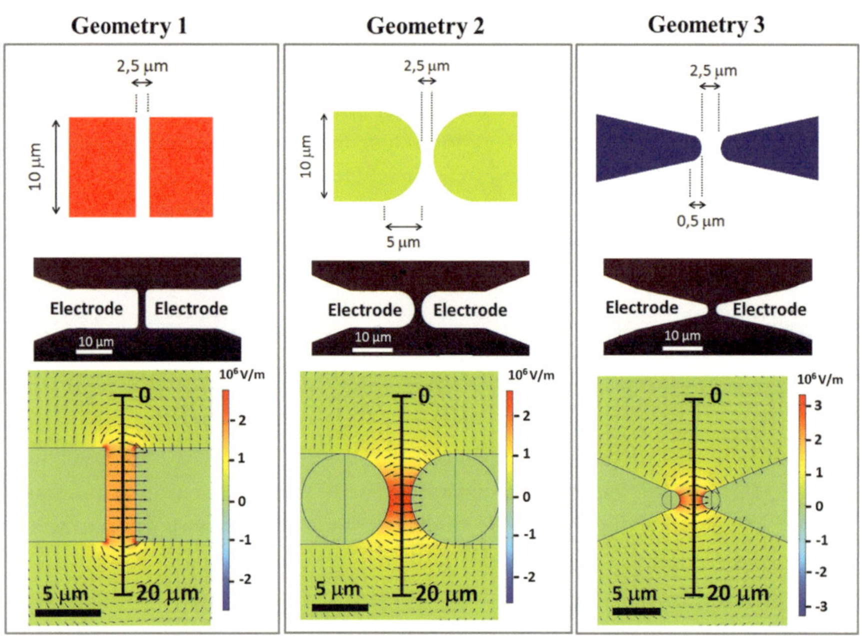

Figure 5. Schematics (top), optical microscopy (centre) and total electric field simulations, using COMSOL Multiphysics tool (below) of the three geometries tested for the Pt electrodes. The 20 µm line was taken for evaluation of the electric field profile for each geometry, as presented in **Figure 8**. Reprinted with permission from [33]. Copyright 2015 by American Vacuum Society.

3.2. NiNWs fabrication and DEP experimental procedure

NiNWs of 4 ± 1 μm-long and 35 ± 5 nm of diameter were fabricated via pulsed electrodeposition into anodized alumina membrane [16, 17]. A 1 M NaOH chemical etching solution at 24°C under agitation was employed to release the NWs from the porous alumina membrane. They were then cleaned with isopropanol and rinsed with deionized water in order to remove organic remains and further dispersed in DMF to avoid NWs clusters formation. The NiNW deposition was performed by DEP (**Figure 4c**), conducted with a HP 8116A pulse/function generator configured with 3 V_{PP} and null offset. The sinusoidal signal was generated for a frequency range between 50 kHz and 1 MHz. Before DEP process, the solution (concentration of ca. 10^8 NiNW mL^{-1}) was sonicated for 120 s at room temperature, in order to uniformly disperse the NiNWs in the DMF. Then, 1 μL solution volume was placed in the gap region and the DEP parameters were applied on each pair of electrodes during 60 s. The DMF excess was immediately rinsed with deionized water and dried with N_2. For each set of DEP parameters, the experiment was repeated ca. 100 times to ensure statistical reliability.

Visual inspection of the gap region by scanning electron microscopy (SEM) was used to evaluate the DEP efficiency for the three electrode geometries and the frequency range used (**Figure 6**). An experiment in which at least one NiNW was deposited—and made electrical contact with a pair of electrodes—was considered as success. For each geometry and frequency, we normalized the number of successes by the total number of experiments. Thus, it was possible to evaluate the efficiency percentage of NiNW deposition (**Figure 7a**) and the average number of deposited NiNW for the successful cases (**Figure 7b**), as a function of the DEP frequency and electrode geometry.

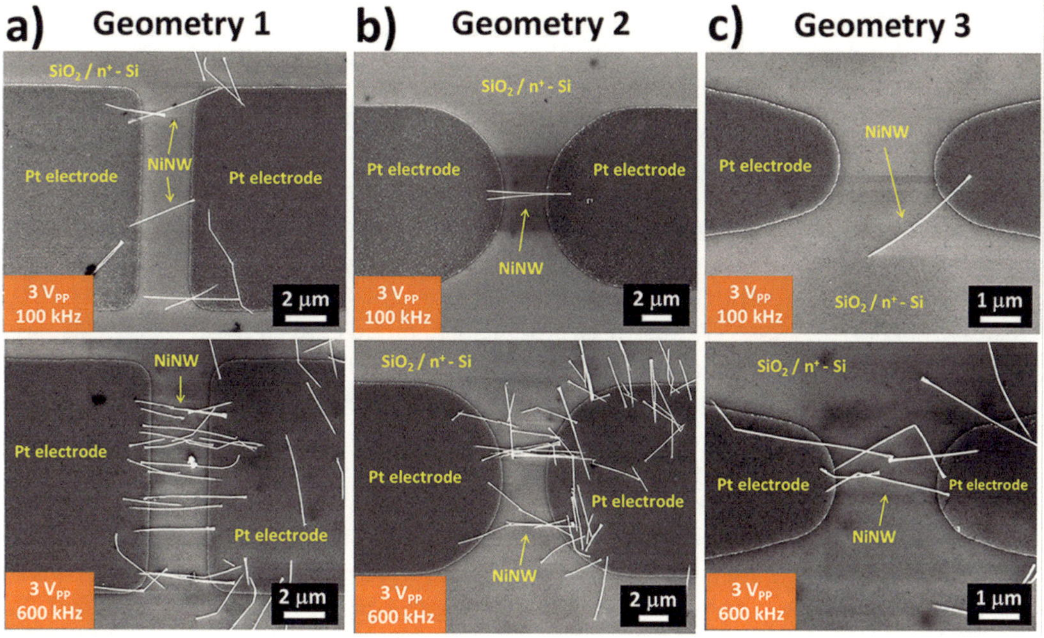

Figure 6. SEM analysis of NiNWs deposited on Pt electrodes for (a) geometry #1, (b) geometry #2 and (c) geometry #3, after DEP experiment (V_{PP} = 3 V, frequency = 100 kHz (upper row) and 600 kHz (lower row)). Reprinted with permission from [33]. Copyright 2015 by American Vacuum Society.

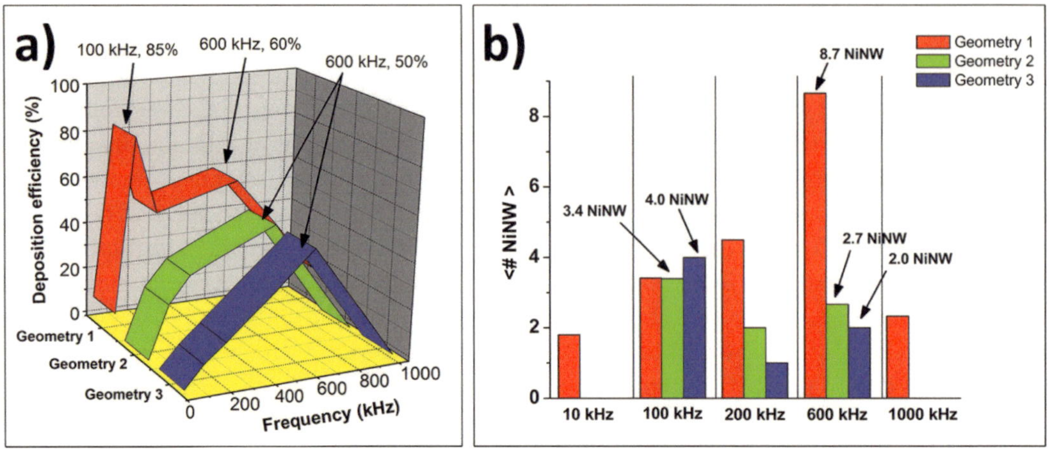

Figure 7. Charts of (a) deposition efficiency and (b) average number of deposited NiNW, obtained for DEP experiment as a function of electric field frequency, for the three electrodes geometries. Reprinted with permission from [33]. Copyright 2015 by American Vacuum Society.

First, it should be noted that DEP and AC electroosmosis might be considered as competing mechanisms for NWs deposition between electrodes. While DEP force is the main mechanism for NW trapping, AC electroosmosis, which occurs simultaneously to DEP, can induce fluid turbulence in the electrodes gap region, thus reducing the probability of success as well as the number of trapped NiNWs. Therefore, as expected from both Eq. (4) and fluid dynamics effects predictions, the DEP efficiency at 10 kHz and 1MHz was almost null, obtaining success only for geometry 1 (8% and 16%, respectively). For frequencies close to 10 kHz, the average AC electroosmosis velocity near the gap is relatively high (**Figure 3b**) and this turbulence prevents the DEP force to trap the NiNW towards the electrodes. On the other hand, for frequencies close to 1 MHz, the DEP force is about two orders of magnitude lower than the one in the range of kHz (**Figure 2a**).

Furthermore, the maximum efficiency obtained for geometry 1 was 85% at 100 kHz, while an efficiency of 60% was reached at 600 kHz. On the other hand, the DEP process was less efficient for geometries 2 and 3, both with the maximum value of 50% obtained for 600 kHz. This discrepancy may be assigned to electric field homogeneity over the electrodes gap, which is larger for geometry 1 than for geometries 2 and 3. The electrode areas are smaller in geometry 2 and 3 cases, which could create inhomogeneities and thus reduce the trapping effect in the gap region. Moreover, the maximum efficiency obtained for geometries 2 and 3 at 600 kHz still represents the frequency region that simultaneously maximizes the DEP force and minimizes the electroosmosis effect. On the other hand, for geometry 1, the larger electrode area increases the probability of success and more NiNWs are captured during DEP process. **Figure 8** presents the simulated total electric field intensity along a 20 μm transversal cross-section in the gap region, indicated in **Figure 5**. We assume that the product between the peak height, h, and its full-width half-maximum, σ, is related to the deposition efficiency. The decreasing product value for geometry 1 to 3 is in agreement with the obtained efficiency results.

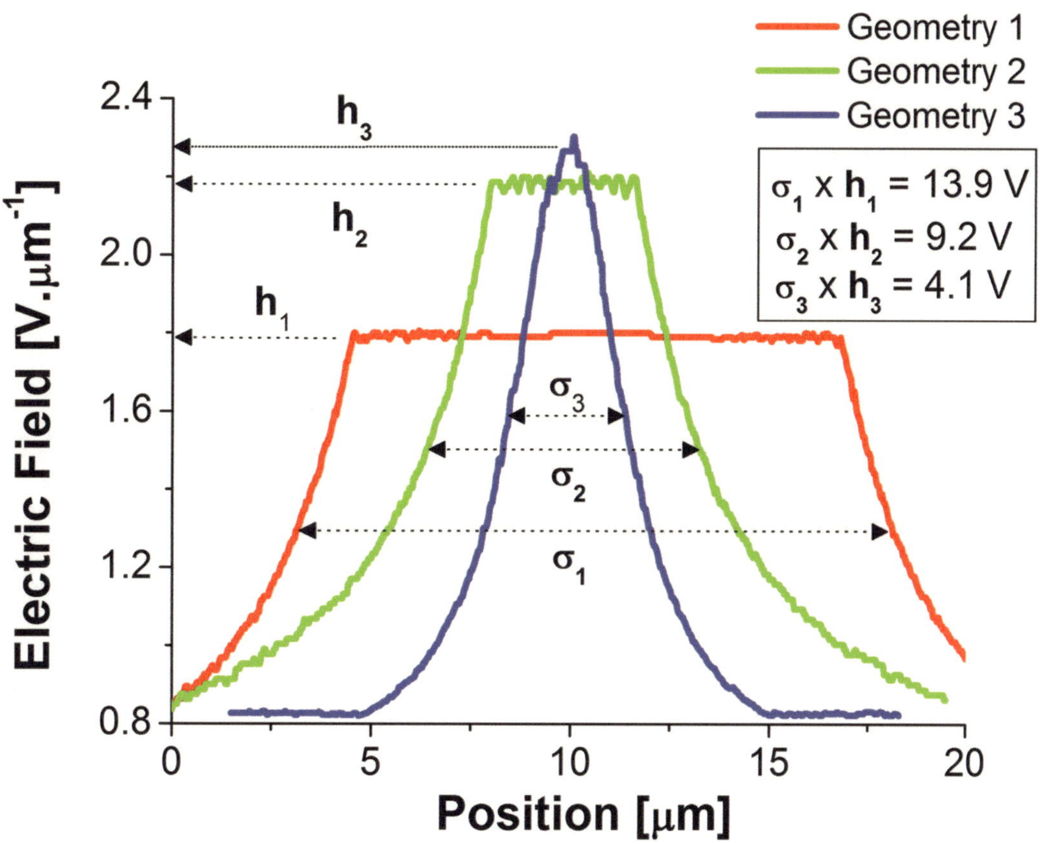

Figure 8. Total simulated electric field amplitude profile for the 20 μm transversal line shown in **Figure 4** for the three geometries, indicating the trapping efficiency to be related to the product between the peak height and its full-width half-maximum. Reprinted with permission from [33]. Copyright 2015 by American Vacuum Society.

Typically, several NiNWs were simultaneously deposited during the successful experiments, with an average number ranging from 1.0 to 8.7. Interestingly, for each investigated geometry, the maximum number of deposited NiNWs was not reached for the frequency yielding the highest efficiency. For geometry 1, only 3.4 NiNWs were deposited at 100 kHz (85% of efficiency), while a peak of 8.7 ones was attained at 600 kHz (60% of efficiency) (**Figure 7**). This result may also be attributed to the crossover between DEP and AC electroosmosis mechanisms. While the DEP force is expected to be maximized at around 100 kHz, the average fluid velocity near the gap is estimated to be close to the maximum peak value (about hundreds of micrometres/s), which limits the number of NiNWs making contact. On the contrary, at a frequency of 600 kHz, the average electroosmosis velocity reduces almost two orders of magnitude—compared to 100 kHz—yielding to less turbulence in the gap region and, therefore, a larger number of NiNWs making electrical contact between electrodes.

The situation is similar for geometries 2 and 3, but inverting the frequencies for which the efficiency and number of deposited NiNWs were maximized. The larger number of deposited NiNWs even for lower efficiency frequency may be attributed to the distortion of the electric field in the electrodes gap created by the first deposited NiNW, reducing the DEP force on the remaining NiNWs dispersed in the DMF solution and thus reducing the number of NiNWs present at higher frequencies. As shown in **Figure 9**, the electric field intensity in the gap region decreases abruptly when the first NiNW makes contact between electrodes, as the electric charges present in the metallic electrodes mainly flows through the NW. However, geometry 1 still presents a larger area, which favours more NiNWs to be present at 600 kHz, as aforementioned.

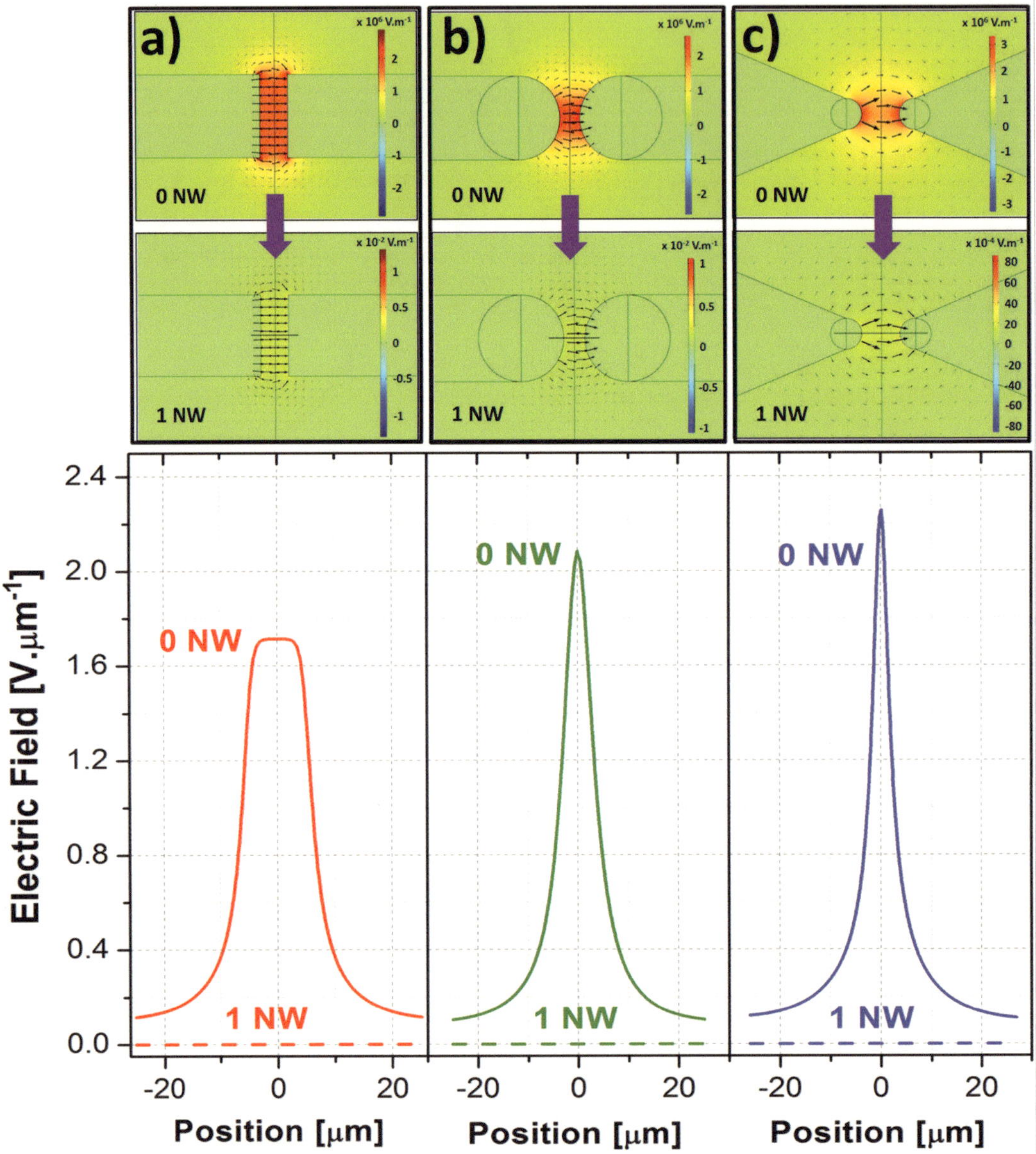

Figure 9. Total electric field simulation (using COMSOL Multiphysics) (top) and curves representing a 50 μm-long cross-section transversal line to the electrodes pair in the gap centre (bottom) for (a) geometry 1, (b) geometry 2 and (c) geometry 3. For all geometries, the total electric field substantially decreases when the first NW makes contact, reducing the probability for the second NW to make contact as well.

Globally, geometry 1 is the most efficient for DEP of NiNWs, as one can obtain efficiency up to 85% (for 3 V_{PP} and 100 kHz). However, our aim, when using DEP process, is to evaluate the transport properties of a single NiNW or only a few of them. Thus, a reasonable result is obtained when only a few NiNWs are present between electrodes. Therefore, geometries 2 and 3 reach ideal average values of NiNW (2.7 and 2.0, respectively, for 600 kHz), still with 50% of efficiency.

3.3. Electrical characterization of a single NiNW isolated by DEP

In order to evaluate the adequacy of DEP protocol for building single NW-based devices or studying fundamental properties of isolated NWs, electrical current-voltage (I-V) measurements were obtained with a Keithley 2400 source meter by applying current (without exceeding 1 nW of power to avoid damaging NiNWs due to heat dissipation) while measuring voltage with a two-wire setup. Parallel equivalent resistance as a function of the number of deposited NWs was measured and presented in a logarithm scale (**Figure 10a**). The linear fit slope of −1.0 ± 0.1 Ω $NiNW^{-1}$ is in good agreement with the ideal case (−1 Ω $NiNW^{-1}$) indicating that all the NWs made proper electrical contact after DEP process. This was accomplished due to metal-like behaviour of both the electrode and the NW. It is worth to mention that the NiNW resistance is about two orders of magnitude larger than the electrodes and cables resistance as well as the contact resistance between the metallic NW and electrodes, making it possible to use a two-wire set-up for this electrical characterization. However, for processes that require DEP of semiconductor and oxide-based NWs as well as carbon-based nanostructures, post-annealing processes are usually required to make proper electrical contact and thus reduce the contact resistance [39, 40]. Alternatively, one may perform DEP of NW using four electrodes in a 4-probes measurement setup for contact resistance subtraction.

In addition, the electric resistivity of one single NiNW, ρ, was measured as a function of temperature, T, using a Physical Property Measurement System (PPMS), in the range of 2–300 K (**Figure 10b**), showing metallic behaviour, as expected. The residual resistivity value of a single NiNW is ρ_0 = 27 μΩ cm, which is in good agreement with the ones with similar dimensions found in literature [17, 33].

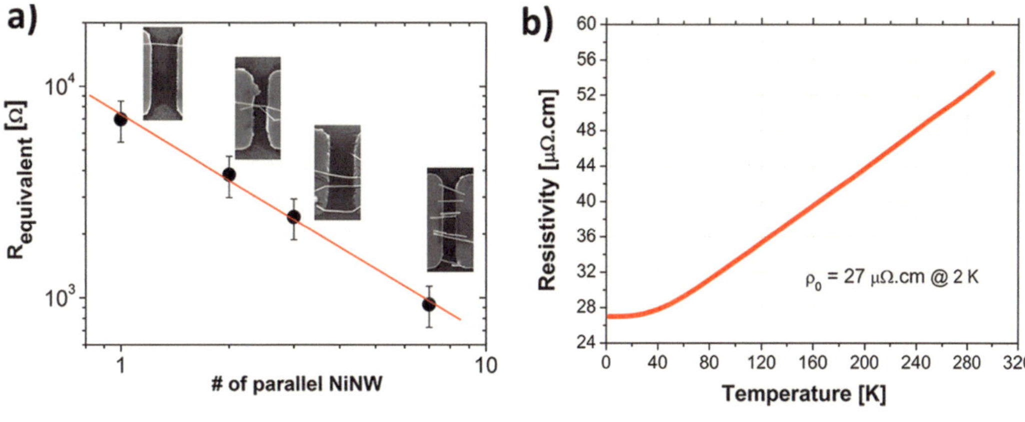

Figure 10. (a) Equivalent parallel resistance versus number of parallel-deposited NiNWs and (b) ρ x T curve of one single NiNW, showing metallic behaviour and a residual resistance of 27 μΩ cm at 2 K. Adapted with permission from [16]. Copyright 2013 by Brazilian Microelectronics Society.

4. Conclusion

This work presented DEP trapping of NiNWs between Pt electrodes defined by photolithography and lift-off. It consists of a powerful protocol for building unique and useful devices. The deposition efficiency and average number of NiNWs were evaluated as a function of the electrodes geometry and DEP frequency. Moreover, the influence of AC electroosmotic fluid flow (as a cross-over mechanism with DEP) could be observed in the mismatch between the frequencies for the largest NiNWs deposition efficiency and the largest number of deposited NiNWs. The maximum deposition efficiencies for square electrodes were 85 and 60% for 100 and 600 kHz, respectively, for averages of 3.4 and 8.7 deposited NiNWs. On the other hand, the efficiency was maximized at 600 kHz for the circular and narrow geometries, with value of 50% and averages of 2.7 and 2.0 NiNWs, respectively. This behaviour can be attributed to electric field inhomogeneities and lower trapping area over the gap present between electrodes in these geometries. For the square electrodes, since it presents a larger electrode area, it captures more NiNWs during DEP process and increases the probability of success, even with electric field intensity slightly lower than circular and narrow geometries. Adequate individual isolated NiNW electrical characterization was allowed by the successful DEP experiment of metallic NWs on the top of metallic electrodes. Additionally, post-annealing processes may be required for improving contact resistance of semiconductor and metal-oxide NWs as well as carbon-based nanostructures to metallic electrodes.

Finally, the DEP process seems to be a promising feature to evaluate fundamental properties of individual NWs as well as to build NW-based sensor devices, since they can be manipulated and isolated with relatively high efficiency. Thus, their individual electrical, thermal and/or optical output signals (in response to the environment stimulus) can be further processed. In addition, NiNWs present ferromagnetic properties, which allow their low current levels to be controlled through magnetic fields. Thus, they can be thought as an alternative for single-NW magnetic sensors as well as a promising alternative to the traditional Si-based MOSFET devices.

Author details

Marcos Vinicius Puydinger dos Santos[1,2,3*], Fanny Béron[1], Kleber Roberto Pirota[1], José Alexandre Diniz[2] and Stanislav Moshkalev[3]

*Address all correspondence to: puyding@ifi.unicamp.br

1 Institute of Physics Gleb Wataghin, University of Campinas, Campinas-SP, Brazil

2 Faculty of Electrical Engineering and Computing, University of Campinas, Campinas-SP, Brazil

3 Center for Semiconductor Components and Nanotechnology, University of Campinas, Campinas-SP, Brazil

References

[1] Wu J, Yin B, Wu F, Myung Y, Banerjee P. Charge transport in single CuO nanowires. Applied Physics Letters. 2014;**105**(18):183506-1-183506-5. DOI: 10.1063/1.4900966

[2] Li M, Li WH, Zhang J, Alici G, Wen W. A review of microfabrication techniques and dielectrophoretic microdevices for particle manipulation and separation. Journal of Physics D: Applied Physics. 2014;**47**(6):63001-1-63001-29. DOI: 10.1088/0022-3727/47/6/063001

[3] Seo K-WW, Lee J-HH, Cho NG, Kang SJ, Kim H-KK, Na S-II, Koo H-WW, Kim T-WW. Simple brush painted Ag nanowire network on graphene sheets for flexible organic solar cells. Journal of Vacuum Science and Technology A. 2014;**32**(6):61201-1-61201-6. DOI: 10.1088/0022-3727/47/6/063001

[4] Ou MN, Yang TJ, Harutyunyan SR, Chen YY, Chen CD, Lai SJ. Electrical and thermal transport in single nickel nanowire. Applied Physics Letters. 2008;**92**(6):063101-1 –063101-3. DOI: 10.1063/1.2839572

[5] Boote JJ, Evans . Dielectrophoretic manipulation and electrical characterization of gold nanowires. Nanotechnology. 2005;**16**(9):1500-1505. DOI: 10.1088/0957-4484/16/9/015.

[6] Huang J, Wan Q. Gas sensors based on semiconducting metal oxide one-dimensional nanostructures. Sensors. 2009;**9**:9903-9924. DOI: 10.3390/s91209903

[7] Cui Y, Wei Q, Park H, Lieber . Nanowire nanosensors for highly sensitive and selective detection of biological and chemical species. Science. 2001;**293**:1289-1292. DOI: 10.1126/science.1062711.

[8] Puydinger dos Santos MV, Lima LPB, Diniz JA, Godoy Filho J. Fabrication of p-type silicon nanowires for 3D FETs using focused ion beam. Journal of Vacuum Science and Technology B. 2013;**31**(6):06FA01-1-06FA01-6. DOI: 10.1116/1.4823763

[9] De Teresa JM, Fernández-Pacheco A. Present and future applications of magnetic nanostructures grown by FEBID. Applied Physics A: Materials Science & Processing. 2014;**117**(4):1645-1658. DOI: 10.1007/s00339-014-8617-7

[10] Svensson CPT, Mårtensson T, Trägårdh J, Larsson C, Rask M, Hessman D, Samuelson L, Ohlsson J. Monolithic GaAs/InGaP nanowire light emitting diodes on silicon. Nanotechnology. 2008;**19**:305201-1-305201-6. DOI: 10.1088/0957-4484/19/30/305201

[11] Zhou H, Wissinger M, Fallert J, Hauschild R, Stelzl F, Klingshirn C, Kalt H. Ordered, uniform-sized ZnO nanolaser arrays. Applied Physics Letters. 2007;**91**:181112-1-181112-3. DOI: 10.1063/1.2805073

[12] Novotny CJ, Yu ET, Yu PKL. InP Nanowire/polymer hybrid photodiode. Nano Letters. 2008;**8**(3):775-779. DOI: 10.1063/1.2805073

[13] Soci C, Zhang A, Xiang B, Dayeh SA, Aplin DPR, Park J, Bao XY, Lo YH, Wang D. ZnO nanowire UV photodetectors with high internal gain. Nano Letters. 2007;**7**(4):1003-1009. DOI: 10.1021/nl070111x

[14] Yoshimura M, Tomioka K, Hiruma K, Hara S, Motohisa J, Fukui T. Growth and characterization of InGaAs nanowires formed on GaAs(111)B by selective-area metal organic vapor phase epitaxy. Japanese Journal of Applied Physics. 2010;**49**:04DH08-1-04DH08-5. DOI: 10.1143/JJAP.49.04DH08

[15] Puydinger dos Santos M V., Velo MF, Domingos RD, Zhang Y, Maeder X, Guerra-Nuñez C, Best JP, Béron F, Pirota KR, Moshkalev S, Diniz JA, Utke I. Annealing-based electrical tuning of cobalt–carbon deposits grown by focused-electron-beam-induced deposition. ACS Applied Materials and Interfaces. 2016;**8**:32496-32503. DOI: 10.1021/acsami.6b12192

[16] Puydinger dos Santos MV, Velo MF, Domingos RD, Bettini J, Diniz JA, Béron F, Pirota KR. Electrodeposited nickel nanowires for magnetic-field effect transistor (MagFET). Journal of Integrated Circuits and Systems. 2016;**11**(1):13-18.

[17] Leitao DC, Sousa CT, Ventura J, Amaral JS, Carpinteiro F, Pirota KR, Vazquez M, Sousa JB, Araujo JP. Characterization of electrodeposited Ni and Ni80Fe20 nanowires. Journal of Non-Crystalline Solids. 2008;**354**:5241-5243. DOI: 10.1016/j.jnoncrysol.2008.05.088

[18] Raychaudhuri S, Dayeh SA, Wang D, Yu ET. Precise semiconductor nanowire placement through dielectrophoresis. Nano Letters. 2009;**9**(6):2260-2266. DOI: 10.1021/nl900423g

[19] Maijenburg AW, Maas MG, Rodijk EJB, Ahmed W, Kooij ES, Carlen ET, Blank DHA, ten Elshof JE. Dielectrophoretic alignment of metal and metal oxide nanowires and nanotubes: A universal set of parameters for bridging prepatterned microelectrodes. Journal of Colloid and Interface Science. 2011;**335**(2):486-493. DOI: 10.1016/j.jcis.2010.12.011

[20] Gierhart BC, Howitt DG, Chen SJ, Smith RL, Collins SD. Frequency dependence of gold nanoparticle superassembly by dielectrophoresis. Langmuir. 2007;**23**(24):12450-12456. DOI: 10.1021/la701472y

[21] Vijayaraghavan A, Sciascia C, Dehm S, Lombardo A, Bonetti A, Ferrari AC, Krupke R. Dielectrophoretic assembly of high-density arrays of individual graphene devices for rapid screening. ACS Nano. 2009;**3**(7):1729-1734. DOI: 10.1021/nn900288d

[22] Decossas S, Mazen F, Baron T, Brémond G, Souifi A. Atomic force microscopy nanomanipulation of silicon nanocrystals for nanodevice fabrication. Nanotechnology. 2003;**14**:1272-1278. DOI: 10.1088/0957-4484/14/12/008

[23] Whang D, Jin S, Wu Y, Lieber CM. Large-scale hierarchical organization of nanowire arrays for integrated nanosystems. Nano Letters. 2003;**3**(9):1255-1259. DOI: 10.1021/nl0345062

[24] Huang Y, Xiangfeng D, Qingqiao W, Lieber C. Directed assembly of one-dimensional nanostructures into functional networks. Science. 2001;**291**:630-633. DOI: 10.1126/science.291.5504.630

[25] Javey A, Nam S, Friedman RS, Yan H, Lieber CM. Layer-by-layer assembly of nanowires for three-dimensional, multifunctional electronics. Nano Letters. 2007;**7**(3):773-777. DOI: 10.1021/nl063056l

[26] Pohl HA. The motion and precipitation of suspensoids in divergent electric fields. Journal of Applied Physics. 1951;**22**(7):869-871. DOI: 10.1063/1.1700065

[27] Xie C, Chen B, Ng CO, Zhou X, Wu J. Numerical study of interactive motion of dielectrophoretic particles. European Journal of Mechanics B/Fluids. 2015;**49**:208-216. DOI: 10.1016/j.euromechflu.2014.08.007

[28] Kang S. Dielectrophoretic motion of two particles with diverse sets of the electric conductivity under a uniform electric field. Computers & Fluids. 2014;**105**:231-243. DOI: 10.1016/j.compfluid.2014.09.029

[29] Schukfeh MI, Storm K, Hansen A, Thelander C, Hinze P, Beyer A, Weimann T, Samuelson L, Tornow M. Formation of nanogaps in InAs nanowires by selectively etching embedded InP segments. Nanotechnology. 2014;**25**:465306-1-465306-6. DOI: 10.1088/0957-4484/25/46/465306

[30] Hamdi FS, Français O, Dufour-Gergam E, Le Pioufle B. How medium osmolarity influences dielectrophoretically assisted on-chip electrofusion. Bioelectrochemistry. 2014;**100**:27-35. DOI: 10.1016/j.bioelechem.2014.05.004

[31] Freer EM, Grachev O, Stumbo DP, Duan X, Martin S, Stumbo DP. High-yield self-limiting single-nanowire assembly with dielectrophoresis. Nature Nanotechnology. 2010;**5**:525-530. DOI: 10.1038/nnano.2010.157

[32] Pohl HA. Dielectrophoresis: The Behavior of Neutral Matter in Non-Uniform Electric Fields. 1st ed. New York: Cambridge University Press; 1978. 579 p.

[33] Puydinger dos Santos M V., Lima LPB, Mayer RA, Béron F, Pirota KR, Diniz JA. Dielectrophoretic manipulation of individual nickel nanowires for electrical transport measurements. Journal of Vacuum Science and Technology B. 2015;**33**(3):031804-1-031804-8. DOI: 10.1116/1.4918732

[34] Dimaki M, Bøggild P. Dielectrophoresis of carbon nanotubes using microelectrodes: a numerical study. Nanotechnology. 2004;**15**:1095-1102. DOI: 10.1088/0957-4484/15/8/039

[35] Smith BD, Mayer TS, Keating CD. Deterministic assembly of functional nanostructures using nonuniform electric fields. Annual Review Physical Chemistry. 2012;**63**:241-263. DOI: 10.1146/annurev-physchem-032210-103346

[36] Liu Y, Chung JH, Liu WK, Ruoff RS. Dielectrophoretic assembly of nanowires. Journal of Physical Chemistry B. 2006;**110**(29):14098-14106. DOI: 10.1021/jp061367e

[37] Green NG, Ramos A, González A, Morgan H, Castellanos A. Fluid flow induced by nonuniform ac electric fields in electrolytes on microelectrodes. I. Experimental measurements. Physical Review E. 2000;**61**(4):4011-4018. DOI: 10.1103/PhysRevE.61.4019

[38] Ramos A, González A, Castellanos A, Green NG, Morgan H. Pumping of liquids with ac voltages applied to asymmetric pairs of microelectrodes. Physical Review E: Statistical Nonlinear Soft Matter Physics. 2003;**67**:056302-1-056302-11. DOI: 10.1103/PhysRevE.67.056302

[39] Rouxinol FP, Gelamo R V., Amici RG, Vaz AR, Moshkalev SA. Low contact resistivity and strain in suspended multilayer graphene. Applied Physics Letters. 2010;**97**:253104-1-253104-4. DOI: 10.1063/1.3528354

[40] Li M, Anderson W, Chokshi N, DeLeon RL, Tompa G. Laser annealing of laser assisted molecular beam deposited ZnO thin films with application to metal-semiconductor-metal photodetectors. Journal of Applied Physics. 2006;**100**:053106-1-053106-4. DOI: 10.1063/1.2344811

Chapter 4

ZnO Nanowires for Dye Sensitized Solar Cells

Simas Rackauskas, Nadia Barbero,
Claudia Barolo and Guido Viscardi

Additional information is available at the end of the chapter

http://dx.doi.org/10.5772/67616

Abstract

This chapter provides a broad review of the latest research activities focused on the synthesis and application of ZnO nanowires (NWs) for dye-sensitized solar cells (DSCs) and composed of three main sections. The first section briefly introduces DSC-working principles and ZnO NW application advantages and stability issues. The next section reviews ZnO NW synthesis methods, demonstrating approaches for controlled synthesis of different ZnO NW morphology and discussing how this effects the overall efficiency of the DSC. In the last section, the methods for ZnO NW interface modification with various materials are discussed, which include ZnO core-shell structures with semiconductive or protective layers, ZnO NW hybrid structures with other materials, such as nanoparticles, quantum dots and carbon nanomaterials and their benefit for charge and light transport in DSCs. The review is concluded with some perspectives and outlook on the future developments in the ZnO nanowire application for DSCs.

Keywords: Nanowires, ZnO nanowire, Synthesis, dye-sensitized solar cell, DSC, Photovoltaics, surface modification, hybrid materials

1. Introduction

World energy demand is rapidly increasing; therefore, a clean and reliable energy source is fundamental to global economy. The environmental impact of the fossil fuel use urges the need for the search of alternative energy resources. Many different renewable energy technologies were developed recently aiming for efficient solutions of clean energy supply; however, a price competitiveness is a biggest drawback. Solar energy is considered as a key solution for environmental challenge, because of its carbon neutral nature and high abundance. The conversion of solar energy to electricity is fulfilled by solar cells based on the photovoltaic effect.

Dye-sensitized solar cells (DSCs) have gained widespread attention because of the ease of fabrication, low production costs and tuneable optical properties, such as colour and transparency [1]. The most attractive properties of DSCs are their low cost and simple manufacturing processes together with their advantageous attributes such as lightweight, low toxicity and good performance in diverse light conditions [2].

ZnO attracted attention as an alternative photoelectrode material for DSCs due to multiple advantages, that is, excellent optoelectronic properties, low cost, easy synthesis, non-toxicity and others. Moreover, ZnO has the diversity of one-dimensional (1D) structures, which suggest attractive approach for photoelectrode scaffolding.

ZnO nanowire (NW) application as DSC photoelectrode shows multiple advantages, such as higher electron mobility and increased optical way because of refraction; moreover, additional functionalities such as flexible DSCs with the help of ZnO NW photoelectrode can also be obtained. The efficiency of ZnO nanowire DSCs is rapidly increasing in recent years; therefore, the interest in ZnO NW application for DSC is rapidly growing. In order to reach higher efficiency of ZnO NW DSCs, further surface modifications with an additional protective layer of TiO_2 or inert materials such as Al_2O_3 are used. Other possibilities lie in tailoring the morphology of NWs, or employing hybrid structures of ZnO nanowires with other materials.

The application of ZnO NW for DSCs is an interesting topic for nanowire community and also to researchers from diverse scientific fields, since much effort was put in this topic, numerous approaches were tried and a definite improvement was reached. Still, there is a need to direct the effort in understanding more deeply the ZnO NW interaction with other materials, processes of light and charge transfer and master the synthesis methods in order to achieve the optimal structure for application in DSCs.

This chapter will provide a comprehensive review of the state-of-the-art research activities focused on the synthesis and application of ZnO nanowires for dye-sensitized solar cells. The first section briefly overviews fundamentals of the DSC and introduces ZnO NW synthesis, divided into vapour and solution phase methods, demonstrating approaches to obtain different ZnO NW morphology. Next, the methods for ZnO NW surface modification are discussed, which include ZnO core-shell structures with semiconductive layers or protective layers and their benefit for charge and light transport in DSCs. In the last part, we will review ZnO NW hybrid structures with other materials, such as nanoparticles, quantum dots or carbon nanomaterials. The chapter will then conclude with the perspectives and the outlook on the future developments in the ZnO nanowire application for DSCs.

2. Fundamentals of dye-sensitized solar cells

The concept of a dye-sensitized solar cell was introduced in 1991, by O'Regan and Grätzel [3]. A schematic diagram showing the operation of a typical DSC is shown in **Figure 1**.

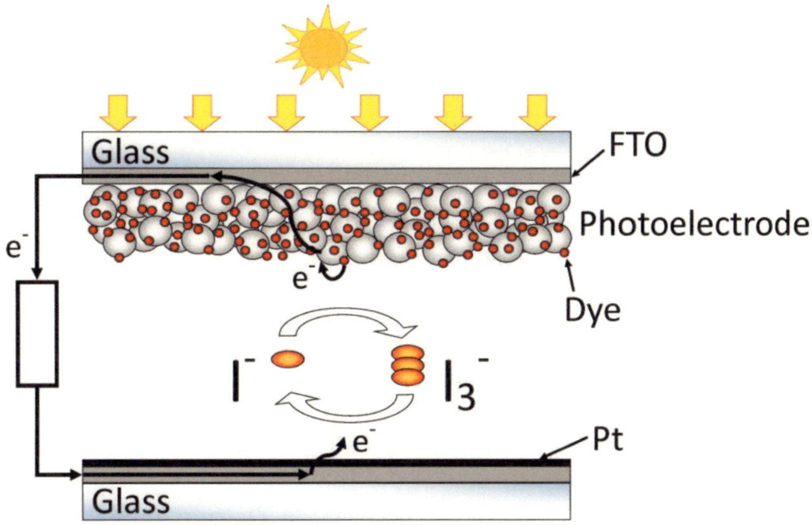

Figure 1. Schematic overview of the typical dye-sensitized solar cell (DSC).

Typical DSCs are composed of a transparent conducting oxide (such as fluorine-doped tin oxide, FTO) on glass, a nanoparticle photoelectrode covered in a monolayer of sensitizing dye, a hole-conducting electrolyte and a platinum-coated, FTO-coated glass back-contact.

Nanoparticles of TiO_2 (anatase) are mostly used as photoelectrode, although alternative wide-band-gap oxides such as ZnO and SnO_2 have also been investigated. A monolayer of the sensitizer is attached to the surface of the nanoparticle photoelectrode. Under illumination, sensitizer photo-excitation results in the injection of an electron into the conduction band of the oxide. The dye is regenerated by electron donation from the electrolyte, mostly from a redox system (iodide/triiodide couple) in an organic solvent. The regeneration of the sensitizer by iodide intercepts the recapture of the conduction band electron by the oxidized dye. The iodide is regenerated, in turn, by the reduction of triiodide at the counter-electrode, and the circuit is completed via electron migration through the external load. The voltage, which is obtained under illumination, corresponds to the difference between the Fermi level of the electron in the solid and the redox potential of the electrolyte. Overall, there are no permanent chemical transformations involved in the generation of electric power from light.

The photoelectrode serves as a support for sensitizer loading and at the same time transport media of photo-excited electrons from sensitizer to the external circuit. Hence, to ensure high dye loading, a large surface area is necessary. Moreover, a fast charge-transport rate is required to ensure high electron collection efficiency. These two properties are the defining characteristics of an ideal photoelectrode [4].

Insight into the factors limiting DSC performance is gained by comparing theoretical cell efficiencies with those of current state-of-the-art cells. The power conversion efficiency (PCE) of a solar cell is given as

$$PCE = \frac{J_{sc} V_{oc} FF}{P_{in}}, \quad (1)$$

where J_{sc} is the absolute value of the current density at short circuit, V_{oc} is the photovoltage at open circuit, *FF* is the fill factor and *P*in is the incident light power density. In principle, the maximum J_{sc} of a DSC is determined by how well the absorption window of its dye overlaps the solar spectrum. Much of the shortfall is due to the poor absorption of low-energy photons by available dyes. Thickening of the photoelectrode to increase its optical density in order to improve the absorption of red and near-infrared light is unsuccessful because the film thickness comes to exceed the electron diffusion length through the nanoparticle network.

The dynamic scale of the processes involved in light to electricity conversion (**Figure 2**) shows that the initial events of electron injection and dye regeneration leading to photoinduced charge separation occur on a femto- to nanosecond or microsecond time scale [6], while the electron transport across the photoelectrode takes place within milliseconds or even seconds [7]. However, for the efficient functioning of the DSC, the diffusion length of the electron should be greater than the thickness of the photoelectrode. Electron diffusion length L_n can be expressed through electron lifetime τ_n and electron diffusion coefficient D_n as

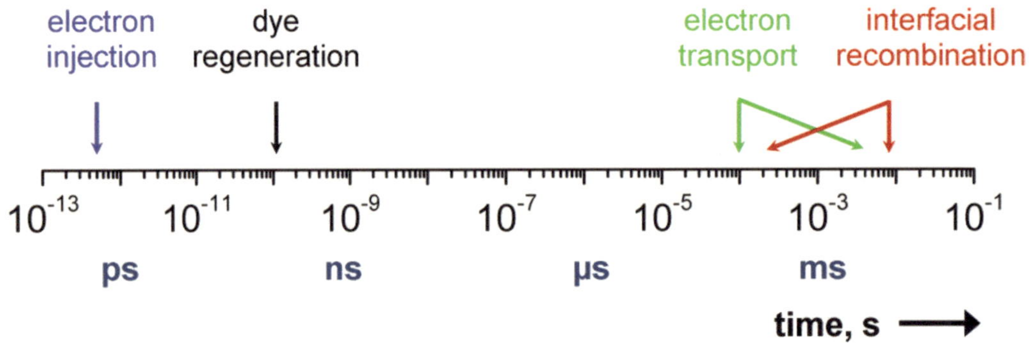

Figure 2. Dynamics competition of the processes involved in the conversion of light to electric power by typical DSCs [5].

$$L_n = \sqrt{D_n \bullet \tau_n}, \quad (2)$$

Disordered network of TiO_2 nanoparticles with numerous grain boundaries weakens electron mobility and results in slow transport and recombination of photo-excited electrons [8]. This greatly hinders the overall PCE of such devices. The mentioned problems associated with the standard architecture of the photoelectrode oblige a search for more effective nanostructured photoelectrode materials and morphologies [9–11].

3. ZnO nanowire synthesis and application in DSCs

The relatively low transport resistance of transparent high-mobility materials, such as ZnO compared to that of anatase TiO_2 nanoparticles, is one of the major advantages for the application in charge injection and collection for DSCs [12–15]. ZnO, which has an energy bandgap similar to that of the ordinarily used TiO_2 but possesses higher electron mobility, is an alternative photoelectrode material for DSCs [16]. Another advantage of ZnO is the diversity of 1D structures. Structurally, ZnO has several fast-growth directions; therefore, various morphologies can be easily obtained.

ZnO nanowires were widely considered as an alternative photoelectrode material (**Figure 3**), since they address many of the mentioned problems; however, up to date the highest power conversion efficiency of ZnO-based DSC reported [18] is still lower than TiO_2-based DSC. The main reason may be that best DSC dyes are designed for TiO_2 photoelectrode, and since ZnO is less stable especially in acid, therefore there is still no efficient dye available for ZnO anode [19, 20]. It was found that electron injection from traditional Ru-based dye to ZnO is much slower than to TiO_2 [21]. However, much research is conducted in order to find modifications for ZnO photoelectrodes and further increase the overall PCE of the DSCs.

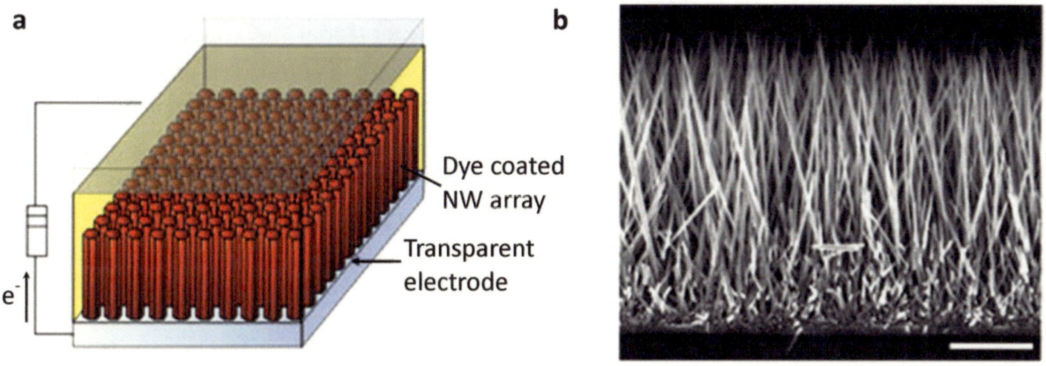

Figure 3. ZnO nanowire-based DSC: (a) a schematic of the cell; (b) SEM image of a nanowire array. Scale bar, 5 μm [17].

3.1. ZnO nanowire DSC stability

ZnO has a low chemical stability in acidic environment. After the DSC assemble, ZnO photoelectrode degrades through the carboxylic groups of the acidic dye attached to it, and thus leads to the formation of Zn^{2+}-dye agglomerates.

On the other hand, the time of dye loading is crucial for the performance of an all-ZnO-based DSC, because longer dye-loading time leads to the formation of more Zn^{2+}-dye agglomerates and shorter immersion time is insufficient for dye adsorption. For this reason, other than traditional Ru-based dyes could be used, such as porphyrin and indoline dyes [22]. Indoline dyes have been found to be a comparatively good match with ZnO because of its lower acidity and the lack of complexing agent. The CR147 dye-sensitized ZnO film is almost free from Zn^{2+}/dye agglomerations. A ZnO DSC with CR147 dye has high PCE of 6.89%, 40% higher than the cell with traditional N719 dye [23]. A metal-free organic dye D149 allows to use lower-sensitizing times, moreover a PCE of 5.14% was reported [24]. A mix of several specially engineered dyes could also be advantageous. Compared to only D149-sensitized cell, the YD2-o-C8-TBA and D149 co-sensitized ZnO DSSC with a wavelength ranging from 475–700 nm exhibited improved photon-to-current conversion efficiencies with cell PCE of 5.6% [25]. Vegetable tannin and their Fe complexes could be used for their low cost and can be obtained from renewable sources, but the efficiencies are up to 0.99% [26].

Other stability problems are associated with a liquid electrolyte used, which tends to leak in time. Device instability and the need for good device packaging have become major problems for commercial application of DSCs. This problem is common to most DSCs; liquid electrolyte could be substituted with quasi-solid-state electrolytes [27]. Metal oxide nanoparticle gelators are applied [28], ZnO nanoparticles gel [29] shows PCE of 4.17%, and the optimized DSC shows a stability of 95% on the PCE value for 150 days. The hole transport is presumably done by Grotthuss-type ion exchange mechanism.

3.2. ZnO nanowire synthesis

As it was earlier stated, competition between the transport and the recombination of photo-excited carriers is one of the main obstacles for developing higher conversion efficiency photoelectrode. To date, numerous morphologies of low-dimensional ZnO nanostructures such as nanowires, nanobelts, nanotubes and nanoflakes have been synthesized and are expected to improve the electron diffusion length in the photoelectrode.

This paragraph reviews the most important approaches of the ZnO NW synthesis with the aim of application in DSCs. In order to obtain an optimal structure for the photoelectrode of DSC, the growth of ZnO NWs should be carefully controlled. ZnO nanowires can be synthesized in a variety of methods; two main synthesis groups can be distinguished: vapour phase and solution phase synthesis.

Vapour phase ZnO NW synthesis involves temperatures typically 400–1100°C, needed to vapourize and transport the precursors. A typical reaction involves ZnO NW growth by vapour-liquid-solid mechanism (VLS) [30, 31] using catalyst particles such as Au, Pt or Ag, or by vapour solid mechanism (VS) [32] without the use of catalysts. Zn vapour is obtained by heating the mixture of ZnO and C [33] or directly from Zn powder [34]. A direct growth on Zn metallic substrate is also interesting for its simplicity [35–37]. ZnO NW growth by vapour phase synthesis typically involves a direct Zn oxidation:

$$Zn + O_2 \rightarrow ZnO \tag{3}$$

Solution phase ZnO NW synthesis is especially interesting for its low-temperature approach and high-output possibilities. Since the typical synthesis conditions apply temperatures lower than 200°C, the variety of materials can be used as substrates. In the solution phase synthesis, the ZnO NW growth process is carried out in aqueous or organic solvents from zinc salts. Normally, aqueous solutions are used and the methods are often referred to as hydrothermal growth [38]. However, other solutions such as ethanol [39] are also used. A typical ZnO NW solution phase growth involves the following reaction:

$$Zn^{2+} + 2\,OH^- \rightarrow Zn\,(OH)_2 \rightarrow ZnO + H_2O \tag{4}$$

where hexamethylenetetramine (HMTA) [40], NH_4OH [41] or NaOH [42] may be used as the hydroxide source. High aspect ratio can be achieved by controlling the preferential growth with catalytic Au nanoparticles [43] or polymers such as polyethylenimine (PEI) [44], or polyethylene glycol (PEG) [45]. ZnO NW aspect ratio can be further improved by sequential growth [46].

Comparing the vapour and liquid phase synthesis, it is important to note that the vapour phase growth of ZnO NWs uses higher temperatures; therefore for low-melting substrates an efficient post synthesis, transferring and attachment methods should be developed, but nonetheless high crystallinity can be obtained without further annealing. On the contrary, most of the solution growth methods use low temperatures, therefore NWs can be directly grown on a variety of low-melting materials and surfaces; however, the synthesis of high aspect ratio structures with high crystallinity is demanding.

Applying the mentioned methods, ZnO NWs could be synthesized in complex forms (**Figure 4**) such as nanoforest [47], tetrapods [49, 50], hierarchical nanowires [48], coral-shaped nanostructures [51], nanocactus [24], flower-like [52] and many others [53–55] by simply controlling the crystal grow direction.

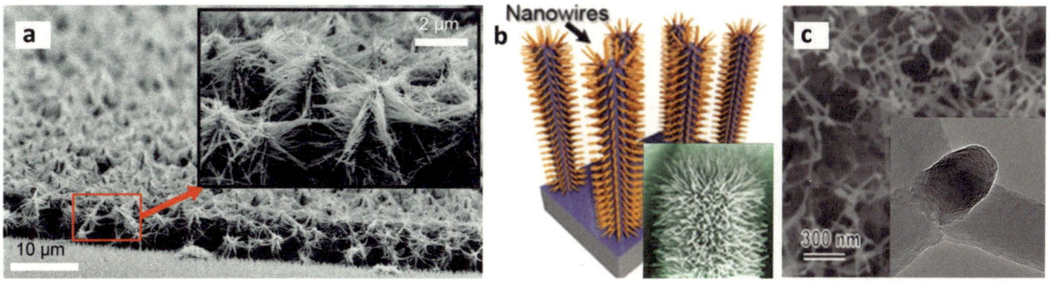

Figure 4. Complex structures of ZnO nanowires: (a) nanoforest [47]; (b) hierarchical nanowires (inset shows SEM figure) [48]; (c) tetrapods [49].

4. ZnO NW interface modifications

The advantage of fast electron transfer is often counterbalanced by faster electron recombination dynamics with the oxidized electrolyte via intrabandgap surface states which ultimately limits the PCE [15]. In order to slow down the recombination and to increase the photovoltage and photocurrent, the ZnO NW interfaces can be covered by a thin layer of various capping materials (Al_2O_3, TiO_2, ZrO_2, etc.). Alternatively, hole transfer can be modified by forming ZnO NW hybrid structures with conductive polymers, or various nanostructures such as quantum dots (QDs), nanocrystals (NCs) and carbon nanomaterials.

4.1. ZnO NW core-shell structure in DSC

The purpose of NW coating with an insulating or semiconducting oxide layer is to diminish the recombination by forming a potential barrier on NW surface, to physically separate electrons from ions and to increase the dye adsorption on the surface. Moreover, band edge can be shifted by using higher bandgap layer to increase the open-circuit voltage. In addition to acting as an effective protection layer for the chemically unstable ZnO against the acidic dye solution, the complete coverage of the NWs with a dense layer can passivate surface traps [12]. ZnO NW core-shell structures with different metal oxides can be obtained with atomic layer deposition (ALD) technique [56, 57]. The coverage of ZnO NWs with an insulating materials results in a low-efficiency DSCs [58–60]; by contrast, ZnO NW core-shell structures with semiconductors have several advantages, especially interesting is TiO_2 shell.

There are several motives why the ZnO-TiO_2 core-shell structure is attractive. TiO_2 is chemically more stable compared to ZnO in acidic dye solutions [61]. Thus, the presence of TiO_2 shell prevents ZnO surface from being dissolved and the formation of Zn^{2+}/dye agglomerates. The TiO_2 shell can also increase the injected electrons and more dye absorption, which lead to a higher light-harvesting efficiency. Moreover, the shell reduces recombination by forming a tunnelling barrier to confine the photoinjected electrons within the core, and by passivating the recombination centres on the core surface [62]. The charge transfer is significantly improved since ZnO has much higher electron mobility compared to TiO_2. DFT ab initio study of ZnO-TiO_2 core-shell structure [63] shows that TiO_2 coating induces changes in surface states and shifts the conduction and valence band edges to higher energies; therefore, an increase in open-circuit voltage and a decrease in short-circuit current are expected.

TiO_2 shell on ZnO NWs can be formed by dip coating in Ti precursors, such as tetrabutyl titanate (TBOT) [64, 65], titanium isopropoxide (TTIP) and titanium tetrachloride ($TiCl_4$). ZnO NW coated with a thin shell layer of TiO_2 (**Figure 5**) showed an increased PCE up to 6% [66]. A comparative study [67] showed that at the same conditions TBOT-treated ZnO DSC showed the best performance with the highest PCE and short-circuit current density of 4.92% and 12.49 mA/cm^2, respectively.

In order to increase the surface area of the photoelectrode and without losing the charge-transport properties, hierarchical structures could be used. A hierarchical structure of ZnO NWs could also increase the efficiency of DSCs by providing a high area for dye absorption

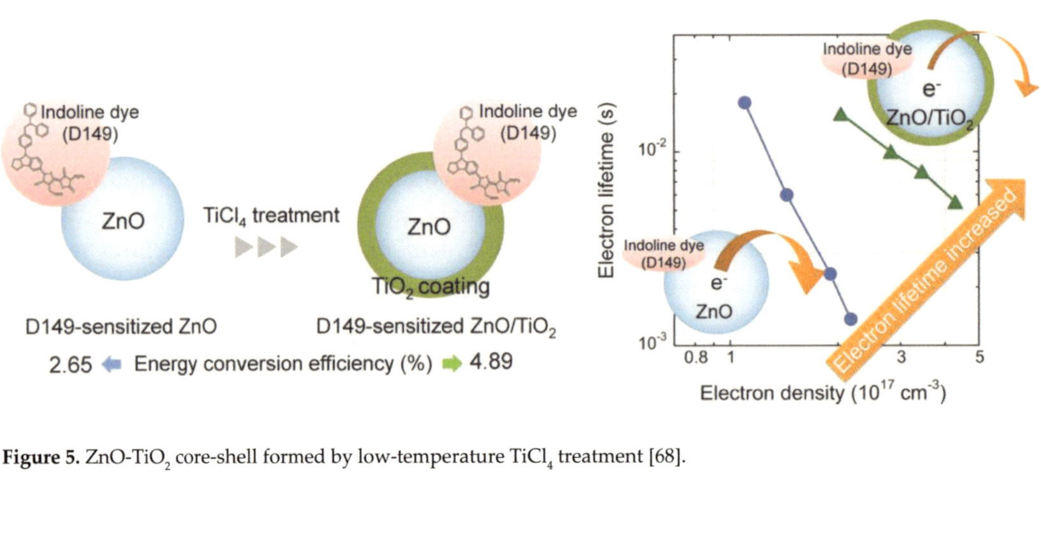

Figure 5. ZnO-TiO_2 core-shell formed by low-temperature $TiCl_4$ treatment [68].

and also effective light scattering. ZnO DSC efficiency can be doubled by decorating ZnO NW photoelectrodes with different nanoparticles of SnO_2 [69], TiO_2 [70, 71] or in situ-synthesized ZnO nanoparticles [72]. Even though the large surface area of nanoparticles empowers a high dye-loading capacity, yet the disordered network with numerous grain boundaries weakens electron mobility and results in slow transport and recombination of photo-excited electrons [8]. The application of ZnO NW with semiconductive nanoparticles has advantages of both high surface area and fast charge transport. A complex structure of hierarchical NWs improves the PCE up to five times compared to the initial NW, the reported efficiencies for nanotrees 2.63% [73], coral-shaped nanostructure gives 4.58% [51], nanocactus 5.14% [24] are encouraging.

ZnO tetrapods used with SnO_2 nanoparticles provided large roughness factors, good charge collection and tunable light-scattering properties. High PCE of 6.31% was attributed to NH_3 treatment, which was believed to create ZnO shell on SnO_2 nanoparticles [74]. A hierarchical core-shell ZnO NW with TiO_2 nanosheets resulted in an outstanding performance with a solid-state electrolyte, showing the conversion efficiencies of up to 7.46% [75].

The prepared ZnO photoelectrode can be further treated to improve the conductance and porosity parameters. Hot-press treatment of the ZnO photoelectrode was demonstrated to improve PCE of the DSC by 45% [76]. Another powerful technique, the room-temperature chemical bath deposition, was used to increase the surface area of the ZnO photoelectrode, a PCE of 5.24% was obtained [77].

4.2. ZnO NW hybrid solar-cell structures

Hybrid polymer solar cells composed of metal oxide nanostructures and polymers have attracted great interest mostly due to the good physical and chemical stability. In such solar cells, p-type donor polymers are combined with n-type acceptor ZnO nanostructures [78] among other oxides. Nanowires are one of the most attractive forms for such solar cells because they provide a direct path for charge transport, high carrier motilities and a high electron affinity necessary for charge injection from the complementary organic donor material [79]. Mostly used conducting polymer poly(3-hexylthiophene) (P3HT) composite with ZnO nanowire shows hybrid solar-cell PCE in the range of 0.2–0.5% [80–82] and can be further increased to 2% by adding dye [83]. Such low-power conversion efficiencies are associated with poor polymer infiltration and therefore low-active interfacial area between polymer and ZnO NWs. An interfacial layer of carbon nanoparticles (CNPs) can be employed to enhance the charge-transfer properties in hybrid ZnO NW solar cells. The energy-level diagram of the solar cells is shown in **Figure 6b**. The LUMO level of the CNPs (-4.00 ± 0.02 eV) is perfectly aligned with both the LUMO level of P3HT (-3.2 eV) and the conduction band edge of TiO_2 (-4.2 eV). This ensures both efficient exciton dissociation at the P3HT/CNP interface and efficient electron extraction. In such a device, ZnO-TiO_2 core-shell works as electron collector and transporter, P3HT acts as electron donor, and the CNP layer acts as electron acceptor [84].

The other possibility is to use quantum dots (QDs) for hole transfer. The use of PbS quantum dots (QDs) is promising because of the tunable, size-dependent bandgap from 0.7 to 1.3 eV [85].

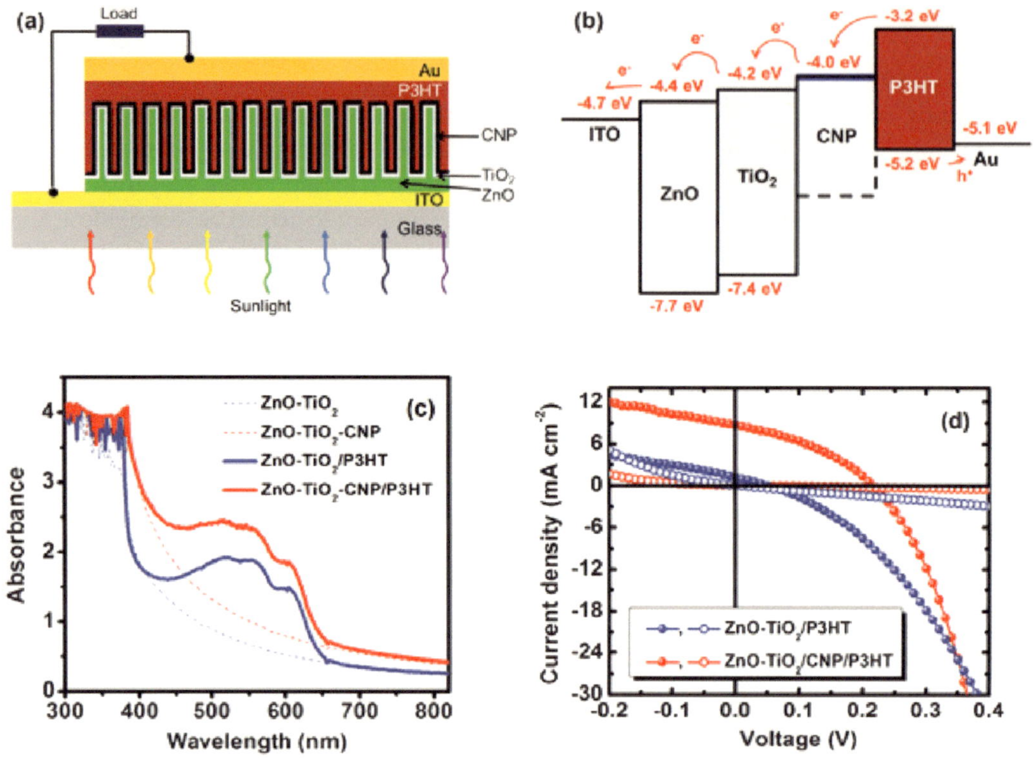

Figure 6. Hybrid solar cell with ZnO-TiO_2 core-shell, carbon nanoparticles (CNP) and conductive polymer P3HT. (a) Device structure; (b) energy-level diagram; (c) absorption spectra of the active films on ITO glass substrates; (d) J-V characteristics (empty: dark; filled: illuminated) of the solar cells with and without CNP [84].

Pairing ZnO with PbS QDs for photovoltaic applications shows high stability in ambient atmosphere. An advantage of using ZnO NWs instead of ZnO planar or nanoparticle layer is demonstrated [86]. Vertically oriented ZnO NW arrays are fully infiltrated with QDs, increasing considerably the light absorption and carrier collection. Ordered interface architecture of ZnO NW arrays can decouple absorption from collection, extending the effective depletion width throughout a thick QD film.

The ZnO NW performance was tested in both earlier mentioned hole conductor configurations: conducting polymer P3HT and PbS QDs [82]. As it is shown in **Figure 7b**, band alignment is similar; however, the PCE of 4.2% and 0.5% for QDs and P3HT, respectively, was obtained. It is worth noting that graphene was used as the conductive electrode instead of the traditional ITO, which demonstrates the flexibility of application for various substrates.

ZnO NW solar-cell absorption in the near-infrared region can be enhanced by using plasmonic nanocrystals (NCs). Ag NC enhances ZnO NW solar-cell (**Figure 8**) PCE from 4.5 to 6.0% [87].

A plasmonic NC traps a photon on the basis of localized surface plasmon resonance, and generates a strong oscillating electric field (i.e. optical near field) that is localized in the vicinity of the NC. The optical near field excites a dye molecule or semiconductor more efficiently than incident far-field light, and therefore photocurrents are enhanced.

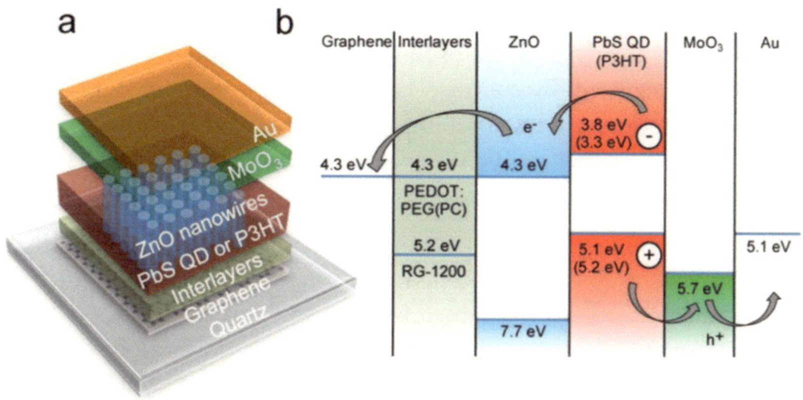

Figure 7. Hybrid ZnO NW solar cells. (a) Schematic diagram of ZnO NWs (400 nm) infiltrated and covered with PbS QDs (300 nm) or P3HT (700 nm); (b) flat-band energy-level diagram of the solar cell [82].

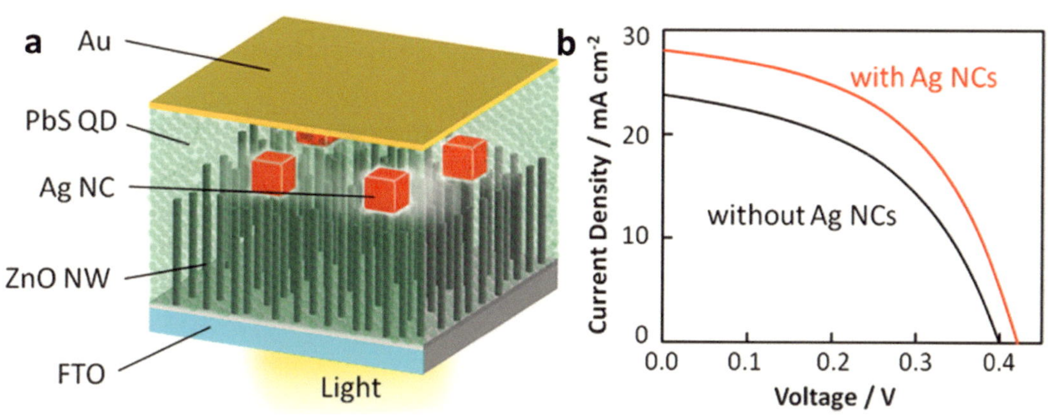

Figure 8. PbS QD/ZnO NW solar cell with embedded Ag NCs. (a) Schematic; (b) J-V characteristics without and with Ag NCs [87].

5. Conclusions

Several conclusions could be made with the proposal for the future trends:

1. First of all, since one of the main parameters for high efficiency of DSC is surface area, more attention should be paid in order to obtain ZnO NW with a structure of high surface, at the same time a virtue of high crystallinity should not be lost. Several possible methods are worth of interest: the obvious trend would be to synthesize nanowires with lower diameter, but the other ways, such as adding NP to the photoelectrode and using NW as charge delivery highway, are also interesting.

2. Non-catalytic growth methods are uncomplicated and much cheaper compared to catalytic ones. Hydrothermal methods have multiple advantages in application since they are low cost and versatile; however, it is problematic to obtain a high aspect ratio structures

with good crystallinity. Nevertheless, these methods are unbeatable for high-yield deposition. A variety of substrates can be used for hydrothermal ZnO NW growth because of the low temperatures used. By contrast, vapour phase growth of ZnO NWs uses higher temperatures, therefore for a low-melting substrates an efficient post-synthesis transferring and attachment methods should be developed; however, high crystallinity can be obtained without further annealing.

3. ZnO NW surface is decomposed in acidic environment of best-to-date dyes, therefore special dyes for ZnO should be developed or alternatively surface modification of ZnO is needed. Best-to-date results are obtained by covering the NW surface with TiO_2 or other semiconductor layer. Alternatively forming a protective layer from dielectric materials such as Al_2O_3 lowers the conductance of photoelectrode, except for an extremely thin layer.

4. Hybrid structures of ZnO NWs are primarily interesting for their high physical and chemical stability, but can also boost other properties such as charge transfer or light absorption; however, more work should be done to understand the underlying mechanisms of ZnO NW interaction with hybrid materials.

Acknowledgements

The work has received funding from the European Union's Seventh Framework programme for research and innovation under the Marie Sklodowska-Curie grant agreement No 609402 – 2020 researchers: Train to Move (T2M).

Author details

Simas Rackauskas*, Nadia Barbero, Claudia Barolo and Guido Viscardi

*Address all correspondence to: simas.rackauskas@gmail.com

Department of Chemistry, NIS Interdepartmental Centre and INSTM Reference Centre, University of Turin, Turin, Italy

References

[1] Mathew S, Yella A, Gao P, Humphry-Baker R, Curchod BFE, Ashari-Astani N, et al. Dye-sensitized solar cells with 13% efficiency achieved through the molecular engineering of porphyrin sensitizers. Nat. Chem. [Internet]. 2014 [cited 2016 Sep 16];6:242-7. Available from: http://www.nature.com/doifinder/10.1038/nchem.1861

[2] Grätzel M. The advent of mesoscopic injection solar cells. Prog. Photovoltaics Res. Appl. [Internet]. John Wiley & Sons, Ltd.; 2006 [cited 2016 Sep 16];14:429-42. Available from: http://doi.wiley.com/10.1002/pip.712

[3] O'regan B, Grätzel M. A low-cost, high-efficiency solar cell based on dye-sensitized colloidal TiO_2 films. Nature. Nature Publishing Group; 1991;353:737-40.

[4] Maçaira J, Andrade L, Mendes A. Review on nanostructured photoelectrodes for next generation dye-sensitized solar cells. Renew. Sustain. Energy Rev. 2013;27:334-49.

[5] Grätzel M. Solar energy conversion by dye-sensitized photovoltaic cells. Inorg. Chem. American Chemical Society. 2005:6841-51.

[6] Bach U, Tachibana Y, Moser J-E, Haque SA, Durrant JR, Grätzel M, et al. Charge separation in solid-state dye-sensitized heterojunction solar cells. J. Am. Chem. Soc. American Chemical Society. 1999;121:7445-6.

[7] Grätzel M. Solar energy conversion by dye-sensitized photovoltaic cells. Inorg. Chem. American Chemical Society. 2005;44:6841-51.

[8] Ye M, Wen X, Wang M, Iocozzia J, Zhang N, Lin C, et al. Recent advances in dye-sensitized solar cells: From photoanodes, sensitizers and electrolytes to counter electrodes. Mater. Today. 2015;18:155-62.

[9] Colodrero S, Forneli A, López-López C, Pellejà L, Míguez H, Palomares E. Efficient transparent thin dye solar cells based on highly porous 1D photonic crystals. Adv. Funct. Mater. [Internet]. WILEY-VCH Verlag; 2012 [cited 2016 Sep 16];22:1303-10. Available from: http://doi.wiley.com/10.1002/adfm.201102159

[10] Wang J, Lin Z. Dye-sensitized TiO_2 nanotube solar cells: Rational structural and surface engineering on TiO_2 nanotubes. Chem. - An Asian J. [Internet]. WILEY-VCH Verlag; 2012 [cited 2016 Sep 16];7:2754-62. Available from: http://doi.wiley.com/10.1002/asia.201200349

[11] Feng X, Zhu K, Frank AJ, Grimes C a, Mallouk TE. Rapid charge transport in dye-sensitized solar cells made from vertically aligned single-crystal rutile TiO_2 nanowires. Angew. Chem. Int. Ed. [Internet]. WILEY-VCH Verlag; 2012 [cited 2016 Sep 16];124:2781-4. Available from: http://doi.wiley.com/10.1002/ange.201108076

[12] Tétreault N, Arsenault É, Heiniger LP, Soheilnia N, Brillet J, Moehl T, et al. High-efficiency dye-sensitized solar cell with three-dimensional photoanode. Nano Lett. [Internet]. American Chemical Society; 2011 [cited 2016 Sep 16];11:4579-84. Available from: http://pubs.acs.org/doi/abs/10.1021/nl201792r

[13] He C, Zheng Z, Tang H, Zhao L, Lu F. Electrochemical impedance spectroscopy characterization of electron transport and recombination in ZnO nanorod dye-sensitized solar cells. J. Phys. Chem. C [Internet]. American Chemical Society; 2009 [cited 2016 Sep 16];113:10322-5. Available from: http://pubs.acs.org/doi/abs/10.1021/jp902523c

[14] Martinson ABF, Goes MS, Fabregat-Santiago F, Bisquert J, Pellin MJ, Hupp JT. Electron transport in dye-sensitized solar cells based on ZnO nanotubes: Evidence for highly efficient charge collection and exceptionally rapid dynamics. J. Phys. Chem. A [Internet]. American Chemical Society; 2009 [cited 2016 Sep 16];113:4015-21. Available from: http://pubs.acs.org/doi/abs/10.1021/jp810406q

[15] Martinson ABF, McGarrah JE, Parpia MOK, Hupp JT. Dynamics of charge transport and recombination in ZnO nanorod array dye-sensitized solar cells. Phys. Chem. Chem. Phys. [Internet]. The Royal Society of Chemistry; 2006 [cited 2016 Sep 16];8:4655-9. Available from: http://www.ncbi.nlm.nih.gov/pubmed/17047762

[16] Zhang Q, Dandeneau CS, Zhou X, Cao G. ZnO nanostructures for dye-sensitized solar cells. Adv. Mater. [Internet]. WILEY-VCH Verlag; 2009 [cited 2016 Sep 16];21: 4087-108. Available from: http://onlinelibrary.wiley.com/doi/10.1002/adma.200803827/abstract%0Ahttp://onlinelibrary.wiley.com/doi/10.1002/adma.200803827/abstract;jsessionid=28F0008F38AEB7367E640662E9AD0B66.f01t04%0Ahttp://onlinelibrary.wiley.com/store/10.1002/adma.200803827/asset/

[17] Law M, Greene LE, Johnson JC, Saykally R, Yang P. Nanowire dye-sensitized solar cells. Nat. Mater. [Internet]. Nature Publishing Group; 2005 [cited 2016 Jul 19];4:455-9. Available from: http://www.nature.com/doifinder/10.1038/nmat1387

[18] Memarian N, Concina I, Braga A, Rozati SM, Vomiero A, Sberveglieri G. Hierarchically assembled ZnO nanocrystallites for high-efficiency dye-sensitized solar cells. Angew. Chem. Int. Ed. [Internet]. WILEY-VCH Verlag; 2011 [cited 2016 Sep 16];50:12321-5. Available from: http://doi.wiley.com/10.1002/anie.201104605

[19] Anta JA, Guille E. ZnO-based dye-sensitized solar cells. J. Phys. Chem. C [Internet]. American Chemical Society; 2012 [cited 2016 Sep 16];116:11413-25. Available from: http://dx.doi.org/10.1021/jp3010025

[20] Jiang WT, Wu C Te, Sung YH, Wu JJ. Room-temperature fast construction of outperformed ZnO nanoarchitectures on nanowire-array templates for dye-sensitized solar cells. ACS Appl. Mater. Interfaces [Internet]. American Chemical Society; 2013 [cited 2016 Sep 16];5:911-7. Available from: http://pubs.acs.org/doi/abs/10.1021/am302570r

[21] Borgwardt M, Wilke M, Kampen T, Mähl S, Xiao M, Spiccia L, et al. Charge transfer dynamics at dye-sensitized ZnO and TiO_2 interfaces studied by ultrafast XUV photoelectron spectroscopy. Sci. Rep. [Internet]. Nature Publishing Group; 2016 [cited 2016 Dec 20];6:24422. Available from: http://www.nature.com/articles/srep24422

[22] Syu Y-K, Tingare Y, Yeh C-Y, Yang J-S, Wu J-J. Panchromatic engineering for efficient zinc oxide flexible dye-sensitized solar cells using porphyrin and indoline dyes. RSC Adv. [Internet]. The Royal Society of Chemistry; 2016 [cited 2016 Sep 30];6:59273-9. Available from: http://xlink.rsc.org/?DOI=C6RA09262D

[23] Lee CP, Chen PW, Li CT, Huang YJ, Li SR, Chang LY, et al. ZnO double layer film with a novel organic sensitizer as an efficient photoelectrode for dye-sensitized solar cells. J. Power Sources. 2016;325:209-19.

[24] Jiang WT, Wu C Te, Sung YH, Wu JJ. Room-temperature fast construction of outperformed ZnO nanoarchitectures on nanowire-array templates for dye-sensitized solar cells. ACS Appl. Mater. Interfaces [Internet]. American Chemical Society; 2013 [cited 2016 Sep 12];5:911-7. Available from: http://pubs.acs.org/doi/abs/10.1021/am302570r

[25] Syu Y-K, Tingare Y, Yeh C-Y, Yang J-S, Wu J-J. Panchromatic engineering for efficient zinc oxide flexible dye-sensitized solar cells using porphyrin and indoline dyes. RSC Adv. [Internet]. Royal Society of Chemistry; 2016 [cited 2016 Sep 12];6:59273-9. Available from: http://xlink.rsc.org/?DOI=C6RA09262D

[26] Çakar S, Güy N, Özacar M, Findik F. Investigation of vegetable tannins and their iron complex dyes for dye sensitized solar cell applications. Electrochim. Acta. Elsevier Ltd. 2016;209:407-22.

[27] Tao L, Huo Z, Ding Y, Li Y, Dai S, Wang L, et al. High-efficiency and stable quasi-solid-state dye-sensitized solar cell based on low molecular mass organogelator electrolyte. J. Mater. Chem. A [Internet]. The Royal Society of Chemistry; 2015 [cited 2016 Dec 20];3:2344-52. Available from: http://xlink.rsc.org/?DOI=C4TA06188H

[28] Dkhissi Y, Huang F, Cheng Y, Caruso RA. Quasi-solid-state dye-sensitized solar cells on plastic substrates. J. Phys. Chem. C [Internet]. American Chemical Society; 2014 [cited 2016 Dec 20];118:16366-74. Available from: http://pubs.acs.org/doi/abs/10.1021/jp408844q

[29] Chang WC, Sie SY, Yu WC, Lin LY, Yu YJ. Preparation of nano-composite gel electrolytes with metal oxide additives for dye-sensitized solar cells. Electrochim. Acta. 2016;212:333-42.

[30] Yang P, Yan H, Mao S, Russo R, Johnson J, Saykally R, et al. Controlled growth of ZnO nanowires and their optical properties. Adv. Funct. Mater. 2002;12:323-31.

[31] Huang MH. Room-temperature ultraviolet nanowire nanolasers. Science (80-.). [Internet]. 2001 [cited 2016 Dec 19];292:1897-9. Available from: http://www.sciencemag.org/cgi/doi/10.1126/science.1060367

[32] Pan ZW, Dai ZR, Wang ZL. Nanobelts of semiconducting oxides. Science (80-.). 2001; 291:1947-9.

[33] Zhang Z, Wang SJ, Yu T, Wu T. Controlling the growth mechanism of ZnO nanowires by selecting catalysts. J. Phys. Chem. C [Internet]. American Chemical Society; 2007 [cited 2014 May 28];111:17500-5. Available from: http://dx.doi.org/10.1021/jp075296a

[34] Sekar A, Kim SH, Umar A, Hahn YB. Catalyst-free synthesis of ZnO nanowires on Si by oxidation of Zn powders. J. Cryst. Growth. 2005;277:471-8.

[35] Rackauskas S, Nasibulin AG, Jiang H, Tian Y, Kleshch VI, Sainio J, et al. A novel method for metal oxide nanowire synthesis. Nanotechnology [Internet]. 2010;20:165603. Available from: http://stacks.iop.org/0957-4484/20/165603

[36] Kleshch VI, Rackauskas S, Nasibulin AG, Kauppinen EI, Obraztsova ED, Obraztsov AN. Field emission properties of metal oxide nanowires. J. Nanoelectron. Optoelectron. [Internet]. 2012 [cited 2016 Nov 28];7:35-40. Available from: http://openurl.ingenta.com/content/xref?genre=article&issn=1555-130X&volume=7&issue=1&spage=35

[37] Rackauskas S, Nasibulin AG, Jiang H, Tian Y, Statkute G, Shandakov SD, et al. Mechanistic investigation of ZnO nanowire growth. Appl. Phys. Lett. [Internet]. 2009 [cited 2013 Oct 3];95:183114. Available from: http://link.aip.org/link/APPLAB/v95/i18/p183114/s1&Agg=doi

[38] Nam D, Park J, Park S, Min Y, Noh Y, Lee D. High-density hydrothermal growth of zinc-oxide nanowires using printed resistive heater. Mater. Lett. 2015.

[39] Wu C, Shen L, Yu H, Huang Q, Zhang YC. Synthesis of Sn-doped ZnO nanorods and their photocatalytic properties. Mater. Res. Bull. 2011;46:1107-12.

[40] Parize R, Garnier J, Chaix-Pluchery O, Verrier C, Appert E, Consonni V. Effects of hexamethylenetetramine on the nucleation and radial growth of ZnO nanowires by chemical bath deposition. J. Phys. Chem. C [Internet]. 2016 [cited 2016 Dec 19];120:5242-50. Available from: http://pubs.acs.org/doi/abs/10.1021/acs.jpcc.6b00479

[41] Cho J, Salleh N, Blanco C, Yang S, Lee C-J, Kim Y-W, et al. Novel synthetic methodology for controlling the orientation of zinc oxide nanowires grown on silicon oxide substrates. Nanoscale [Internet]. 2014 [cited 2016 Dec 19];6:3861-7. Available from: http://www.ncbi.nlm.nih.gov/pubmed/24584438

[42] Xu S, Wang ZL. One-dimensional ZnO nanostructures: Solution growth and functional properties. Nano Res. 2011;4:1013-98.

[43] Gu C, Shanshan L, Huang J, Shi C, Liu J. Preferential growth of long ZnO nanowires and its application in gas sensor. Sens. Actuators, B Chem. 2013;177:453-9.

[44] Hu X, Shen X, Li H, Masuda Y, Ohji T, Kato K. Polyethylenimine-assisted synthesis of transparent ZnO nanowhiskers at ambient temperatures. Thin Solid Films. 2014;558:134-9.

[45] Panigrahy B, Aslam M, Misra DS, Bahadur D. Polymer-mediated shape-selective synthesis of ZnO nanostructures using a single-step aqueous approach. CrystEngComm [Internet]. Royal Society of Chemistry; 2009 [cited 2016 Dec 19];11:1920. Available from: http://xlink.rsc.org/?DOI=b904833m

[46] Caicedo N, Thomann J-S, Leturcq R, Lenoble D. Aspect ratio improvement of ZnO nanowires grown in liquid phase by using step-by-step sequential growth. CrystEngComm [Internet]. The Royal Society of Chemistry; 2016 [cited 2016 Dec 19];18:5502-11. Available from: http://xlink.rsc.org/?DOI=C6CE00904B

[47] Ko SH, Lee D, Kang HW, Nam KH, Yeo JY, Hong SJ, et al. Nanoforest of hydrothermally grown hierarchical ZnO nanowires for a high efficiency dye-sensitized solar cell. Nano Lett. [Internet]. American Chemical Society; 2011 [cited 2016 May 17];11:666-71. Available from: http://dx.doi.org/10.1021/nl1037962

[48] Bielinski AR, Kazyak E, Schlepütz CM, Jung HJ, Wood KN, Dasgupta NP. Hierarchical ZnO nanowire growth with tunable orientations on versatile substrates using atomic

layer deposition seeding. Chem. Mater. [Internet]. 2015 [cited 2016 Dec 18];27:4799-807. Available from: http://pubs.acs.org/doi/abs/10.1021/acs.chemmater.5b01624

[49] Rackauskas S, Klimova O, Jiang H, Nikitenko A, Chernenko KA, Shandakov SD, et al. A novel method for continuous synthesis of ZnO tetrapods. J. Phys. Chem. C [Internet]. American Chemical Society; 2015; Available from: http://dx.doi.org/10.1021/acs.jpcc.5b03702

[50] Rackauskas S, Mustonen K, Järvinen T, Mattila M, Klimova O, Jiang H, et al. Synthesis of ZnO tetrapods for flexible and transparent UV sensors. Nanotechnology [Internet]. 2012;23:95502. Available from: http://www.ncbi.nlm.nih.gov/pubmed/22327417

[51] Lamberti A, Gazia R, Sacco A, Bianco S, Quaglio M, Chiodoni A, et al. Coral-shaped ZnO nanostructures for dye-sensitized solar cell photoanodes. Prog. Photovoltaics Res. Appl. [Internet]. 2014 [cited 2016 Jul 18];22:189-97. Available from: http://doi.wiley.com/10.1002/pip.2251

[52] Pugliese D, Bella F, Cauda V, Lamberti A, Sacco A, Tresso E, et al. A chemometric approach for the sensitization procedure of ZnO flowerlike microstructures for dye-sensitized solar cells. ACS Appl. Mater. Interfaces [Internet]. American Chemical Society; 2013 [cited 2016 Dec 19];5:11288-95. Available from: http://pubs.acs.org/doi/abs/10.1021/am403527m

[53] Kuo S-Y, Yang J-F, Lai F-I. Improved dye-sensitized solar cell with a ZnO nanotree photoanode by hydrothermal method. Nanoscale Res. Lett. [Internet]. 2014 [cited 2016 Dec 19];9:206. Available from: http://www.pubmedcentral.nih.gov/articlerender.fcgi?artid=4022993&tool=pmcentrez&rendertype=abstract

[54] Zhang S, Chen H-S, Matras-Postolek K, Yang P. ZnO nanoflowers with single crystal structure towards enhanced gas sensing and photocatalysis. Phys. Chem. Chem. Phys. [Internet]. The Royal Society of Chemistry; 2015 [cited 2016 Dec 19];17:30300-6. Available from: http://xlink.rsc.org/?DOI=C5CP04860E

[55] Gao R, Tian J, Liang Z, Zhang Q, Wang L, Cao G. Nanorod-nanosheet hierarchically structured ZnO crystals on zinc foil as flexible photoanodes for dye-sensitized solar cells. Nanoscale [Internet]. Royal Society of Chemistry; 2013 [cited 2016 Dec 19];5:1894-901. Available from: http://dx.doi.org/10.1039/C2NR33599A%5Cnhttp://pubs.rsc.org/en/content/articlepdf/2013/nr/c2nr33599a

[56] Law M, Greene LE, Radenovic A, Kuykendall T, Liphardt J, Yang P. ZnO-Al_2O_3 and ZnO-TiO_2 core-shell nanowire dye-sensitized solar cells. J. Phys. Chem. B. American Chemical Society ; 2006;110:22652-63.

[57] Zhang Y, Lu H-L, Wang T, Ren Q-H, Chen H-Y, Zhang H, et al. Photoluminescence enhancement of ZnO nanowire arrays by atomic layer deposition of ZrO_2 layers and thermal annealing. Phys. Chem. Chem. Phys. [Internet]. The Royal Society of Chemistry; 2016 [cited 2016 Dec 19];18:16377-85. Available from: http://xlink.rsc.org/?DOI=C6CP01900E

[58] Bu IYY. Effects of post annealing temperature on formation of alumina core shell on zinc oxide nanowire: optoelectronic properties and dye sensitized solar cell applications. J. Mater. Sci. Mater. Electron. [Internet]. Springer US; 2014 [cited 2016 Dec 19];25:4458-65. Available from: http://link.springer.com/10.1007/s10854-014-2188-4

[59] Li TC, Góes MS, Fabregat-Santiago F, Bisquert J, Bueno PR, Prasittichai C, et al. Surface passivation of nanoporous TiO_2 via atomic layer deposition of ZrO_2 for solid-state dye-sensitized solar cell applications. J. Phys. Chem. C [Internet]. American Chemical Society; 2009 [cited 2016 Jul 19];113:18385-90. Available from: http://pubs.acs.org/doi/abs/10.1021/jp906573w

[60] Loh L, Dunn S. Recent progress in ZnO-based nanostructured ceramics in solar cell applications. J. Nanosci. Nanotechnol. American Scientific Publishers; 2012;12:8215-30.

[61] Greene LE, Yuhas BD, Law M, Zitoun D, Yang P. Solution-grown zinc oxide nanowires [Internet]. Inorg. Chem. 2006 [cited 2016 Jul 20]. p. 7535-43. Available from: http://pubs.acs.org/doi/abs/10.1021/ic0601900

[62] Law M, Greene LE, Radenovic A, Kuykendall T, Liphardt J, Yang P. ZnO-Al_2O_3 and ZnO-TiO_2 core-shell nanowire dye-sensitized solar cells. J. Phys. Chem. B [Internet]. American Chemical Society; 2006 [cited 2016 Jul 19];110:22652-63. Available from: http://pubs.acs.org/doi/abs/10.1021/jp0648644

[63] Pazoki M, Nafari N, Taghavinia N, O'Regan B, Gratzel M, Hagfeldt A, et al. Ab initio study of electronic effects in the ZnO/TiO_2 core/shell interface: application in dye sensitized solar cells. RSC Adv. [Internet]. The Royal Society of Chemistry; 2014 [cited 2016 Jul 19];4:301-7. Available from: http://xlink.rsc.org/?DOI=C3RA45973J

[64] Feng Y, Ji X, Duan J, Zhu J, Jiang J, Ding H, et al. Synthesis of ZnO@TiO_2 core–shell long nanowire arrays and their application on dye-sensitized solar cells. J. Solid State Chem. [Internet]. 2012 [cited 2016 Jul 19];190:303-8. Available from: http://dx.doi.org/10.1016/j.jssc.2012.02.026

[65] Ji X, Liu W, Leng Y, Wang A, Ji X, Liu W, et al. Facile synthesis of ZnO@TiO_2 core-shell nanorod thin films for dye-sensitized solar cells. J. Nanomater. [Internet]. Hindawi Publishing Corporation; 2015 [cited 2016 Jul 19];2015:1-5. Available from: http://www.hindawi.com/journals/jnm/2015/647089/

[66] Gan X, Li X, Gao X, Zhuge F, Yu W. ZnO nanowire/TiO_2 nanoparticle photoanodes prepared by the ultrasonic irradiation assisted dip-coating method. Thin Solid Films. 2010;518:4809-12.

[67] Wang D, Wang W, Ma X, Zhang C, Zhao J, Zhang X. Comparative study on the influence of TiO_2 precursors on ZnO-based dye-sensitized solar cells. Ind. Eng. Chem. Res. [Internet]. American Chemical Society; 2015 [cited 2016 Dec 20];54:12639-45. Available from: http://pubs.acs.org/doi/abs/10.1021/acs.iecr.5b03627

[68] Sakai N, Miyasaka T, Murakami TN. Efficiency enhancement of ZnO-based dye-sensitized solar cells by low-temperature $TiCl_4$ treatment and dye optimization. J. Phys.

Chem. C [Internet]. American Chemical Society; 2013 [cited 2016 Dec 6];117:10949-56. Available from: http://pubs.acs.org/doi/abs/10.1021/jp401106u

[69] Chen W, Qiu Y, Yang S. A new ZnO nanotetrapods/SnO_2 nanoparticles composite photoanode for high efficiency flexible dye-sensitized solar cells. Phys. Chem. Chem. Phys. [Internet]. The Royal Society of Chemistry; 2010 [cited 2016 Dec 18];12:9494-501. Available from: http://xlink.rsc.org/?DOI=c000584c

[70] Liu J, Wei A, Zhao Y, Lin K, Luo F. Dye-sensitized solar cells based on ZnO nanoflowers and TiO_2 nanoparticles composite photoanodes. J. Mater. Sci. Mater. Electron. [Internet]. Springer US; 2014 [cited 2016 Dec 19];25:1122-6. Available from: http://link.springer.com/10.1007/s10854-013-1698-9

[71] Yang J, Lin Y, Meng Y, Lin Y. Oriented ZnO nanotubes arrays decorated with TiO_2 nanoparticles for dye-sensitized solar cell applications. Appl. Phys. A Mater. Sci. Process. 2014;114:1195-9.

[72] Gao R, Cui Y, Liu X, Wang L, Cao G. A ZnO nanorod/nanoparticle hierarchical structure synthesized through a facile in situ method for dye-sensitized solar cells. J. Mater. Chem. A [Internet]. The Royal Society of Chemistry; 2014 [cited 2016 Dec 18];2:4765. Available from: http://www.scopus.com/inward/record.url?eid=2-s2.0-84897711028&partnerID=tZOtx3y1

[73] Ko SH, Lee D, Kang HW, Nam KH, Yeo JY, Hong SJ, et al. Nanoforest of hydrothermally grown hierarchical ZnO nanowires for a high efficiency dye-sensitized solar cell. Nano Lett. [Internet]. American Chemical Society; 2011 [cited 2016 Dec 19];11:666-71. Available from: http://pubs.acs.org/doi/abs/10.1021/nl1037962

[74] Chen W, Qiu Y, Zhong Y, Wong KS, Yang S. High-efficiency dye-sensitized solar cells based on the composite photoanodes of SnO_2 nanoparticles/ZnO nanotetrapods. J. Phys. Chem. A. American Chemical Society; 2010;114:3127-38.

[75] Miles DO, Lee CS, Cameron PJ, Mattia D, Kim JH. Hierarchical growth of TiO_2 nanosheets on anodic ZnO nanowires for high efficiency dye-sensitized solar cells. J. Power Sources. 2016;325:365-74.

[76] Haque Choudhury MS, Kishi N, Soga T. Hot-compress: A new postdeposition treatment for ZnO-based flexible dye-sensitized solar cells. Mater. Res. Bull. [Internet]. 2016 [cited 2016 Dec 18];80:135-8. Available from: https://www.scopus.com/inward/record.uri?eid=2-s2.0-84964068595&partnerID=40&md5=50ce3ec4267cd7360fa5c41dc52d7463

[77] Chang G-J, Lin S-Y, Wu J-J. Room-temperature chemical integration of ZnO nanoarchitectures on plastic substrates for flexible dye-sensitized solar cells. Nanoscale [Internet]. The Royal Society of Chemistry; 2014 [cited 2016 Dec 19];6:1329-34. Available from: http://www.ncbi.nlm.nih.gov/pubmed/24362771

[78] Oosterhout SD, Wienk MM, van Bavel SS, Thiedmann R, Koster LJA, Gilot J, et al. The effect of three-dimensional morphology on the efficiency of hybrid polymer solar cells.

Nat. Mater. [Internet]. Nature Publishing Group; 2009 [cited 2016 Dec 20];8:818-24. Available from: http://dx.doi.org/10.1038/nmat2533

[79] Briseno AL, Holcombe TW, Boukai AI, Garnett EC, Shelton SW, Fréchet JJM, et al. Oligo- and polythiophene/ZnO hybrid nanowire solar cells. Nano Lett. [Internet]. American Chemical Society; 2010 [cited 2016 Dec 20];10:334-40. Available from: http://pubs.acs.org/doi/abs/10.1021/nl9036752

[80] Ravirajan P, Peiró AM, Nazeeruddin MK, Graetzel M, Bradley DDC, Durrant JR, et al. Hybrid polymer/zinc oxide photovoltaic devices with vertically oriented ZnO nanorods and an amphiphilic molecular interface layer. J. Phys. Chem. B. American Chemical Society ; 2006;110:7635-9.

[81] Olson DC, Lee YJ, White MS, Kopidakis N, Shaheen SE, Ginley DS, et al. Effect of polymer processing on the performance of poly(3-hexylthiophene)/ZnO nanorod photovoltaic devices. J. Phys. Chem. B [Internet]. American Chemical Society ; 2007 [cited 2016 Dec 20];111:16640-5. Available from: http://pubs.acs.org/cgi-bin/doilookup/?10.1021/jp0757816

[82] Park H, Chang S, Jean J, Cheng JJ, Araujo PT, Wang M, et al. Graphene cathode-based ZnO nanowire hybrid solar cells. Nano Lett. [Internet]. American Chemical Society; 2013 [cited 2016 Dec 20];13:233-9. Available from: http://pubs.acs.org/doi/abs/10.1021/nl303920b

[83] Thitima R, Patcharee C, Takashi S, Susumu Y. Efficient electron transfers in ZnO nanorod arrays with N719 dye for hybrid solar cells. Solid. State. Electron. 2009;53:176-80.

[84] Li Y, Li S, Jin L, Murowchick JB, Peng Z. Carbon nanoparticles as an interfacial layer between TiO_2-coated ZnO nanorod arrays and conjugated polymers for high-photocurrent hybrid solar cells. RSC Adv. [Internet]. Royal Society of Chemistry; 2013 [cited 2016 Dec 20];3:16308. Available from: http://xlink.rsc.org/?DOI=c3ra42614a

[85] Moreels I, Lambert K, Smeets D, De Muynck D, Nollet T, Martins JC, et al. Size-dependent optical properties of colloidal PbS quantum dots. ACS Nano [Internet]. American Chemical Society; 2009 [cited 2016 Dec 21];3:3023-30. Available from: http://pubs.acs.org/doi/abs/10.1021/nn900863a

[86] Jean J, Chang S, Brown PR, Cheng JJ, Rekemeyer PH, Bawendi MG, et al. ZnO nanowire arrays for enhanced photocurrent in PbS quantum dot solar cells. Adv. Mater. [Internet]. WILEY-VCH Verlag; 2013 [cited 2016 Dec 20];25:2790-6. Available from: http://doi.wiley.com/10.1002/adma.201204192

[87] Kawawaki T, Wang H, Kubo T, Saito K, Nakazaki J, Segawa H, et al. Efficiency enhancement of PbS quantum Dot/ZnO nanowire bulk-heterojunction solar cells by plasmonic silver nanocubes. ACS Nano [Internet]. American Chemical Society; 2015 [cited 2016 Dec 20];9:4165-72. Available from: http://pubs.acs.org/doi/abs/10.1021/acsnano.5b00321

Chapter 5

Solution-Processable Nanowire Field-Effect Transistors

Maxim Shkunov, Grigorios Rigas and
Marios Constantinou

Additional information is available at the end of the chapter

http://dx.doi.org/10.5772/intechopen.68800

Abstract

Solution-processable single-crystalline inorganic semiconducting nanowires are excellent building blocks for printable electronics requiring high performance of semiconducting components. Excellent charge carrier mobilities of crystalline nanowires combined with solvent-based nanowire processing open up possibilities for low-cost nanowire electronics targeting a variety of applications ranging from flexible circuits to chemical and biological sensors. Nanowire field-effect transistors are key devices for most of such applications. Recent developments in controllable nanowire positioning and orientation on the substrates and electrical property selection provide the necessary technological breakthroughs enabling the fabrications of reproducible nanowire transistors. In this chapter, we discuss the nanowire assembly methods and high-spatial-resolution scanning probe microscopy techniques towards scalable fabrication of high-performance printable nanowire field-effect transistors.

Keywords: inorganic semiconducting nanowires, printed electronics, field-effect transistors, solution processability

1. Introduction

The growing interest for printed electronics using functional nanomaterials is due to a plethora of conventional and novel applications that can be generated at a fraction of the cost as compared to conventional microelectronic device fabrication involving multiple photolithographic and etching steps. These applications include flexible circuits [1], biological and chemical sensors [2–4], photovoltaic and piezo energy harvesting [5–7], low-dissipation power electronics [8], terahertz detectors [9, 10], memory storage devices [11], artificial e-skin sensors [12], flexible displays [13] and optoelectronic devices [14] including lasers [15, 16]. One-dimensional inorganic semiconductors in the form of single-crystalline nanowires

(NWs) hold most of the promise to act as a backbone for these applications. This is due to nanowires' excellent charge transport properties, the availability of both direct and indirect bandgap semiconducting materials, high aspect ratios allowing nanowires to bridge typical device electrode gaps, solution processability, chemical stability, room temperature deposition capabilities and compatibility with a variety of substrates including both rigid and flexible substrates such as plastics and paper. Separation of nanowire growth from the device assembly allows to employ low-temperature, bottom-up device fabrication approach utilising solution processing of nanowire devices. General nanowire device fabrication process flow is illustrated in **Figure 1**, where synthesised nanowires are dispersed in organic solvents for form functional inks, and then solution-based deposition is used to assemble nanowires on substrates, ideally with a good degree of nanowire orientation and positioning in respect to device electrodes. Density of nanowires after solvent-based disposition is defined by the type of the device and can vary significantly from monolayer coverage for transistor devices to very dense layers intended as energy storage electrodes in, for example, Li-ion batteries [17]. The devices are then finished by depositing the required counter-electrodes, dielectric

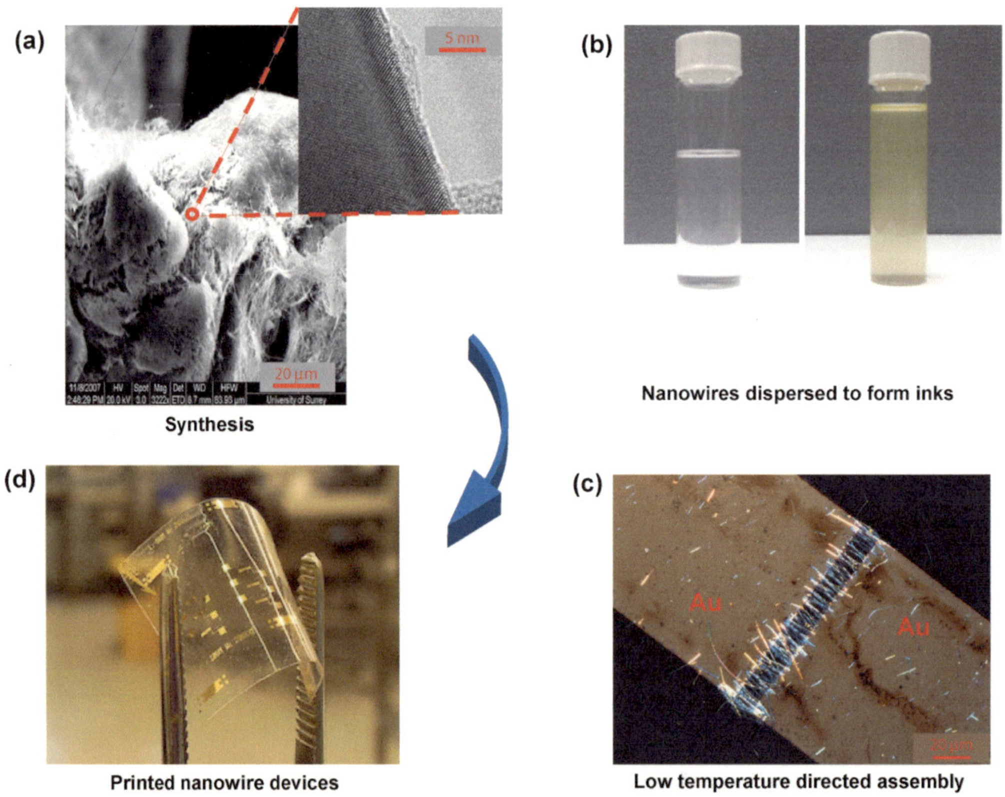

Figure 1. A concept of printable nanowire electronics on flexible substrates, where NW powder (a) is dispersed to form semiconducting nanowire inks (b), followed by aligned deposition of nanowires between device electrodes (c), and finished device can be fabricated at low-temperature process compatible with plastic substrates (d). (b) shows example of ZnO NW ink on the left and Si NW ink on the right.

layers and conducting tracks and also encapsulation layers, if devices need protection from the environment.

Reproducible positioning and alignment of nanowires with respect to device electrodes have been one of the key challenges in solution-processed nanowire electronics and has attracted a significant research effort. Advances toward nanowire assembly from liquid-based 'inks' onto device substrates are discussed in Section 3, paying attention to positioning, alignment and orientation. Recent progress toward selective nanowire deposition using dielectrophoresis is discussed in relation to separation of nanowires with different physical dimensions, such as lengths, and also selection based on the their electrical properties and types of nanowires.

Nanowire surface states play an important role in the corresponding device electrical characteristics, and large surface-to-volume ratio of nanowires opens, on one hand, enormous opportunities for surface receptor functionalisation toward chemical and biological sensors and, on the other hand, can lead to undesirable electrical hysteresis effects due to the presence of surface trap states that need to be eliminated. Further discussion of nanowire surface functionalisation and passivation methods towards stable, hysteresis-free device operation is described in Section 4.

Multiple nanowire channel devices raise particular characterisation challenges associated with both good contact formation between the nanowires and electrodes and also the need to identify nanowires which fully connect the device electrodes and those that do not participate in the conduction mechanism. Characterisation methods based on conducting scanning probe microscopy techniques are discussed in Section 5 in relation to conducting and semiconducting nanowires to probe their positioning, morphology, conductivities and other physical parameters.

2. Nanowire transistor device structures

Field-effect transistors (FETs) are key building blocks for modern electronics as they provide a variety of functions essential for analogue and digital electronics. Printable electronic fabrication approach is highly beneficial for a variety of senor transistor applications, including chemical and biological sensing and tensile and optical sensing, where extremely dense transistor integration is not required, and in many cases, sensors need to be positioned some distance apart from each other to allow, for example, transistor functionalisation with chemical or biological receptors.

Traditional microelectronic fabrication involves mainly top-down processes, such as masking; etching; stripping; ion implantation, with multiple photolithographic steps; and high-temperature annealing (often around 900°C), resulting in high amount of chemical waste required for the processing and very significant amount of energy use.

Printable electronics is based around the concept of bottom-up fabrications with additive manufacturing steps aimed at building the devices by functional inks printing and deposition techniques involving consecutive printing of conducting, semiconducting, and dielectric layers and components. These functional inks thus need to contain nanoparticles and/or material providing the necessary properties. Examples include semiconducting nanowire inks, metal nanoparticle inks and polymer dielectric inks. Printable nanowire FETs can be fabricated by various deposition techniques, with specifics related to the particular conductor, semiconductor, or dielectric materials. Metal nanoparticle ink deposition for nanowire FET electrodes, for example, can be performed by ink-jet printing [18], whereas popular gate dielectric deposition for NW FETs is by spin-coating polymer insulator materials [19]. Semiconductor nanowire transistor channel plays a defining role in the FET operation, so more attention will be devoted to the NW assembly as discussed in the next section.

FET device structures that are fully compatible with solution-based printing fabrication belong to thin-film transistor family and are represented by bottom-gate FETs and top-gate FETs. **Figure 2** is showing schematics for both types of NW FETs.

Top- and bottom-gate NW FET structures each have particular advantages: Bottom-gate structure is ideal for sensing applications, where NW channel area is fully exposed on the top to chemical or biological analytes. Additionally, bottom-gate design offers straightforward device prototyping when using commercially available oxidised silicon (Si/SiO_2) wafers, which can be directly used as the substrate, incorporating conducting common gate, highly doped Si and gate dielectric SiO_2. Such substrates are also highly amenable for ink-jet printing for electrode deposition.

Top-gate transistor architecture allows better gate control over nanowire channel due to wrap-around gate dielectric geometry, naturally occurring after the dielectric deposition. Importantly, top-gate FETs demonstrate improved environmental stability due to nanowire channel encapsulation provided by the gate dielectric together with metal gate electrode covering the channel, thus protecting nanowires from atmospheric exposure [20].

Main challenges in printable NW FET fabrication (Section 3) are due to deposition and controllable alignment of nanowires, which need to bridge the device electrodes; reproducible number of nanowires in FETs; and solvent-based deposition and alignment compatible with large-scale printing.

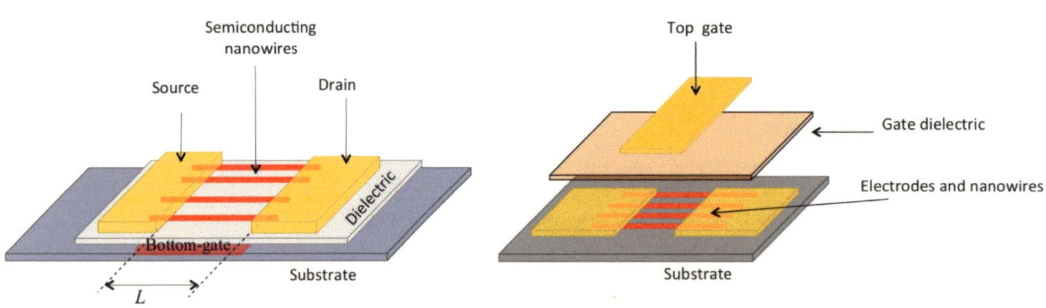

Figure 2. Schematics of bottom-gate (a) and top-gate (b) nanowire FETs.

Surface states are always present in nanomaterials due to their high surface area, and they play an important role in nanowire device performance. Thus, interface properties of nanowires need to be controlled to eliminate unwanted electrical hysteresis effects as discussed in Section 4.

3. Nanowire assembly

The bottom-up fabrication approach for nanowire transistors relies on the availability of grown nanowire materials and also on the efficient assembly methods for nanowire solvent-based deposition. The NW synthesis has been demonstrated using various methods, including supercritical fluid-liquid-solid growth (SFLS) for Si and Ge NWs [21], vapour liquid solid (VLS) growth for Si nanowires and the laser ablation for SnO_2 nanowires [22]. For the preparation of the NW formulations, the as-synthesised NWs are typically dispersed in various organic solvents using low-power sonication process for short periods of time to gently remove nanowires from the growth substrates and to promote homogeneous distribution of nanowire materials in the solvent.

Nanowire deposition and alignment on device substrates represent the next fabrication step. Several self-assembly techniques have been developed in the past for transferring the NWs onto the microelectrode structures to form the conducting channels of the FET devices. The main challenges in self-assembly are the precise alignment and positioning of the nanowires on predefined electrode areas and the purification and selection of nanowires with desired electrical properties. Most of the reported assembly techniques, apart from dielectrophoresis (DEP) provide no or very limited control of NW lateral alignment and no electrical properties selection. In this section, various NW deposition methods are discussed, and more attention is paid to the DEP process enabling electrical and morphological properties' selectivity.

3.1. Nanowire alignment techniques

A number of techniques have been developed for the alignment and deposition of NWs on the substrates, including Langmuir-Blodgett (LB), blown-bubble films (BBF), flow-directed assembly, electrostatic interactions, mechanical shear forces and electric field-assisted assembly with dielectrophoresis.

The Langmuir-Blodgett deposition is a very powerful technique for the alignment of high-density arrays of parallel nanostructures over large areas with a precise spacing between nanoparticles from μm down to nm scale. The method was first reported for barium chromate nanorods [23], with inspiration taken from centuries-old technique for timber-floating on rivers [24]. The technique has been adapted and developed to assemble a variety of nanowires from metallic silver NWs [25] to semiconducting nanowires, including silicon [26]. In the LB technique, the surfactant-wrapped NWs (e.g. 1-octadecylamine) are dispersed on the water surface of a Langmuir-Blodgett trough. Due to the impartial surfactant's solubility in water, the NWs are floating on the surface of the trough. Then, computer-controlled barriers, positioned at the edges of the LB trough, slowly compress the NWs to a higher density at the surface. The density of the monolayer, which depends on NW-to-NW stacking, is mainly limited by the amount of the suspended NWs, and their direction of alignment is dictated by the barrier orientation of the LB trough. Barrier compression causes NWs to reorient and to align parallel to the trough barrier edge, forming a highly dense monolayer. Then the NW layer can be transferred to a substrate, such as silicon wafer; however, some degree of misalignment can occur during the transfer of nanowire from the surface of the trough onto the solid substrate. This LB technique is compatible with large-area electronic fabrication; however, it is difficult to control the NW alignment yield and reproducibility of the alignment on top of the FET source-drain electrode structures [25, 26], and no lateral nanowire alignment is typically obtained.

The blown-bubble film assembly technique is able to provide uniformly aligned and controlled density of NW by utilising a bubble expansion process [27]. The surface of nanomaterials is chemically functionalised, and then, nanowires are mixed with epoxy resin. For example, Si NWs are modified using 5,6-epoxyhexyltriethoxysilane, dispersed into tetrahydrofuran and then homogenously mixed with a known mass of epoxy resin. Afterward, a bubble is expanded from the epoxy-nanomaterial viscous formulation at a controlled direction and speed, and then bubble surface is touched by a substrate to transfer the nanowires resulting in well-defined nanomaterial-incorporated thin films. The alignment of the nanomaterials is mainly attributed to the shear stress present in the epoxy fluid during the bubble expansion

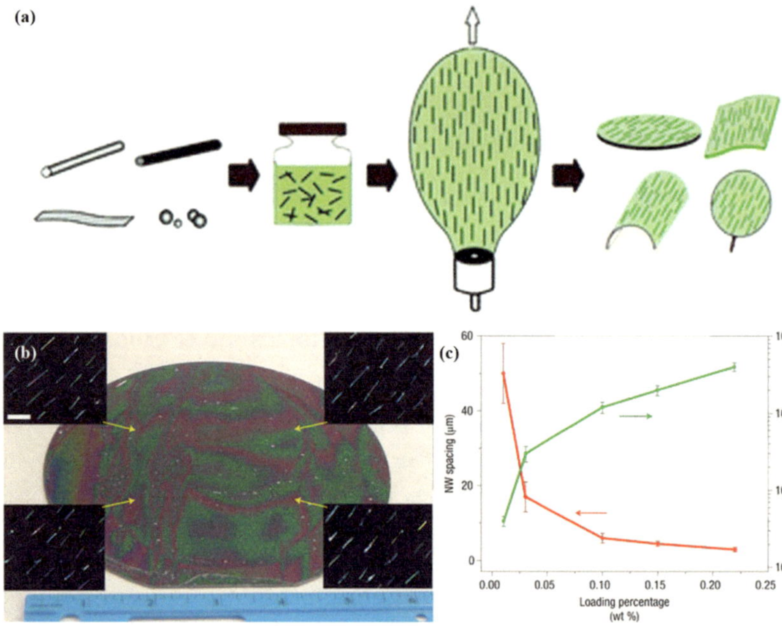

Figure 3. (a) Schematic of the BBF technique. The nanomaterials are dispersed in polymer solution, the solution is expanded as a bubble using a die, and then the uniformly aligned nanomaterials are transferred to different substrates including plastics. (b) Optical image of 6 inch Si wafer with aligned Si NWs (scale bar, 10 μm). (c) Density vs. concentration curve, indicating the density can be controlled by loading a percentage of nanomaterials [27]. © The Royal Society of Chemistry 2008.

process (**Figure 3a**). The viscosity of the homogeneous solution is the key parameter for the uniform distribution of nanomaterials in the resulting bubble films (**Figure 3b–c**). This technique provides a dense and uniform NW distribution suitable for large-area applications, and it is applicable to both rigid and flexible substrates. Although, it requires a large sample volume of NWs, the residual epoxy resin needs to be removed in order to define electrical contacts for NW FET applications, and it is difficult to control the number of the aligned NWs across the source-drain FET electrodes [27]

Highly aligned and dense nanowire packing was demonstrated by flow-directed assembly of nanowires in a microfluidic channel structure formed between poly(dimethylsiloxane) (PDMS) mould and a flat substrate (Si/SiO_2 modified with amine functionality). Assembled arrays of NWs, formed by passing the suspended NWs through the microfluidic channels, were aligned along the flow direction. The shear flow near the surface of the substrate immobilised the NWs and aligned them along the flow. At high flow rates, larger shear forces were produced and led to better alignment. This technique provided a controlled NW density and a flexibility to meet complex device configurations. However, the alignment process was attainable only for NWs with diameter <15 nm, it required the fabrication of microfluidic channels, it was challenging to scale up the process to large wafers, it was difficult to control the number of the aligned NWs across the source-drain FET electrodes and no lateral nanowire alignment was achievable [28, 29].

The electrostatic interaction method for nanowire assembly was developed by Heo et al. [30], and it relied on the electrostatic interactions of the positively charged surfaces of the semiconducting nanomaterials with the negatively charged surface of the substrate. Si NWs were functionalised with amine-terminated aminopropyltriethoxysilane (APTS) self-assembled monolayer (SAM) and well dispersed in aqueous solution. Then Si NWs were deposited on negatively charged substrates (i.e. SiO_2) with hydrophobic octadecyltrichlorosilane patterns. Therefore, the positively charged NWs were attracted and aligned by the negatively charged APTS surface regions of the substrate. This technique was able to prepare arbitrarily shaped patterns using conventional microfabrication facilities; nonetheless it required the removal of the APTS SAM on the Si NWs with buffer oxide etching process.

A nanowire contact printing method was developed by et al. [31] which utilises the shear forces generated by mechanically sliding two solid surfaces, giving a high degree of order of assembled NWs. In this method two substrates are needed, namely, the donor and the receiver substrate. The donor substrate is covered with a dense "lawn" of NWs, while the receiver is coated with lithographically patterned resist. The method is based on a contact printing process by sliding the surfaces of the two substrates against each other under gentle pressure (**Figure 4a**). Also, lubricant is used to minimise the NW-NW mechanical friction and their uncontrolled breakage during the sliding process. During printing, NWs are transferred from the donor substrate to the receiver substrate by the van der Waals interactions, resulting in the direct alignment of a dense monolayer array of NWs (**Figure 4b–e**). This technique offers the ability to transfer monolayer arrays of NWs on both rigid and flexible substrates with a controllable density. The disadvantage of this technique is the lack of control of the number of the aligned NWs across the source-drain FET electrodes, and NWs tend to break during the transfer process [29, 31].

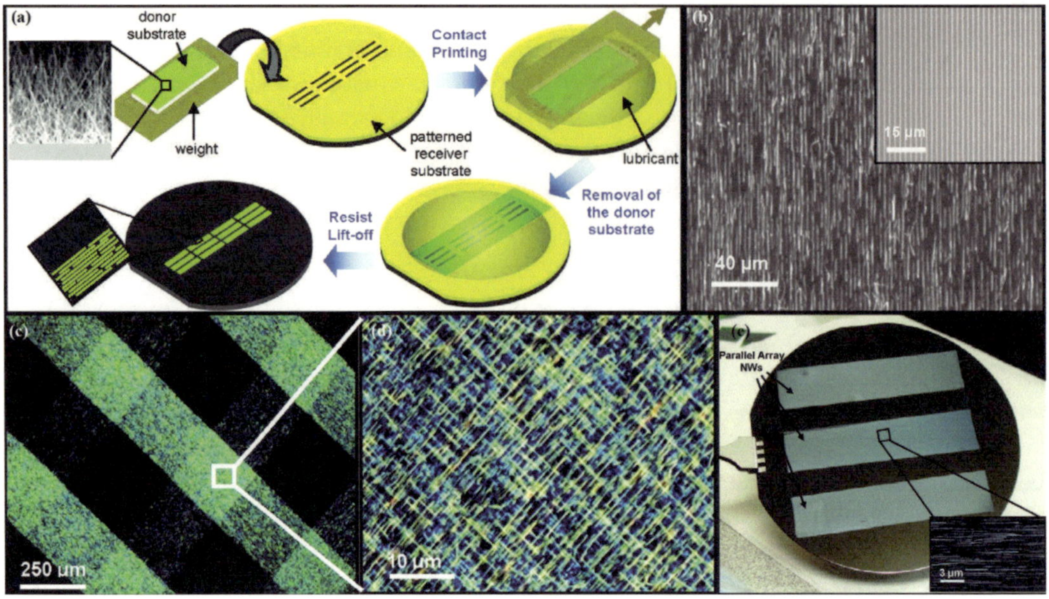

Figure 4. (a) A schematic illustration of the contact printing method. (b) Optical image of the assembly of Ge NWs on a Si/SiO_2 substrate. (c–d) Optical images of a cross assembly printing of Si NWs. (e) Printed Ge NWs on a 4 inch Si/SiO_2 wafer, including an inset SEM image of the printed NWs [31]. © 2008 American Chemical Society.

The electric field-assisted assembly known as dielectrophoresis (DEP) is currently the most promising nanowire assembly technique for the self-alignment of nanomaterials. During a DEP process, a polarisable nanoparticle, suspended in a liquid, is subjected to a non-uniform alternating electric field (AC), and its charges separate and accumulate at the surface, forming a dipole, which experiences the force dependent on the gradient of the electric field resulting in NW self-assembly across the electrode gap [32, 33]. The working principle of DEP is based on the balancing of NW-NW interaction and dielectrophoretic and opposing hydrodynamic drag forces. When the NW "ink" (nanowire dispersion) flows over the electrode gap with applied AC electric field, it gets polarised and attracted to the regions with the strongest electric field gradient, resulting in nanowire deposition across the electrode gap by means of dielectrophoretic forces (F_{DEP}), as illustrated in **Figure 5a**. The deposition of NWs across the FET source-drain electrodes mainly depends on the difference in the dielectric properties (ε) of the nanomaterial and the medium, based on the Clausius-Mossotti factor (CMF) and the electric field gradient. As opposed to previously mentioned nanowire assembly techniques, the DEP method requires very small volumes of NW dispersion (a few µL per DEP alignment), and it offers a precise, dense and controllable deposition of NW monolayer arrays on predefined electrode structures with a high degree of orientation control as shown in **Figure 5b–d**. The NW density can be controlled by tailoring the NW concentration in the formulation and the applied DEP signal voltage [33, 34].

3.2. Self-assembly and purification of nanowires

Some of the key challenges in solution-processed NW applications are the simultaneous purification, selection, and alignment of NWs on the desired substrate areas, which have attracted

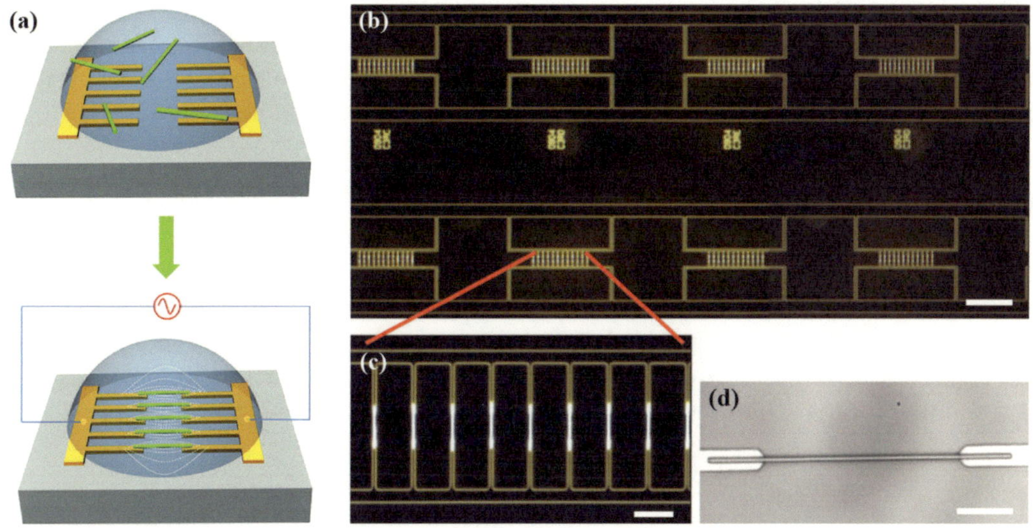

Figure 5. (a) Graphical illustration of dielectrophoretic assembly [1]. © 2012 American Chemical Society. (b) Optical image of NW assembly on eight multi-finger electrode arrays of total 96 sites (scale bar, 200 μm). (c) Precision alignment of NWs on a single sense array (scale bar, 20 μm). (d) Deep ultraviolet image of a NW aligned on a single electrode (scale bar, 4 μm) [34]. © 2010 Macmillan Publishers Limited.

a significant research effort. Despite the substantial progress in NW assembly demonstrated with LB, BBF, flow-directed assembly, electrostatic interactions and mechanical shear forces, the simultaneous precise deposition of NWs with respect to device electrodes and the selective deposition of NWs based on morphological or electrical properties received little attention, and NWs were typically deposited as-synthesised, which affected device reproducibility, performance and reliability.

The first precise orientation positioning, selective deposition and separation of single Si NWs based on their electrical quality from a mixture of other types of nanowires were demonstrated using DEP with capillary assembly by Collet et al. [35]. The combination of both DEP and capillary assembly involved dragging the liquid meniscus of the NW solution on top of the substrate, with a controlled velocity and temperature which led to the trapping of single NWs by the DEP forces on predefined locations coupled with the capillary force acting to align the nanowires. The DEP force trapped the NWs and also adjusted their orientation at the surface of the electrodes followed by the liquid evaporation. A mixture of monodispersed Si, ZnO and InAs nanowires dispersed in isopropanol was used in this study.

The precise orientation of the NWs was achieved by employing stronger electronic polarisation along the long NW axis as compared to the short axis under the application of the AC electric field. In the case of Si NWs, the optimum capture frequency was obtained at 50 kHz and for InAs and ZnO were 500 kHz and 2 MHz, respectively. The authors argued that each NW exhibits its own specific signature based on their electrical properties, which leads to the assembly of different nanostructures at different frequencies. As a proof of concept, the authors provided separation and precise alignment of single Si and InAs nanowires from NW mixture containing both types of nanowires, into alternating columns on the substrate, by switching the applied frequency between 50 and 500 kHz in each column of DEP electrodes, once the meniscus of the NW solution reached the desired set of electrodes. This work has successfully demonstrated a separation of nanowires with different electrical conductivities by examining their optimum capture frequency, which is influenced by the conductivity and level of doping.

Key progress in DEP nanowire selection and purification was demonstrated by Constantinou et al. [33] by developing a scalable one-step solution process for the selection, collection and ordered assembly of Si NW arrays (**Figure 6a**) with desired electrical properties from a poly-disperse as-synthesised NW dispersions. The method was applied to SFLS-grown Si NWs with various lengths, diameters and electrical conductivities. The technique utilised the fluidic shear forces coupled with DEP self-alignment to position high-quality NWs with the lowest level of traps across the electrode gap and to remove weakly interacting NWs and

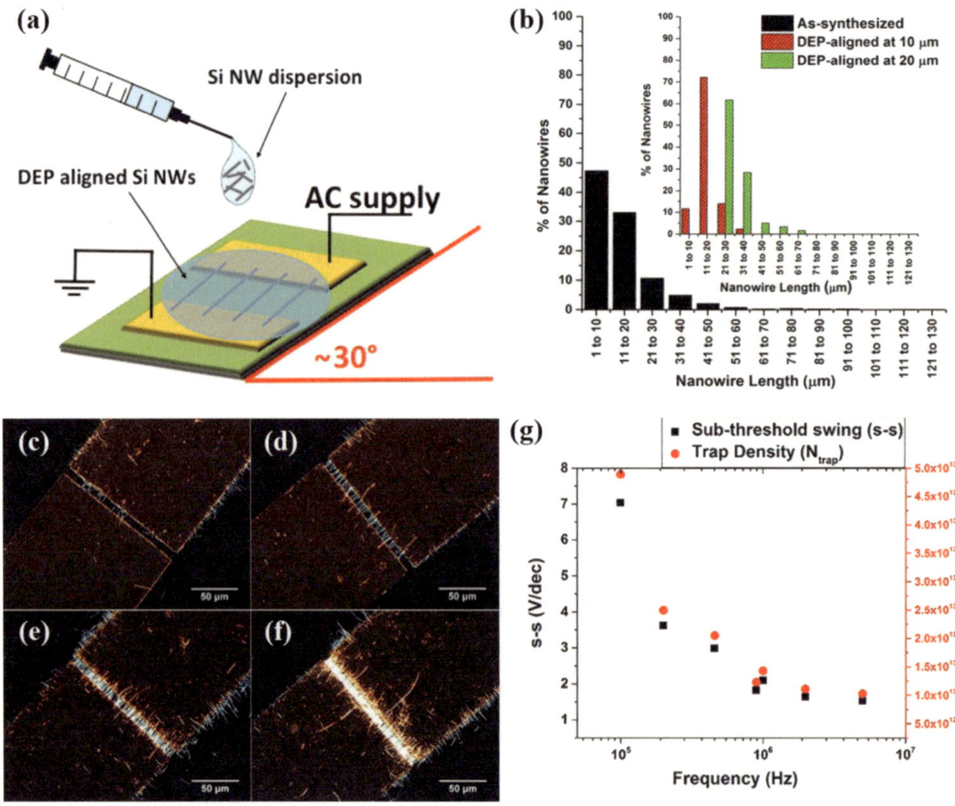

Figure 6. (a) Graphical illustration of the fluidic shear force with DEP technique used for the NW deposition and DEP alignment. The substrate was tilted at approximately 30° angle (versus horizontal plane) to provide gravity-assisted flow of NW dispersion. (b) Length histogram of NWs collected by DEP across two different electrode gaps, 10 and 20 μm. Controllable DEP alignment of SFLS Si NWs across parallel electrode bas with 10 μm spacing at (c) 5 V, (d) 10 V, (e) 15 V and (f) 20 V. (g) Subthreshold swing (s-s) and NW trap density (N_{trap}) data extracted from transfer characteristics of FETs for NWs collected from 100 kHz to 5 MHz [33]. © 2016 American Chemical Society.

impurities from the device channel region. The DEP assembly process employed a drop-cast method on an inclined substrate to provide a gravity-assisted slow flow of NW formulations perpendicular to the pre-patterned DEP electrode gap after the DEP sinusoidal signal was applied (**Figure 6c–f**). The DEP selectivity based on the NW electrical parameters was controlled by altering the frequency dependence of the DEP force. The DEP method coupled with fluidic shear forces offered a high degree of flexibility for NW positioning with NWs being oriented along the channel and perpendicular to the electrode edges. Importantly, the DEP also served as a nanowire length selection tool by providing the strongest interactions for the NWs with lengths at least the same size as the electrode gap or slightly longer opening the possibility for a sequential separation of NWs by length (**Figure 6b**). Main advancement of the method was in systematical examination of the electrical response of the Si NWs aligned at various DEP signal frequencies. It was highlighted that the F_{DEP} is frequency dependent and the level of polarisability is related to the conductivity of the nanostructure, allowing the selection of highly conducting NWs by altering the applied DEP frequency. Thus, lower frequency range collects NWs with various qualities including poor quality that may contain defects and associated traps, while high frequencies induce the collection of only the highest quality, low-defect semiconducting Si NWs, with higher conductivity. The electrical properties of the NWs aligned at various frequencies were compared using FET devices by extracting both subthreshold swing (s-s) and trap density (Ntrap) data as shown in **Figure 6g** and via conducting atomic force microscopy analysis. As a proof of concept, high-performance Si NW FETs with excellent performance were demonstrated with high FET on-currents in the mA range and low subthreshold swing (s-s) and trap density.

The proposed methodology enables the effective selection and assembly of heterogeneous nanowires of any type and provides a printed electronic fabrication approach that can be scaled up for industrial applications.

4. Surface modification approaches for nanowire FETs

The undesirable electrical hysteresis effect in NW FETs is one of the main sources for performance instabilities, which affects the reproducibility and potential applications of NW FETs such as sensors, circuits elements and logic gates, that causes a shift in the threshold voltage (V_{TH}) during normal device operation under gate bias stress. In most FET applications, besides memory devices, even a small degree of hysteresis is unwanted.

The hysteresis behaviour mainly originates from electron or hole trapping/detrapping at the semiconductor/insulator interface due to dielectric surface functionalities and adsorbed small molecules (e.g. oxygen and water). Typically such interface trapping is very strongly pronounced in bottom-gate NW FETs with SiO_2 gate dielectric [36], where silanol groups trap charges at the interfacial layer between the semiconductor material and the gate dielectric [37]. In NW FETs, the hysteresis also originates from the NW surface traps, and it is observed in moisture-free environments and ambient air conditions, highlighting the demand for NW passivation. In this section, we describe NW surface passivation methods with an oxide shell for minimising hysteresis effects.

4.1. Silicon nanowire surface passivation methods

The majority of the NW surface passivation methods include high-temperature annealing which is not suitable for plastic electronic applications. However, successful passivation results for Chemical Vapour Deposition grown Si NWs have been demonstrated by Kawashima et al. [38] and Wang et al. [39]. Both approaches focused on the synthesis of core-shell Si NWs with the core acting as a channel and the shell acting as a gate dielectric.

Kawashima et al. [38] investigated the growth of a 24-nm thermal oxide layer on the surface of Si NWs (forming a core-shell structure) using rapid thermal processing in an oxygen atmosphere at 1100°C for 2–6 min. The authors compared the I–V characteristics of bare and core-shell Si NW FETs, although hysteresis appeared in both cases with a much smaller effect on the core-shell Si NW FETs than the bare Si NW FETs. The results proved that the hysteresis is drastically reduced by the formation of the oxide shell around the Si NW core, effectively passivating the Si NW surface and preventing the adsorption of oxygen and moisture which cause hysteresis. In a similar approach by Wang et al. [39], a silicon nanowire core was covered with a 10-nm-thick SiO_2 shell at lower annealing temperature (700°C) but for a much longer time duration (4 h), also resulting in near-zero hysteresis in devices.

A different approach [40] demonstrated thermal oxidation using ozone for forming a good quality 5-nm-thick SiO_2 shell layer obtained at 600°C processing temperature. However, such high-temperature annealing is not fully suitable for printed electronic applications, and a low-temperature process is still required.

A very efficient method for nanowire passivation that results in a dramatic reduction of surface trap states on Si NWs was demonstrated by Constantinou et al. [41]. This method is ideally suited for nanomaterials intended for printed electronic applications, as it is quick, fully solution based and conducted at room temperature and normal pressure. The passivation phenomenon was directly linked to the interfacial mild oxidation effect induced by the processing solvent.

Dimethylformamide (DMF) solvent treatment was used to directly reduce the hysteresis effect in solution-processed SFLS-grown Si NW FET devices [41]. The choice of solvents for the NW dispersion had a dramatic impact on the NW surface trap density and on the hysteresis effect. The undoped Si NWs were dispersed in both anisole and DMF, with DMF solvent-processed FETs exhibiting significantly low hysteresis of ~3 V, with the lowest reported hysteresis value being 0.1 V (**Figure 7a**), whereas the average hysteresis values of FETs based on nanowires dispersed in anisole were 25–32 V. Thus, DMF passivation effect of the Si NW shell resulted in a drastically suppressed NW FET hysteresis with up to 300 times (from 32 to 0.1 V) reduction as compared to anisole dispersion solvent [41]. For the investigation of the NW surface passivation, high-resolution X-ray photoelectron spectroscopy (XPS) technique was used to analyse the surface of the DMF- and anisole-treated Si NWs. It was observed that DMF-treated Si NWs showed a significant 35% increase of SiO_2 composition of NW shell around the core compared to the reference sample (dry NWs) as shown in **Figure 7b**. The formation of a SiO_2 -rich shell passivates the Si core channel area and provides stability against environmental ageing and prevents the adsorption of water molecules that are responsible for the hysteretic behaviour of the Si NW FETs (**Figure 7c**). As a further study, the DMF-treated devices were exposed to

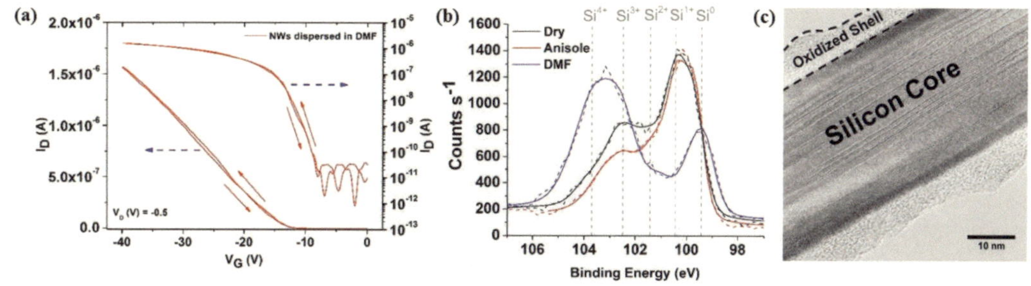

Figure 7. (a) I–V characteristics of DMF-treated Si NW FET showing a very low hysteresis of 0.1 V. (b) XPS high-resolution spectrum of the Si2p region of SFLS-grown Si NWs dispersed in anisole and DMF. (c) HRTEM image of Si NW dispersed in DMF showing a single crystal silicon core with 30 nm diameter and 5–8-nm-thick amorphous oxidised polyphenylsilane shell [41]. © 2015 American Chemical Society.

ambient conditions for 1000 h showing environmental stability, with a marginal increase of the hysteresis. In summary, the simplicity and effectiveness of solution-based mild oxidation methods offer fast passivation technological solution for printable nanowire-based electronics.

5. Scanning probe microscopy techniques for the characterisation of nanowires and devices

5.1. Introduction to conducting atomic force microscopy

Scanning probe microscopy (SPM) techniques evolved quite rapidly over the last two decades for incorporating additional modes beyond conventional topography [42, 43] and were able to provide additional characterisation information in comparison with traditional nanomaterial electron microscopy techniques such scanning electron microscopy. Driven by the introduction of novel one-dimensional conducting and semiconducting nanomaterials, SPM coupled with electrical modes became an integral part for the development and further understanding of these materials due to the unique capability of extracting critical information about the electrical properties of nanoparticles with a spatial resolution comparable to their nanoscale size [44, 45]. By combining this new level of information with conventional topography data [46], a correlation between structural and electrical properties can be established, thus enabling further optimisation of nanoparticles' growth and device integration mechanisms. In this section, the discussion will focus on recent advances in the field of conductive atomic force microscopy (c-AFM) for nanowire characterisation.

c-AFM is a current-based sensing technique for mapping the conductivity variations in samples. It utilises a conductive AFM tip which is brought in contact with a biased sample. During the scan, any current flowing through the sample and into the tip is recorded, forming a current map of the scanned surface (**Figure 8**). c-AFM can simultaneously map the topography and current distribution of a sample (**Figure 9**), thus making it ideal for evaluating defects formed during the fabrication of functional devices made with conducting or semiconducting NW networks.

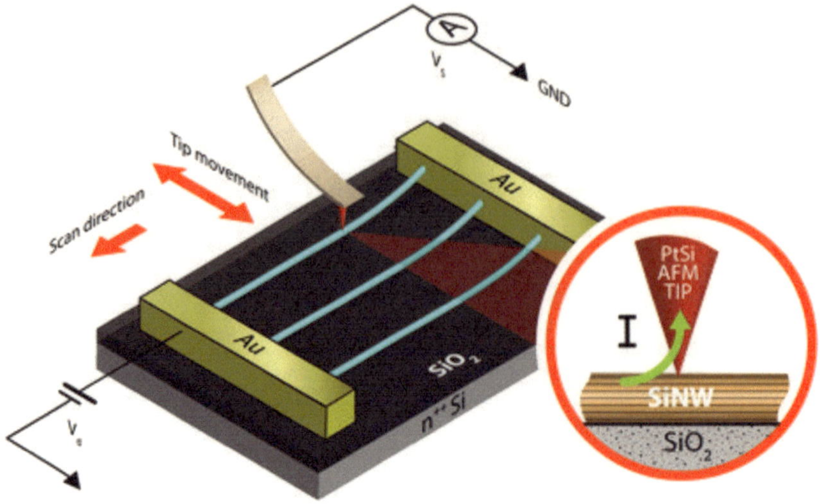

Figure 8. Schematic of c-AFM operating principle [33]. © 2016 American Chemical Society.

In the pursue of solution processed, high conductivity and transparent alternatives to the ITO electrodes, commonly used in optoelectronic devices, metallic NWs were extensively studied [47–49]. One of the main fabrication challenges was the ability of the entire NW network to conduct and thus to minimise the effective sheet resistivity. Bridging the naturally occurring gaps between isolated nanowires during deposition (**Figure 9**) was considered of high importance, and several optimisation approaches were suggested [48, 50, 51]. While conventional measurement techniques can be employed for evaluating the conductivity improvement of these methods on the macroscopic scale, nanoscopic evaluation of the nanowire-to-nanowire connections requires a special approach. One of recent examples was based on using reduced graphene oxide (rGO) as a way of connecting discontinuities in silver nanowire (AgNW) networks [51]. c-AFM was successfully used for imaging these junctions and for proving the bridging effect induced by rGO flakes.

Example of metallic NWs acting as the electrode layer represents one of the most straightforward systems to be examined with c-AFM. Solution-processed nanowire electronics relies heavily on the active layers of the devices consisting of semiconducting NWs. Thus, efficient characterisation of nanowires, including FET device components and any defects on nanowires or contact areas, is essential for the advancement of printed electronics.

5.2. c-AFM characterisation of nanowires with high spatial resolution

As discussed in Section 3, the purification of NWs based on their electrical properties is essential for their efficient incorporation in functional devices. Establishing a relation between the nanowire deposition methods and the properties of assembled nanowires, with very high spatial resolution, is key for high-performance printable nanowire-based FETs. Recently, c-AFM technique was employed for probing the electrical properties of polydispersed Si NWs, after selection processes using different dielectrophoretic signal frequencies [33]. In order to

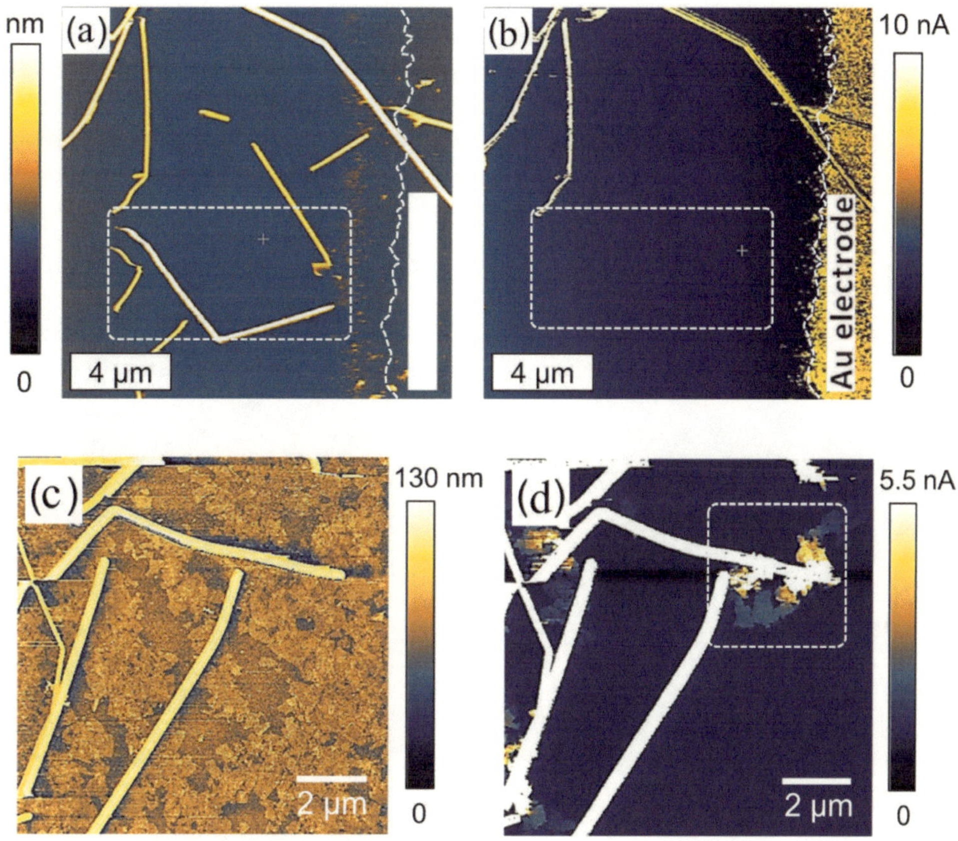

Figure 9. (a–b) c-AFM images including topography (a) and current maps (b) obtained for an AgNW network at a tip bias of 1.6 V. As deposited NWs may result in discontinuities which affect the sheet resistance of the layer. (c–d) Topography (c) and current maps (d) of isolated AgNWs, electrically connected using rGO [51]. © 2016 AIP Publishing.

investigate the conductivity of the low- and high-DEP frequency-selected NWs, c-AFM measurements were conducted with Si NW channels in real FETs. The devices were prepared with Si NWs assembled at DEP frequencies of 220 kHz and 5 MHz. The samples were biased at the drain electrode of the structure, as shown in **Figure 8**, with the conductive platinum-silicide AFM tip being in contact with the Si NWs to act as a ground electrode. Such two-terminal current measurement allows probing the NW conductivity directly, as a function of position. The material for the tip was chosen in order to reduce the contact resistance with the underlying semiconducting nanostructures, without increasing the cantilever deflection and thus allowing soft contact with the sample. By keeping the bias constant at 0.5 V and maintaining constant force applied to the AFM tip, the current maps for NW samples deposited at various DEP signal frequencies were obtained by scanning the samples in contact mode in the direction perpendicular to the NWs. Both conducting AFM current data and AFM topological data were collected at the same time allowing to differentiate the NWs deposited in the channel, but not electrically connected to the contacts, from NWs that were fully electrically connected.

Figure 10 demonstrates typical images obtained for both height (AFM mode) and current (conducting mode) measurements of Si NWs aligned at different DEP frequencies. During the measurement, a common trend was observed that NWs collected using a high, 5 MHz, DEP signal showed ~14 times higher current than the ones collected at low, 220 kHz, frequency. Current levels up to 233 pA were obtained for the NWs aligned at 5 MHz (**Figure 10b**), while peak

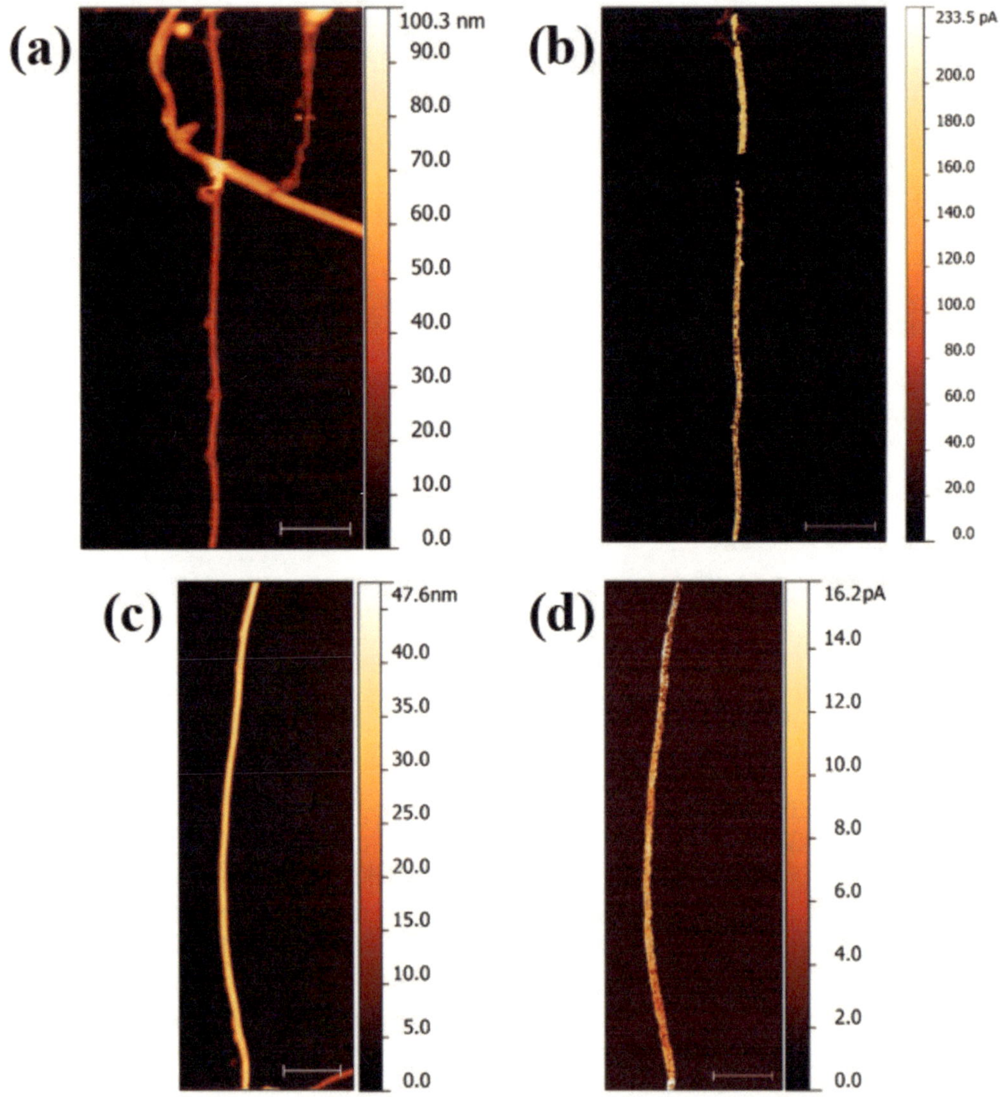

Figure 10. c-AFM data. (a) Height map of Si NWs aligned at 5 MHz and (b) the corresponding current map. Only one straight, fully electrically connected NW is visible in the image. (c) Height map of a NW aligned at 220 kHz and (d) its current map. Scale bars are 1 μm. Conducting Si NWs have similar diameters, (a) 33 nm and (c) 34 nm. All the images were obtained in the trace mode with a pixel size of 5.9 nm. NWs aligned at higher frequency showed higher peak currents up to 233 pA (b), compared to the 16-pA peak current found in the lower frequency assembled NWs (d) [33]. © 2016 American Chemical Society.

current values of 16 pA were measured for the ones assembled at 220 kHz (**Figure 10d**). These current differences between the NWs collected at high and low DEP signal frequencies were fully consistent with the transistor measurements, demonstrating that selective nanoparticle assembly at various dielectrophoretic frequencies is directly correlated with the conductivity of the NWs.

Incorporation of semiconducting NWs into functional devices requires to consider surface oxide properties as well, to enable the efficient coupling of these nanowires with the rest of the device components [52]. In Si NWs, for example, native or intentionally grown surface oxide can act as an insulator or as a poor charge transport medium with limited conduction [53]. In such cases, the uniformity of this oxide layer can be of fundamental importance for the overall functionality of the device. Conventional approaches for imaging any defects in the oxide shell, such as transmission electron microscopy, rely on the observed structural uniformity only and may lead to incorrect assumptions regarding its electrical characteristics.

In a work published in 2008, Stratakis et al. [54] compared the differences between oxide grown on planar Si and Si NWs, in an attempt to evaluate the reliability of these nanostructures for future applications. By examining several nanowires, it was found that the interface barrier for electron injection from silicon core into the oxide shell was similar to planar oxides. Trap density however was susceptible to the non-uniformities of the oxide shell, which, despite appearing to be planar under conventional topography or TEM, had localised weaknesses (**Figure 11**). These weakness areas did not appear to lower substantially the threshold for current leakage through the oxide, but they were potential reliability hazards in nanowire-based devices. This work demonstrated quite successfully that electrical characteristics in NWs may vary significantly depending on the growth conditions of the shell layer. It can also serve as an example of the level of information that can be extracted using advanced SPM techniques.

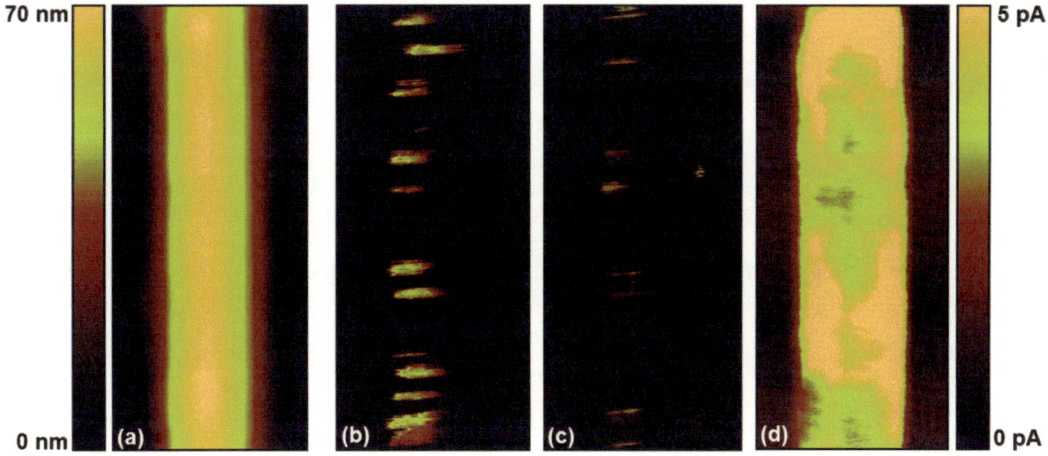

Figure 11. c-AFM current through NW oxide. (a) AFM topography image of part of a NW. Images b, c and d are current maps of the same part of the wire. At the pristine state of this wire, current was first detected at a bias of 13 V (b). Emission is inhomogeneous. Reduced average current in the second scan (c) is attributed to electron trapping in the oxide. At a higher voltage (16 V), increased current density appears from the entire surface of the nanowire (d) [54]. © 2008 American Chemical Society.

6. Conclusions

Solution-processed nanowire field-effect transistors have reached the threshold of being commercially viable due to the developments in nanowires controllable assembly. The main criticism for nanowire-based electronics was linked in the past to the lack of control of a number of nanowires between the device electrodes, thus limiting the devices reproducibility. Recent developments in the field, focusing on solvent-based reproducible and controllable positioning, orientation and placement of nanowires on predefined substrate locations, also coupled with nanowires' purification and selection methods, have provided the necessary breakthroughs to enable scalable fabrication of nanowire devices, including FETs.

For some time, the main target in NW assembly was to reach a high deposition yield and well-aligned nanostructures over large areas as an attempt to achieve superior device performances and reproducibility. However, more precise control of nanowire properties in fabricated devices can only be achieved by full integration of separation and purification processes for nanomaterials, based on their electrical and morphological properties. Currently, the only assembly technique offering all of these features is electric-field–assisted dielectrophoresis. This method is scalable and can be applied in large-area device deposition. Scanning probe microscopy characterisation techniques, such as conductive AFM, capable of simultaneous topological and conductivity measurements of nanowire active channels in electronic devices with high special resolution, have helped to optimise the nanowire properties and the corresponding selection and deposition methods.

In summary, solution-processed nanowire field-effect transistors represent a powerful technological platform with high potential for low-cost large-area electronics with applications targeting chemical, biological and tensile sensing; flexible circuits; memory; displays; and photonic devices.

Author details

Maxim Shkunov*, Grigorios Rigas and Marios Constantinou

*Address all correspondence to: m.shkunov@surrey.ac.uk

Advanced Technology Institute, Electrical and Electronic Engineering, University of Surrey, Guildford, UK

References

[1] Liu X, Long Y-Z, Liao L, Duan X, Fan Z. Large-scale integration of semiconductor nano wires for high-performance flexible electronics. ACS Nano. 2012;**6**(3):1888-1900. DOI:10.1021/nn204848r

[2] Cui Y, Wei QQ, Park HK, Lieber CM. Nanowire nanosensors for highly sensitive and selective detection of biological and chemical species. Science. Aug 2001;**293** (5533):1289-1292. DOI: 10.1126/science.1062711

[3] Patolsky F, Lieber CM. Nanowire nanosensors. Materials Today. Apr 2005;**8**(4):20-28. DOI: 10.1016/S1369-7021(05)00791-1

[4] Gao AR, et al. Enhanced sensing of nucleic acids with silicon nanowire field effect transistor biosensors. Nano Letters. Oct 2012;**12**(10):5262-5268. DOI: 10.1021/nl302476h

[5] Tian B, Kempa TJ, Lieber CM. Single nanowire photovoltaics. Chemical Society Reviews. 2009;**38**(1):16-24. DOI: 10.1039/B718703N

[6] Song JH, Zhou J, Wang ZL. Piezoelectric and semiconducting coupled power generating process of a single ZnO belt/wire. A technology for harvesting electricity from the environment. Nano Letters. Aug 2006;**6**(8):1656-1662. DOI:10.1021/nl060820v

[7] Seo M-H, et al. Versatile transfer of an ultralong and seamless nanowire array crystallized at high temperature for use in high-performance flexible devices. ACS Nano. 2017;**11**(2):1520-1529. DOI: 10.1021/acsnano.6b06842

[8] Pregl S, Heinzig A, Baraban L, Cuniberti G, Mikolajick T, Weber WM. Printable parallel arrays of Si nanowire Schottky-Barrier-FETs with tunable polarity for complementary logic, IEEE Transactions on Nanotechnology. May 2016;**15**(3):549-556. DOI:10.1109/TNANO.2016.2542525

[9] Peng K, et al. Single nanowire photoconductive terahertz detectors. Nano Letters. Jan 2015;**15**(1):206-210. DOI:10.1021/nl5033843

[10] Peng K, Parkinson P, Gao Q, Boland JL, Li Z, Wang F, Mokkapati S, Fu L, Johnston MB, Tan HH, Jagadish C., Single n^+ -i-n^+ InP nanowires for highly sensitive terahertz detection. Nanotechnology. 2017;**28**(12):125202. DOI:10.1088/1361-6528/aa5d80

[11] Yang B, Buddharaju KD, Teo SHG, Singh N, Lo GQ, Kwong DL, Vertical silicon-nanowire formation and gate-all-around MOSFET. Ieee Electron Device Letters. Jul 2008;**29**(7):791-794. DOI:10.1109/LED.2008.2000617

[12] Takei K, et al. Nanowire active-matrix circuitry for low-voltage macroscale artificial skin. Nature Materials. Oct 2010;**9**(10):821-826. DOI:10.1038/nmat2835

[13] Ju SY, et al. Fabrication of fully transparent nanowire transistors for transparent and flexible electronics. Nature Nanotechnology. Jun 2007;**2**(6):378-384. DOI:10.1038/nmat2835

[14] Gudiksen MS, Lauhon LJ, Wang J, Smith DC, Lieber CM. Growth of nanowire superlattice structures for nanoscale photonics and electronics. Nature. Feb 7 2002;**415**(6872):617-620. DOI: 10.1038/415617a

[15] Duan XF, Huang Y, Agarwal R, Lieber CM. Single-nanowire electrically driven lasers. Nature. Jan 2003;**421**(6920):241-245. DOI: 10.1038/nature01353

[16] Feng C, Xu Z, Wang X, Yang H, Zheng L, Fu H. Organic-nanowire–SiO2 core–shell microlasers with highly polarized and narrow emissions for biological imaging. ACS Applied Materials & Interfaces. 2017;**9**(8):7385-7391. DOI:10.1021/acsami.6b13387

[17] Kennedy T, Brandon M, Ryan KM. Advances in the application of silicon and germanium nanowires for high-performance lithium-ion batteries. Advanced Materials. Jul 2016;**28**(27):5696-5704. DOI:10.1002/adma.201503978

[18] Noh YY, Cheng XY, Sirringhaus H, Sohn JI, Welland ME, Kang DJ. Ink-jet printed ZnO nanowire field effect transistors. Applied Physics Letters. Jul 2007;**91**(4):p. 3. DOI: 10.1063/1.2760041

[19] Opoku C, Chen L, Meyer F, Shkunov M. Solution processable nanowire field-effect transistors. MRS Proceedings. 2011;**1287**:69-74

[20] Opoku C, Hoettges KF, Hughes MP, Stolojan V, Silva SRP, Shkunov M. Solution processable multi-channel ZnO nanowire field-effect transistors with organic gate dielectric. Nanotechnology. Oct 2013;**24**(40). Art. no. 405203. DOI: 10.1088/0957-4484/24/40/405203

[21] Hanrath T, Korgel BA. Supercritical fluid–liquid–solid (SFLS) synthesis of Si and Ge nanowires seeded by colloidal metal nanocrystals. Advanced Materials. 2003;**15**(5):437-440. DOI: 10.1002/adma.200390101

[22] Hu J, Bando Y, Liu Q, Golberg D, Laser-ablation growth and optical properties of wide and long single-crystal SnO2 ribbons. Advanced Functional Materials. 2003;**13**(6):493-496. DOI: 10.1002/adfm.200304327

[23] Kim F, Kwan S, Akana J, Yang P. Langmuir-Blodgett nanorod assembly. Journal of the American Chemical Society. 2001;**123**(18):4360-4361. DOI: 10.1021/ja0059138

[24] Yang P, Nanotechnology: Wires on water. Nature. 2003;**425**(6955):243-244. DOI: 10.1038/425243a

[25] Tao A, et al. Langmuir-Blodgett silver nanowire monolayers for molecular sensing using surface-enhanced Raman spectroscopy. Nano Letters. 2003;**3**(9):1229-1233. DOI: 10.1021/nl0344209

[26] Whang D, Jin S, Wu Y, Lieber CM. Large-scale hierarchical organization of nanowire arrays for integrated nanosystems. Nano Letters. 2003;**3**(9):1255-1259. DOI: 10.1021/nl0345062

[27] Yu G, Li X, Lieber CM, Cao A. Nanomaterial-incorporated blown bubble films for large-area, aligned nanostructures. Journal of Materials Chemistry. 2008;**18**(7):728-734. DOI: 10.1039/B713697H

[28] Huang Y, Duan X, Wei Q, Lieber CM. Directed assembly of one-dimensional nanostructures into functional networks. Science. 2001;**291**(5504):630-633. DOI: 10.1126/science.291.5504.630

[29] Yu G, Lieber CM. Assembly and integration of semiconductor nanowires for functional nanosystems. Pure and Applied Chemistry. 2010;**82**(12):2295-2314. DOI: 10.1351/PAC-CON-10-07-06

[30] Heo K, et al. Large-scale assembly of silicon nanowire network-based devices using conventional microfabrication facilities. Nano Letters. 2008;**8**(12):4523-4527. DOI: 10.1021/nl802570m

[31] Fan Z, et al. Wafer-scale assembly of highly ordered semiconductor nanowire arrays by contact printing. Nano Letters. 2008;**8**(1):20-25. DOI: 10.1021/nl071626r

[32] Hughes MP, Nanoelectromechanics in Engineering and Biology. CRC Press; 2002, Boca Raton, FL. ISBN 9780849311833

[33] Constantinou M, et al. Simultaneous tunable selection and self-assembly of Si nanowires from heterogeneous feedstock. ACS Nano. Apr 2016;**10**(4):4384-4394. DOI: 10.1021/acsnano.6b00005

[34] Freer EM, Grachev O, Duan X, Martin S, Stumbo DP. High-yield self-limiting single-nanowire assembly with dielectrophoresis. Nature Nanotechnology. 2010;**5**(7):525-530. DOI: 10.1038/nnano.2010.106

[35] Collet M, et al. Large-scale assembly of single nanowires through capillary-assisted dielectrophoresis. Advanced Materials. 2015;**27**(7):1268-1273. DOI: 10.1002/adma.201403039

[36] Paska Y, Haick H. Interactive effect of hysteresis and surface chemistry on gated silicon nanowire gas sensors. ACS Applied Materials & Interfaces. May 2012;**4**(5):2604-2617. DOI: 10.1021/am300288z

[37] Kim SH, et al. Hysteresis behaviour of low-voltage organic field-effect transistors employing high dielectric constant polymer gate dielectrics. Journal of Physics D: Applied Physics. 2010;**43**(46):465102. DOI:10.1088/0022-3727/43/46/465102

[38] Kawashima T, Saitoh T, Komori K, Fujii M. Synthesis of Si nanowires with a thermally oxidized shell and effects of the shell on transistor characteristics. Thin Solid Films. 2009;**517**(16):4520-4526. DOI: 10.1016/j.tsf.2008.12.042

[39] Wang Y, Lew K-K, Mattzela J, Redwing J, Mayer T. Top-gated field effect devices using oxidized silicon nanowires. In: Device Research Conference Digest, 2005. DRC'05. 63rd;**1**: IEEE; 2005. pp. 159-160. DOI: 10.1109/DRC.2005.1553101

[40] Fukata N, Kaminaga J, Takiguchi R, Rurali R, Dutta M, Murakami K. Interaction of boron and phosphorus impurities in silicon nanowires during low-temperature ozone oxidation. The Journal of Physical Chemistry C. 2013;**117**(39):20300-20307. DOI: 10.1021/jp406713p

[41] Constantinou M, et al. Interface passivation and trap reduction via a solution-based method for near-zero hysteresis nanowire field-effect transistors. ACS Applied Materials & Interfaces. 2015;**7**(40):22115-22120. DOI: 10.1021/acsami.5b07140

[42] Giridharagopal R, Cox PA, Ginger DS. Functional scanning probe imaging of nanostructured solar energy materials. Accounts of Chemical Research. Sep 2016;**49**(9):1769-1776. DOI: 10.1021/acs.accounts.6b00255

[43] Tsoi WC, et al. Surface and subsurface morphology of operating nanowire: fullerene solar cells revealed by photoconductive-AFM. Energy & Environmental Science. Sep 2011;**4**(9):3646-3651. DOI: 10.1039/C1EE01944A

[44] Kopanski JJ. Electrical scanning probe microscopes to address industrial nano-metrology needs of integrated circuits and nanoelectronic devices. In 29th Conference on Precision Electromagnetic Measurements (CPEM 2014); 2014. pp. 214-215. DOI: 10.1109/CPEM.2014.6898335

[45] Berger R, Butt HJ, Retschke MB, Weber SAL. Electrical modes in scanning probe microscopy. Macromolecular Rapid Communications. Jul 2009;**30**(14):1167-1178. DOI: 10.1002/marc.200900220

[46] Kalinin SV, et al. Big, deep, and smart data in scanning probe microscopy. ACS Nano. Oct 2016;**10**(10):9068-9086. DOI: 10.1021/acsnano.6b04212

[47] Leem DS, Edwards A, Faist M, Nelson J, Bradley DDC, de Mello JC. Efficient organic solar cells with solution-processed silver nanowire electrodes. Advanced Materials. Oct 2011;**23**(38):4371. DOI: 10.1002/adma.201100871

[48] Sannicolo T, Lagrange M, Cabos A, Celle C, Simonato JP, Bellet D. Metallic nanowire-based transparent electrodes for next generation flexible devices: A review. Small. Nov 2016;**12**(44:6052-6075. DOI: 10.1002/smll.201602581

[49] Langley D, Giusti G, Mayousse C, Celle C, Bellet D, Simonato J-P. Flexible transparent conductive materials based on silver nanowire networks: A reviewNanotechnology. Nov 15 2013;**24**(45). Art. no. 452001. DOI: 10.1088/0957-4484/24/45/452001

[50] Kim SH, Choi WI, Kim KH, Yang DJ, Heo S, Yun DJ. Nanoscale chemical and electrical stabilities of graphene-covered silver nanowire networks for transparent conducting electrodes, Scientific Reports. Sep 2016;**6**. Art. no. 33074. DOI: 10.1038/srep33074

[51] Shaw JE, Perumal A, Bradley DDC, Stavrinou PN, Anthopoulos TD. Nanoscale current spreading analysis in solution-processed graphene oxide/silver nanowire transparent electrodes via conductive atomic force microscopy. Journal of Applied Physics. May 2016;**119**(19). Art. no. 195501. DOI: 10.1063/1.4949502

[52] Li Y, Qian F, Xiang J, Lieber CM. Nanowire electronic and optoelectronic devices. Materials Today. Oct 2006;**9**(10):18-27. DOI: 10.1016/S1369-7021(06)71650-9

[53] Vogel EM. Technology and metrology of new electronic materials and devices. Nature Nanotechnology. Jan 2007;**2**(1):25-32. DOI: 10.1038/nnano.2006.142

[54] Stratakis E, et al. Imaging dielectric properties of Si nanowire oxide with conductive atomic force microscopy complemented with femtosecond laser illumination. Nano Letters. Jul 2008;**8**(7):1949-1953. DOI: 10.1021/nl0807171

Chapter 6

Semiconductor Nanowire MOSFETs and Applications

Hao Zhu

Additional information is available at the end of the chapter

http://dx.doi.org/10.5772/67446

Abstract

Semiconductor nanowires have aroused a lot of scientific interest and have been regarded as one of the most promising candidates that would make possible building blocks in future nanoscale devices and integrated circuits. Employing nanowire as metal-oxide-semiconductor field-effect transistor (MOSFET) channel can enable a gate-surrounding structure allowing an excellent electrostatic gate control over the channel for reducing the short-channel effects. This chapter introduces the basic physics of semiconductor nanowires and addresses the problem of how to synthesize semiconductor nanowires with low-cost, high-efficiency and bottom-up approaches. Effective integration of nanowires in modern complementary metal-oxide-semiconductor (CMOS) technology, specifically in MOSFET devices, and non-volatile memory applications is also reviewed. By extending the nanowire MOSFET structure into a universal device architecture, various novel semiconductor materials can be investigated. Semiconductor nanowire MOSFETs have been proved to be a strong and useful platform to study the physical and electrical properties of the novel material. In this chapter, we will also review the investigations on topological insulator materials by employing the nanowire field-effect transistor (FET) device structure.

Keywords: semiconductor nanowire, gate-surrounding, MOSFET, self-assembly, flash-like non-volatile memory, topological insulator nanowire FETs

1. Introduction

Since the invention of integrated circuit in the 1950s, the scaling of metal-oxide-semiconductor field-effect transistor (MOSFET) continues with the emergence of new technologies to extend complementary metal-oxide-semiconductor (CMOS) down to ever smaller technology node. However, the CMOS scaling has deviated from the trends predicted by Moore and the scaling rules set forth by Dennard et al. due to fundamental physical and technical limitations [1, 2]. Limitations such as heat dissipation, leakage current and channel length modulation have become a major concern that will inevitably lead to the slowing down in CMOS scaling when approaching atomic dimension. Thus, there is an urgent need to develop new semiconductor technology to solve the issues including cost, speed, density, reliability and power dissipation. A great deal of efforts has been made, and among all the candidates, nanoelectronics involved with replacing existing silicon-based technology has risen to be one of the most promising solutions towards continuous CMOS scaling.

During the past decades, semiconductor nanowires synthesized by bottom-up techniques and derived field-effect transistor (FET) devices have been intensively studied as the fundamental building blocks for nanoelectronic devices and circuit technologies [3–7]. Compared to planar devices based on bulk materials, the nanowires have a smaller channel and large surface-to-volume ratio. In addition, the gate-surrounding or gate-all-around (GAA) structure that can be formed in the nanowire FET allows excellent electrostatic gate control over the nanowire channel. Moreover, the GAA nanowire transistors enable ultimate CMOS device scaling with the best possible short-channel control considering the quantum confinement effects and the scattering at atomic dimensions. The GAA horizontal nanowire FET architecture exhibits high similarities to the FinFET, which is the predominant technology in the current 10 nm or even 7 nm process node. Thus, GAA FETs are very promising candidates in the sub-7 nm nodes to extend the scalability beyond the limits imposed by the FinFET technology with much less complexity compared to the alternative scaling approaches [8].

Till date, nanowire FETs have mainly been fabricated using the "top-down" approaches based on advanced lithography with nanowire prepared by dry/wet etching which usually yields well-oriented nanowire arrays [9, 10]. However, it has been well recognized that nanowires synthesized by "bottom-up" techniques such as chemical vapour deposition (CVD) can have lower cost and higher quality compared to the"top-down" methods [11]. Thus, it would be very attractive to develop an approach to manufacture such nanowire FETs with an excellent performance. Nevertheless, the technique to fabricate FETs from CVD grown silicon nanowires remains a barrier to the development of devices with optimized performance. Current approaches are primarily based on the harvesting and positioning the as-grown nanowires using aligning methods such as fluidic alignment, dielectrophoresis or nanoscale probe methods [12–17]. Such methods will inevitably introduce contaminants to the nanowire surfaces, which will adversely influence the device interface state density and possibly the nanowire surface roughness, and the device performance will be deteriorated as a result.

In this chapter, we first review the fundamentals of semiconductor nanowires and synthetic strategies. Then we introduce a novel self-alignment fabrication process for nanowire FET applications. The high-quality self-aligned nanowire channels possess clean surface and the fabricated FET devices exhibit excellent performance including large on/off ratio, small subthreshold slope and small leakage current. Such an effective nanowire integration scenario is very attractive for different materials and device investigations. We focus on the flash memory based on the self-aligned Si nanowire FETs and the study of topological insulator nanowire FETs using the self-alignment techniques. The results demonstrate that by employing the one-dimensional nanowire as the active component in the electronic devices, the flash memory performance has been significantly improved such as lower operation voltage and better

http://dx.doi.org/10.5772/67446

reliability, and the basic principles behind the topological insulators have been manifested strongly enabling more delicate characterizations to understand the physics of the specific material. We believe that nanowire MOSFETs will open up a suit of potential applications in future sub-10 nm semiconductor technology as well as solutions towards emerging micro and nanoelectronic device challenges.

2. Review of physics of semiconductor nanowires

The theory of nanowires will be briefly reviewed with a focus on silicon nanowires since the features discussed here can be extended to other types of semiconductor nanowires. The structural and mechanical properties of silicon nanowires will be reviewed first, and then the electronic properties will be discussed in which we will also consider the device-related issues and limitations.

2.1. Structural and mechanical properties

The Si nanowires are intriguing mostly due to the extra high surface-to-volume ratio and the well-defined single crystalline orientation. The nanowire growth direction has been widely studied, and the connection between the diameter and the favoured crystal direction has been established [18]. It has been reported both experimentally and theoretically that the catalyst-assisted Si nanowires with smallest diameter prefer the <110> direction, while the nanowires with larger diameter favours the <111> direction [19–21]. The growth direction determines the nanowire cross section to some extent. A pentagonal cross section has been observed from the ground-state structure for Si nanowires up to 5 nm [22, 23]. This cross section due to the joint of five prisms cut of a Si (110) plane has rarely been reported. A more well-known structure in good agreement with experimental work is the hexagonal cross section for <110> Si nanowires with a bulk core [18, 24].

Mechanical properties of Si become quite different when reaching nanoscale with lower dimension. Due to the compressive surface stress, the Young's modulus of <100> Si nanowires softens as the surface-to-volume ratio increases and a steep decrease has been detected on shrinking of the nanowire diameter to 2–2.5 nm [25]. It was further discovered that the Young's modulus of Si nanowires is strongly anisotropic [26]. Different from the <100> nanowires, which exhibit lowest values, the wires grown along <110> direction give the highest values. These results are also in good agreement with the experimental work [27]. Currently, the mechanical properties of nanowires have become a promising research direction. Static and dynamic nanowire bending with atomistic simulations, which relies on periodic boundary conditions, reveals the influence of surface stress more clearly and provides valuable information for nanowire-based device designs.

2.2. Electronic properties

It is widely recognized that bulk Si has an indirect band gap, with the valence band maximum at the Γ point and the conduction minimum at about 85% along the Γ and X direction.

However, Si nanowires grown along most orientations have a direct band gap, with the maximum of the valence band and the minimum of the conduction band aligned at the same point in k-space. As a result, such nanowire materials have become encouraging candidates in optoelectronics or photonics applications [28].

The electronic properties of nanowires depend on the nanowire growth direction, cross section and diameter. However, it is interesting to find that the band gap of Si nanowire is insensitive to the shape of cross section. The band gap of a 1 nm nanowire was found constant with change less than 0.09 eV by simulating cross sections that are even utterly different, as long as providing the same surface-to-volume ratio [29, 30].

Different types of Si nanowire based electronic devices have been fabricated and studied for a variety of application purposes. An important cause of electrical conductance degradation in such devices is the scattering occurring at the surface with the presence of surface defects or surface roughness [31, 32]. Generally, the surface roughness disorder induces irregularities in the density of states along the nanowire axis, resulting in the carrier reflection and reduction in conductance. A more detailed study shows that the backscattering due to the surface roughness strongly depends on the nanowire orientation, and the anisotropy comes from the difference in band structure [33, 34]. In particular, electrons are less sensitive to surface roughness in <110> nanowires, and the transfer of holes is more smooth in <111> nanowires. Other electronic transport parameters such as the mean free path, length and the localization length can be better explained with the above incorporated differences.

Despite the defects or roughness from the nanowire surface, the existence of impurities is another major source of scattering, especially when the nanowire size is scaled down below 10 nm. Unlike the bulk in which the carriers are just slowed down by the impurities, the trajectories of the carriers in an extremely thin one-dimensional medium can be entirely backscattered because the scattering potential often extends throughout most of the cross sections of the thin wire [34]. Impurities can be originated from imperfect growth or intentional doping for targeted electrical properties when referred to semiconductors. For the dopants in Si nanowires, impurities like donor impurities, either segregate to the surface forming electrically inactive components or stay in the wire producing a strong backscattering, both will lead to the decreased conductance.

3. Silicon nanowire synthesis and field-effect transistors

3.1. Silicon nanowire synthesis

There has been a large number of reported works on the nanowire synthesis of a wide spectrum of semiconductor materials during the past decades. Both chemical and physical methods are intensively studied. Understanding the growth mechanism of these synthesis approaches is helpful in developing one-dimensional nanostructures of the desired materials and derivative electrical devices.

http://dx.doi.org/10.5772/67446

Solution phase synthesis of nanowires is a low cost yet high yield method in which selective capping agents are often employed, especially in an anisotropic growth, for the kinetic control of the evolving nanostructure to allow elongation along certain axis by preferentially adsorbing to specific crystal facets [35–37]. The major limitation of such solution phase synthesis lies in the empirical trial-and-error methodology for capping agents. Template-assisted electrochemical synthesis is another popular approach to grow one-dimensional nanostructures. Synthesis of metals, semiconductors and conductive polymers and oxides has been reported by using templates such as anodic aluminium oxide (AAO) and polycarbonate membranes [38, 39]. This method shares the advantages such as low-cost, well-orientation, ambient temperature, pressure operation and feasibility of batch fabrication of nanowire array. However, complete template filling still remains a challenge for the nanowire synthesis through electrodeposition. Template-assisted synthesis has been more utilized in the formation of hollow nanotube and core-shell nanowire structures in combination with atomic layer deposition (ALD) technique [40, 41].

Vapour phase synthesis of one-dimensional nanostructure is probably the most extensively explored approach. Numerous techniques have been developed to grow nanowire from gas precursors. Among all the vapour phase methods, the most widely studied and successful approach in generating high-quality single-crystal nanowires in large quantities is the vapour-liquid-solid (VLS) mechanism [4]. We will be focusing on this method in this section, and the following contents reviewed in this chapter are based on the nanowires synthesized by using the VLS approach.

A typical VLS process starts from the dissolution of vapour reactants into a catalytic alloy phase, followed by the crystal nucleation at the liquid-solid interface. As illustrated in **Figure 1**, gold nanoparticles are commonly used as catalyst for Si nanowire growth by the

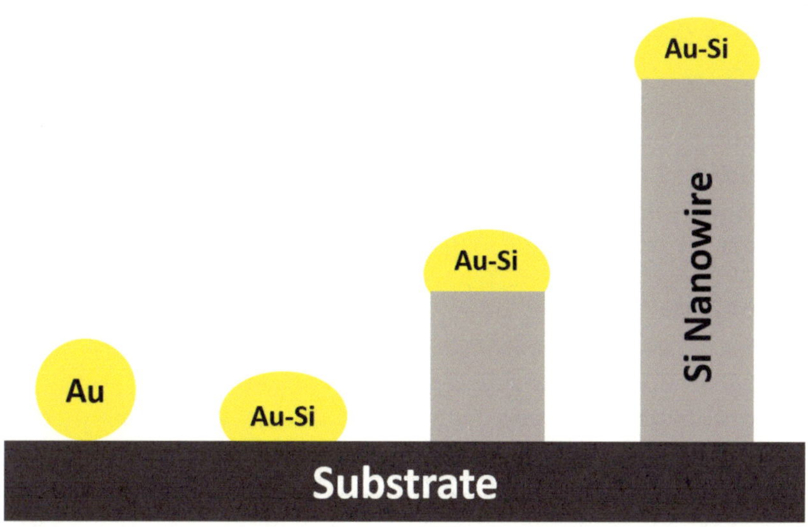

Figure 1. Schematic illustration of the VLS mechanism for Si nanowire synthesis using Au nanoparticle as catalyst.

VLS mechanism. The Au particles catalyze the gas precursor (e.g. SiH_4) and reacts with the Si atoms from the vapour phase, forming Au-Si eutectic droplets. Si was introduced from the vapour and adsorbed onto the liquid surface, then diffused into the droplet leading to a supersaturated state where the Si atoms precipitate and the nanowire starts nucleating. In the growth, the droplet serves as the "seeds" limiting the lateral growth of individual nanowires, and the droplet size remains unchanged during the growth which determines the diameter of the subsequent nanowires.

The VLS process has become a widely used growth method for one-dimensional nanostructures of a broad variety of materials. In addition to elemental semiconductors such as Si and Ge, binary compounds including III-V semiconductors (e.g. GaN, GaAs and InAs), oxides (e.g. ZnO, SiO_2 and ITO) and chalcogenides (e.g. Bi_2Se_3, In_2Se_3 and CdS) have also been fabricated and studied in nanowire morphology following the VLS mechanism.

3.2. Silicon nanowire field-effect transistors

In this section, we review the fabrication and characterization of Si nanowire FETs prepared by using a self-alignment method based on VLS synthesis mechanism [42]. Unlike the traditional nanowire harvesting and alignment methods, the self-alignment approach not only enables simultaneous batch fabrication of reproducible and homogeneous nanowire devices of high quality, but also limits the contaminations of the nanowire during the fabrication process. The fabricated self-aligned Si nanowire FETs exhibit excellent current-voltage (I-V) characteristics, high on/off current ratio and small subthreshold slope providing an excellent platform for other devices and material investigations and applications.

3.2.1. FET fabrication

The Si nanowire FETs are fabricated using the directed self-alignment process as shown in **Figure 2**. The main concept of this fabrication approach is that Si nanowires are synthesized from Au catalysts on predefined locations as on wafers and well aligned with the source/drain and gate electrodes by photolithographic processes without harvesting the nanowires. At the first step of the fabrication process, a thin layer of Au catalyst (~1 nm) was deposited on the SiO_2/Si substrate and patterned by photolithography and lift-off processes. The Si nanowires were grown from the catalyst at the defined locations in a low-pressure chemical vapour deposition (LPCVD) furnace at 440°C for 2 hours with an ambient SiH_4 stream under a pressure of 500 mTorr. The nanowires are typically 20 μm in length and 20 nm in diameter. Immediately after the VLS growth, the Si nanowires were loaded in a dry oxidation furnace and oxidized at 750°C for 30 minutes in O_2 to form a ≈ 3 nm thick SiO_2 which was expected to provide a good interface between the nanowire and the subsequent top gate dielectric stack [42]. Photolithographic and lift-off processes were then performed to form the source and drain contacts. To facilitate proper contact formation, a 2% HF wet etch was applied to remove the oxides from the Si nanowires at the source/drain region before electrode metal (Al in this case) deposition. The next step is to deposit top gate dielectric (25 nm HfO_2) by ALD at 250°C, followed by a deposition of 5 nm of Al_2O_3 to improve the interface with the

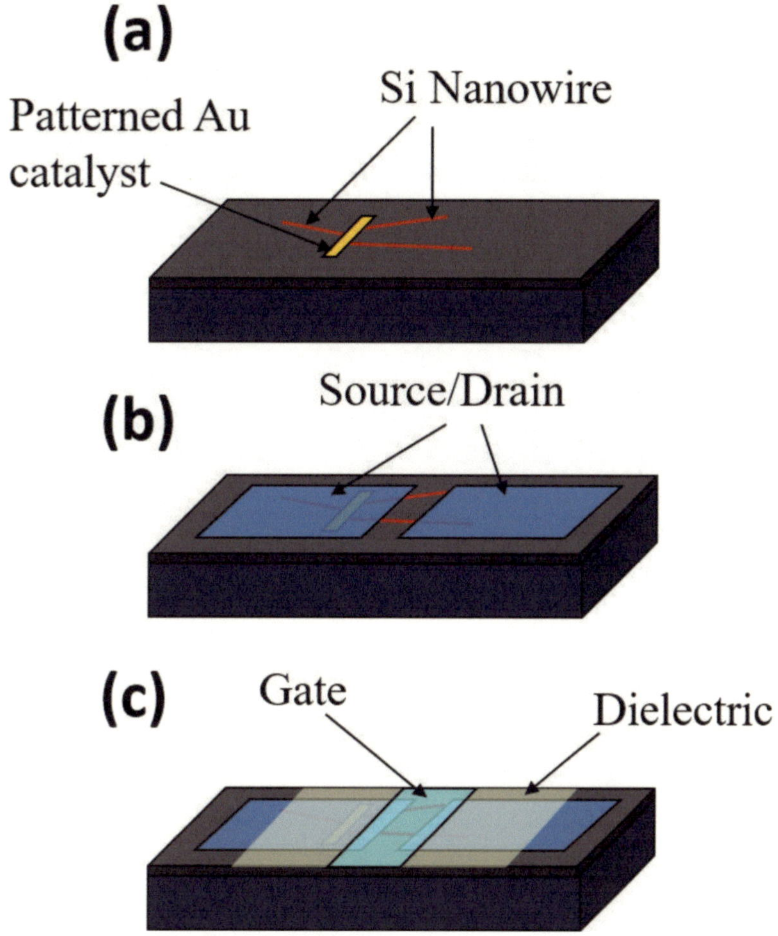

Figure 2. Self-alignment fabrication process: (a) patterned Au catalyst on SiO_2; (b) synthesis of Si nanowire from the Au catalyst, nanowire oxidation and alignment of source/drain contacts; (c) deposition of gate dielectric and pattern of top gate electrode [42].

Al top gate which was formed by the same lift-off process as of the source/drain electrodes. The final devices were annealed in forming gas (5% H_2 in N_2) at 325°C for 5 minutes to reduce the interface trap density between the nanowire and dielectric layer, as well as to improve the contact between the Al metal to the Si nanowire and HfO_2. **Figure 3** shows the top-view scanning electron microscopy (SEM) image of a finished self-aligned Si nanowire FET with gate length of 2 μm.

3.2.2. Electrical characterizations

Si nanowire FET with single or multiple nanowire channels is expected to exhibit better performance than planer bulk Si. The Gate-surrounding structure can be enabled allowing excellent electrostatic gate control over the channel for reduced short-channel effects.

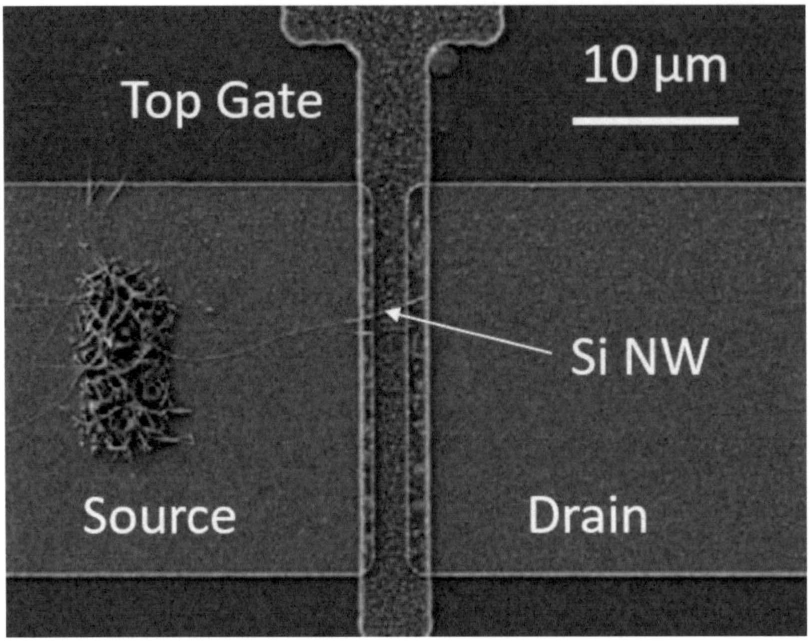

Figure 3. Top-view SEM image of a typical self-aligned Si nanowire FET with gate length of 2 μm [42].

The following equations define the minimum gate length to avoid short-channel effects for single-gate, double-gate and surrounding-gate structures:

$$\lambda_{\text{single-gate}} = \sqrt{\frac{\varepsilon_{Si}\, t_{ox}\, t_{Si}}{\varepsilon_{ox}}} \tag{1}$$

$$\lambda_{\text{double-gate}} = \sqrt{\frac{\varepsilon_{Si}\, t_{ox}\, t_{Si}}{2\,\varepsilon_{ox}}} \tag{2}$$

$$\lambda_{\text{surrounding-gate}} = \sqrt{\frac{2\,\varepsilon_{Si}\, t_{Si}^{\,2} \ln\left(1 + \frac{2\,t_{ox}}{t_{Si}}\right) + \varepsilon_{ox}\, t_{Si}^{\,2}}{16\,\varepsilon_{ox}}} \tag{3}$$

where t_{Si}, ε_{Si}, t_{ox} and ε_{ox} are the thickness and permittivity of Si and gate oxide, respectively [43]. The above equations demonstrate that the surrounding gate structure offers the best characteristics for controlling the short-channel effects.

Figure 4a and **b** shows the output characteristics of a self-aligned Si nanowire FET in both linear and logarithmic scale. From the drain current (I_{DS}) versus drain voltage (V_{DS}) curves, the leakage-affected region, weak, moderate and strong inversion operation regions of the FET are clearly shown in **Figure 4b**. In the weak inversion region, I_{DS} increases exponentially with gate voltage (V_{GS}) due to the diffusion of carriers (holes) and is saturated at 3 ϕ_t (~78 mV at room temperature, ϕ_t = kT/q is the thermal voltage). In moderate inversion region, I_{DS} varies by a couple of orders of magnitude, changing gradually from one form of functional dependence to the other. In strong inversion region, I_{DS} approximately follows the quadratic law

http://dx.doi.org/10.5772/67446

and is proportional to $(V_{GS} - V_{TH})^2$ with saturation at $V_{DS} = V_{GS} - V_{TH}$. These curves demonstrate that the self-aligned Si nanowire FET has similar electrical behaviours to those of conventional MOSFETs, even though it has much simpler device structure and no source/drain junction doping. It should be noted that the I_{DS}–V_{DS} curves increase sharply in the linear region, indicating a small source and drain contact resistance.

Due to the Schottky contacts between the Al contacts and the intrinsic Si nanowire, Schottky-barrier pMOSFET characteristics are expected for these FETs. **Figure 5** shows typical transfer characteristics of self-aligned Si nanowire FET with V_{DS} of −50, −100 and −150 mV, respectively.

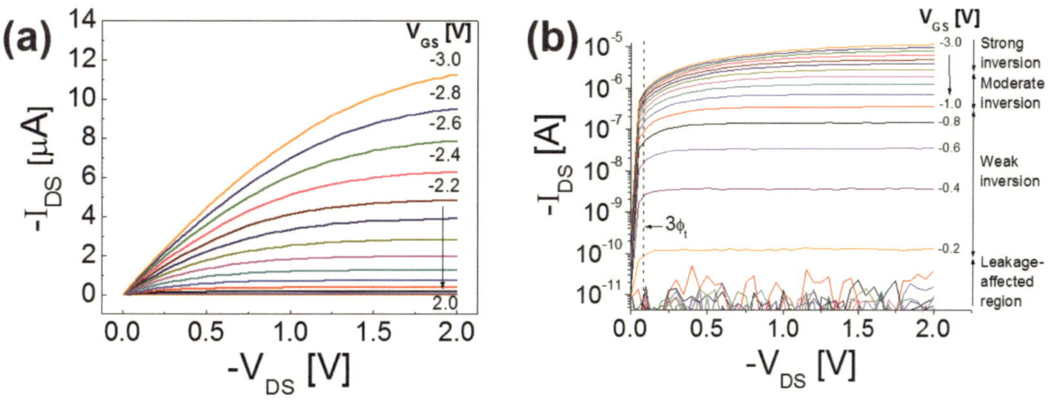

Figure 4. Output characteristics of the self-aligned Si nanowire FET in (a) linear and (b) logarithmic scale [42].

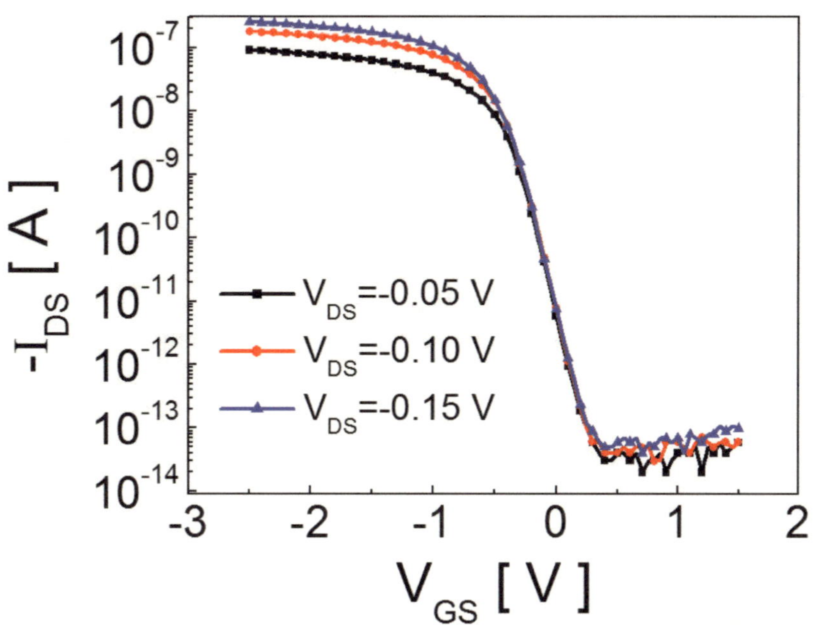

Figure 5. Transfer characteristics of the Si nanowire FET, with V_{DS} value of −50, −100 and −150 mV, respectively [42].

The on/off current ratio is ~10^8 within a 1.2 V V_{GS} window and no ambipolar behaviour is observed. The carrier mobility can be calculated from the linear region of the transconductance when V_{DS} = −50 mV with a presumed diameter of 20 nm for the nanowire channel. The calculated results show that the fabricated Si nanowire FETs have a relatively consistent hole mobility around 100 cm^2/Vs. The subthreshold swing (SS) can be extracted from the subthreshold region of I_{DS} in the log-scale I_{DS}–V_{GS} curves. The SS values of the Si nanowire FETs are as low as 75 mV/dec, which is quite small as compared to most reported results on nanowire FETs and poly-Si thin-film transistors [42].

The Si nanowire FETs fabricated through the self-alignment approach exhibit excellent performance as indicated by a high on/off current ratio (~10^8), small leakage current (<10^{-14} A), good carrier mobility (~100 cm^2/Vs) and small subthreshold slope (75 mV/dec). These excellent characteristics are due to the clean interfaces formed in the self-alignment fabrication process, and such Si nanowire FETs are very attractive for future nanoelectronic device applications.

4. Silicon nanowire-based flash-like non-volatile memory

4.1. Introduction of flash-like non-volatile memory

Today, computing architectures and electronic systems built on CMOS components are still pursuing without any sign of slowing down of requirements for low power, fast speed and high density alternatives. Up to now, electronic systems whose main function is to focus on date computing and storage take up more than half of the semiconductor market, and the demand is still growing explosively in areas such as portable electronic devices and systems. Solid-state mass storage occupies a large portion of this market, due to their compatibility with CMOS scaling technology, suitability for harsh environment without mechanical parts and the fact that most types of memory are non-volatile.

Non-volatile memory is typically employed for the task of secondary storage or long-term storage, which usually does not require fast operation speed or integration density. The current primary storage or on-chip memory still relies upon volatile forms of random-access memory. During the past decades, the size of cache memory in the central processing unit (CPU), which is also known as the static random-access memory (SRAM) has been doubled several times as a feasible strategy to increase the CPU capability. However, increasing SRAM will decrease CPU net information throughput because it is volatile and occupies a large chip floor space. Thus, it will be a revolutionary breakthrough in microelectronics if a truly non-volatile memory can be implemented as the on-chip memory in CPUs replacing SRAM.

4.1.1. Basic memory concepts and scaling challenges

Among all the non-volatile memory candidates for primary storage applications, flash memory is the most widely studied and electrically accessible form and is the most promising non-volatile memory in the electronics market. Flash memory has fast read access times, good retention and reliability and CMOS compatible fabrication process [44–46].

Also known as the floating-gate memory (shown in **Figure 6**), a flash memory device stores the trapped electrons in the floating gate until they are removed by another application of electrical field. Because the floating gate is insulated from the control gate and the MOSFET channel by a relatively thick blocking oxide and a thin tunnelling oxide, respectively, the trapped electrons can be retained for many years, and logical "0" and "1" states can be defined according to the presence or absence of electrons trapped in the floating gate. In the past years, the flash capacity, integrated density and performance have been continued to increase with lower manufacturing cost. Due to its compatibility with conventional CMOS process, it is easier and more reliable to integrate flash memory than other forms of non-volatile memory in logic and analog devices with increasing embedded and stand-alone memory to achieve higher chip performance.

However, current flash memory also exhibit disadvantages such as relatively slow write/erase speed and medium endurance, which make it far below the standards of on-chip memory applications. In a conventional floating-gate memory, the tunnelling oxide can hardly be scaled below 7 nm due to the requirement of data retention. Moreover, ultra-thin tunnelling oxide will lead to severe stress-induced leakage current (SILC) issue. On the other hand, the conventional floating-gate materials, poly Si and oxynitride operate at large programming voltages and endure only 10^5 operation cycles. Even though some new flash technologies are promising for low-voltage operations, the voltage supply is in excess of the working voltage standard of the advanced processes, and the gap between the operation voltage of memory device and CMOS logic continues to be broadened.

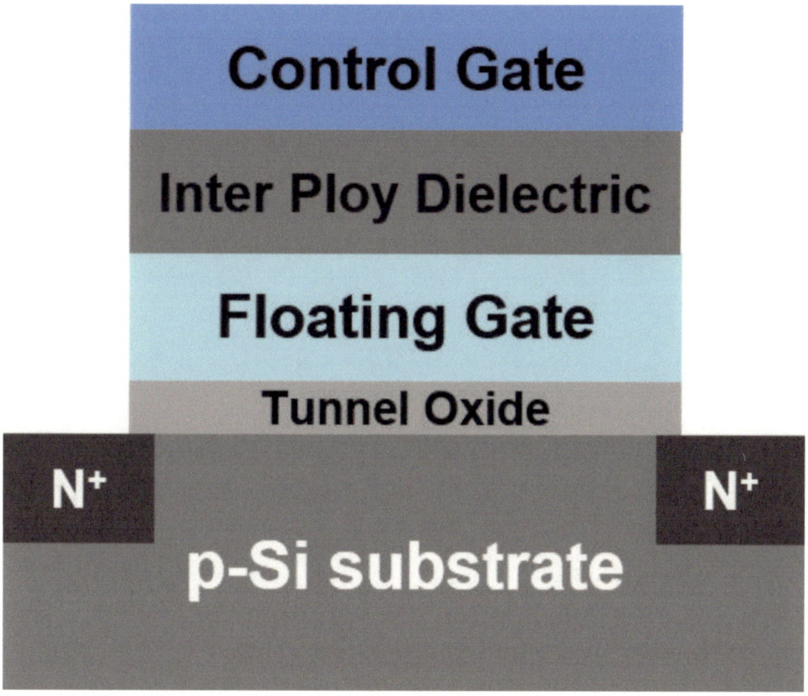

Figure 6. Schematic diagram of a floating-gate flash memory.

4.1.2. SONOS type charge-trapping memory

In recent decades, polysilicon-oxide-nitride-oxide-silicon (SONOS) type charge-trapping non-volatile memory has attracted intensive attention to replace the conventional floating-gate memory due to their advantages such as lower consumption, better reliability and scalability, and simpler structure and fabrication process [47–49]. **Figure 7** demonstrates the schematic structure of a SONOS charge-trapping memory, in which the electrons tunnel through the tunnelling layer and stored in the nitride layer. However, the conventional charge-trapping materials in SONOS memory are not compatible with the dimensional scaling and lead to poor performance in speed, retention and endurance. Furthermore, to achieve faster speed, the tunnelling oxide must be shrinking to enhance the electric field across it while the blocking oxide should be thicker to suppress the leakage current. But thinner tunnelling layer may in turn cause poor stability and reliability, and thicker blocking layer requires larger working voltages. Great efforts have been made for good memory characteristics via various approaches to gain a trade-off between speed and data retention. An effective strategy is to use non-planar channel such as nanotubes and nanowires, to enhance the gate control over the channel. Compared to planar devices based on bulk materials, the nanowires have larger surface-to-volume ratio, therefore, it requires less stored charges to induce the same memory window for nanowire-based flash memory.

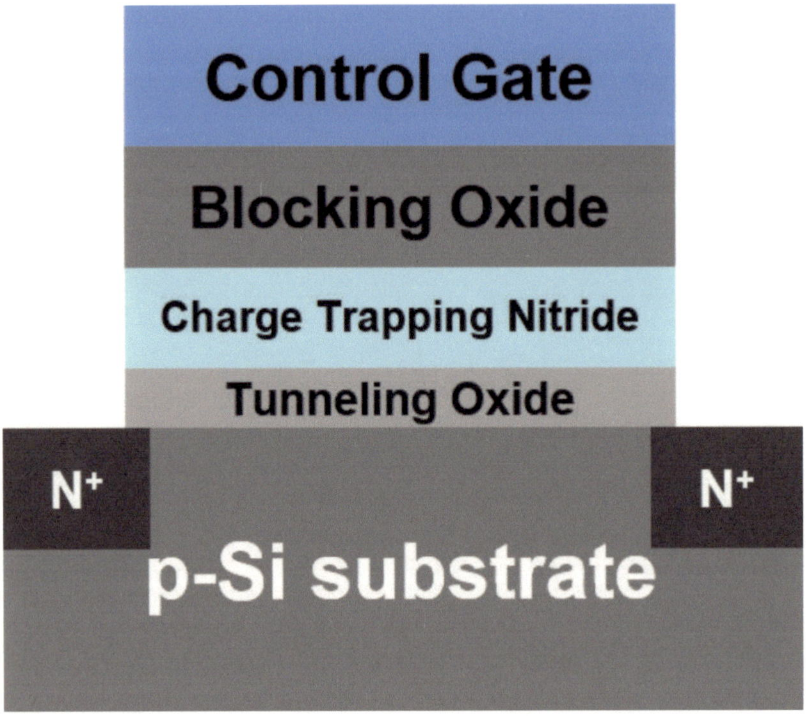

Figure 7. Schematic diagram of a SONOS-type charge-trapping flash memory.

The following in this section reviews the work on the device engineering of flash memory with novel dielectric materials and redox molecules as charge trapping medium based on the Si nanowire FETs. As compared with planar structure memory devise, much better performances have been achieved by utilizing the one-dimensional nanowire channel in the memory devices.

4.2. Silicon nanowire-based dielectric charge-trapping memory

4.2.1. History of one-dimensional nanostructure based charge-trapping memory

The first non-volatile memory cell was demonstrated in 2002 by Fuhrer et al. with one-dimensional nanostructure as the active channel component based on carbon nanotube [50]. Sufficient memory window was witnessed, and Fuhrer et al. predicted that the charge detection of such narrow and high mobility nanotube FETs will outperform planar FET cells [50]. In 2003, Choi et al. reported a carbon nanotube based non-volatile memory with oxide-nitride-oxide (ONO) dielectric stack [51]. A high electric field was generated by 5 V top gate voltage surrounding the surface of 3 nm nanotube, which was sufficiently high to produce Fowler-Nordheim tunnelling.

In 2006, Cha et al. reported a non-volatile flash memory having GaN nanowires with SiO_2 as a charge trapping material, and the operation mechanism was also investigated [52]. It was reported that the electric field distribution in the gate dielectric was different at the centre and the edge beneath the gate. The field is uniform under the centre of the gate across the oxide, and the field distribution at the edge changes along the distance. Non-volatile memory cells based on Si nanowire with CMOS compatible gate dielectric were reported by Zhu et al. in 2011 [53]. HfO_2 was used for charge storage, the thickness of which does not affect the electric field across the tunnelling oxide. The programming speed and retention time have been significantly improved through the surrounding gate structure.

Various high-k materials such as Y_2O_3, Gd_2O_3 and ZrO_2 have been proposed and studied as charge-trapping layer for better reliability and speed [54–56]. Using thicker high-k dielectric with relatively larger band gap as the blocking oxide can be expected to lower the electric field across the blocking layer. Integration of novel dielectric materials as the gate stack in the nanowire FET architecture is therefore very promising to realize practical applications of flash memory with faster speed, higher density, smaller size and better reliability.

4.2.2. Device fabrication

The Si nanowire FET based dielectric charge-trapping flash memory is prepared by following the self-alignment method [57]. Typically, a 300 nm SiO_2 was first grown on Si substrate by dry oxidation. Then a thin film of Au catalyst (1–3 nm) was deposited on patterned area pre-defined by photolithography. Then the Si nanowires were grown from the catalyst following the VLS mechanism and were oxidized at 750°C for 30 minutes to form a ≈ 3 nm SiO_2 which will function as the tunnelling layer. The 3/100 (unit: nm) Ti/Pt source and drain electrodes were then patterned with photolithography and deposited by e-beam evaporation. The channel length between the source/drain electrodes was controlled to be 5 μm. The following process is the deposition of charge-trapping layer and blocking oxide dielectric. Different charge-trapping layer and stacks were fabricated—Ta_2O_5, $Ta_2O_5/Al_2O_3/Ta_2O_5$ (TAT) and a reference HfO_2, with thickness of 20, 6/8/6, 20 nm, respectively [57]. The blocking oxide for all memory devices was selected to be 25 nm for Al_2O_3. Ta_2O_5 was deposited by sputtering, whereas Al_2O_3 and HfO_2 layers were deposited by ALD with TMA and TEMAH as precursors, respectively. Finally, a 100 nm Pd top gate was formed by the same photolithographic and lift-off processes as the source/drain electrodes.

4.2.3. Characterization and performance discussion

Three SONOS-like charge-trapping flash memory devices with structures of metal/Al_2O_3/Ta_2O_5/Al_2O_3/Ta_2O_5/SiO_2/Si (MATATOS), metal/Al_2O_3/Ta_2O_5/SiO_2/Si (MATOS) and metal/Al_2O_3/HfO_2/SiO_2/Si (MAHOS) have been fabricated and electrically characterized [57]. **Figure 8** shows the schematic top view and a cross-sectional transmission electron microscope (TEM) image of a MATATOS device.

Typical output characteristics (I_{DS}–V_{DS}) of the self-aligned Si nanowire FET based flash memory are shown in **Figure 9a** (MATATOS) with smooth and well-saturated drain current curves and a small source and drain contact resistance. **Figure 9b–d** shows the transfer characteristics of the three memory devices with counterclockwise hysteresis loops suggesting the charging and discharging behaviour. By comparing the memory window, the MATATOS and MATOS devices show larger values than the MAHOS device [57].

The programming and erasing (P/E) operations on the MATATOS memory were measured by studying the threshold voltage shift (ΔV_{Th}) under different P/E operations and were

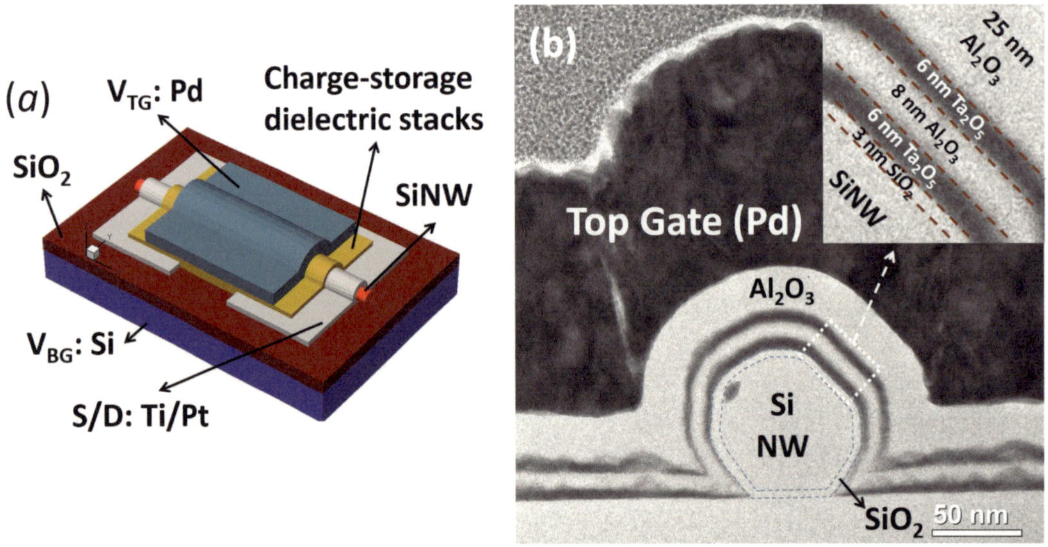

Figure 8. (a) Schematic of top-view of the dielectric charge-trapping flash memory based on Si nanowire FET; (b) cross-sectional TEM image of the dielectric flash memory with TAT charge-trapping layer [57].

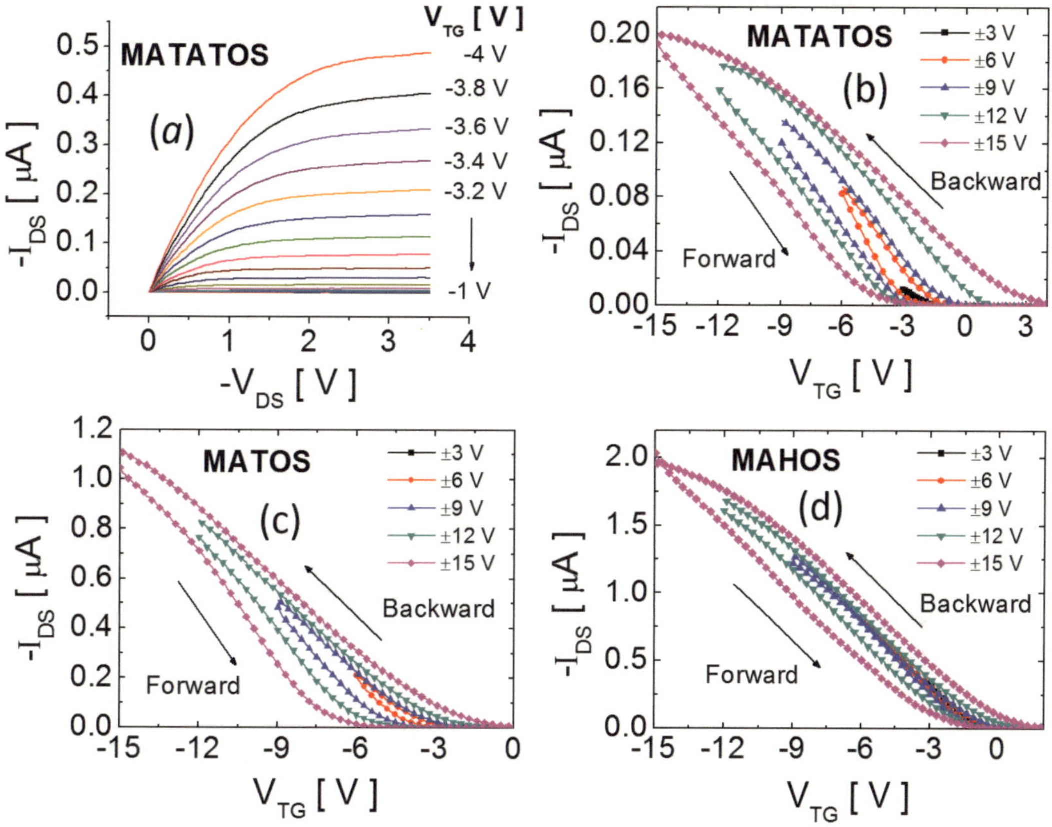

Figure 9. (a) Output characteristics of MATATOS device for V_{GS} from −4 to −1 V with 0.2 V step; transfer characteristics of (b) MATATOS, (c) MATOS, and (d) MAHOS devices, showing counterclockwise hysteresis loops under different V_{GS} sweep ranges [57].

illustrated in **Figure 10**. The I_{DS}–V_{GS} curves show clear ΔV_{Th} under accumulative rectangular gate voltage pulses but exhibiting same subthreshold slope (~300 mV/dec) before and after the programming or erasing operations. This indicates that the ΔV_{Th} is due to the fixed charges in the charge-trapping layers instead of the interface states. By stressing a positive gate voltage, the electrons will tunnel through the tunnelling oxide from the Si nanowire and get trapped in the charge-trapping layers. This corresponds to the programming operations, which will result into a V_{Th} shift towards the positive direction. Negative gate voltage will reversely shift the V_{Th} in the opposite direction by removing the pre-trapped electrons from the charge-trapping layer to the Si nanowire. This is defined as the erasing operations. By comparing the ΔV_{Th} under different P/E gate voltages, it was found that both MATATOS and MATOS devices showed faster speed than the MAHOS device, and MATOS device showed a larger ΔV_{Th} for a long P/E time due to the thicker charge-trapping layer.

Figure 11a shows the charge retention properties of the devices. Less than 25% charge loss was observed for both MATATOS and MATOS devices. **Figure 11b** demonstrated the

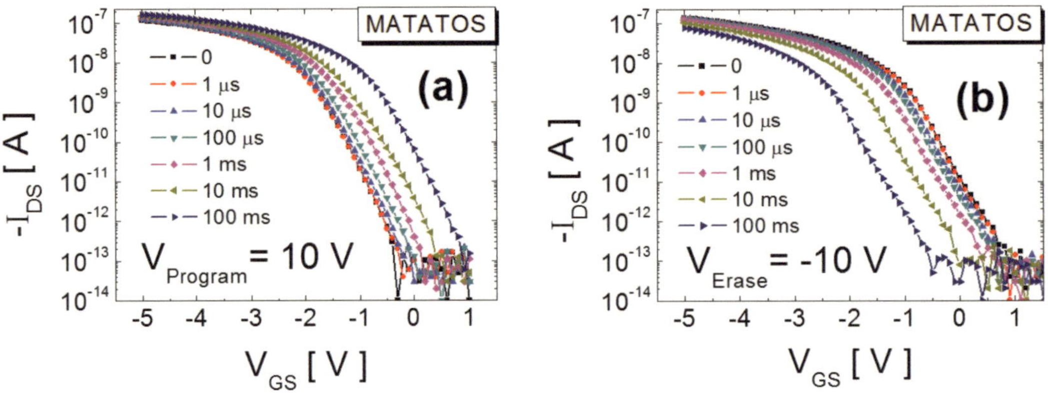

Figure 10. (a) Programming and (b) erasing operations with accumulated P/E time at ±10 V gate voltage pulses on the MATATOS flash memory device [57].

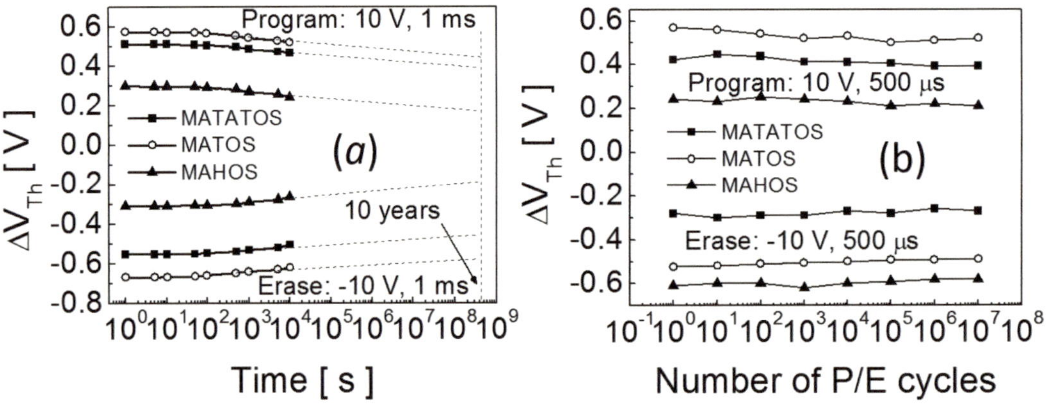

Figure 11. (a) Retention and (b) endurance characteristics of three flash memory devices [57].

endurance properties of the three devices. Very small memory window degradations were observed for all the devices after 10^7 P/E cycles. Such good endurance behaviour arises from the good interface between the Si nanowire and the high-k stacks formed in the self-alignment fabrication process.

The design of the TAT charging-trapping stack with an Al_2O_3 layer sandwiched between two Ta_2O_5 layers is for the multiple charge storage, in which the Al_2O_3 layer functions as the blocking oxide for the first Ta_2O_5 layer and tunnelling oxide for the second Ta_2O_5 layer. As shown in **Figure 12**, clear two-step charging behaviour is successfully observed in the MATATOS device. The first step around 10 V gate voltage is due to the charge storage in the first Ta_2O_5 layer, and the second step at around 20 V is observed indicating the charge-trapping centres in both Ta_2O_5 layers have been filled with electrons. The charge density of the two Ta_2O_5 layers are calculated to be 1.75×10^{19} and 4.98×10^{19} cm^{-3}, respectively.

http://dx.doi.org/10.5772/67446

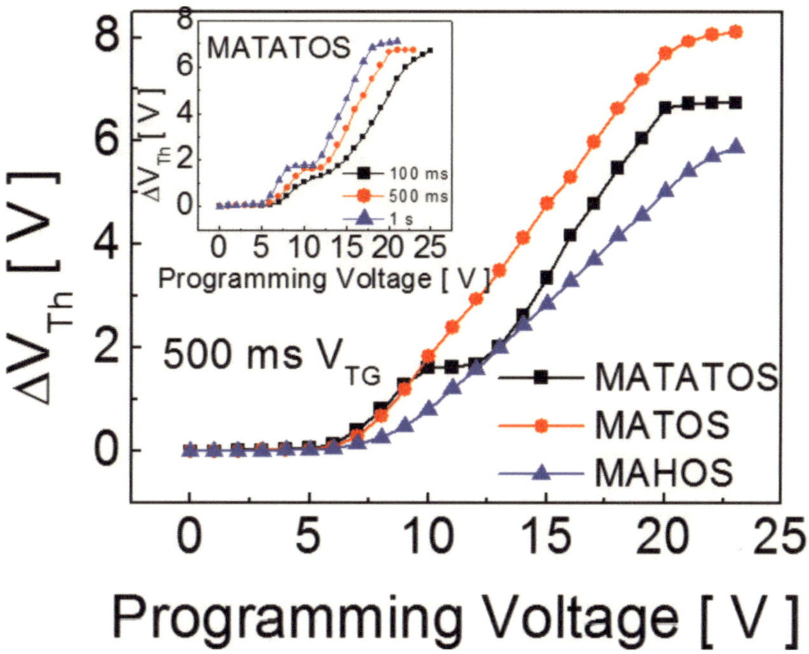

Figure 12. ΔV_{Th} as a function of programming voltage of the three devices. MATATOS device shows a two-step charging storage behaviour. Inset: ΔV_{Th} versus programming voltage of the MATATOS device under gate voltage pulse with different width [57].

By comparing the above results from the flash memory based on Si nanowire FET to the electrical performance of planar capacitor structure with same gate dielectric stack, the nanowire FET-based flash memory exhibits faster speed, better endurance and more remarkable discrete multi-bit memory storage at lower operation voltages [58]. The scaling from planar Si to nanoscale Si nanowire channel effectively enhance the gate electric field introduced by the gate-surrounding structure, enabling faster speed and lower operation voltage. The device reliability is improved with the clean nanowire surface and dielectric interfaces formed by using the self-alignment technique. Such a high-performance and CMOS compatible flash memory is very attractive for future multi-bit non-volatile memory applications.

4.3. Silicon nanowire-based molecular charge-trapping memory

4.3.1. Introduction of redox-active molecules

CMOS and semiconductor non-volatile memory scaling have generated various approaches towards building memory devices with higher scalability and better performance. The hybrid silicon/molecular approach is very attractive as a technology that leverages advantages afforded by a molecule-based active medium with the vast infrastructure of traditional MOS technology.

In reduction-oxidation, redox-active molecules can be attached on various surfaces such as Si and SiO_2 by forming a self-assembled monolayer (SAM) or multiple layers with simple and low-cost processes. Due to the inherent reduction and oxidation of the redox centres, such molecules can exhibit distinct charged and discharged states which can represent the logic on and off states. It has been demonstrated that the redox-active molecules attached on Si structures are stable and can endure more than 10^{12} P/E cycles [59]. Such an excellent reliability is derived from the intrinsic properties of redox molecules. Thus, incorporating redox-active molecules as charge-storage medium in a Si-based flash memory is quite interesting. By taking advantage of the high-quality thin oxide surrounding the Si nanowire, which is readily feasible in various chemical functionalizations, a molecular flash memory with redox molecules attached on the Si nanowire surface serving as the charge storage medium can have even better memory performance including lower operation voltage, faster speed, better device scalability and better reliability.

4.3.2. Redox-active molecules attachment and memory device fabrication

Two redox-active molecules were integrated and studied in the molecular flash memory: α-ferrocenylethanol (referred as ferrocene) and $Ru_2(ap)_4(C_2C_6H_4P(O)(OH)_2)$ (referred as Ru_2), in which ap = 2-anilinopyridinate, with the molecular structures shown in **Figure 13** [60].

The molecular flash memory device fabrication follows the self-alignment Si nanowire FET process as well. After the catalyst patterning, nanowire growth and oxidation and source/drain electrode formation, the molecules SAM attachment on the nanowire was performed by placing droplets of a solution of dichloromethane with 3-mM ferrocene and 2 mM Ru_2 on the active areas separately [60]. Each drop was in place for 3–4 minutes and the samples were held at 100°C in an N_2 environment during the attachment. Saturated SAM will be formed after ~30 minutes. After the self-assembly process, dichloromethane was used to rinse the substrates to remove any residual molecules that are not bonded to the SiO_2 surface. Then, the samples were immediately loaded into the ALD chamber for a deposition of 25 nm Al_2O_3 with TMA and H_2O as precursors at 100°C. Finally, a 100 nm Pd top gate was formed with photolithographic and lift-off processes. A reference sample without molecules was fabricated for comparative study.

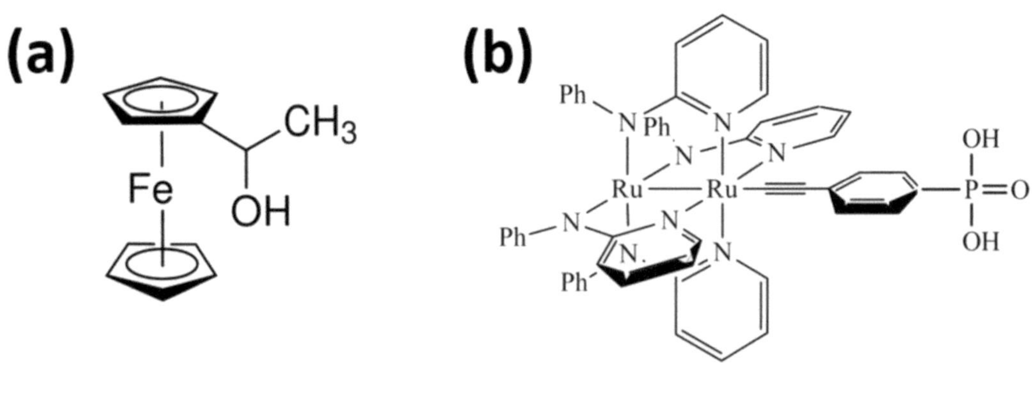

Figure 13. Molecular structure of (a) α-ferrocenylethanol and (b) $Ru_2(ap)_4(C_2C_6H_4P(O)(OH)_2)$ [60].

4.3.3. Electrical characterization and memory performance

Schematic structure of a completed molecular flash memory device was shown in **Figure 14a**. **Figure 14b** shows the TEM image of the cross-section of a ferrocene-attached molecular flash memory. A clear gate surrounding structure has been obtained, with an "intermixed" region observed (indicated by the red dash line). Schottky-barrier p-type MOSFET characteristics have been observed for the Si nanowire based molecular flash memory cells as the source/drain was engineered as Schottky junction. **Figure 15a** and **b** shows the output characteristics of a typical ferrocene molecular flash memory. Smooth and well saturated I_{DS}–V_{DS} curves have been observed with clear leakage-affected region and the weak, moderate and strong inversion operation regions. From the I_{DS}–V_{GS} curves shown in **Figure 15c** and **d**, counterclockwise hysteresis loops were obtained for both devices, suggesting the charge trapping mechanism. The log-scale transfer characteristics shown in the insets demonstrated an on/off ratio as high as ~10^7. The inset (ii) in **Figure 15c** shows the I_{DS}–V_{GS} curves of the reference sample (without molecules), and a negligible memory window was observed, ruling out the possibility of charge storage in the Al_2O_3 dielectric traps.

The P/E speed characterizations of the ferrocene flash memory were shown in **Figure 16a** and **b**. Threshold voltage shift towards the positive (negative) direction was observed during the programming (erasing) operations, indicating the electrons (holes) were injected from the Si nanowire through the SiO_2 and stored in the centres of the molecules. From **Figure 16c** and **d**, both molecular memory devices showed fast P/E speed, which arises from the intrinsic fast speed of the charging behaviour of the molecules and the strong electric field induced through the top gate control over the channel in the gate-surrounding structure.

Similar to the multiple charging behaviour demonstrated in the previous dielectric flash memory, the Ru_2 molecular flash memory is also designed and expected for the application

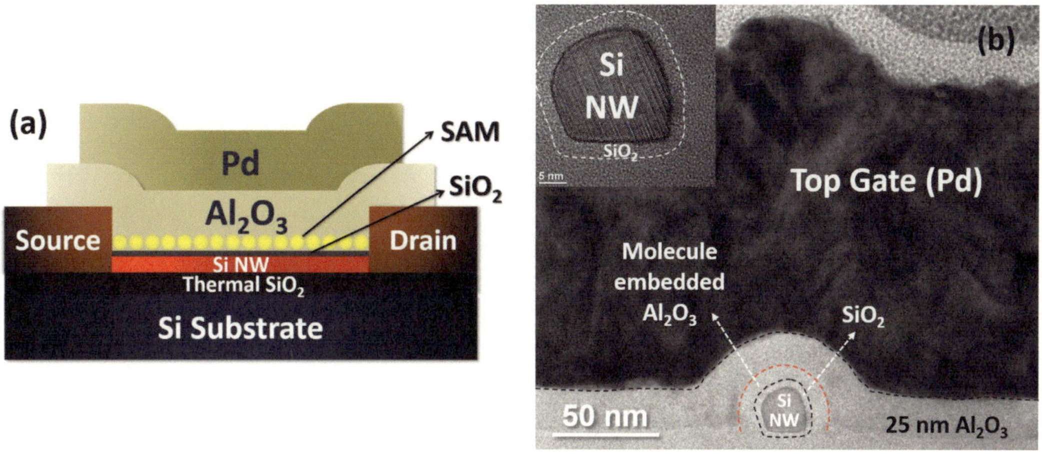

Figure 14. (a) Schematic structure of a completed molecular flash memory device based on Si nanowire FET; (b) TEM image of the cross section of a ferrocene-attached molecular flash memory. The red dashed line indicates the ferrocene-embedded Al_2O_3 region. Inset: Cross section of the nanowire channel, with SiO_2 layer indicated by the dashed line [60].

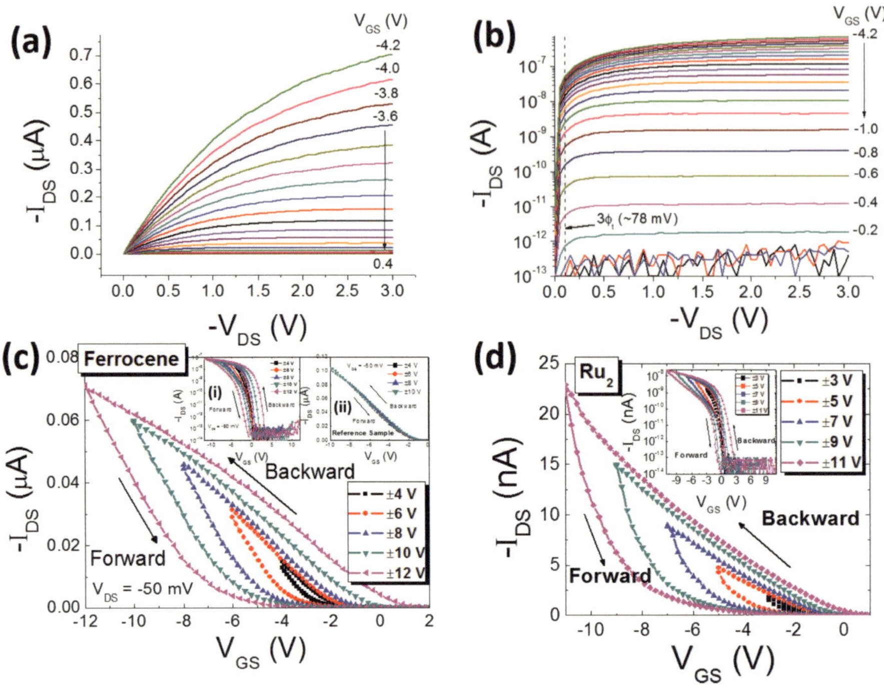

Figure 15. (a) Linear and (b) log-scale output characteristics of ferrocene molecular flash memory; I_{DS}–V_{GS} of the (c) ferrocene and (d) Ru_2 molecular flash memory with the insets showing the log-scale transfer curves. Inset (ii) in (c) shows the I_{DS}–V_{GS} curves of the reference sample with negligible hysteresis observed [60].

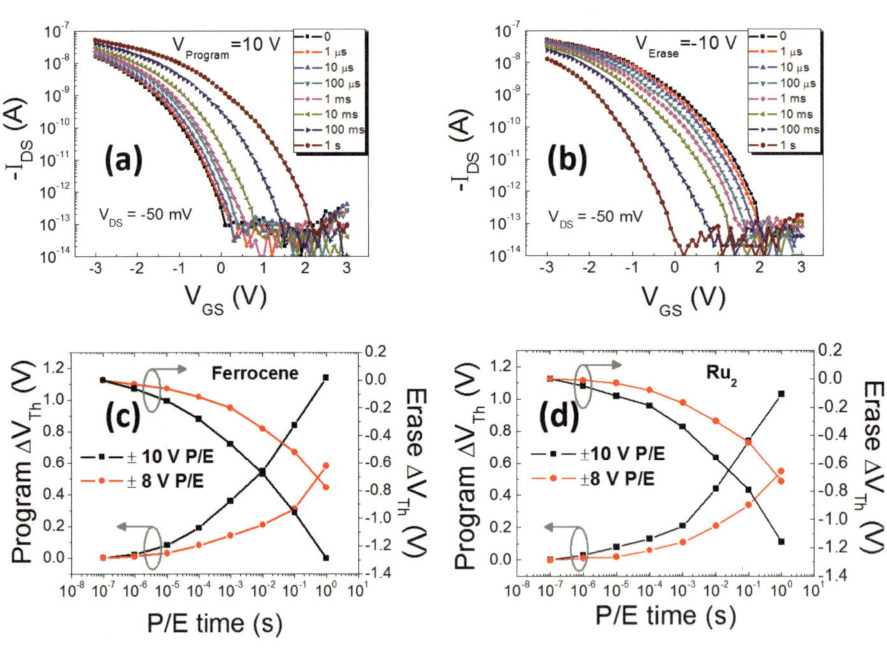

Figure 16. (a) Programming and (b) erasing operations of the ferrocene molecular memory under accumulative rectangular P/E gate voltage pulses. Speed characterizations of the (c) ferrocene and (d) Ru_2 molecular flash memory [60].

of multi-bit memory storage due to the two redox centres, which can exhibit stable and distinct charged states at different voltage levels [57, 60]. As shown in **Figure 17b**, two charged steps were observed at around 10 and 14 V, respectively. The overall charging density of the Ru_2 SAM was calculated to be 1.12×10^{13} cm^{-2}, which is sufficiently high for discrete multi-bit memory applications. One of the most intriguing features of a molecular flash memory is the reliability. **Figure 17c–f** shows the date retention and the endurance properties of the

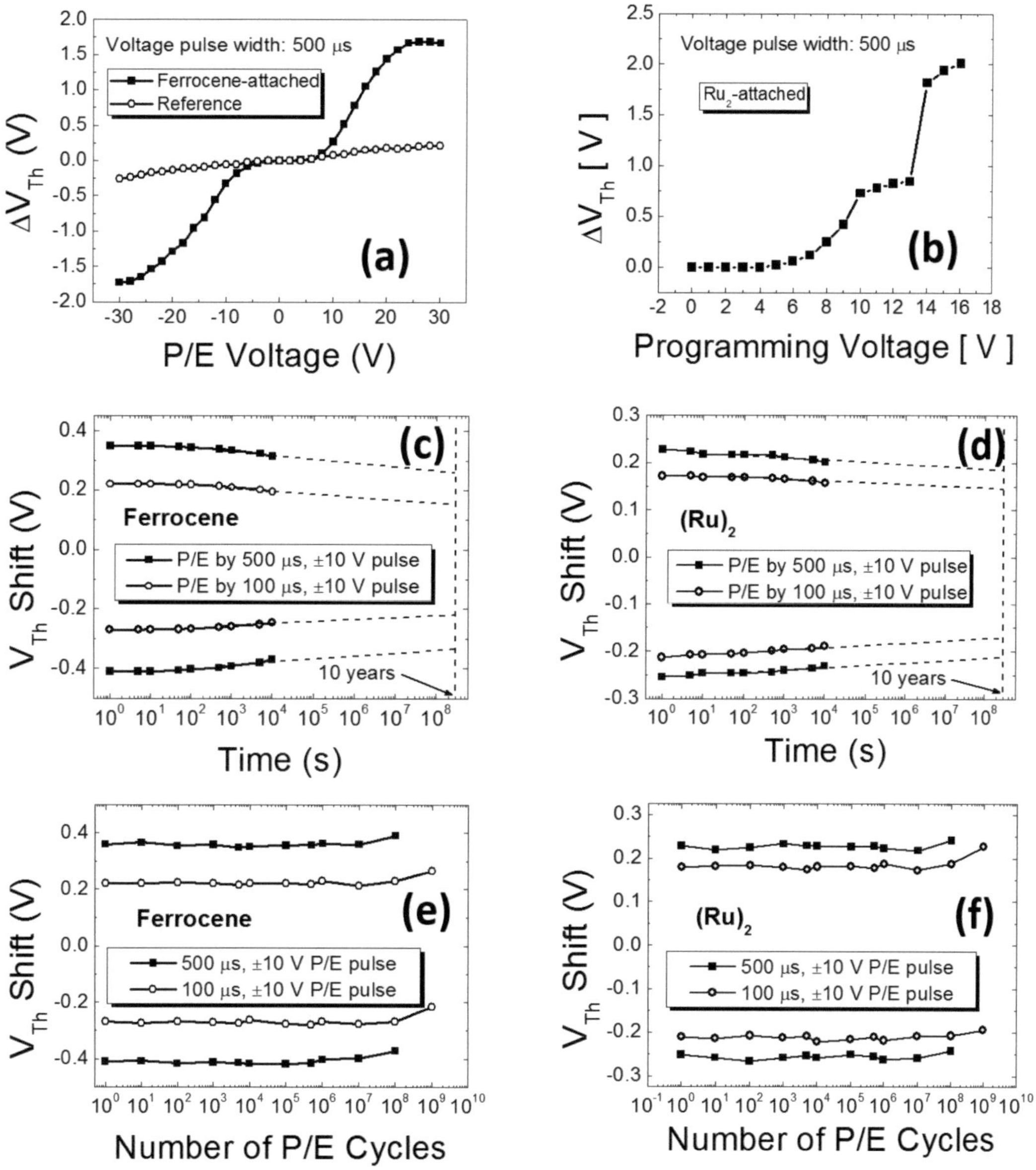

Figure 17. ΔV_{Th} of (a) ferrocene and reference sample and (b) Ru_2 flash memory as a function of P/E voltage. Room temperature retention characteristics of (c) ferrocene and (d) Ru_2 flash memory. Endurance properties of (e) ferrocene and (f) Ru_2 flash memory [60].

ferrocene and Ru_2 flash memory devices, respectively. The projected 10 year memory window showed a charge loss of only 20%, and the excellent endurance characteristics were demonstrated by the negligible memory window degradation after 10^9 P/E cycles, which is about 10,000 times better than that of the conventional floating gate memory [60]. Such a good reliability is due to the intrinsic stable redox behaviour of the molecules and the high-quality tunnelling oxide with clean solid/molecule and dielectric interfaces by using the self-aligned nanowire FET fabrication process. The nanowire FET based molecular flash memory is thus very attractive for future fast speed, high-endurance and high-density on-chip non-volatile memory applications.

5. Nanowire field-effect transistor as a platform for novel materials research

Semiconductor nanowires have shown unique properties in the manner of both physics and technology. The significance of nanowires over planar materials has been more and more discovered and focussed through various interesting and fundamental phenomena, when nanowires have nanoscale diameter at or even below the characteristic length scale of such basic parameters as phonon mean free path, exciton Bohr radius, magnetic domain size, exciton diffusion length and so forth [61]. Many physical properties of semiconductor nanowire are utterly different from the planar bulk materials due to the confines of nanowire surface. Moreover, the large surface-to-volume ratio of nanowires allows for distinct structural, electrical and transport behaviours, as well as advanced technological applications.

This section focuses on the applications of nanowire MOSFETs as a platform for novel materials research. Topological insulator materials and devices are discussed here as an example illustrating the significance of semiconductor with nanowire morphology in the understanding and implementation of fundamental physics and properties behind the materials.

5.1. Introduction to topological insulators

Topological insulators are characterized as a new class of materials having insulating band gaps in the bulk but gapless surface states topologically protected by time-reversal symmetry [62, 63]. Recently discovered three-dimensional topological insulators such as Bi_2Se_3, Bi_2Te_3 and Sb_2Te_3 have been intensively investigated both theoretically and experimentally [64, 65]. Most current experimental research focuses on the surface states of thin films grown by a molecular beam epitaxy (MBE) or mechanically exfoliated from bulk materials. For example, the gapless surface states featuring helical Dirac electrons have been observed by angle-resolved photoemission spectroscopy (ARPES) and scanning tunnelling microscopy (STEM) techniques. A few groups have reported the modification of surface conduction of such materials by doping, electrical gating or polarized light [66–68]. But there has rarely been reported of high-performance nanoelectronic devices based on topological insulators such as the analog of MOSFET. For conventional CMOS devices, the Si surface conduction is protected by thermal SiO_2 to optimize the inversion properties for good transistor performance. For topological insulators, the gapless surface state is derived from the inherent material properties and maintains a robust surface conduction. Therefore, it will be very attractive to integrate the topological insulators as the active conducting channel in MOSFETs.

However, despite the significant efforts made in engineering of materials and devices based on bulk topological insulators, it is always a challenge to modulate the surface conduction due to the dominant contribution from bulk conduction. Topological insulator nanowires can be expected to greatly enhance the surface conduction due to their extra-high surface-to-volume ratio. Here, we will review the work on MOSFET devices based on topological insulator nanowires and the effective separation of surface conduction from the bulk conduction by an external electrical means [69].

5.2. VLS nanowire synthesis of topological insulator Bi_2Se_3 nanowires

The Bi_2Se_3 topological insulator nanowire FETs were fabricated by following the self-alignment process, similar to that of previous Si nanowire FETs [69]. The essential steps are as follows: the SiO_2/Si substrates with patterned Au catalyst were loaded in the downstream end in a horizontal tube furnace while Bi_2Se_3 source powder was located at the heating centre. The furnace was heated to a temperature in a range of 500–550°C and kept for 2 hours with a 50 sccm flow of Ar gas as a carrier gas. The Bi_2Se_3 nanowires were grown following the VLS route at pre-defined locations with typical length of 20 μm and 50 nm in diameter. Ti (3 nm)/Pt (100 nm) source/drain electrodes were formed by photolithography. A layer of 30 nm HfO_2 was then deposited at 250°C by ALD covering the nanowire channel and also a part of source/drain electrodes. The final step is the formation of 100 nm Pd top gate electrode.

Figure 18a shows the SEM image of the as-synthesized Bi_2Se_3 nanowires. Au nanoparticles were found at the top end of each wire, indicating the VLS mechanism. The high-resolution TEM (HRTEM) image shown in **Figure 18** demonstrates that the Bi_2Se_3 nanowires are in a well-defined single-crystal rhombohedral phase with growth direction close to $[11\bar{2}0]$. From the cross-sectional TEM image (**Figure 18d**); it is clear that the hexagonal nanowire core is surrounded by the insulating HfO_2 and the Omega-shaped top gate [69].

5.3. Characterizations of Bi_2Se_3 nanowire FET

The Bi_2Se_3 nanowire FET shows excellent transfer characteristics, as shown in **Figure 19**, such as close-to-zero cutoff current, strong-inversion-like on state current and over 10^8 on/off ratio within 1.0 V gate voltage. The nanowire FET has unipolar current dominated by electron conduction, which is similar to a conventional long-channel Schottky-barrier MOSFET with either electron or hole conduction determined by the unipolar Schottky junctions at the source and drain. **Figure 19b** and **c** shows well-saturated and smooth I_{DS}–V_{DS} curves. I_{DS} saturates roughly at $V_{DS} = V_{GS} - V_{Th}$ in the highly conductive region but does not saturate at $V_{DS} \approx 3\phi_t$ in the weak/moderate conductive region. Instead, I_{DS} keeps increasing significantly after $3\phi_t$, suggesting the fact that the Bi_2Se_3 nanowire FET does not follow the diffusion current model as described for the conventional MOSFETs. It is believed that I_{DS} in the weak/moderate conductive regions is also dominated by drift current [69].

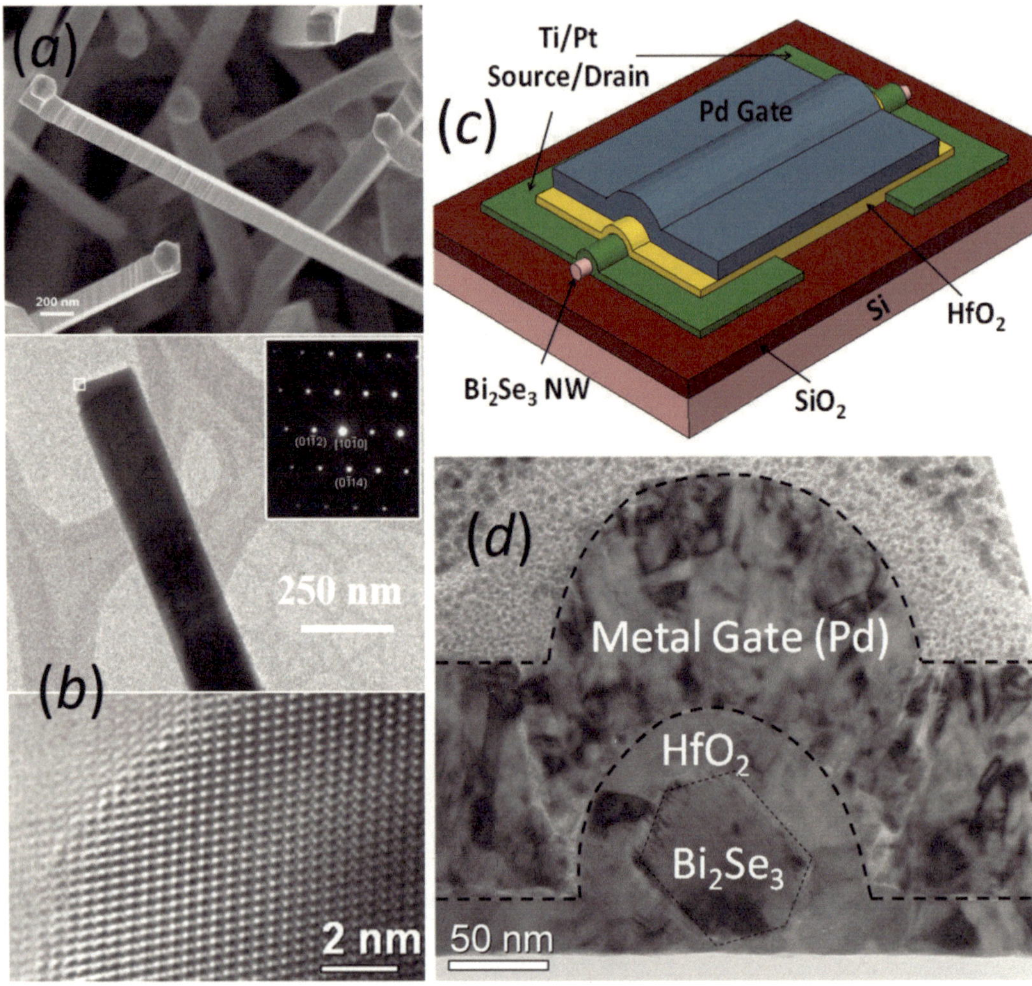

Figure 18. (a) SEM image of Bi_2Se_3 nanowires synthesized by following the VLS mechanism; (b) HRTEM image of Bi_2Se_3 nanowire showing the $[11\bar{2}0]$ growth direction. Inset: Magnified region of the nanowire; (c) schematic structure of a Bi_2Se_3 nanowire FET; (d) cross-sectional TEM image of a Bi_2Se_3 nanowire FET [69].

Further characterizations at different temperatures confirm that the saturation of I_{DS} is due to the electron velocity saturation at the source end instead of the pinch-off at the drain end of the nanowire channel, because a linear relationship was observed between the saturation drain current and the over-threshold voltage (V_{GS}–V_{Th}) as shown in **Figure 19**. By comparing the transfer characteristics at different temperatures (**Figure 20a**), it was found that I_{DS}–V_{GS} curves obtained at temperatures lower than 240 K show a clear cutoff region with higher on/off ratio. The off state current for temperatures greater than 240 K flattens and saturates at negative voltages much below V_{Th}. Such temperature dependence indicates metallic conduction in the on state and insulating behaviour in the off state. By fitting the strongly activated temperature-dependent current to $I_{DS,Off} = I_0 e^{-E_a/kT}$ where E_a is the activation energy, k is

http://dx.doi.org/10.5772/67446

Boltzmann's constant and I_0 is a constant prefactor. E_a value of about 0.33 eV was obtained which is very close to the reported band gap value of bulk Bi_2Se_3 [69].

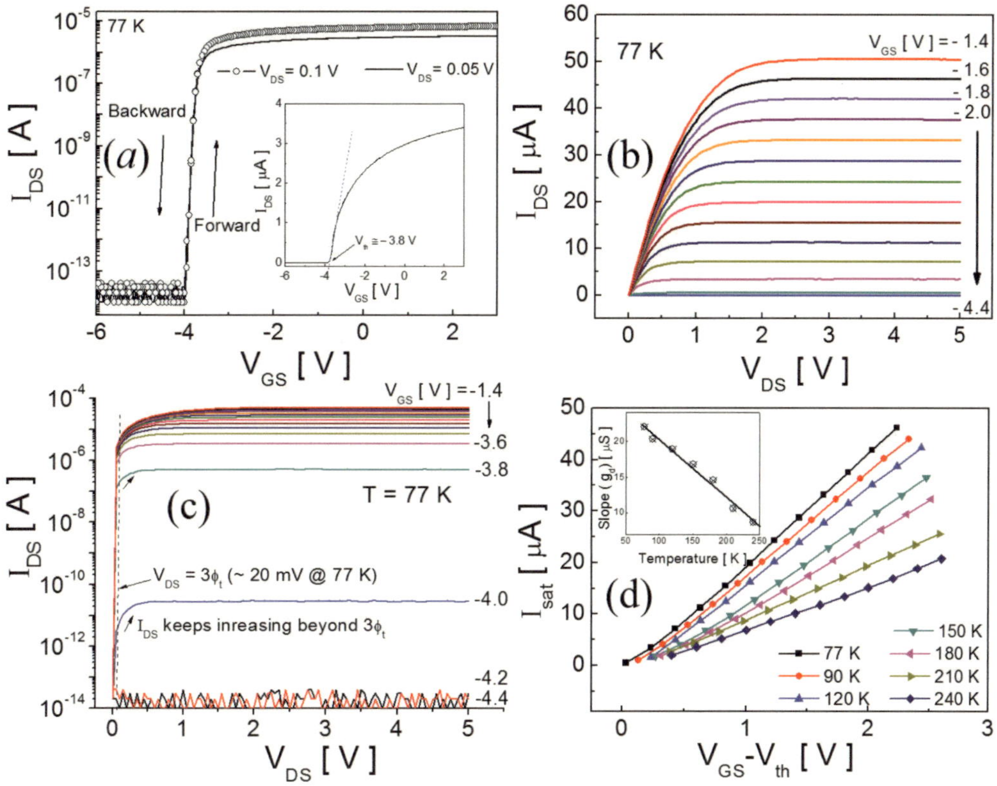

Figure 19. (a) Transfer characteristics and (b and c) output characteristics of the Bi_2Se_3 nanowire FET at 77 K. (d) I_{DS} as a function of over-threshold voltage. Inset: Linear fit slope versus temperature [69].

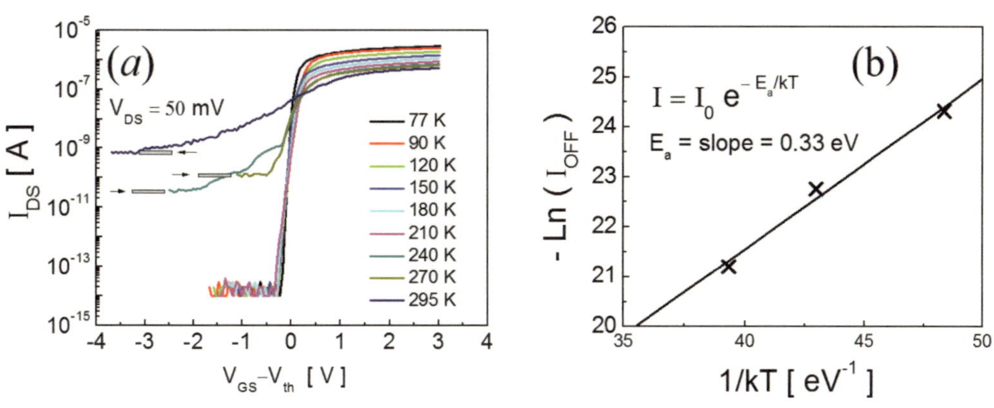

Figure 20. (a) I_{DS}–V_{GS} curves at different temperature; (b) $\ln(I_{DS})$ at off state versus 1/kT above 240 K and its fit to $I_{DS,Off} = I_0 e^{-E_a/kT}$. [69].

5.4. Performance discussion

In the off state, the gate voltage is large enough to fully deplete the electrons from the nanowire, and the small temperature-dependent off-state current is due to the thermal excitations across the energy band gap of the bulk of Bi_2Se_3 nanowire. This further indicates the fact that the electric field generated surround the gate by the gate voltage below the threshold is strong enough to modify the spectrum of the nanowire and destroy the surface conduction channels [69]. Different with the conventional semiconductor nanowires, the saturated current of the Bi_2Se_3 nanowire FET in the on state is linear in gate voltage, indicating metallic conduction and is most likely flowing at the nanowire surface. Such interpretation also agrees with the temperature dependence shown in **Figure 20a**. The most significant achievement is that the surface metallic conduction and the insulating switch-off can be controlled by a surprisingly small gate voltage, resulting from the excellent gate control by the surrounding gate nanowire FET structure.

The above electrical performance obtained from the Bi_2Se_3 topological insulator nanowire FET is very impressive by taking advantages of the nanowire surrounding gate structure, leading to an enhanced gate control over the nanowire channel to realize electrical behaviour that have not been observed on planar counterparts. For example, the sharp switching from cutoff to surface conduction and saturation current by a gate voltage of a few volts is neither expected nor has been previously reported. Since the spin and momentum are locked in the surface states of topological insulators, possibilities of electric manipulation of spin current using gate voltage as well as novel circuit applications may be opened up using such one-dimensional topological insulator materials.

6. Conclusions

The physics and operation principles of nanowire materials and device have been systematically studied. Semiconductor nanowires enable the surrounding-gate structures, which significantly enhance the gate control over the channel in the electrical devices, leading to quite distinct and interesting device behaviours compared with the planar or bulk materials. The "self-alignment" approach enables simultaneous batch fabrication of large numbers of nanowire devices, while effectively reducing the processing steps in which nanowire surfaces might be contaminated. Si nanowire FETs with excellent electrical performances have been used as the platform to fabricate novel flash non-volatile memory devices. Both high-k dielectric and molecular charge-trapping memory demonstrate excellent memory behaviour and are very interesting for future on-chip non-volatile memory applications. The "self-alignment" method and the nanowire FET device architecture have been proved to be an effective platform and approach to be implemented and study other novel materials. High-performance topological insulator Bi_2Se_3 nanowire FET has been fabricated and investigated. The surface states are successfully separated from the bulk conduction within a small range of gate voltage due to the strong electric field induced through the surrounding-gate structure formed in the self-alignment fabrication process. Such a high-performance nanoelectronic device and the analysis on surface conduction have never been previously reported. Therefore, the nanowire MOSFETs not only exhibit their potential in future CMOS scaling at advanced technology nodes but also provide an excellent approach for novel materials research towards next-generation micro and nanoelectronic devices.

Author details

Hao Zhu

Address all correspondence to: hao_zhu@fudan.edu.cn

School of Microelectronics, Fudan University, Shanghai, P.R. China

References

[1] Dennard R.H., Gaensslen F.H., Yu H.N., Rideout V.L., Bassous E., Leblanc A.R. Proceedings of the IEEE. 1999;**87**:668-678.

[2] Moore G.E. IEDM Technical Digest. 1975;11-13.

[3] Tans S.J., Verschueren A.R.M., Dekker C. Nature. 1998;**393**:49-52.

[4] Wagner R.S., Ellis W.C. Applied Physics Letters. 1964;**4**:89-90.

[5] Ucjikoga S. MRS Bulletin. 2002;**27**:881-886.

[6] Duan X..F, Niu C.M., Sahi V., Chen J., Parce J.W., Empedocles S., et al. Nature. 2003;**425**:274-278.

[7] Wu Y.C., Chang T.C., Liu P.T., Chou C.W., Wu Y.C., Tu C.H., et al. IEEE Transactions on Nanotechnology. 2006;**5**:157-162.

[8] Mertens H., Ritzenthaler R., Chasin A., Schram T., Kunnen E., Hikavyy A., et al. IEDM Technical Digest. 2016;524-527.

[9] Gunawan O., Sekaric L., Majumdar A., Rooks M., Appenzeller J., Sleight J.W., et al. Nano Letters. 2008;**8**:1566-1571.

[10] Zervas M., Sacchetto D., Micheli G.D., Leblebici Y. Microelectron Engineering. 2011;**88**:3127-3132.

[11] Westwater J., Gosain D.P., Tomiya S., Usui S., Ruda H. Journal of Vacuum Science & Technology B. 1997;**15**:554-557.

[12] Huang Y., Duan X.F., Wei Q.Q., Lieber C.M. Science. 2001;**291**:630-633.

[13] Salalha W., Zussman E. Physics Fluids. 2005;**17**:063301.

[14] Evoy S., DiLello N., Deshpande V., Narayanan A., Liu H., Riegelman M., et al. Microelectron Engineering. 2004;**75**:31-42.

[15] Papadakis S.J., Hoffmann J.A., Deglau D., Chen A., Tyagi P., Gracias D.H. Nanoscale. 2011;**3**:1059-1065.

[16] Li Q.L., Koo S.M., Richter C.A., Edelstein M.D., Bonevich J.E., Kopanski J.J., et al. IEEE Transactions on Nanotechnology. 2007;**6**:256-262.

[17] Hertel T., Martel R., Avouris P. Journal of Physical Chemistry B. 1998;**102**:910-915.

[18] Wu Y., Cui Y., Huynh L., Barrelet C.J., Bell D.C., Lieber C.M. Nano Letters. 2004;**4**:433-436.

[19] Schmidt, V., Senz S., Gosele U. Nano Letters. 2005;**5**:931-935.

[20] Wang C.X., Hirano M., Hosono H. Nano Letters. 2006;**6**:1552-1555.

[21] Akiyama T., Nakamura K., Tomonori I. Physical Review B. 2006;**74**:033307.

[22] Zhao Y., Yakobson B.I. Physical Review Letters. 2003;**91**:035501.

[23] Marks L.D. Reports on Progress in Physics. 1994;**57**:603-649.

[24] Ma D.D.D., Lee C.S., Au F.C.K., Tong S.Y., Lee S.T. Science. 2003;**299**:1874-1877.

[25] Lee B., Rudd R.E. Physical Review B. 2007;**75**:041305.

[26] Ma L., Wang J., Zhao J., Wang G. Chemical Physics Letters. 2008;**452**:183-187.

[27] Leu P.W., Svizhenko A., Cho K. Physical Review B. 2008;**77**:235305.

[28] Canham L.T. Applied Physics Letters. 1990;**57**:1046-1048.

[29] Ng M.F., Zhou L., Yang S.W., Sim L.Y., Tan V.B.C., Wu P. Physical Review B. 2007;**76**:155435.

[30] Yao D., Zhang G., Li B. Nano Letters. 2008;**8**:4557-4561.

[31] Wang J., Polizzi E., Ghosh A., Datta S., Lundstrom M. Applied Physics Letters. 2005;**87**:043101.

[32] Luisier M., Schenk A., Fichtner W. Applied Physics Letters. 2007;**90**:102103.

[33] Persson M.P., Lherbier A., Niquet Y.M., Triozon F., Roche S. Nano Letters. 2008;**8**:4146-4150.

[34] Lherbier A., Persson M.P., Niquet Y.M., Triozon F., Roche S. Physical Review B. 2008;**77**:085301.

[35] Geaney H., Mullane E., Ryan K.M. Journal of Materials Chemistry C. 2013;**1**:4996-5007.

[36] Kornienko N., Whitmore D.D., Yu Y., Leone S.R., Yang P. ACS Nano. 2015;**9**:3951-3960.

[37] Zhang D., Eaton S.W., Yu Y., Dou L., Yang P. Journal of the American Chemical Society. 2015;**137**:9230-9233.

[38] Furneaux R.C., Rigby W.R., Davidson A.P. Nature. 1989;**337**:147-149.

[39] Fleisher R.L., Price P.B., Walker R.M. Nuclear Tracks in Solids. Berkeley: University of California Press; 1975.

[40] Wang C.C., Kei C.C., Yu Y.W., Perng T.P. Nano Letters. 2007;**7**:1566-1569.

[41] Kemell M., Pore V., Tupala J., Ritala M., Leskela M. Chemistry of Materials. 2007;**19**:1816-1820.

[42] Zhu H., Li Q., Yuan H., Baumgart H., Ioannou D.E., Richter C.A. Solid-State Electronics. 2012;**78**:92-96.

[43] Colinge J.P. Solid-State Electronics. 2004;**48**:897-905.

[44] Atwood G. IEEE Transactions on Device and Materials Reliability. 2004;**4**:301-305.

[45] Lai S.K. IBM Journal of Research and Development. 2008;**52**:529-535.

[46] Pavan P., Bez R., Olivo P., Zanoni E. Proceedings of the IEEE. 1997;**85**:1248-1271.

[47] Wu K.H., Chien H.C., Chan C.C., Chen T.S., Kao C.H. IEEE Transactions on Electron Devices. 2005;**52**:987-992.

[48] Honda K., Hashimoto S., Cho Y. Applied Physics Letters. 2005;**86**:063515.

[49] Wrazien S.J., Zhao Y.J., Krayer J.D., White M.H. Solid-State Electronics. 2003;**47**:885-891.

[50] Fuhrer M.S., Kim B.M., Brintlinger T. Nano Letters. 2002;**2**:755-759.

[51] Choi W.B., Chae S., Bae E., Lee J.W., Cheong B.H., Kim J.R., Kim J.J. Applied Physics Letters. 2003;**82**:275-277.

[52] Cha H.Y., Wu H., Chae S., Spencer M.G. Journal of Applied Physics. 2006;**100**:024307.

[53] Zhu X., Li Q., Ioannou D.E., Gu D., Bonevich J.E., Baumgart H., et al. Nanotechnology. 2011;**22**:254020.

[54] Wang Y.Q., Hwang W.S., Zhang G., Samudra G., Yeo Y.C., Yoo W.J. IEEE Transactions on Electron Devices. 2007;**54**:2699-2705.

[55] Huang X.D., Lai P.T., Liu L., Xu J.P. Applied Physics Letters. 2011;**98**:242905.

[56] Huang X.D., Liu L., Xu J.P., Lai P.T. Applied Physics Letters. 2011;**99**:112903.

[57] Zhu H., Bonevich J.E., Li H., Richter C.A., Yuan H., Kirillov O., et al. Applied Physics Letters. 2014;**104**:233504.

[58] Zhu H., Yuan H., Li H., Richter C.A., Kirillov O., Ioannou D.E., Li Q. IEEE Transaction on Nanotechnology. 2013;**12**:1151-1157.

[59] Liu Z.M., Yasseri A.A., Lindsey J.S., Bocian D.F. Science. 2003;**302**:1543-1545.

[60] Zhu H., Pookpanratana S.J., Bonevich J.E., Natoli S.N., Hacker C.A., Ren T., et al. ACS Applied Material and Interfaces. 2015;**7**:27306-27313.

[61] Hochbaum A.I., Yang P. Chemical Reviews. 2010;**110**:527-546.

[62] Bernevig B.A., Hughe, T.L., Zhang S.C. Science. 2006;**314**:1757-1761.

[63] Zhang H., Liu C.X., Qi X.L., Dai X., Fang Z., Zhang S.C. Nature Physics. 2009;**5**:438-442.

[64] Hasan M.Z., Kane C.L. Reviews of Modern Physics. 2010;**82**:3045-3067.

[65] Qi X.L., Zhang S.C. Reviews of Modern Physics. 2011;**83**:1057-1110.

[66] Kong D., Chen Y., Cha J.J., Zhang Q., Analytis J.G., Lai K., et al. Nature Nanotechnology. 2011;**6**:705-709.

[67] Steinberg H., Gardner D.R., Lee Y.S., Jarillo-Herrero P. Nano Letters. 2010;**10**:5032-5036.

[68] Cho S., Butch N.P., Paglione J., Fuhrer M.S. Nano Letters. 2011;**11**:1925-1927.

[69] Zhu H., Richter C.A., Zhao E., Bonevich J.E., Kimes W.A., Jang H.J., et al. Scientific Reports. 2013;**3**:1757.

Silicon Quasi-One-Dimensional Nanostructures for Photovoltaic Applications

Rosaria Anna Puglisi, Valentina Lombardo and Sebastiano Caccamo

Additional information is available at the end of the chapter

http://dx.doi.org/10.5772/67749

Abstract

Thanks to the silicon abundance, stability, non-toxicity and well known electronic properties, Si based solar cells have represented the leading actors in the photovoltaic market and future projections confirm this predominance. However, half of the module cost is due to the material consumption and processing. In order to decrease the costs, a cut in the Si consumption must be operated, with consequent decrement in the optical absorption, generated current and device efficiency. To keep the performance level, a proper Si surface design with the objective to trap the light, has been developed. One of the most popular approaches is to use silicon nanowires embedded in the solar cell emitter where they play the role of optically and electrically active layer, thanks to their excellent optical absorption properties. However, also another material has been the terminus of the light-trapping materials, the silicon nanoholes. Their mechanical robustness is superior, making their integration inside the cell easier and cost-effective. The review will bring about all of the most common methods to fabricate these two types of nanostructures when used for solar cells applications, their optical properties and some critical aspects related to their high surface to volume ratio which modify the recombination processes.

Keywords: silicon, nanowires, nanoholes, synthesis, solar cells

1. Introduction

Silicon nanowires (SiNWs) are very popular in the third-generation solar cells because of their outstanding electrooptical properties [1–4]. Once integrated within the cell active area, they: (i) increase the solar spectrum optical absorption, thanks to light-trapping mechanisms, thus increasing the electron-hole generation and consequently, the device short circuit current [2, 5]; (ii) due to spatial confinement, improve the coupling with certain wavelengths [6, 7]; (iii) their radial symmetry allows for the formation of core-shell junctions, which decouple the light path from carrier path, improving independently the absorption and the electricity collection [8, 9]. Currently, another quasi-one-dimensional (1D) material is overlooking the panorama of the solar cells latest generation: nanoholes [10, 11]. These nanowire "mirror" structures offer all their advantages but are more robust and then easily integrable. The review aims to give an overview of all the synthesis methods and electrooptical properties of both structures, for the first time, presently proposed in the literature giving a space to the issues still open. The two complementary materials, SiNWs and silicon nanoholes (SiNHs), are first described in terms of the synthesis methods. As far as SiNWs are concerned, various techniques are present in the literature, mostly based on growth or etch, both assisted by metals. The vapor-liquid-solid (VLS) method [12, 13] is among the most popular methods to grow the nanowires (NWs). In this process, a liquid eutectic droplet is formed by a metal catalyst particle and the solid Si substrate brought at high temperature. A vapor Si precursor reacts with the liquid eutectic by dissociating the Si adatom, which then diffuses inside it to precipitate and form the 1D NWs. Many aspects dramatically controlling the NW morphology like the precursor flow gas [14], the substrate type, and crystallographic orientation [8, 12, 15]; the SiNWs growth direction [12] and substrate surface chemical conditions [16] will be reviewed and discussed. When SiNWs are deposited by using plasma enhanced chemical vapor deposition (CVD) systems, also the plasma power plays an important role in the formation of SiNWs [17]. The most common metal exploited as a catalyst is gold, thanks to its low eutectic temperature. A part of the literature recognizes the importance to remove the Au catalyst droplets, residual after the growth, because they create deep level traps in Si and increase carrier recombination [18, 19]. To avoid the electronic defects introduced by Au, also other catalysts will be discussed [20]. The review will also address the metal-assisted chemical etching (MAE), another very popular method to obtain SiNW arrays [6, 21–27]. MAE is based on a redox reaction in which a metal with a suitable reduction potential, e.g. Ag, catalyzes the Si oxidation through a highly oxidant agent, such as peroxide aqueous solution. The so-formed silicon oxide is thus removed by the etchant agent. Again the substrate properties such as the synthesis process conditions, the doping type, and concentration will be examined [22, 24]. Nanosphere lithography (NSL) is usually applied together with the MAE process to obtain the ordered SiNWs. It exploits the self-assembly of water-dispersed polystyrene (PS) nanoparticles onto a substrate, where they spontaneously arrange in a hexagonal pattern. A metal layer is then thermally evaporated, acting as a catalyst for successive MAE etch in the regions not protected by the PS particles, which leads to the formation of the nanostructures [26]. SiNWs are excellent platforms to build core-shell architectures, where the inner part of the 1D Si nanostructure is doped of one type, while the external shell is doped of the opposite type [8]. This geometry allows to separate the electrical carrier path, which proceeds along the NW radial direction, from the optical absorption, which takes place along the axial one. In this way, the carrier collection distance is smaller or close to the minority carrier diffusion length. A few approaches presented in the literature propose to add a layer of intrinsic Si between the inner and outer regions, forming a p-i-n coaxial junction [9], in order to increase the effect of the electric field of the depleted region during the charge separation process. Regarding the doping methods for the SiNWs, conventional routes, such as gas source-based techniques or ion implantation [18], will be described, as well as the new cost-effective approach called molecular monolayer doping (MD), consisting in immersing the Si to be doped inside a liquid solution with dopant chemical precursors [28]. The techniques of formation SiNHs will be then considered [10, 29]. Etching in this case is the only key process, typically assisted by metal nanoparticles or in a more standard approach through dry etch machines. The lithographic masks are used if ordered NHs are targeted. As for the NW formation, the metal-assisted etching is a very popular method to create NHs for its versatility and affordability. The size of the holes obtained is strictly related to the metal amount, substrate orientation, and etching duration [11, 21, 30]. The roughness left by the etching process is well recognized in the literature as an issue for the application of solar cells because of its impact on the carrier recombination, thus different surface passivation procedures are developed [11, 31]. UV lithography merged with the MAE etch is proposed in the literature as one of the most conventional strategies for the synthesis of an ordered nanoporous pattern [10]. As alternative strategies, many interesting nonconventional lithographies proposed in the literature will be reviewed. Among the others, nanoimprinting lithography (NIL) is a high-throughput and low-cost technique and not limited by the effects of wave diffraction or scattering which instead occur in the photolithographic methods. For these reasons, this technique can be considered as a promising candidate for mass production [32, 33]. Block copolymer (BCP) lithography is another useful alternative, thanks to its simplicity, low cost, and suitability for the semiconductor industry [34]. The basic principle is the self-assembly of block-copolymers: when a BCP solution is spun onto a solid substrate, it can lead to the formation of a specific template over the whole sample area. The realized pattern can act as a soft mask for the successive pattern transfer to underlying substrates. A less known method to fabricate NHs will be also described, the focused ion beam (FIB) which lets to obtain diameters below few nm [33]. After the synthesis the SiNHs, optical properties [35–38] will be presented, highlighting for the first time the pros and cons of both structures.

2. Silicon nanowires

The VLS method is one of the predominant techniques to form SiNWs. In the VLS method, the Si precursor gas, typically silane, is introduced into a CVD or a physical vapor deposition chamber where liquid metal catalyst particles, forming an eutectic with the substrate, react with the Si precursor atoms. Inside the liquid metal droplet, the Si atoms undergo to supersaturation and diffuse toward the crystalline Si growing interface. The nucleation of SiNWs depends on the deposition temperature and the partial pressure of the precursor gas. At low pressures, the chemical potential of the wire is higher than that of the vapor phase due to the high surface-to-volume ratio, and this prevents the wire nucleation in small eutectic droplets

[39]. When the pressure increases, the steam chemical potential increases as well and the growth of small nanowires becomes possible [40, 41]. SiNWs grown by the VLS method can change dramatically their morphology by changing also the precursor gas flow [14]. SiNWs have been grown directly on a p-type <111> Si substrate. Gold catalysts were deposited by sputtering in Ar plasma. The size of Au nanoparticles was controlled by the sputtering conditions and the thickness of Au film. After the metal deposition, SiNWs were grown using SiH_4 and N_2 as a carrier gas and changing flow gas ratio (SiH_4:N_2) as in the following: 1:3.2, 1:6, and 1:8. Scanning electron microscopy (SEM) and X-ray diffraction (XRD) results showed that with the second dilution (1:6) the density obtained was the highest, the size of NWs more uniform and the crystalline structure more pronounced than in the other cases. With respect to thermal CVD, plasma-enhanced CVD systems present the advantage to decompose the precursor gas by using the plasma energy instead of the high temperature. Moreover, in an inductively coupled plasma (ICP) CVD reactor, the high plasma density and the absence of acceleration of ions toward the substrate grant the absence of structural substrate damage. In reference [17], different plasma power values between 0 and 1000 W were explored to study its role on the Si NWs growth. Morphological results showed that three growth regimes can be identified: a low plasma region from 0 to 60 W, where only 1D SiNW structures were grown, an intermediate region from 60 to 100 W where both 1D and 2D films were obtained, and a third zone at high plasma power (from 100 to 1000 W) where SiNWs were absent and only continuous uncatalyzed Si layer was obtained. When deposited by VLS onto thermally oxidized Si substrates [12], a percentage of 44% of SiNWs with radius smaller than 10 nm showed by high-resolution Transmission Electron Microscopy (HRTEM) a diamond-like structure, while this characteristic was not present in those with larger diameters, which presented the wurtzite structure confirmed also by Raman measurements. In order to explain this result, a thermodynamic analysis was carried out considering the difference in standard free energy between diamond and wurtzite structure in NWs and the contribution due to the surface energy and stress. They found that the final structure is produced by the large stress in the surface curvature developed during the Si precipitation and nucleation with the catalyst. The surface pressure during the successive growth is reduced by the development of facets. VLS is also used to fabricate hexagonally faceted SiNWs as large as 30 μm in diameter [13]. SiNWs were grown on Si <111> substrates by Au catalyzed VLS process by using $SiCl_4$ as a precursor and H_2 as a carrier gas. The metal catalyst was first deposited onto the substrate by thermal evaporation and then annealed. The SiNWs obtained by this process present hexagonal facets and have diameters ranging between 30 nm and 30 μm, with lengths larger than 50 μm. After the deposition, they dispersed the SiNWs by sonication on an n-type Si wafer and doped the NW by p-type Al diffusion. Then deposited by lithography four metallic contacts to perform four-point-probe measurements on a single NW. They demonstrate ohmic n-type behavior and measure the contact resistance, the effective resistivity and the doping level. They found that bulk resistivities ranged from 0.02 to 0.08 Ω cm, corresponding to 1×10^{18} to 1×10^{17} cm^{-3} electrically active dopants. P- and n-type doped SiNWs have also been obtained by VLS growth in CVD with SiH_4 [15]. The doping process took place during the growth by adding PH_3 or B_2H_6 to the process gas. B-doped Si <111> substrates or borosilicate glass substrates with a thin B-doped multicrystalline Si layer on top were used. The NWs were grown and doped under identical conditions for both substrate types. Morphological characterization showed no differences in

the SiNWs despite the different substrates, with lengths of about 3–6 μm and diameters of about 20–100 nm with a 2-nm thick SiO_2 shell. SiNWs can be grown by VLS also on stainless steel foil substrates [8]. In this case, the experiment started with a 5-nm thick Au film deposited as a catalyst layer over the stainless steel substrate. Then p-type SiNWs have been synthesized in a CVD chamber at 650°C by using silane, hydrogen, hydrochloric acid, and trimethylboron as the dopant precursor gas. The obtained SiNWs show diameters of about 100 nm and a length of about 16 μm. The array was first oxidized and then dipped in HF to remove the grown oxide, in order to expose a part of the SiNWs to the deposition of a thin n-type a-Si conformal layer by plasma enhanced CVD (PECVD) using PH_3 as a doping gas. In such a way, a core-shell structure is formed. It presents the advantage to separate the electrical carrier path, which travels along the NW radius, from the optical absorption, which takes place along the NW length. The deposition of the amorphous Si layer is proposed by the authors as an effective strategy to passivate the NWs through the minimization of the nonradiative surface recombination. The introduction of an intrinsic layer in the p-i-n coaxial SiNWs array for fabrication of solar cells has also been suggested in the literature [9], with the advantage to increase the effect of the electric field in the depleted region during the charge separation process. The p-type core of SiNW was fabricated by the VLS method, while the intrinsic and n-type shells were produced by chemical vapor deposition using in the external shell phosphine as a precursor. The outer portions were grown at temperatures higher than the core to suppress the eutectic mediated growth and allow the deposition the NW surface. TEM analysis showed that the p core of SiNW is monocrystalline, and the shells are polycrystalline. An average value of SiNW diameter was measured to be 300 nm. Then p-i-n coaxial SiNW solar cells were fabricated and characterized under air mass 1.5 global (AM 1.5G) illumination. I–V showed an open circuit voltage V_{oc} of 0.260 V, a short circuit current I_{sc} of 0.503 nA, a fill factor FF of 55.0%, and an efficiency of 3.4%. Much more complicated structures are proposed in reference [42]. In this chapter, the authors fabricated four different coaxial SiNW architectures, respectively, p/n, p/in, p/pn, and p/pin. The SiNWs core was grown by the Au-catalyzed VLS method, while the shell was grown by the vapor solid (VS) method using SiH_4, B_2H_6, and PH_3. The core diameters were 100 nm. High-resolution transmission electron microscopy (HRTEM) is used to understand a crystal orientation <112> of the core along the length of the NW. The external NW surfaces were instead assigned to two {111} planes, two {011} planes, and four {113} planes. Energy Dispersive X-Ray Spectroscopy (EDX) was used to characterize the phosphorous dopant and oxygen profiles in the p/pin sample. The O and P profiles and the spatial maps of the P X-ray counts exhibited signals at the edges of the outer n-shell, with the P peaks shifted by 10–15 nm inward with respect to the O ones, thus indicating that the phosphorus atoms are localized at the interface with the oxidized shell. The location of dopant active atoms in SiNWs fabricated by VLS was also studied after one or more steps of chemical oxidation and etching which reduced controllably the SiNWs diameter [43]. The n-type (p-type) SiNWs were synthesized using SiH_4, PH_3 (B_2H_6), and H_2 (He and Ar). Subsequent cycles of chemical wet oxidation/etching were carried out in order to remove 3 nm of material for each cycle. The SiNWs were integrated inside a field effect transistor (FET) device and the electrical characterization showed that, after the removal of the surface layer in SiNWs with diameter smaller than 22 nm, the electrical characteristics strongly changed, revealing the segregation of the dopant atoms on the surface of the SiNWs. Conversely, for diameters larger than 22 nm, the characterization showed that the core of the NWs instead contained dopant atoms. In relation to the *ex-situ* methods to dope the SiNWs, reference [18] demonstrates how the doping of SiNWs is possible by ion implantation, and from this system, a possible functional device is fabricated. This study is realized on three different SiNWs: (1) n doped, (2) n doped at the top and p doped at the bottom, and (3) p doped at the top and n doped at the bottom. The SiNWs were grown by Au catalyzed VLS in an electron beam evaporation system. NWs were 250–500 nm long and 150–400 nm in diameter. The implantation was performed by using B and P ions. The energy was chosen in order for the ion depth to match the NWs length. The NWs were tilted of a few degrees to avoid channeling effects. After ion implantation, a thermal process was necessary in order to remove the implantation damage. In reference [44], the authors present an innovative approach to dope SiNWs in order to fabricate solar cells. SiNWs have been fabricated by the VLS method in an ICP-CVD chamber and showed lengths of about 500 nm and diameters between 2.5 and 70 nm. After the NWs synthesis, the native SiO_2 was removed by HF and the sample immersed in aqua regia in order to etch away the gold agglomerates. Morphological characterization showed the successful elimination of the residual gold agglomerates. SiNWs were doped by molecular monolayer doping, an innovative method that consists in immersing the Si substrate to be doped inside a liquid solution composed of the chemical precursor of the dopant at high temperature [28]. This process leads to the formation of a monolayer of molecules containing the dopant atoms. The sample was then covered with a SiO_2 layer and annealed to diffuse dopant atoms into the substrate. The cell was completed with front and back contacts realized with Ag paste and sputtered Al. I-V measurements in dark showed a rectifying behavior of the diodes but with a high value of the reverse saturation current. This was attributed to the presence of Au residuals, from the VLS process, inside the bulk of the wafer. The high temperature process used in a group of samples for the dopant diffusion let increase the leakage current, probably due to the structural damage induced by the high temperature on the SiNWs. Under solar simulator illumination, the cells showed a photovoltaic effect. I_{sc} presented higher values in the SiNW cells with respect to the planar reference cell, as expected thanks to the improved optical absorption of the nanostructures. However, the increment was also related to the better coupling between the nanostructured surface and the Ag paste than with the planar Si case. Another study is focused on the removal of gold catalyst [19]. In this chapter, SiNW arrays were fabricated by the VLS method from a <111> Si wafer patterned by lithography to define the regions where the Au catalyst was deposited. The SiNWs obtained under these conditions were vertically oriented and 50 μm long, with an average diameter of 2.6 μm. The SiNW array was HF etched and Au was then removed using a commercial etchant. Energy Dispersive X-Ray Spectroscopy (EDX) measurement was performed and Au was detected on the NWs tip and within the Si matrix. To remove these inclusions, multiple cycles of thermal oxidation and etching steps were performed. As it is clear that the literature underlines the importance of the residual Au removal, because it creates deep level traps in the Si band gap, producing phenomena of charge recombination and hopping conductivity, thus deteriorating the carriers transport. Besides gold, other catalysts have also been explored. In or Sn has been used for the SiNWs synthesis on the glass substrate by the VLS method [20] using SiH_4 plasma at 300–600°C. The morphology, density, and properties of the array during the growth were studied as a function of the growth parameters, such as temperature and dilution of Si source in H_2. Energy dispersive X-ray (EDX) analysis was also performed to investigate the residual metal catalyst concentration after the growth. The authors highlighted that the solubility of In is poor in Si, indeed the In content in the NWs was found to be below the EDX limit of detection, while on the top of the nanowire reaches concentrations of 5%. This behavior shows the advantages of these metal catalysts when compared with gold. The role of the interface between In and the substrate was also investigated. In one case, In nanodots (NDs) were thermally evaporated over a Si <111> substrate and then transferred to the growth chamber, while in a second case, they were deposited by Radio frequency (RF) magnetron sputtering and the NWs growth process followed without air breaking [16]. Morphological characterization of the catalyst particles revealed a spherical shape, in the first case, thus indicating a large contact angle with the substrate. The main reason was attributed to the presence of a thin oxide layer at the interface between the In catalyst particles and the Si substrate. This layer hampered the precipitation of Si atoms through the eutectic liquid, causing a low density of SiNWs.

In the second experiment, as a result of elimination of the air breaking, no oxide layer formed at the In/Si interface. This optimization produced a decrease of the In particles contact angle from 140 to 80°. As a result, the NWs density increased with respect to the first process. **Figure 1(a)** shows the results of this last experiment, where the contact angle is measured by SEM observation in tilted view of the In catalyst particles deposited over the Si substrate. Vertically aligned <111> oriented SiNWs were obtained (**Figure 1b**) with a uniform diameter of about 18 nm and a length of approximately 100 nm, as observed by TEM (**Figure 1c**). The diffraction analysis (**Figure 1e** and **f**) confirmed their epitaxial orientation with respect to the

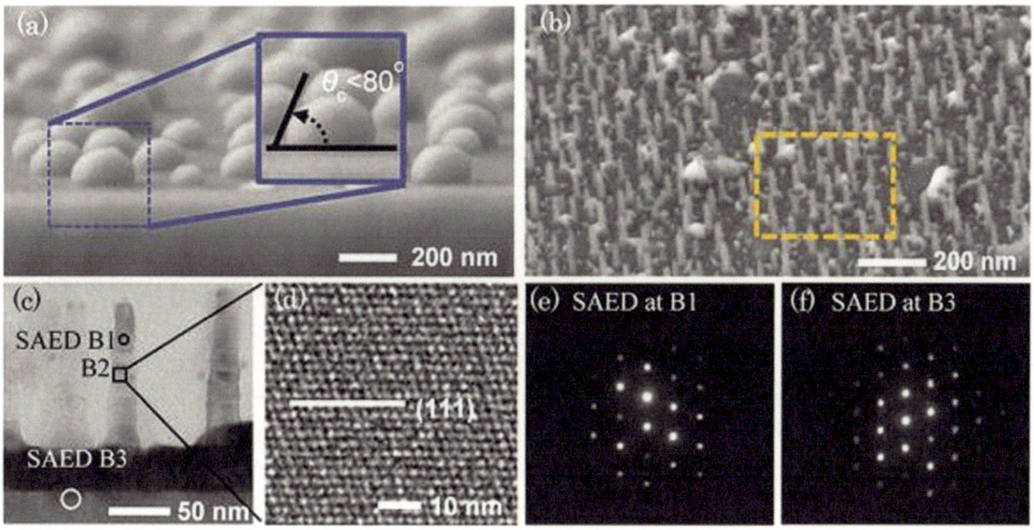

Figure 1. (a) SEM images of In NDs grown by In sputtering, (b) SEM image of Si NWs grown at TS = 600 1C for 60 min. The magnified image in the dotted rectangular area is shown in **Figure 4b**. (c) TEM micrograph of the vertically aligned Si NWs fabricated using In NDs, (d) HR-TEM micrograph of the as-grown (111) oriented Si NWs taken at point "B1" of (c), (e), and (f) shows the SAED patterns taken at points"B1" and "B3," respectively, of (c) [16].

substrate. One of the alternative methods proposed in the literature to synthesize SiNWs is the combination of VLS and laser ablation [45]. The authors compare the method to the one used for the synthesis of carbon nanotubes [46]: a vacuum furnace with a solid source of Si mixed with Fe and a cold finger were used for the SiNWs deposition. The target was kept at 1200°C and the laser pulsed for tens of nanoseconds. The growth took place by evaporating the material from the target, which in vacuum combined to form the liquid catalyst-Si phase and then the supersaturated solid SiNWs. The nanostructures deposited as a net on the furnace inner walls close to the cold finger. This approach then provides template-free SiNWs. TEM images showed that SiNWs were curved, with length of few hundreds of μm and external diameters ranging from 3 to 43 nm with a peak distribution at 16 nm. A 5-nm amorphous shell over the crystalline core was also found. Raman spectra at 521 cm^{-1} of the nanowires showed a tail at lower wavelength numbers, attributed by the authors, to the presence of internal defects besides the SiNWs small size.

MAE is another widespread method to obtain SiNW array [6, 21–27]. A metal catalyst and an etchant solution of HF/H_2O_2 in various ratios are used. MAE has two evidence sources: the first one is that the chemical etching occurs only under the portion covered by the metal, and second is that the "holes" dug into the substrate have the same diameter of the catalyst particles. Peng et al. [20] describe the MAE working principles as a "catalytic conversion of chemical free energy into propulsive mechanical power." Indeed, they explain that the metal basically acts as not just a catalyst but also as a galvanic cell. When deposited onto a Si substrate, we can distinguish two interfaces: the first one is the Ag/etchant solution, which is constituted by the oxidant agent (H_2O_2) and the etchant, HF. This interface works as the cathode of the cell, thus promoting the reduction of the peroxide to water through the consumption of both protons and electrons. Vice versa, the second interface, Si substrate/Ag, works as the anode of the galvanic cell. Here, two reactions play: one is the oxidation of Si, which turns in SiO_2, and the latter one is the etching of Si oxide by the HF, which forms the soluble H_2SiF_6 species. This process depends on both aqueous HF and H_2O_2 species: indeed the electrokinetic model proposed by the authors effectively works only in the presence of both chemicals, and etching rate increases with increasing the concentration of both species. Alternatively, a solution of silver nitrate, fluoridric acid, and water can be applied [6, 22–27]. The particularity of the MAE samples is that the SiNWs are epitaxial, i.e., show single crystal cores and the crystallographic orientation is the same of the used wafer, and maintain the same type and doping level as the starting wafer. As a counterpart, the NW surface is usually very rough due to the fact that it is produced by a chemical etch. The following redox reaction: $Si + 2H_2O_2 + 6HF \rightarrow H_2SiF_6 + 4H_2O$ is explicitly indicated as the base of the NWs formation [22]. A systematic study on the dependence of the MAE SiNW formation process has been performed [22]. They explored six different substrates, p- and n-type, with several degrees of doping levels. A 0.4 M (0.15 M) solution of H_2O_2 for lightly and heavily doped wafers was used. The etching process was performed in dark and at room temperature. The samples were cleaned to dissolve the Ag catalyst and the grown oxide layer. The length and diameter of SiNWs demonstrate a dependence on the doping level of the original substrate: the NWs in heavily doped wafer were smaller than the lightly doped ones and in both cases were several μm in length and 50–200 nm in diameter. The reason whereby heavily doped Si was more active than the lightly doped one was ascribed to the high carrier concentration favoring the electrons transfer [22]. Oblique SiNWs can also be obtained by MAE [23]. A p-type Si <111> wafer previously cleaned was subjected to the evaporation of a net-like layer of 10–50 nm of silver nanoparticles. Successively, the sample was immersed in a solution of DI water, HF (4.6 M) and H_2O_2 (0.44 M) from room temperature to 80°C and various times to modulate the NWs length. The final SiNWs present oblique orientations with two preferential directions, the <100> and <110>. The authors attribute this morphology to the anisotropic etching of silicon and to the competition existing during the catalytic etching along different directions. Besides oblique NWs, zigzag SiNWs can also be produced by immersing cleaned Si wafers in aqueous HF solution (4.6 M) with 0.01 or 0.04 M silver nitrate ($AgNO_3$) solution for 40 min at different temperatures [24]. The array formation using the same solution 0.01 M $AgNO_3$ at two different temperatures (25 and at 55°C) was first focalized. Results showed that at lower temperatures, Ag dots created arrays of straight SiNWs, while at 55°C, the Ag dots produced curved SiNWs. Then a 0.04 M solution at 45–55°C was used and three types of curved SiNWs were obtained (**Figure 2**): I type with a turning angle of 150°, longitudinal orientations on <111> and <113>; II type with 125°, <100> and <111> orientation; III 90° and a longitudinal orientation that alternates between two orthogonal <100> directions (**Figure 2c**).

Two possible etching mechanisms were proposed to explain the results: (1) single-particle etching mechanism in which the Ag nanodots can move, as a result of perturbations, from the <111> direction to the <113> direction (I group). Since the last direction is not energetically favorite, the dots switch back to <111>; (2) a multiparticle etching mechanism in which the higher etching temperature/silver ion concentration enhances the etching activity, so the Ag

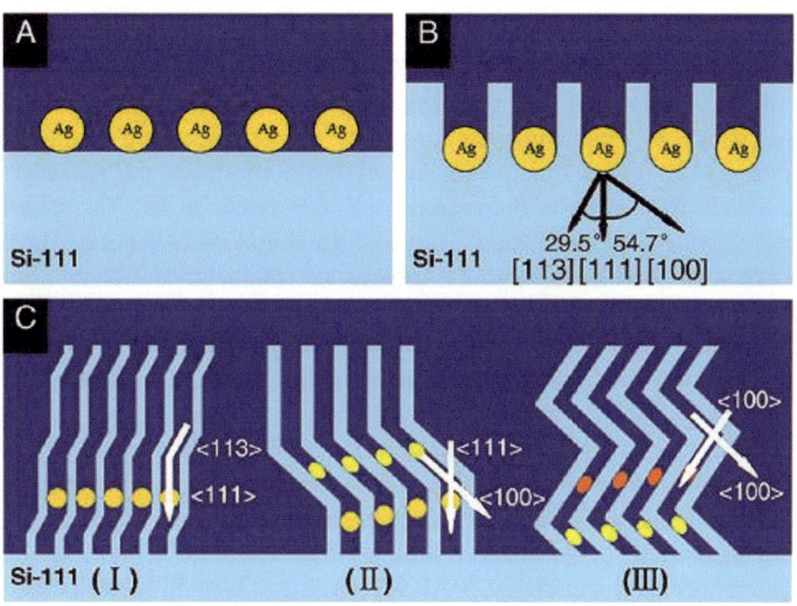

Figure 2. The proposed growth or etching process of the zigzag Si nanowires [24].

particles can move from the <111> direction to the three different <100> directions (groups II and III).

The MAE process can be coupled to a mask of anodic aluminum oxide (AAO) to obtain an ordered array of vertical aligned SiNWs [25]. From SEM characterization performed on AAO masks, the authors estimated the mask pore diameter, interpore distance, and pore density to be 20 nm, 60 nm and 2.9×10^{10} cm^{-2}, respectively. SEM characterization was used also to study the array after silver removal, and in this case, it was evident that the SiNWs had a different diameter at the top and the bottom parts. This difference was assigned to the dissolution of metal during the etching or to the isotropic etching of Si in HF/H_2O_2 and the different exposure times of the NW parts. The evolution of the NWs diameter was also studied as a function of thickness of the metal film, and the results show that the trend is inversely proportional to the thickness. This behavior was explained by considering the decrease of the original pore diameter when the metal thickness increases, due to the conformality of the deposition. Nano sphere lithography (NSL) exploits the self-assembly of polymeric nanospheres in order to create a soft mask for patterning the surface. When water-dispersed PS nanoparticles are spun onto a substrate, they spontaneously arrange in a hexagonal pattern in which they leave triangular voids between each sphere. A successive dry etch leads to a shrinkage of the PS spheres, which thus exposes a major portion of the substrate surface. A metal layer is then thermally evaporated, covering the unprotected substrate, while the area under each PS particle remains metal-free. Therefore, the polymeric nanospheres are removed by dissolving them into a specific solvent or through a lift-off method thus leaving a metal layer with circular voids. This metallic film acts as a catalyst for successive processes such as the MAE etch, which leads to the formation of SiNWs. The final structure characteristics strictly depend on some geometrical parameters such as the "in solution"-diameter of the polymeric particles, the effective diameter related to the polymer/substrate interaction and finally the space between two neighbor particles after the deposition. In reference [26], the NSL obtained by spreading a monodisperse suspension of PS microspheres onto the substrate was used. The size of the PS spheres was then reduced by reactive ion etching process (RIE). The samples were then subjected to the deposition of a gold thin film, thus producing an anti-dot pattern, and subsequently etched in HF/H_2O_2, which acted only on the regions of Si covered with Au. The sample was then immersed into toluene to dissolve the PS spheres, dried and characterized. The SiNWs have size similar to the ones of the reduced PS spheres, i.e., diameter of 400 nm and a length of about 1 μm. Among the others, they studied the influence of the substrate doping type and concentration. As for reference [22], it is found that the doping type influences the etching rate of Si: the substrate with the lower doping concentration showed deeper etching than highly doped one. Axial junctions can be also obtained if using properly doped substrates [6]. The synthesis of SiNWs started with three evaporation steps of amorphous Si layers on borosilicate glass, respectively, doped by: (1) boron (p+), (2) phosphorus (n), and (3) phosphorus highly doped (n+), followed by crystallization with a diode laser at different wavelengths (depending on the doping types). The recrystallized multilayers showed large Si grains (of the order of a few tens of micrometers). Successively, MAE wet etching was performed in order to obtain a SiNW array, presenting an axial junction p+/n/n+ along their length, with the different doping types as the original multilayers. Optical properties showed a reflectance less than 5% in the wavelength between 300 and 1000 nm. The SiNWs were integrated inside a solar cell with the highest efficiency of 4.4%. In reference [27], the authors have synthesized SiNW arrays by using MAE with silver nitrate, fluoridric acid, and water of an n-type Si substrate. The Si oxide was removed by HF and the SiNWs were then covered with a shell of p-type amorphous Si deposited by low pressure CVD and finally the amorphous layer was crystallized by rapid thermal annealing (RTA). SEM analysis showed vertically aligned core-shell SiNWs with a length of 18 μm, total diameter of 350–400 nm, and a coverage of 50%. TEM results showed that the shell was constituted by nanocrystalline domains of about 5 nm. As recognized by the authors, the large surface roughness typical of the MAE etched SiNWs can lead to enhanced depletion region traps, causing strong carrier recombination effects. SiNWs obtained by MAE have been integrated in a solar cell [1]. The nanostructures have been etched by a p-type Si <100> substrates. The n+ emitter was obtained by $POCl_3$ diffusion at 930°C. Aluminum and silver films were evaporated on the rear contact of the cell, the first to remove the parasitic junction and the second to form the back electrode. The front contact was made by Ti/Pd/Ag grid by mask evaporation. The solar cell presented a V_{oc} of 548.5 mV, an I_{sc} of 26.06 mA, and a FF of 0.6512 and a power conversion efficiency of 9.31%. The photovoltic (PV) conversion was not high if compared to other geometries such as the devices based on slantingly aligned SiNW arrays that have been demonstrated in reference [23]. The difference in efficiency was correlated to the low current collection yield or to enhanced surface electron-hole recombination velocity. VLS grown SiNWs was doped to obtain axial p-i-n junctions [2]. The solar cell was fabricated with a single SiNW and the intrinsic region length was varied from 0 to 4 μm. The I-V data for the intrinsic region of 0, 2, and 4 μm show V_{oc} and I_{sc} values of: 0.12 V and 3.5 pA, 0.24 V and 14.0 pA, and 0.29 V and 31.1 pA, respectively.

The results show a systematic improvement in both V_{oc} and I_{sc} with increasing i-segment length, where the largest increase is observed in moving from the p-n to p-i-n structure. The best value of FF (51%) was shown in p-i-n device with i-region of 4 μm and it yielded a maximum power output per NW of 4.6 pW. The efficiency in the single axial NW cell is calculated by considering the projected active area and results 0.5% less than the coaxial geometry which produced 3.4% [9]. The authors also monitored I_{sc} and V_{oc} as a function of the temperature and found that V_{oc} showed a linear dependence while I_{sc} slightly increased. A tandem structure p-i-n+-p+-i-n (**Figure 3A** and **B**) was also fabricated in order to investigate the behavior of series of single-junction SiNW device. Tandem device was created under the same conditions of the single junction NWs. They finally compared the results of p-i-n axial NW to tandem NW device with the same intrinsic length of 2 μm, the values of V_{oc} and I_{sc} are 0.23 V and 10.2 pA, and 0.39 V and 8.2 pA, respectively. These data showed an increment of 57% of V_{oc} (**Figure 3C**) due to the presence of the tunnel junction n+/p+. However, the increment is not large as expected due to the parasitic series resistances at the tunnel interface [2]. A study on the current collection and generated photovoltage for several NW diameters and consequent junction areas with respect to the minority carrier diffusion length has been also pursued [13]. The p-n junction was formed by depositing and diffusing Al over the eutectic temperature on a single NW. The work results suggest that a trade-off between small NWs (enhancing current collection) and large ones (enhancing photovoltage) can be reached when the NW radius is comparable to the minority carrier diffusion length. One of the processing steps more critical

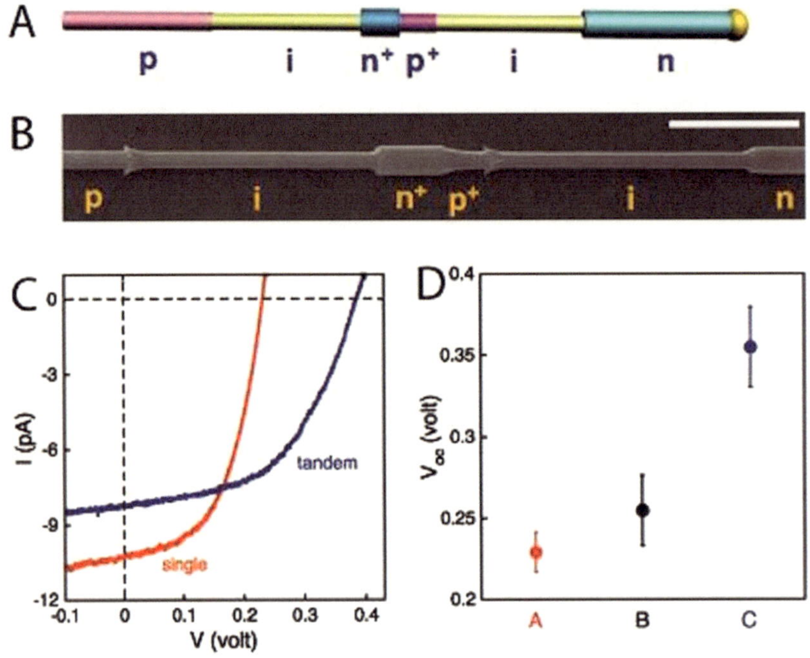

Figure 3. Tandem axial SiNW photovoltaic devices. (A) Schematic of two p-i-n diodes integrated in series on a single NW. (B) SEM images of a selectively etched tandem p-i-n+-p+-i-n SiNW; scale bar is 1 μm. (C) *I-V* responses recorded on p-i (2 μm) -n (single) and p-i-n+-p+-i-n, *i*) 2 μm (tandem) SiNW devices under AM 1.5G illumination. (D) *V*oc for p-i (2 μm) -n (case A), p-i (4 μm) -n (case B), and p-i-n+-p+-i-n, *i*) 2 μm (case C) axial SiNW devices. Error bars are ±1 standard deviation [2].

for the integration of SiNWs inside the solar cells is the passivation process. Indeed, the presence of surface states on the NWs, due to the high surface to volume ratio of these 1D nanostructures, leads to copious carrier recombination events detrimental for the device efficiency. Conductance of SiNWs was investigated before and after a passivation treatment with organic molecules (4-nitrophenyl octadecanoato or tetraethylammonium bromide) [47]. The SiNWs were integrated inside a FET transistor and characterized in terms of the passivation process effects. The treatments with both molecules improved similarly the device on/off ratio, the transconductance and the mobility. The same results were obtained also by changing the first molecule chain lengths. The explanation was based on the fact that the molecules lead to a stable and relatively nonpolar Si-O-C bond, thus reducing the availability of the surface to water hydrolysis, and as a consequence the density of polar surface sites, such as SiO^-.

3. Silicon nanoholes

SiNHs are emerging as promising materials for the fabrication of performing solar cells because of their interesting properties and easy integration in the device processing. Since their first proposal in the solar cells field [10], several examples have appeared in the literature, referring to them as "pores," or "black-Si," "holey structures," and "honeycomb" [48, 49]. Regarding their synthesis, NHs can be realized in Si through electrochemical anodization and oxidation carried in an aqueous electrolyte of HF and H_2O_2 [50]. The process is based on the localization of the electric field on specific substrate sites, due to the depletion layer formed on its surface. The experiments are run onto n-type Si and the surfaces are texturized by applying different electrolyte concentrations and current densities. The pore density can be tuned with the current and HF concentration, while it does not show any clear dependence on the H_2O_2 content. These data have been explained by assuming that the bottleneck process is the removal of the oxidized surfaces rather than their oxidation. The pore depth does not strongly depend on the peroxide concentration, while it increases with the HF. Depths as large as tens of microns can be reached. The oxidation/etching process exploited in the previous case is also employed in MAE, which as for the NW formation is very frequently used to create NHs. The size of the holes obtained is strictly related to both the duration of the metal deposition and etching steps [21]. If compared to the electrochemical anodization, MAE is assisted by a metallic catalyst while the first one is boosted by an external current which supplies the carriers to the Si surface [50]. The fabrication of NHs by MAE is affected by different factors [30]. As seen in the last reference, the samples are immersed in $AgNO_3$ aqueous solution and the amount of deposited Ag, in terms of dot density and size, increases with the immersion time. The Ag particle density will be higher onto more reactive surfaces (e.g., <111> Si than <100> Si), but for very long deposition times the area covered from metal seems to be independent of the substrate orientation. Moreover, this parameter can have a strong effect on the NH morphology: for metal coverage lower than 70%, in the <100> Si substrate the etching path changes abruptly from the vertical to the horizontal direction along the (100) axis, thus showing a clear lateral etching component. While in the case of the <111> Si substrates, the etching process follows a much more irregular path. However, any lateral etching is suppressed for both Si <100> and Si <111> samples when the Ag coverage is higher than 70%. Indeed under these conditions, the nanoparticles sink all together at the bottom of holes and they all follow the equivalent <100> directions, thus forming long, vertical, and aligned structures in the <100> substrate, and oblique pores in the <111> ones [30]. Many interesting approaches found in the literature merge the MAE process with different lithographic techniques to realize ordered patterns. Among these, UV lithography is one of the most conventional strategies proposed [10]. The lithographic steps with the exposure to deep UV light are followed by silver evaporation onto a mask, which undergoes to a lift-off process, thus leaving Ag dots on the Si surface. A MAE wet etching is run by an aqueous oxidant mixture of H_2O_2 and HF, which molds the substrate. After that silver dots are removed with nitric acid from the holes bottom. The morphology shows a regular square-packed NHs array with precise features, such as diameter (800 nm), pitch ($\approx$1 μm), edge-to-edge distance ($\approx$200 nm), and depth ($\approx$1.4 μm) (**Figure 4**). According to this synthetic approach, the lateral dimensions of NHs are strictly conditioned by the specific size of the shadow mask and by the wavelength of the light source used, which can give severe resolution limits when very small sizes are required. As a consequence, the smallest features achievable by the typical UV lithography are on the order of hundreds of nanometers. Differently, the third dimension of the structures can be easily controlled by calibrating the etching time.

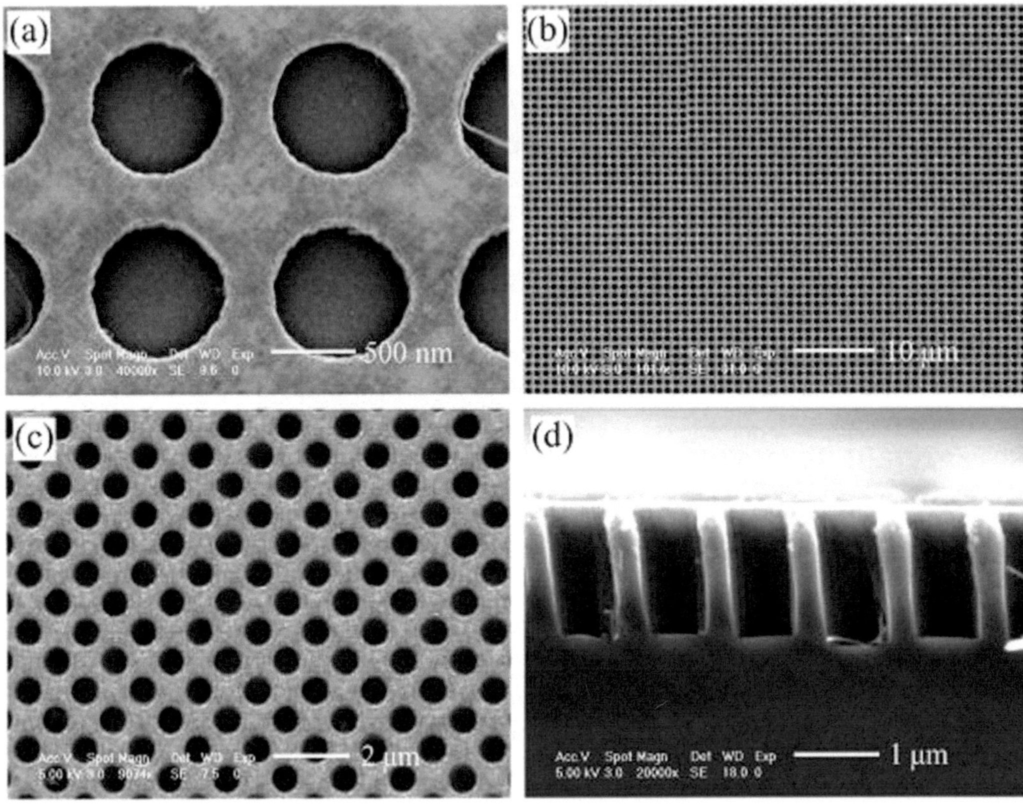

Figure 4. SEM images of silicon nanoholes produced in an 8–12 Ω cm p-Si(100) wafer. (a) Top-view SEM image of ordered, shallow, flat-bottomed pits in Si formed during the initial etching stage. Silver is seen at the bottom of these pits. Marker: 500 nm (b and c): top-view SEM images of ordered silicon nanoholes with large depths after prolonged immersion in HF and H_2O_2 solution. Marker: 10 and 2 μm figure (b) and (c), respectively, (d) cross-sectional view of silicon nanoholes; the hole channels are cylindrical and vertical with respect to the Si surface. Marker: 1 μm [10].

Nanoimprinting lithography (NIL) guarantees the realization of structures with vertical and smooth sidewalls [32, 33]. It consists in the pattern transfer from a hard mold to a resist through mechanical pressure. The replication is then run from the resist to an underlying substrate by successive treatments, such as dry etching. This approach can be exploited for the fabrication of a SiNHs array [51] by using a polydimethylsiloxane (PDMS) mold and poly(methyl methacrylate) (PMMA) resist in order to pattern the surface. The pores are finally realized by etching the Si substrate with SF_6/O_2 gaseous mixture. This approach leads to a regular NH array in which holes have 200 nm of diameter and the aspect ratio >1. BCP lithography allows the formation of nanotemplates by exploiting self-assembling mechanisms. For the sake of simplicity, it will be now considered the behavior of a binary species, which means a linear di-block copolymer containing A and B as repeating units. If A and B blocks have a poor chemical affinity in their molten state, which can be estimated by the Flory-Huggins interaction parameter, an unfavorable excess in the mixing free energy takes place. The energy excess does not entropically support the mixing process between the blocks thus promoting a mesoscopic phase separation between A and B. The shape and size of the phases have a strong dependence on the relative fraction of each polymeric block in the chain and the global molecular weight of the polymer, respectively. An ordered pattern through the BCP self-assembly onto a SiO_2/Si substrate can be realized. The di-block copolymer PS-b-PMMA with a 70:30 ratio, polydispersity 1.1, and molecular weight 67,000 gr/mol is typically used. Before proceeding with the deposition of the PS-b-PMMA, the SiO_2 substrate surface with an analogous random polymer is neutralized. Then the PS-b-PMMA is spun from 1 w/w% toluene solution onto the random layer, and this sample is annealed in a range of 160–190°C for a couple of hours. The annealing promotes the separation of the two polymeric phases thus forming a template in which PMMA cylindrical microdomains are organized in a hexagonal array and embedded into a PS matrix. The exposition of the annealed film to the UV light lets cure the PS chains and thus allows the successive development of the PMMA portions in acetic acid. Therefore, this step leaves a regular porous PS film, which can be exploited by using the standard lithography concepts, taking into account the necessary technical measures. The literature results demonstrated that the thickness of the copolymer layer plays a major role in the self-assembling process [52]. Indeed, it is noted that too thick films lead to a quasi-lamellar morphology, while too thin layers do not reproduce the hexagonal pattern, thus losing the planar order. A suitable range of spin speed for the optimal self-assembly is then identified. Both the annealing temperature and its duration also have a strong role for determining the planar order of the film. Indeed, the annealing must be run over the glass transition temperature of both polymeric blocks in order to give the right mobility for allowing the movements of the chains. However, there is small range of temperatures that can guarantee the lateral order of the domains, thus allowing the diffusion of the molecules and avoiding the order-to-disorder transition. Inside this range of temperature, the self-assembling process can be tuned in order to obtain ordered hexagonal macrodomains as large as microns [53]. The template replication down to the SiO_2 and then to the Si substrate to form NHs can be carried by a CHF_3/Ar gas mixture followed by cycling SF_6/O_2 and CHF_3/Ar according to the procedure of the Bosch process [54]. The choice about the gases mixture and the general approach for both the etching cycles is made in order to promote the anisotropic etching which removes material from the bottom of the structures realized while it protects the sidewall and preserves them from the etching. Indeed, the passivation layer realized by the CHF_3 prevents the enlargement of the structures due to any isotropic etching effect, while SF_6/O_2 acts as a reagent for the etching of the Si substrate. This procedure leads to a hexagonal array of 30 nm-deep SiNHs and an aspect ratio ≈1. NSL is also used for the fabrication of SiNHs. Unlike the process used for the SiNWs formation, here the metallic film acts as a hard mask for successive processes such as the reactive ion etching (RIE) which leads to the formation of the Si holes. PS spheres are spun from a liquid solution on a Si substrates, realizing a monolayer of polymeric spheres with a high hexagonal packed array. The PS spheres diameter is shrunk with a RIE etch which thus increases the separation between them. A metallic film is deposited through a thermal evaporation onto the PS layer while a successive lift-off in a sonication bath removes PS, thus leaving the porous metallic layer that works as a hard mask for the successive dry etching of the Si. Finally, the metallic layer is removed. The PS spheres have also been embedded inside a monomer matrix [55]. A suspension of silica particles in a solution of a specific monomer (ETPTA) is spun onto the Si substrate. After monomer curing, the polymeric mask with silica particles embedded and positioned in hexagonal configuration is obtained. The vertical pressures exerted by the spin coating and the polymerization constrict the PS particles on the substrate to form a nonclose packed array. The polymeric mask is then removed by plasma oxygen, thus leaving a 2D film of silica nanoparticles. A metallic film is deposited onto the colloidal pattern via e-beam technique and it acts as a hard mask after the silica lift-off and during the successive RIE etching. Finally, the mask is removed via etching. SEM characterization shows that the diameter of the SiNHs is equal to the silica particles diameter, $D \approx 320$ nm and it is correlated to the center-to-center distance, d, since $d = 1.4$ D. In the work of Jiang et al., SiNHs are five times smaller with respect to those ones obtained by Hulteen et al. [56]. It is found that by using particles with $D = 264$ nm, the in-plane particle diameter, a, is ≈0.23 D, while the interparticles spacing, $d \approx 0.6$ D, and it does not depend on the substrate [56]. SiNHs can also be realized by two other uncommon techniques, the fast atoms bombardment (FAB) and FIB. In the first approach, SF_6 beam is used to erode the Si surface [57]. This method first requires a partial protection of the Si surface with a mask, for example with the alumina AAO mask: the sample is soaked in an oxalic solution and electrically oxidized at a constant voltage to obtain an ordered porous array with a periodicity of approximately 100 nm. FAB is thus performed using SF_6 gas under an acceleration voltage of 2.0 kV, to transfer the porous texture from alumina to the underlying Si, keeping both the directionality and in-plane hexagonal array of the alumina mask. FIB is also exploited in order to manufacture NHs with diameters below 5 nm. The approach is also called the "ion beam sculpting technique" [32]. Nanopores of about 100 nm diameter are drilled by the focused beam of the FIB onto a low-stress SiN layer supported by a Si substrate. After multiple beam scans on the pore and its surrounding areas, a strong reduction of the diameter of the NHs is observed, down to 5 nm. The authors speculate that the holes shrink because of a mass flow finalized to the minimization of the surface tension energy, when temperature and entropy are locally increased by the high-energy electron beam, although the explanation is not complete. If pore diameter smaller than 5 nm is desired, the above process becomes uncontrolled and the authors suggest to use a 200-keV beam of a TEM equipment, as a more controllable sculpting process. Using this second strategy, pores with less than 1 nm diameter can be fabricated. Despite the well-known behavior for which high-energy species usually tend to enlarge structures, surprisingly this does not occur in this case. It must be highlighted that this mechanism works as here described when the initial pore diameter is ≤50 nm. In fact, for pores which have ≥80 nm, different dynamics occur [58].

As for the SiNWs, the passivation process is a critical step and it has been addressed in the literature with several solutions. The SiNHs surface can be passivated with a defects removal etching (DRE) process, with the objective to remove the ion bombardment defects [31]. The DRE process is performed onto the sample using an oxidant aqueous solution of $HF:HNO_3$ which works through two different steps: (1) nitric acid oxides Si to SiO_2 and (2) HF removes the oxide formed. DRE duration strongly influences the depth of the SiNHs since the longer the etching, the deeper the holes obtained. Oh et al. [11] also recognize the strategic importance of surface passivation and treat the NHs sample by immersing it into an aqueous solution of 1% of tetramethylammonium hydroxide (TMAH), to decrease the area of the realized texture and the structural defects due to the MAE wet etching [11]. Despite the

http://dx.doi.org/10.5772/67749

passivation issues not yet fully overcome, the electrooptical properties of the NH arrays are extremely interesting. Indeed as reported in the literature, a periodic array with subwavelength dimensions, such as the SiNHs, manifests antireflecting behaviors. When an electromagnetic radiation passes from a lower refractive index medium (e.g., air) to another one that has a higher optical density (i.e., Si), an abrupt variation of the optical parameter occurs. In this case, the transmitted "power" of the incident light is partially lost through the reflections. Differently, when the refractive index changes gradually, the losses due to external reflections can be minimized since the incident light sees the surface as a continuous system in which the power is transferred from the air to the bulk "layer by layer." This effect is more evident in deep surface textures, for which the reflectance reduction is also explained by the light-trapping effect, which means that multiple internal reflections increase the light path inside the structures and lead to the absorption enhancement. Koynov et al. investigate the optical response of a textured surface throughout the entire spectrum range of interest for photovoltaics from 350 to 1000 nm [35]. They find a very strong reduction of the reflectance that passes from 35 to 45% of the flat reference to very low values (5%) for the random textured sample. At the same time, the total transmittance, T, does not change within 350–1000 nm. Since $A + T + R = 1$, where A is the absorption and R is the total reflectance, these results imply that the absorption of a textured Si surface must be higher than that of the untreated sample. R decreases with increasing the pore depth [36] and, for each value of depth, R is about constant into the spectral range from 250 to 1000 nm. After 1000 nm, the reflectance increases and this occurs for each structure and independently of the depth, also in accordance with other experimental data [37]. From these results, the authors provide the universal law correlating the reflectance to the depth of nanostructures and to the wavelength of the incident light, $R\,(d, \lambda)$. The universal scaling-law is an exponential decay: $R(d, \lambda) = R_0(\lambda)\, e^{-C(\frac{d}{\lambda})}$ where $R_0(\lambda)$ is the wavelength depending-reflectance of the flat reference, d is the depth of the NHs, λ is the wavelength of the incident light, and C is a constant. Such subwavelength structures can be designed in order to increase the optical absorption of materials [38]. For instance, the indirect band gap of Si is the main cause of its low absorption in the IR region, which can be thus improved by the surface texturing. About this, geometrical features of the nanopattern employed can play a role in the reflectance reduction. For example, when the optical properties of SiNWs and SiNHs with the same dimensions are compared, it is possible to see that the latter ones show superior optical properties, thanks to a more efficient light-trapping capability. These results are deduced by the reflectance measurements of both SiNHs and SiNWs arrays (**Figure 5**).

The reflectance of a shallow NHs array can decrease until 5% within 300–1100 nm wavelength and it is always lower for NHs than for NWs of comparable size. The reflectance can be further reduced by increasing depth/length of the structures and it reaches the minimum value in the UV region while its maximum is always achieved in the IR range. Similar to SiNWs, SiNHs can represent the base for radial p-n junctions, with the carrier channel being placed in between the hole interspacing, rather than in the inner core of the nanostructure. The doping can be performed only by *ex-situ* methods, by using the standard gas source [10] or by MD [29]. The last one allows the optimization of the doping step since it avoids the typical stochastic spatial distribution or the crystal damage, leading instead to a high conformality. Authors in

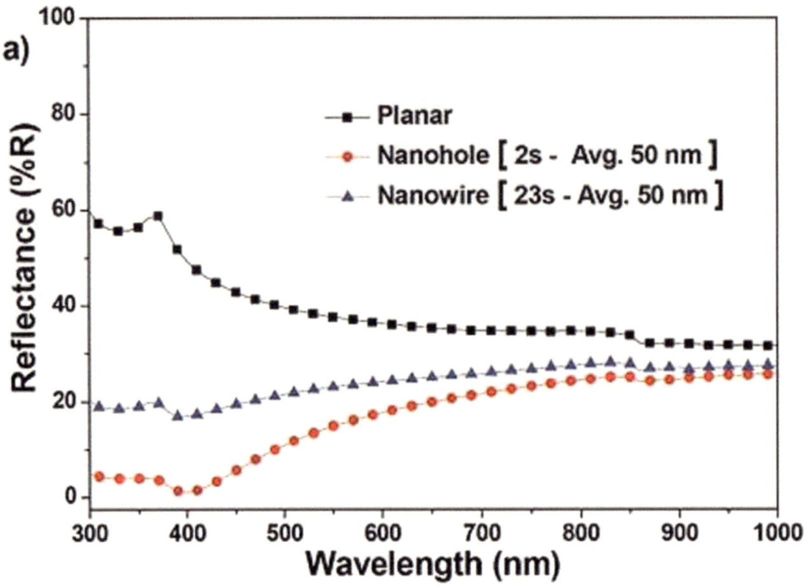

Figure 5. a) The optical reflectance of SiNW and SiNH arrays measured by the integrating sphere, including the reflectance average lengths of 50 nm. All the figures are show here with the reflectance of referenced polished n-type Si wafer [38].

reference [29] investigate the doping profile by spreading resistance measurements for two different annealing temperatures. They find a boron concentration peak of 10^{19} cm^{-3} and a junction depth of about 20 nm for a diffusion process at 950°C, while, when the annealing temperature is 1050°C, the carrier peak concentration slightly decreases and a junction depth of 70 nm is obtained. In the case of small SiNHs, a smaller thermal budget could modulate the diffusion depth in order to properly design the carrier channel width.

4. Conclusions

In this work, various SiNWs synthesis methods have been reviewed, the most common based on the VLS catalyzed growth. The advantages are control of external impurities by use of vacuum instrumentation, low surface roughness, control over structural and morphological properties, crystallinity. When the growth takes place in a plasma-based CVD chamber, the additional benefits consist in the low temperatures and high growth rate. The substrate crystallographic orientation determines the NW directions. Among the methods examined those based on the combination of the catalyzed growth and the laser induction do not allow the manufacture of the cell, as the NWs obtained are free-standing and have no contact with the substrate. The NWs have been exploited as building blocks for the formation of core-shell metallurgical junctions. A large variety of designs have been proposed, where the intrinsic layer plays the relevant role to improve the cell efficiency. One of the issues largely debated is about the efficient removal of the metallic catalyst from the SiNWs, because it is related to the electrical performance deterioration. Regarding the optical properties, light trapping by multiple reflections in the case of forest-like configuration, and increased absorption in the UV in the case of structures smaller than 5 nm have emerged. SiNHs are frequently fabricated directly by MAE which leads to random nanostructures or by lithographic approaches and successive etching to obtain recurring patterns. A wide range of diameter sizes going from hundreds of nm, attainable by common lithographies, down to 30 nm, by BCP, can be achieved. Smaller SiNHs can be synthesized by FIB with drawbacks for the wafer scalability of the process. The most interesting SiNHs optical properties manifest drop in reflectance down to 5%, demonstrated for randomic structures with 50 nm of diameter. Even for these mirror structures, it is possible to build core-shell p-n layers and innovative *ex-situ* techniques are proposed providing conformality and cost-effective solutions.

Acknowledgements

R.A.P. wish to mention the National Program PON R&C 2007–2013, project "Tecnologie per l'ENERGia e l'Efficienza energETICa (ENERGETIC)" (PON02_00355_3391233) for financial support.

Author details

Rosaria Anna Puglisi[1*], Valentina Lombardo[1,2] and Sebastiano Caccamo[1,2]

*Address all correspondence to: rosaria.puglisi@imm.cnr.it

1 Consiglio Nazionale delle Ricerche – Istituto per la Microelettronica e Microsistemi Strada Ottava, Catania, Italy

2 Dipartimento di Fisica, università degli Studi di Catania, Catania, Italy

References

[1] Peng K, Xu Y, Wu Y, et al. Alligned single-crystalline Si nanowire array for photovoltaic applications. Small. 2005;**1**(11):1062-1067. DOI: 10.1002/smll.200500137

[2] Kempa TJ, Tian B, Kim DR, et al. Single and tandem axial p-i-n nanowire photovoltaic devices. Nano Letters. 2008;**8**(10):3456-3460. DOI: 10.1021/nl8023438

[3] Schmidt BV, Wittemann JV, Senz S, Go U. Silicon nanowires: A review on aspects of their growth and their electrical properties. Advanced Materials. 2009;**21**:2681-2702. DOI: 10.1002/adma.200803754

[4] Ramanujam J, Shiri D, Verma A. Silicon nanowire growth and properties: A review. Materials Express. 2011;**1**(2):105-126. DOI: 10.1166/mex.2011.1013

[5] Kelzenberg MD, Boettcher SW, Petykiewicz JA, et al. Enhanced absorption and carrier collection in Si wire arrays for photovoltaic applications. Nature Materials. 2010;**9**(3):239-244. DOI: 10.1038/nmat2635

[6] Sivakov V, Andrä G, Gawlik A, et al. Silicon nanowire-based solar cells on glass: Synthesis, optical properties, and cell parameters. Nano Letters. 2009;**9**(4):1549-1554. DOI: 10.1021/nl803641f

[7] Zhang RQ, Hou C, Gao N, et al. Multi-field effect on the electronic properties of silicon nanowires. ChemPhysChem. 2011;**12**(7):1302-1309. DOI: 10.1002/cphc.201100030

[8] Tsakalakos L, Balch J, Fronheiser J, et al. Silicon nanowire solar cells. Applied Physics Letters. 2007;**91**:233117-1. DOI: 10.1063/1.2821113

[9] Tian B, Zheng X, Kempa TJ, et al. Coaxial silicon nanowires as solar cells and nanoelectronic power sources. Nature. 2007;449:885-889. DOI: 10.1038/nature06181

[10] Peng K, Wang X, Li L, et al. High-performance silicon nanohole solar cells. Journal of the American Chemical Society. 2010;**132**:6872-6873. DOI: 10.1021/ja910082y

[11] Oh J, Yuan H, Branz HM. An 18.2%-efficient black-silicon solar cell achieved through control of carrier recombination in nanostructures. Nature Nanotechnology. 2012;**7**:743-748. DOI: 10.1038/nnano.2012.166

[12] Fontcuberta i Morral A, Arbiol J, Prades JD, et al. Synthesis of silicon nanowires with wurtzite crystalline structure by using standard chemical vapor deposition. Advanced Materials. 2007;**19**:1347-1351. DOI: 10.1002/adma.200602318

[13] Kelzenberg MD, Turner-Evans DB, Kayes BM, et al. Photovoltaic measurements in single-nanowire Si solar cells. Nano Letters. 2008;**8**:710-714. DOI: 10.1021/nl072622p

[14] Yung Kuo C, Gau C, Tong Dai B. Photovoltaic characteristics of silicon nanowire arrays synthesized by vapor liquid solid process. Solar Energy Materials & Solar Cells. 2011;**95**(1):154-157. DOI: 10.1016/j.solmat.2010.04.028

[15] Yu L, Rigutti L, Tchernycheva M, et al. Silicon nanowire-based solar cells. Nanotechnology. 2008;**19**:295203. DOI: 10.1088/0957-4484/19/29/295203

[16] Ajmal Khan M, Ishikawa Y, Kita I, et al. Control of verticality and (111) orientation of In-catalyzed silicon nanowires grown in the vapour–liquid–solid mode for nanoscale device applications. Journal of Materials Chemistry C. 2015;**3**:11577-11580. DOI: 10.1039/C5TC01338K

[17] Garozzo C, La Magna A, Mannino G, et al. Competition between uncatalyzed and catalyzed growth during the plasma synthesis of Si nanowires and its role on their optical properties. Journal of Applied Physics. 2013;**113**(21):214311-214317. DOI: 10.1063/1.4809557

[18] Hoffmann S, Bauer J, Ronning C, et al. Axial p-n junctions realized in silicon nanowires by ion implantation. Nano Letters. 2009;**9**(4):1341-1344. DOI: 10.1021/nl802977m

[19] Kendrick CE, Yoon HP, Yuwen YA, et al. Radial junction silicon wire array solar cells fabricated by gold-catalyzed vapor-liquid-solid growth. Applied Physics Letters. 2010;**97**(14):42-45. DOI: 10.1063/1.3496044

[20] Yu L, O'Donnell B, Alet PJ, et al. All-in-situ fabrication and characterization of silicon nanowires on TCO/glass substrates for photovoltaic application. Solar Energy Materials & Solar Cells. 2010;**94**(11):1855-1859. DOI: 10.1016/j.solmat.2010.06.021

[21] Peng BK, Lu A, Zhang R, et al. Motility of metal nanoparticles in silicon and induced anisotropic silicon etching. Advanced Functional Materials. 2008;**18**:3026-3035. DOI: 10.1002/adfm.200800371

[22] Zhang ML, Peng KQ, Fan X, et al. Preparation of large-area uniform silicon nanowires arrays through metal-assisted chemical etching. Journal of Physical Chemistry C. 2008;**112**(12):4444-4450. DOI: 10.1021/jp077053o

[23] Fang H, Li X, Song S, et al. Fabrication of slantingly-aligned silicon nanowire arrays for solar cell applications. Nanotechnology. 2008;**19**(25):255703. DOI: 10.1088/0957-4484/19/25/255703

[24] Chen H, Wang H, Zhang XH, et al. Wafer-scale synthesis of single-crystal zigzag silicon nanowire arrays with controlled turning angles. Nano Letters. 2010;**10**(3):864-868. DOI: 10.1021/nl903391x

[25] Huang Z, Zhang X, Reiche M, et al. Extended arrays of vertically aligned Sub-10 nm diameter [100] Si nanowires by metal-assisted chemical etching. Nano Letters. 2008;**8**(9):3046-3051. DOI: 10.1021/nl802324y

[26] Mikhael B, Elise B, Xavier M, et al. New silicon architectures by gold-assisted chemical etching. ACS Applied Materials & Interfaces. 2011;**3**(10):3866-3873. DOI: 10.1021/am200948p

[27] Garnett EC, Yang P, Garnett EC, et al. Silicon nanowire radial p-n junction solar cells. Journal of the American Chemical Society. 2008;**130**:9224-9225. DOI: 10.1021/ja8032907

[28] Puglisi RA, Caccamo S, D'Urso L, et al. A comprehensive Study on the physicochemical and electrical properties of Si doped with the molecular doping method. Physics of the Status Solid A. 2015;**1694**(8):1685-1694. DOI: 10.1002/pssa.201532030

[29] Garozzo C, Giannazzo F, Italia M, et al. Radial junctions formed by conformal chemical doping for innovative hole-based solar cells. Materials Science and Engineering: B. 2013;**178**:686-690. DOI: 10.1016/j.mseb.2012.11.019

[30] Milazzo RG, Arrigo GD, Spinella C, et al. Ag-Assisted chemical etching of (100) and (111) n-type silicon substrates by varying the amount of deposited metal. Journal of the Electrochemical Society. 2012;**159**(9):521-525. DOI: 10.1149/2.008209jes

[31] Chen T, Yu P, Chen S, et al. Characteristics of large-scale nanohole arrays for thin-silicon photovoltaics. Progress in Photovoltaics: Research and Applications. 2014;**22**:452-461. DOI: 10.1002/pip.2291

[32] Lo CJ, Aref T, Bezryadin A. Fabrication of symmetric sub-5 nm nanopores using focused ion and electron beams. Nanotechnology. 2006;**17**:3264-3267. DOI: 10.1088/0957-4484/17/13/031

[33] Chen J, Shi J, Decanini D, et al. Gold nanohole arrays for biochemical sensing fabricated by soft UV nanoimprint lithography. Microelectronic Engineering. 2009;**86**:632-635. DOI: 10.1016/j.mee.2008.12.093

[34] Bates FS, Fredrickson GH. Block copolymers—Designer soft materials. Physics Today. 1999;**52**(2):32. DOI: 10.1063/1.882522

[35] Koynov S, Brandt MS, Stutzmann M. Black nonreflecting silicon surfaces for solar cells. Applied Physics Letters. 2006;**88**:203107(1-3). DOI: 10.1063/1.2204573

[36] Branz HM, Yost VE, Ward S, et al. Nanostructured black silicon and the optical reflectance of graded-density surfaces. Applied Physics Letters. 2009;**94**(23):1-4. DOI: 10.1063/1.3152244

[37] Hadobas K, Kirsch S, Carl A, et al. Reflection properties of nanostructure-arrayed silicon. Nanotechnology. 2000;**11**:161-164. DOI: 0957-4484/00/030161

[38] Thiyagu S, Syu H, Hsueh C, et al. Optical trapping enhancement from high density silicon nanohole and nanowire arrays for efficient hybrid organic–inorganic solar cells. RSC Advances. 2015;**5**:13224-13233. DOI: 10.1039/C4RA13536A

[39] Wagner RS and Ellis WC. Vapor liquid solid mechanism of single crystal growth. Applied Physics Letters. 1964;**4**(5):89-90. DOI: 10.1063/1.1753975

[40] Westwater J, Gosain DP, Tomiya S, et al. Growth of silicon nanowires via gold/silane vapor–liquid–solid reaction. Journal of Vacuum Science & Technology. B. 1997;**15**(3):554-557. DOI: 10.1116/1.589291

[41] Puglisi RA, Mannino G, Scalese S, et al. Silicon nanowires obtained by low temperature plasma-based chemical vapor deposition. Materials Research Society Symposium Proceedings. 2012;**1408**:139-144. DOI: 10.1557/opl.2012.718

[42] Kempa TJ, Cahoon JF, Kim S-K, et al. Coaxial multishell nanowires with high-quality electronic interfaces and tunable optical cavities for ultrathin photovoltaics. Proceedings of the National Academy of Sciences. 2012;**109**(5):1407-1412. DOI: 10.1073/pnas.1120415109

[43] Xie P, Hu Y, Fang Y, Huang J, Lieber CM. Diameter-dependent dopant location in silicon and germanium nanowires. Proceedings of the National Academy of Sciences. 2009;**106**(36):15254-15258. DOI: 10.1073/pnas.0906943106

[44] Puglisi RA, Garozzo C, Bongiorno C, et al. Molecular doping applied to Si nanowires array based solar cells. Solar Energy Materials & Solar Cells. 2015;**132**:118-122. DOI: 10.1016/j.solmat.2014.08.040

[45] Zhang YF, Tang YH, Wang N, et al. Silicon nanowires prepared by laser ablation at high temperature. Applied Physics Letters. 1998;**72**(15): 1835-1837. DOI: 10.1063/1.121199

[46] Yu DP, Sun XS, Lee CS, et al. Synthesis of boron nitride nanotubes by means of excimer laser ablation at high temperature. Applied Physics Letters. 1998;**1966**:1-4. DOI: 10.1063/1.121236

[47] Cui Y, Zhong ZH, Wang DL, et al. High performance silicon nanowire field effect transistors. Nano Letters. 2003;**3**(2):149-152. DOI: 10.1021/nl025875l

[48] Zhao J, Wang A, Ferrazza MAG, Zhao J, et al. Silicon solar cells monocrystalline silicon solar cells. Applied Physics Letters. 1998;**73**(14):1991-1994. DOI: 10.1063/1.122345

[49] Hauser H, Michl B, Schwarzkopf S, et al. Honeycomb texturing of silicon via nanoimprint lithography for solar cell applications. IEEE Journal of Photovoltaics. 2012;**2**(2):114-122. DOI: 10.1109/JPHOTOV.2012.2184265

[50] Martins GF, Thompson AJR, Goller B, et al. Fabrication of "finger-geometry" silicon solar cells by electrochemical anodisation. Journal of Materials Science. 2013;**48**:2977-2985. DOI: 10.1007/s10853-012-7075-x

[51] Hamouda F, Barbillon G, Held S, et al. Nanoholes by soft UV nanoimprint lithography applied to study of membrane proteins. Microelectronic Engineering. 2009;**86**:583-585. DOI: 10.1016/j.mee.2008.11.086

[52] Guarini KW, Black CT, Milkove KR, et al. Nanoscale patterning using self-assembled polymers for semiconductor applications. Journal of Vacuum Science & Technology B. 2001;**19**(6):2784-2788. DOI: 10.1116/1.1421551

[53] Puglisi RA, La Fata P, Lombardo S. Tailoring the long-range order of block copolymer based nanomasks on flat substrates. Applied Physics Letters. 2007;**91**:0531041-3. DOI: 10.1063/1.2766694

[54] Puglisi RA. Towards ordered silicon nanostructures through self-assembling mechanisms and processes. Journal of Nanomaterials. 2015;**2015**:1-20. DOI: 10.1155/2015/586458

[55] Jiang P, McFarland MJ. Wafer-Scale periodic nanohole arrays templated from two-dimensional nonclose-packed colloidal crystals. Journal of the American Chemical Society. 2005:**127**;3710-3711. DOI: 10.1021/ja042789

[56] Hulteen JC, Duyne RP Van, Hulteen JC, et al. Nanosphere lithography: A materials general fabrication process for periodic particle array surfaces. Journal of Vacuum Science and Technology A. 1995;**2011**:1553-1558. DOI: 10.1116/1.579726

[57] Sai H, Fujii H, Arafune K, et al. Antireflective subwavelength structures on crystalline Si fabricated using directly formed anodic porous alumina masks. Applied Physics Letters. 2006;**88**:201116(1-4). DOI: 10.1063/1.2205173

[58] Storm AJ, Chen JH, Ling XS, et al. Fabrication of solid-state nanopores with single-nanometre precision. Nature Materials. 2003;**2**:537-541. DOI: 10.1038/nmat941

Chapter 8

Opportunities of Scanning Probe Microscopy for Electrical, Mechanical and Electromechanical Research of Semiconductor Nanowires

Pavel Geydt, Mikhail S. Dunaevskiy and
Erkki Lähderanta

Additional information is available at the end of the chapter

http://dx.doi.org/10.5772/intechopen.68162

Abstract

In this chapter, three types of phenomena (electrical, mechanical, and electromechanical) that can be investigated in individual III–V semiconductor nanowires with scanning probe microscope are presented. Transport measurements in GaAs nanowires based on stable electric connection provided opportunity to study individual vertical freestanding nanowires under gentle precisely controlled force. Latter approach appears superior to studies of horizontally fixed nanowires because studying vertical as-grown nanowires avoids charge leakage into the substrate and impact of defects caused by breakage of nanowires. Principles of thermionic emission theory are used to characterize electrical effects in individual as-grown nanowires. Effects of SiO_2 protective layer, surface passivation layers, illumination, and influence of sweeping rate of current-voltage recording are analyzed. Elastic studies are performed for individual InP nanowires affixed at one end. Bending of the tapered nanowires with diameters of a narrow free end either 10 or 20 nm was performed under different loading forces. It allowed calculation of flexibility coefficient profiles along the nanowires' axes. Improved numerical model for tapered nanowires leads to the finding of Young's modulus of wurtzite InP material in nanowires. Piezoelectric measurements permitting registration of reverse piezo effect with opportunities of direct piezo response recording for individual wurtzite GaAs nanowires are briefly described.

Keywords: scanning probe microscopy, AFM, PFM, current-voltage characteristics, Young's modulus

1. Introduction

Recent technological challenges of nanotechnology and industry are lying in two main fields: characterization and production of nanostructures. In order to enhance the quality of nanodevices, it is vital to know properties of the nanostructures involved and specific features that are attributed to their scaling. Since nanowires (NWs) are one of the most prominent objects of nanoscience, significant attention is being paid to their properties and throughput investigation [1–7].

However, already choosing an appropriate method for studying NWs is hard due to their small size and fragility. NWs are quasi one-dimensional structures with nanoscale width possessing properties that dramatically change due to the size effects [8]. Diameter of NWs is approximately 10–100 nm with considerable value of length-to-diameter ratio. In addition, surface-to-volume ratio of NWs defines their strong dependence from the surface effects [1]. Typically, well-established massive methods of investigation are used to characterize the entire arrays of NWs [2, 5]: photoluminescence (PL), X-ray diffraction (XRD), scanning electron microscopy (SEM), transmission electron microscopy (TEM), optical methods, etc. All above-mentioned techniques have limitations on spatial resolution and type of information provided by each certain technique. Definitely, a precise and accurate method focused onto single-object study and revealing features of individual nanostructures should be proposed to support inductive reasoning over properties of NWs inside arrays. Furthermore, if properties of an individual object will be revealed in details, it seems achievable to tune the object's properties and then replicate improved structures into a regular array. Later, such "perfect" array of NWs can be used to create advanced devices based on NWs. Thus, detailed understanding of structure's parameters and its material's properties is expected to benefit innovative technologies. Nanowire-based devices already attract significant attention as units establishing new energy sources [3, 9–11], sensors [12–14], transistors [7, 15], NEMS [16, 17], etc.

Due to the challenges associated with NWs, a novel multifunctional helping hand needs to be utilized for their systematic study. Exactly scanning probe microscopy (SPM) as a family of research methods provides visualization, examination, and even changing of properties of individual nanostructures [18–20]. SPM-based methods were successfully performing this challenging role for NWs as reported previously [21–26]. For example, exactly SPMs are enthusiastically used in studies of new materials and assemblies in modern field of piezotronics [10, 27, 28]. However, description of prospects of electromechanical SPM methods for NWs seems not properly structured yet. Moreover, SPM is an intensively progressing tool, which regularly establishes new and enhanced opportunities for an advanced research [29–31]. We will try to represent accumulated knowledge and our experience in NWs and SPM to help in further studies of NWs. This is done with aspiration that adequate description of SPM methods will stimulate designing and R&D of NW-based devices. Thus, we will represent our results of studies of an individual III-V semiconductor NWs within an enhanced experimental methodology.

Primary issue in such work is establishing of classification under which a comprehensive study will be presented. Among different variants: (1) by composition of material used in NWs (GaAs, InP, GaN); (2) by microscopic method (force spectroscopy, quantitative nanomechanical mapping (QNM), conductive atomic force microscopy (C-AFM), Kelvin probe force microscopy (KPFM), piezoresponse force microscopy (PFM), etc.); (3) by geometrical orientation of the NWs (vertical, horizontal, inclined, or stumps rest after breakage of NWs); and (4) by properties of interest; we preferred the latter class. Therefore, the chapter consists of three following sections: electrical, mechanical, and electromechanical properties of NWs and represents methods of their investigation by modern SPM approaches. Latter means that magnetic [32–35], optical [29, 36–39], acoustic [10, 40–45], catalytic [46], and other properties [47–49] will not be reflected in recent chapter due to format limitations but are proposed to be accessed from the denoted comprehensive works.

NWs described in Sections 2 and 3 were grown by vapor-liquid-solid (VLS) mechanism [4] in metaloxide vapor phase epitaxy (MOVPE) chamber. Catalytic golden droplets were sputtered on substrate under the temperature ~450°C followed by the initiation of a volatile precursor's flow in a predefined protocol [50]. Diameters of NWs were ~100 nm in accordance with width of the Au droplets. Lengths were controlled by duration of the precursor's flow: ~1 μm for vertical structures, ~5 μm for horizontal NWs, and ~3 μm for inclined NWs. NWs for electrical studies were highly p-doped (~10^{19} cm^{-3}) by diethylzinc flow. Comprehensive details about fabrication setup and technique can be accessed from [50]. Samples with horizontal NWs in Section 2 were prepared by a procedure similar to the above described, although after fabrication, the NWs were broken from the substrate and dispersed in distilled water. Acquired suspension was placed onto clean SiO_2 substrate. When water has evaporated, NWs appeared fixed to the substrate by adhesive force. This step was followed by optical lithography, which finally connected separate NWs into electric circuit connected with macro-sized electrodes. Circuiting has been initially checked by SEM and verified by SPM performed afterward. Inclined InP NWs were fabricated on silicon wafers without intentional doping to preserve the pure wurtzite crystal structure of InP.

Subsequent studies were carried out on standard SPM Multimode 8 (Bruker, USA) equipped with additional modules, e.g., either PeakForce KPFM or PeakForce TUNA [51]. Both regimes were accompanied by quantitative nanomechanical mapping (QNM). Separately, few types of spectroscopies were executed. This means that individual single-point-oriented studies were performed with SPM, e.g., force-distance spectroscopy $F(z)$, current-voltage spectroscopy (I–V curves), or current-time spectroscopy $I(t)$.

PFM studies presented in Section 4 were done for wurtzite GaAs NWs fabricated with molecular beam epitaxy (MBE) method (details can be found in [52]). Measurements were performed in standard protocol, where specific AC voltage (frequency ~30 kHz, amplitude 10 V) was applied to the substrate. This has induced mechanical oscillations of the piezo responsive phase in studied NWs, which was registered by high-frequency lock-in of the SPM controller. In a modified version of piezo experiment, we observed current instead of mechanical oscillations. NWs were stressed, but not broken and piezo potential was induced in them. Therefore, mapping of current flowing through the area with NWs was performed under the gentle force. In such approach, piezoelectric NWs are bended and can generate an observable electric current.

2. Electrical measurements

Industrial importance of electrical characterization of NWs defined that electrical measurements of NWs are broadly studied and reported nowadays [7, 53–55]. Indeed, GaAs and InP NWs are proposed for next-generation NW-based solar cells [9, 25, 56–58], while their efficiency is characterized from current-voltage characteristics of an array of NWs (rarely from single individual NW [11]). SPM electrical measurements can be classified into two main types. First one is registration of electric current and its dependence from voltage applied between contacting probe and sample [7] by C-AFM or PeakForce Tuna methods. Secondly, registration of surface electric potential of a sample (i.e. work function of a NW's surface combined with visualization of injected electrostatic charge), is achievable in lift scan of KPFM.

For proper electrical measurements of NWs, substrate must be highly conductive (**Figure 1a**). Obviously, resistance of a wafer should be significantly smaller than resistance of the NW connected in series with substrate (not more than few Ohms). Therefore, substrates' resistance can be neglected [7, 25] (**Figure 1b**). While probe-NW contact resistance is also neglected and shunting is avoided, the I–V characteristics indicate Schottky barrier near zero voltage [7] and linear resistance of a NW (when Schottky barrier becomes open as will be discussed further in 2.1.1.). Connection to a bias source should be direct and not leading to any significant RLC elaboration. Additionally, SPM probes should be highly conductive with uniform thickness of stable rigid conductive coating. It is important to know the shape of a probe's tip apex and utilize probes with height of the tip pyramid applicable for micrometer-tall NWs. For example, KPFM studies are routinely done by probes with height of a pyramidal tip $\sim$2.5 μm. Such probes must not be used for studies of NWs with heights above $\sim$2.0 μm, because a cantilever's beam would strike NW instead of scanning done appropriately with the tip's apex. Morphology of a sample should be considered from established growth procedure or with subsequent visualization techniques, e.g., SEM or TEM. Moreover, following less articulated phenomena should be kept in mind for accurate measurement.

- "Plowing effect" [59], i.e., sliding of the location of probe-sample contact in opposite direction from the cantilever's holder after probe is approaching onto a sample. Referred contact is arranged between the end of tip's apex and a sample. Sliding between them emerges due to the lateral force acting onto the probe's tip mounted on a cantilever affixed at on end. Such sliding (i.e. plowing) occurs every time when the SPM probe approaches the surface, e.g. during initial engage onto the sample, and in each tapping cycle. Plowing/sliding length can reach tens of nanometers depending on setpoint force and is able to damage the surface of flat soft samples. This phenomenon must be considered to understand where exactly the current-voltage spectrum is being registered.
- Influence of setpoint force of push by SPM probe onto stability of contact and forcing, i.e., penetrating, through a surface oxide and chemisorbed water layers.
- Flexibility of a cantilever with related possibility that both vertical and torsional bends will be represented under dissimilar setpoint forces.
- Geometry of a contact between the round top part of an NW (**Figure 1c**) and rounded SPM probe's tip elongated into a pyramid.

Aforementioned features should be taken into account with environmental conditions of specific experiment, i.e., pressure of ambient gas or vacuum level, composition and humidity of this surrounding medium, temperature of the system, and illumination of certain individual NW under the study. Consideration of these factors is important because environment, e.g. air

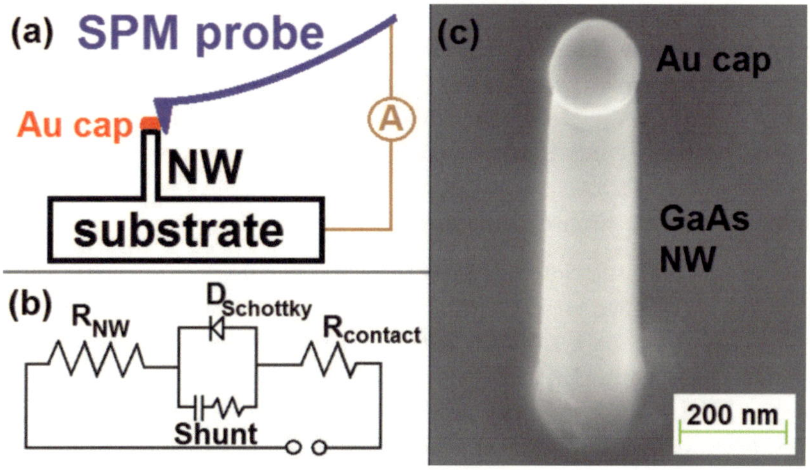

Figure 1. (a) Scheme of the electroconductive experiment by SPM probe in contact with an NW's metal cap. (b) Equivalent electric circuit proposed for the electrical conductivity of a semiconductor NW with a metal cap. (c) Single vertical p^{+}-GaAs NW on p^{++}-GaAs substrate.

in an AFM laboratory room, leads to surface oxidation and covering of NWs by a layer of water. Therefore, electrical, mechanical, and other properties of NWs experience aging. In practice, aging in GaAs is more intensive than for InP, but deterioration of NW can be decelerated by passivation with AlGaAs, GaN, GaP, or other thin coatings [56, 60].

2.1. Electrical conductivity of vertical and horizontal nanowires

Electrical resistance of the conductive channel [7, 56] in a semiconductor NW can be measured when bias is arranged between SPM's probe and a NW positioned in direct contact. Later evaluation of transconductance can be done in terms of the thermionic emission theory [7]. It should be noted that high amount of surface states are one of the most widely known disadvantages of GaAs as a material [55]. Indeed, charges can be trapped by NW's surface due to the defective nature of surface of nanostructures. These trapped carriers induce electric charge that hampers transport of mobile carriers through the conduction channel. Moreover, layer of a nonstoichiometric oxide grows on bulk semiconductor material of the NW. This oxide layer similarly acts as a charge-trapping agent blocking the carrier's transport in the NW, i.e., decreasing width of the conductive channel [61].

SPM controller allows measuring small current flowing through the system when contact is arranged with single individual NW. It should be noted that since SPM controller is solely measuring the electric current signal, an adequate model should be used to interpret the obtained signal (**Figure 1b**). Reason is that shunt or parasitic conductance through undesired coatings, e.g. chemisorbed water layer (from moisture content), nonstoichiometric oxide, or passivation layer can influence the obtained I–V characteristics. Capability to measure the as-grown vertical NWs is the main advantage of SPM techniques in comparison to methods requiring breakage and postgrown processing, which affect properties of these nanostructures. Additionally, SPM is preferential to massive methods, where parameters are registered for an array containing hundreds to millions of NWs, and further experimental values are averaged. Indeed, massive approaches do not allow qualifying the growth quality of NWs and examination of their local features, e.g., properties of Schottky barrier located between metal cap and a semiconductor NW. Thus, the prominent advantage of SPM in research of NWs is exactly in studying of local properties of individual structures.

2.1.1. Obtaining the current-voltage characteristics

Approach of taking I–V curves initiates from scanning the area covered by NWs (**Figure 2a**), while DC bias ~1 V is applied between SPM probe and a substrate. In modern microscopes, it is possible to record few channels of data simultaneously during a single scan. This provides obtaining of maps demarcating independent records of properties of a studied material when SPM data scan file is analyzed. Later, it becomes possible to correlate features from different channels, e.g., *topography* and *current*, and understand details of certain features observed in overall morphology. Vertical NW appears as pyramids in *topography* (**Figure 2b**) due to convolution effect of pyramidal tip and cylindrical or hexagonal NW. Preliminary map of currents in close proximity to a conductive NW typically represents an ensemble of conductive points or distinctive regions with regular shape indicating currents responsible for different facets of the SPM probe tip's pyramid (**Figure 2b** and **c**).

C-AFM study is considerably less expensive and more automated than similar studies performed on SEM with full metal probe or practice of STM with visualization of movements by the station's SEM [62–64]. Noteworthy that in C-AFM, it is only necessary to choose the location where current is maximum and to perform spectroscopy of current at this specified spot (**Figure 3**). Typically, position of the spot is situated at a distance approximately tens of nanometers from the top part of a vertical NW's metal cap visible from *topography* (**Figure 2b**). Reason here is that the most conductive area only indicates position where contact resistance $R_{CONTACT}$ between probe and NW is the smallest.

Capability to observe conductance of long vertical NWs (**Figure 2b** and **3**) highlights benefits of C-AFM technique in comparison to measurements of those done on horizontal NWs connected by lithography. Advantages include simplicity of procedure, availability of standard conductive probes, and absence of necessity in costly additional visualizing SEM. Moreover, utilization of such SEM is associated with bombardment of the nanostructures with focused high-energy electron beam. It can affect conductivity of NWs both when beam is switched on and off. When SEM is performing visualization, the amount of charge carriers within the certain observed NW increases by injection of electrons in SEM observation procedure. Moreover, if SEM beam is switched off, localized charges can persist in surface oxide and spoil the entire sample and measurement. On the contrary, C-AFM is based on standard AFM. A flexible cantilever with highly responsive piezo scanner and bias control system is used in this regime. These features allow measuring morphology and conductivity of fragile nanoobjects with heights more than 5 μm without their breakage. It is possible to arrange experiments with electric bias applied by SPM, although unfortunately, the aforementioned deterioration of NWs by localized charges can still exist.

Nevertheless, few shortcomings arise in SPM method of I–V spectroscopy. As we have noticed from experimental work, drift of the sample occurs during the scanning, so that the spot of measurements moves toward hardly predictable location. It leads to decline of reproducibility

http://dx.doi.org/10.5772/intechopen.68162

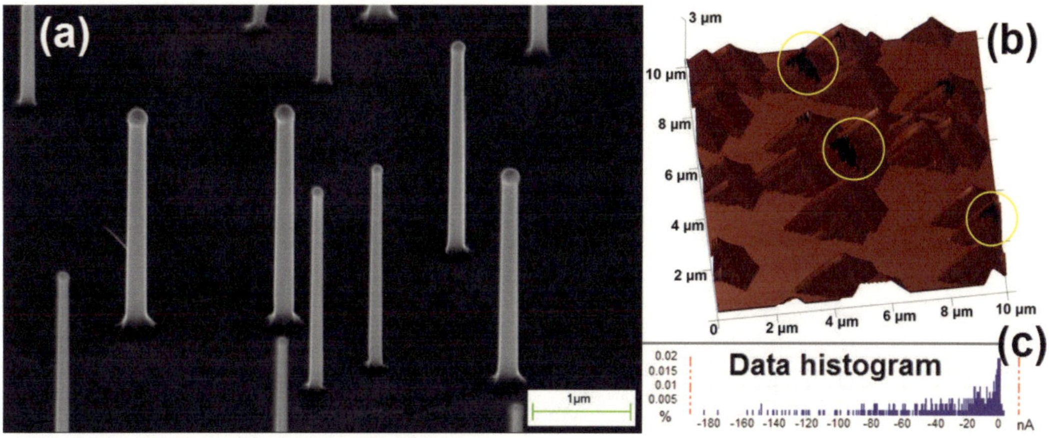

Figure 2. (a) SEM image of an array of vertical GaAs NWs. (b) 3D model of topography of an array of vertical GaAs NWs, where NWs are seen as pyramids due to shape convolution. Electric current map is overplayed on 3D topography indicating regions with high current density near the NW tops as black spots marked with circles. (c) Distribution of values of electric currents in an array of NWs obtained by PeakForce Tuna module, where highest absolute values indicate stable electric contact.

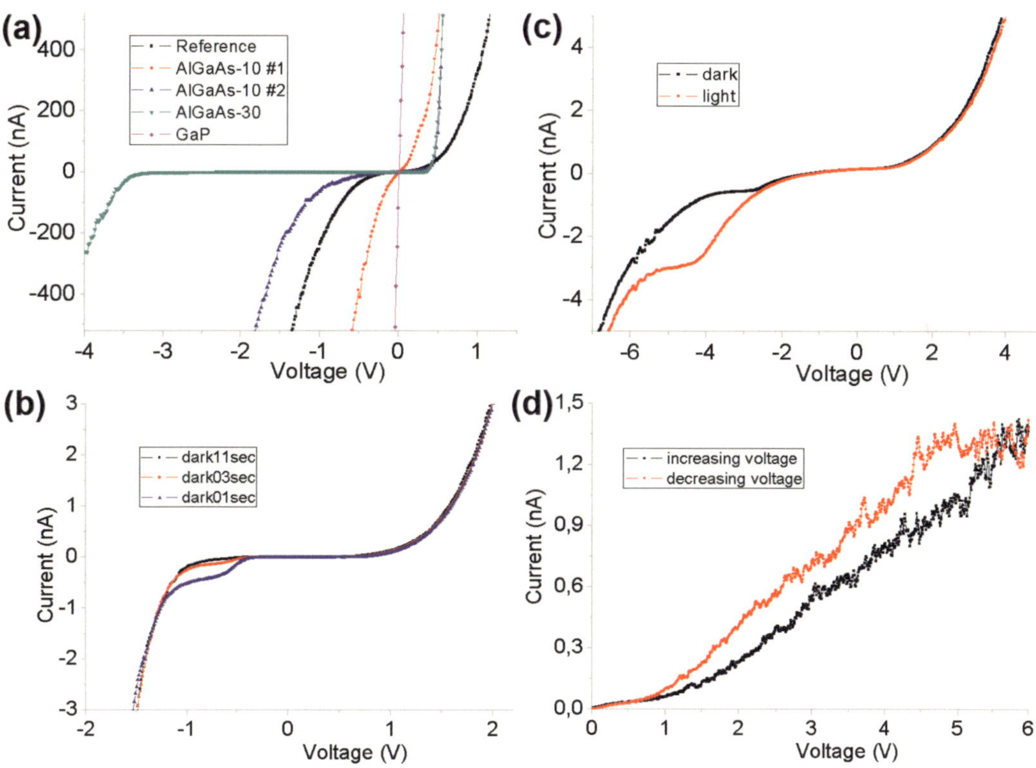

Figure 3. I–V curves for vertical (a)–(c) and horizontal (d) GaAs NWs taken by SPM. (a) Comparison of different passivating shells. (b) Indication of effects of sweeping duration onto accumulation of charges in NWs for 1 s, 3 s, and 11 s. (c) Comparison between electric output of the same pGaAs NW on nGaAs substrate in light and dark conditions. (d) Hysteresis of electric current for GaAs NW.

of I–V curves for same NW after long series of measurements. This can be solved by obtaining more representative statistics of measurements, repeating the tests by fresh probes and with significant suppression of a setpoint force acting during I–V spectroscopy acquisition.

Analysis of the I–V spectra [7] (**Figure 3**) acquired by SPM controller for certain individual NWs in a series of samples prepared with different treatments can be done. It is possible to evaluate effects of passivating shell onto conductive behavior of NWs. For example, in **Figure 3a**, two passivating compounds are compared for vertical GaAs NWs, i.e., GaP and AlGaAs. It is noticeable that current through the NW in region of forward bias increases for both of them in comparison with unpassivated reference NWs. We studied these passivating shells in details previously [65, 66]. GaP appeared to be the most advantageous passivating material from experimental measurements of GaAs NWs. This conclusion is obvious from I–V curves taken on vertical NWs since I–V spectra for sample with GaP ultrathin shell demonstrate neither a specific knee in region of forward bias nor the junction voltage parameter. Reason for that is decrease of concentration of surface states at the surface of vertical NWs, which widens the conduction channel in NWs. Practically, I–V curves for GaP passivated NWs were almost linear, so that quasi-Ohmic contact was formed between Au cap and highly doped GaAs crystal part of the NW.

Indeed, latter contact area should arrange a Schottky junction, and NW system should resemble the rectifying shape of I–V curve for a Schottky diode [7, 67] in accordance with the diode law

$$I = I_0\left(e^{\frac{eV}{kT}} - 1\right) \tag{1}$$

where I_0 is a reverse-bias leakage current (A), q is an elementary charge 1.6×10^{-19} C, k is the Boltzmann constant 1.23×10^{-23} J/K, and T is an absolute temperature (K).

It is considered that explanation for phenomenon of negligence of Schottky input is that Fermi-level pinning is reduced in highly doped NW [68]. In fact, level of doping can be quantified from the linear serial resistance region of I–V curves. The slope equals to resistance of a system by Ohm's law and following equation

$$R = \frac{l}{eN_A \mu \pi r_{NW}^2} \tag{2}$$

where length of the NW, L, is approximately 1 μm, elementary charge $q = 1.6 \times 10^{-19}$ C, mobility of carriers $\mu = 400$ cm^2/V s, and NW's radius r_{NW} is approximately 50 nm, while carrier concentration is N_A. Since resistance of GaP-passivated p-GaAs NWs is ~100 kΩ from the slope, the doping level N_A is approximately 10^{19} cm^{-3}, verifying the value anticipated from fabrication protocol of VLS growth.

Passivation by GaP [69] or AlGaAs [70] was previously found among the most advantageous methods to decrease the density of surface states in GaAs NWs by other authors. Interestingly, it is possible to distinguish instabilities of quality of passivation for NWs over the single sample as visible for two I–V curves taken on two separate GaAs NWs with equal 10-nm AlGaAs-passivating shells. In fact, sizes of NWs have specific distribution as can be seen in **Figure 2a**. From AlGaAs passivation results obtained for NW#1 (red curve) and NW#2 (blue curve), it is possible to conclude that thin AlGaAs passivation is also advantageous for GaAs NWs, although effect of passivation is less pronounced comparing with NWs shelled by GaP. However, summarized effect of the surface states (fast traps and slow traps) of native oxide significantly depends on NW's diameter. I–V curve for NW with AlGaAs shell passivation layer with thickness 30 nm (cyan curve) already resembles the shape characteristic for a diode.

Such NWs have forward threshold voltage (junction voltage) approximately 0.2 V and indicate leakage current and reverse breakdown voltage lower than −3 V. Therefore, such thick shell passivation approach is detrimental for NWs. Reason for that is accumulation of electric charge at the shell and its capacitive adverse input blocking the conduction channel of the NW. Moreover, inclination of all curves in regions of linear resistance is equal because it represents the concentration of charge carriers, i.e., same doping level [65, 66].

Furthermore, analysis of curves recorded with different sweeping rates indicates that charge accumulation and redistribution effects can be visualized for GaP-passivated pGaAs NWs on nGaAs substrates (**Figure 3b**). Corresponding curves represent both Schottky barrier input and p–n junction barrier. It should be noted that detailed shape of I–V curves depends not only on sweeping rate (duration of registration of a spectra) but also on starting voltage of such registration and cycle. Latter means that hysteresis effects can be observed from I–V curves of NWs. Therefore, all curves in **Figure 3a–c** were shown for sweeping from maximum voltage, i.e. while bias was consistently shifted toward negative potential. Additionally, illumination effects can be observed for certain NW systems with p-n barrier. As it is seen from **Figure 3c** for unpassivated pGaAs NW on nGaAs substrate, over-lighted conditions led to intensive rise in current through NW. In addition, the serial linear resistance regions for unpassivated NW (**Figure 3c**) are significantly further from 0 V comparing with passivated NWs (**Figure 3b**). Partly, the observed increase of current can be explained by the stimulation of photo-induced conductivity. Availability to register I–V curves is more impressive for vertical NWs but can be done also for horizontal NWs. In this respect, for example, resistivity and charge accumulation near the Schottky barrier (near the interface between the NW and each electrode) can be (**Figure 3d**).

2.1.2. Mapping of electric current (C-AFM in contact mode, PeakForce TUNA in semicontact mode)

Another option is performing the C-AFM scanning without further spectroscopy, while DC bias between probe and sample is being manually or programmably changed. This possesses registering the map of values of current for each manually specified bias. The approach is time-consuming, but can allow plotting of considerable amount of data points on I–V plot and later reconstruction of detailed shape of I–V curves. Advantage of the mapping approach is elimination of the adverse drift, because it is possible to reconstruct the location of a NW's cap independently in every scan. This location of maximum current would be defined to consider and plot exactly current values from the locations of maximum currents. It should be noted that high-aspect ratio, i.e., so-called lent-shaped scans are proposed to be done. These rectangular shape scans with low amount of scan lines along the slow scan axis reduce duration of

SPM measurements. Another advantage is that mapping approach allows obtaining statistics to characterize the distribution and repeatability of conductance of NWs in an array. Unfortunately, mapping approach accelerates degradation of the SPM probe by wearing its conductive coating and can result into damage or breakage of the NW. This is due to significant forces (in the order of hundreds to thousands of nanonewtons) acting onto NWs in contact mode of C-AFM. Stability of conductive coating of SPM probes from wearing can be verified as unfailing of maximum values of registered electric current with time during a series of scans.

Therefore, the most advantageous procedure of measurements of NW's resistance should be based on method, where force of interaction is minimized and controlled. Most of recent SPM setups provide this opportunity, which appeared initially as PeakForce™ mode (Bruker). PeakForce is based on ScanAsyst feature allowing automated calculation of optimal setpoint force and feedback gain. Here, it should be noted that the regime combining both accurate measurement of topography and measurement of current is called PeakForce TUNA mode. The regime is realized as additional mode of our standard SPM microscope. It is beneficial in comparison to C-AFM because considerably smaller force of interaction can be established and controlled, i.e., force is controlled with precision ~0.01 nN instead of few nN in C-AFM. Simultaneously, current is being registered in the range up to 500 nA with outstanding accuracy ~65 fA, which is also two order of magnitude more accurate than C-AFM with its high noise level. Practically, measurements of morphology of vertical NWs are done in such a way that force of interaction onto flexible few micrometer-tall NWs can be controlled to be approximately 0.1 nN.

Since PeakForce Tuna is a semicontact regime, electric current is being measured in every act of tapping. Direct contact established during a single tap lasts approximately tens of microseconds (depending on setpoint force and tapping frequency), but this duration is enough for modern electronics to register current precisely with picoampere accuracy. This approach allows measurements of long and thin NWs (length ~4 μm and diameter ~10 nm) without damaging the NWs as was shown in our previous works [65, 66]. Last important detail is exactly the composition of conductive coating on SPM probes utilized for electric measurements. Industrial techniques of fabrication of SPM probes are consistently developing. Old types of coating materials, e.g., so-called diamond-like coated probes (DCP), are becoming more rigid and reliable. Simultaneously, noble metals are being used for coatings, e.g., alloy of Pt/Ir (Bruker) or Au. These coatings have thickness approximately tens of nanometers and tend to be defaced when significant forces or electrical biases above 2 V are applied. It leads to decline in repeatability of results for excessive biases or insufficient scale of applied electric potential for I–V curves. We consider that most reliable results can be obtained with modern DCP probes (NT-MDT), or AFM probes made entirely of highly conductive inert metal, e.g., full metal platinum probes (Rocky mountain nanotechnology) or with SPM probes covered with tungsten semicarbide W_2C (ScanSens). All of these types of probes are available in market representing various resonant frequencies and spring constants. Still unfortunately, progress of SPM probes seems less significant than predicted by corresponding roadmaps for technologies [71]. Hence, a holdback in one of the nanotechnologies decelerates research of others as happens in the study of NWs.

2.2. Electrostatic surface potential and charging of horizontal NWs

KPFM is a method of registration of surface electric potential originated from a classical experimental technique invented in late nineteenth century. Its principle of operation utilizes control of electric potential field between measuring probe and studied sample [72]. SPM probe moves in predetermined pattern above the surface with constant lift height and performs mapping of surface's electric potential with millivolt-level accuracy. The advanced SPM version of Kelvin probe setup is KPFM operated in frequency modulation regime. Lateral resolution of imaging is in the order of radius of the SPM tip, while most recent conductive SPM tips have radius approximately 20 nm. Indeed, KPFM can be acclaimed as one of the most important and informative tools to study electrostatic effects in NWs. In a forefront KPFM experimental procedure, scanning is being done in a single pass, simultaneously mapping *topography* and *potential*. This feature is called PeakForce KPFM FM, and it reduces the duration of experiment twice in comparison to two-pass method [73]. Thus, major achievements in this method were reached during latest two decades as a result of advanced nanotechnology and modern computing. It should be noted that resolution of KPFM is strongly dependent on quality factor of SPM's probe oscillations Q. Parameter Q increases in vacuum due to depressed damping of induced mechanical vibrations of SPM cantilever that significantly improve signal-to-noise ratio. Common resolution in ambient air conditions is limited by tens of nanometers partly due to the moisture and water layer covering surfaces of both sample and SPM probes.

KPFM experiments with horizontal NWs fixed in metal-semiconductor-metal (MSM) geometry are well established and were described in previous works [74]. However, it seems worth adding few vital previously unreported details to disclose the procedure of surface potential microscopy in even more comprehensive way. Optimal procedure of KPFM measurements assumes that horizontally lying NW (see marked area in **Figure 4a**) should be oriented perpendicularly to the plane of the vertical deflection of an SPM cantilever (see scheme in **Figure 4b**). Typically, this is achieved by manual rotating of the sample itself or sample's holder in desired direction before the measurements. Scanning should be done with fast scanning direction coincident with long axis of an NW, which is done by insertion of scan angle of 90°. These methodical improvements result into elimination of influence of shape of the pyramidal SPM tip because front and back angles of the tip pyramid are typically different, while the two other side angles are identical. Another benefit is a significant suppression of parasitic capacitive influence of massive SPM tip onto lateral resolution [75]. Moreover, such arrangement routinely eliminates spikes related to topography roughness of NWs visible in lift pass of the double pass KPFM mode. This is because typical surface profile of the NW in the direction of fast scanning is a smooth flat line. Latter allows decreasing the lift scan distance below 20 nm without any risk of accidental contacting the NW in a noncontact pass. Cantilever and the pyramidal tip should be as small as allowed by features of topology and roughness. In addition, tip radius need to be as small as possible on probe with stable conductive coating. Consideration of all aforementioned factors can allow sub-nanometer resolution of KPFM imaging [72].

Injection of electric charge with its subsequent registration by SPM involves the following steps. Electric bias should be switched on and adjusted to zero with feedback system recording

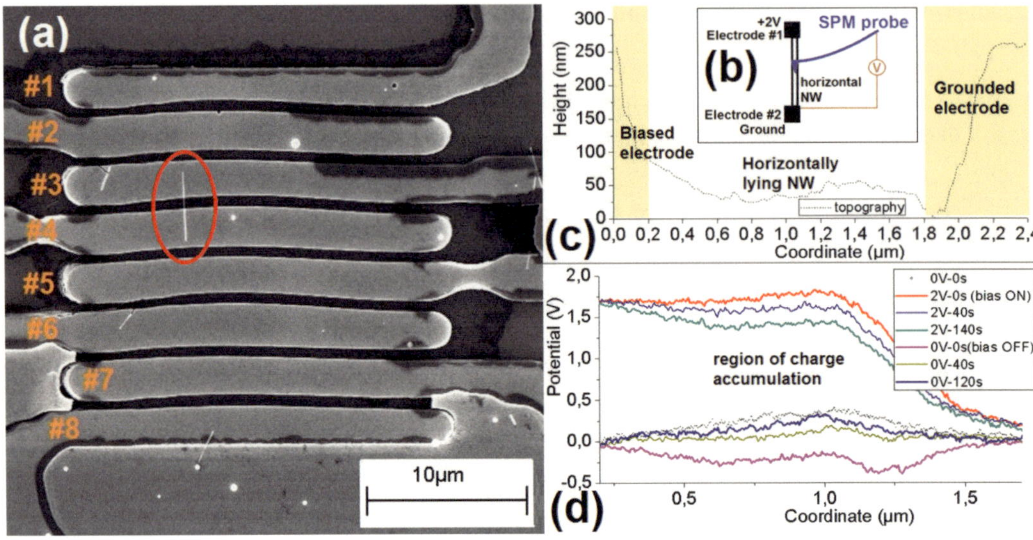

Figure 4. (a) SEM image of a microscheme with pGaAs NW covered by Zn-Au micropads grown by optical lithography. Red mark indicates an NW circuiting the pads #3 and #4, which were used for biasing and grounding. (b) Scheme of the surface potential experiment by SPM probe lifted above the horizontally lying NW. NW is rotated on scheme to represent perpendicular direction of its axis to bending plane of the SPM's probe. (c) Topography profile of the horizontal pGaAs NW. (d) Surface electrical profiles for bias 2 V applied followed by 0 V applied to the left electrode, i.e., to electric micropad #3.

electric *potential* of a sample. First, surface *potential* map should clearly indicate position and orientation of an NW on a substrate in *topography* (**Figure 4c**). It is expected that *potential* profile of the NW should be smooth and flat (**Figure 4d**). If NW is properly connected to microelectrodes and further massive electric feed source, then *potential* of the electrodes should immediately reflect the established parameter of DC bias (i.e., zero if grounded). It should be highlighted that surface *potential* profile of the NW would follow the set value of electric potential of an electrode. It is remarkable that the profile can change in time due to accumulation of electric charges so that shape of profile evolves depending on barrier and NW (**Figure 4d**). Moreover, shape of the profile qualifies this energy barrier at the border between two materials, which indicates certain type of electric contact, i.e., Ohmic [65–67] or Schottky (**Figure 4d**).

Furthermore, potential of both electrodes immediately returns to initial value when bias is switched off, but the distribution of surface electric potential profile of an NW in KPFM occurs with finite observable rate (**Figure 4d**). It means that exactly the charge injected into the NW is visible in profile. Reason is that accumulated charge whose highest concentration is located at coordinate of maximum $U_{MAX}(x)$ is distributed according to (1) diffusive transport model, (2) tunneling into substrate, and (3) considering Coulomb repulsion [76]. In other words, electric charges become redistributed and dynamics of this process are restricted by certain major distribution mechanism. The major mechanism of charge distribution in semiconductor NWs should be diffusion. If charge is leaking, then transistors and energy sources produced on base of such NWs would be ineffective. This means that for properly arranged transport measurements with SPM $dU(t)$ should be proportional to square root of time. From $U_{MAX}(t)$ dependence and full-width at half-maximum of $U_{MAX}(t)$ plot, it becomes possible to characterize the mobility of charge carriers and location of defects in NWs. It should be noted that experiments in horizontal NWs and recording their surface potential profiles are preferable to be done for NWs fixed at two ends in a "bridge geometry" [77, 78]. The "bridge geometry" arrangement is laborious in fabrication, although superior to the horizontal MSM arrangement. Reason is that in bridge geometry, the NW is not in electrical contact with a substrate, which eradicates possibility of charge tunneling into the substrate. Similar KPFM investigation over the length of the NW can be proposed also for vertical or inclined NWs, although the substrate with such NWs must be cleaved and then tilted so that SPM probe would have access to the side surface of the NW instead of solely the NW's metal cap. Potentially, electrical studies of NWs can be done by an SPM probe oriented horizontally [79] but not with vertical standard AFM and STM probe tips. In addition, multiprobe techniques already allow advanced operation with vertical NWs [80], but such measuring stations are nowadays expensive and rare.

3. Evaluation of mechanical strength for inclined NWs

Evaluation of strength characteristics of nanoscale objects is a nontrivial task, particularly because it is impeded by properties of their surface, geometry of a measurement, and physical effects of scaling. In this respect, studies of NWs by SPM are challenging because these structures are highly bendable, oscillate with own frequency due to thermal vibrations, and because properties along the NW can be unevenly distributed. Another challenge is that NWs kept in atmospheric environment become partly oxidized, and their properties tend to resemble features of nanocomposites. One of the major details that are widely neglected is that NWs represent specific degree of tapering (**Figures 1c**, **2a**, and **5a**). Reason for that is parasitical input of vapor-solid input of crystal growth [52], which occurs concurrently with VLS mechanism. Tapering of group III–V semiconductor NWs is more pronounced under high-temperature fabrication and for excessive values for relation of V/III precursor sources inside a MOVPE chamber [50]. We will pay special attention to the impact of tapering onto measurement of properties of NWs in this section.

Few methods possessing nanoscale accuracy enable measurements of 1D nanostructures. One of the noncontact approaches makes use of in-situ TEM in order to measure amplitude of free and induced mechanical oscillations of an NW fixed at one end [81]. To certain degree of accuracy, it is possible to calculate elastic parameters with such TEM or high-resolution SEM visualization. Typically, it is only needed to calculate elastic modulus from abovementioned amplitude and geometrical parameters with known density of the NW's material. However, first, certain degree of inaccuracy arises from inexactness of measured geometry and vibrational output. Second, the major reason for error can be that no adequate models were proposed for tapered structures and their mechanics. It should be noted that flexibility coefficient of such tapered NWs is variable along the length.

Methods based on SPM employ direct mechanical contact between structure and a probe and have advantage in their locality. In fact, the only SPM approach presented in literature is when the

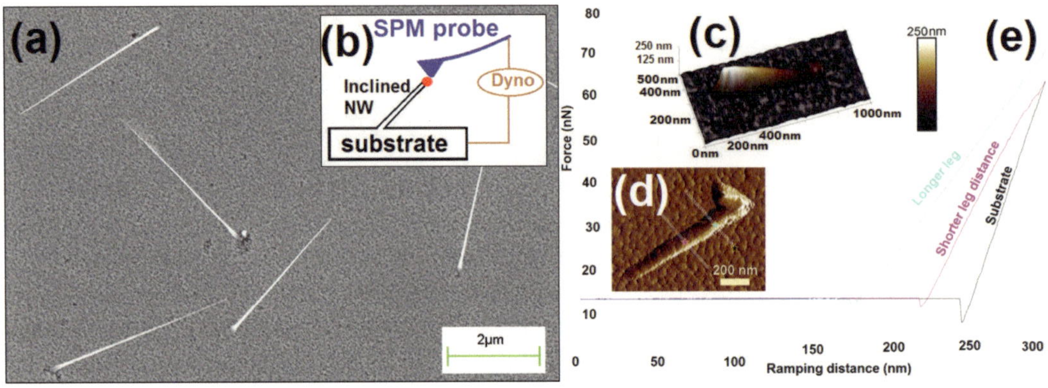

Figure 5. (a) SEM image of an array of inclined tapered wurtzite InP NWs on Si substrate. (b) Scheme of the bending experiment by SPM probe brought in contact with inclined NW. (c) 3D topography model of an inclined NW representing angle 22° for visualization of the NW under small setpoint force. (d) PeakForce error channel representing NW's fixed end at lower left and free end at upper right. Force spectroscopy was performed according to the coordinates seen by color marks. (e) force-load curves *F*(*z*) for substrate, middle of NW, and free end of NW.

probe is bending an NW in a definite leg distance point [82, 83]. Leg distance and NW's diameter should be accurately measured in order to evaluate the Young's modulus. We will describe this simple approach in detail at first and then represent our original approach based on mapping of elastic parameters. Latter allows precise control of setpoint force that provides significantly more accurate and precise results for tapered NWs, also entitled as conical NWs or nanocones [84].

3.1. Method of force-load curves F(z)

The simplest approach is based on bending of an NW by SPM probe in exactly defined position [85]. This experiment can be performed on inclined cylindrical NWs (**Figure 5a–c**). Procedure involves measurement of *topography* and defining the location, where stable contact for bending can be established at farthest distance from the NW's fixed end (**Figure 5d**). Longer leg distance provides higher values of displacement and thereof bigger amount of data points plotting the *F*(*z*) curve without breaking the NW.

SPM probe for such experiments needs to be flexible and dull, i.e., tip radius is desirable to be considerably more than 30 nm. Initially, cantilever becomes bended on any flat rigid substrate (see black curve on **Figure 5e**), e.g., in proximity to the NW. When spring constant of SPM cantilever becomes precisely recognized, the measurement is called calibrated conferring the force-distant curve *F*(*z*) and its slope. Afterward, bending of the desired structure is performed. It is clearly visible that magenta and turquoise curves (**Figure 5e**) taken correspondingly for location near the middle length of the NW and closer to the free end indicate two different slopes. For a highly bendable structure, measured force-load curve *F*(*z*) will consist of two components, i.e,. bending of a cantilever and own bend of the NW [85]. For finest evaluation of the NW's spring constant, it is preferable to choose a cantilever with spring constant similar to the anticipated stiffness of the NW. Since spring constant of cantilever k_{CANT} is known from calibration, the other component is exactly spring constant of the NW k_{NW} [85].

First difficulty of this one-point spectroscopy approach is that NWs can be tapered to specific extent. Their diameter d_{NW} is depending on leg distance x, but since $d_{NW}(x)$ is changing along the length of the NW, then their k_{NW} is nonlinearly dependent from leg distance as $k_{NW}(x)$, where $x \in [0;L]$. Despite that, k_{NW} characterizing elastic properties of a studied NW is adequately obtained only for one coordinate situated considerably far from NW's end. Second, it does not seem achievable to recognize this leg distance x precisely, because inclined NWs are partly merged with the substrate at the fixed end, which is an artifact of fabrication procedure (**Figure 5d**). However, it is obvious that SPM probe is unable to visualize any area beneath the structure (**Figure 5c**). Third, since diameter strongly affects the value of elastic modulus, inaccuracy of a diameter is crucial. Finally, such measurement can hardly be qualified as stable, because sliding effect is significant, so repeatability over time is also a significant drawback. Nonetheless, with this technique authors of [82, 83, 85] have successfully measured and reported elastic moduli of 1D nanostructures with reasonable accuracy. In was achievable because method of force-load curves is adequate for thick long nanostructures.

3.2. Advanced QNM for bendable nanostructures

We have generalized an advanced procedure allowing measurements of inclined NWs to be carried out with significant accuracy [86]. Measurements are proposed to be performed with the help of SPM equipped by a module registering setpoint force at each moment, e.g., QNM. In fact, major SPM manufacturers had promoted concurrent methods analogous to QNM (HybriD from NT-MDT, pulsed force mode from WITec and jumping mode from Nanotec Electronica) independently, while Bruker firstly introduced their QNM™ in 2009. As a result, this advanced experimental technique is now attainable by plenty of research teams with various microscopes worldwide. Let us describe our experimental approach with theoretical background and associated limitations, which will be formulated in terminology of the Bruker's QNM mode.

Initially, it is possible to find individual NWs by built-in optical microscope providing magnification 200x. In fact, even location of thin NWs having diameters 10 nm can be visualized by optical microscope, because they scatter the light. Such NWs become visible as single grey pixel routinely resolvable by optical matrix (**Figure 6a**). The scanning is done by the blunt flexible SPM probe moving along the long axis of an inclined NW. It should be noted that specific probes can have a tip shape allowing to settle on the nanostructure and arrange a stable mechanical contact. For example, manufactured NP-STT10 probes (Bruker) with two tip apexes are proposed for stabilization of contact between NW and SPM probe and to avoid sliding effect (see two tip apexes in **Figure 6b**). Separately, ordinary flexible one-apex probes should be used for surface visualization. QNM measurements of bendable NWs are proposed to be performed under different setpoint forces (**Figure 6c** and **d**) in order to obtain significant statistics to calculate flexibility coefficients (**Figure 6e**) and model the Young's modulus. Interestingly, such remarkably precise scanning requires not more than 10 min with rectangular scans described in Section 2.1.2. Entire profile of a flexibility coefficient is obtained instead of low representative single data point in the method of force-load curves (**Figure 5e**). These data are reproducible during the scan, for different forces and by probes with different spring constants.

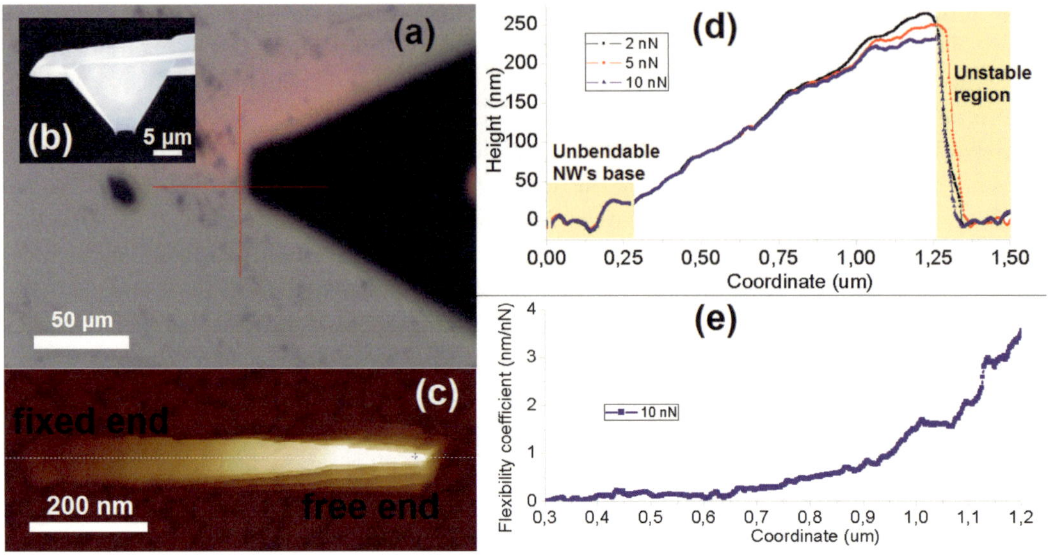

Figure 6. (a) Optical image 125 × 80 µm representing individual NWs as grey spots near the triangle-shaped SPM cantilever. (b) SEM image of BRUKER NP-STT10 probe with two apexes at the end of the tip and an arc between them. (c) High resolution rectangular shaped 2D topography image of an inclined InP NW under significant force setpoint 10 nN. (d) Comparison of topography profiles for bending of inclined InP NW under different forces. (e) Bending profile representing flexibility spectra of a single bendable NW along its length, where fixed end is on left and free end is on right side.

Our approach is based on Hooke's law and Euler-Bernoulli theory. The body compression law (seventeenth century Hooke's law) states that

$$F = k\delta \tag{3}$$

where k is a linear stiffness coefficient, i.e., spring constant, of a structure and δ is a linear displacement, i.e., deformation, by the force F acting on same axis with deformation.

In case of a NW, which is bended down by specific setpoint force F until the NW's "bended position", the equation for deflection ω should be written as follows

$$F = \mathrm{k}_{NW}(x) \cdot \omega_{NW,\ F}(x) \tag{4}$$

Noteworthy that the "spring constant" k_{NW} is not invariable along the NW structure in bending experiment, but depends from a distance to the fixed end $x \in [0,L]$, where L is length of NW.

Deflection of NW $\omega_{\mathrm{NW,F}}$ under certain applied force F is a parameter measurable from SPM topography data (**Figure 6d**)

$$\omega_{NW,F}(x) = \Delta\text{height} = h_F(x) - h_0(x) \tag{5}$$

where h is a registered height, so that $h_{\mathrm{10nN}}(x)$ is the height observed by SPM when the probe pushes with a setpoint force 10 nN at a certain coordinate $x \in [0,L]$.

It should be noted that F is a fixed parameter and this setpoint force is established with piconewton accuracy in QNM. The spring constant k is known as dependent from the leg distance x, so that relation between k and x is cubic for a cylindrical beam $k \sim x^3$ similar to experimental plot of flexibility profile f(x) in **Figure 6e**. We will later calculate how exactly k is related with x for a tapered beam with the cone angle 2α and diameter in the middle of the NW's length D_{MID}. Height $h_0(x)$ is typically considered as the height of a structure scanned by SPM. Despite that, the topography is always measured under specific setpoint force applied by SPM. Therefore, the force for $h_0(x)$ is neglected. SPM setpoint force is ~10 pN for a gentle scanning, which is affordably small.

Euler-Bernoulli theory introduced in eighteenth century defines bending of an isotropic beam [87], depending on composition (by elastic modulus) and geometrical shape of a beam (e.g., by length L and diameter D)

$$\frac{d^2}{dx^2}\left(EI\frac{d^2\omega}{dx^2}\right) = q \tag{6}$$

where $I(x)$ is second moment of inertia of a beam.

The equation for deflection of an NW by a point contact established by an SPM probe along the location in NW $x \in [0,L]$

$$E\frac{d^2}{dx^2}\left(I(x)\frac{d^2\omega}{dx^2}\right) = 0 \tag{7}$$

The boundary conditions appropriate for our model are

$$\omega(0) = 0;\ \omega'(0) = 0;\ \omega''(L) = 0;\ \omega'''(L) = -\frac{F}{EI(X)} \tag{8}$$

Accounting these boundary conditions, the generalized form of a solution of Eq. (7) is

$$\omega_{NW,F}(x) = \frac{F}{6EI(x)}(3Lx^2 - x^3) \tag{9}$$

The second moment of inertia for a cross-section of a conical beam is described as

$$I(x) = \frac{\pi D_{MID}^4}{64}(1 + \frac{\alpha}{D_{MID}}(2x - L)) \tag{10}$$

We now can formulate the dependence of flexibility of a tapered NW:

$$f_{NW}(x) = \frac{1}{k_{NW}(x)} = \frac{\text{Deflection}}{\text{Force}} = \frac{\omega_{NW,F}(x)}{F} = \frac{1}{E_{mat}} \cdot [\text{coeff}(x, L, D_{MID}, \alpha)] \tag{11}$$

where flexibility coefficient $f_{NW}(x)$ is presented as a function of material (E), location of applied force (x), and the shape of a structure (L, D_{MID}, α).

Finally, the formula for calculation of the Young's modulus of a tapered NW in our model is the following

$$E_{\text{mat}} = \frac{64}{3\pi} \frac{1}{f_{NW}(x)} \frac{1}{D_{\text{mid}}^4} x^3 \frac{1}{(1 + \frac{\alpha}{D_{\text{mid}}}(2x - L))(1 - \frac{\alpha L}{D_{\text{mid}}})^3} \tag{12}$$

It must be noted that for appropriate application of any model, it is vital to consider not only the generalized form of relation between variables, but also assumptions of the model, defining its limits of use. Two of such assumptions for Euler-Bernoulli theory are isotropy of the material and elastic regime of bending. Material of the NW is considered isotropic across its entire length, discarding e.g., surface effects, possible defects in crystal structure, etc. In reality, defects act as deformation accelerators and result into spikes visible on bending profile. Elastic regime of bending means that object's deformation should be very small with the force applied. In addition, elastic modulus in our model is considered independent from temperature and speed of bending. In fact, Young's modulus E descends with heating [88], although it is not considered in our model. Indeed, any physical body becomes heated when the work is applied upon it. Therefore, bending by SPM probe performed in a regular manner 2000 times per second can be seem to add heat into the studied NW. Reason for ignoring the possible thermal input is that NWs have considerably high thermal conductivity and small length, so that heat is rapidly distributed into the substrate and dissipated in atmosphere.

Remarkably, the value of Young's modulus is obtained for one-single NW from a substantial series of data points $\omega_{\text{NW,F}}(x)$ recorded by SPM. Indeed, each of these values is already appropriate to define E, but the series of data values contains ~400 informative data records providing high reproducibility of a result. Furthermore, each setpoint force provides specific bending to verify that deflection is within the linearity range. If multiplied by amount of data values for each applied setpoint force, then significant approximation of flexibility coefficient is achieved, providing unprecedented accuracy of this approach.

Unfortunately, the model involves few geometrical parameters of NWs, which are difficult for estimation with significant accuracy. Visualization techniques, e.g., SEM or SPM topography measurement, were used to provide L, D_{MID}, and α. Due to limitations of lateral resolution of SEM scanning (~2 nm) inaccuracy of measurement of diameters of the studied NWs was significant. Moreover, relative error of diameter evaluation increases for thin NWs, while we have studied NWs with diameter of top part as small as 10 nm. Therefore, a special protocol of estimation of diameter of the NWs was utilized to correct the values. Initially, sizes were controlled during fabrication with accuracy approximately ±3 nm. After the fabrication, diameters were controlled by SEM. After that we performed breakage of few NWs from an array to estimate their height, while NWs were lying on flat substrate. As a result, for NWs with $D_{\text{MID}} = 40$ nm, relative error of estimation of Young's modulus was ±40%. Therefore, we performed the HR-SEM investigation and numerical modeling that decreases the relative error until approximately 20%.

This level of accuracy allowed defining that the Young's modulus of InP in thinner NWs increased from 60±7 GPa until 140±30 GPa. Remarkably, 62 GPa is exactly the value of Young's modulus of zinc blende InP. The result was obtained from a series of 16 inclined

NWs. This means that Young's modulus of wurtzite InP is approximately twice bigger than that of zinc blende InP. Wurtzite structure of NWs was verified by TEM and microPL analysis. Already this, i.e., estimation of Young's moduli of wurtzite InP, is a challenging achievement, which can be acquired only in thin NWs. Reason is that wurtzite InP phase is unstable with increasing of the sample's size, so that bulk material transforms into zinc blende structure while the NWs under our studies were affordably thin. Own scaling effects of Young's modulus of wurtzite phase can be separately investigated when size of the material becomes lower than 50 nm in diameter. Another issue to highlight is that a proper model of tapered NWs was proposed and used [86], which was not performed by other authors. The results were achieved for NWs with upper diameter (width of Au caps) $\sim$10 nm.

In order to understand the unlikeness of experiment with bendable objects, it is essentially important to differentiate between "stiffness of the structure" and "rigidity of the material within a structure". Such distinguishing seems necessary because *deformation* channel data are calculated only in the case when both force applied and material's stress occur at the same location. These calculations for SPM in QNM are done in accordance with Derjaguin-Muller-Toporov (DMT) elastic model [24], used as built-in in QNM mode. Indeed, DMT model was introduced in SPM to be mainly used for characterization of flat surfaces and the method succeeded in this pursue. However, for bendable objects, DMT channel data and underlying model are inapplicable. This is explained considering that the stress occurs solely at the fixed end of a bendable structure possessing one fixed end. Therefore, the location of stress (fixed end) is situated micrometers away from location of force applied by SPM probe. In QNM, while DMT model is used, the deformation is quantified from adhesion data, while setpoint force is precisely controlled by feedback system. Topography is being registered routinely, but highly bendable NWs are twisting down during their scanning. Thus, bending reflects on registered topography. Similar conclusion that morphology depends on setpoint force applied by SPM probe was previously demonstrated in studies of exceedingly soft materials [89]. Force needs to be kept in such small range that deformation is consistently $\sim$1 nm across the entire flat surface of a sample. Lastly, topography must be measured in appropriate time moment of PeakForce procedure. The sync distance parameter needs to be manually adjusted to be approximately 6 μs (for PeakForce frequency 2 kHz) later [86] than established automatically during the first engaging onto a sample prior to scanning. Autofitting of time moment of the PeakForce value for each tapping cycle can solve this problem and probably will be introduced in forthcoming versions of SPM software.

Considering other types of experiments possible to be realized with SPM, classical indentation can be mentioned. For indentation measurements, sharp and extremely stiff SPM probes are required. Unfortunately, elastic deformation and indentation models consider different geometries and types of stress and should not be compared. Elastic models include stretching, squeezing, and torsional stress, while indentation results depend on a shape of the indenter, etc. Interesting summary of results obtained by these various methods of ZnO NWs was prepared by authors of [90]. They have shown that elastic modulus of ZnO measured by different groups with different methods varied by almost twice.

Summarizing that the above-described elastic profile mapping approach is significantly more advanced in comparison with spectroscopic method of investigation of Force-Load curves. It is first because diameter D is not considered constant due to tapering of NWs. Secondly, because $k_{NW}(L)$ is being evaluated at plenty of definitely established positions, but not at one single location (typically at the oscillating free end of NW and in disfavoring semicontact tapping mode). Our approach allows precise definition of coordinates of the fixed end necessary for the evaluation of the leg distance for each definite data point from hundreds of those measured experimentally. Even though defining these coordinates is not possible for SPM visualization, it is reachable for fitting procedure in our numerical model. Accuracy of evaluation of Young's modulus seems to be approximately 5 GPa for NWs with radius above 30 nm. Finally, our approach is easily applicable for conical NWs. The presented comparatively easy SPM procedure of measurements of elastic characteristics is proposed to be used for various one-dimensional nanostructures and conical nanoobjects.

4. Studying piezo response

Piezoelectricity in NWs is an electromechanical phenomenon [91], so we left it to last section of this chapter. Piezophototronic devices based on NWs are widely discussed as sources that would replace both solar cells and windmills in the distant future. Therefore, such nanogenerators based on piezo responsive NWs represent a new generation of sustainable energy sources and must be studied in details. Interestingly, SPM is exactly the instrument that is used by plethora of research teams in the field of nanogenerators of electricity. Previously, we reflected two types of experiments, i.e., spectroscopy and mapping. Being consistent with this methodological assemblage, two self-determined PFM approaches will be discussed in this section for wurtzite GaAs NWs (**Figure 7a** and **b**). First type is a standard PFM oriented toward detection of reverse piezo coefficient. Second type is registration of piezo current and indirect evaluation of piezo coefficient, although reasons for generation of current in this case are not entirely understood up to recent moment.

4.1. Reverse piezo coefficient studies of horizontal NWs and nanowire stumps

Experiment requires reliable contact to be arranged between a sample and a dull probe mounted on a stiff cantilever. This is one of the methodical differences comparing to all other experiments described in this chapter where flexible levers were much preferable. Hard contact commonly leads to destruction of vertical and inclined NWs due to significant forces applied to the sample and fragility of NWs. Thus, the method is applicable either for horizontal NWs or for "stumps" (**Figure 7b** insets (i) and (iii), correspondingly). Stump structure is a residue of NW left after its upper part was broken due to application of excessive force (nonelastic stress). The part left is lower than initial NW, so its thermal oscillations are less intensive in lateral direction. Unfortunately, it seems impossible to predict existence of piezo responsive phase in the stump after the breakage. Nevertheless, this becomes explicit from PFM imaging, where studies are performed in C-AFM contact regime.

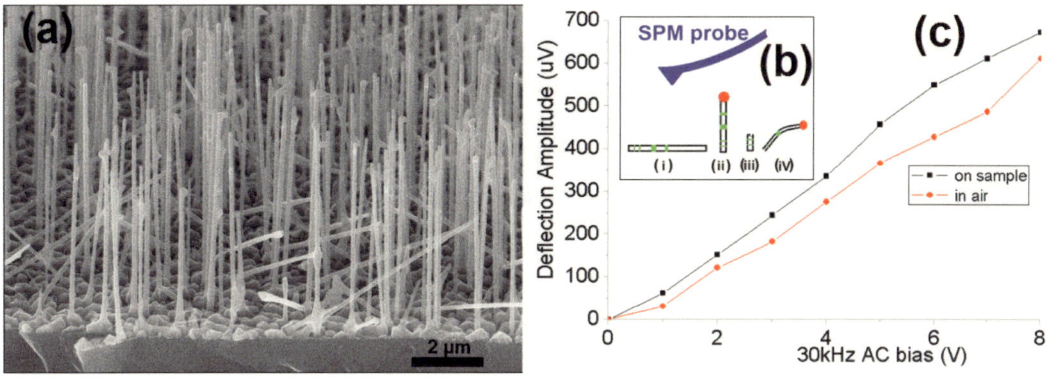

Figure 7. (a) SEM image of an array of vertical wurtzite GaAs NWs grown by MBE on Si substrate. (b) Schemes of the PFM experiment by SPM probe for horizontal (i), vertical (ii) and inclined (iv) NWs and NW stump (iii). (c) Comparison between PFM amplitude output from a stump of wurtzite GaAs NW and air 5 μm above the NW.

4.1.1. Spectroscopy of surface oscillations in comparison with non-piezoelectric phase and air

In spectroscopic PFM method (**Figure 7c**), it is necessary to establish a point-like contact (area of contact is few tens of square nanometers) with NW's segment containing piezo responsive phase. Then, high-frequency variable bias V_{AC} is being applied to the substrate. Frequency of the voltage applied f_{AC} should be different from the resonant frequency of the SPM probe contacting sample under certain force. After that, mechanical oscillations [$\mu V_{DER}/V_{AC}$] are registered by high-frequency feedback amplifier of photodetector's signal. Indeed, all signals obtained in SPM are processed as relation of *derivative microvolts* μV_{DER} of signal, but if measurement is properly calibrated considering sensitivity of photodetector as [μV_{DER}/nm], then result would be represented in [nm/V_{AC}]. In addition, flexibility of the SPM cantilever can be easily calibrated in [nN/nm] (see probe calibration protocol in Section 3.1), which would provide a value of inverse piezo coefficient of a studied material in [nN/V_{AC}]. Unfortunately, given force in [nN] does not represent the value of elastic response correctly in contact mode AFM, but is used only as guiding parameter of force. However, if standard test sample with known piezo coefficient is used for comparison of its piezo response with data recorded for unknown material, then result for sample would be recorded in standard layout [nN/V_{AC}]. In such a way, it becomes possible to compare properties of new materials with references and tabulated data for d_{33} vertical piezo coefficient for stumps or d_{13} and d_{15} coefficients for horizontal NWs.

It should be noted that measurements of both piezo active phase and nonpiezo active phase in same conditions should be compared to eliminate the influence of background noise onto derived signal and evaluate solely input of piezo response (**Figure 7c**). This is due to capacitive effects influencing the compelled resonant mechanical oscillations of an SPM probe situated in micrometer distance from the sample. Obtained dependence of mechanical oscillations from applied potential can be wrongly interpreted as pure piezoresponse signal (**Figure 7c**). The logic here should be that if signal in certain NW is different from measurements of nonpiezoelectric material with same geometry and composition, then piezo phase exists in the NW. Fortunately, single experimental measurement requires few minutes, which permits quick finding the NW stumps with piezo phase. Significant drawback of this approach is that it is almost impossible to characterize thickness of the piezo active inclusion in the stump, while all remaining part of NW cannot consist of piezo active material. Therefore, additional tools are required to characterize its structure and composition.

4.1.2. Mapping of mechanical oscillations induced by AC voltage

Method of mapping allows avoiding the sample's drift and builds up a map of piezo response (**Figure 8**). Such map indicates more intensive contrast for areas with stronger piezo signal.

If a sample is homogeneous or provides areas with homogeneous piezo response, e.g., on surface of horizontal NWs, then such areas provide significantly representative array of data. Thus, measurements in PFM mapping approach are more accurate comparing with PFM spectroscopy. Drawbacks of this approach are that sample and probe are more intensively worn and torsional movements are strengthened due to friction. Moreover, triboelectric effect arising from friction between SPM probe and surface can influence the result [92]. Triboelectric effect can generate localized charges of opposite signs in the NW and SPM probe after each tap due to friction and difference in work functions between the sample and the probe. These charges can shift surface potential of a sample and lead to generation of adverse current. However, such current is detrimental solely for evaluation of piezo coefficient. Intriguing new field of NW technology was formulated during last decade focused on production of current with triboelectric effect. For comprehensive analysis of prospects of triboelectric nanogenerators, reader should access [92].

4.2. Piezo current visualization for vertical and inclined NWs

Instead of reverse piezo effect, it can be more beneficial for applied science and industry to evaluate useful electric current produced by NWs containing piezo responsive phase. We will briefly describe its phenomenological value. Visualization of electric current for thick, vertical, and inclined NWs (see insets (ii) and (iv) in **Figure 7b**, correspondingly) is considered uncomplicated for SPM with force control (see Section 3.2). Latter can be crucial for testing of NWs fabricated from new materials with the following two techniques.

4.2.1. Recording of current during mechanical bending in local point contact geometry

In the method of force-load curves (discussed in Section 3.1), it is possible to perform spectroscopy of current in time $I(t)$. If registration of current flowing through the system is switched on, then it becomes possible to measure electric current flowing through an individual NW in every moment of its bending (see Section 2.1.2) (**Figure 9**). It must be admitted that current spikes were registered while no bias was applied to the sample. However, the electric current observed by SPM can be associated with increase of concentration of photo induced carriers [93, 94]. Parasitic electrically induced potential can also lead to adverse stimulation of carriers. Since laser beam falling onto the SPM cantilever near the probe can over-illuminate the measured NW then both of these mechanisms can lead to observation of electrical spots during C-AFM spectroscopy or mapping. These influences can be overcomed with adequate grounding and performing the experiments in dark conditions [93, 94].

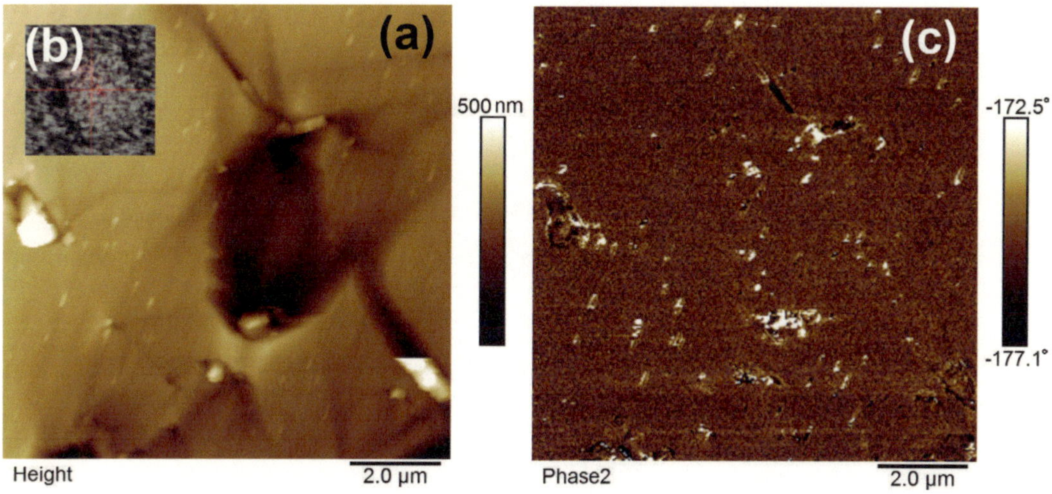

Figure 8. (a) 2D topography of a field with stumps left from an array of wurtzite GaAs NWs. (b) Optical image 100 × 100 µm representing square shaped field left after the breakage of vertical wurtzite GaAs NWs. (c) 2D image with a map of phase of the oscillations in lateral direction for a field with wurtzite GaAs stumps seen on (a) image. White spots indicate areas with higher amplitude indicating individual stumps.

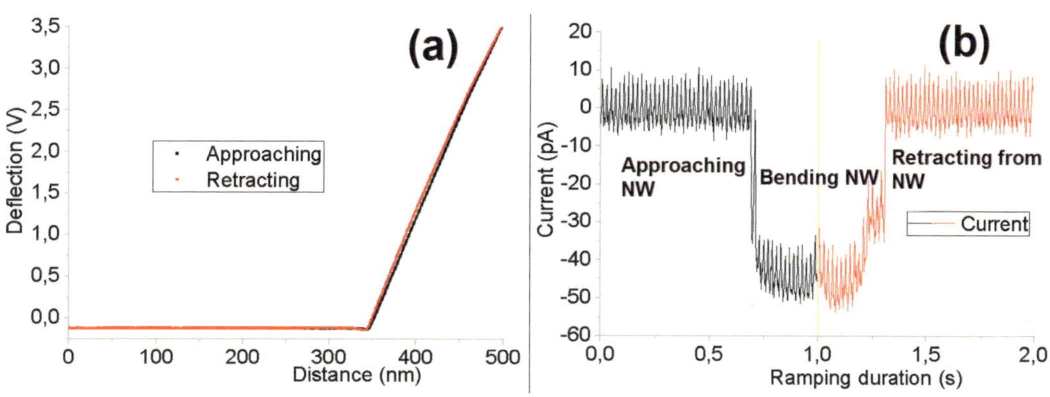

Figure 9. (a) Plot of F(z) for approaching and retracting from the sample where wurtzite NW was bended. (b) Current-time spectra $I(t)$ registered during the bending of wurtzite GaAs NW.

Additionally, since these measurements are performed during almost seconds (while the value of current is considerably stable), it is potentially possible to evaluate the charge transferred $q[C]$ and power $P[pW]$. This gives path to evaluation of performance of piezotronic devices with help of SPM by assessment of single NWs.

4.2.2. Mapping of electrical current in NWs induced by controlled mechanical bending

Mapping of electric current in principle can help to visualize certain NWs producing higher values of electric current See **Figure 10** taken under 0 V applied bias on WZ GaAs, where black spots are supposedly representing the current spikes measured in contact by SPM probe and generated due

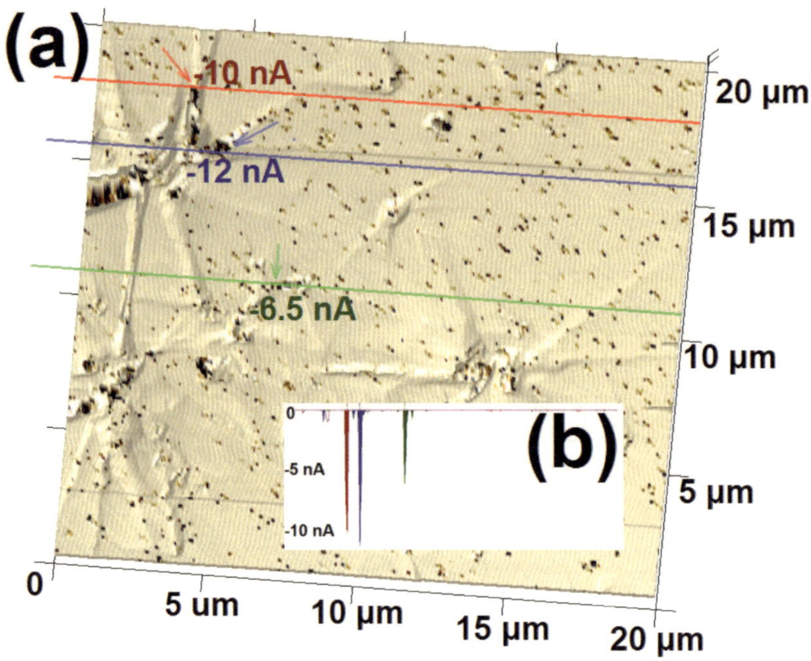

Figure 10. (a) 3D topography model of surface covered by stumps of wurtzite GaAs NWs seen as black spots due to overlaying of electric current map onto 3D topography. (b) Comparison of profiles of electric current taken for spots with highest electric current on from (a) image.

to mechanical stress of piezo responsive WZ GaAs material. This might indicate statistics about level of fabrication defects and overall quality of an array. Unfortunately, this approach is highly dependent from definite recognition of properties of energy barriers involved. Both Schottky barrier at the top part of NW and diffusion barrier at the bottom of NW (existing if materials of NW and substrate are different, e.g., Si and GaAs) can affect the result. In addition, triboelectricity can be an adverse factor during mapping of mechanical oscillations and it is aggravated for mapping in comparison with spectroscopy. Thus, triboelectric effect elaborates calculation of the input of piezo current and further evaluation of direct piezo coefficient of NW in nonvertical axis.

Shift of electric potential, which is registered as piezo response of the NWs' material, is suggested to be the result of redistribution of charge carriers inside NWs during their scanning/bending. Correspondingly, electric current that can be measured for piezoelectric NWs in PeakForce Tuna microscopy should indicate transfer of above-mentioned electric charges, i.e., electric current. Order of magnitude for such current is approximately few nanoamperes (**Figure 10b**). In addition, it can be possible to recalculate the current into potential, if resistance of the system is known. Unfortunately, heterojunction at the contact area between NW and a substrate obscures the transport mechanisms in such NWs. Revealing of mechanism of generation of electric current requires a separate comprehensive methodical study. This study should be performed on NWs with negligible energy barrier at its bottom contact with a substrate.

When aforementioned methodical difficulties will be overcomed, a map of current density generated from each individual NW from an array would be visualized. In this respect, it should be considered that force of push onto the NWs affects the value of electric potential induced in NWs. At the same time, altering rate of a setpoint force regulates the value of electric current produced by NW-based nanogenerators. Integer of current from single NW in time by its physical meaning defines the charge transferred into the SPM probe and recorded by SPM controller. Moreover, current from an array of NWs can be summed. Anticipated current output from an array containing plenty of NWs can be substantial for functioning of NW-based devices and can be characterized by current per certain area of an energy source [A/cm^2]. Applicability of latter approach needs to be approved by experiments with properly arranged massive electrode grown above an array. Therefore, we can conclude that local methods still need to be strengthened by massive methods of research of NWs. The proposed concept of registering the piezo current presented here has debatable arguments but provides a room for improvements. If it would be realized in near future, it seems prolific for quick and accurate estimation of electricity generated by piezoelectric materials and NWs.

5. Conclusion

To summarize, nine types of experiments were described in this chapter with associated advantages and drawbacks of the SPM techniques. Different types of SPM probes paired as sharp/dull and flexible/stiff were specified for certain experiments with NWs. Congregation of methodical information was motivated toward better understanding of opportunities of microscopic techniques for studies of NWs. Comprehensive usage of these techniques depends on measuring systems available for certain research groups interested in study of NWs. While SPM systems continuously develop in latest decades, we emphasize that SPM is essentially useful, precise, and versatile tool for abovementioned purposes when properly operated. SPM allows studying electrical, mechanical, piezo properties, in addition to magnetic and optical properties, biocompatibility for medicine, chemical reactivity, etc.

In the course of our SPM research work, few significant findings were obtained in materials science. GaP should be preferably chosen as passivation material for GaAs NWs, while AlGaAs appeared also beneficial. Second, transparent protective medium of SiO_2 has a complex influence onto conductive properties of NWs, i.e., energy barrier is reduced for unpassivated NWs, while it is strengthened for GaP-passivated NWs. These results were first obtained due to possibilities of SPM force control. Additionally, few methodical improvements were proposed. In the field of mechanics, the original mapping procedure of studies of tapered 10 nm thin NWs fixed at one end has been established. This allowed to first measurement of Young's modulus of wurtzite InP material, which exists recently only in nanothin state, i.e., solely in NWs. Therefore, it is a remarkable example of gathering a promising exotic nanomaterial and the only technique that can correctly describe its properties. Moreover, few original SPM experiments were proposed for comprehensive characterization of piezo properties of piezoelectric NWs in the field of electromechanics. Individual wurtzite GaAs NWs were experimentally investigated on the basis of proposed methodical electromechanical concepts. Lastly, studies of electrical conductivity of as-grown NWs are possible only with SPM. We developed the methodology of examination of as-grown vertical NWs and analysis of their I–V curves. The presented approach doe not require an objectionable stage of the breakage of the

NWs followed by their fixation in horizontal lithographic MSM pattern for analyses as was done previously by other authors.

Considering the further outlook of SPM methods in studies of NWs, described approaches can be partly adapted to nonsemiconductor classes of materials, e.g., metals, polymers, insulators, ceramics, carbon ribbons, etc. Modern computational tools with proper procedures can assist standard microscope stations in realization of fast measurements and data recognition. Furthermore, components of SPM and experimental practices tend to be developed to an automated stage, resulting in even more powerful technique for studying versatile one-dimensional structures. Utilization of SPM allows obtaining of vital information about NWs, their individual and even local properties, which helps to better understanding of properties of an array of NWs and to design future devices based on NWs.

Acknowledgements

Authors wish to thank a research group in Department of Micro and Nanosciences, Micronova, operating in Aalto University, Finland for fabrication and SEM visualization of vertical, horizontal and inclined NWs presented in Sections 2 and 3. Additionally, we thank the group of Laboratory of Nanoelectronics from St. Petersburg Academic University, St. Petersburg, Russia for growth and TEM visualization of piezo responsive GaAs NWs. PG thanks Dr. Prokhor Alekseev for thorough discussions providing significant contribution to the final version of the manuscript, Alina Lomenkova for help with graphical data analysis and Andrey Shubin from ScanSens Gmbh for assistance with tungsten semicarbide probes. This work was supported by EU project Horizon2020-MSCA-RISE-691010 Hunter.

Author details

Pavel Geydt[1*], Mikhail S. Dunaevskiy[2] and Erkki Lähderanta[1]

*Address all correspondence to: pavel.geydt@lut.fi

1 Lappeenranta University of Technology, Lappeenranta, Finland

2 Ioffe Institute, St. Petersburg, Russia

References

[1] Xia Y, Yang P, Sun Y, Wu Y, Mayers B, Gates B, Yin Y, Kim F, Yan H. One-dimensional nanostructures: Synthesis, characterization, and applications. Advances Materials. 2003;**15**(5):353–389. DOI: 10.1002/adma.200390087

[2] Dasgupta NP, Sun J, Liu C, Brittman S, Andrews SC, Lim J, Gao H, Yan R, Yang P. 25th anniversary article: Semiconductor nanowires – synthesis, characterization, and applications. Advances Materials. 2004;**26**(14):2137–2184. DOI: 10.1002/adma.201305929

[3] Hochbaum AI, Yang P. Semiconductor nanowires for energy conversion. Chemical Reviews. 2010;**110**(1):527–546. DOI: 10.1021/cr900075v

[4] Lu W, Xiang J, editors. Semiconductor Nanowires: From Next-Generation Electronics to Sustainable Energy. London, United Kingdom: RSC Publishing; 2015. p. 448. DOI: 10.1039/9781782625209

[5] Arbiol J, Xiong Q, editors. Semiconductor Nanowires: Materials, Synthesis, Characterization and Applications. 1st ed. United Kingdom: Woodhead Publishing; 2015. p. 572. DOI: 10.1016/B978-1-78242-253-2.01001-X

[6] Prete P, editor. Nanowires. Croatia: InTech; 2010. p. 428. DOI: 10.5772/3457

[7] Zhang Z, Yao K, Liu Y, Jin C, Liang X, Chen Q, Peng L-M. Quantitative analysis of current–voltage characteristics of semiconducting nanowires: Decoupling of contact effects. Advanced Functional Materials. 2007;**17**:2478–2489. DOI: 10.1002/adfm.200600475

[8] Arzt E. Size effects in materials due to microstructural and dimensional constraints: A comparative review. Acta Materials. 1998;**46**(16):5611–5626. DOI: 10.1016/S1359-6454(98)00231-6

[9] Duan X, Huang Y, Cui Y, Wang J, Lieber CM. Indium phosphide nanowires as building blocks for nanoscale electronic and optoelectronic devices. Nature. 2001;**409**:66–69. DOI: 10.1038/35051047

[10] Wang ZL, Song J. Piezoelectric nanogenerators based on zinc oxide nanowire arrays. Science. 2006;**312**(5771):242–246. DOI: 10.1126/science.1124005

[11] Holm JV, Jørgensen HI, Krogstrup P, Nygård J, Liu H, Aagesen M. Surface-passivated GaAsP single-nanowire solar cells exceeding 10% efficiency grown on silicon. Nature communications. 2013;**4**(1498):1–5. DOI: 10.1038/ncomms2510

[12] Cui Y, Wei Q, Park H, Lieber CM. Nanowire nanosensors for highly sensitive and selective detection of biological and chemical species. Science. 2001;**293**(5533):1289–1292. DOI: 10.1126/science.1062711

[13] Wan Q, Li QH, Chen YJ, Wang TH, He XL, Li JP, Lin CL. Fabrication and ethanol sensing characteristics of ZnO nanowire gas sensors. Applied Physics Latters. 2004;**84**(18):3654–3656. DOI: 10.1063/1.1738932

[14] Patolsky F, Lieber CM. Nanowire nanosensors. Materials Today. 2005;**8**(4):20–28. DOI: 10.1016/S1369-7021(05)00791-1

[15] Lu W, Xie P, Lieber CM. Nanowire transistor performance limits and applications. IEEE Transactions on Electron Devices. 2008;**55**(11):2859–2876. DOI: 10.1109/TED.2008.2005158

[16] Husain A, Hone J, Postma HWC, Huang XMH, Drake T, Barbic M, Scherer A, Roukes ML. Nanowire-based very-high-frequency electromechanical resonator. Applied Physics Latters. 2003;**83**(6):1240–1242. DOI: 10.1063/1.1601311

[17] Feng XL, He R, Yang P, Roukes ML. Very high frequency silicon nanowire electromechanical resonators. Nano Letters. 2007;**7**(7):1953–1959. DOI: 10.1021/nl0706695

[18] Binnig G, Quate CF, Gerber C. Atomic force microscope. Physical Review Letters. 1986;**56**:930–933. DOI: 10.1103/PhysRevLett.56.930

[19] Giessibl FJ. Advances in atomic force microscopy. Reviews of Modern Physics. 2003;**75**:949–983. DOI: 10.1103/RevModPhys.75.949

[20] Bhushan B, Fuchs H, editors. Applied Scanning Probe Methods II—Scanning Probe Microscopy Techniques. 1st ed. Germany: Springer-Verlag Berlin Heidelberg; 2006. p. 420. DOI: 10.1007/b139097

[21] Marszalek PE, Greenleaf WJ, Li H, Oberhauser AF, Fernandez JM. Atomic force microscopy captures quantized plastic deformation in gold nanowires. Proceedings of the National Academy of Sciences USA. 2000;**97**(12):6782–6786. DOI: https://www.ncbi.nlm.nih.gov/pmc/articles/PMC18594/

[22] Dobrokhotov VV, Yazdanpanah MM, Pabba S, Safir A, Cohn RW. Visual force sensing with flexible nanowire buckling springs. Nanotechnology. 2007;**19**(3):035502. DOI: 10.1088/0957-4484/19/03/035502

[23] Wu C-H, Yeh N. Electrical properties and photoresponses of silicon nanowires with selective anchored gold nanoparticles via scanning probe bond breaking nanolithography. Japanese Journal of Applied Physics. 2009;**48**(4S): 04C152. DOI: 10.1143/JJAP.48.04C152

[24] Sahin O. Harmonic Force Microscope: A New Tool for Biomolecular Identification and Material Characterization Based on Nanomechanical Measurements [dissertation]. USA: Stanford University; 2005. p. 113. Available from: https://searchworks.stanford.edu/view/6212100

[25] Beinik I. Electrical Characterization of Semiconductor Nanostructures by Conductive Probe Based Atomic Force Microscopy Techniques [dissertation]. Austria: Montanuniversität Leoben; 2011. p. 118. Available from: http://www.unileoben.ac.at/images/stories/Bibliothek/edoc/AC08511205n01vt.pdf

[26] Sadewasser S, Glatzel T, editors. Kelvin Probe Force Microscopy—Measuring and Compensating Electrostatic Forces. Germany: Springer; 2012. 334 p. DOI: 10.1007/978-3-642-22566-6

[27] Teichert C, Beinik I. Conductive atomic-force microscopy investigation of nanostructures in microelectronics. In: Bhushan B, editor. Scanning Probe Microscopy in Nanoscience and Nanotechnology 2. Germany: Springer Berlin Heidelberg; 2010. pp. 691–721. DOI: 10.1007/978-3-642-10497-8_23

[28] Wang ZL. Nanopiezotronics. Advanced Materials. 2007;**19**(6):889–892. DOI: 10.1002/adma.200602918

[29] Nakayama Y, Pauzauskie PJ, Radenovic A, Onorato RM, Saykally RJ, Liphardt J, Yang P. Tunable nanowire nonlinear optical probe. Nature. 2007;**447**:1098–1101. DOI: 10.1038/nature05921

[30] Rossi N, Braakman FR, Cadeddu D, Vasyukov D, Tütüncüoglu G, i Morral AF, Poggio M. Vectorial scanning force microscopy using a nanowire sensor. Nature Nanotechnology. 2017;**12**(2):150–155. DOI: 10.1038/nnano.2016.189

[31] US Patent 8484756 B2. Bertness KA, Sanford NA, Kabos P, Wallis TM. Tip-Mounted Nanowire Light Source Instrumentation [Internet]. 2011. Available from: https://www.google.com/patents/US8484756 [Accessed: 01.02.2017]

[32] Yang G, Tang J, Kato S, Zhang Q, Qin LC, Woodson M, Liu J, Kim JW, Littlehei PT, Park C, Zhouless O. Magnetic nanowire based high resolution magnetic force microscope probes. Applied Physics Latters. 2005;**87**(12):123507. DOI: 10.1063/1.2043237

[33] Park JJ, Reddy M, Stadler BJH, Flatau AB. Hysteresis measurement of individual multilayered Fe-Ga/Cu nanowires using magnetic force microscopy. Journal of Applied Physics. 2013;**113**(17):17A331. DOI: 10.1063/1.4795818

[34] Tabasum MR, Zighem F, Medina JDLT, Encinas A, Piraux L, Nysten B. Magnetic force microscopy investigation of arrays of nickel nanowires and nanotubes. Nanotechnology. 2014;**25**(24):245707. DOI: 10.1088/0957-4484/25/24/245707

[35] Béron F, dos Santos MVP, de Carvalho PG, Moura KO, Arzuza LCC, Pirota KR. How to characterize cylindrical magnetic nanowires. In: Khan M, editor. Magnetic Materials. Croatia: InTech; 2016. pp. 41–69. DOI: 10.5772/63482

[36] Dresselhaus MS, Lin Y-M, Rabin O, Black MR, Kong J, Dresselhaus G. Nanowires. In: Bhushan B, editor. Springer Handbook of Nanotechnology. Germany: Springer Berlin Heidelberg; 2010. pp. 119–167. DOI: 10.1007/978-3-642-02525-9_4

[37] Yang P, Yan H, Mao S, Russo R, Johnson J, Saykally R, Morris N, Pham J, He R, Choi H-J. Controlled growth of ZnO nanowires and their optical properties. Advanced Functional Materials. 2002;**12**(5):323–331. DOI: 10.1002/1616-3028(20020517)12:5<323::AID-ADFM323>3.0.CO;2-G

[38] Kim DC, Dheeraj DL, Fimland BO, Weman H. Polarization dependent photocurrent spectroscopy of single wurtzite GaAs/AlGaAs core-shell nanowires. Applied Physics Latters. 2013;**102**(14):142107. DOI: 10.1063/1.4801865

[39] Zhang Y, Wu J, Aagesen M, Liu H. III–V nanowires and nanowire optoelectronic devices. Journal Physics D: Applied Physics. 2015;**48**(46):463001. DOI: 10.1088/0022-3727/48/46/463001

[40] Mukdadi OM, Datta SK, Dunn ML. Acoustic-phonon dispersion in nanowires. Journal of Applied Physics. 2005;**97**(7):0743123. DOI: 10.1063/1.1871333

[41] McGary PD, Tan L, Zou J, Stadler BJH, Downey PR, Flatau AB. Magnetic nanowires for acoustic sensors (invited). Journal of Applied Physics. 2006;**99**(8):08B310. DOI: 10.1063/1.2167332

[42] Young ESK, Bouravleuv AD, Cirlin GE, Dhaka V, Lipsanen H, Tchernycheva M, Scherbakov AV, Platonov AV, Akimov AV, Kent AJ. Electrical detection of picosecond acoustic pulses in vertical transport devices with nanowires. Applied Physics Latters. 2014;**104**(6):062102. DOI: 10.1063/1.4864637

[43] Xie Q-Y, Ju Z-Y, Tian H, Xue Q-T, Chen Y-Q, Tao L-Q, Mohammad MA, Zhang X-Y, Yangac Y, Ren T-L. A point acoustic device based on aluminum nanowires. Nanoscale. 2016;**8**:5516–5525. DOI: 10.1039/C5NR06999H

[44] Hernández-Mínguez A, Möller M, Breuer S, Pfüller C, Somaschini C, Lazić S, Brandt O, García-Cristóbal A, de Lima Jr. MM, Cantarero A, Geelhaar L, Riechert H, Santos PV. Acoustically driven photon antibunching in nanowires. Nano Letters. 2012;**12**(1):252–258. DOI: 10.1021/nl203461m

[45] Kargar F, Debnath B, Kakko J-P, Säynätjoki A, Lipsanen H, Nika DL, Lake RK, Balandin AA. Direct observation of confined acoustic phonon polarization branches in free-standing semiconductor nanowires. Nature Communications. 2016;**7**:13400. DOI: doi:10.1038/ncomms13400

[46] Burt DP, Wilson NR, Weaver JMR, Dobson PS, Macpherson JV. Nanowire probes for high resolution combined Scanning electrochemical microscopy—atomic force microscopy. Nano Letters. 2005;**5**(4):639–643. DOI: 10.1021/nl050018d

[47] Hersam MC, Hoole ACF, O'Shea SJ, Welland ME. Potentiometry and repair of electrically stressed nanowires using atomic force microscopy. Applied Physics Latters. 1998;**72**(8):915. DOI: 10.1063/1.120872

[48] Trukhin VN, Buyskikh AS, Kaliteevskaya NA, Bourauleuv AD, Samoilov LL, Samsonenko YB, Cirlin GE, Kaliteevski MA, Gallant AJ. Terahertz generation by GaAs nanowires. Applied Physics Latters. 2013;**103**(7):072108. DOI: 10.1063/1.4818719

[49] Miao X, Chabak K, Zhang C, Mohseni PK, Walker Jr D, Li X. High-speed planar GaAs nanowire arrays with f_{max}> 75 GHz by wafer-scale bottom-up growth. Nano Letters. 2015;**15**(5):2780–2786. DOI: 10.1021/nl503596j

[50] Haggrén T. Nanowire Technology for Optoelectronic Applications [dissertation]. Finland: Aalto University; 2016. p. 76. DOI: https://aaltodoc.aalto.fi/handle/123456789/21918

[51] Bruker Inc. Application Note #132—Simultaneous Electrical and Mechanical Property Mapping at the nanoscale with PeakForce TUNA [Internet]. 2011. Available from: https://www.bruker.com/fileadmin/user_upload/8-PDF-Docs/SurfaceAnalysis/AFM/ApplicationNotes/AN132-RevA0-Simultaneous_Electrical_Mechanical_Property_Mapping_with_PeakForce_TUNA-AppNote.pdf [Accessed: 01.02.2017]

[52] Dubrovskii VG, Cirlin GE, Ustinov VM. Semiconductor nanowhiskers: Synthesis, properties, and applications. Semiconductors+. 2009;**43**(1539):1585–1628. DOI: 10.1134/S106378260912001X

[53] Rojo MM, Calero OC, Lopeandia AF, Rodriguez-Viejob J, Martín-Gonzalez M. Review on measurement techniques of transport properties of nanowires. Nanoscale. 2013;**5**:11526–11544. DOI: 10.1039/C3NR03242F

[54] Kaja K. Development of nano-probe techniques for work function assessment and application to materials for microelectronics [dissertation]. France: Université Joseph-Fourier-Grenoble; 2010. p. 221. DOI: https://tel.archives-ouvertes.fr/tel-00515370/en

[55] Chia ACE. Electrical Characterization and Optimization of Gallium Arsenide Nanowire Ensemble Devices [dissertation]. Canada: McMaster University; 2013. p. 155. DOI: https://macsphere.mcmaster.ca/handle/11375/13335

[56] LaPierre RR, Chia ACE, Gibson SJ, Haapamaki CM, Boulanger J, Yee R, Kuyanov P, Zhang J, Tajik N, Jewell N, Rahman KMA. III–V nanowire photovoltaics: Review of design for high efficiency. Physica Status Solidi Rapid Research Letters. 2013;**7**(10):815–830. DOI: 10.1002/pssr.201307109

[57] van Dam D, van Hoof NJJ, Cui Y, van Veldhoven PJ, Bakkers EPAM, Gómez Rivas J, Haverkort JEM. High-efficiency nanowire solar cells with omnidirectionally enhanced absorption due to self-aligned indium-tin-oxide Mie scatterers. American Chemical Society Nano 2016;**10**:11414−11419. DOI: 10.1021/acsnano.6b06874

[58] Malloroqui AD. Nanowire-Based Solar Cells: Device Design and Implementation [dissertation]. EPFL Lausanne: Switzerland; 2014. p. 108. DOI: https://infoscience.epfl.ch/record/196411

[59] Giordano MA, Schmidt SR. Applications of contact mode AFM to manufacturing processes. In: Bhushan B, editor. Scanning Probe Microscopy in Nanoscience and Nanotechnology 2. Germany: Springer Berlin Heidelberg; 2011. pp. 867–914. DOI: 10.1007/978-3-642-03535-7_25

[60] Tajik N. Sulfur Passivation of III–V Semiconductor Nanowires [dissertation]. Canada: McMaster University; 2013. p. 132. DOI: https://macsphere.mcmaster.ca/handle/11375/12834

[61] Werner F, Limbach F, Carsten M, Denker C, Malindretos J, Rizzi A. Electrical conductivity of InN nanowires and the influence of the native indium oxide formed at their surface. Nano Letters. 2009;**9**(4):1567–1571. DOI: 10.1021/nl8036799

[62] Lin X, He X, Lu J, Gao L, Huan Q, Deng Z, Cheng Z, Shi D, Gao H. Manipulation and four-probe analysis of nanowires in UHV by application of four tunneling microscope tips: a new method for the investigation of electrical transport through nanowires. Surface and Interface Analysis. 2006;**38**(6):1096–1102. DOI: 10.1002/sia.2333

[63] Timm R, Persson O, Engberg DLJ, Fian A, Webb JL, Wallentin J, Jönsson A, Borgström MT, Samuelson, Mikkelsen A. Current–voltage characterization of individual as-grown nanowires using a scanning tunneling microscope. Nano Letters. 2013;**13**(11):5182–5189. DOI: 10.1021/nl402570u

[64] Lord AM, Walton AS, Maffeis TG, Ward MB, Davies P, Wilks SP. ZnO nanowires with Au contacts characterised in the as-grown real device configuration using a local multi-probe method. Nanotechnology. 2014;**25**(42):425706. DOI: 10.1088/0957-4484/25/42/425706

[65] Geydt P, Alekseev PA, Dunaevskiy M, Lähderanta E, Haggrén T, Kakko J-P, Lipsanen H. Observation of linear I–V curves on vertical GaAs nanowires with atomic force microscope. Journal of Physics: Conference Series. 2015;**661**(1):012031. DOI: 10.1088/1742-6596/661/1/012031P

[66] Geydt P, Alekseev PA, Dunaevskiy MS, Haggrén T, Kakko J-P, Lähderanta E, Lipsanen H. Influence of surface passivation on electric properties of individual GaAs nanowires studied by current–voltage AFM measurements. Lithuanian Journal of Physics. 2016;**46**(2):92–101. DOI: 10.3952/physics.v56i2.3305

[67] Ellis JA, Barnes PA. Current–voltage characteristics of a GaAs Schottky diode accounting for leakage paths. Applied Physics Latters. 1999;**76**(1):124. DOI: 10.1063/1.125677

[68] Suyatin DB, Jain V, Nebol'sin VA, Trägårdh J, Messing ME, Wagner JB, Persson O, Timm R, Mikkelsen A, Maximov I, Samuelson L, Pettersson H. Strong Schottky barrier reduction at Au-catalyst/GaAs-nanowire interfaces by electric dipole formation and fermi-level unpinning. Nature Communications. 2014;**5**(3221):1–8. DOI: 10.1038/ncomms4221

[69] Darbandi A, Salehzadeh O, Kuyanov P, La Pierre RR, Watkins SP. Surface passivation of tellurium-doped GaAs nanowires by GaP: Effect on electrical conduction. Journal of Applied Physics. 2014;**115**(23):234305. DOI: 10.1063/1.4883960

[70] Dementyev PA, Dunaevskii MS, Samsonenko YB, Cirlin GE, Titkov AN. Current-voltage characteristics of silicon-doped GaAs nanowhiskers with a protecting AlGaAs coating overgrown with an undoped GaAs layer. Semiconductors+. 2010;**44**(5):610–615. DOI: 10.1134/S1063782610050118

[71] Morita S. Roadmap of Scanning Probe microscopy. 1st ed. Germany: Springer-Verlag Berlin Heidelberg; 2007. p. 201. DOI: 10.1007/978-3-540-34315-8

[72] Melitz W, Shen J, Kummel AC, Lee S. Kelvin probe force microscopy and its application. Surface Science Reports. 2011;**66**(1):1–27. DOI: 10.1016/j.surfrep.2010.10.001

[73] Li G, Mao B, Lan F, Liu L. Practical aspects of single-pass scan Kelvin probe force microscopy. Review of Scientific Instruments 2012;**83**(11):113701. DOI: 10.1063/1.4761922

[74] Prokhor AA. Investigation of Charge and Electric Field Distribution in Nanostructures by Scanning Probe Microscopy (in Russian) [dissertation]. Russian Federation: LETI; 2013. p. 159. Available from: http://fizmathim.com/issledovanie-raspredeleniya-zaryadov-i-

elektricheskih-poley-v-pribornyh-nanostrukturah-metodami-skaniruyuschey-zondovoy-m DOI: http://search.rsl.ru/ru/record/01005535363

[75] Dunaevskiy M, Alekseev P, Girard P, Lashkul A, Lahderanta E, Titkov A. Analysis of the lateral resolution of electrostatic force gradient microscopy. Journal of Applied Physics. 2009;**112**(6):064112. DOI: 10.1063/1.4752430

[76] Hong KM, Noolandi J, Street RA. Theory of radiative recombination by diffusion and tunneling in amorphous Si: H. Physical Review B. 1981;**23**(6):2967. DOI: 10.1103/PhysRevB.23.2967

[77] He R, Gao D, Fan R, Hochbaum AI, Carraro C, Maboudian R, Yang P. Si Nanowire bridges in microtrenches: Integration of growth into device fabrication. Advanced Materials. 2005;**17**(17):2098–2102. DOI: 10.1002/adma.200401959

[78] Yong Oh J, Park J-T, Jang H-J, Cho W-J, Islam MS. 3D-transistor array based on horizontally suspended silicon nano-bridges grown via a bottom-up technique. Advanced Materials. 2014;**26**(14):1929–1934. DOI: 10.1002/adma.201304245

[79] Shi C, Luu DK, Yang Q, Liu J, Chen J, Ru C, Xie S, Luo J, Ge J, Sun Y. Recent advances in nanorobotic manipulation inside scanning electron microscopes. Microsystems & Nanoengineering. 2016;**2**(16024):1–16. DOI: 10.1038/micronano.2016.24

[80] Qin S, Kim T-H, Wang Z, Li A-P. Nanomanipulation and nanofabrication with multi-probe scanning tunneling microscope: From individual atoms to nanowires. Review of Scientific Instruments. 2012;**83**(6):063704. DOI: 10.1063/1.4727878

[81] Wang ZL, Gao RP, Pan ZW, Dai ZR. Nano-scale mechanics of nanotubes, nanowires, and nanobelts. Advanced Engineering Materials. 2001;**3**(9):657–661. DOI: 10.1002/1527-2648(200109)3:9<657::AID-ADEM657>3.0.CO;2-0

[82] Barth S, Harnagea C, Mathur S, Rosei F. The elastic moduli of oriented tin oxide nanowires. Nanotechnology. 2009;**20**(11):115705. DOI: 10.1088/0957-4484/20/11/115705

[83] Chen CQ, Shi Y, Zhang YS, Zhu J, Yan YJ. Size dependence of young's modulus in ZnO nanowires. Physical Review Letters. 2006;**96**:075505. DOI: 10.1103/PhysRevLett.96.075505

[84] Wang B, Stevens E, Leu PW. Strong broadband absorption in GaAs nanocone and nanowire arrays for solar cells. Optics Express. 2014;**22**(S2):A386–395. DOI: 10.1364/OE.22.00A386

[85] Alekseev PA, Dunaevskii MS, Stovpyaga AV, Lepsa M, Titkov AN. Measurement of Young's modulus of GaAs nanowires growing obliquely on a substrate. Semiconductors+. 2012;**46**(5):641–646. DOI: 10.1134/S106378261205003X

[86] Geydt P, Dunaevskiy M, Alekseev P, Kakko J-P, Haggrén T, Lähderanta E, Lipsanen H. Direct measurement of elastic modulus of InP nanowires with Scanning probe microscopy in PeakForce QNM mode. Journal of Physical: Conference Series. 2016;**768**(1): 012029. DOI: 10.1088/1742-6596/769/1/012029

[87] Gere JM, Goodno BJ. Mechanics of Materials. 8th ed. Stamford, USA: Cengage Learning; 2013. p. 1098. DOI: http://trove.nla.gov.au/version/185463313

[88] Wachtman Jr JB, Tefft WE, Lam Jr DG, Apstein CS. Exponential temperature dependence of Young's modulus for several oxides. Physical Reviews. 1961;**122**(6):1754. DOI: 10.1103/PhysRev.122.1754

[89] Dokukin ME, Sokolov I. Quantitative mapping of the elastic modulus of soft materials with Harmonix and PeakForce QNM AFM modes. Langmuir. 2012;**8**(46):16060–16071. DOI: 10.1021/la302706b

[90] Xu F, Qin Q, Mishra A, Gu Y, Zhu Y. Mechanical properties of ZnO nanowires under different loading modes. Nano Research. 2010;**3**(4):271–280. DOI: 10.1007/s12274-010-1030-4

[91] Espinosa HD, Bernal RA, Minary-Jolandan M. A review of mechanical and electromechanical properties of piezoelectric nanowires. Advanced Materials. 2012;**24**(34):4656–4675. DOI: 10.1002/adma.201104810

[92] Wang ZL. Triboelectric nanogenerators as new energy technology and self-powered sensors—principles, problems and perspectives. Faraday Discuss. 2014;**176**:447–458. DOI: 10.1039/C4FD00159A

[93] Lin Y-F, Song J, Ding Y, Lu S-Y, Wang ZL. Alternating the output of a CdS nanowire nanogenerator by a white-light-stimulated optoelectronic effect. Advanced Materials. 2008;**20**:3127–3130. DOI: 10.1002/adma.200703236

[94] Song J, Wang X, Liu J, Liu H, Li Y, Wang ZL. Piezoelectric potential output from ZnO nanowire functionalized with p-type oligomer. Nano Letters. 2008;**8**(1):203–207. DOI: 10.1021/nl072440v

Chapter 9

Diluted Magnetic DNA Nanowires

Caner Değer, Vahap Eldem and İzzet Paruğ Duru

Additional information is available at the end of the chapter

http://dx.doi.org/10.5772/67921

Abstract

DNA, as a natural biological nanowire, could be modified by inorganic atoms, either conducting or semi-conducting ones, to render feasible for technological applications and sequencing. Magnetic phase transition of modified DNA (M-DNA) nanowire, similar to diluted magnetic semiconductor (DMS) materials, occurs by changing parameters such as temperature and doping ratio having critical values. Tuning such parameters provide a pleasant control about determination of the magnetic property of M-DNA, particularly room-temperature soft ferromagnetism. In this chapter, a fundamental theory for anti-ferromagnetic Cr^{3+}-doped M-DNA nanowire named as diluted magnetic organic structures (DMOS) is tried to figure out the interactions representing the exotic behaviour of these type of organometallics. However, authors detailed preparation of nanowire as a global input and overall procedure of simulation based on Markov chain Monte Carlo (MCMC) method.

Keywords: DMOS, M-DNA, dipolar Interaction, hysteresis, organic nanowire, MCMC

1. Introduction

With unique structural features and self-assembly nature, deoxyribonucleic acid (DNA) has long been considered as ubiquitous bionanomaterials. The unique structural patterns and dynamic behaviours of DNA mainly arise from hydrogen bonds between purine and pyrimidine bases which are fundamental building blocks in double helix. Moreover, the editable, modifiable, scalable, and controllable natures of polynucleotides make synthetic DNA an attractive material for several applications at the nanoscale fields, including optoelectronics [1], spintronics [2], nanophotonics [3], nanobiosensors [4, 5] and even data storages such as nucleic acid memory. On the other hand, despite all these advantages, optical, electronic and magnetic properties of bare DNA are not suitable for such applications. Therefore, by doping DNA with divalent metal ions (Cu^{2+}, Ni^{2+}, Zn^{2+}, and Co^{2+}), called "modified DNA (M-DNA)", it is possible to tailor or control the magnetic and electrical properties of DNA. Besides, changing magnetic property of DNA via selective incorporation of metal ions between purine-pyrimidine pairs and hydrogen bonds not only provides a tunable material but also gives a clue about the location of bases. However, Hu et al. [6] investigated the electronic and magnetic properties of graphene and Fe and probe the DNA bases adsorbed by first-principal calculations theoretically to detect DNA bases. As a matter of fact, researchers recently focused on conceiving magnetic properties of M-DNA nanowires even though Azbel [7] derived an explicit formula that provides the consistency of theory and experiments while studying phase transitions in DNA. Alivisatos et al. [8], Mirkin et al. [9], and Braun et al. [10] were the first researchers who studied DNA-templated self-assembly, followed by Richter et al. [11], Park et al. [12] and Keren et al. [13]. Fabrication of DNA nanowires extend the field of workspace generated by organic structures. Park et al. [12] studied detection of nucleic acids with high sensitivity and specificity. Besides fabrication of DNA nanowires, band-gap effects on temperature-dependent magnetism of DNA molecules have been evaluated by Yi [14]. On the other hand, Savin et al. [15] described the thermal properties of DNA, especially heat conductivity of the DNA double helix, constructing a 3D coarse-grain (CG) model. Therefore, some studies have been carried out to reveal magnetic properties of M-DNA, both theoretically and experimentally. Electronic states of metal ion-doped M-DNA (m is Mn, Ni, Mg, Co and Fe) are figured out by optical absorption concluding charge transfer from Fe^{2+} to DNA and transferred charges should be positioned close to bases [16]. Computational models are properly developed to investigate the electrical conduction in DNA nanowires by Panahi and Chitsazanmoghaddam [17] and Behnia and Fathizadeh [18]. The studies performed by Dugasani et al. [19] and Nikiforov et al. [20] discovered that divalent metal ion–doped DNA exhibits soft ferromagnetic (FM) behaviour. Both of the studies claim that the origin of the soft ferromagnetic behaviour is the interaction between doped ions. Long-range interactions have also been proposed as the cause of room-temperature ferromagnetism in M-DNA. On the other hand, a comprehensive theoretical study proposed a new type of Hamiltonian approach suggesting both inorganic and organic structures including M-DNA nanowires [21]. One of the significant terms of Hamiltonian indicates exchange interaction, the ferromagnetic or anti-ferromagnetic interaction between atoms. The effect of ion concentration on the system was investigated to determine the proper ratio for technological applications. The result of the study is also compatible with experimental studies.

2. Constructing modified DNA nanowires

In this section, a simple workflow is introduced for construction of the modified DNA structure doped by a metal ion (Cr^{3+}) to investigate magnetic properties, especially soft ferromagnetism related to doping ratio. Firstly, a set of gene sequences should be designated as a sample group considering the guanine-cytosine (G-C)% content and base composition herewith sequence length. It is important to know the base composition of DNA and the number of hydrogen bounds since dopant ion is replaced with hydrogen atoms. After the determination of sample group (base sequences), base pairs containing organic atoms will be converted to a 3D structure by assigning atomistic coordinates of atoms in Euclidian space. van Dijk and Bonvin [22] developed a model to produce atomistic coordinates with connections of atoms from the singe base sequence of a gene in the mentioned coordinate space using reasonable parameters. The key parameters are roll, tilt and twist during the modelling procedure combined with each other. By the usage of 3D DART modelling tool [23], resulting DNA helical was ready to be modified by a dopant ion. Doping process is required to start after the successful finish of former work. Dugasani et al. [19] suggested the most probable locations of subsequent dopant ions between base pairs and PO4, and at last, the random displacement of dopants instead of hydrogen atoms of N-H-N and N-H-O.

2.1. Preparation of the sample group

G-C base pair has three electrostatic hydrogen bonds between nitrogen and oxygen atoms, while adenine-thymine (A-T) base pairs are connected to each other via two hydrogen bonds in the Watson-Crick model. Duru et al. [21] doped a number of genes with length of 1000 bases whose G-C ratio ranged from 40.8 to 69.5%. Sample group prepared has hydrogen bonds between the G and C bases ranging from 1124 to 2085 and the ratio of A to T content ranging from 610 to 1184. Authors prepared a sample group having different genes and G-C contents to re-simulate the M-DNA. The sample sequences of DNA (1000 bases) in length were obtained from Ensembl Genome browser (www.ensembl.org). Detailed information about the base composition of selected genes is presented in **Table 1**.

2.2. Atomistic structure of DNA

Polynucleotide chains constitute DNA by helical circling which is either right handed or left handed. As the most common form, B-DNA is right handed. There are a major groove and a minor groove through the outer region, respectively. Yet another right-handed helical A-DNA is a more congested molecule. Contrary to A-DNA and B-DNA, Z-DNA is a left-handed helical DNA. Base pairs are seen as zigzags. Moreover, helical structure forms a deep indentation instead of two. The Watson-Crick model, B-DNA, is considered as the normal form. The model is for the damp form due to the high water activity in vivo. That model has 10 base pairs per turn of the double helix, with a rise of 3.4 Å/bp [24], and A-DNA has 11 base pairs per turn and a rise of 2.6 Å/bp [25]. **Table 2** compares some properties of A-DNA, B-DNA and Z-DNA.

Sequences of all genes of the sample group should be used to visualize the atomic coordinates using U-GENE (free open-source cross-platform) separately (can be downloaded from http://ugene.net/). All of the atoms have the same spherical geometry. First, distances between the first 10 nearest neighbours of all atoms in DNA are calculated to provide a reasonable process time during simulation. van Dijk et al. [22] proposed a method to generate a canonical structure including atomic positions of A-DNA and B-DNA. The modelling procedure is clearly detailed in their work. B-DNA should be preferred for simulation process due to the most observed structural form of bare DNA in vivo.

Gene name	Length (bp)	A (%)	T (%)	G (%)	C (%)	%GC	M.W. (Da)
ADGB[a]	1000	32.8	26.4	17.1	23.7	40.8	312,051
AGRN[b]	1000	15.1	15.4	34.7	34.8	69.5	312,177
AK1[c]	1000	25	18.2	24.2	32.6	56.8	313,148
ALB[d]	1000	29.2	28.2	19.8	22.8	42.6	311,340
FOXA1[e]	1000	19.7	13	35.5	31.8	67.3	311,793
MFN1[f]	1000	30.3	27.4	17.6	24.7	42.3	312,046
PODN[g]	1000	21	17.3	33.5	28.2	61.7	311,130
SMOX[h]	1000	21.9	18.8	29.2	30.1	59.3	311,944
SMTN[j]	1000	21.2	14.1	33.4	31.3	64.7	311,929
TESK1[k]	1000	18.6	18.8	31.1	31.5	62.6	311,883

[a] androglobin
[b] agrin
[c] adenylate kinase 1
[d] albumin
[e] forkhead box A1
[f] mitofusin 1
[g] podocan
[h] spermine oxidase
[j] smoothelin
[k] testis-specifc kinase 1 [21].

Table 1. List of genes with different base compositions and molecular weights (M.W.).

	A-DNA	B-DNA	Z-DNA
Helix sense	right-handed	right-handed	left-handed
Repeating unit	1	1	2
Rotation/bp	32.7°	34.3°	60°/2
Inclination of bp to axis	+19°	−1.2°	−9°
Rise/bp along axis	0.26 nm	0.34 nm	0.37 nm
Rise/turn of helix	2.86 nm	3.57 nm	4.56 nm
Mean propeller twist	+18°	+16°	0°
Nucleotide phosphate to phosphate distance	0.59 nm	0.70 nm	C: 0.57 nm, G: 0.61 nm
Diameter	2.3 nm	2.0 nm	1.8 nm

Table 2. Comparing geometries of A-DNA, B-DNA and Z-DNA.

Figure 1 shows A-T and G-C base pairs, including base atoms having spherical geometry and bounds roughly. Furthermore, the flow diagram of the distance calculated between nearest neighbours is given by **Figure 2**. Finalization process of preparing input data would be followed by the doping process explained in the next section.

http://dx.doi.org/10.5772/67921

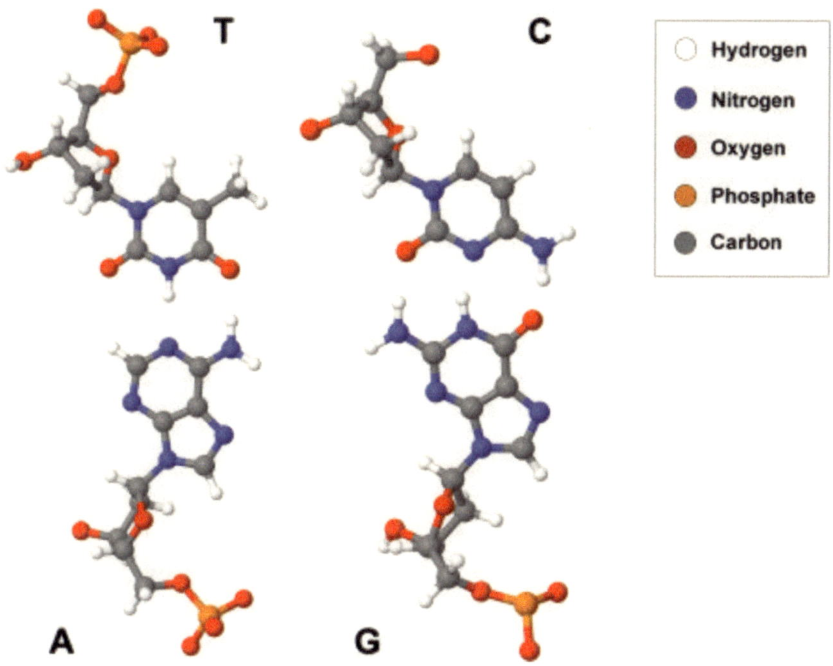

Figure 1. A molecular structure of A-T and C-G base pairs.

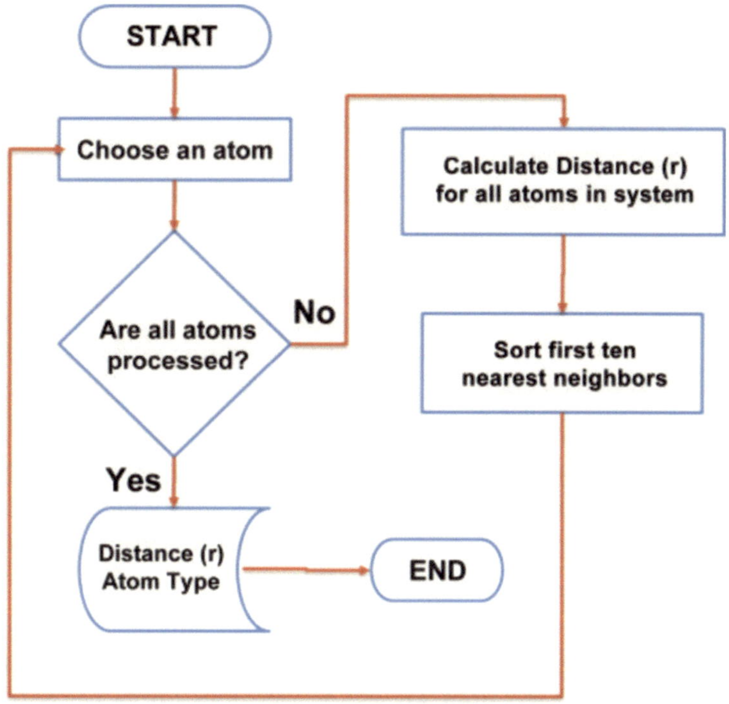

Figure 2. A flow diagram of first 10 nearest neighbours' distance calculation.

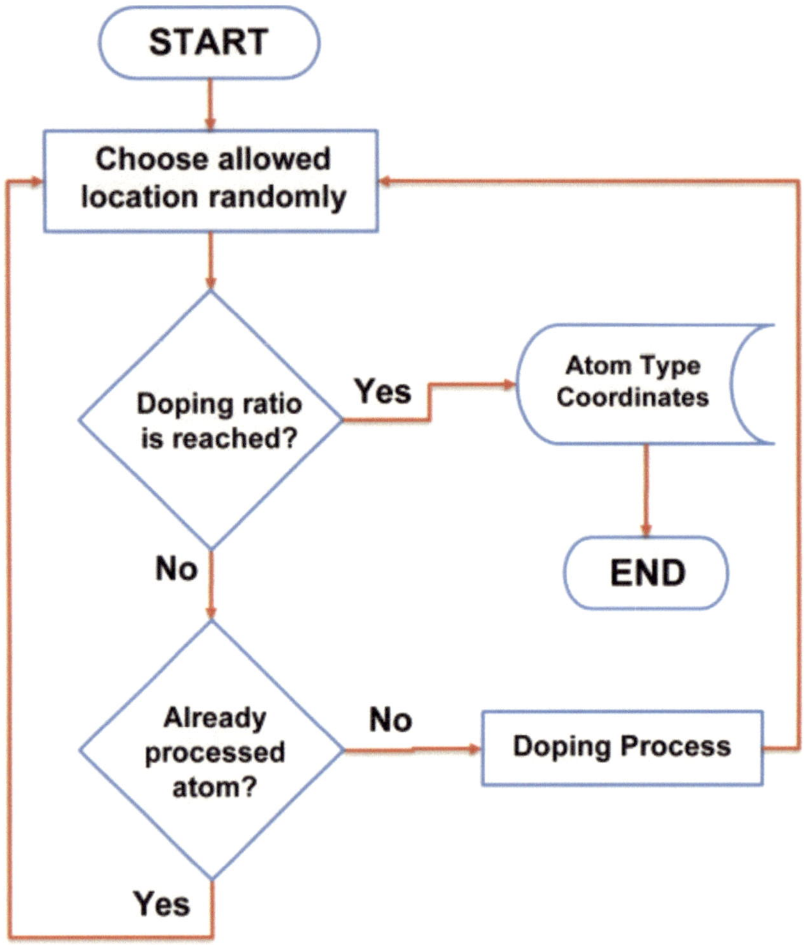

Figure 3. A flow diagram of doping process.

2.3. Doping process

The dopant ions are distributed to DNA almost randomly assuming that the binding probabilities of the dopant to G-C, A-T and $(PO^4)^-$ are similar (33%) in contrast to Dugasani et al. [19].

The number of hydrogen atoms that connects base pairs affects the magnetic phase of M-DNA by changing concentration of metal ions doped to B-DNA [26, 27]. Selection of DNA bases will be investigated and is based on G-C ratio since it includes three hydrogen atoms, while A-T has two bounds of hydrogen. Therefore, G-C ratio of selected DNA is changing from 40.8 to 69.5% to reveal the effect of G-C ratio on doping process and magnetic behaviour.

In the light of recent experimental studies [26–30], dopant ions have different probabilities on where they will bind. **Figures 4** and **5** illustrate the highest probabilities of Cr^{3+} positioning through the DNA helical structure. Hydrogen bind N-N or N-O or N-C atoms existed on the inner edge of pairs of helical structure and have critical importance since researchers mainly say that dopant ion is actually replaced by hydrogen. So, Cr^{3+}ions are located instead of hydrogen binding to a base pair randomly during doping process for a desired value of doping ratio. Flow diagram in **Figure 3** shows the doping process precisely. In **Figure 4**, dash circles represent the probable positions of doped ion, removing relevant hydrogen in G-C and A-T pair, respectively. Also, dopant ion Cr^{3+} binds to phosphate, as illustrated in **Figure 5**. There are probable positions where dopant ions will be located. It is thought that unbounded oxygen of phosphate holds on to the dopant ion according to Dugasani et al. [19]. Other transition metals (Mn, Fe, Co, Ni and Cu) can be doped to DNA, and they can show ferromagnetic behaviour [19, 26].

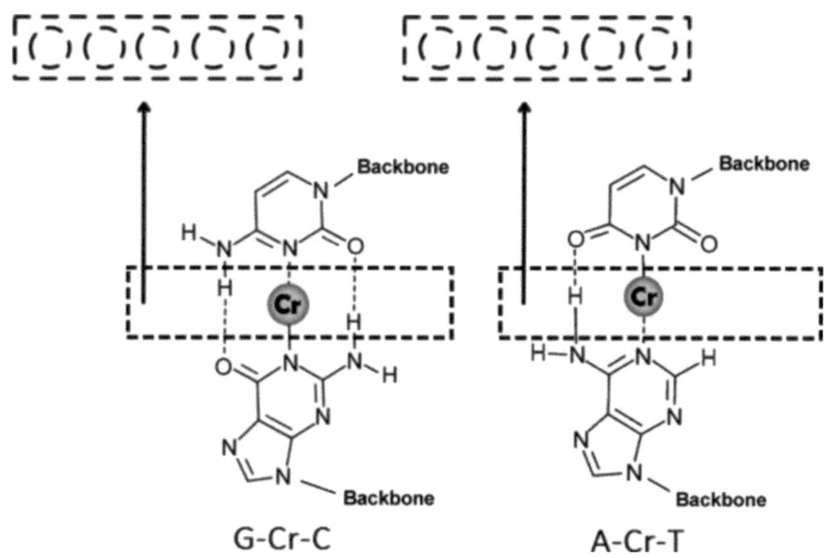

Figure 4. An illustration of probable positions of dopant Cr located instead of hydrogen atom. Dashed circles indicate the possible locations, while grey circle is the most probable one between G-C and A-T base pairs.

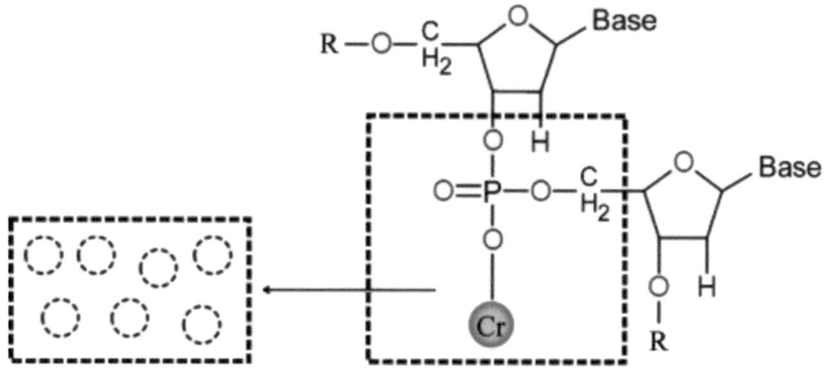

Figure 5. An illustration of probable positions of dopant Cr binding to $(PO_4)^-$. Dashed circles indicate the possible locations, while the grey circle is the most probable one.

3. Simulating the hysteresis

The relationship between applied magnetic field and magnetization is generally explained by using hysteresis loops in solid state physics. A detailed investigation of the hysteresis loop gives immense information about magnetic properties of the material. Saturation magnetization (M_S), described as the maximum value of the magnetization achieved in a sufficiently large magnetic field, can be determined from hysteresis loop. Also, remanent magnetization (M_R), the magnetization that remains in a material when the magnetizing force is zero, and coercive field (H_C), the amount of reverse magnetic field that has to be applied to make the magnetization return to zero, are indicated by hysteresis loop.

Initially, every domain in a ferromagnetic material has an intense magnetization but the whole material is generally unmagnetized because all of the domains are randomly oriented with respect to one another in the absence of magnetic field. The only need is a small external magnetic field to observe ferromagnetism phenomenon. This small magnetic field can cause the magnetic domains to line up with each other and parallel to external magnetic field, the material is now said to be magnetized. In contrast to paramagnets, ferromagnets tend to stay magnetized even if external magnetic field is removed. This tendency is called as "hysteresis". Saturation magnetization, remanent magnetization and coercive field can be viewed in a "*hysteresis loop*" as in **Figure 6.**

3.1. Dipolar Heisenberg Hamiltonian

DNA nanowire is fixed on a glass substrate, and bases adenine, thymine, cytosine and guanine are formed by H, C, O, N, P and dopant Cr^{3+} repeated through z direction in cylindrical coordinates during the simulation process. Glass is chosen because of its non-magnetic property to prevent any external magnetic contribution from another structure except the M-DNA structure since precision in measurement is important. However, doping process affected positions to be changed with little shifts by some perturbations to preserve considerable symmetry and prevent the fall to pieces of the including atoms. The Hamiltonian of the system is described by:

$$\mathcal{H} = -\sum_{<i,j>} J_{\alpha\beta}\left(r_j - r_i\right) S_i^{\alpha} S_j^{\beta} + D \sum_{<i,k>} \left[\frac{S_i^{\alpha} S_k^{\beta}}{r_{ik}^3} - \frac{3\left(S_i^{\alpha} r_{ik}\right)\left(S_k^{\beta} r_{ik}\right)}{r_{ik}^5}\right] - \mu_B g \sum_i B_{\alpha} S_i^{\alpha} \tag{1}$$

where $J_{\alpha\beta}$ is the exchange coupling constant, B_{α} ($\alpha = 0$) is the magnetic field applied along the z-direction, μ_B is the Bohr magneton, and g is the gyromagnetic ratio. α is set to zero to apply external uniform magnetic field through –z direction. The exchange term due to uniqueness and "sole" importance while explaining the strange nature of diluted magnetic semiconductor (DMS) structures is required to be assimilated. This term dipolar Heisenberg Hamiltonian represents the exchange interaction and is related to the distance and type of atoms. It is calculated by summing up nearest neighbours over the whole existed atoms in the system. Theory of diluted magnetic organic structures (DMOS) will be detailed in the next section. The second term factorized by D represents the magnetic dipolar interaction. Even though it can be considered as irrelevant to the subject of chapter, dipole-dipole interaction is also critical to the understanding of magnetic dipoles in optical lattices. The significance of dipolar interac-

http://dx.doi.org/10.5772/67921

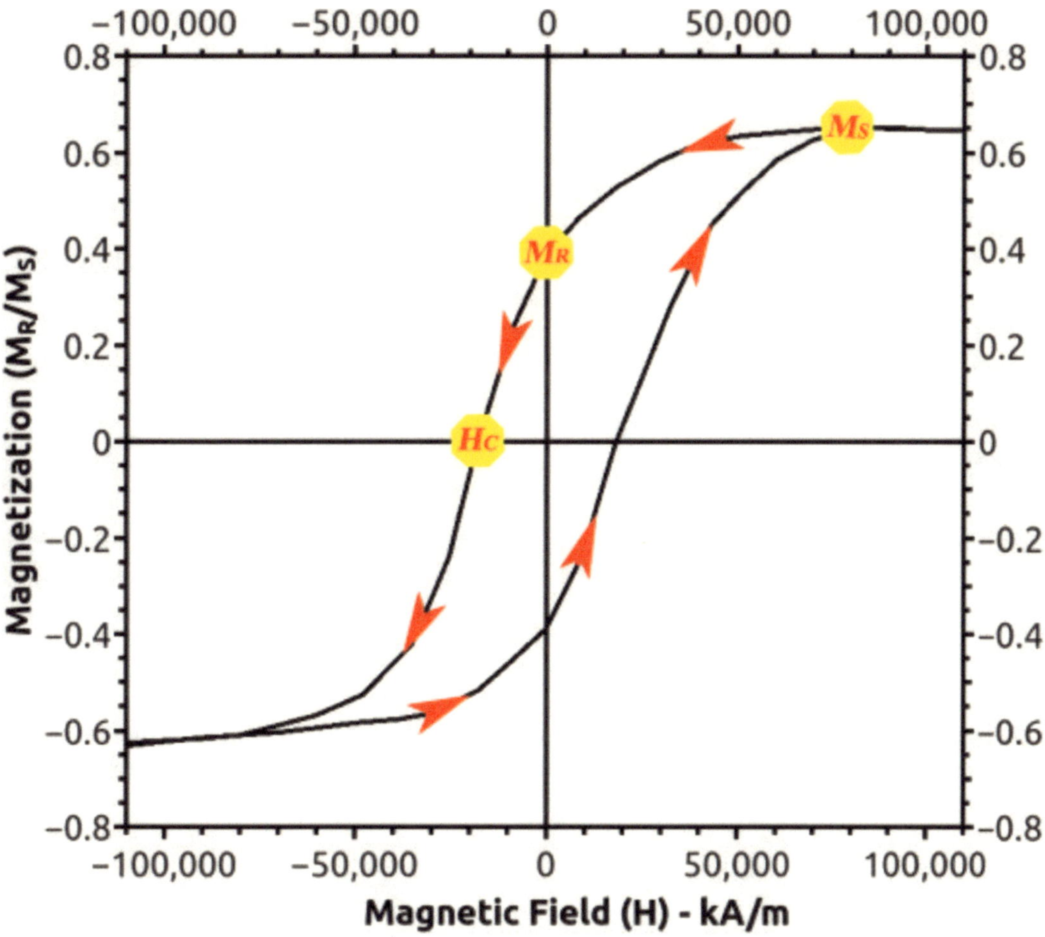

Figure 6. A schematic representation of the hysteresis loop belonging to a ferromagnetic material at a certain temperature (T). H_c= 20,000 (kA/m) and normalized remanent magnetization M_R=0.4.

tion in this theory is intrinsically balancing exchange energy to keep the essence of DMOS dialectic. Finally, the time comes up to talk about the last term, Zeeman energy. The effect of the external field is implicated as the Zeeman energy term by choosing an easy axis for the applied magnetic field. Heisenberg Hamiltonian, as a well-known Hamiltonian, is widely and greatly studied not only in modelling magnetic systems but also there is a distinct scientific "accent" in this theory named diluted magnetic organic structures based on the DMS theory mentioned in [31].

3.2. Theory of diluted magnetic organic structures

Theory of DMOS consists of two basic emphasis on the base of exchange interaction between first 10 nearest neighbours of A, T, C and G atoms in M-DNA helical. First is about the distance-dependent behaviour which means that effect of $J_{\alpha\beta}$ will decrease when distance between interacting neighbours increase. $J_{\alpha\beta} \propto \frac{\delta}{r}$; δ is any positive scalar. The other is that the major assumption of the theory provides a fundamental essence diverging from superexchange (Kramers-Anderson superexchange), Ruderman–Kittel–Kasuya–Yosida (RKKY) and similar ones. Even if the former can be easily understood, one can deal with the second to perceive it properly. The coupling strength of the atoms between the ions and organic atoms is scaled to the ion-ion coupling strength. That was so fast. Let us start again. Consider an exchange interaction that exists between a random determined X (C, H, O, N, P and Cr^{3+}) and Y (C, H, O, N, P and Cr^{3+}). If X and Y are both Cr^{3+}, the interaction type will be anti-ferromagnetic. However, if X and Y are both organic atoms, $J_{\alpha\beta}$ will vanish because extra soft magnetic contribution by organic atoms should be neglected. Lastly, if X (Y) is organic and Y (X) is Cr^{3+}, they will interact as ferromagnetic. Eq. (2) formulizes the mentioned assumption clearly.

$$J_{\alpha\beta}(r) = \begin{cases} -1 & \alpha = \beta = Cr \\ 0 < J(r) \ll 1 & \alpha = Cr,\ \beta = organic \\ 0 & \alpha = organic,\ \beta = organic \end{cases} . \tag{2}$$

According to a theory, the FM interaction between AFM Cr^{3+} ions and organic atoms, besides $J_{\alpha\beta,}$ vanishes for the organic type of atoms. Furthermore, the system will be aware of the anti-ferromagnetic interaction between Cr^{3+}ions. Actually, Duru et al. [21, 31] proposed a very similar method for the interacting atoms of diluted magnetic semiconductors.

In order to investigate effect of interactions between Cr-Cr and Cr-X (X = C, P, H, N and O) ions on magnetic behaviour, the authors have tried different J-coupling constants. During the simulation process, a reasonable exchange coupling constant between the above-mentioned atoms is traced. The aim is providing consistent characteristic results of the experimental studies.

3.3. MCMC simulation method

Markov chain Monte Carlo (MCMC) simulation method based on the metropolis algorithm is explained by defining an organic lattice introduced is Section 2. Metropolis algorithm [32] uses a transition probability related to energy difference between the first and trial state generated randomly. The ordered structured is formed by Monte Carlo method and also is stochastic due to the nature of the methodology. This time-dependent behaviour is defined by a basic equation in stochastic models like Ising model [33]. Eq. (3) shows:

$$\frac{\partial P_n(t)}{\partial t} = -\sum_{n \neq m} \left[P_n(t) W_{n \to m} - P_m(t) W_{m \to n} \right] \tag{3}$$

Actually, P_n is the probability of the state n at a known finite temperature, t and energy E_n.

$$P_n(t) = \frac{e^{-\frac{E_n}{k_B T}}}{Z} \tag{4}$$

Z denotes the partition function. $W_{n \to m}$ represents the transition probability from state n to state m. In the balance condition, $P_n(t) W_{n \to m} = P_m(t) W_{m \to n}$ will be valid since $\frac{\partial P_n(t)}{\partial t} = 0$. It is known as "detailed balance" [34]. Since it is hard to defeat the problem of partition function Z, one should construct a Markov chain [35]: The new state will be generated by the former state.

http://dx.doi.org/10.5772/67921

Therefore, Z became unnecessary and

$$\frac{P_n(t)}{P_m(t)} = \frac{e^{-\frac{E_n}{k_B T}}}{e^{-\frac{E_m}{k_B T}}} = e^{-\frac{E_n - E_m}{k_B T}} \tag{5}$$

$$\Delta E = E_n - E_m \tag{6}$$

Any transition probability agreed with detailed balance should be accepted. The first selection probability in statistical physics is the metropolis algorithm.

$$W_{m\to n} = \begin{cases} \frac{1}{\tau} e^{-\frac{\Delta E}{k_B T}}, & \Delta E > 0 \\ \frac{1}{\tau}, & \Delta E < 0 \end{cases} \tag{7}$$

τ, denotes the time which is required for a spin flip. Besides the ordinary Monte Carlo, it is transformed to a transition probability to avoid the partition function and in a natural way to take us to a well-known form, Markov chain. Expectation value of an observable A should be calculated by $< A > = \sum_n^N A_n P_n$. However, mean value of magnetization can be easily calculated by $< M > = \frac{1}{N} \sum_i^N S_i$. Flow diagram details the simulation process explicitly, which is shown in **Figure 7**. First, a random atom of the M-DNA nanowire is selected to calculate its energy as described by dipolar Hamiltonian. Second, this step is followed by determining a random magnetic moment of the relevant atom in 3D vector space. Simulation world is calling this state as a trial state. Once again, an energy calculation via to trial state is performed to determine the difference between the trial and former one. Then, the system can decide the

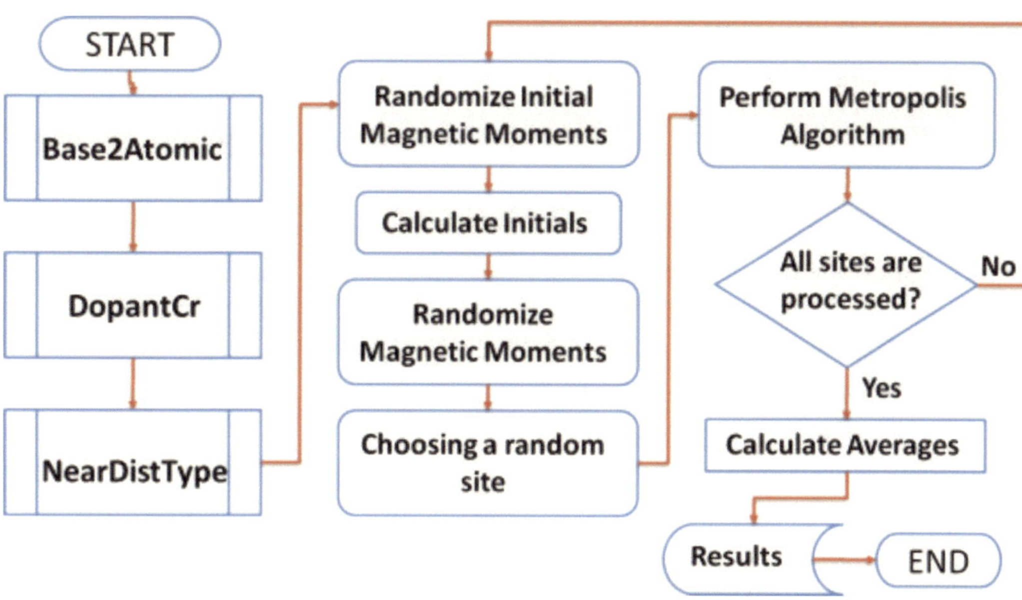

Figure 7. A flow diagram of the simulation process.

new state that will be exist according to the minimization of the energy ($\Delta E < 0$). What if the energy difference is bigger than zero? The answer is Boltzmann statistics. If $\Delta E > 0$, a pseudo random number will be generated and compared with $e^{\frac{\Delta E}{k_B T}}$. If random number is bigger than $e^{\frac{\Delta E}{k_B T}}$, trial state will be the new state. In this equation, beta is the inverse temperature included in Boltzmann factor k_B.

4. Magnetic measurements

Magnetic behaviour of materials strongly depends on temperature. Even if magnetic moments are coupled to each other in ferromagnetic materials, thermal agitations can cause random alignment of the magnetic moments in certain temperatures. The thermal agitations tend to keep the atomic moments pointed at random. The result is only partial alignment in the field direction and therefore a small positive susceptibility. The effect of an increase in temperature is an increase of the randomizing effect of thermal agitation that decreases the susceptibility. There is a maximum temperature for all ferromagnets for the disappearance of ferromagnetic property as a result of thermal energy. This critical temperature is called as "Curie temperature". Thermal agitation becomes dominant compared to molecular field upward of Curie temperature. For example, Curie temperatures of iron and cobalt are 1043 and 1400 Kelvin, respectively. Thermal stability of a magnetic material is required to use the material in technological applications. Therefore, determining magnetic properties of a material in different temperatures is the crucial point for the material being a candidate for technological purposes. For our case, room temperature ferromagnetism is the subject so hysteresis loops are measured at room temperature. Magnetization-temperature (M-T) graphs and hysteresis curves of M-DNA can be found in this topic.

4.1. Magnetization-temperature

Two modes are possible for M-T measurements: Zero field cooling (ZFC) and field cooling (FC). In ZFC, the sample is cooled without any applied magnetic field to the desired temperature. Then, the data are collected while heating with certain value of applied magnetic field. In FC, the sample is cooled with some applied magnetic field to the desired temperature. Then, the temperature is increased in the same applied magnetic field. The data can be collected while either the cooling or heating process. The field cooling magnetization-temperature measurement of various samples of M-DNA can be found in **Figures 8** and **9**.

Duru et al. [31] proposed a decrement in the remanent magnetization of the system at temperature $k_B T = 0.01$, from $x = 0.15$ to 0.30, which can be seen in **Figure 8**. The magnetic behaviour of the system strongly depends on the doping ratio (x) because of changing the number of ferromagnetic interactions between the atoms. When a doped atom has all host atomic neighbours, the amount of ferromagnetic coupling for the interacting doped host atoms reaches a maximum. In contrast, decreasing the number of doped-host neighbours and increasing the number of doped-doped neighbours cause the disappearance of ferromagnetism. At the same time, thermal agitations play a well-known role in magnetic systems, especially providing disordered states. A detailed discussion about the effect of doping ratio will be carried out with the information provided by hysteresis curves in the next topic.

http://dx.doi.org/10.5772/67921

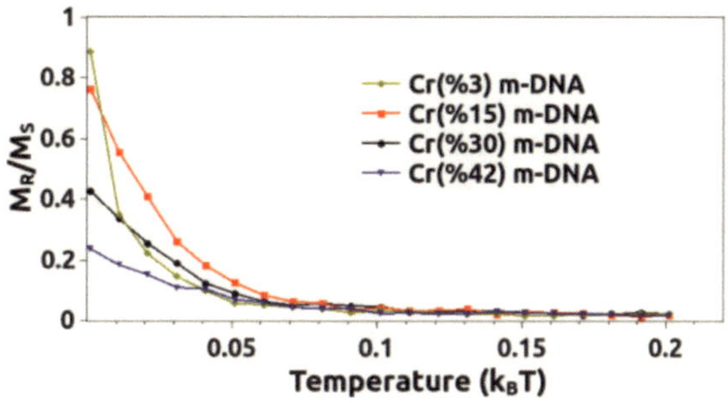

Figure 8. FC of Cr-doped DNA with different doping ratios: 3, 15, 30 and 42%. M_R/M_S ratio of 15%-doped M-DNA is larger when compared to 3, 30 and 42% at 0.02 k_BT.

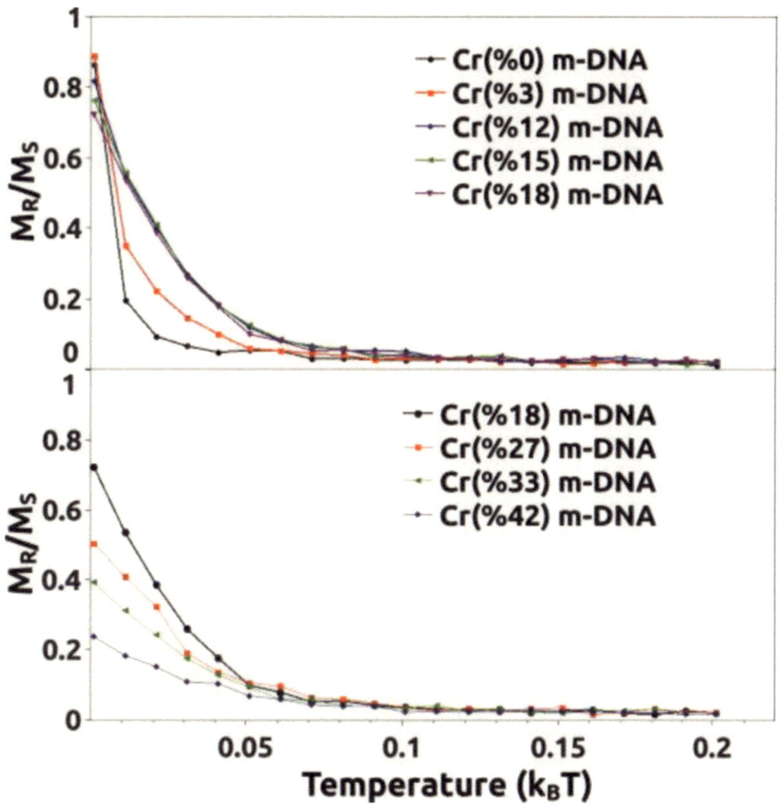

Figure 9. FC of M-DNA doped by Cr **(a)** 0, 3, 12, 15 and 18% **(b)** 18, 27, 33 and 42%.

4.2. Hysteresis curves

The idea of using DNA, a natural nanowire, in magnetism-based technological applications arose when the first paramagnetic resonance signal was observed by Bliumenfeld and Bendersky [36]. The signal explicitly revealed that B-DNA is a paramagnetic material. A hysteresis curve of B-DNA verifies its paramagnetic phase as shown in **Figure 10**. The outputs of the simulation process for B-DNA were also used to calibrate the main parameters of the doped-DNA systems [31].

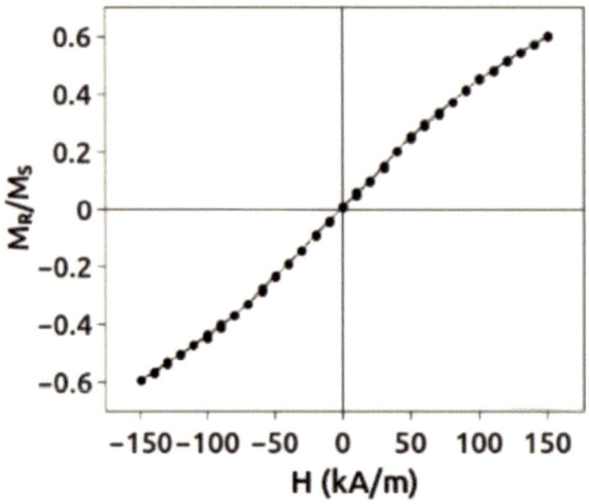

Figure 10. Hysteresis loop of B-DNA at k_BT/J=0.02.

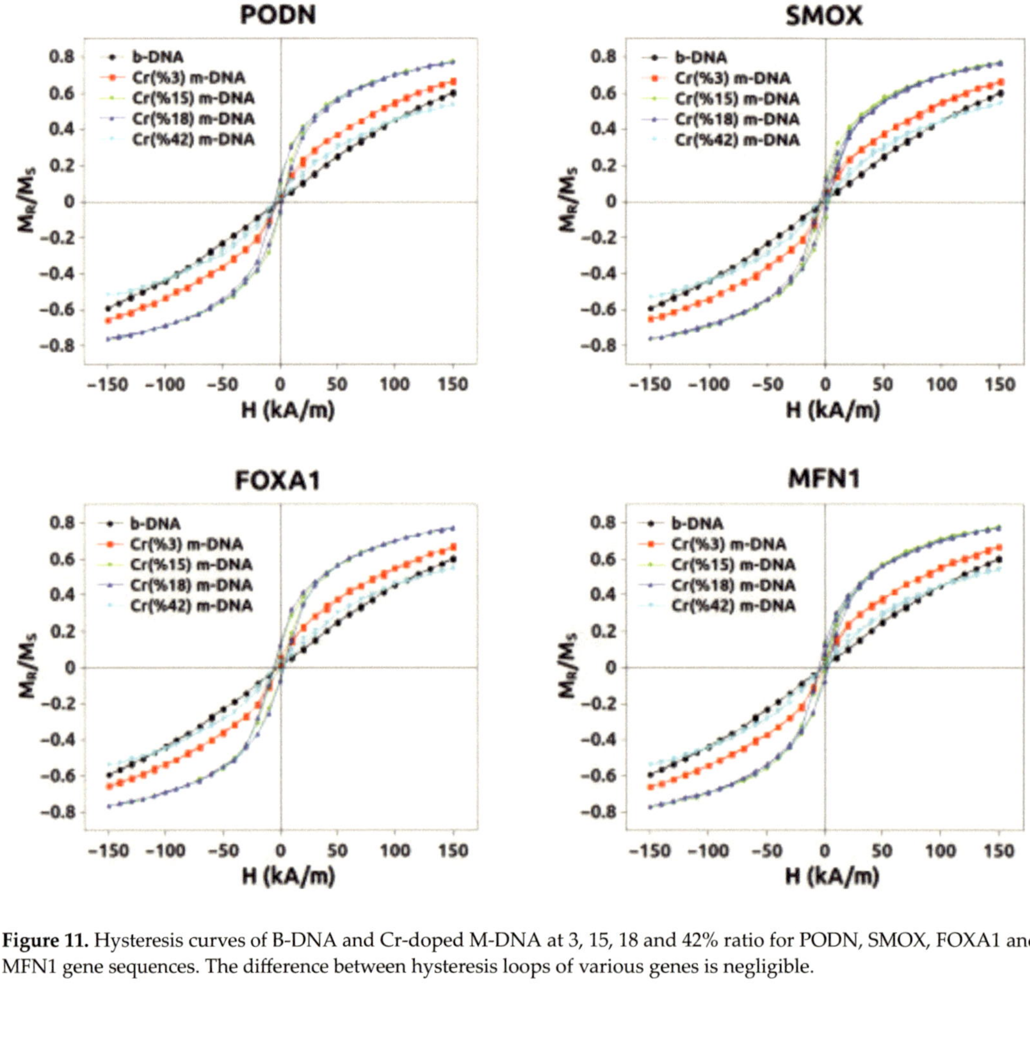

Figure 11. Hysteresis curves of B-DNA and Cr-doped M-DNA at 3, 15, 18 and 42% ratio for PODN, SMOX, FOXA1 and MFN1 gene sequences. The difference between hysteresis loops of various genes is negligible.

Although the Curie temperature of M-DNA is not so far from room temperature, **Figure 11** makes it clear that Cr-doped M-DNA has a relatively higher coercivity for $x = 0.15$ against other organomagnetic materials [37]. It is supposed that higher Cr concentrations larger than 15% lead the system to a state of disorder. Decrement in the M_R and coercive field indicates a change in the magnetic phase of DNA with increasing the doping ratio after $x = 0.15$ [31]. Briefly, the system has nonlinear phase transition behaviour with increasing the doping ratio from 0 to 42%: first, it is paramagnetic as in B-DNA and then gains soft ferromagnetic and antiferromagnetic properties, respectively.

Particularly, residual magnetism in system vanishes for $x = 0.42$, indicating that the ferromagnetic behaviour of the system completely vanishes due to AFM ion coupling, while there is no significant difference between $x = 0.15$ and 0.18 because of the random distribution of the Cr ions.

When **Figure 12** is roughly skimmed, DMOS behaviour should be understood easily. Remanent magnetization normalized by saturation magnetization has the bigger value near to 15% doping ratio and decreases with increasing doping ratio. Hysteresis and remanent magnetization figures of only four different genes of the sample group are implemented to this section since it is sufficient to verify the theory.

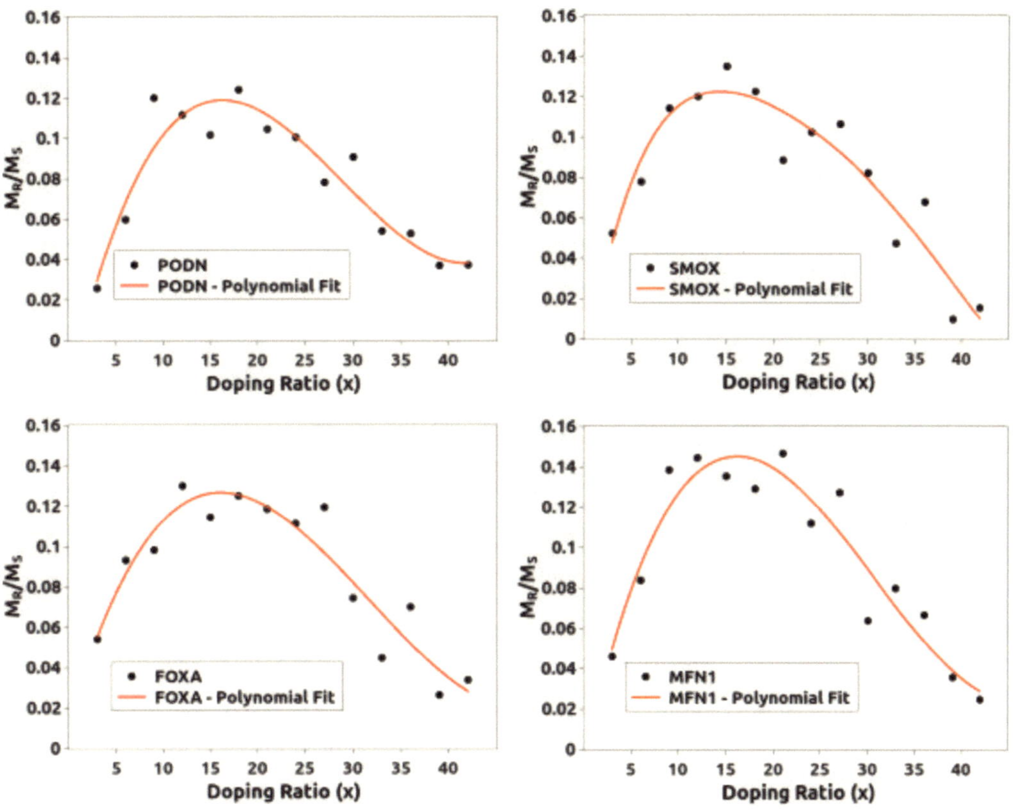

Figure 12. M_R/M_S MR/MS versus doping ratio of Cr-doped M-DNA for PODN, SMOX, FOXA1 and MFN1 gene sequences. The presentation of changing magnetic phase is supported by second-order polynomial fitting to have a clear understanding.

5. Summary

Geometrical structure of B-DNA nanowire in Euclidian space is generated by a given sample set of genes, including non-dispersive base sequences that consist of certain G-C ratio. The modifying process of B-DNA nanowire by doping Cr^{3+} ions is explained clearly and detailed by flow diagram as made from the simulation process of the whole system. However, MCMC method based on metropolis algorithm is used to mimic M-DNA nanowire to discover a proper Hamiltonian in the case of DMOS theory. Existence of a ferromagnetic interaction between Cr-X (X = N, H, O, P and C) and an anti-ferromagnetic interaction between Cr-Cr neighbouring atoms underlies the theory with Heisenberg Hamiltonian-included dipolar interaction. Moreover, hysteresis curves and remanent magnetization versus doping ratio of Cr is presented after critical temperature is determined by simulating of FC. $k_B T/J$ ratio is set to ~0.1 after several trials to calibrate parameters according to the paramagnetic behaviour. Duru et al. proposed a decrement in the remanent magnetization of the system at temperature $k_B T = 0.01$ from $x = 0.15$ to 0.30. Moreover, hysteresis curves reveal the DMS characteristic of M-DNA by Cr^{3+}, and it is not crucially dependent to type of atom. Different type of atom should only change coercive field or doping ratios that system gains ferromagnetic property. One must not forget that the dopants must have conducting property such as metals or semi-conductors.

Acknowledgements

We would like to thank the Scientific Research Commission (BAPKO) of Marmara University supporting this chapter under the research project numbered FEN-A-100616-0275.

Author details

Caner Değer[1], Vahap Eldem[2] and İzzet Paruğ Duru[1]*

*Address all correspondence to: izzetparug.duru@marun.edu.tr

1 Department of Physics, Marmara University, İstanbul, Turkey

2 Department of Biology, İstanbul University, İstanbul, Turkey

References

[1] Dugasani SR, Park B, Gnapareddy B, Pamanji SR, Yoo S, Lee KW, Lee S, Jun SC, Kim JH, Kim C, Park SH. Tunable near white light photoluminescence of lanthanide ion (Dy^{3+}, Eu^{3+} and Tb^{3+}) doped DNA lattices. RSC Advances. 2015;**5**(69):55839-55846. DOI: 10.1039/C5RA07360J

[2] Göhler B, Hamelbeck V, Markus TZ, Kettner M, Hanne GF, Vager Z, Naaman R, Zacharias H. Spin selectivity in electron transmission through self-assembled monolayers of double-stranded DNA. Science. 2011;**331**(6019):894-897. DOI: 10.1126/science.1199339

[3] Steckl AJ. DNA—a new material for photonics? Nature Photonics. 2007;**1**:3-5. DOI: 10.1038/nphoton.2006.56

[4] Nam J-M, Thaxton CS, Mirkin CA. Nanoparticle-based bio-bar codes for the ultrasensitive detection of proteins. Science. 2003;**301**(5641):1884-1886. DOI: 10.1126/science.1088755

[5] Lu Y, Goldsmith BR, Kybert NJ, Johnson ATC. DNA-decorated graphene chemical sensors. Applied Physics Letters. 2010;**97**(8). DOI: 10.1063/1.3483128

[6] Hu J-F (胡建芬), Feng L (冯琳), Zhang W-X (张文星), Li Y (李勇), Lu Y-X (卢亚鑫). Detection of DNA bases using Fe atoms and graphene. Chinese Physics Letters. 2016;**33**(1). DOI: 10.1088/0256-307X

[7] Azbel MY. Phase transitions in DNA. Physical Review A. 1979;**20**(4):1671-1684. DOI: 10.1103/PhysRevA.20.1671

[8] Alivisatos AP, Johnsson KP, Peng X, Wilson TE, Loweth CJ, Bruchez MP Jr, Schultz PG. Organization of "nanocrystal molecules" using DNA. Nature. 1996;**382**:609-611. DOI: 10.1038/382609a0

[9] Mirkin CA, Letsinger RL, Mucic RC, Storhoff JJ. A DNA-based method for rationally assembling nanoparticles into macroscopic materials. Nature. 1996;**382**:607-609. DOI: 10.1038/382607a0

[10] Braun E, Eichen Y, Sivan U, Ben-Yoseph G. DNA-templated assembly and electrode attachment of a conducting silver wire. Nature. 1998;**391**:775-778. DOI: 10.1038/35826

[11] Richter J, Mertig M, Pompe W, Mönch I, Schackert HK. Construction of highly conductive nanowires on a DNA template. Applied Physics Letters. 2001;**78**(4). DOI: 10.1063/1.1338967

[12] Park SJ, Taton TA, Mirkin CA. Array-based electrical detection of DNA with nanoparticle probes. Science. 2002;**295**(5559):1503-1506. DOI: 10.1126/science.1067003

[13] Keren K, Berman RS, Buchstab E, Sivan U, Braun E. DNA-templated carbon nanotube field-effect transistor. Science. 2003;**302**(5649):1380-1382. DOI: 10.1126/science.1091022

[14] Yi J. Emergent paramagnetism of DNA molecules. Physical Review B. 2006;**74**(21):212406. DOI: 10.1103/PhysRevB.74.212406

[15] Savin AV, Mazo MA, Kikot IP, Manevitch LI, Makoto AVO. Heat conductivity of the DNA double. Physical Review B. 2011;**83**(24):245406. DOI: 10.1103/PhysRevB.83.245406

[16] Sakamoto H, Mizoguchi K. Electronic states of DNA and M-DNA studied by optical absorption. Physical Review E. 2014;**89**(2):022719. DOI: 10.1103/PhysRevE.89.022719

[17] Panahi M, Chitsazanmoghaddam M. Comment on "modeling the electrical conduction in DNA nanowires: Charge transfer and lattice fluctuation theories". Physical Review E. 2016;**93**(4):046401. DOI: 10.1103/PhysRevE.93.046401

[18] Behnia S, Fathizadeh S. Modeling the electrical conduction in DNA nanowires: Charge transfer and lattice fluctuation theories. Physical Review E. 2015;**91**(2):022719. DOI: 10.1103/PhysRevE.91.022719

[19] Dugasani SR, Lee N, Lee J, Kim B, Hwang SU, Lee KW, Kang WN, Park SH. Magnetic characteristics of copper ion-modified DNA thin films. Scientific Reports. 2013;**3**(1819). DOI: 10.1038/srep01819

[20] Nikiforov VN, Kuznetsov VD, Yevdokimov YM, Irkhind VY. Magnetic properties of Gd^{3+} ions in the spatially distributed DNA molecules. Journal of Magnetism and Magnetic Materials. 2014;**368**:338-341. DOI: 10.1016/j.jmmm.2014.06.008

[21] Duru IP, Değer C, Eldem V, Kalayci T, Aktas S. Theory of room temperature ferromagnetism in Cr modified DNA nanowire. Journal of Physics D: Applied Physics. 2016;**49**:135004. DOI: 10.1088/0022-3727/49/13/135004

[22] van Dijk M, Bonvin AMJJ. 3D-DART: A DNA structure modelling server. Nucleic Acid Research. 2009;**37**(2). DOI: 10.1093/nar/gkp287

[23] Utrecht Biomolecular Interaction Web Portal. 3DDART@BonvinLab [Internet]. Available from: http://milou.science.uu.nl/services/3DDART/

[24] Watson CD, Crick FHC. Molecular structure of nucleic acids. Nature. 1953;**4356**:737

[25] Franklin R, Gosling RG. The structure of sodium thymonucleate fibres. II. The cylindrically symmetrical Patterson function. Acta Crystallographica. 1953b;**6**:678-685. DOI: 10.1107/S0365110X53001940

[26] Dugasani SR, Kim M, Lee I-y, Kim JA, Gnapareddy B, Lee KW, Kim T, Huh N, Kim G-H, Park SC. Construction and characterization of Cu^{2+}, Ni^{2+}, Zn^{2+}, and Co^{2+} modified-DNA crystals. Nanotechnology. 2015;**26**(27):275604. DOI: 10.1088/0957-4484/26/27/275604

[27] Mizoguchi K. Physical properties of natural DNA and metal ion inserted M-DNA. In: Heckman EM, Singh TB, Yoshida J, editors. Proceedings of SPIE. Nanobiosystems: Processing, Characterization, and Applications; 2008. DOI: 10.1117/12.801478

[28] Dong R, Yan X, Li K, Ban G, Wang M, Cui S, Yang B. Effects of metal ions on conductivity and structure of single DNA molecule in different environmental conditions. Nanoscale Research Letters. 2010;**5**:1431. DOI: 10.1007/s11671-010-9657-3

[29] Park H-Y, Dugasani SR, Kang D-H, Jeon J, Jang SK, Lee S, Roh Y, Park SH, Park J-H. n- and p-type doping phenomenon by artificial DNA and M-DNA on two-dimensional transition metal dichalcogenides. ACS Nano. 2014;**8**(11):11603-11613. DOI: 10.1021/nn5048712

[30] Joseph J, Schuster GB. Long-distance radical cation hopping in DNA: The effect of thymine–Hg(II)–thymine base pairs. Organic Letters. 2007;**9**(10):1843-1846. DOI: 10.1021/ol070135a

[31] Duru İP, Değer C, Kalaycı T, Arucu M. A computational study on magnetic effects of Zn Cr O 1−x x type diluted magnetic semiconductor. Journal of Magnetism and Magnetic Materials. 2015;**396**:268-274. DOI: 10.1016/j.jmmm.2015.08.031

[32] Metropolis N, et al. Journal of Chemical Physics. 1953;**21**:1087

[33] Kawasaki K. In: Domb C, Green MS, editors. Chapter 4.Mode Coupling and Critical Dynamics. Phase Transitions and Critical Phenomena. Vol. 5. London: Academic Press; 1976

[34] Boltzmann L. Lectures on Gas Theory. Berkeley, CA, USA: University of California Press; 1964

[35] Hastings WK. Monte Carlo sampling methods using Markov chains and their applications. Biometrika. 1970;**57**(1):97-109. DOI: 10.1093/biomet/57.1.97. JSTOR 2334940. Zbl 0219.65008

[36] Bliumenfeld LA, Bendersky VA. Magnetic and Dielectric Properties of High-ordered Macromolecular Structures. Doklady Akademii Nauk, SSSR. 1960;133:1451-1454

[37] Jain R, Kabir K, Gilroy JB, Mitchell KAR, Wong K, Hicks RG. High-temperature metal–organic magnets. Nature. 2007;**445**:291-294

Chapter 10

Recent Trends in Plasmonic Nanowire Solar Cells

Mohamed Hussein,
Mohamed Farhat Othman Hameed and
Salah S. A. Obayya

Additional information is available at the end of the chapter

http://dx.doi.org/10.5772/67617

Abstract

Light trapping is crucial for low-cost and highly efficient nanowire (NW) solar cells (SCs). In order to increase the light absorption through the NWSCs, plasmonic materials can be incorporated inside or above the NW design. In this regard, two novel designs of plasmonic NWSCs are reported and analyzed using 3D finite difference time domain method. The geometrical parameters of the reported designs are studied to improve their electrical and optical efficiencies. The ultimate and power conversion efficiencies (PCE) are used to quantify the conversion efficiency of the light into electricity. The first design relies on funnel shaped SiNWs with plasmonic core while the cylindrical NWs of the second design are decorated by Ag diamond shaped. The calculated ultimate efficiency and PCE of the plasmonic funnel design are equal to 44% and 18.9%, respectively with an enhancement of 43.3 % over its cylindrical NWs counterpart. This enhancement can be explained by the coupling between the three optical modes, supported by the upper cylinder, lower cone and plasmonic material. Moreover, the cylindrical SiNWs decorated by Ag diamond offer an ultimate efficiency and short-circuit current density of 25.7%, and 21.03 mA/cm^2, respectively with an improvement of 63% over the conventional cylindrical SiNWs.

Keywords: light trapping, plasmonic solar cells, nanowires, finite-difference time-domain

1. Introduction

It has become obvious that global stability, economic prosperity, and quality of life are linked intimately with sufficient supplies of clean energy. Searching for energy resources to meet the world's growing demand is a major challenge for the next half century. Traditional energy sources are currently depleted at the rate of approximately 4.1 × 10^{20} J/yr, which is equivalent to continuous power consumption of 13 trillion watt [1]. An increase in the global population to 9 billion people, along with the rapid worldwide technology development and economic growth, leads to an inevitable increase of more than double the demand for energy (to 30 TW) by 2050 and above the triple demand (to 40 TW) by the end of this century [2]. Because of the dramatic hydrocarbon-based power consumption, the harmful effects (such as global warming, acid precipitation, air pollution, ozone depletion, and forest destruction) are increasingly in catastrophic forms. In the pursuit of limiting these drawbacks, suitable actions aimed at reducing the dependence on the fossil fuels are mandatory. The most urgent challenge is in the search for clean and renewable alternative energy resources required for the sustainable development of human civilization. The Sun is the primary source for clean and abundant energy, with nearly 120,000 TW of incident radiations on the earth's surface which has the potential for covering the human needs [3, 4]. This contains 1.5×10^{18} KWh of energy; which is larger than that of the traditional reserves of oil, coal, and gas which form 1.75×10^{15}, 1.4×10^{15}, and 5.5×10^{15} KWh, respectively [5, 6]. Hence, solar radiation annually provides over 100 times the energy supplied by the entire known fossil fuel sources. Harvesting solar power has become the modern trend in worldwide research projects reported in the scientific community [5, 6]. The photovoltaics (PVs) or solar cell (SC) devices are considered a promising technology that converts the sunlight directly into electricity on a very large scale without causing pollution [7]. The PVs are considered attractive and a promising candidate that would significantly contribute in the future global renewable energy sources. The main challenge in the PV devices is the cost reduction per watt with respect to the fossil fuel technologies. The cost/watt ratio of the photovoltaic devices is at least 1.5 times higher than the electricity generated from the fossil fuels [8]. The main factor influencing the cost/watt ratio is the size of the active material, mostly crystalline silicon (c-Si), and the manufacturing process. The 30–40% of cost/watt in the c-Si SCs is mainly due to the silicon substrate.

Until now, over 90% of the photovoltaic market is currently dominated by the c-Si solar cells. A maximum power conversion efficiency of 25.6% has been achieved for the c-Si solar cell which is approaching the Shockley-Queisser theoretical limit of 29.4%. Therefore, further improvement in the Si solar cell efficiency is required to compete with the fossil fuels and other alternative sources of energy [9]. Different efficient and reliable avenues have been used to enhance the solar cell efficiency and reduce the cost/watt ratio [9]. In the attempt to do so, different materials have been used, such as crystalline silicon (c-Si) [10], amorphous silicon (a-Si) [11], gallium arsenide (GaAs) [12], copper indium gallium selenide (CIGS) cadmium telluride (CdTe) [9, 13], GaSb/GaAs quantum rings [14], organic material [15], and tandem solar cells by combining several materials [16]. The c-Si material is commercially used because it is cheap and abundant compared to other materials. However, the absorption of the c-Si is poor, and a very large thickness is needed to absorb the solar energy. The bandgap of silicon (1.12 eV) is not at the perfect bandgap for solar spectrum, which severely limits the efficiency of the solar cell. Alternatively, a-Si has a better absorption, and it is usually used for thin-film solar cell materials; however, its efficiency is not higher than the c-Si solar cells. The GaAs has high conversion efficiency up to 30% because its bandgap is close to the optimized bandgap for single absorbed material on the expenses of the high total cost. Also, the cadmium telluride (CdTe) has good conversion efficiency, and the cost can be minimized due to the rapid technology development based on this material. However, the Cd is a highly polluted material, and it has several drawbacks. Recently, different materials can be used to absorb various spectra of solar energy with high efficiency [17, 18]. However, such multi-junction designs need complicated fabrication techniques and expensive manufacturing processes. Thus, the multi-junction solar cells are difficult to be commercialized in the near future.

Currently, the thicknesses of the crystalline silicon wafers used in solar cell market is about 180–300 μm which affects the total cell price due to the silicon materials and processing. Therefore, thin film solar cells attract a great attention, with film thickness in the range of 1–2 μm, as it can be deposited on cheap substrates such as glass, plastic, or stainless steel. However, thin-film solar cell suffers from its weak absorption near the material bandgap especially when using indirect bandgap materials (Si). One of the most alternative important approaches to increase the thin-film solar cell efficiency is the light-trapping techniques to enhance the light absorption into the solar cell active materials. These techniques can increase the light path of the trapped photons in the solar cell materials to generate the electron-hole pair under light illumination. Consequently, the material cost can be minimized by using cheap substrate and solar cell materials with high efficiency. Further, semiconductor nanowires with a few nanometers in diameters can reduce SC material cost with high efficiency. The antireflection layers can be also used to improve the light transmission efficiency [19]. The refractive index of the antireflection layer is chosen between the refractive indices of the active material and surrounding air. The most widely used antireflection materials include SiO_2, SiNx, TiO_2, Al_2O_3, CeO_2, etc. [20]. The antireflection layer can be merged with the textured structures for better light-trapping improvements. The textured silicon surface structures can be etched by various etchants, such as NaOH [21], TMAH [22, 23], and KOH [24]. Different textured structures have been introduced, including the pyramidal textured surfaces [25], inverted pyramidal [26], honeycomb [27], nanowire [28, 29], gratings [30], V-grooved [31], and other types of structures [32–34]. Most of these structures have large size and are not suitable for thin-film SCs with micrometer or sub-micrometer ranges [30]. Recently, the plasmonic SCs attract sharp attention because it can be used for efficient light trapping through the active layer of the SC. Further, it is simple to merge the plasmonic materials into nanowire and thin-film SC designs [35]. Therefore, the plasmonic nanoparticles are considered as pivotal cornerstone in increasing the light trapping in thin-film and nanowire SCs.

Following this introduction, a review of plasmonic light-trapping techniques for solar cell applications will be discussed. In addition, the recent trends in NW SCs are introduced thoroughly. Further, two novel designs of plasmonic NW SCs are presented and analyzed using 3D full-vectorial finite-difference time-domain method (FDTD). In this regard, plasmonic funnel-shaped SiNWs and plasmonic diamond SiNWs are investigated to present the effect of adding plasmonic materials through or above the SiNWs. The performance of the suggested NWSCs has been investigated in terms of the ultimate efficiency and short-circuit current density, open-circuit voltage, power conversion efficiency, and absorbed field profiles. The reported funnel-shaped SiNWs with plasmonic core show an ultimate efficiency of 36.9% which is higher than cylindrical and conical counterparts by 36.7 and 22.2%, respectively. Further, the plasmonic funnel-shaped design has short-circuit current density of 30.2 mA/cm^2 which is greater than 24.5 and 22.12 mA/cm^2 of the conical and cylindrical designs. The enhancement in the ultimate efficiency and hence short-circuit current density apparently can be attributed to the combination between three types of optical modes that are supported by the upper cylinder, the lower tapered cone, and the plasmonic material. On the other hand, the SiNW decorated by Ag diamond shows an ultimate efficiency of 25.7% which is higher than conventional SiNWs without plasmonic material by 63%. Additionally, the reported design with Ag diamond shows short-circuit current density of 21.03% mA/cm^2 which is greater than the conventional counterpart SiNWs with an improvement of 32.6%.

2. Review of plasmonic light trapping for solar cell applications

At the emergence of the eighteenth century, metallic nanoparticles have gained much interest in the research community for their optical properties. Years later, the Raman scattering was enhanced through the utilization of metallic structures [36] which attracted more researchers to the field of plasmonic nanoparticles. Since then, plasmonic resonance, with its fundamental properties, has been introduced to various applications such as integrated optics, biosensing [37, 38], and energy harvesting [39–41]. In metallic nanoparticles, plasmon resonances give rise to a resonance scattering which reduces the reflection over the spectral range. This anti-reflection effect, caused by random and periodic arrays of metallic nanoparticles, has been extensively studied, theoretically and experimentally, for photovoltaic applications [42, 43]. Stuart and Hall have led the research work in the field of light-sensitive devices enhanced by plasmonic resonance where an enhancement factor of 18 has been reported for a silicon-on-insulator photodetector at 800 nm wavelength. This device has a thickness of 165 nm and is covered by silver nanoparticles on its surface [44]. Additionally, Schaadt et al. [45] proposed a highly doped wafer-based solar cells coated by gold nanoparticles on its surface which achieved an enhancement of 80% for the photocurrent at 500 nm wavelength. Gold nanoparticles have been also used in amorphous thin-film silicon SCs, by Derkacs et al. [43], which improved the overall power conversion efficiency (PCE) by 8%. Pillai et al. [46] achieved 19 and 33% increase in the overall photocurrent for wafer-based SCs and 1.25 μm thick silicon-on-insulator design, respectively, using silver nanoparticles on the cells' surface. Ouyang et al. [47] employed self-assembled Ag nanoparticles on the surface of thin-film silicon SCs which enhanced the short-circuit current density by 27%. A relative photocurrent enhancement of 10% was reported by Beck et al. [48] for silicon cell of thickness 22 μm covered by an array of Ag nanoparticles. In recent years, Tan et al. [49] used Ag nanoparticles as a plasmonic back reflector which offered a net short-circuit current density gain of 2 mA/cm^2. Moreover, different types of solar cells have achieved a performance enhancement based on the plasmonic effect of metallic nanoparticles such as organic and dye-sensitized SCs [50, 51]. In a pioneering breakthrough by Ding et al. [52], plasmonic effect was used to enhance the light absorption and hence, the efficiency of dye-sensitized SCs which achieved a PCE of 5.9% [52]. For a dye-sensitized TiO_2 film, reported by Hägglund et al. [53], an enhanced carrier generation was achieved by employing gold nanodisc. Rand et al. [54] also introduced organic SCs based on ultrathin film using tiny silver nanoparticles which enhanced the cell efficiency. Furthermore, an efficiency enhancement of 1.7 in magnitude was reported by Morfa et al.

for organic bulk heterojunction solar cells [55]. Another organic SCs proposed by Chen et al. achieved an improved PCE using Au nanoparticles incorporated into anodic buffer layer [56].

3. State of the art of semiconductor nanowire solar cells

The current crystalline Si cells are widely used in the PV market with efficiency in the proximity of 25% [57]. A major dilemma lies in the high material cost. The use of cheaper materials, such as polycrystalline thin films or organic semiconductors, decreases the material cost. The amount of required absorbing material can be further decreased by increasing the concentration of the light through a small area [58]. A dramatic improvement in the fabrication techniques can further minimize the loss mechanism and hence increase the SC efficiency. The use of semiconductor nanoantennas (NAs) or NWs readily fulfills all these requirements. In this regard, the NW SCs with structures of a few nanometers in diameters have been considered for PV power generation. The semiconductor NWs have advantages in terms of optical absorption enhancement [59]. This is mainly due to the small reflectivity of the NW array (NWA) with large open area on the frontal surface. The light is trapped by multiple total internal reflections through each nanoscale cylindrical NW resonator [60]. In 2007, Hu and Chen [59] introduced the cylindrical SiNWs with an ultimate efficiency of 15.5%. The behavior of the semiconductor NWs strongly depends on their structural geometrical parameters [59]. Consequently, intense theoretical studies of the optical absorption are required to maximize the ultimate efficiency of the NWA-based SCs. The NWAs with optimized parameters apparently offer higher broadband absorption than solid films [61]. As the NW diameter, period, and wavelength are of the same order, strong light scattering and further light trapping occur [61]. The modified design presented by C. Lin and Povinelli [62] have enhanced the ultimate efficiency up to 23.84% through increasing the lattice constant to 600 nm [62]. The ultimate efficiency of 20.44% was obtained using partially aperiodic vertically SiNWs [63]. Li et al. [64] accomplished an ultimate efficiency of 30.5% by tuning silicon nanowire parameters. The NWAs with random orientation, length, diameter, and position have been experimentally investigated in 2008 [65]. It has been shown that reflection or backscattering was detected with slight effects on the absorption. The optical properties of vertically aligned silicon NWAs with three types of structural randomness, i.e., random position, diameters, and lengths, were numerically investigated [66]. The NWAs with random position displayed slight absorption enhancement, while those with random diameter or length exhibited a significant absorption enhancement. Therefore, the structural randomness in vertical nanowire arrays can further enhance the optical absorption compared to ordered nanowire arrays. The substantial reflectance and absorption of slanting silicon NWAs attracted attention due to their established applications in SCs [67]. The slanting NWAs achieve a significant suppression of light reflection and an enhancement of light-harvesting ability than the conventional vertical structure. The slanted NWAs offer an ultimate efficiency of 33.45%, which exceeds the 30.5% of optimum vertical SiNWs [67]. Wanget al. [68] also reported an ultimate efficiency of 29.1% by using elliptical SiNWAs. On top of that, an ultimate efficiency of 39.3% is achieved using SiNWs arranged in a decagonal lattice [28]. Additionally, periodic vertical cone arrays form a promising family of structures that are currently under examination for PV technology. Wang et al. proved that the light absorption can be considerably enhanced by using nanocones [69, 70]. In addition, Ko et al. [71] demonstrated that an ultimate efficiency of 28.9% can be obtained by decreasing the cone bottom diameter to 20 nm. Further, the SiNWs funnel shape achieved an ultimate efficiency of 41.8% [29].

4. Simulation methodology and numerical results

4.1. Plasmonic funnel SiNWs

The optical simulation of the suggested designs is carried out using 3D finite-difference time-domain (FDTD) method via Lumerical FDTD solution software package [72]. The FDTD method is widely used in solar cell applications due to its ability to make the simulation physically realistic through using the multi-coefficient material model. Moreover, the FDTD has the capacity to simulate broadband frequency ranges in a single simulation run. In this study, the 3D FDTD method is employed for investigating the optical properties of the SiNWs over the main region of the solar spectrum [72]. **Figure 1(a)** shows the utilized computational domain for the proposed plasmonic funnel-shaped SiNW unit cell in x-z plane. However, **Figure 1(b)** illustrates the x-y view of the suggested design. The top and bottom of the unit cell are surrounded by perfectly matched layers (PML) boundary conditions, while the unit cell is surrounded by periodic boundary conditions (BCs) along x and y directions to simulate periodic square SiNW array. The width and length of the unit cell are equal to 500 nm, while the height is taken as 4000 nm. In order to achieve high accuracy, fine meshing of $\Delta x = \Delta y = \Delta z = 5$ nm is used, which is less than

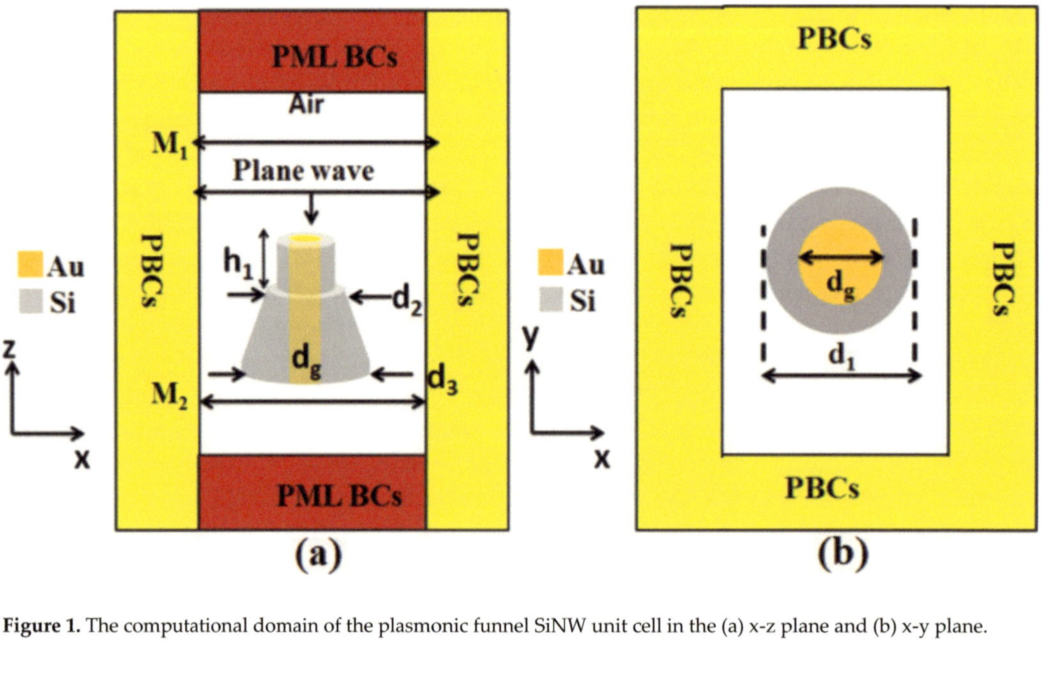

Figure 1. The computational domain of the plasmonic funnel SiNW unit cell in the (a) x-z plane and (b) x-y plane.

1/60 of the shortest wavelength. The proposed plasmonic design is specifically illuminated from the top by a plane wave ranging from 300 to 1100 nm, which covers the main part of the solar spectrum. The incident wave is normal to the x-y plane with the electric field polarized along the x-axis as shown in **Figure 1(a)**. In this study, we only consider lightly doped silicon material. At frequencies above the bandgap, the difference in the optical constants between lightly doped and intrinsic silicon is negligible, which gives us the convenience of using the same optical constants for both p and n regions of the core-shell structure in silicon NW SC [59]. The refractive indices of the SiNWs are taken from experimentally measured optical constants from Ref. [73]. The plasmonic funnel SiNW design shown in **Figure 1** consists of silicon cylinder of height h_1 and a cone of height h_2. The total thickness is equal to h_1+h_2 which is fixed to 2.33 μm that is comparable to the thickness of thin-film solar cell. The Si cylinder has a diameter d_1 with a gold core of diameter d_g, while the cone has upper and lower base diameters of d_2 and d_3, respectively. In this investigation, the NWs are arranged in a square lattice with a periodicity of $\Lambda = 500$ nm.

The amount of the total absorbed power by the suggested design can be calculated as a function of the wavelength λ using the total reflection R (λ) and total transmission T (λ) according to the following relation:

$$A(\lambda) = 1\text{-}R(\lambda)\text{-}T(\lambda) \tag{1}$$

Further, the absorbed power by the SiNWs only is calculated as

$$\alpha(\lambda) = \mathbf{A}(\lambda) - \mathbf{Ag}(\lambda) \tag{2}$$

where Ag(λ) is the absorbed power by the plasmonic material. The wavelength-dependent reflection and transmission are calculated by using two frequency domain monitors (M_1) and (M_2), placed above and below the unit cell, respectively, as shown in **Figure 1(a)**. In order to have a convergent absorption, the source and reflection monitors are placed above the SiNWs at a distance 400 and 600 nm, respectively. However, the transmission monitor is placed below the structure by 300 nm. The ultimate efficiency (η) defined as a figure of merit is used to evaluate the absorption capability of the proposed structures for solar cell applications. The ultimate efficiency is calculated from the following equation [4, 8]:

$$\eta = \frac{\int_{300\,\mathrm{nm}}^{\lambda_g} F_s(\lambda)\, \alpha(\lambda) \frac{\lambda}{\lambda_g}\, d\lambda}{\int_{300\,\mathrm{nm}}^{4000\,\mathrm{nm}} F_s(\lambda)\, d\lambda} \tag{3}$$

where λ is the wavelength of the incident light, E_g is the bandgap wavelength of the Si, and $F_s(\lambda)$ is the photon flux density in the ASTM AM 1.5 solar spectrum [74]. In this study, λ_g is taken as 1100 nm, corresponding to the energy gap of the silicon material. The short-circuit current density (J_{sc}) is related to the ultimate efficiency by assuming perfect carrier collection efficiency, i.e., each photon absorbed with energy higher than the bandgap produces one electron-hole pair, and all generated carriers are collected to produce current without recombination losses [28, 29]. The short-circuit current density J_{sc} of the proposed design can be expressed as [29]

$$J_{sc} = \eta \frac{q\lambda_g}{hc} \int_{300\,\mathrm{nm}}^{4000\,\mathrm{nm}} F_s(\lambda)\, d\lambda = 81.83\eta (mA/cm^2) \tag{4}$$

where q is the electron charge, h is Planck's constant, η is ultimate efficiency, and c is the speed of light.

The open-circuit voltage can be related to the maximum short-circuit current density by [75]:

$$V_{oc} = \frac{K_B T_c}{q} \ln\left(1 + \frac{J_{sc}}{J_0}\right), \tag{5}$$

where K_B (ev/K) is the Boltzmann constant, T_c is the ambient temperature of the solar cell temperature that is assumed to be equal to the ambient temperature (i.e., 300 K), and J_0 is the dark saturation current density. The semiempirical expression for the dark saturation current density is given by [75]:

$$J_o = 1.5 \times 10^9 \ln\left(-\frac{E_g}{K_B T_c}\right), \tag{6}$$

The fill factor (FF) is defined as [75]

$$FF = \frac{v_{oc} - \ln(v_{oc} + 0.72)}{v_{oc} + 1}, \tag{7}$$

where $v_{oc} = \frac{V_{oc}}{(KT/q)}$,

To quantify the electrical performance of the proposed design, the power conversion efficiency (PCE) is calculated from the following equation:

$$PCE = \frac{v_{oc} J_{sc} FF}{I_{in}}, \tag{8}$$

where I_{in} is the incident power density at AM 1.5 and is equal to 900.14 w/m^2 [76].

Figure 2(a) depicts the absorption spectra of the cylindrical, conical, and proposed plasmonic funnel NWs versus the wavelength. It should be noted that the absorption of the suggested plasmonic design $\alpha(\lambda)$ shown in this figure is due to the Si only. In this study, d_1 = 300 nm, d_2 = 400 nm, d_g = 120 nm, d_3 = 500 nm, h_1 = 500 nm, and h_2 = 1830 nm. Further, the studied three shapes have the same total height of 2330 nm. It is revealed from this figure that the plasmonic funnel shape shows superior absorption than conical and cylindrical counterparts over the

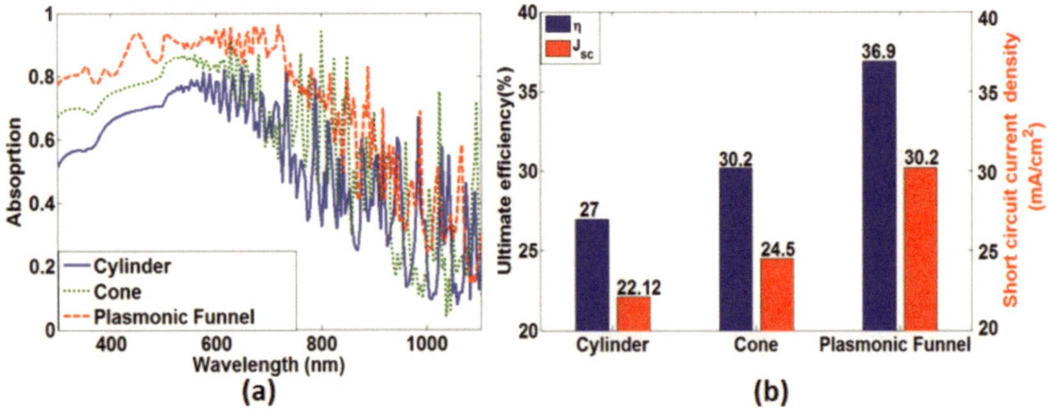

Figure 2. (a) Absorption spectra and (b) ultimate efficiency and short-circuit current density of the cylindrical, conical, and plasmonic funnel NWs.

entire wavelength band. **Figure 2(b)** shows the ultimate efficiency and short-circuit current density of the investigated NWs. The ultimate efficiencies of these structures are equal to 27, 30.2, and 36.9%, respectively. Therefore, it may be noted that the plasmonic funnel-shaped NWs offer an overall enhancement of 36.7 and 22.2% over the cylindrical and conical counterparts. Further, the funnel-shaped design has short-circuit current density of 30.2 mA/cm^2 which is greater than 24.5 and 22.12 mA/cm^2 of the conical and cylindrical designs. The enhancement in the ultimate efficiency and hence short-circuit current density apparently can be attributed to the combination between the modes supported by the upper cylinder and the lower tapered cone along with the plasmonic modes generated by the gold core. Additionally, the proposed design offers continuous reflections between the SiNWs until the incident angle becomes smaller than the critical angle resulting in high light trapping and broadband absorption. Furthermore, the absorption enhancement can be also explained from the waveguide modal dispersion of the NWs [29]. The leaky mode resonance relies on the NW diameter. Therefore, the use of plasmonic funnel-shaped NW with multiple diameters will increase the number of leaky mode resonances. The cylindrical part with small diameter d_1 can absorb short wavelength spectrum. However, the long wavelength spectrum is absorbed by the conical part with upper and lower base diameters d_2 and d_3, respectively. Additionally, the gold core generates plasmonic modes which play an important role for the broadband absorption.

To further confirm the underlying mechanism of the absorption enhancement, the absorbed field profiles inside the cylindrical, conical, and plasmonic funnel shapes have been presented in **Figure 3(a)–(c)**, respectively in x-z plane at λ = 688 nm. It is reveled from these figures that the incident light has more penetration and confinement through the proposed funnel shape with plasmonic core than the conventional conical and cylindrical Si structures. Therefore, the absorption and consequently the ultimate efficiency are enhanced.

In order to ensure that the absorption enhancement of the suggested design is due to the enhanced light trapping in the Si material, the absorption through the plasmonic core and the surrounding Si material are calculated and shown in **Figure 4**. The absorption of the Si NW funnel design [29] without plasmonic material is also plotted in **Figure 4**. It may be noted from this figure that the absorption through the plasmonic metal is small. Further, the plasmonic

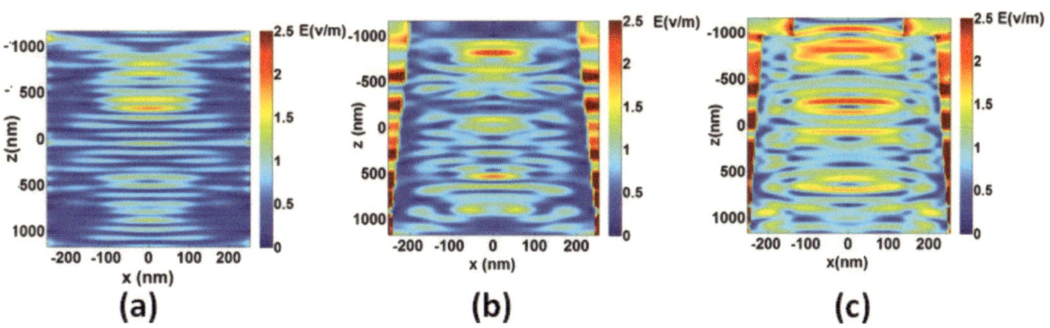

Figure 3. Absorbed field profiles of the (a) cylindrical, (b) conical, and (c) plasmonic funnel SiNWs in x-z plane at λ = 688 nm.

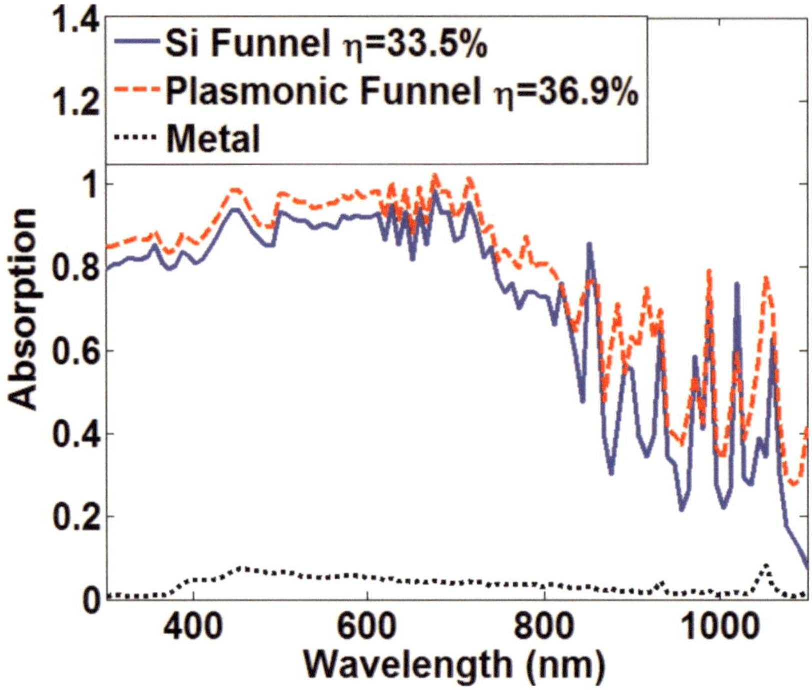

Figure 4. Absorption spectra through the Si funnel [29], metallic layer, and plasmonic funnel NWs.

funnel design has better absorption than that of the Si funnel design [29]. The simulation results also reveal that the plasmonic funnel-shaped NWs achieve an ultimate efficiency of 36.9% which is higher than that of the Si funnel counterpart of ultimate efficiency of 33.5%. The metallic core supports plasmonic modes which increase the light confinement through the silicon NWs. Therefore, the hybrid funnel nanowires improve the light trapping and hence the ultimate efficiency.

To understand the electrical enhancement in the reported design, the ultimate efficiency (η), short-circuit current density (J_{sc}), open-circuit voltage (V_{oc}), fill factor (FF), and power conversion efficiency (PCE) are calculated and shown in **Table 1**. In this table, the calculated values of η, J_{sc}, V_{oc}, FF, and PCE of the cylindrical [66], conical [70], Si funnel [29], and proposed plasmonic funnel SiNWs are listed in detail. It may be noted from this table that the V_{oc} and PCE of the plasmonic funnel shape are higher than that reported in Refs. [29, 66, 70]. Additionally, the PCE of the reported design is higher than that of the cylindrical SiNWs [66] and Si funnel [29] counterparts by 9.95 and 38.2%, respectively. This enhancement is attributed to the better absorption in the plasmonic funnel design which is responsible for enhancing the conversion efficiency. Therefore, the suggested design has a better conversion from optical to electrical power than the previously published design [29, 66, 70].

In order to enhance the absorption capabilities of the proposed funnel-shaped SiNWs with plasmonic core, the effects of the structure geometrical parameters are investigated. **Figure 5(a)** shows the effect of the top cylindrical diameter (d_1) on the ultimate efficiency

Structure	η	J_{sc} (mA/cm²)	V_{oc} (v)	FF	PCE
Cylindrical SiNWs [66]	27%	22.0941	0.6538	0.8380	13.6405
Conical SiNWs [70]	30.2%	24.7127	0.6567	0.8385	15.3247
Funnel SiNWs [29]	33.5%	27.4131	0.6594	0.8390	17.0687
Plasmonic funnel NWs	36.9%	30.1953	0.6619	0.8395	18.8723

Table 1. The ultimate efficiency (η), short-circuit current density (J_{sc}), open-circuit voltage (V_{oc}), fill factor (FF), and power conversion efficiency (PCE) of different structures.

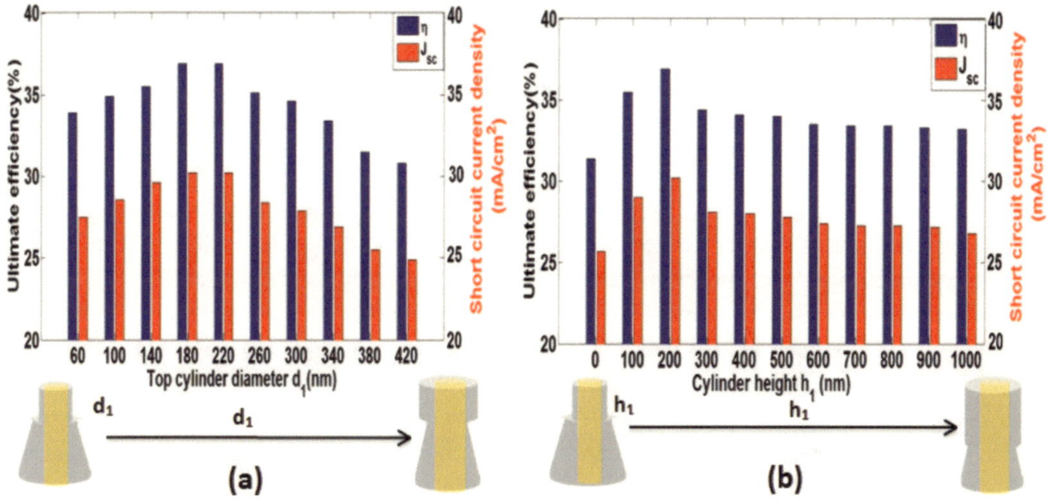

Figure 5. Variation of the short-circuit current density and ultimate efficiency with the (a) cylinder diameter d_1 and (b) cylinder height h_1.

and short-circuit current density for the suggested structure. In this investigation, the other parameters Λ, h_1, d_2, d_3, and h_2 are kept constants at 500, 500, 400, 500, and 1830 nm, respectively. As the diameter (d_1) increases from 60 to 180 nm, the absorption through the SiNWs and hence the ultimate efficiency increase. The ultimate efficiency and short-circuit current density have maximum values of 36.9% and 30.2 mA/cm² at d_1 = 180 nm, respectively. As d_1 is further increased, the reflection will be increased, and hence the absorption and ultimate efficiency are decreased. As a result, the short-circuit current density decreases from 30.2 to 24.9 mA/cm² by increasing d_1 from 180 to 420 nm, respectively. Therefore, the tuned d_1 = 180 nm will be used for the subsequent simulations. Next, the effect of the cylinder height h_1 of the suggested design is also investigated. As h_1 increases from 0 to 200 nm, the ultimate efficiency and short-circuit current density are increased. Therefore, it is concluded that the funnel shape has a better absorption capability than the conical design at $h_1 = 0$. It may be also seen from **Figure 5(b)** that the cylinder height has a slight effect on the short-circuit current density of the proposed design when $h_1 \geq 300$ nm. This is due to the fixed cylindrical diameter d_1 which faces the incident light. Therefore, the amount of trapped light inside the proposed design is slightly changed with h_1 variation. The maximum ultimate efficiency of 36.9% is obtained at $h_1 = 200$ nm.

Figure 6(a) displays the relation between the short-circuit current density and ultimate efficiency against the cone upper base diameter (d_2) for the suggested design. In this study, the tuned geometrical parameters of the plasmonic design are taken as $d_1 = 180$ nm, $d_3 = 500$, and $h_1 = 200$ nm. As d_2 increases from 160 to 360 nm, the active material increases, and hence higher order optical modes are generated. Therefore, the ultimate efficiency is increased from 33.3 to 36.96%. The tapered shape of the plasmonic design with $d_2 = 360$ nm has maximum ultimate efficiency of 36.9% and J_{sc} of 30.2 mA/cm^2. The effect of the cone base diameter (d_3) is also studied and shown in **Figure 6(b)**. It is revealed from **Figure 6(b)** that as the cone base diameter (d_3) of the plasmonic funnel design increases, the active material will be increased which decreases the transmission through the proposed design. Therefore, the absorption and ultimate efficiency are increased. An optimum ultimate efficiency of 36.9% and J_{sc} of 30.2 mA/cm^2 are obtained at 500 nm cone base diameter.

From the practical point of view, the absorption spectra of the funnel SiNWs with plasmonic core and cylindrical SiNWs decorated on 2 μm Si substrate and Ag back reflector of thickness 200 nm are calculated and shown in **Figure 7**. Compared with the cylindrical SiNWs of ultimate efficiency = 30.6%, the proposed plasmonic design shows an ultimate efficiency of 44% with an enhancements of 43.3% [7]. Additionally, short-circuit current densities of 36 mA/cm^2 and 25.1 mA/cm^2 are achieved by the plasmonic funnel design and cylindrical SiNWs, respectively. This is due to the coupling between the nanowire arrays and underlying substrate which leads to a strong absorption enhancement in the lower energy region as well as good light-trapping improvement in the longer wavelength as displayed in **Figure 7(a)**.

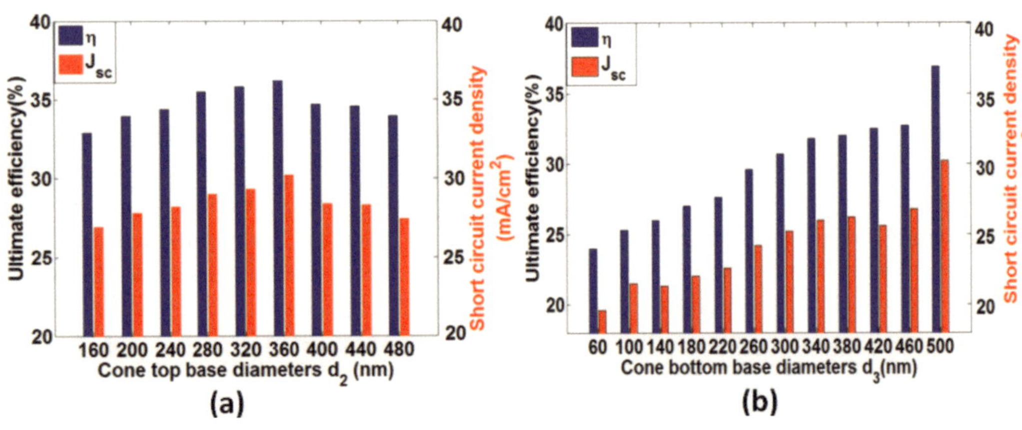

Figure 6. Variation of the short-circuit current density and ultimate efficiency with the cone (a) top diameter d_2 and (b) bottom diameter d_3.

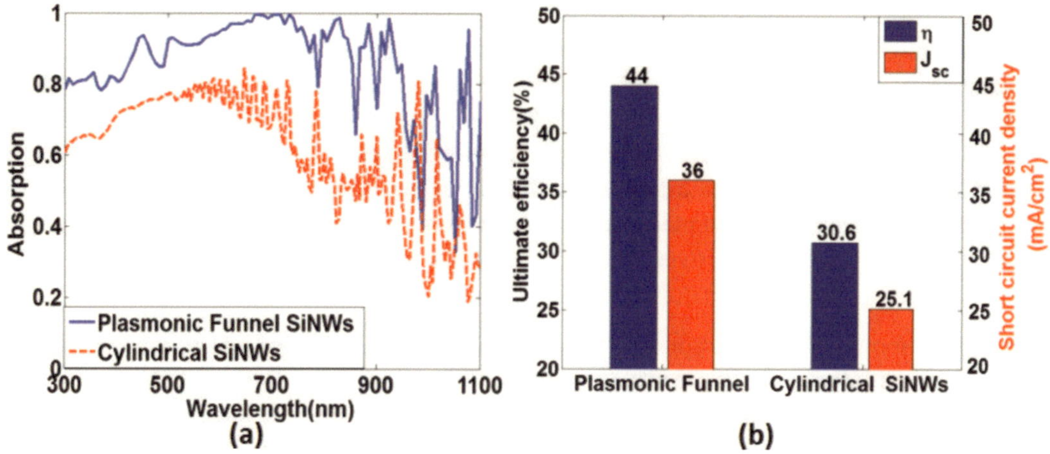

Figure 7. (a) Absorption spectra and (b) ultimate efficiency and short-circuit current density of the proposed plasmonic funnel SiNWs and cylindrical SiNWs with Si substrate and Ag back reflector.

4.2. SiNWs decorated by plasmonic diamond

In order to test the effect of plasmonic material over the NW structure, the cylindrical SiNWs decorated by Ag diamond is studied. **Figure 8(a)** illustrates the computational domain of the proposed plasmonic unit cell in the x-z plane, while **Figure 8(b)** shows the top view of the suggested design. The positive and negative z directions of the unit cell are surrounded by PML boundary conditions, while the unit cell is surrounded by periodic BCs along x

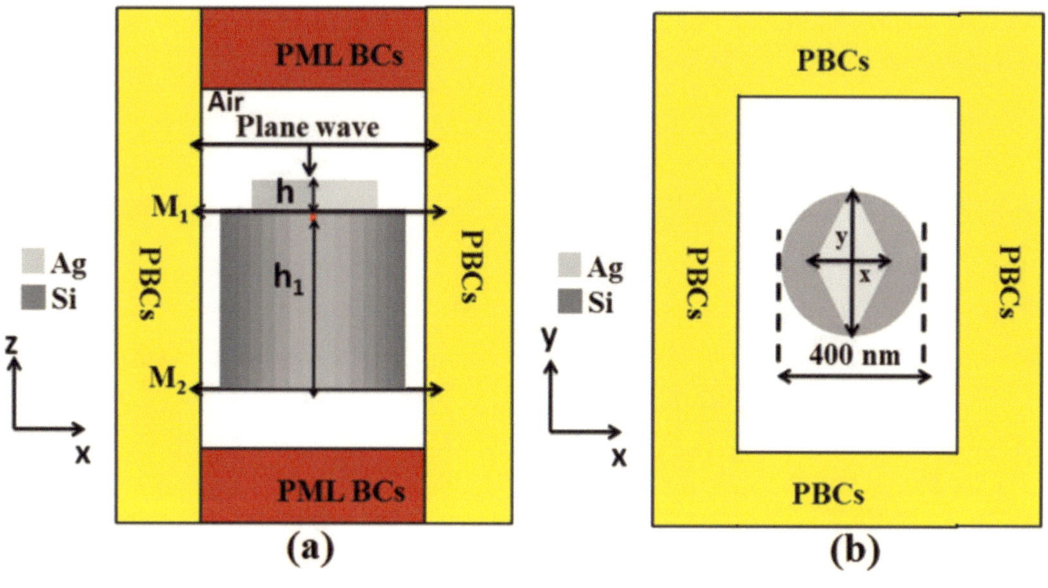

Figure 8. Computational domain of the proposed cylindrical SiNWs decorated by Ag diamond in the (a) x-z plane and (b) x-y plane.

and y directions to simulate periodic square SiNW array. The reported design of the silicon NWs decorated by Ag diamond is specifically illuminated from the top by a plane wave ranging from λ = 300 to 1100 nm, which covers the main part of the solar spectrum. Two frequency-domain field monitors M_1 and M_2 are placed at the top and bottom of the cell, respectively, to calculate the absorption in the SiNW which is defined by $A_{Si} = M_1 - M_2$. In this study, the height and the diameter of the Si nanowire are equal to 500 and 400 nm, respectively. Furthermore, the major and minor axes of the Ag diamond shape are taken as x and y, respectively. In this investigation, x = y = 320 nm, while the diamond thickness h is equal to 80 nm.

Figure 9(a) shows the absorption spectra of the cylindrical SiNWs and cylindrical SiNWs decorated by Ag diamond. It is revealed from this figure that the cylindrical SiNWs with Ag diamond show superior absorption than cylindrical counterpart over the entire wavelength band. Additionally, **Figure 9(b)** shows the ultimate efficiency and short-circuit current density of the studied designs. The ultimate efficiency of the plasmonic design is equal to 21.1% which is greater than the conventional cylindrical design of ultimate efficiency = 16%. Therefore, the cylindrical-shaped design with Ag diamond offers an overall enhancement of 31.9% over the cylindrical SiNWs counterpart. Further, the plasmonic design has a short-circuit current density of 17.3 mA/cm^2 which is greater than that of the conventional cylindrical design of 13.1 mA/cm^2. The enhancement in the ultimate efficiency and hence short-circuit current density can be attributed to the generated plasmonic mode by the upper Ag diamond which is responsible for improving the absorption through the SiNWs.

Table 2 shows the calculated η, J_{sc}, V_{oc}, FF, and PCE of the cylindrical SiNWs and cylindrical SiNWs decorated by Ag diamond. It is found that the electrical parameters are also enhanced due to the existing plasmonic materials above the active layer. Additionally, the V_{oc} is increased from 640 to 647 mV, and the PCE is also improved from 7.9 to 10.6 with 34.1% enhancement over the cylindrical SiNW counterpart.

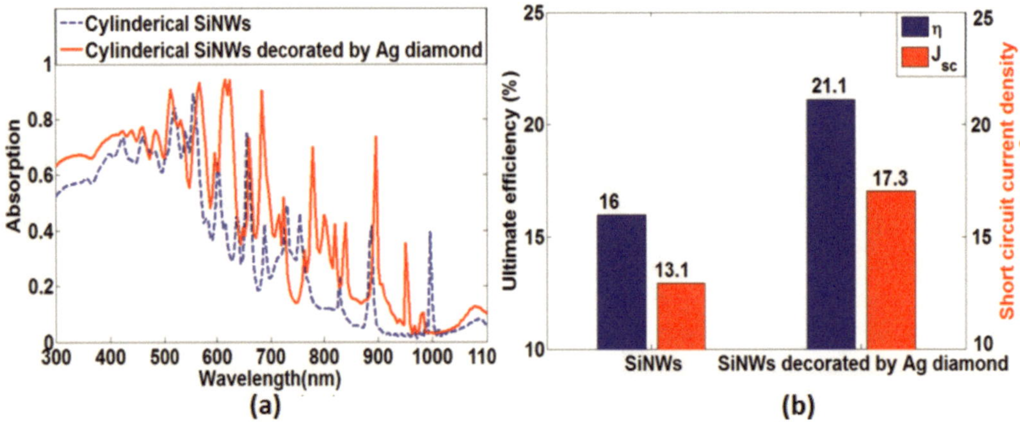

Figure 9. (a) Absorption spectra and (b) ultimate efficiency and short-circuit current density of cylindrical SiNWs and cylindrical SiNWs decorated by Ag diamond.

Structure	η	J_{sc} (mA/cm²)	V_{oc} (v)	FF	PCE
Cylindrical SiNWs	16%	13.1	0.64	0.835	7.9
Cylindrical SiNWs decorated with Ag diamond	21.1%	17.3	0.647	0.8369	10.6

Table 2. The calculated values of η, J_{sc}, V_{oc}, FF, and PCE for cylindrical SiNWs and cylindrical SiNWs decorated by Ag diamond.

To further explore the underlying mechanism of the absorption enhancement in the proposed design, the field profiles in x-y and x-z planes at $\lambda = 616$ nm are shown in **Figure 10**. As shown from **Figure 10(b) and (d)**, the field penetration through the proposed design is greater than that of the conventional SiNWs. This is due to the plasmonic particles that have increased the light confinement around the plasmonic material and, thus, allowed more waveguide modes to couple into the SiNWs which is responsible for the absorption enhancement. Additionally, the figures show high field intensity penetration inside the reported SiNWs which is defined by the field hot spots in the figures.

In order to further enhance the absorption capabilities of the proposed SiNWs decorated by Ag diamond, the effects of the structure geometrical parameters are studied. **Figure 11(a)** shows the impact of the diamond minor axis (x) on the ultimate efficiency and short-circuit current density. In this investigation, the other parameters Λ, h_1, and y are kept constant at 500, 500, 400, and 320 nm, respectively. It is concluded that as the minor axis (x) increases to 280 nm, the absorption and hence the ultimate efficiency and short-circuit current density also increase. A maximum ultimate efficiency and short-circuit current density of 21.4% and 17.5 mA/cm², receptively, are obtained at x = 280 nm. If the minor axis is further increased, the ultimate efficiency and short-circuit current density will be decreased. Therefore, the tuned value of 280 nm for the minor axis will be used for the subsequent simulations. Next, the effect of the major axis y of the Ag diamond is also investigated and shown in **Figure 11(b)**. It may be seen from this figure that the ultimate efficiency and short-circuit current density have maximum values of 21.9% and 17.1% at y = 360 nm, respectively. If the major axis y is further increased, the reflection will be increased which decreases the absorption and ultimate efficiency as shown in **Figure 11(b)**.

Next, the effect of the diamond thickness (h) on the ultimate efficiency and short-circuit density is also investigated and shown in **Figure 12(a)**. It is revealed from this figure that the diamond height has a slight effect on the absorption and hence short-circuit current density and ultimate efficiency. A maximum ultimate and short-circuit current density of 22.5% and 18.4 mA/cm², respectively, occur at h = 120 nm. **Figure 12(b)** shows the variation of ultimate efficiency and short-circuit density with the periodicity of the NW, while the other parameters are kept constant at h = 120 nm, h_1 = 500 nm, x = 280 nm, and y = 360 nm. As the periodicity decreases, the ultimate efficiency and short-circuit current density increase. The maximum ultimate efficiency of 25.7% is obtained at periodicity of 400 nm which is higher than that of the conventional SiNWs without plasmonic material by 63% at the same periodicity. This enhancement is due to the change of the periodicity which leads to a change in the filling ratio (at fixed diamond size) which in turn influences the generated localized surface plasmon resonance.

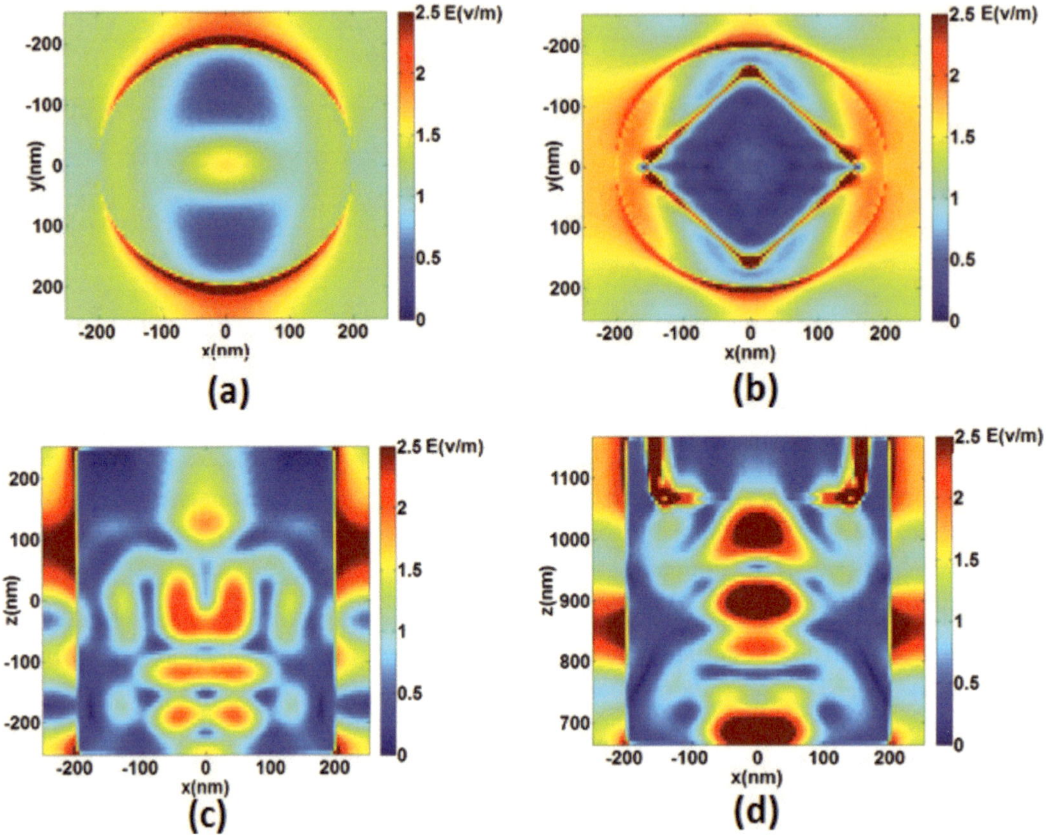

Figure 10. Absorbed field profiles for cylindrical SiNWs without and with Ag diamond particle in the (a, b) x-y plane and in the (c, d) x-z plane at $\lambda = 616$ nm.

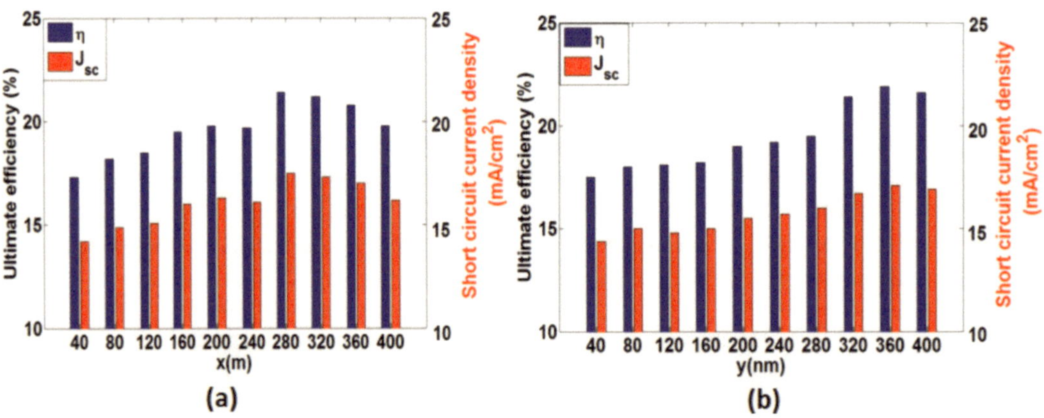

Figure 11. Variation of the short-circuit current density and ultimate efficiency of the conventional SiNWs and SiNWs decorated by Ag diamond with the (a) minor axis x and (b) major axis y.

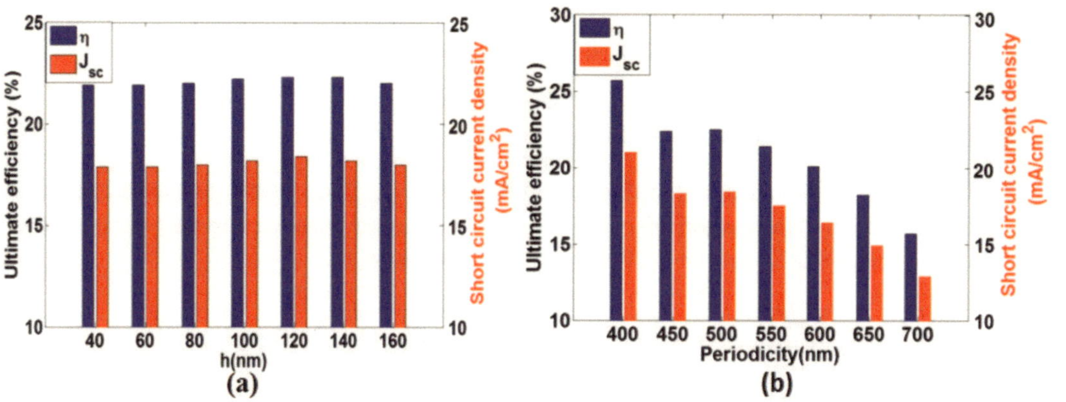

Figure 12. Variation of the short-circuit current density and ultimate efficiency for the conventional SiNWs and SiNWs decorated by Ag diamond with the (a) diamond height and (b) periodicity.

5. Conclusion

In this chapter, the possibility of improving the light trapping through the SiNWs by using plasmonic materials is discussed and studied. In this regard, the effects of using plasmonic materials inside or above the SiNWs are introduced. In these investigations, two novel designs of plasmonic SiNW solar cells are presented and analyzed using the 3D FDTD method. A full study for optical and analytical electrical efficiencies has been carried out. The ultimate efficiency and power conversion efficiency have been calculated to quantify the overall performance of the proposed designs. Further, the designed geometrical parameters are tuned to maximize the optical and electrical efficiencies. The obtained ultimate efficiency and power conversion efficiency of the plasmonic funnel design are equal to 44 and 18.9% with an enhancement of 43.3% over its cylindrical NW counterpart. Additionally, the cylindrical SiNWs decorated by Ag diamond offer an ultimate efficiency and short-circuit current density of 25.7% and 21.03 mA/cm^2, respectively, with an improvement of 63% over the conventional SiNWs. The proposed plasmonic solar cell designs could be very crucial to present low-cost, high-absorption efficiency, and shape-controlled nanopillars for energy-harvesting applications.

Author details

Mohamed Hussein[1,2], Mohamed Farhat Othman Hameed[1,3] and Salah S. A. Obayya[1,3]*

*Address all correspondence to: sobayya@zewailcity.edu.eg

1 Centre of Photonics and Smart Materials, Zewail City of Science and Technology, Giza, Egypt

2 Faculty of Science, Ain Shams University, Cairo, Egypt

3 Faculty of Engineering, Mansoura University, Mansoura, Egypt

References

[1] G. Sadashivappa, and N. Sharvari, "Nanoantenna-a review," International Journal of Renewable Energy Technology Research, Vol. 4, pp. 1-9, 2015.

[2] M. A. Khan, P. Sichanugrist, S. Kato, and Y. Ishikawa, "Theoretical investigation about the optical characterization of cone-shaped pin-Si nanowire for top cell application," Energy Science & Engineering, Vol. 4, pp. 383-393, 2016.

[3] L. Mescia, and A. Massaro, "New trends in energy harvesting from earth long-wave infrared emission," Advances in Materials Science and Engineering, Vol. 2014, p. 10, 2014.

[4] F. Manzano-Agugliaro, A. Alcayde, F. G. Montoya, A. Zapata-Sierra, and C. Gil, "Scientific production of renewable energies worldwide: an overview," Renewable and Sustainable Energy Reviews, Vol. 18, pp. 134-143, 2//2013.

[5] A. Arigliano, P. Caricato, A. Grieco, and E. Guerriero, "Producing, storing, using and selling renewable energy: the best mix for the small medium industry," Computers in Industry, Vol. 65, pp. 408-418, 4//2014.

[6] O. D. Miller, "Photonic Design: From Fundamental Solar Cell Physics to Computational Inverse Design," ed, 2012 arXiv:1308.0212 [physics.optics].

[7] W. Wang, J. Zhang, X. Che, and G. Qin, "Large absorption enhancement in ultrathin solar cells patterned by metallic nanocavity arrays," Scientific Reports, Vol. 6, p. 34219, 10/05/online 2016.

[8] W. Raja, A. Bozzola, P. Zilio, E. Miele, S. Panaro, H. Wang ,A. Toma, A. Alabastri, F. Angelis and R. Zaccaria, et al., "Broadband absorption enhancement in plasmonic nanoshells-based ultrathin microcrystalline-Si solar cells," Scientific Reports, Vol. 6, p. 24539, 04/15 11/15/received 03/31/accepted 2016.

[9] S. Foster, and S. John, "Light-trapping design for thin-film silicon-perovskite tandem solar cells," Journal of Applied Physics, Vol. 120, p. 103103, 2016/09/14 2016.

[10] D. M. Powell, M. T. Winkler, H. J. Choi, C. B. Simmons, D. B. Needleman, and T. Buonassisi, "Crystalline silicon photovoltaics: a cost analysis framework for determining technology pathways to reach baseload electricity costs," Energy & Environmental Science, Vol. 5, pp. 5874-5883, 2012.

[11] O. El Daif, E. Drouard, G. Gomard, A. Kaminski, A. Fave, M. Lemiti, et al., "Absorbing one-dimensional planar photonic crystal for amorphous silicon solar cell," Optics Express, Vol. 18, pp. A293–A299, 2010/09/13 2010.

[12] J. Grandidier, D. M. Callahan, J. N. Munday, and H. A. Atwater, "Gallium arsenide solar cell absorption enhancement using whispering gallery modes of dielectric nanospheres," IEEE Journal of Photovoltaics, Vol. 2, pp. 123-128, 2012.

[13] B. Li, J. Liu, G. Xu, R. Lu, L. Feng, and J. Wu, "Development of pulsed laser deposition for CdS/CdTe thin film solar cells," Applied Physics Letters, Vol. 101, p. 153903, 2012/10/08 2012.

[14] P. J. Carrington, M. C. Wagener, J. R. Botha, A. M. Sanchez, and A. Krier, "Enhanced infrared photo-response from GaSb/GaAs quantum ring solar cells," Applied Physics Letters, Vol. 101, p. 231101, 2012/12/03 2012.

[15] J. Cao, Z. Zhan, L. Hou, Y. Long, P. Liu, and W. Mai, "Optical modeling of organic solar cells based on rubrene and C70," Applied Optics, Vol. 51, pp. 5718-5723, 2012/08/10 2012.

[16] D. Shahrjerdi, S. W. Bedell, C. Ebert, C. Bayram, B. Hekmatshoar, K. Fogel, et al., "High-efficiency thin-film InGaP/InGaAs/Ge tandem solar cells enabled by controlled spalling technology," Applied Physics Letters, Vol. 100, p. 053901, 2012/01/30 2012.

[17] A. G. Imenes, and D. R. Mills, "Spectral beam splitting technology for increased conversion efficiency in solar concentrating systems: a review," Solar Energy Materials and Solar Cells, Vol. 84, pp. 19-69, 10//2004.

[18] A. Polman, and H. A. Atwater, "Photonic design principles for ultrahigh-efficiency photovoltaics," Nature Materials, Vol. 11, pp. 174-177, 03//print 2012.

[19] J. A. Dobrowolski, D. Poitras, P. Ma, H. Vakil, and M. Acree, "Toward perfect antireflection coatings: numerical investigation," Applied Optics, Vol. 41, pp. 3075-3083, 2002/06/01 2002.

[20] P. Singh, S. N. Sharma, and N. M. Ravindra, "Applications of porous silicon thin films in solar cells and biosensors," JOM, Vol. 62, pp. 15-24, 2010.

[21] Z. Xi, D. Yang, and D. Que, "Texturization of monocrystalline silicon with tribasic sodium phosphate," Solar Energy Materials and Solar Cells, Vol. 77, pp. 255-263, 5/30/2003.

[22] D. Iencinella, E. Centurioni, R. Rizzoli, and F. Zignani, "An optimized texturing process for silicon solar cell substrates using TMAH," Solar Energy Materials and Solar Cells, Vol. 87, pp. 725-732, 5//2005.

[23] P. Papet, O. Nichiporuk, A. Kaminski, Y. Rozier, J. Kraiem, J. F. Lelievre, et al., "Pyramidal texturing of silicon solar cell with TMAH chemical anisotropic etching," Solar Energy Materials and Solar Cells, Vol. 90, pp. 2319-2328, 9/22/2006.

[24] S. Hayashi, T. Minemoto, H. Takakura, and Y. Hamakawa, "Influence of texture feature size on spherical silicon solar cells," Rare Metals, Vol. 25, pp. 115-120, 2006/10/01 2006.

[25] P. Campbell, and M. A. Green, "Light trapping properties of pyramidally textured surfaces," Journal of Applied Physics, Vol. 62, pp. 243-249, 1987/07/01 1987.

[26] S. C. Baker-Finch, and K. R. McIntosh, "Reflection of normally incident light from silicon solar cells with pyramidal texture," Progress in Photovoltaics: Research and Applications, Vol. 19, pp. 406-416, 2011.

[27] J. Zhao, A. Wang, M. A. Green, and F. Ferrazza, "19.8% efficient "honeycomb" textured multicrystalline and 24.4% monocrystalline silicon solar cells," Applied Physics Letters, Vol. 73, pp. 1991-1993, 1998/10/05 1998.

[28] M. Hussein, M. F. O. Hameed, N. F. F. Areed, and S. S. A. Obayya, "Ultra-high efficient solar cell based on decagonal arrays of silicon nanowires," Optical Engineering, Vol. 53, pp. 117105-117105, 2014.

[29] M. Hussein, M. F. O. Hameed, N. F. F. Areed, A. Yahia, and S. S. A. Obayya, "Funnel-shaped silicon nanowire for highly efficient light trapping," Optics Letters, Vol. 41, pp. 1010-1013, 2016/03/01 2016.

[30] Y.-C. Lee, C.-F. Huang, J.-Y. Chang, and M.-L. Wu, "Enhanced light trapping based on guided mode resonance effect for thin-film silicon solar cells with two filling-factor gratings," Optics Express, Vol. 16, pp. 7969-7975, 2008/05/26 2008.

[31] H. Nakaya, M. Nishida, Y. Takeda, S. Moriuchi, T. Tonegawa, T. Machida, et al., "Polycrystalline silicon solar cells with V-grooved surface," Solar Energy Materials and Solar Cells, Vol. 34, pp. 219-225, 1994/09/01 1994.

[32] X. D. Wang, E. Graugnard, J. S. King, Z. L. Wang, and C. J. Summers, "Large-scale fabrication of ordered nanobowl arrays," Nano Letters, Vol. 4, pp. 2223-2226, 2004/11/01 2004.

[33] J. W. Leem, D. H. Joo, and J. S. Yu, "Biomimetic parabola-shaped AZO subwavelength grating structures for efficient antireflection of Si-based solar cells," Solar Energy Materials and Solar Cells, Vol. 95, pp. 2221-2227, 8//2011.

[34] J. W. Leem, J. S. Yu, Y. M. Song, and Y. T. Lee, "Antireflective characteristics of disordered GaAs subwavelength structures by thermally dewetted Au nanoparticles," Solar Energy Materials and Solar Cells, Vol. 95, pp. 669-676, 2//2011.

[35] H. A. Atwater, and A. Polman, "Plasmonics for improved photovoltaic devices," Nature Materials, Vol. 9, pp. 205-213, 03//print 2010.

[36] S. Nie, and S. R. Emory, "Probing single molecules and single nanoparticles by surface-enhanced Raman scattering," Science, Vol. 275, p. 1102, 1997.

[37] S. A. Maier, M. L. Brongersma, P. G. Kik, S. Meltzer, A. A. G. Requicha, and H. A. Atwater, "Plasmonics—a route to nanoscale optical devices," Advanced Materials, Vol. 13, pp. 1501-1505, 2001.

[38] X. D. Hoa, A. G. Kirk, and M. Tabrizian, "Towards integrated and sensitive surface plasmon resonance biosensors: a review of recent progress," Biosensors and Bioelectronics, Vol. 23, pp. 151-160, 9/30/2007.

[39] M. Hussein, N. F. F. Areed, M. F. O. Hameed, and S. S. A. Obayya, "Design of flower-shaped dipole nano-antenna for energy harvesting," IET Optoelectronics, Vol. 8, pp. 167-173, 2014.

[40] O. Salah, A. Nihal Fayez Fahmy, O. H. Mohamed Farhat, and A. Mohamed Hussein, "Optical nano-antennas for energy harvesting," in Innovative Materials and Systems for Energy Harvesting Applications, M. Luciano, L. Onofrio, and P. Francesco, Eds., Hershey, PA, USA: IGI Global, 2015, pp. 26-62.

[41] Y. M. El-Toukhy, M. Hussein, M. F. O. Hameed, A. M. Heikal, M. M. Abd-Elrazzak, and S. S. A. Obayya, "Optimized tapered dipole nanoantenna as efficient energy harvester," Optics Express, Vol. 24, pp. A1107–A1122, 2016/07/11 2016.

[42] K. Nakayama, K. Tanabe, and H. A. Atwater, "Plasmonic nanoparticle enhanced light absorption in GaAs solar cells," Applied Physics Letters, Vol. 93, p. 121904, 2008.

[43] D. Derkacs, S. H. Lim, P. Matheu, W. Mar, and E. T. Yu, "Improved performance of amorphous silicon solar cells via scattering from surface plasmon polaritons in nearby metallic nanoparticles," Applied Physics Letters, Vol. 89, p. 093103, 2006.

[44] H. R. Stuart, and D. G. Hall, "Island size effects in nanoparticle-enhanced photodetectors," Applied Physics Letters, Vol. 73, pp. 3815-3817, 1998.

[45] D. M. Schaadt, B. Feng, and E. T. Yu, "Enhanced semiconductor optical absorption via surface plasmon excitation in metal nanoparticles," Applied Physics Letters, Vol. 86, p. 063106, 2005.

[46] S. Pillai, K. R. Catchpole, T. Trupke, and M. A. Green, "Surface plasmon enhanced silicon solar cells," Journal of Applied Physics, Vol. 101, p. 093105, 2007.

[47] Z. Ouyang, X. Zhao, S. Varlamov, Y. Tao, J. Wong, and S. Pillai, "Nanoparticle-enhanced light trapping in thin-film silicon solar cells," Progress in Photovoltaics: Research and Applications, Vol. 19, pp. 917-926, 2011.

[48] F. J. Beck, S. Mokkapati, and K. R. Catchpole, "Plasmonic light-trapping for Si solar cells using self-assembled, Ag nanoparticles," Progress in Photovoltaics: Research and Applications, Vol. 18, pp. 500-504, 2010.

[49] H. Tan, R. Santbergen, A. H. Smets, and M. Zeman, "Plasmonic light trapping in thin-film silicon solar cells with improved self-assembled silver nanoparticles," Nano Letters, Vol. 12, pp. 4070-4076, 2012.

[50] L. Lu, Z. Luo, T. Xu, and L. Yu, "Cooperative plasmonic effect of Ag and Au nanoparticles on enhancing performance of polymer solar cells," Nano Letters, Vol. 13, pp. 59-64, 2012.

[51] A. P. Kulkarni, K. M. Noone, K. Munechika, S. R. Guyer, and D. S. Ginger, "Plasmon-enhanced charge carrier generation in organic photovoltaic films using silver nanoprisms," Nano Letters, Vol. 10, pp. 1501-1505, 2010.

[52] I. Ding, J. Zhu, W. Cai, S. J. Moon, N. Cai, P. Wang, et al., "Plasmonic dye-sensitized solar cells," Advanced Energy Materials, Vol. 1, pp. 52-57, 2011.

[53] C. Hägglund, M. Zäch, and B. Kasemo, "Enhanced charge carrier generation in dye sensitized solar cells by nanoparticle plasmons," Applied Physics Letters, Vol. 92, p. 013113, 2008.

[54] B. P. Rand, P. Peumans, and S. R. Forrest, "Long-range absorption enhancement in organic tandem thin-film solar cells containing silver nanoclusters," Journal of Applied Physics, Vol. 96, pp. 7519-7526, 2004.

[55] A. J. Morfa, K. L. Rowlen, T. H. R. III, M. J. Romero, and J. V. D. Lagemaat, "Plasmon-enhanced solar energy conversion in organic bulk heterojunction photovoltaics," Applied Physics Letters, Vol. 92, p. 013504, 2008.

[56] F.-C. Chen, J.-L. Wu, C.-L. Lee, Y. Hong, C.-H. Kuo, and M. H. Huang, "Plasmonic-enhanced polymer photovoltaic devices incorporating solution-processable metal nanoparticles," Applied Physics Letters, Vol. 95, p. 013305, 2009.

[57] R. M. Ferdous, A. W. Reza, and M. F. Siddiqui, "Renewable energy harvesting for wireless sensors using passive RFID tag technology: a review," Renewable and Sustainable Energy Reviews, Vol. 58, pp. 1114-1128, 2016.

[58] J. Nelson, The Physics of Solar Cells. Vol. 1: World Scientific, Imperial College, UK, 2003.

[59] L. Hu, and G. Chen, "Analysis of optical absorption in silicon nanowire arrays for photovoltaic applications," Nano Letters, Vol. 7, pp. 3249-3252, 2007.

[60] L. Cao, P. Fan, A. P. Vasudev, J. S. White, Z. Yu, W. Cai, et al., "Semiconductor nanowire optical antenna solar absorbers," Nano Letters, Vol. 10, pp. 439-445, 2010.

[61] C. Lin, and M. L. Povinelli, "Optical absorption enhancement in silicon nanowire and nanohole arrays for photovoltaic applications," in Proc. SPIE 7772, Next Generation (Nano) Photonic and Cell Technologies for Solar Energy Conversion, 77721G (August 24, 2010); doi:10.1117/12.860254.

[62] C. Lin, and M. L. Povinelli, "Optical absorption enhancement in silicon nanowire arrays with a large lattice constant for photovoltaic applications," Optics Express, Vol. 17, pp. 19371-19381, 2009.

[63] C. Lin, and M. L. Povinelli, "Optimal design of aperiodic, vertical silicon nanowire structures for photovoltaics," Optics Express, Vol. 19, pp. A1148–A1154, 2011.

[64] J. Li, H. Yu, S. M. Wong, X. Li, G. Zhang, P. G.-Q. Lo, et al., "Design guidelines of periodic Si nanowire arrays for solar cell application," Applied Physics Letters, Vol. 95, p. 243113, 2009.

[65] O. L. Muskens, J. G. Rivas, R. E. Algra, E. P. Bakkers, and A. Lagendijk, "Design of light scattering in nanowire materials for photovoltaic applications," Nano Letters, Vol. 8, pp. 2638-2642, 2008.

[66] H. Bao, and X. Ruan, "Optical absorption enhancement in disordered vertical silicon nanowire arrays for photovoltaic applications," Optics Letters, Vol. 35, pp. 3378-3380, 2010.

[67] L. Hong, Rusli, X. Wang, H. Zheng, H. Wang, and H. Yu, "Design guidelines for slanting silicon nanowire arrays for solar cell application," Journal of Applied Physics, Vol. 114, p. 084303, 2013.

[68] R. Ren, Y.-X. Guo, and R.-H. Zhu, "Enhanced absorption in elliptical silicon nanowire arrays for solar energy harvesting," Optical Engineering, Vol. 53, pp. 027102-027102, 2014.

[69] B. Wang, E. Stevens, and P. W. Leu, "Strong broadband absorption in GaAs nanocone and nanowire arrays for solar cells," Optics Express, Vol. 22, pp. A386–A395, 2014.

[70] B. Wang, and P. W. Leu, "Enhanced absorption in silicon nanocone arrays for photovoltaics," Nanotechnology, Vol. 23, p. 194003, 2012.

[71] M.-D. Ko, C.-K. Baek, T. Rim, S. Park, and Y.-H. Jeong, "Optical and electrical characteristics of asymmetric nanowire solar cells," Journal of Applied Physics, Vol. 111, p. 073102, 2012.

[72] F. Lumerical, "Solution, FDTD Solutions 6.5," ed. http://www.lumerical.com/tcad-products/fdtd/

[73] E. D. Palik, Handbook of Optical Constants of Solids Vol. 3: Academic press, Elsevier Inc. 1998.

[74] NREL. Reference Solar Spectral Irradiance: ASTM G-173. Available: http://rredc.nrel.gov/solar/spectra/am1.5/astmg173/astmg173.html. 2016.

[75] Z. Jia, Q. Cheng, J. Song, M. Si, and Z. Luo, "Optical properties of a grating-nanorod assembly structure for solar cells," Optics Communications, Vol. 376, pp. 14-20, 2016.

[76] P. R. Pudasaini, and A. A. Ayon, "Nanostructured thin film silicon solar cells efficiency improvement using gold nanoparticles," Physica Status Solidi (a), Vol. 209, pp. 1475-1480, 2012.

Chapter 11

A Comparative Study of Nanowire Arrays for Maximum Power Transmission

Hasan Aykut Şatana, Barışcan Karaosmanoğlu and Özgür Ergül

Additional information is available at the end of the chapter

http://dx.doi.org/10.5772/67447

Abstract

In this chapter, we present a comparative study of nanowire arrays for the purpose of transmitting electromagnetic energy to long distances with minimum loss. Silver nanowires at an infrared frequency are considered as a case study, using an accurate simulation environment based on the surface-integral equations and the multilevel fast multipole algorithm (MLFMA). Reliable numerical results are obtained by considering arrays as three-dimensional structures with finite sizes in all dimensions. Nanowires with different cross sections are compared in alternative array configurations to assess and compare their transmission properties. While there are no significantly varying performances for the arrays of few elements, large discrepancies occur as the number of nanowires increases. We show that arrays involving nanowires with hexagonal cross sections in hexagonal arrangements can provide significantly better power transmissions in comparison to others. We also present the superiority of a particular case, where the nanowires and gaps between them, both with hexagonal cross sections, are equally distributed, leading to a honeycomb structure. This type of structures demonstrates high-quality transmissions that can be useful in diverse application areas, such as optical coupling, subwavelength imaging, and energy harvesting.

Keywords: nanowire arrays, optical power transmission, surface integral equations, multilevel fast multiple algorithm

1. Introduction

Nanowires are excellent tools for guiding electromagnetic waves with minimum loss at optical frequencies [1–4] particularly as an alternative to dielectric structures at nanometer scales. Plasmonic properties of metals, such as silver (Ag) and gold (Au), enable high-quality power transmissions along nanowires via surface plasmon polaritons [5]. In addition to their favorable characteristics that fully exploit the benefits of the surface plasmons to transfer electromagnetic energy to long distances [6, 7], nanowires are relatively easier to fabricate, making them popular in diverse optics applications, such as optical coupling [4], subwavelength imaging [8–10], microscopy and sensing [11–13], and energy harvesting [14, 15]. While the first samples of fabricated nanowires in the literature are often deformed with undesired defects on their surfaces, rapid advances in nanotechnology have enabled the fabrication of progressively improved structures with desired geometric features. Recent studies demonstrate not only isolated nanowires with smooth surfaces but also their periodic arrangements in regular and highly ordered forms [16, 17].

As in all subareas of nanotechnology, experimental studies on nanowires are supported by numerical and computational analysis [18, 19]. The plasmonic properties of metals at optical frequencies can be modeled using tabulated or formulated permittivity values [20] inserted into the conventional solvers for penetrable bodies [21]. Accurate simulation environments based on various methods and techniques can provide an ability to investigate the effects of the cross-sectional shapes [22–24] and the choice of the material [22] on the nanowire characteristics before their fabrications. For plasmonic problems, not only limited to nanowires, computational methods based on the differential equations are very popular in the literature due to their availability as commercial and noncommercial software. On the other hand, for nanowire simulations, these methods may not enable a rigorous analysis, as the nanowire arrays are typically long in terms of wavelength, while they must be modeled as finite to accurately quantify their electromagnetic responses in most cases. In addition, in array configurations, several nanowires must be modeled together by considering all interactions and coupling between them, without employing any periodicity and/or infinity. For this purpose, integral-equation formulations, particularly the surface-integral equations [25–31] that need discretizations only on the surfaces of the structures, are very suitable for the computational analysis of nanowires, as well as similar structures.

In parallel to the improved capabilities in nanotechnology, recent studies focus on geometric properties (e.g., cross section and length) of nanowires to maximize their performances in the related applications [32, 33]. On the other hand, the best geometric properties of the nanowires, especially their cross sections, and their arrangements in array configurations for maximizing the power output in transmission lines are not well known. In this chapter, we present a comparative study of nanowire arrays in terms of power transmission, that is, transfer of the electromagnetic energy with minimum loss. For numerical tests, we particularly consider transmission properties of Ag nanowires at an infrared frequency (250 THz) and compare various cross-sectional geometries in alternative arrangements. An accurate solver based on surface-integral equations and the multilevel fast multipole algorithm (MLFMA) [34, 35] is used to analyze nanowires as three-dimensional finite structures with full parametric modeling of metals using measured values of the electrical permittivity. Surface formulations lead to dense matrix solutions, whose solutions can be challenging due to the ill-conditioned nature of the problems, large negative permittivity values at optical frequencies, and large dimensions in terms of the operation wavelength. These challenges are solved by using MLFMA in an inner/outer iterative scheme. With the high efficiency and accuracy of the developed solver, we analyze realistic structures by including all mutual couplings between the interacting nanowires. The results demonstrate interesting properties of hexagonal arrangements of hexagonal cells, leading to honeycomb structures, which demonstrate much better transmission capabilities in comparison to others when the arrays are relatively large. In the literature, it is shown that such hexagonal arrangements of nanowires can indeed be fabricated [13, 17], while we also show that the periodicity and the size of the nanowires (that may bring additional challenges in the fabrication) are also critical on the transmission performances of the arrays.

This chapter is organized as follows. First, in Section 2, we briefly present the simulation environment used in this study with some details of the major components. Parameters and variables of the nanowire simulations are discussed in Section 3. Then, Section 4 includes numerical results and comparisons, followed by the discussion and concluding remarks in Sections 5 and 6.

2. Simulation environment

The developed simulation environment for the analysis of plasmonic problems, such as nanowires at optical frequencies, is extensively discussed in Ref. [36]. In this section, we consider the major components of the solver for accurate and efficient analysis of nanowires and their arrays.

2.1. Modified combined tangential formulation

Consider a homogeneous object (nanowire or nanowire array) located in free space (vacuum) and illuminated by time-harmonic sources described by incident electric and magnetic fields ($\boldsymbol{E}^{\text{inc}}$ and $\boldsymbol{H}^{\text{inc}}$). The object is assumed to be nonmagnetic with permeability $\mu_o = 4\pi \times 10^{-7}$ H/m and characterized by a permittivity $\varepsilon_p = (-\varepsilon_R + i\varepsilon_I)\varepsilon_o$, where $\varepsilon_o = 1/(c^2\mu_o)$ and $c = 299792458$ m/s. The equivalence theorem allows us to formulate the problem with equivalent currents $\boldsymbol{J} = \hat{n} \times \boldsymbol{H}$ and $\boldsymbol{M} = -\hat{n} \times \boldsymbol{E}$ defined on the surface of the object with outward normal $\hat{n}$. In phasor domain with the $\exp(-i\omega t)$ time convention, these currents radiate in homogeneous spaces using the impulse (Green's) functions $g_{o,p}(\boldsymbol{r},\boldsymbol{r}') = \exp(ik_{o,p}|\boldsymbol{r} - \boldsymbol{r}'|)/(4\pi|\boldsymbol{r} - \boldsymbol{r}'|)$, where $k_o = \omega\sqrt{\mu_o}\sqrt{\varepsilon_o}$ and $k_p = \omega\sqrt{\mu_o}\sqrt{\varepsilon_p}$ are the wavenumbers for the angular frequency $\omega = 2\pi f = 2\pi c/\lambda_o$ (hence, λ_o is the wavelength in free space). Outer and inner problems are superimposed using the boundary conditions, that is, continuity of the tangential electric and magnetic field intensities, leading to a set of equations that are discretized and converted into matrix equations for numerical solutions.

The method to enforce the boundary conditions leads to alternative ways to construct formulations. Among many, we prefer a recently developed modified combined tangential formulation (MCTF), which can be written as

$$\begin{bmatrix} \overline{\overline{Z}}_{11}^{\text{MCTF}} & \overline{\overline{Z}}_{12}^{\text{MCTF}} \\ \overline{\overline{Z}}_{21}^{\text{MCTF}} & \overline{\overline{Z}}_{22}^{\text{MCTF}} \end{bmatrix} \cdot \begin{bmatrix} \boldsymbol{a}_J \\ \boldsymbol{a}_M \end{bmatrix} = \begin{bmatrix} \boldsymbol{w}_1^{\text{MCTF}} \\ \boldsymbol{w}_2^{\text{MCTF}} \end{bmatrix}, \tag{1}$$

where

$$\overline{\boldsymbol{Z}}_{11}^{\mathrm{MCTF}}[m,n] = i\omega\mu_o \int_{S_m} d\boldsymbol{r}\boldsymbol{t}_m(\boldsymbol{r}) \cdot \int_{S_n} d\boldsymbol{r}' \boldsymbol{b}_n(\boldsymbol{r}')[g_o(\boldsymbol{r},\boldsymbol{r}') + g_p(\boldsymbol{r},\boldsymbol{r}')] \\ + \frac{i}{\omega} \int_{S_m} d\boldsymbol{r}\boldsymbol{t}_m(\boldsymbol{r}) \cdot \int_{S_n} d\boldsymbol{r}' \nabla' \cdot \boldsymbol{b}_n(\boldsymbol{r}') \left[\frac{1}{\varepsilon_o} \nabla g_o(\boldsymbol{r},\boldsymbol{r}') + \frac{1}{\varepsilon_p} \nabla g_p(\boldsymbol{r},\boldsymbol{r}')\right] \tag{2}$$

$$\overline{\boldsymbol{Z}}_{12}^{\mathrm{MCTF}}[m,n] = -\int_{S_m} d\boldsymbol{r}\boldsymbol{t}_m(\boldsymbol{r}) \cdot \int_{PV,S_n} d\boldsymbol{r}' \boldsymbol{b}_n(\boldsymbol{r}') \times [\nabla' g_o(\boldsymbol{r},\boldsymbol{r}') + \nabla' g_p(\boldsymbol{r},\boldsymbol{r}')] \tag{3}$$

$$\overline{\boldsymbol{Z}}_{21}^{\mathrm{MCTF}}[m,n] = -\frac{\mu_o}{\sqrt{\varepsilon_o}\sqrt{\varepsilon_p}} \overline{\boldsymbol{Z}}_{12}^{\mathrm{MCTF}}[m,n] \tag{4}$$

$$\overline{\boldsymbol{Z}}_{22}^{\mathrm{MCTF}}[m,n] = i\omega\mu_o \int_{S_m} d\boldsymbol{r}\boldsymbol{t}_m(\boldsymbol{r}) \cdot \int_{S_n} d\boldsymbol{r}' \boldsymbol{b}_n(\boldsymbol{r}') \left[\frac{\sqrt{\varepsilon_o}}{\sqrt{\varepsilon_p}} g_o(\boldsymbol{r},\boldsymbol{r}') + \frac{\sqrt{\varepsilon_p}}{\sqrt{\varepsilon_o}} g_p(\boldsymbol{r},\boldsymbol{r}')\right] \\ + \frac{i}{\omega\sqrt{\varepsilon_o}\sqrt{\varepsilon_p}} \int_{S_m} d\boldsymbol{r}\boldsymbol{t}_m(\boldsymbol{r}) \cdot \int_{S_n} d\boldsymbol{r}' \nabla' \cdot \boldsymbol{b}_n(\boldsymbol{r}')[\nabla g_o(\boldsymbol{r},\boldsymbol{r}') + \nabla g_p(\boldsymbol{r},\boldsymbol{r}')] \tag{5}$$

and

$$\boldsymbol{w}_1^{\mathrm{MCTF}}[m] = -\int_{S_m} d\boldsymbol{r}\boldsymbol{t}_m(\boldsymbol{r}) \cdot \boldsymbol{E}^{\mathrm{inc}}(\boldsymbol{r}) \tag{6}$$

$$\boldsymbol{w}_2^{\mathrm{MCTF}}[m] = -\frac{\mu_o}{\sqrt{\varepsilon_o}\sqrt{\varepsilon_p}} \int_{S_m} d\boldsymbol{r}\boldsymbol{t}_m(\boldsymbol{r}) \cdot \boldsymbol{H}^{\mathrm{inc}}(\boldsymbol{r}). \tag{7}$$

In the above, $\boldsymbol{b}_n$ and $\boldsymbol{t}_m$ are the basis and testing functions for $\{n,m\} = \{1,2,\ldots,N\}$, and PV in Eq. (3) represents the principal value of the integral. In the numerical solutions in this chapter, we consider the Rao-Wilton-Glisson (RWG) functions defined on pairs of discretization triangles, that is, $\boldsymbol{b}_n(\boldsymbol{r}) = 0$ if $\boldsymbol{r} \notin S_n$ and $\boldsymbol{t}_m(\boldsymbol{r}) = 0$ if $\boldsymbol{r} \notin S_m$. Solution of the matrix equation in Eq. (1) provides the coefficients expanding the current densities, that is, $\boldsymbol{a}_J$ and $\boldsymbol{a}_M$ in $\boldsymbol{J}(\boldsymbol{r}) = \sum \boldsymbol{a}_J[n]\boldsymbol{b}_n(\boldsymbol{r})$ and $\boldsymbol{M}(\boldsymbol{r}) = \sum \boldsymbol{a}_M[n]\boldsymbol{b}_n(\boldsymbol{r})$.

It is remarkable that MCTF is stable for arbitrarily large values of ε_R (as opposed to the conventional formulations [37]). As $\varepsilon_R \to \infty$ and $\varepsilon_p \to -\infty$, we have $\boldsymbol{a}_M \to 0$, which is enforced by the second row that reduces into

$$\overline{\boldsymbol{G}} \cdot \boldsymbol{a}_M = 0. \tag{8}$$

In the above, $\overline{\boldsymbol{G}}$ is the Gram matrix, that is,

$$\overline{\boldsymbol{G}}[m,n] = \int_{S_m} d\boldsymbol{r}\boldsymbol{t}_m(\boldsymbol{r}) \cdot \boldsymbol{b}_n(\boldsymbol{r}), \tag{9}$$

which contains only five nonzero elements (corresponding to the intersections of the RWG functions) per row or column. While the second row leads to vanishing magnetic currents, the first row of MCTF becomes

$$i\omega\mu_o \int_{S_m} d\boldsymbol{r}\boldsymbol{t}_m(\boldsymbol{r}) \cdot \sum_{n=1}^{N} \boldsymbol{a}_J[n] \int_{S_n} d\boldsymbol{r}' g_o(\boldsymbol{r},\boldsymbol{r}') \boldsymbol{b}_n(\boldsymbol{r}') = -\int_{S_m} d\boldsymbol{r}\boldsymbol{t}_m(\boldsymbol{r}) \cdot \boldsymbol{E}^{\rm inc}(\boldsymbol{r}) \tag{10}$$

for $m = 1, 2, \ldots, N$, which is the standard electric-field integral equation for perfectly conducting objects.

2.2. Accelerated matrix-vector multiplications with MLFMA

Matrix equations derived from MCTF can be solved iteratively, where the matrix-vector multiplications can be performed efficiently as

$$\begin{bmatrix} y_1 \\ y_2 \end{bmatrix} = \begin{bmatrix} \overline{\boldsymbol{Y}}_{11}^{\rm MCTF} & \overline{\boldsymbol{Y}}_{12}^{\rm MCTF} \\ \overline{\boldsymbol{Y}}_{21}^{\rm MCTF} & \overline{\boldsymbol{Y}}_{22}^{\rm MCTF} \end{bmatrix} \cdot \begin{bmatrix} \boldsymbol{x}_J \\ \boldsymbol{x}_M \end{bmatrix} + \begin{bmatrix} \overline{\boldsymbol{U}}_{11}^{\rm MCTF} & \overline{\boldsymbol{U}}_{12}^{\rm MCTF} \\ \overline{\boldsymbol{U}}_{21}^{\rm MCTF} & \overline{\boldsymbol{U}}_{22}^{\rm MCTF} \end{bmatrix} \cdot \begin{bmatrix} \boldsymbol{x}_J \\ \boldsymbol{x}_M \end{bmatrix}, \tag{11}$$

where $\overline{\boldsymbol{Y}}_{ab}^{\rm MCTF}$ for $\{a,b\} = \{1, 2\}$ represents the near-field interactions that are between the basis and testing functions close to each other for a given distance threshold. There are only $\mathcal{O}(N)$ nonzero elements in each $\overline{\boldsymbol{Y}}_{ab}^{\rm MCTF}$ with an identical sparsity pattern. Most of the interactions (computations in matrix-vector multiplications) corresponding to the elements in the far-field matrices $\overline{\boldsymbol{U}}_{ab}^{\rm MCTF} \approx \overline{\boldsymbol{Z}}_{ab}^{\rm MCTF} - \overline{\boldsymbol{Y}}_{ab}^{\rm MCTF}$ are calculated on the fly using MLFMA. These matrix elements are never computed directly or stored in memory, while their multiplications with given input vectors $\boldsymbol{x}_J$ and $\boldsymbol{x}_M$ can still be performed with controllable accuracy, as described in [36]. Using MLFMA, a matrix equation can efficiently be solved via Krylov-subspace algorithms with $\mathcal{O}(N\log N)$ complexity per matrix-vector multiplication or iteration.

In the solutions of plasmonic problems, inner interactions are localized as the negative permittivity increases. Since MCTF involves the combinations of interactions from the inner and outer media, the contributions from the inner medium become very small for some long-distance interactions. This allows us to omit such contributions without deteriorating the accuracy [38]. As an example, consider the first term for $\overline{\boldsymbol{Z}}_{11}^{\rm MCTF}$ in Eq. (2). The integrand involves the combination of Green's functions as

$$g_o(\boldsymbol{r},\boldsymbol{r}') + g_p(\boldsymbol{r},\boldsymbol{r}') = \frac{\exp(ik_o|\boldsymbol{r}-\boldsymbol{r}'|)}{4\pi|\boldsymbol{r}-\boldsymbol{r}'|} - \frac{\exp(ik_p|\boldsymbol{r}-\boldsymbol{r}'|)}{4\pi|\boldsymbol{r}-\boldsymbol{r}'|} = \frac{\exp(ik_o|\boldsymbol{r}-\boldsymbol{r}'|) + \exp(ik_p|\boldsymbol{r}-\boldsymbol{r}'|)}{4\pi|\boldsymbol{r}-\boldsymbol{r}'|}. \tag{12}$$

In a normalized form, the contribution of the inner medium becomes more obvious as

$$\frac{g_o(\boldsymbol{r},\boldsymbol{r}') + g_p(\boldsymbol{r},\boldsymbol{r}')}{g_o(\boldsymbol{r},\boldsymbol{r}')} = 1 + \frac{\exp(ik_p|\boldsymbol{r}-\boldsymbol{r}'|)}{\exp(ik_o|\boldsymbol{r}-\boldsymbol{r}'|)}. \tag{13}$$

If the interactions in MLFMA are to be calculated with an error threshold of Δe, which is determined by the sampling and truncation used in the diagonalization of Green's function [34], it is possible to omit the contribution from the inner medium if

$$|\exp(ik_p|\boldsymbol{r}-\boldsymbol{r}'|)| \leq \Delta e|\exp(ik_o|\boldsymbol{r}-\boldsymbol{r}'|)| = \Delta e. \quad (14)$$

For large values of ε_R, this inequality can be approximated as

$$|\boldsymbol{r}-\boldsymbol{r}'| \geq \frac{-\ln(\Delta e)}{k_o\sqrt{\varepsilon_R}} \quad (15)$$

as the condition for the distance to omit the related inner interaction. For example, for two digits of accuracy ($\Delta e = 0.01$) and $\varepsilon_R = 100$ at 250 THz, it is safe to omit the interaction if $|\boldsymbol{r}-\boldsymbol{r}'| \geq 88$ nm. This significantly accelerates the matrix-vector multiplications with MLFMA. We emphasize that omitting interactions as described above has no negative effect in the accuracy of the results. In fact, as the frequency drops and ε_R increases, we expect that inner interactions decay quickly such that the inner matrix becomes a Gram matrix and the formulation turns into a perfectly conducting version.

3. Nanowire simulations

In this section, we present the major parameters of the nanowire simulations considered in this chapter using the implementation described in Section 2.

3.1. Nanowire models

We consider Ag nanowires of length 5 μm at a fixed frequency of 250 THz, at which the nanowire length corresponds to approximately $4.17\lambda_o$. At this frequency, the permittivity of Ag can be found as approximately $-60.76 + 4.31i$ [20]. For the cross sections of the nanowires, we use square, circular (circular-like, more precisely, depending on the discretization size), and hexagonal shapes. For all shapes, the cross-sectional area of each nanowire is selected as 0.01 μm^2 (e.g., a square nanowire has 0.1×0.1 μm cross section) other than scaling trials for the hexagonal cases. For numerical solutions, nanowires are discretized with 0.05–0.075 μm triangles ($\lambda_o/24$–$\lambda_o/16$), on which the RWG functions are defined. As shown in the numerical results, we consider arrays of nanowires in 2×1, 2×2, 4×4, and 6×6 arrangements, while the main focus is the comparison of the 6×6 arrays. In the regular arrays, the distance between the centers of two nanowires is 0.2 μm. In the hexagonal arrangements involving hexagonal cells, the center-to-center distance between two nearby nanowires is approximately 0.186 μm. Discretization of each nanowire requires 2772–5376 unknowns. Then, for the 6×6 arrays, these discretizations lead to matrix equations involving 99,792–193,536 unknowns, which need a fast and efficient solver, such as MLFMA, to complete all simulations in hours.

3.2. Excitation

Each nanowire array is excited by a pair of Hertzian dipoles oriented in the opposite directions. The electric and magnetic field intensities created by a single dipole can be written as

http://dx.doi.org/10.5772/67447

$$\boldsymbol{E}^{\text{inc}}(\boldsymbol{r}) = i\omega\mu_o \frac{\exp(ik_oR)}{4\pi R} \left\{ \boldsymbol{I}_D \left(1 + \frac{i}{k_oR} - \frac{1}{k_o^2R^2} \right) - \boldsymbol{I}_D \cdot \hat{\boldsymbol{R}}\hat{\boldsymbol{R}} \left(1 + \frac{3i}{k_oR} - \frac{3}{k_o^2R^2} \right) \right\} \tag{16}$$

$$\boldsymbol{H}^{\text{inc}}(\boldsymbol{r}) = \boldsymbol{I}_D \times \hat{\boldsymbol{R}} \frac{\exp(ik_oR)}{4\pi R} \left(\frac{1}{R} - ik_o \right), \tag{17}$$

where $\boldsymbol{R} = \boldsymbol{r} - \boldsymbol{r}' = \hat{\boldsymbol{R}}R$ and $\boldsymbol{I}_D$ is the dipole moment. The distance between the dipoles is 0.2 μm and they are located symmetrically with 0.2 μm distance from the nanowires. An example to the localization of the dipoles with respect to a 4×4 array of square nanowires is depicted in **Figure 1**. Except for the 2×1 arrays, the transverse locations of the dipoles correspond to gaps between the nanowires, leading to efficient dipole-to-array coupling. For the same reason, the power flow is mainly confined inside these arrays (on the interfaces between the nanowires and gaps), while for the 2×1 arrays (and somewhat for the 2×2 arrays), the transmission occurs mostly on the outer sides.

3.3. Field and power calculations

After the current coefficients are found through iterative solutions employing MLFMA, the near-zone electric and magnetic fields can be found anywhere using the discretized form of

$$\boldsymbol{E}^{\text{sec}}(\boldsymbol{r}) = i\omega\mu_o \int d\boldsymbol{r}' \boldsymbol{J}(\boldsymbol{r}') g_u(\boldsymbol{r},\boldsymbol{r}') + \frac{i}{\omega\varepsilon_u} \int d\boldsymbol{r}' \nabla' \cdot \boldsymbol{J}(\boldsymbol{r}') \nabla g_u(\boldsymbol{r},\boldsymbol{r}') - \int d\boldsymbol{r}' \nabla g_u(\boldsymbol{r},\boldsymbol{r}') \times \boldsymbol{M}(\boldsymbol{r}') \tag{18}$$

$$\boldsymbol{H}^{\text{sec}}(\boldsymbol{r}) = i\omega\varepsilon_u \int d\boldsymbol{r}' \boldsymbol{M}(\boldsymbol{r}') g_u(\boldsymbol{r},\boldsymbol{r}') + \frac{i}{\omega\mu_o} \int d\boldsymbol{r}' \nabla' \cdot \boldsymbol{M}(\boldsymbol{r}') \nabla g_u(\boldsymbol{r},\boldsymbol{r}') + \int d\boldsymbol{r}' \boldsymbol{J}(\boldsymbol{r}') \times \nabla' g_u(\boldsymbol{r},\boldsymbol{r}'), \tag{19}$$

where the integrals are over the surfaces of the structure. The calculations are carried out for the outer ($u = o$) and inner ($u = p$) problems, leading to vanishingly zero fields for the inner and outer regions, respectively. This also allows us to check the accuracy of solutions by testing the equivalence theorem. We note that, for the outer problem, incident fields $\boldsymbol{E}^{\text{inc}}$ and $\boldsymbol{H}^{\text{inc}}$ must be added to the secondary fields $\boldsymbol{E}^{\text{sec}}$ and $\boldsymbol{H}^{\text{sec}}$ to arrive at the total values. Then, by superimposing the results from inner and outer calculations, overall graphics demonstrating the electric and magnetic field intensities in the vicinity of the object are obtained. We

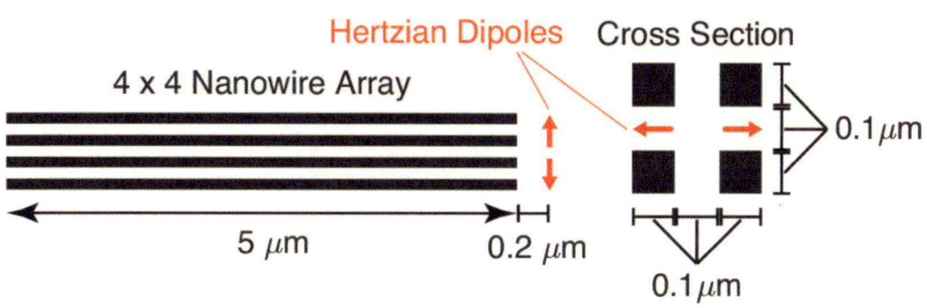

Figure 1. Positions of Hertzian dipoles with respect to a 4×4 nanowire array. The length of all arrays considered in this chapter is 5 μm, while the dipoles oriented in the opposite directions are located symmetrically with 0.2 μm distance from the nanowires.

particularly focus on the magnitudes of the electric field intensity and the magnetic field intensity with units V/m and A/m, respectively, in dB scales. In addition, as a major quantity in the analysis of nanowires, we calculate the Poynting vector (W/m^2), simply called the power density in this chapter, as

$$\boldsymbol{S}(\boldsymbol{r}) = \frac{1}{2}\boldsymbol{E}(\boldsymbol{r}) \times \boldsymbol{H}^*(\boldsymbol{r}), \tag{20}$$

where * is the conjugate operation. In this equation, $\boldsymbol{E}$ and $\boldsymbol{H}$ represent the field intensities, corresponding to only secondary fields inside the object and the sum of secondary and incident fields outside the object. We note that the time-average power density is the real part of the expression in Eq. (20). Similar to the field intensities, we present the magnitude of the power density in the results. Considering the continuity of the tangential fields on surfaces, the normal component of the power density is continuous across nanowire/air interfaces, while the main power flow is in the tangential directions (along nanowires). As an example to field intensity and power density distributions, **Figure 2** depicts the excitation of a pair of nanowires with square cross sections. Electric field intensity, magnetic fields intensity, and the real part of the power density are plotted in the vicinity of the dipoles and the tips of the nanowires, where the coupling of the electromagnetic energy to the nanowire system is clearly visible.

3.4. Iterative solutions and parameters

Nanowire problems are challenging in terms of iterative solutions, partially due to MCTF that provides accurate solutions of plasmonic problems at the cost of worse conditioning of matrix equations in comparison to those derived from the second-kind integral equations. Therefore, iterative solutions are performed using a flexible generalized minimal residual (FGMRES) algorithm, which allows for a nested strategy involving inner solutions through GMRES. Specifically, FGMRES employs MLFMA (with 1% maximum error in all solutions) for the matrix-vector multiplications, while it is preconditioned by inner solutions via GMRES. In the inner layer, GMRES employs an approximate form of MLFMA (AMLFMA) [39] that is derived from MLFMA by reducing the number of harmonics. GMRES is further preconditioned via

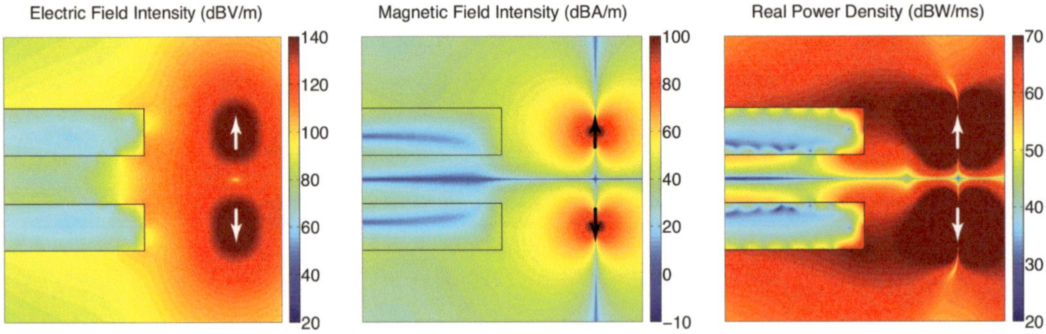

Figure 2. Excitation of a pair of nanowires with square cross sections through a pair of Hertzian dipoles. The electric field intensity, magnetic field intensity, and the real part of the power density are plotted in the vicinity of the dipoles (shown with arrows) and the tips of the nanowires.

http://dx.doi.org/10.5772/67447

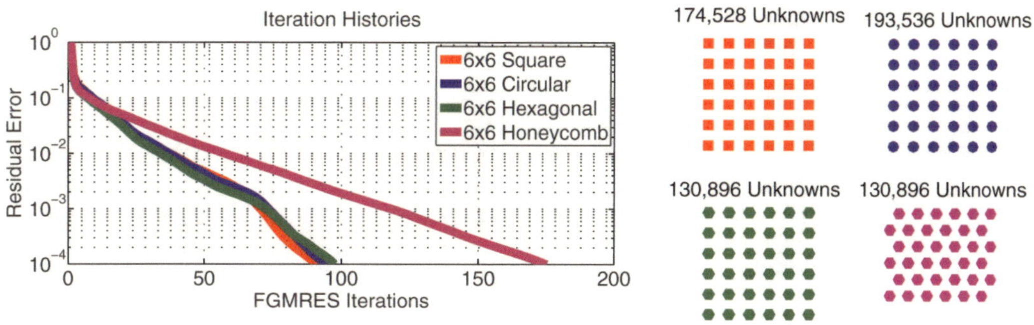

Figure 3. Iteration histories for the solutions of 6×6 nanowire arrays.

algebraic preconditioners based on the near-field interactions. As an example to iterative solutions, **Figure 3** depicts the iteration histories for the analysis of 6×6 nanowire arrays (with number of unknowns shown in **Figure 3**). The residual error is reduced to 10^{-4} (as in all solutions presented in this chapter), requiring nearly 100 iterations for the circular, square, and hexagonal cross sections. For the honeycomb structure, the number of iterations increases to more than 150 to achieve the desired residual error. The reason for larger number of iterations for the honeycomb structure seems to be related to very high transmission capabilities of the nanowires in this configuration, as shown in the simulation results.

4. Nanowire arrays for power transfer

Using the developed solver, we consider nanowires and their arrangements in alternative configurations. Near-zone field intensity and power density values, which are computed accurately with given error thresholds as described above, are used to precisely assess the efficiency of power transmission levels, leading to a rigorous study of nanowire arrays as transmission lines at optical frequencies.

4.1. Double nanowires

First, we consider transmission by pairs of nanowires with different cross sections. **Figure 4** presents the electric field intensity (dBV/m), magnetic field intensity (dBA/m), and power density (dBW/m^2) in the vicinity of the nanowires (on a plane containing dipoles). In each case, a total of $200 \times 800 = 160{,}000$ samples are used for high-quality plots. The excitation dipoles are located on the right-hand side, as clearly observed by large intensity and density values on this side. We observe that the power transmission is almost identical for different cross sections, that is, circular, hexagonal, and square. We note that the nanowires have the same cross-sectional area and the distance between them is the same for all cases. In general, there is a high-quality transmission of the electromagnetic energy along the nanowires, independent of the cross section. In each case, the variations in the electric field intensity, magnetic field intensity, and power density are related to the optical length of the nanowires, which is 5 μm corresponding to approximately $4.17\lambda_0$. While not clearly visible in these plots,

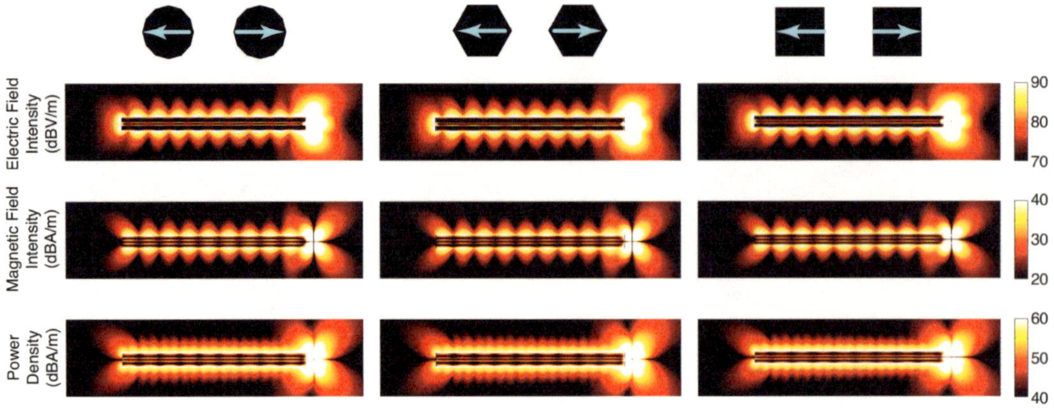

Figure 4. The electric field intensity, magnetic field intensity, and power density values in the vicinity of 2×1 nanowire arrays with different cross sections. The transverse locations of the dipoles (arrows) are also shown.

plasmonic oscillations with much higher rates in the power values exist on the air/metal interfaces, as depicted in **Figure 2** for the nanowires with square cross sections. **Figure 4** also shows that when the transmission line involves only two nanowires, the transmission occurs mostly along the outer surfaces.

4.2. Nanowires in array configurations

Next, we focus on 2×2 systems of nanowires, as depicted in **Figure 5**, again with different cross sections. In this case, the near-field samples (that are on the plane containing the dipoles) do not coincide with the nanowires. Comparing the results for different cross sections, we observe similar characteristics with high-quality transmissions. Specifically, none of the cross-sectional shapes seems to have an advantage over others. Considering the outputs of the

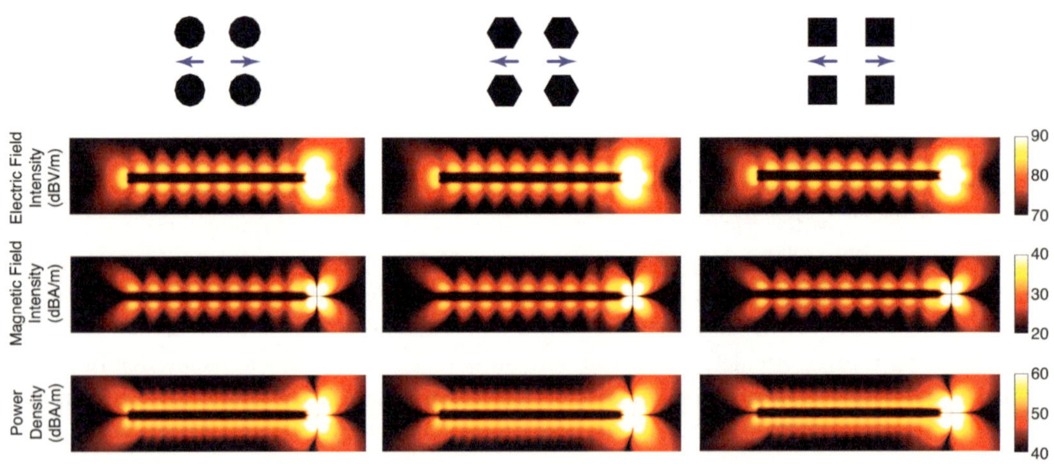

Figure 5. The electric field intensity, magnetic field intensity, and power density values in the vicinity of 2×2 nanowire arrays with different cross sections. The transverse locations of the dipoles (arrows) are also shown.

transmission lines, where the electromagnetic energy is coupled to air, 2×2 arrays seem to be slightly better than the nanowire pairs shown in **Figure 4**. For example, considering the power density distributions, larger beams occur on the left-hand sides of the plots, indicating better transmissions with the 2×2 arrays. Values along the nanowires seem to decrease in comparison to 2×1 systems, but this is (misleadingly) due to the sampling plane that is separated from the nanowires for the 2×2 arrangements. Therefore, for a fair comparison, we need to consider the outputs of the nanowires (on the left-hand sides of the plots).

Figure 6 presents similar results for 4×4 arrays. In this case, in addition to circular, square, and hexagonal nanowires that are regularly arranged, we consider a small-scale honeycomb structure involving hexagonal nanowires in a hexagonal arrangement. We note that the term honeycomb is used to describe the overall structure with hexagonal gaps between hexagonal nanowires with the same cross-sectional areas. In comparison to smaller arrays, the transmissions through 4×4 arrays, which occur in two parallel paths for the regular arrangements, are mostly confined inside the arrays. It is remarkable that the transmission properties are almost the same for the regular arrangements of the nanowires with circular, square, and hexagonal cross sections, while the transmission along the honeycomb structure occurs differently as distributed into more pathways. Nevertheless, considering again the outputs of the nanowires, there is not a clear advantage of this structure over others.

4.3. Comparisons of larger arrays

Dramatic changes occur when the nanowire arrays get larger. In **Figure 7**, we consider arrays of 6×6 nanowires that are arranged similar to the 4×4 arrays. For the regular arrays of nanowires with circular, square, and hexagonal cross sections, the electromagnetic energy is transmitted again in two main lines close to the transverse locations of the dipoles. The transmission performance of these arrays is very similar to each other. On the other hand, a significant improvement is observed when the arrangement is also made hexagonal, leading to a honeycomb structure. This structure clearly outperforms the others with higher quality field and power transmissions. This is further verified in **Figure 8**, where the intensity and density values are sampled at the outputs of the arrays. In these plots, the samples are selected on the transverse plane located at 0.2 μm distance from the nanowire tips (at the output side). In the power density values, more than 5 dB (more than three times) enhancement is observed when the honeycomb structure is used instead of the others.

4.4. Investigation of cross-sectional area

After demonstrating the significantly higher transmission capabilities of the honeycomb structure, it is questioned whether the cross-sectional areas of the nanowires are optimal or not. For this purpose, we consider hexagonal arrangements of 6×6 nanowires with only hexagonal cross sections. As depicted in **Figure 9**, we scale the cross-sectional dimensions by 0.5, 0.75, 1.25, and 1.375, when 1.0 corresponds to the original honeycomb structure. It can be observed that scaling cross-sectional dimensions both to smaller and to larger values leads to deteriorations in the transmission properties of the array. This is again verified by the output plots in **Figure 10**, where the results for a scaling with 1.125 ratio is included as well as for the original

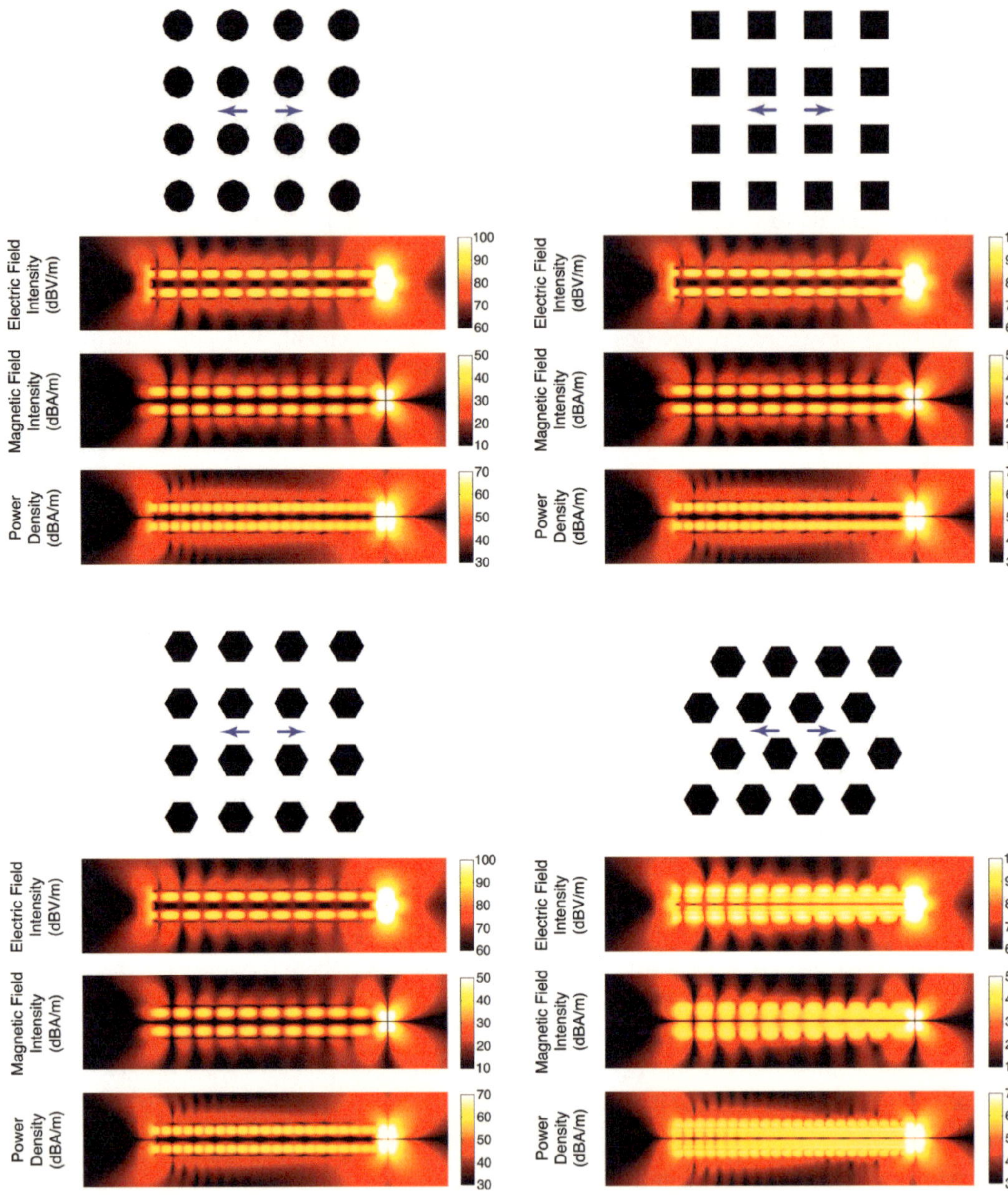

Figure 6. The electric field intensity, magnetic field intensity, and power density values in the vicinity of 4×4 nanowire arrays with different cross sections. The transverse locations of the dipoles (arrows) are also shown. We note that, in the longitudinal direction, the distance between the dipoles and the arrays is 0.2 μm in all cases.

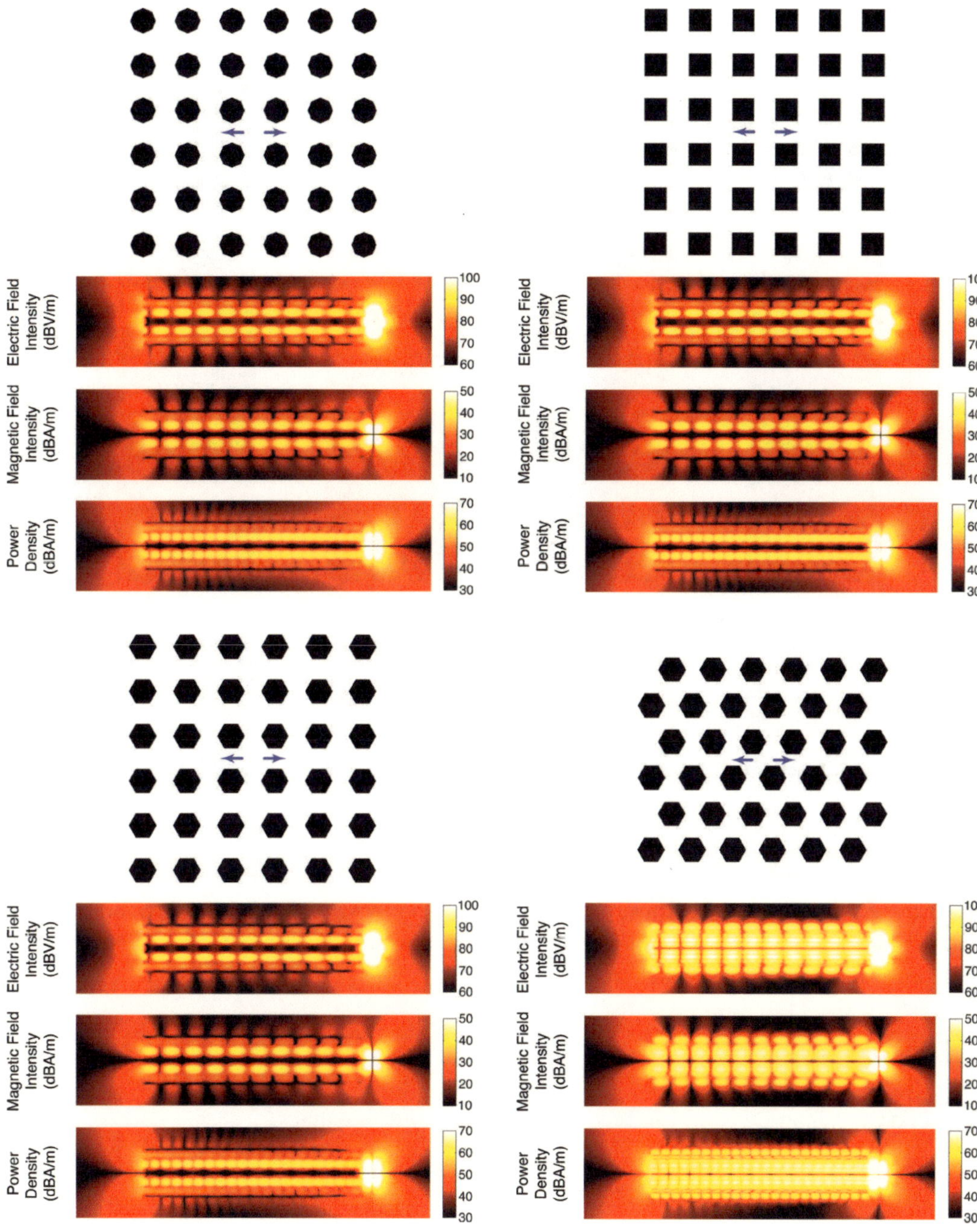

Figure 7. The electric field intensity, magnetic field intensity, and power density values in the vicinity of 6×6 nanowire arrays with different cross sections. The transverse locations of the dipoles (arrows) are also shown. We note that, in the longitudinal direction, the distance between the dipoles and the arrays is 0.2 μm in all cases.

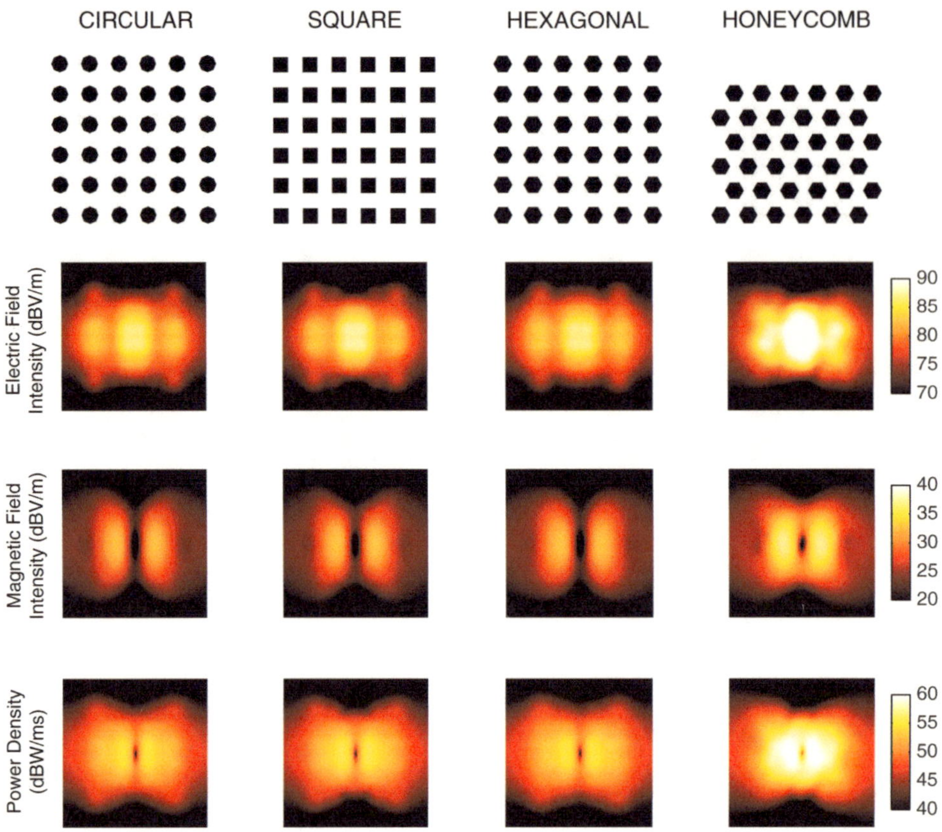

Figure 8. The electric field intensity, magnetic field intensity, and power density values at the outputs of 6×6 nanowire arrays with different cross sections.

structure. As demonstrated in these plots, even small changes in the cross-sectional areas of the nanowires deteriorate the performance of the array, showing the optimality of the honeycomb structure with equally filled and empty spaces.

5. Discussion

Based on the numerical results in Section 4, one may conclude that

- the 6×6 honeycomb structure demonstrates very good transmission properties, in comparison to other arrays with the same number of nanowires under the same conditions,
- closely packing the hexagonal nanowires (scaling up) deteriorates the transmission quality, and
- reducing the dimensions (scaling down) also has a negative effect.

In order to explain the results, we now consider the electric current density ($\boldsymbol{J} = \hat{\boldsymbol{n}} \times \boldsymbol{H}$) on the nanowire surfaces. **Figure 11** depicts the dominant electric current (in dB scale with 20 dB

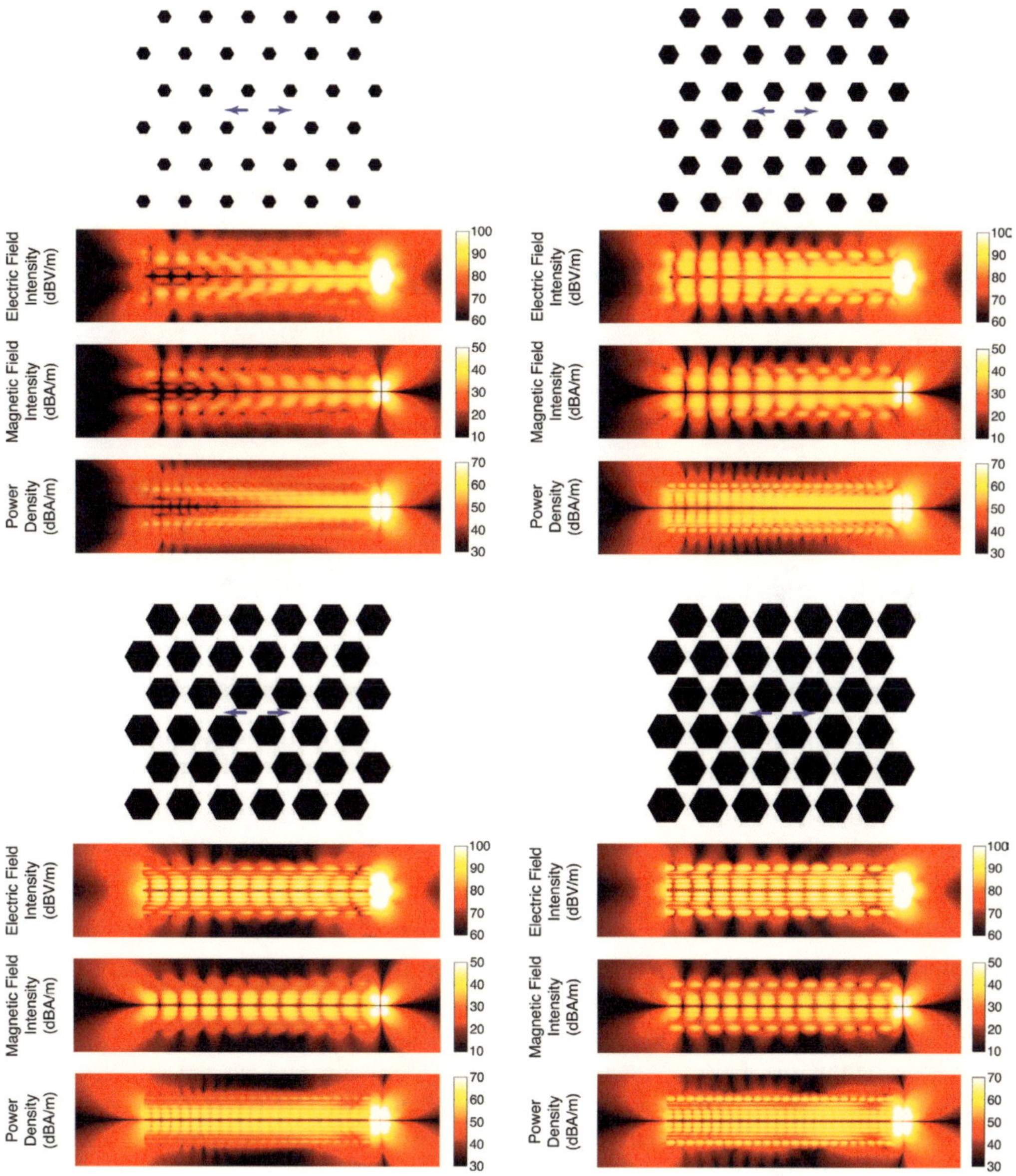

Figure 9. The electric field intensity, magnetic field intensity, and power density values in the vicinity of 6×6 nanowire arrays with hexagonal cross sections of different areas. We note that, in the longitudinal direction, the distance between the dipoles and the arrays is 0.2 μm in all cases.

dynamic range) along the 6×6 arrays of nanowires with circular, square, and hexagonal cross sections, as well as along the honeycomb structure. We also consider the hexagonal arrangement of hexagonal nanowires scaled by the factors of 0.5 and 1.4. It is evident that the current density is much higher for the honeycomb structure. In addition to three-dimensional views, we present a side view when the dipoles are oriented outward/inward. Comparison of current distributions reveals significant differences on the behaviors of the arrays. First, considering

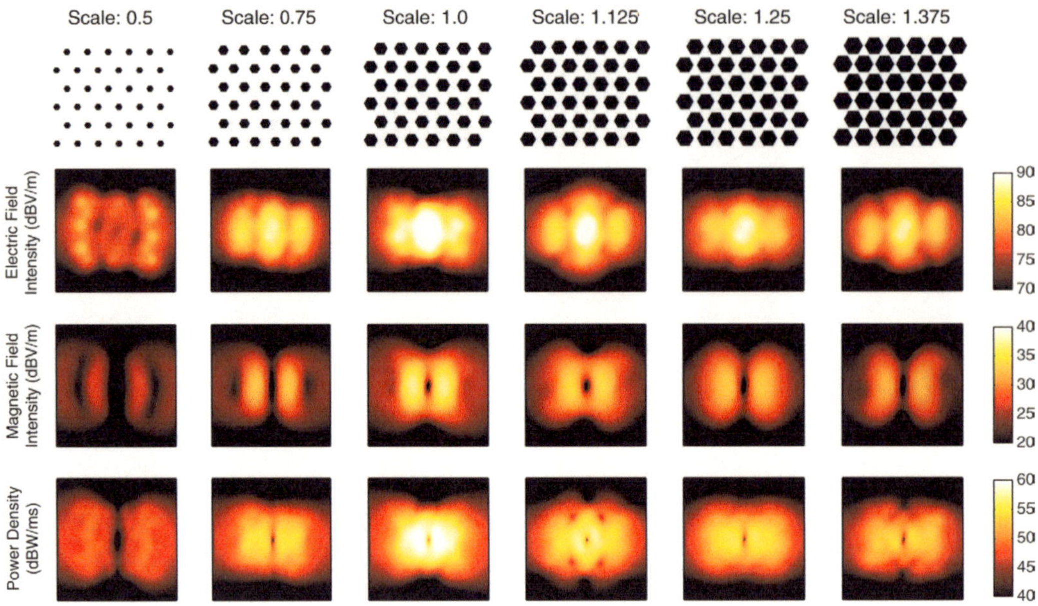

Figure 10. The electric field intensity, magnetic field intensity, and power density values at the outputs of 6×6 nanowire arrays with hexagonal cross sections of different areas. The scale of 1.0 corresponds to the original honeycomb structure.

the array with circular cross sections, it can be observed that the currents along the nanowires are out of phase. Specifically, the current density has maximum locations at different longitudinal positions for the inner two rows and the outer rows. The same situation is also observed for the arrays with square and hexagonal cross sections. All these observations are consistent with the near-zone plots depicted in **Figure 7**.

The out-of-phase currents along the regular arrays can be explained by the adverse effects of neighboring nanowires. Specifically, considering Ampere's circuital law, each nanowire creates a magnetic field that induces current in the opposite direction (to the main current flow) on the neighboring nanowires. Due to the cancellations of the main and secondary currents, the overall currents along the nanowires become out of phase, reducing the transmission capability of the arrays. From this perspective, nanowires in a honeycomb structure are located optimally such that the adverse interactions between nanowires are minimized. Considering how the nanowires are located, it can be observed that nearby surfaces of neighboring nanowires are not aligned and the magnetic coupling between the nanowires is weaker than those in the other arrangements. There are surfaces with identical alignments but these surfaces (on the same sides of the nanowires) are far away from each other. As a result, the currents along the nanowires are almost in phase, as depicted in **Figure 11**.

Scaling the nanowires in the honeycomb structures always has an adverse effect. Decreasing cross-sectional area reduces the overall area of the surfaces for the propagation of plasmonic waves. In addition, increasing distance between the surfaces of the neighboring nanowires adds extra phase in the magnetic coupling, deteriorating the synchronization of the currents. Obviously, increasing the cross-sectional area also has adverse effect as a result of increasing

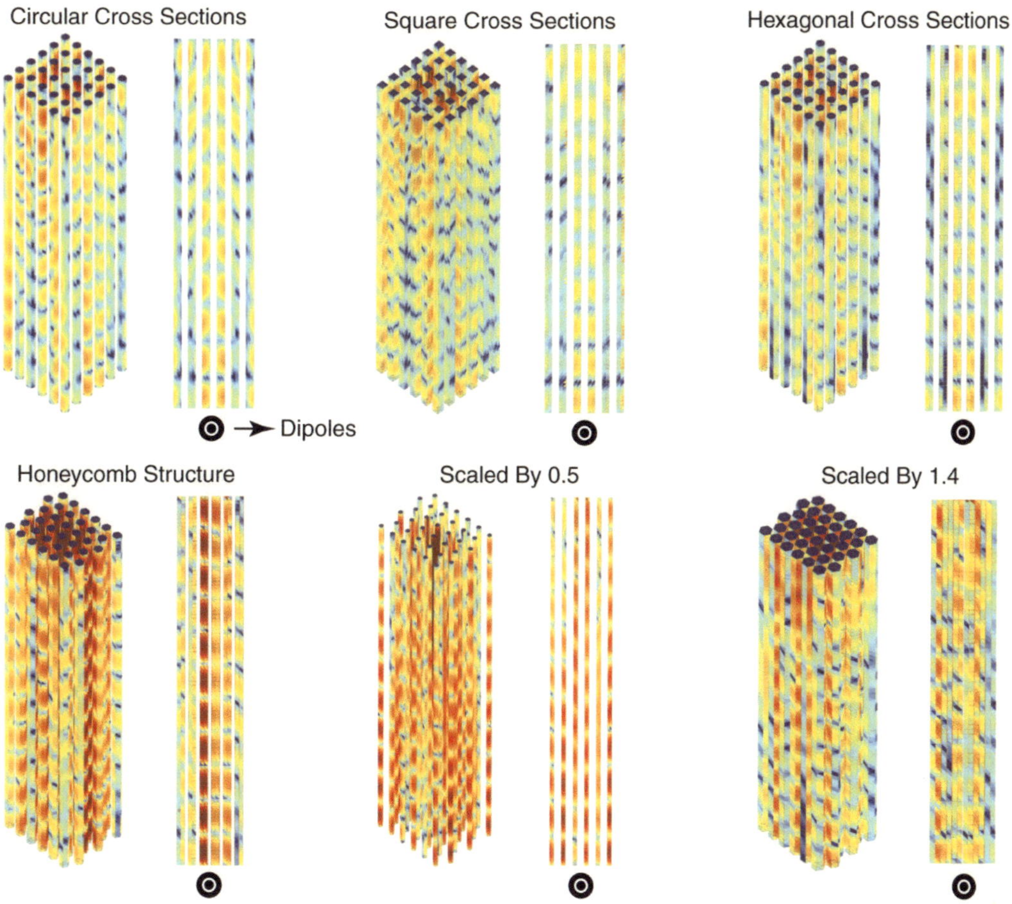

Figure 11. The magnitude of the electric current density induced on the nanowire surfaces for different array configurations. The location and orientation (outward and inward) of the dipole pair are shown in side views.

coupling between the nanowires. It appears that the honeycomb structure is a kind of optimal way to transmit power.

6. Concluding remarks

An investigation of nanowire transmission lines may lead to optimal structures that can be used in nano-optical devices. In this chapter, we present the effects of cross-sectional geometry and arrangement of array elements in different configurations. Using an accurate full-wave simulation environment, we obtain reliable results of three-dimensional finite arrays without resorting to approximation methods and asymptotic optical techniques that may have significant accuracy problems. Based on a comparative study of Ag nanowires at an infrared frequency, we show that power transmission becomes significantly higher for nanowires with hexagonal cross sections and their hexagonal arrangements, leading to honeycomb structures. We also explain the favorable electromagnetic characteristics of these structures, which may pave a way for improved nanowire systems for optimal power transmission.

Acknowledgements

This work was supported by the Scientific and Technical Research Council of Turkey (TUBITAK) under the Research Grant 113E129 and by the Turkish Academy of Sciences (TUBA) in the framework of the Young Scientist Award Program.

Authors contributions

Ö.E. developed the theory and formulations, B.K. and Ö.E. implemented the solvers, and H.A.Ş. conducted the numerical experiments.

Author details

Hasan Aykut Şatana, Barışcan Karaosmanoğlu and Özgür Ergül*

*Address all correspondence to: ozgur.ergul@eee.metu.edu.tr

Department of Electrical and Electronics Engineering, Middle East Technical University, Ankara, Turkey

References

[1] Ditlbacher H, Hohenau A, Wagner D, Kreibig U, Rogers M, Hofer F, Aussenegg FR, Krenn JR. Silver nanowires as surface plasmon resonators. Phys. Rev. Lett. 2005;**95**: 257403.

[2] Sanders AW, Routenberg DA, Wiley BJ, Xia Y, Dufresne ER, Reed MA. Observation of plasmon propagation, redirection, and fan-out in silver nanowires. Nano Lett. 2006;**6**: 1822–1826.

[3] Akimov AV, Mukherjee A, Yu CL, Chang DE, Zibrov AS, Hemmer PR, Park H, Lukin MD. Generation of single optical plasmons in metallic nanowires coupled to quantum dots. Nature. 2007;**450**:402–406.

[4] Guo X, Qiu M, Bao J, Wiley BJ, Yang Q, Zhang X, Ma Y, Yu H, Tong L. Direct coupling of plasmonic and photonic nanowires for hybrid nanophotonic components and circuits. Nano Lett. 2009;**9**:4515–4519.

[5] Wang W, Yang Q, Fan F, Xu H, Wang ZL. Light propagation in curved silver nanowire plasmonic waveguides. Nano Lett. 2011;**11**:1603–1608.

[6] Li Z, Bao K, Fang Y, Guan Z, Halas NJ, Nordlander P, Xu H. Effects of a proximal substrate on plasmon propagation in silver nanowires. Phys. Rev. B. 2010;**82**:241402.

[7] Ma Y, Li X, Yu H, Tong L, Gu Y, Gong Q. Direct measurement of propagation losses in silver nanowires. Opt. Lett. 2010;**35**:1160–1162.

[8] Yao J, Liu Z, Liu Y, Wang Y, Sun C, Bartal G, Stacy AM, Zhang X. Optical negative refraction in bulk metamaterials of nanowires. Science. 2008;**321**:930.

[9] Liu Y, Bartal G, Zhang X. All-angle negative refraction and imaging in a bulk medium made of metallic nanowires in the visible region. Opt. Exp. 2008;**16**:15439–15448.

[10] Casse BDF, Lu WT, Huang YJ, Gultepe E, Menon L. Super-resolution imaging using a three-dimensional metamaterials nanolens. Appl. Phys. Lett. 2010;**96**:023114.

[11] Sun M, Zhang Z, Wang P, Li Q, Ma F, Xu H. Remotely excited Raman optical activity using chiral plasmon propagation in Ag nanowires. Light Sci Appl. 2013;**2**:112.

[12] Huang Y, Fang Y, Zhang Z, Zhu L, Sun M. Nanowire-supported plasmonic waveguide for remote excitation of surface-enhanced Raman scattering. Light Sci Appl. 2014;**3**:199.

[13] Wang X, Summers CJ, Wang ZL. Large-scale hexagonal-patterned growth of aligned ZnO nanorods for nano-optoelectronics and nano-sensor arrays. Nano Lett. 2004;**4**:423–426.

[14] Rockstuhl C, Fahr S, Lederer F. Absorption enhancement in solar cells by localized plasmon polaritons. J. Appl. Phys. 2008;**104**:123102.

[15] Bergin SM, Chen Y, Rathmell AR, Charbonneau P, Lib Z, Wiley BJ. The effect of nanowire length and diameter on the properties of transparent, conducting nanowire films. Nanoscale. 2012;**4**:1996–2004.

[16] Yin AJ, Li J, Jian W, Bennett AJ, Xu JM. Fabrication of highly ordered metallic nanowire arrays by electrodeposition. Appl. Phys. Lett. 2001;**79**:1039–1041.

[17] Wang RC, Liu CP, Huang JL. ZnO hexagonal arrays of nanowires grown on nanorods. Appl. Phys. Lett. 2005;**86**:251104.

[18] Kottmann JP, Martin OJF. Plasmon resonances of silver nanowires with a non-regular cross section. Phys. Rev. B. 2001;**64**:235402.

[19] Yılmaz A, Karaosmanoğlu B, Ergül Ö. Computational electromagnetic analysis of deformed nanowires using the multilevel fast multipole algorithm. Sci. Rep. 2015;**5**:8469.

[20] Johnson PB, Christy RW. Optical constants of the noble metals. Phys. Rev. B. 1972;**6**:4370–4379.

[21] Ergül Ö. Solutions of large-scale electromagnetics problems involving dielectric objects with the parallel multilevel fast multipole algorithm. J. Opt. Soc. Am. A. 2011;**28**:2261–2268.

[22] Kottmann JP, Martin OJF. Influence of the cross section and the permittivity on the plasmon-resonance spectrum of silver nanowires. Appl. Phys. B. 2001;**73**,299–304.

[23] Futamata M, Maruyama Y, Ishikawa M. Local electric field and scattering cross section of Ag nanoparticles under surface plasmon resonance by finite difference time domain method. J. Phys. Chem. B. 2003;**107**:7607–7617.

[24] Giannini V, Rodriguez-Oliveros R, Sanchez-Gil JA. Surface plasmon resonances of metallic nanostars/nanoflowers for surface-enhanced Raman scattering. Plasmonics. 2010;**5**:99–104.

[25] Hohenester U, Krenn J. Surface plasmon resonances of single and coupled metallic nanoparticles: a boundary integral method approach. Phys. Rev. B. 2005;**72**:195429.

[26] Sondergaard T. Modeling of plasmonic nanostructures: green's function integral equation methods. Phys. Stat. Sol. B. 2007;**244**:3448–3462.

[27] Kern AM, Martin OFJ. Surface integral formulation for 3D simulations of plasmonic and high permittivity nanostructures. J. Opt. Soc. Am. A. 2009;**26**:732–740.

[28] Gallinet B, Martin OFJ. Scattering on plasmonic nanostructures arrays modeled with a surface integral formulation. Photon. Nanostruct. Fund. Appl. 2010;**8**:278–284.

[29] Rodriguez-Oliveros R, Sanchez-Gil JA. Localized surface-plasmon resonances on single and coupled nanoparticles through surface integral equations for flexible surfaces. Opt. Exp. 2011;**16**:12208–12219.

[30] Solis DM, Taboada JM, Obelleiro F. Surface integral equation method of moments with multiregion basis functions applied to plasmonics. IEEE Trans. Antennas Propag. 2015;**63**:2141–2152.

[31] Karaosmanoğlu B, Gür UM, Ergül Ö. Investigation of nanoantennas using surface integral equations and the multilevel fast multipole algorithm. In Proceedings of the Progress in Electromagnetics Research Symposium (PIERS); Prague, 2015, pp. 2026–2030.

[32] Bora M, Fasenfest BJ, Behymer EM, Chang ASP, Nguyen HT, Britten JA, Larson CC, Chan JW, Miles RR, Bond TC. Plasmon resonant cavities in vertical nanowire arrays. Nano Lett. 2010;**10**:2832–2837.

[33] Nauert S, Paul A, Zhen Y, Solis D, Vigderman L, Chang WS, Zubarev ER, Nordlander P, Link S. Influence of cross sectional geometry on surface plasmon polariton propagation in gold nanowires. ACS Nano. 2014;**8**:572–580.

[34] Chew WC, Jin JM, Michielssen E, Song J. Fast and efficient algorithms in computational electromagnetics. Boston: Artech House; 2001.

[35] Ergül Ö, Gürel L. The multilevel fast multipole algorithm (MLFMA) for solving large-scale computational electromagnetics problems. Chichester, West Sussex, UK: Wiley-IEEE; 2014.

[36] Çekinmez A, Karaosmanoğlu B, Ergül Ö. Integral-equation formulations of plasmonic problems in the visible spectrum and beyond. In Reyhanoğlu M, editor. Dynamical systems—analytical and computational techniques. InTech; 2017.

http://dx.doi.org/10.5772/67447

[37] Karaosmanoğlu B, Yımaz A, Ergül Ö. On the accuracy and efficiency of surface formulations in fast analysis of plasmonic structures via MLFMA. In Proceedings of the Progress in Electromagnetics Research Symposium (PIERS); Shanghai, 2016, pp. 2629–2633.

[38] Karaosmanoğlu B, Yılmaz A, Gür UM, Ergül Ö. Solutions of plasmonic structures using the multilevel fast multipole algorithm. Int J RF Microw Comput. Aided Eng. 2016;**26**:335–341.

[39] Önol C, Karaosmanoğlu B, Ergül Ö. Efficient and accurate electromagnetic optimizations based on approximate forms of the multilevel fast multipole algorithm. IEEE Antennas Wirel Propag. Lett. 2016;**15**:1113–1115.

MODERN FUZZY CONTROL SYSTEMS AND ITS APPLICATIONS

Edited by **S. Ramakrishnan**

Modern Fuzzy Control Systems and Its Applications
http://dx.doi.org/10.5772/65834
Edited by S. Ramakrishnan

Contributors

Danil Dintsis, Makoto Katoh, Mao Li, Tawanda Mushiri, Liviu Dinca, Jenica-Ileana Corcau, Beyda Tasar, Arif Gulten, Longzhi Yang, Yanpeng Qu, Fei Chao, Zheming Zuo, Bahadır Ergun, Cumhur Sahin, Ugur Kaplan, Cheng-Lun Chen, Daniela Perdukova, Pavol Fedor, Peter Hofmann, Junmin Li, Guo Zhang, Jinsha Li, Wafa Gritli, Hajer Gharsallaoui, Mohamed Benrejeb, Hidehiro Ikeda, Hamzah Ahmad, Nur Aqilah Othman, Zulkifli Musa, Saifudin Razali, Jose Navarro-Pedreño, Ignacio Melendez-Pastor, Ignacio Gómez, Antonis A. Zorpas, Dus?an Krokavec, Anna Filasova? Stefan Breban, Ioana Gros, Calin Marginean, Petre Dorel Teodosescu, Brandon Matthew Cook, Timothy Arnett, Kelly Cohen, Anatoly Khomonenko, Sergey Adadurov, Aleksandr Krasnovidov, Teodor Lucian Grigorie, Ruxandra Botez

Published by InTech
Janeza Trdine 9, 51000 Rijeka, Croatia

Publishing Process Manager Romina Rovan
Technical Editor SPi Global
Cover InTech Design team

First published August, 2017
Printed in Croatia
Legal deposit, Croatia: National and University Library in Zagreb

Additional hard copies can be obtained from orders@intechopen.com

Modern Fuzzy Control Systems and Its Applications, Edited by S. Ramakrishnan
p. cm.
Print ISBN 978-953-51-3389-6
Online ISBN 978-953-51-3390-2

For EU product safety concerns:
IN TECH d.o.o., Prolaz Marije Krucifikse Kozulić 3, 51000 Rijeka, Croatia,
info@intechopen.com or visit our website at intechopen.com.

Meet the editor

Dr. S. Ramakrishnan is a professor and the head of the Department of Information Technology, Dr. Mahalingam College of Engineering and Technology, Pollachi, India. Dr. Ramakrishnan is a reviewer of 25 international journals such as IEEE transactions, IET journals, *ACM Computing Reviews*, Elsevier science journals, Springer journals, Wiley journals, etc. He is in the editorial board of seven international journals. He is a guest editor of special issues in three international journals including the *Telecommunication Systems Journal* of Springer. He has published 139 papers and 6 books. Dr. S. Ramakrishnan has published a book on wireless sensor networks for CRC Press, USA; four books on speech processing and pattern recognition for InTech Publisher, Croatia; and a book on computational techniques for Lambert Academic Publishing, Germany.

Contents

Preface

Control systems play an important role in engineering. Fuzzy logic is the natural choice for designing control applications and is the most popular and appropriate for the control of home and industrial appliances. Academic and industrial experts are constantly researching and proposing innovative and effective fuzzy control systems. This book is an edited volume and has twenty-one innovative chapters arranged into five sections covering applications of fuzzy control systems in energy and power systems, navigation systems, imaging, and industrial engineering.

Section 1 is on modern fuzzy control systems and has five valuable chapters written by leading researchers. These five chapters provide strong theoretical-cum-application-orientated presentations. Improvisations of conventional models such as Takagi-Sugeno and Mamdani models along with numerical illustrations and applications are provided in these five chapters.

Section 2 is on energy and power system applications and has four well-written chapters by scholarly authors. All these four chapters provide requisite mathematical models, illustrations, and experimental analysis.

Section 3 is on navigation systems and has four rich chapters written by eminent authors. Out of these four chapters, two chapters focus on mobile navigations, and the other two chapters focus on inertial and aerial navigations. These four chapters provide not only theoretical stuffs but also experimental results.

Section 4 is on imaging and healthcare applications and comprises three chapters written by renowned authors. This section covers object-based image analysis, camera calibration, and prosthetic hand.

The last section is on industrial engineering applications containing five potential chapters written by distinguished authors. The applications are presented with mathematical models and simulation results. The five applications include control systems for fruit juice-producing company, MIMO systems in mechatronics, vibration control, wastewater treatment plant and soil vulnerability testing.

Overall, this book provides a rich set of modern fuzzy control systems and their applications and will be a useful resource for the graduate students, researchers, and practicing engineers in the field of electrical engineering.

I would like to express my sincere thanks to all authors for their contribution and effort to bring in this wonderful book. My earnest gratitude and appreciation are extended to the InTe-chOpen Publisher, in particular Ms. Romina Rovan. I would like to express my heartfelt thanks to the management, secretary, and principal of my institute. Finally, I would like to extend my dearest thanks to my family members and in particular to my cute daughter Abira-mi.

Dr. S. Ramakrishnan
Professor and Head, Department of Information Technology
Dr. Mahalingam College of Engineering and Technology

Pollachi, India

Section 1

Modern Fuzzy Control Systems

Chapter 1

Stabilizing Fuzzy Control via Output Feedback

Dušan Krokavec and Anna Filasová

Additional information is available at the end of the chapter

http://dx.doi.org/10.5772/68129

Abstract

The chapter presents new conditions suitable in design of stabilizing static as well as dynamic output controllers for a class of continuous-time nonlinear systems represented by Takagi-Sugeno models. Taking into account the affine properties of the TS model structure, and applying the fuzzy control scheme relating to the parallel-distributed output compensators, the sufficient design conditions are outlined in the terms of linear matrix inequalities. Depending on the proposed procedures, the Lyapunov matrix can be decoupled from the system parameter matrices using linear matrix inequality techniques or a fuzzy-relaxed approach can be applied to make closed-loop dynamics faster. Numerical examples illustrate the design procedures and demonstrate the performances of the proposed design methods.

Keywords: continuous-time nonlinear systems, Takagi-Sugeno fuzzy systems, linear matrix inequality approach, parallel-distributed compensation, output feedback

1. Introduction

Contrarily to the linear framework, nonlinear systems are too complex to be represented by unified mathematical resources and so, a generic method has not been developed yet to design a controller valid for all types of nonlinear systems. An alternative to nonlinear system models is Takagi-Sugeno (TS) fuzzy approach [1], which benefits from the advantages of suitable linear approximation of sector nonlinearities. Using the TS fuzzy model, each rule utilizes the local system dynamics by a linear model and the nonlinear system is represented by a collection of fuzzy rules. Recently, TS model-based fuzzy control approaches are being fast and successfully used in nonlinear control frameworks. As a result, a range of stability analysis conditions [2–5] as well as control design methods have been developed for TS fuzzy systems [6–9], relying mostly on the feasibility of an associated set of linear matrix inequalities

(LMI) [10]. An important fact is that the design problem is a standard feasibility problem with several LMIs, potentially combined with one matrix equality to overcome the problem of bilinearity. In consequence, the state and output feedback control based on fuzzy TS systems models is mostly realized in such structures, which can be designed using numerical techniques based on LMIs.

The TS fuzzy model-based state control is based on an implicit assumption that all states are available for measurement. If it is impossible, TS fuzzy observers are used to estimate the unmeasurable state variables, and the state controller exploits the system state variable estimate values [11–14]. The nonlinear output feedback design is so formulated as two LMI set problem, and treated by the two-stage procedure using the separation principle, that is, dealing with a set of LMIs for the observer parameters at first and then solving another set of LMIs for the controller parameters [15]. Since, the fuzzy output control does not require the measurement of system state variables and can be formulated as a one LMI set problem, such structure of feedback control is preferred, of course, if the system is stabilizable.

From a relatively wide range of problems associated with the fuzzy output feedback control design for the continuous-time nonlinear MIMO systems approximated by a TS model, the chapter deals with the techniques incorporating the slack matrix application and fuzzy membership-relaxed approaches. The central idea of the TS fuzzy model-based control design, that is, to derive control rules so as to compensate each rule of a fuzzy system and construct the control strategy based on the parallel-distributed compensators (PDC), is reflected in the approach of output control. Motivated by the above mentioned observations, the proposed design method respects the results presented in Refs. [16, 17], and is constructed on an enhanced form of quadratic Lyapunov function. Comparing with the approaches based only on quadratic Lyapunov matrix [18], which are particular in the case of large number of rules, that are very conservative as a common symmetric positive definite matrix, is used to verify all Lyapunov inequalities, presented principle naturally extends the affine TS model properties using slack matrix variables to decouple Lyapunov matrix and the system matrices in LMIs, and gives substantial reducing of conservativeness. Moreover, extra quadratic constraints are included to incorporate fuzzy membership functions relaxes [19, 20] and applied for static as well as dynamic TS fuzzy output controllers design. Note, other constraints with respect to, for example, to decay rate and closed-loop pole clustering can be utilized to extend the proposed design procedures.

The remainder of this chapter is organized as follows. In Section 2, the structure of TS model for considered class of nonlinear systems is briefly described, and some of its properties are outlined. The output feedback control design problem for systems with measurable promise variables is given in Section 3, where the design conditions that guarantees global quadratic stability are formulated and proven. To complete the solutions, Section 4 formulate the static decoupling principle in static TS fuzzy output control, and the method is reformulated in Section 5 in defined criteria for TS fuzzy dynamic output feedback control design. Section 6 gives the numerical examples to illustrate the effectiveness of the proposed approach, and to confirm the validity of the control scheme. The last section, Section 7, draws conclusions and some future directions.

Throughout the chapter, the following notations are used: $\boldsymbol{x}^T$, $\boldsymbol{X}^T$ denotes the transpose of the vector $\boldsymbol{x}$ and matrix $\boldsymbol{X}$, respectively, for a square matrix $\boldsymbol{X} = \boldsymbol{X}^T > 0$ (respectively, $\boldsymbol{X} = \boldsymbol{X}^T < 0$)

http://dx.doi.org/10.5772/68129

means that $\boldsymbol{X}$ is a symmetric positive definite matrix (respectively, negative definite matrix), the symbol $\boldsymbol{I}_n$ represents the n-th order unit matrix, $\mathbb{R}$ denotes the set of real numbers, and $\mathbb{R}^{n \times r}$ denotes the set of all $n \times r$ real matrices.

2. Takagi-Sugeno fuzzy models

The systems under consideration are from one class of multi-input and multi-output (MIMO) dynamic systems, which are nonlinear in sectors and represented by TS fuzzy model. Constructing the set of membership functions $h_i\ (\boldsymbol{\theta}(t))$, $i = 1, 2, \ldots, s$, where

$$\boldsymbol{\theta}(t) = \begin{bmatrix} \theta_1(t) & \theta_2(t) & \cdots & \theta_q(t) \end{bmatrix}, \tag{1}$$

is the vector of premise variables, the final states of the systems are inferred in the TS fuzzy system model as follows

$$\dot{\boldsymbol{q}}(t) = \sum_{i=1}^{s} h_i(\boldsymbol{\theta}(t))(\boldsymbol{A}_i\boldsymbol{q}(t) + \boldsymbol{B}_i\boldsymbol{u}(t)), \tag{2}$$

with the output given by the relation

$$\boldsymbol{y}(t) = \boldsymbol{C}\boldsymbol{q}(t), \tag{3}$$

where $\boldsymbol{q}(t) \in \mathbb{R}^n$, $\boldsymbol{u}(t) \in \mathbb{R}^r$, $\boldsymbol{y}(t) \in \mathbb{R}^m$ are vectors of the state, input, and output variables, $\boldsymbol{A}_i \in \mathbb{R}^{n \times n}$, $\boldsymbol{B}_i \in \mathbb{R}^{n \times r}$, $\boldsymbol{C} \in \mathbb{R}^{m \times n}$ are real finite values matrix, and where $h_i(\boldsymbol{\theta}(t))$ is the averaging weight for the i-th rule, representing the normalized grade of membership (membership function).

By definition, the membership functions satisfy the following convex sum properties.

$$0 \le h_i(\boldsymbol{\theta}(t)) \le 1, \quad \sum_{i=1}^{s} h_i(\boldsymbol{\theta}(t)) = 1 \quad \forall i \in \langle 1, \ldots, s \rangle. \tag{4}$$

It is assumed that the premise variable is a system state variable or a measurable external variable, and none of the premise variables depends on the inputs $\boldsymbol{u}(t)$.

It is evident that a general fuzzy model is achieved by fuzzy amalgamation of the linear system models. Using a TS model, the conclusion part of a single rule consists no longer of a fuzzy set [21], but determines a function with state variables as arguments, and the corresponding function is a local function for the fuzzy region that is described by the premise part of the rule. Thus, using linear functions, a system state is described locally (in fuzzy regions) by linear models, and at the boundaries between regions an interpolation is used between the corresponding local models.

Note, the models, Eqs. (2) and (3), are mostly considered for analysis, control, and state estimation of nonlinear systems.

Assumption 1 *Each triplet* ($\boldsymbol{A}_i$, $\boldsymbol{B}_i$, $\boldsymbol{C}$) *is locally controllable and observable, the matrix* $\boldsymbol{C}$ *is the same for all local models.*

It is supposed in the next that the aforementioned model does not include parameter uncertainties or external disturbances, and the premise variables are measured.

3. Static fuzzy output controller

In the next, the fuzzy output controller is designed using the concept of parallel-distributed compensation, in which the fuzzy controller shares the same sets of normalized membership functions like the TS fuzzy system model.

Definition 1 *Considering Eqs. (2) and (3), and using the same set of normalized membership function Eq. (4), the fuzzy static output controller is defined as*

$$\boldsymbol{u}(t) = \sum_{j=1}^{s} h_j(\boldsymbol{\theta}(t))\boldsymbol{K}_j\boldsymbol{y}(t) = \sum_{j=1}^{s} h_j(\boldsymbol{\theta}(t))\boldsymbol{K}_j\boldsymbol{C}\boldsymbol{q}(t). \tag{5}$$

Note that the fuzzy controller Eq. (5) is in general nonlinear.

Considering the system, Eqs. (2) and (3), and the control law, Eq. (5), yields

$$\dot{\boldsymbol{q}}(t) = \sum_{i=1}^{s}\sum_{j=1}^{s} h_i(\boldsymbol{\theta}(t))h_j(\boldsymbol{\theta}(t))\left(\boldsymbol{A}_i + \boldsymbol{B}_i\boldsymbol{K}_j\boldsymbol{C}\right)\boldsymbol{q}(t) = \sum_{i=1}^{s}\sum_{j=1}^{s} h_i(\boldsymbol{\theta}(t))h_j(\boldsymbol{\theta}(t))\boldsymbol{A}_{cij}\boldsymbol{q}(t), \tag{6}$$

$$\boldsymbol{A}_{cij} = \boldsymbol{A}_i + \boldsymbol{B}_i\boldsymbol{K}_j\boldsymbol{C}, \quad \boldsymbol{A}_{cji} = \boldsymbol{A}_j + \boldsymbol{B}_j\boldsymbol{K}_i\boldsymbol{C}. \tag{7}$$

Proposition 1 (*standard design conditions*). *The equilibrium of the fuzzy system Eqs.* (2) *and* (3), *controlled by the fuzzy controller Eq.* (5), *is global asymptotically stable if there exist a positive definite symmetric matrix* $\boldsymbol{W} \in I\!R^{n \times n}$ *and matrices* $\boldsymbol{Y}_j \in I\!R^{r \times m}$, $\boldsymbol{H} \in I\!R^{m \times m}$ *such that*

$$\boldsymbol{W} = \boldsymbol{W}^T > 0, \tag{8}$$

$$\boldsymbol{A}_i\boldsymbol{W} + \boldsymbol{W}\boldsymbol{A}_i^T + \boldsymbol{B}_i\boldsymbol{Y}_i\boldsymbol{C} + \boldsymbol{C}^T\boldsymbol{Y}_i^T\boldsymbol{B}_i^T < 0, \tag{9}$$

$$\frac{\boldsymbol{A}_i\boldsymbol{W} + \boldsymbol{W}\boldsymbol{A}_i^T}{2} + \frac{\boldsymbol{A}_j\boldsymbol{W} + \boldsymbol{W}\boldsymbol{A}_j^T}{2} + \frac{\boldsymbol{B}_i\boldsymbol{Y}_j\boldsymbol{C} + \boldsymbol{C}^T\boldsymbol{Y}_j^T\boldsymbol{B}_i^T}{2} + \frac{\boldsymbol{B}_j\boldsymbol{Y}_i\boldsymbol{C} + \boldsymbol{C}^T\boldsymbol{Y}_i^T\boldsymbol{B}_j^T}{2} < 0, \tag{10}$$

$$\boldsymbol{C}\boldsymbol{W} = \boldsymbol{H}\boldsymbol{C} \tag{11}$$

for $i = 1, 2, \ldots, s$ as well as $i = 1, 2, \ldots, s-1$, $j = i+1, i+2, \ldots, s$, and $h_i(\boldsymbol{\theta}(t))h_j(\boldsymbol{\theta}(t)) \neq 0$.

When the above conditions hold, the control law gain matrices are given as

$$\boldsymbol{K}_i = \boldsymbol{Y}_i\boldsymbol{H}^{-1}. \tag{12}$$

Proof. (compare, for example, Ref. [16]) Prescribing the Lyapunov function candidate of the form

$$v(\boldsymbol{q}(t)) = \boldsymbol{q}^T(t)\boldsymbol{P}\boldsymbol{q}(t) > 0, \tag{13}$$

where $\boldsymbol{P} \in \mathbb{R}^{n \times n}$ is a symmetric positive definite matrix, the time derivative of Eq. (13) along the system trajectory is

$$\dot{v}(\boldsymbol{q}(t)) = \dot{\boldsymbol{q}}^T(t)\boldsymbol{P}\boldsymbol{q}(t) + \boldsymbol{q}^T(t)\boldsymbol{P}\dot{\boldsymbol{q}}(t) < 0. \tag{14}$$

Inserting Eq. (6) into Eq. (14), it has to be satisfied

$$\dot{v}(\boldsymbol{q}(t)) = \sum_{i=1}^{s}\sum_{j=1}^{s} h_i(\boldsymbol{\theta}(t))h_j(\boldsymbol{\theta}(t))\boldsymbol{q}^T(t)\boldsymbol{P}_{cij}\boldsymbol{q}(t) < 0, \tag{15}$$

$$\boldsymbol{P}_{cij} = \boldsymbol{P}\boldsymbol{A}_{cij} + \boldsymbol{A}_{cij}^T\boldsymbol{P}. \tag{16}$$

Since $\boldsymbol{P}$ is positive definite, the state coordinate transform can be defined as

$$\boldsymbol{q}(t) = \boldsymbol{W}\boldsymbol{p}(t), \quad \boldsymbol{W} = \boldsymbol{P}^{-1}, \tag{17}$$

and subsequently, Eqs. (15) and (16) can be rewritten as

$$\dot{v}(\boldsymbol{p}(t)) = \sum_{i=1}^{s}\sum_{j=1}^{s} h_i(\boldsymbol{\theta}(t))h_j(\boldsymbol{\theta}(t))\boldsymbol{p}^T(t)\boldsymbol{W}_{cij}\boldsymbol{p}(t) < 0, \tag{18}$$

$$\boldsymbol{W}_{cij} = \boldsymbol{A}_{cij}\boldsymbol{W} + \boldsymbol{W}\boldsymbol{A}_{cij}^T. \tag{19}$$

Permuting the subscripts i and j in Eq. (18), also it can write

$$\dot{v}(\boldsymbol{q}(t)) = \sum_{i=1}^{s}\sum_{j=1}^{s} h_i(\boldsymbol{\theta}(t))h_j(\boldsymbol{\theta}(t))\boldsymbol{p}^T(t)\boldsymbol{W}_{cji}\boldsymbol{p}(t) < 0, \tag{20}$$

$$\boldsymbol{W}_{cji} = \boldsymbol{A}_{cji}\boldsymbol{W} + \boldsymbol{W}\boldsymbol{A}_{cji}^T. \tag{21}$$

Thus, adding Eqs. (17) and (19), it yields

$$2\dot{v}(\boldsymbol{p}(t)) = \sum_{i=1}^{s}\sum_{j=1}^{s} h_i(\boldsymbol{\theta}(t))h_j(\boldsymbol{\theta}(t))\boldsymbol{p}^T(t)\big(\boldsymbol{W}_{cij} + \boldsymbol{W}_{cji}\big)\boldsymbol{p}(t) < 0 \tag{22}$$

and subsequently,

$$\dot{v}(\boldsymbol{p}(t)) = \sum_{i=1}^{s} h_i^2(\boldsymbol{\theta}(t))\boldsymbol{p}^T(t)\boldsymbol{W}_{cii}\boldsymbol{p}(t) + 2\sum_{i=1}^{s-1}\sum_{j=i+1}^{s} h_i(\boldsymbol{\theta}(t))h_j(\boldsymbol{\theta}(t))\boldsymbol{p}^T(t)\frac{\boldsymbol{W}_{cij}+\boldsymbol{W}_{cji}}{2}\boldsymbol{p}(t) < 0, \quad (23)$$

which leads to the set of inequalities.

$$(\boldsymbol{A}_i + \boldsymbol{B}_i\boldsymbol{K}_i\boldsymbol{C})\boldsymbol{W} + \boldsymbol{W}(\boldsymbol{A}_i + \boldsymbol{B}_i\boldsymbol{K}_i\boldsymbol{C})^T < 0, \quad (24)$$

$$\frac{(\boldsymbol{A}_i + \boldsymbol{B}_i\boldsymbol{K}_j\boldsymbol{C})\boldsymbol{W}}{2} + \frac{(\boldsymbol{A}_j + \boldsymbol{B}_j\boldsymbol{K}_i\boldsymbol{C})\boldsymbol{W}}{2} + \frac{\boldsymbol{W}(\boldsymbol{A}_i + \boldsymbol{B}_i\boldsymbol{K}_j\boldsymbol{C})^T}{2} + \frac{\boldsymbol{W}(\boldsymbol{A}_j + \boldsymbol{B}_j\boldsymbol{K}_i\boldsymbol{C})^T}{2} < 0 \quad (25)$$

for $i = 1, 2, \ldots, s$ as well as $i = 1, 2, \ldots, s - 1$, $j = 1 + 1, i + 2, \ldots, s$ and $h_i(\boldsymbol{\theta}(t))h_j(\boldsymbol{\theta}(t)) \neq 0$.

Thus, setting here

$$\boldsymbol{K}_j\boldsymbol{C}\boldsymbol{W} = \boldsymbol{K}_j\boldsymbol{H}\boldsymbol{H}^{-1}\boldsymbol{C}\boldsymbol{W}, \quad (26)$$

where $\boldsymbol{H}$ is a regular square matrix of appropriate dimension and defining

$$\boldsymbol{H}^{-1}\boldsymbol{C} = \boldsymbol{C}\boldsymbol{W}^{-1}, \quad \boldsymbol{Y}_j = \boldsymbol{K}_j\boldsymbol{H}, \quad (27)$$

the LMI forms of Eqs. (9) and (10) are obtained from Eqs. (24) and (25), respectively, and Eq. (27) implies Eq. (11). This concludes the proof.

Trying to minimize the number of LMIs owing to the limitation of solvers, Proposition 1 is presented in the structure, in which the number of stabilization conditions, used in fuzzy controller design, is equal to $N = (s^2 + s)/2 + 1$. Evidently, the number of stabilization conditions is substantially reduced if s is large.

Proposition 2 (*enhanced design conditions*). *The equilibrium of the fuzzy system Eqs.* (2) *and* (3), *controlled by the fuzzy controller Eq.* (5), *is global asymptotically stable if for given a positive* $\delta \in \mathbb{R}$, *there exist positive definite symmetric matrices* $\boldsymbol{V}, \boldsymbol{S} \in \mathbb{R}^{n \times n}$, *and matrices* $\boldsymbol{Y}_j \in \mathbb{R}^{r \times m}$, $\boldsymbol{H} \in \mathbb{R}^{m \times m}$ *such that*

$$\boldsymbol{S} = \boldsymbol{S}^T > 0, \quad \boldsymbol{V} = \boldsymbol{V}^T > 0, \quad (28)$$

$$\begin{bmatrix} \boldsymbol{A}_i\boldsymbol{S} + \boldsymbol{S}\boldsymbol{A}_i^T + \boldsymbol{B}_i\boldsymbol{Y}_i\boldsymbol{C} + \boldsymbol{C}^T\boldsymbol{Y}_i^T\boldsymbol{B}_i^T & * \\ \boldsymbol{V} - \boldsymbol{S} + \delta\boldsymbol{A}_i\boldsymbol{S} + \delta\boldsymbol{B}_i\boldsymbol{Y}_i\boldsymbol{C} & -2\delta\boldsymbol{S} \end{bmatrix} < 0, \quad (29)$$

$$\begin{bmatrix} \boldsymbol{\Phi}_{ij} & * \\ \boldsymbol{V} - \boldsymbol{S} + \delta\frac{\boldsymbol{A}_i\boldsymbol{S} + \boldsymbol{A}_j\boldsymbol{S}}{2} + \delta\frac{\boldsymbol{B}_i\boldsymbol{Y}_j + \boldsymbol{B}_j\boldsymbol{Y}_i}{2}\boldsymbol{C} & -2\delta\boldsymbol{S} \end{bmatrix} < 0, \quad (30)$$

$$\boldsymbol{C}\boldsymbol{S} = \boldsymbol{H}\boldsymbol{C}, \quad (31)$$

for $i = 1, 2, \ldots, s$, *as well as* $i = 1, 2, \ldots, s - 1$, $j = 1 + 1, i + 2, \ldots, s$, $h_i(\boldsymbol{\theta}(t))h_j(\boldsymbol{\theta}(t)) \neq 0$, *and*

$$\boldsymbol{\Phi}_{ij} = \frac{\boldsymbol{A}_i\boldsymbol{S} + \boldsymbol{S}\boldsymbol{A}_i^T}{2} + \frac{\boldsymbol{A}_j\boldsymbol{S} + \boldsymbol{S}\boldsymbol{A}_j^T}{2} + \frac{\boldsymbol{B}_i\boldsymbol{Y}_j\boldsymbol{C} + \boldsymbol{C}^T\boldsymbol{Y}_j^T\boldsymbol{B}_i^T}{2} + \frac{\boldsymbol{B}_j\boldsymbol{Y}_i\boldsymbol{C} + \boldsymbol{C}^T\boldsymbol{Y}_i^T\boldsymbol{B}_j^T}{2}. \quad (32)$$

When the above conditions hold, the control law gain matrices are given as

$$\boldsymbol{K}_i = \boldsymbol{Y}_i\boldsymbol{H}^{-1}. \quad (33)$$

Here and hereafter, $*$ denotes the symmetric item in a symmetric matrix.

Proof. Writing Eq. (6) in the form

$$\sum_{i=1}^{s}\sum_{j=1}^{s} h_i(\boldsymbol{\theta}(t))h_j(\boldsymbol{\theta}(t))\big(\boldsymbol{A}_{cij}\boldsymbol{q}(t) - \dot{\boldsymbol{q}}(t)\big) = \boldsymbol{0}, \quad (34)$$

then with an arbitrary symmetric positive definite matrix $\boldsymbol{S} \in \mathbb{R}^{n \times n}$ and a positive scalar $\delta \in \mathbb{R}$, it yields

$$\sum_{i=1}^{s}\sum_{j=1}^{s} \mathrm{h}_i(\boldsymbol{\theta}(t))\mathrm{h}_j(\boldsymbol{\theta}(t))\big(\boldsymbol{q}^T(t)\boldsymbol{S} + \delta\dot{\boldsymbol{q}}^T(t)\boldsymbol{S}\big)\big(\boldsymbol{A}_{cij}\boldsymbol{q}(t) - \dot{\boldsymbol{q}}(t)\big) = 0. \quad (35)$$

Since $\boldsymbol{S}$ is positive definite, the new state variable coordinate system can be introduced so that

$$\boldsymbol{p}(t) = \boldsymbol{S}\boldsymbol{q}(t), \quad \dot{\boldsymbol{p}}(t) = \boldsymbol{S}\dot{\boldsymbol{q}}(t), \quad \boldsymbol{V} = \boldsymbol{S}^{-1}\boldsymbol{P}\boldsymbol{S}^{-1}. \quad (36)$$

Therefore, Eq. (14) can be rewritten as

$$\dot{v}(\boldsymbol{p}(t)) = \dot{\boldsymbol{p}}^T(t)\boldsymbol{V}\boldsymbol{p}(t) + \boldsymbol{p}^T(t)\boldsymbol{V}\dot{\boldsymbol{p}}(t) < 0 \quad (37)$$

and Eq. (35) takes the form

$$\sum_{i=1}^{s}\sum_{j=1}^{s} \mathrm{h}_i(\boldsymbol{\theta}(t))\mathrm{h}_j(\boldsymbol{\theta}(t))\big(\boldsymbol{p}^T(t) + \delta\dot{\boldsymbol{p}}^T(t)\big)\big(\boldsymbol{A}_{cij}\boldsymbol{S}\boldsymbol{p}(t) - \boldsymbol{S}\dot{\boldsymbol{p}}(t)\big) = 0. \quad (38)$$

Thus, adding Eq. (38) as well as the transposition of Eq. (38) to Eq. (37), it yields

$$\begin{aligned}\dot{v}(\boldsymbol{p}(i)) &= \dot{\boldsymbol{p}}^T(t)\boldsymbol{V}\boldsymbol{p}(t) + \boldsymbol{p}^T(t)\boldsymbol{V}\dot{\boldsymbol{p}}(t) \\ &\quad + \sum_{i=1}^{s}\sum_{j=1}^{s} h_i(\boldsymbol{\theta}(t))\mathrm{h}_j(\boldsymbol{\theta}(t))\big(\boldsymbol{p}^T(t) + \delta\dot{\boldsymbol{p}}^T(t)\big)\big(\boldsymbol{A}_{cij}\boldsymbol{S}\boldsymbol{p}(t) - \boldsymbol{S}\dot{\boldsymbol{p}}(t)\big) \\ &\quad + \sum_{i=1}^{s}\sum_{j=1}^{s} h_i(\boldsymbol{\theta}(t))\mathrm{h}_j(\boldsymbol{\theta}(t))\Big(\boldsymbol{A}_{cij}\boldsymbol{S}\boldsymbol{p}(t) - \boldsymbol{S}\dot{\boldsymbol{p}}(t)\big)^T\big(\boldsymbol{p}(t) + \delta\dot{\boldsymbol{p}}(t)\big) < 0.\end{aligned} \quad (39)$$

Using the notation

$$\boldsymbol{p}_c^T(t) = \left[\boldsymbol{p}^T(t) \quad \dot{\boldsymbol{p}}^T(t)\right], \tag{40}$$

the inequality Eq. (39) can be written as

$$\dot{v}(\boldsymbol{p}_c(t)) = \sum_{i=1}^{s}\sum_{j=1}^{s} h_i(\boldsymbol{\theta}(t))\mathrm{h}_j(\boldsymbol{\theta}(t))\boldsymbol{p}_c^T(t)\boldsymbol{S}_{cij}\boldsymbol{p}_c(t) < 0, \tag{41}$$

$$\boldsymbol{S}_{cij} = \begin{bmatrix} (\boldsymbol{A}_i + \boldsymbol{B}_i\boldsymbol{K}_j\boldsymbol{C})\boldsymbol{S} + \boldsymbol{S}(\boldsymbol{A}_i + \boldsymbol{B}_i\boldsymbol{K}_j\boldsymbol{C})^T & * \\ \boldsymbol{V} - \boldsymbol{S} + \delta(\boldsymbol{A}_i + \boldsymbol{B}_i\boldsymbol{K}_j\boldsymbol{C})\boldsymbol{S} & -2\delta\boldsymbol{S} \end{bmatrix} < 0. \tag{42}$$

Permuting the subscripts i and j in Eq. (41), and following the way used above, analogously it can obtain

$$\dot{v}(\boldsymbol{p}_c(t)) = \sum_{i=1}^{s}\mathrm{h}_i^2(\boldsymbol{\theta}(t))\boldsymbol{p}_c^T(t)\boldsymbol{S}_{cii}\boldsymbol{p}_c(t) + 2\sum_{i=1}^{s-1}\sum_{j=i+1}^{s}\mathrm{h}_i(\boldsymbol{\theta}(t))\mathrm{h}_j(\boldsymbol{\theta}(t))\boldsymbol{p}_c^T(t)\frac{\boldsymbol{S}_{cij} + \boldsymbol{S}_{cji}}{2}\boldsymbol{p}_c(t) < 0. \tag{43}$$

Since $r = m$, it is now possible to set

$$\boldsymbol{K}_j\boldsymbol{C}\boldsymbol{S} = \boldsymbol{K}_j\boldsymbol{H}\boldsymbol{H}^{-1}\boldsymbol{C}\boldsymbol{S}, \tag{44}$$

where $\boldsymbol{H}$ is a regular square matrix of appropriate dimension and introducing

$$\boldsymbol{H}^{-1}\boldsymbol{C} = \boldsymbol{C}\boldsymbol{S}^{-1}, \quad \boldsymbol{Y}_j = \boldsymbol{K}_j\boldsymbol{H} \tag{45}$$

then Eqs. (42) and (45) imply Eqs. (29)–(31). This concludes the proof.

Note, Eq. (42) leads to the set of LMIs only if δ is a prescribed constant. (δ can be considered as a tuning parameter). Considering δ as a LMI variable, Eq. (42) represents the set of bilinear matrix inequalities (BMI).

Theorem 1 (*enhanced relaxed design conditions*). *The equilibrium of the fuzzy system Eqs.* (2) *and* (3), *controlled by the fuzzy controller Eq.* (5), *is global asymptotically stable if for given a positive* $\delta \in \mathbb{R}$ *there exist positive definite symmetric matrices* $\boldsymbol{V}, \boldsymbol{S} \in \mathbb{R}^{n \times n}$, *the matrices* $\boldsymbol{X}_{ij} = \boldsymbol{X}_{ji}^T \in \mathbb{R}^{r\times n}$, *and* $\boldsymbol{Y}_j \in \mathbb{R}^{r \times m}$, $\boldsymbol{H} \in \mathbb{R}^{m \times m}$ *such that*

$$\boldsymbol{S} = \boldsymbol{S}^T > 0, \quad \boldsymbol{V} = \boldsymbol{V}^T > 0, \quad \begin{bmatrix} \boldsymbol{X}_{11} & \boldsymbol{X}_{12} & \cdots & \boldsymbol{X}_{1s} \\ \boldsymbol{X}_{21} & \boldsymbol{X}_{22} & \cdots & \boldsymbol{X}_{2s} \\ \vdots & \vdots & \ddots & \vdots \\ \boldsymbol{X}_{s1} & \boldsymbol{X}_{s2} & \cdots & \boldsymbol{X}_{ss} \end{bmatrix} > 0, \tag{46}$$

$$\begin{bmatrix} \boldsymbol{A}_i\boldsymbol{S} + \boldsymbol{S}\boldsymbol{A}_i^T + \boldsymbol{B}_i\boldsymbol{Y}_i\boldsymbol{C} + \boldsymbol{C}^T\boldsymbol{Y}_i^T\boldsymbol{B}_i^T + \boldsymbol{X}_{ii} & * \\ \boldsymbol{V} - \boldsymbol{S} + \delta\boldsymbol{A}_i\boldsymbol{S} + \delta\boldsymbol{B}_i\boldsymbol{Y}_i\boldsymbol{C} & -2\delta\boldsymbol{S} \end{bmatrix} < 0, \tag{47}$$

$$\begin{bmatrix} \boldsymbol{\Phi}_{ij} & * \\ \boldsymbol{V} - \boldsymbol{S} + \delta\frac{\boldsymbol{A}_i\boldsymbol{S} + \boldsymbol{A}_j\boldsymbol{S}}{2} + \delta\frac{\boldsymbol{B}_i\boldsymbol{Y}_j + \boldsymbol{B}_j\boldsymbol{Y}_i}{2}\boldsymbol{C} & -2\delta\boldsymbol{S} \end{bmatrix} < 0, \tag{48}$$

$$CS = HC, \tag{49}$$

for $i = 1, 2, \ldots, s$, *as well as* $i = 1, 2, \ldots, s-1$, $j = 1+1, i+2, \ldots, s$, $h_i(\boldsymbol{\theta}(t))h_j(\boldsymbol{\theta}(t)) \neq 0$ *and*

$$\boldsymbol{\Phi}_{ij} = \frac{\boldsymbol{A}_i\boldsymbol{S} + \boldsymbol{S}\boldsymbol{A}_i^T}{2} + \frac{\boldsymbol{A}_j\boldsymbol{S} + \boldsymbol{S}\boldsymbol{A}_j^T}{2} + \frac{\boldsymbol{B}_i\boldsymbol{Y}_j\boldsymbol{C} + \boldsymbol{C}^T\boldsymbol{Y}_j^T\boldsymbol{B}_i^T}{2} + \frac{\boldsymbol{B}_j\boldsymbol{Y}_i\boldsymbol{C} + \boldsymbol{C}^T\boldsymbol{Y}_i^T\boldsymbol{B}_j^T}{2} + \frac{\boldsymbol{X}_{ij} + \boldsymbol{X}_{ji}}{2}. \tag{50}$$

When the above conditions hold, the control law gain matrices are given as

$$\boldsymbol{K}_i = \boldsymbol{Y}_i\boldsymbol{H}^{-1}. \tag{51}$$

Proof. Introducing the positive real term

$$v_v(\boldsymbol{\theta}(t)) = \boldsymbol{q}^T(t)\boldsymbol{Z}(\boldsymbol{\theta}(t))\boldsymbol{q}(t) > 0, \tag{52}$$

$$\boldsymbol{Z}(\boldsymbol{\theta}(t)) = \boldsymbol{Z}^T(\boldsymbol{\theta}(t)) = \sum_{i=1}^{s}\sum_{j=1}^{s} \mathrm{h}_i(\boldsymbol{\theta}(t))\mathrm{h}_j(\boldsymbol{\theta}(t))\boldsymbol{Z}_{ij} > 0, \tag{53}$$

where $\boldsymbol{Z}_{ij} = \boldsymbol{Z}_{ji}^T \in \mathbb{R}^{n\times n}$, $i, j = 1, 2, \ldots, s$ is the set of associated matrices and using the state coordinate transform Eq. (36), then Eq. (53) can be rewritten as

$$v_v(\boldsymbol{p}(t)) = \sum_{i=1}^{s}\sum_{j=1}^{s} \mathrm{h}_i(\boldsymbol{\theta}(t))\mathrm{h}_j(\boldsymbol{\theta}(t))\boldsymbol{p}^T(t)\boldsymbol{X}_{ij}\boldsymbol{p}(t) > 0, \quad \boldsymbol{X}_{ij} = \boldsymbol{S}^{-1}\boldsymbol{Z}_{ij}\boldsymbol{S}^{-1} = \boldsymbol{X}_{ji}^T, \tag{54}$$

where

$$\boldsymbol{Z}(\boldsymbol{\theta}(t)) = \left[\mathrm{h}_1(\boldsymbol{\theta}(t)\boldsymbol{p}(t) \quad \mathrm{h}_2(\boldsymbol{\theta}(t)\boldsymbol{p}(t) \quad \cdots \quad \mathrm{h}_s(\boldsymbol{\theta}(t)\boldsymbol{p}(t)\right] \begin{bmatrix} \boldsymbol{X}_{11} & \boldsymbol{X}_{12} & \cdots & \boldsymbol{X}_{1s} \\ \boldsymbol{X}_{21} & \boldsymbol{X}_{22} & \cdots & \boldsymbol{X}_{2s} \\ \vdots & \vdots & \ddots & \vdots \\ \boldsymbol{X}_{s1} & \boldsymbol{X}_{s2} & \cdots & \boldsymbol{X}_{ss} \end{bmatrix} \begin{bmatrix} \mathrm{h}_1(\theta(t)p(t) \\ \mathrm{h}_2(\theta(t)p(t) \\ \vdots \\ \mathrm{h}_s(\theta(t)p(t) \end{bmatrix} \tag{55}$$

is symmetric, an positive definite if Eq. (46) is satisfied. Then, in the sense of the Krasovskii theorem (see, for example, Ref. [22]), it can be set up in Eq. (39)

$$\begin{aligned} \dot{v}(\boldsymbol{p}(i)) &= \dot{\boldsymbol{p}}^T(t)\boldsymbol{V}\boldsymbol{p}(t) + \boldsymbol{p}^T(t)\boldsymbol{V}\dot{\boldsymbol{p}}(t) \\ &\quad + \sum_{i=1}^{s}\sum_{j=1}^{s} \mathrm{h}_i(\boldsymbol{\theta}(t))\mathrm{h}_j(\boldsymbol{\theta}(t))\left(\boldsymbol{p}^T(t) + \delta\dot{\boldsymbol{p}}^T(t)\right)\left(\boldsymbol{A}_{cij}\boldsymbol{S}\boldsymbol{p}(t) - \boldsymbol{S}\dot{\boldsymbol{p}}(t)\right) \\ &\quad + \sum_{i=1}^{s}\sum_{j=1}^{s} \mathrm{h}_i(\boldsymbol{\theta}(t))\mathrm{h}_j(\boldsymbol{\theta}(t))(\boldsymbol{A}_{cij}\boldsymbol{S}\boldsymbol{p}(t) - \boldsymbol{S}\dot{\boldsymbol{p}}(t))^T(\boldsymbol{p}(t) + \delta\dot{\boldsymbol{p}}(t)) \\ &< -\sum_{i=1}^{s}\sum_{j=1}^{s} \mathrm{h}_i(\boldsymbol{\theta}(t))\mathrm{h}_j(\boldsymbol{\theta}(t))\boldsymbol{p}^T(t)\boldsymbol{X}_{ij}\boldsymbol{p}(t) \\ &< 0, \end{aligned} \tag{56}$$

which in the consequence, modifies Eq. (42) as follows

$$S_{cij} = \begin{bmatrix} (A_i + B_iK_jC)S + S(A_i + B_iK_jC)^T + X_{ij} & * \\ V - S + \delta(A_i + B_iK_jC)S & -2\delta S \end{bmatrix} < 0. \quad (57)$$

Following the same way as in the proof of Proposition 2, then Eqs. (47) and (48) can be derived from Eq. (57), while Eq. (55) implies Eq. (46). This concludes the proof.

This principle naturally exploits the affine TS model properties. Introducing the slack matrix variable S into the LMIs, the system matrices are decoupled from the equivalent Lyapunov matrix V. Note, to respect the conditions $X_{1j} = X_{ji}^T$, the set of inequalities Eqs. (47) and (48) have to be constructed. In the opposite case, constructing a set on s^2 LMIs, the constraint conditions have to be set as $X_{1j} = X_{ij}^T > 0$, that is, the weighting matrices have to be symmetric positive definite.

Corollary 1 *Prescribing $S = V$ and using the Schur complement property, then Eq.* (57) *implies*

$$A_{cij}S + SA_{cij}^T + X_{ij} + 0.5\delta SA_{cij}^T\delta^{-1}S^{-1}\delta A_{cij}S < 0 \quad (58)$$

and for $\delta = 0$ evidently, it has to be

$$A_{cij}S + SA_{cij}^T + X_{ij} < 0. \quad (59)$$

Evidently, then Eqs. (47) and (48) imply

$$S(A_i + B_iK_iC)^T + (A_i + B_iK_iC)S + X_{ii} < 0, \quad (60)$$

$$\frac{(A_i + B_iK_jC)S}{2} + \frac{(A_j + B_jK_iC)S}{2} + \frac{S(A_i + B_iK_jC)^T}{2} + \frac{S(A_j + B_jK_iC)^T}{2} + \frac{X_{ij} + X_{ji}}{2} < 0. \quad (61)$$

Considering $S = W$ and comparing with Eqs. (23) and (24), then Eqs. (60) and (61) are the extended set of inequalities Eqs. (23) and (24). The result is that the equilibrium of the fuzzy system Eqs. (2) and (3), controlled by the fuzzy controller Eq. (5), is global asymptotically stable if there exist a positive definite symmetric matrices $S \in \mathbb{R}^{n \times n}$, the matrices $X_{1j} = X_{ji}^T \in \mathbb{R}^{r\times n}$, and $Y_j \in \mathbb{R}^{r \times n}$, $H \in \mathbb{R}^{m \times m}$ such that

$$S = S^T > 0, \quad \begin{bmatrix} X_{11} & X_{12} & \cdots & X_{1s} \\ X_{21} & X_{22} & \cdots & X_{2s} \\ \vdots & \vdots & \ddots & \vdots \\ X_{s1} & X_{s2} & \cdots & X_{ss} \end{bmatrix} > 0, \quad (62)$$

$$A_iS + SA_i^T + B_iY_iC + C^TY_i^TB_i^T + X_{ii} < 0, \quad (63)$$

$$\frac{A_iS + SA_i^T}{2} + \frac{A_jS + SA_j^T}{2} + \frac{B_iY_jC + C^TY_j^TB_i^T}{2} + \frac{B_jY_iC + C^TY_i^TB_j^T}{2}\frac{X_{ij} + X_{ji}}{2} < 0, \quad (64)$$

$$CS = HC, \quad (65)$$

http://dx.doi.org/10.5772/68129

for $i = 1, 2, \ldots, s$, as well as $i = 1, 2, \ldots, s-1$, $j = 1+1, i+2, \ldots, s$, and $h_i(\boldsymbol{\theta}(t))h_j(\boldsymbol{\theta}(t)) \neq 0$. Subsequently, if this set of LMIs is satisfied, the set of control law gain matrices is given as

$$\boldsymbol{K}_i = \boldsymbol{Y}_i \boldsymbol{H}^{-1}. \tag{66}$$

These LMIs form relaxed design conditions.

Note the derived results are linked to some existing finding when the design problem involves additive performance requirements and the relaxed quadratic stability conditions of fuzzy control systems (see, e.g., Refs. [11, 19]) are equivalently steered.

4. Forced mode in static output control

In practice, the plant with $r = m$ (square plants) is often encountered, since in this case, it is possible to associate with each output signal as a reference signal, which is expected to influence this wanted output. Such mode, reflecting nonzero set working points, is called the forced regime.

Definition 2 *A forced regime for the TS fuzzy system Eqs.* (2) *and* (3) *with the TS fuzzy static output controller Eq.* (5) *is foisted by the control policy*

$$\boldsymbol{u}(t) = \sum_{j=1}^{s} h_j(\boldsymbol{\theta}(t))\boldsymbol{K}_j \boldsymbol{y}(t) + \sum_{i=1}^{s}\sum_{j=1}^{s} h_i(\boldsymbol{\theta}(t))h_j(\boldsymbol{\theta}(t))\boldsymbol{W}_{ij}\boldsymbol{w}(t), \tag{67}$$

where $r = m$, $\boldsymbol{w}(i) \in \mathbb{R}^m$ is desired output signal vector, and $\boldsymbol{W}_{ij} \in \mathbb{R}^{m \times m}$, $i, j = 1, 2, \ldots s$ is the set of the signal gain matrices.

Lemma 1. *The static decoupling challenge is solvable if* $(\boldsymbol{A}_i, \boldsymbol{B}_i)$ *is stabilizable and*

$$rank\begin{bmatrix} \boldsymbol{A}_i & \boldsymbol{B}_i \\ \boldsymbol{C} & \boldsymbol{0} \end{bmatrix} = n + m. \tag{68}$$

Proof. If $(\boldsymbol{A}_i, \boldsymbol{B}_i)$ is stabilizable, it is possible to find $\boldsymbol{K}_j$ such that matrices $\boldsymbol{A}_{cij} = \boldsymbol{A}_i + \boldsymbol{B}_i\boldsymbol{K}_j\boldsymbol{C}$ are Hurwitz. Assuming that for such $\boldsymbol{K}_j$, it yields

$$\text{rank}\begin{bmatrix} \boldsymbol{A}_i & \boldsymbol{B}_i \\ \boldsymbol{C} & \boldsymbol{0} \end{bmatrix} = \text{rank}\begin{bmatrix} \boldsymbol{A}_i & \boldsymbol{B}_i \\ \boldsymbol{C} & \boldsymbol{0} \end{bmatrix}\begin{bmatrix} \boldsymbol{I}_n & \boldsymbol{0} \\ \boldsymbol{K}_j\boldsymbol{C} & \boldsymbol{I}_m \end{bmatrix} = \text{rank}\begin{bmatrix} \boldsymbol{A}_i + \boldsymbol{B}_i\boldsymbol{K}_j\boldsymbol{C} & \boldsymbol{B}_i \\ \boldsymbol{C} & \boldsymbol{0} \end{bmatrix}, \tag{69}$$

$$\text{rank}\begin{bmatrix} \boldsymbol{A}_i + \boldsymbol{B}_i\boldsymbol{K}_j\boldsymbol{C} & \boldsymbol{B}_i \\ \boldsymbol{C} & \boldsymbol{0} \end{bmatrix} = \text{rank}\begin{bmatrix} \boldsymbol{I}_n & \boldsymbol{0} \\ -\boldsymbol{C}(\boldsymbol{A}_i + \boldsymbol{B}_i\boldsymbol{K}_j\boldsymbol{C})^{-1} & \boldsymbol{I}_m \end{bmatrix}\begin{bmatrix} \boldsymbol{A}_i + \boldsymbol{B}_i\boldsymbol{K}_j\boldsymbol{C} & \boldsymbol{B}_i \\ \boldsymbol{C} & \boldsymbol{0} \end{bmatrix}, \tag{70}$$

respectively, then

$$\text{rank}\begin{bmatrix} \boldsymbol{A}_i & \boldsymbol{B}_i \\ \boldsymbol{C} & \boldsymbol{0} \end{bmatrix} = \text{rank}\begin{bmatrix} \boldsymbol{A}_i + \boldsymbol{B}_i\boldsymbol{K}_j\boldsymbol{C} & \boldsymbol{B}_i \\ \boldsymbol{0} & -\boldsymbol{C}(\boldsymbol{A}_i + \boldsymbol{B}_i\boldsymbol{K}_j\boldsymbol{C})^{-1}\boldsymbol{B}_i \end{bmatrix} = n + m, \tag{71}$$

since $\text{rank}(\boldsymbol{A}_i + \boldsymbol{B}_i\boldsymbol{K}_j\boldsymbol{C}) = n$, and $\text{rank}\boldsymbol{B}_i = m$.

Thus, evidently, it has to be satisfied

$$\mathrm{rank}\left(\boldsymbol{C}(\boldsymbol{A}_i + \boldsymbol{B}_i\boldsymbol{K}_j\boldsymbol{C})^{-1}\boldsymbol{B}_i\right) = m. \tag{72}$$

This concludes the proof.

Theorem 2. *To reach a forced regime for the TS fuzzy system Eqs.* (2) *and* (3) *with the TS fuzzy control policy Eq.* (67), *the signal gain matrices have to take the forms*

$$\boldsymbol{W}_{ij} = \left(\boldsymbol{C}(\boldsymbol{A}_i + \boldsymbol{B}_i\boldsymbol{K}_j\boldsymbol{C})^{-1}\boldsymbol{B}_i\right)^{-1}, \tag{73}$$

where $\boldsymbol{W}_{ij} \in \mathbb{R}^{m \times m}$, $i, j = 1, 2, \ldots s$.

Proof. In a steady state, which corresponds to $\dot{\boldsymbol{q}}(t) = \boldsymbol{0}$, the equality $\boldsymbol{y}_o = \boldsymbol{w}_o$ must hold, where $\boldsymbol{q}_o \in \mathbb{R}^n$, $\boldsymbol{\theta}_o \in \mathbb{R}^q$, $\boldsymbol{y}_o$, $\boldsymbol{w}_o \in \mathbb{R}^m$ are the vectors of steady state values of $\boldsymbol{q}(t)$, $\boldsymbol{\theta}(t)$, $\boldsymbol{y}(t)$, $\boldsymbol{w}(t)$, respectively.

Substituting Eq. (67) in Eq. (2) yields the expression

$$\sum_{i=1}^{s}\sum_{j=1}^{s}\mathrm{h}_i(\boldsymbol{\theta}_o)\mathrm{h}_j(\boldsymbol{\theta}_o)\left((\boldsymbol{A}_i + \boldsymbol{B}_i\boldsymbol{K}_j\boldsymbol{C})\boldsymbol{q}_o + \boldsymbol{B}_i\boldsymbol{W}_{ij}\boldsymbol{w}_o\right) = \boldsymbol{0}, \tag{74}$$

$$-\sum_{i=1}^{s}\sum_{j=1}^{s}\mathrm{h}_i(\boldsymbol{\theta}_o)\mathrm{h}_j(\boldsymbol{\theta}_o)\boldsymbol{q}_o = -\boldsymbol{q}_o = \sum_{i=1}^{s}\sum_{j=1}^{s}\mathrm{h}_i(\boldsymbol{\theta}_o)\mathrm{h}_j(\boldsymbol{\theta}_o)(\boldsymbol{A}_i + \boldsymbol{B}_i\boldsymbol{K}_j\boldsymbol{C})^{-1}\boldsymbol{B}_i\boldsymbol{W}_{ij}\boldsymbol{w}_o, \tag{75}$$

respectively, and it can be set

$$\boldsymbol{y}_o = \boldsymbol{C}\boldsymbol{q}_o = -\sum_{i=1}^{s}\sum_{j=1}^{s}\mathrm{h}_i(\boldsymbol{\theta}_o)\mathrm{h}_j(\boldsymbol{\theta}_o)\boldsymbol{C}(\boldsymbol{A}_i + \boldsymbol{B}_i\boldsymbol{K}_j\boldsymbol{C})^{-1}\boldsymbol{B}_i\boldsymbol{W}_{ij}\boldsymbol{w}_o = \boldsymbol{I}_m\boldsymbol{w}_o. \tag{76}$$

Thus, Eq. (76) gives the solution

$$\boldsymbol{W}_{ij}^{-1} = -\boldsymbol{C}(\boldsymbol{A}_i + \boldsymbol{B}_i\boldsymbol{K}_j\boldsymbol{C})^{-1}\boldsymbol{B}_i, \tag{77}$$

which implies Eq. (68). Hence, declaredly,

$$\mathrm{rank}\boldsymbol{W}_j = \mathrm{rank}\left(\boldsymbol{C}(\boldsymbol{A}_i + \boldsymbol{B}_i\boldsymbol{K}_j\boldsymbol{C})^{-1}\boldsymbol{B}_i\right) = m. \tag{78}$$

This concludes the proof.

The forced regime is basically designed for constant references and is very closely related to shift of origin. If the command value $\boldsymbol{w}(t)$ is changed "slowly enough," the above scheme can do a reasonable job of tracking, that is, making $\boldsymbol{y}(t)$ follow $\boldsymbol{w}(t)$ [23].

5. Bi-proper dynamic output controller

The full order biproper dynamic output controller is defined by the equation

$$\dot{\boldsymbol{p}}(t) = \sum_{j=1}^{s} \mathrm{h}_j(\boldsymbol{\theta}(t))\left(\boldsymbol{J}_j\boldsymbol{p}(t) + \boldsymbol{L}_j\boldsymbol{y}(t)\right), \tag{79}$$

$$\boldsymbol{u}(t) = \sum_{j=1}^{s} \mathrm{h}_j(\boldsymbol{\theta}(t))\left(\boldsymbol{M}_j\boldsymbol{p}(t) + \boldsymbol{N}_j\boldsymbol{y}(t)\right), \tag{80}$$

where $\boldsymbol{p}(t) \in \mathbb{R}^h$ is the vector of the controller state variables and the parameter matrix

$$\boldsymbol{K}_j^\circ = \begin{bmatrix} \boldsymbol{J}_j & \boldsymbol{L}_j \\ \boldsymbol{M}_j & \boldsymbol{N}_j \end{bmatrix}, \tag{81}$$

$\boldsymbol{K}_j^\circ \in \mathbb{R}^{(n+r)\times(h+m)}$, is considered in this block matrix structure with respect to the matrices $\boldsymbol{J}_j \in \mathbb{R}^{h\times h}$, $\boldsymbol{L}_j \in \mathbb{R}^{h\times m}$, $\boldsymbol{M}_j \in \mathbb{R}^{r\times h}$, and $\boldsymbol{N}_j \in \mathbb{R}^{r\times m}$. For simplicity, the full order $p = n$ controller is considered in the following.

To analyze the stability of the closed-loop system structure with the dynamic output controller, the closed-loop system description implies the following form

$$\dot{\boldsymbol{q}}^\circ(t) = \sum_{i=1}^{s}\sum_{j=1}^{s} h_i(\boldsymbol{\theta}(t))h_j(\boldsymbol{\theta}(t))\boldsymbol{A}_{cij}^\circ\boldsymbol{q}^\circ(t), \tag{82}$$

$$\boldsymbol{y}^\circ(t) = \boldsymbol{I}^\circ\boldsymbol{C}^\circ\boldsymbol{q}^\circ(t), \tag{83}$$

where

$$\boldsymbol{q}^{\circ T}(t) = \begin{bmatrix} \boldsymbol{q}^T(t) & \boldsymbol{p}^T(t) \end{bmatrix}, \tag{84}$$

$$\boldsymbol{A}_{cij}^\circ = \begin{bmatrix} \boldsymbol{A}_i + \boldsymbol{B}_i\boldsymbol{N}_j\boldsymbol{C} & \boldsymbol{B}_i\boldsymbol{M}_j \\ \boldsymbol{L}_j\boldsymbol{C} & \boldsymbol{N}_j \end{bmatrix}, \quad \boldsymbol{I}^\circ = [\boldsymbol{0} \quad \boldsymbol{I}_m], \quad \boldsymbol{C}^\circ = \begin{bmatrix} \boldsymbol{0} & \boldsymbol{I}_n \\ \boldsymbol{C} & \boldsymbol{0} \end{bmatrix} \tag{85}$$

and $\boldsymbol{A}_{cij}^\circ \in \mathbb{R}^{2n\times 2n}$, $\boldsymbol{I}^\circ \in \mathbb{R}^{m\times(n+m)}$, $\boldsymbol{C}^\circ \in \mathbb{R}^{(n+m)\times 2n}$.

Introducing the notations

$$\boldsymbol{A}_i^\circ = \begin{bmatrix} \boldsymbol{A}_i & \boldsymbol{0} \\ \boldsymbol{0} & \boldsymbol{0} \end{bmatrix}, \quad \boldsymbol{B}_i^\circ = \begin{bmatrix} \boldsymbol{0} & \boldsymbol{B}_i \\ \boldsymbol{I}_n & \boldsymbol{0} \end{bmatrix}, \tag{86}$$

where $\boldsymbol{A}_i^\circ \in \mathbb{R}^{2n\times 2n}$, $\boldsymbol{B}_i^\circ \in \mathbb{R}^{2n\times(n+r)}$, the closed-loop system matrices take the equivalent forms

$$\boldsymbol{A}^{\circ}_{cij} = \boldsymbol{A}^{\circ}_{i} + \boldsymbol{B}^{\circ}_{i}\boldsymbol{K}^{\circ}_{j}\boldsymbol{C}^{\circ}. \tag{87}$$

In the sequel, it is supposed that $(\boldsymbol{A}^{\circ}_{i}, \boldsymbol{B}^{\circ}_{i})$ is stabilizable, $(\boldsymbol{A}^{\circ}_{i}, \boldsymbol{C}^{\circ}_{i})$ is detectable [24].

Note this kind of controllers can be preferred in fault tolerant control (FTC) structures with virtual actuators [25].

Theorem 3 (*relaxed design conditions*). *The equilibrium of the fuzzy system Eqs.* (2) *and* (3) *controlled by the fuzzy dynamic output controller Eqs.* (79) *and* (80) *is global asymptotically stable if there exist a positive definite symmetric matrix* $\boldsymbol{S}^{\circ} \in \mathbb{R}^{2n \times 2n}$, *symmetric matrices* $\boldsymbol{X}^{\circ}_{ij} = \boldsymbol{X}^{\circ}_{ji} \in \mathbb{R}^{2n\times 2n}$, *a regular matrix* $\boldsymbol{H}^{\circ} \in \mathbb{R}^{(n+m)\times(n+m)}$, *and matrices* $\boldsymbol{Y}^{\circ}_{j} \in \mathbb{R}^{(n+r)\times(n+m)}$ *such that*

$$\boldsymbol{S}^{\circ} = \boldsymbol{S}^{\circ T} > 0, \quad \begin{bmatrix} \boldsymbol{X}^{\circ}_{11} & \boldsymbol{X}^{\circ}_{12} & \cdots & \boldsymbol{X}^{\circ}_{1s} \\ \boldsymbol{X}^{\circ}_{21} & \boldsymbol{X}^{\circ}_{22} & \cdots & \boldsymbol{X}^{\circ}_{2s} \\ \vdots & \vdots & \ddots & \vdots \\ \boldsymbol{X}^{\circ}_{s1} & \boldsymbol{X}^{\circ}_{s2} & \cdots & \boldsymbol{X}^{\circ}_{ss} \end{bmatrix} > 0, \tag{88}$$

$$\boldsymbol{A}^{\circ}_{i}\boldsymbol{S}^{\circ} + \boldsymbol{S}^{\circ}\boldsymbol{A}^{\circ T}_{i} + \boldsymbol{B}^{\circ}_{i}\boldsymbol{Y}^{\circ}_{i}\boldsymbol{C}^{\circ} + \boldsymbol{C}^{\circ T}\boldsymbol{Y}^{\circ T}_{i}\boldsymbol{B}^{\circ T}_{i} + \boldsymbol{X}^{\circ}_{ii} < 0, \tag{89}$$

$$\frac{\boldsymbol{A}^{\circ}_{i} + \boldsymbol{A}^{\circ}_{j}}{2}\boldsymbol{S}^{\circ} + \boldsymbol{S}^{\circ}\frac{\boldsymbol{A}^{\circ T}_{i} + \boldsymbol{A}^{\circ T}_{j}}{2} + \frac{\boldsymbol{B}^{\circ}_{i}\boldsymbol{Y}^{\circ}_{j} + \boldsymbol{B}^{\circ}_{j}\boldsymbol{Y}^{\circ}_{i}}{2}\boldsymbol{C}^{\circ} + \boldsymbol{C}^{\circ T}\frac{\boldsymbol{Y}^{\circ T}_{j}\boldsymbol{B}^{\circ T}_{i} + \boldsymbol{Y}^{\circ T}_{i}\boldsymbol{B}^{\circ T}_{j}}{2} + \frac{\boldsymbol{X}^{\circ}_{ij} + \boldsymbol{X}^{\circ}_{ji}}{2} < 0, \tag{90}$$

$$\boldsymbol{C}^{\circ}\boldsymbol{S}^{\circ} = \boldsymbol{H}^{\circ}\boldsymbol{C}^{\circ}, \tag{91}$$

for all $i \in \langle 1, 2, \dots s\rangle$, $i < j \le s$, $i, j \in \langle 1, 2, \dots s\rangle$, *respectively, and* $h_i(\boldsymbol{\theta}(t))h_j(\boldsymbol{\theta}(t)) \neq 0$.

When the above conditions hold, the set of control law gain matrices are given as

$$\boldsymbol{K}^{\circ}_{j} = \boldsymbol{Y}^{\circ}_{j}(\boldsymbol{H}^{\circ})^{-1}, \quad j = 1, 2, \dots, s \tag{92}$$

Proof. Defining the Lyapunov function as follows

$$v(\boldsymbol{q}^{\circ}(t)) = \boldsymbol{q}^{\circ T}(t)\boldsymbol{P}^{\circ}\boldsymbol{q}^{\circ}(t) > 0, \tag{93}$$

where $\boldsymbol{P}^{\circ} \in \mathbb{R}^{2n \times 2n}$ is a positive definite matrix, then the time derivative of $v(\boldsymbol{q}(t))$ along a closed-loop system trajectory is

$$\dot{v}(\boldsymbol{q}^{\circ}(t)) = \dot{\boldsymbol{q}}^{\circ T}(t)\boldsymbol{P}^{\circ}\boldsymbol{q}^{\circ}(t) + \boldsymbol{q}^{\circ T}(t)\boldsymbol{P}^{\circ}\dot{\boldsymbol{q}}^{\circ}(t) < 0. \tag{94}$$

Substituting Eq. (87), then Eq. (94) implies

$$\dot{v}(\boldsymbol{q}^{\circ}(t)) = \sum_{i=1}^{s}\sum_{j=1}^{s} h_i(\boldsymbol{\theta}(t))h_j(\boldsymbol{\theta}(t))\boldsymbol{q}^{\circ T}(t)\boldsymbol{P}^{\circ}_{cij}\boldsymbol{q}^{\circ}(t) < 0, \tag{95}$$

$$\boldsymbol{P}^{\circ}_{cij} = \boldsymbol{P}^{\circ}\boldsymbol{A}^{\circ}_{cij} + \boldsymbol{A}^{\circ T}_{cij}\boldsymbol{P}^{\circ}. \tag{96}$$

Since $\boldsymbol{P}^{\circ}$ is positive definite, the state coordinate transform can now be defined as

http://dx.doi.org/10.5772/68129

$$\boldsymbol{q}^{\circ}(t)=\boldsymbol{S}^{\circ}\boldsymbol{p}^{\circ}(t),\quad \boldsymbol{S}^{\circ}=(\boldsymbol{P}^{\circ})^{-1}, \tag{97}$$

and subsequently Eqs. (95) and (96) can be rewritten as

$$\dot{v}(\boldsymbol{p}^{\circ}(t))=\sum_{i=1}^{s}\sum_{j=1}^{s}h_i(\boldsymbol{\theta}(t))h_j(\boldsymbol{\theta}(t))\boldsymbol{p}^{\circ T}(t)\boldsymbol{S}^{\diamond}_{cij}\boldsymbol{p}^{\circ}(t)<0, \tag{98}$$

$$\boldsymbol{S}^{\diamond}_{cij}=\boldsymbol{A}^{\circ}_{cij}\boldsymbol{S}^{\circ}+\boldsymbol{S}^{\circ}\boldsymbol{A}^{\circ T}_{cij}. \tag{99}$$

Introducing, analogously to Eqs. (54) and (55), the positive term

$$v_v(\boldsymbol{p}^{\circ}(t))=\boldsymbol{p}^{\circ T}(t)\boldsymbol{Z}^{\circ}(\boldsymbol{\theta}(t))\boldsymbol{p}^{\circ}(t)>0, \tag{100}$$

defined by the set of matrices $\left\{\boldsymbol{X}^{\circ}_{ij}=\boldsymbol{X}^{\circ T}_{ji}\in\mathbb{R}^{n\times n},\ i,j=1,2,\dots,s\right\}$ in the structure Eq. (88) such that

$$\boldsymbol{Z}^{\circ}(\boldsymbol{\theta}(t))=\boldsymbol{Z}^{\circ T}(\boldsymbol{\theta}(t))=\sum_{i=1}^{s}\sum_{j=1}^{s}h_i(\boldsymbol{\theta}(t))h_j(\boldsymbol{\theta}(t))\boldsymbol{X}^{\circ}_{ij}>0, \tag{101}$$

then, in the sense of Krasovskii theorem, it can be set up

$$\dot{v}(\boldsymbol{p}^{\circ}(t))=\sum_{i=1}^{s}\sum_{j=1}^{s}h_i(\boldsymbol{\theta}(t))h_j(\boldsymbol{\theta}(t))\boldsymbol{p}^{\circ T}(t)\boldsymbol{S}^{\circ}_{cij}\boldsymbol{p}^{\circ}(t)<0, \tag{102}$$

where

$$\boldsymbol{S}^{\circ}_{cij}=\boldsymbol{A}^{\circ}_{cij}\boldsymbol{S}^{\circ}+\boldsymbol{S}^{\circ}\boldsymbol{A}^{\circ T}_{cij}+\boldsymbol{X}^{\circ}_{ij}. \tag{103}$$

Therefore, Eq. (102) can be factorized as follows

$$\dot{v}(\boldsymbol{p}^{\circ}(t))=\sum_{i=1}^{s}h_i^2(\boldsymbol{\theta}(t))\boldsymbol{p}^{\circ T}(t)\boldsymbol{S}^{\circ}_{cii}\boldsymbol{p}^{\circ}(t)+2\sum_{i=1}^{s-1}\sum_{j=i+1}^{s}h_i(\boldsymbol{\theta}(t))h_j(\boldsymbol{\theta}(t))\boldsymbol{p}^{\circ T}(t)\frac{\boldsymbol{S}^{\circ}_{cij}+\boldsymbol{S}^{\circ}_{cji}}{2}\boldsymbol{p}^{\circ}(t)<0, \tag{104}$$

which, using Eq. (87), leads to the following sets of inequalities

$$\boldsymbol{A}^{\circ}_{i}\boldsymbol{S}^{\circ}+\boldsymbol{S}^{\circ}\boldsymbol{A}^{\circ T}_{i}+\boldsymbol{B}^{\circ}_{i}\boldsymbol{K}^{\circ}_{j}\boldsymbol{C}^{\circ}\boldsymbol{S}^{\circ}+\boldsymbol{S}^{\circ}\boldsymbol{C}^{\circ T}\boldsymbol{K}^{\circ T}_{j}\boldsymbol{B}^{\circ T}_{i}+\boldsymbol{X}^{\circ}_{ij}<0, \tag{105}$$

$$\begin{aligned}&\frac{\left(\boldsymbol{A}^{\circ}_{i}+\boldsymbol{B}^{\circ}_{i}\boldsymbol{K}^{\circ}_{j}\boldsymbol{C}^{\circ}\right)\boldsymbol{S}^{\circ}}{2}+\frac{\left(\boldsymbol{A}^{\circ}_{j}+\boldsymbol{B}^{\circ}_{j}\boldsymbol{K}^{\circ}_{i}\boldsymbol{C}^{\circ}\right)\boldsymbol{S}^{\circ}}{2}+\frac{\boldsymbol{S}^{\circ}\left(\boldsymbol{A}^{\circ}_{i}+\boldsymbol{B}^{\circ}_{i}\boldsymbol{K}^{\circ}_{j}\boldsymbol{C}^{\circ}\right)^{T}}{2}\\&+\frac{\boldsymbol{S}^{\circ}\left(\boldsymbol{A}^{\circ}_{j}+\boldsymbol{B}^{\circ}_{j}\boldsymbol{K}^{\circ}_{i}\boldsymbol{C}^{\circ}\right)^{T}}{2}+\frac{\boldsymbol{X}^{\circ}_{ij}+\boldsymbol{X}^{\circ}_{ji}}{2}<0,\end{aligned} \tag{106}$$

for $i=1,2,\dots,s$, as well as $i=1,2,\dots,s-1$, $j=1+1,i+2,\dots,s$, and $h_i(\boldsymbol{\theta}(t))h_j(\boldsymbol{\theta}(t))\neq 0$.

Analyzing the product $B_i^\circ K_j^\circ C^\circ S^\circ$, it can set

$$B_i^\circ K_j^\circ C^\circ S^\circ = B_i^\circ K_j^\circ H^\circ (H^\circ)^{-1} C^\circ S^\circ = B_i^\circ Y_j^\circ C^\circ, \tag{107}$$

where

$$K_j^\circ H^\circ = Y_j^\circ, \quad (H^\circ)^{-1} C^\circ = C^\circ (S^\circ)^{-1} \tag{108}$$

and $H^\circ \in \mathbb{R}^{(m+n)\times(m+n)}$ is a regular square matrix. Thus, with Eq. (108), then Eqs. (105) and (106) implies Eqs. (89) and (90) and Eq. (108) gives Eq. (91). This concludes the proof.

This theorem provides the sufficient condition under LMIs and LME formulations for the synthesis of the dynamic output controller reflecting the membership function properties.

For the same reasons as in Theorem 1, the following theorem is proven.

Theorem 4 (*enhanced relaxed design conditions*). *The equilibrium of the fuzzy system Eqs.* (2) *and* (3) *controlled by the fuzzy dynamic output controller Eqs.* (79) *and* (80) *is global asymptotically stable if for given a positive* $\delta \in \mathbb{R}$ *there exist positive definite symmetric matrices* $V^\circ, S^\circ \in \mathbb{R}^{n\times n}$, *and matrices* $Y_j^\circ \in \mathbb{R}^{r\times n}$, $H^\circ \in \mathbb{R}^{m\times m}$ *such that*

$$S^\circ = S^{\circ T} > 0, \quad V^\circ = V^{\circ T} > 0, \quad \begin{bmatrix} X_{11}^\circ & X_{12}^\circ & \cdots & X_{1s}^\circ \\ X_{21}^\circ & X_{22}^\circ & \cdots & X_{2s}^\circ \\ \vdots & \vdots & \ddots & \vdots \\ X_{s1}^\circ & X_{s2}^\circ & \cdots & X_{ss}^\circ \end{bmatrix} > 0, \tag{109}$$

$$\begin{bmatrix} A_i^\circ S^\circ + S^\circ A_i^{\circ T} + B_i^\circ Y_i^\circ C^\circ + C^{\circ T} Y_i^{\circ T} B_i^{\circ T} & * \\ V^\circ - S^\circ + \delta A_i^\circ S^\circ + \delta B_i^\circ Y_i^\circ C^\circ & -2\delta S^\circ \end{bmatrix} < 0, \tag{110}$$

$$\begin{bmatrix} \Phi_{ij}^\circ & * \\ V^\circ - S^\circ + \delta \dfrac{A_i^\circ S^\circ + A_j^\circ S^\circ}{2} + \delta \dfrac{B_i^\circ Y_j^\circ + B_j^\circ Y_i^\circ}{2} C^\circ & -2\delta S^\circ \end{bmatrix} < 0, \tag{111}$$

$$C^\circ S^\circ = H^\circ C^\circ, \tag{112}$$

for $i = 1, 2, \ldots, s$, *as well as* $i = 1, 2, \ldots, s-1$, $j = 1+1, i+2, \ldots, s$, $h_i(\boldsymbol{\theta}(t))h_j(\boldsymbol{\theta}(t)) \neq 0$, *and*

$$\Phi_{ij}^\circ = \frac{A_i^\circ S^\circ + S^\circ A_i^{\circ T}}{2} + \frac{A_j^\circ S^\circ + S^\circ A_j^{\circ T}}{2} + \frac{B_i^\circ Y_j^\circ C^\circ + C^{\circ T} Y_j^{\circ T} B_i^{\circ T}}{2} + \frac{B_j^\circ Y_i^\circ C^\circ + C^{\circ T} Y_i^{\circ T} B_j^{\circ T}}{2}. \tag{113}$$

When the above conditions hold, the control law gain matrices are given as

$$K_i^\circ = Y_i^\circ (H^\circ)^{-1}. \tag{114}$$

Proof. Since Eq. (82), Eq. (87) takes formally the same structure as Eqs. (6) and (7), following the same way as in the proof of Theorem 1, the conditions given in Theorem 4 can be obtained. From this reason, the proof is omitted. Compare, for example, Ref. [17].

Following the presented results, it is evident that the standard as well as the enhanced conditions for biproper dynamic output controller design can be derived from Theorem 3 and Theorem 4 in a simple way.

http://dx.doi.org/10.5772/68129

6. Illustrative example

The nonlinear dynamics of the system is represented by TS model with s = 3 and the system model parameters [20]

$$A_1 = \begin{bmatrix} -1.0522 & -1.8666 & 0.5102 \\ -0.4380 & -5.4335 & 0.9205 \\ -0.5522 & 0.1334 & -0.4898 \end{bmatrix}, A_2 = \begin{bmatrix} -1.0565 & -1.8661 & 0.5116 \\ -0.4380 & -5.4359 & 0.9214 \\ -0.5565 & 0.1339 & -0.4884 \end{bmatrix},$$

$$A_3 = \begin{bmatrix} -1.0602 & -1.8657 & 0.5133 \\ -0.4381 & -5.4353 & 0.9216 \\ -0.5602 & 0.1343 & -0.4867 \end{bmatrix}, B = \begin{bmatrix} -0.1765 & 0.0000 \\ 0.0000 & 0.0000 \\ 0.1176 & 0.4721 \end{bmatrix}, C = \begin{bmatrix} 1 & 0 & 0 \\ 0 & 1 & 0 \end{bmatrix}.$$

To the state vector $\boldsymbol{q}(t)$ are associated the premise variables and the membership functions as follows

$$\boldsymbol{\theta}(t) = \begin{bmatrix} \theta_1(t) \\ \theta_2(t) \\ \theta_3(t) \end{bmatrix}, \theta_i(t) = \begin{cases} \theta_1(t) \text{ if } q_1(t) \text{ is about } 2.5, & h_1(\theta_2(t)) = 1 - \frac{1}{2.5}|\theta_2(t) - 2.5| \\ \theta_2(t) \text{ if } q_1(t) \text{ is about } 0, & h_2(\theta_1(t)) = 1 - \frac{1}{2.5}|\theta_1(t)| \\ \theta_3(t) \text{ if } q_1(t) \text{ is about } -2.5, & h_3(\theta_3(t)) = 1 - \frac{1}{2.5}|\theta_3(t) + 2.5| \end{cases}$$

while the generalized premise variable is $\theta(t) = q_1(t)$.

Thus, solving Eqs. (46)–(49) for prescribed $\delta = 1.2$ with respect to the LMI matrix variables $\boldsymbol{S}, \boldsymbol{V}$ $\boldsymbol{H}, \boldsymbol{Y}_i$, j = 1, 2, 3, and $\boldsymbol{X}_{ij}$, i, j = 1, 2, 3 using Self–Dual–Minimization (SeDuMi) package for Matlab [26], then the feedback gain matrix design problem was feasible with the results

$$\boldsymbol{S} = \begin{bmatrix} 0.3899 & -0.0102 & -0.0000 \\ -0.0102 & 0.1596 & -0.0000 \\ -0.0000 & -0.0000 & 0.4099 \end{bmatrix}, \boldsymbol{V} = \begin{bmatrix} 0.9280 & 0.1235 & -0.1525 \\ 0.1235 & 1.1533 & -0.3979 \\ -0.1525 & -0.3979 & 0.7574 \end{bmatrix},$$

$$\boldsymbol{H} = \begin{bmatrix} 0.3899 & -0.0102 \\ -0.0102 & 0.1596 \end{bmatrix},$$

$$\boldsymbol{X} = \begin{bmatrix} 0.4567 & 0.0983 & -0.0517 & 0.0694 & 0.0463 & -0.0174 & 0.0694 & 0.0463 & -0.0174 \\ 0.0983 & 0.7153 & -0.1118 & 0.0463 & 0.1906 & -0.0441 & 0.0463 & 0.1905 & -0.0440 \\ -0.0517 & -0.1118 & 0.1883 & -0.0175 & -0.0442 & 0.0143 & -0.0175 & -0.0442 & 0.0142 \\ 0.0694 & 0.0463 & -0.0175 & 0.4573 & 0.0981 & -0.0515 & 0.0695 & 0.0463 & -0.0174 \\ 0.0463 & 0.1906 & -0.0442 & 0.0981 & 0.7154 & -0.1115 & 0.0463 & 0.1905 & -0.0440 \\ -0.0174 & -0.0441 & 0.0143 & -0.0515 & -0.1115 & 0.1876 & -0.0175 & -0.0441 & 0.0142 \\ 0.0694 & 0.0463 & -0.0175 & 0.0695 & 0.0463 & -0.0175 & 0.4578 & 0.0978 & -0.0514 \\ 0.0463 & 0.1905 & -0.0442 & 0.0463 & 0.1905 & -0.0441 & 0.0978 & 0.7152 & -0.1111 \\ -0.0174 & -0.0440 & 0.0142 & -0.0174 & -0.0440 & 0.0142 & -0.0514 & -0.1111 & 0.1868 \end{bmatrix},$$

$$\boldsymbol{Y}_1 = \begin{bmatrix} 0.5607 & -0.4590 \\ 0.1544 & -0.1191 \end{bmatrix}, \boldsymbol{Y}_2 = \begin{bmatrix} 0.5558 & -0.4577 \\ 0.1579 & -0.1207 \end{bmatrix}, \boldsymbol{Y}_3 = \begin{bmatrix} 0.5518 & -0.4566 \\ 0.1606 & -0.1222 \end{bmatrix},$$

Substituting the above parameters into Eq. (51) to solve the controller parameters, the following gain matrices are obtained

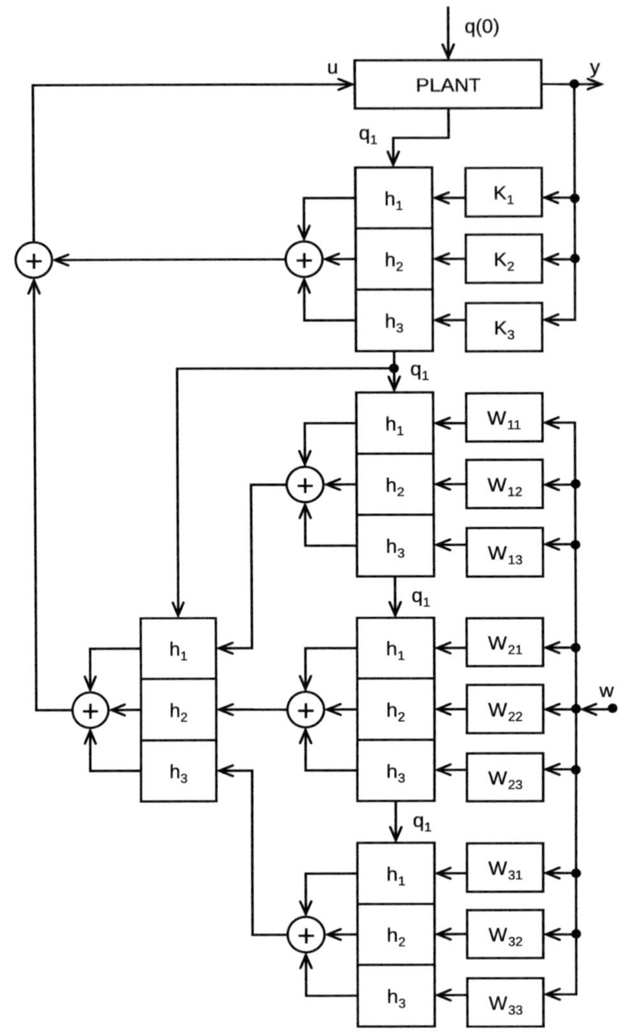

Figure 1. TS fuzzy static output control structure in a forced mode.

$$K_1 = \begin{bmatrix} 1.3653 & -2.7895 \\ 0.3772 & -0.7224 \end{bmatrix}, K_2 = \begin{bmatrix} 1.3530 & -2.7823 \\ 0.3860 & -0.7318 \end{bmatrix}, K_3 = \begin{bmatrix} 1.3428 & -2.7761 \\ 0.3925 & -0.7406 \end{bmatrix},$$

$$A_{c22} = \begin{bmatrix} -1.2953 & -1.3750 & 0.5116 \\ -0.4380 & -5.4359 & 0.9214 \\ -0.2152 & -0.5388 & -0.4884 \end{bmatrix}, A_{c31} = \begin{bmatrix} -1.3012 & -1.3734 & 0.5133 \\ -0.4381 & -5.4353 & 0.9216 \\ -0.2216 & -0.5348 & -0.4867 \end{bmatrix},$$

For simplicity, other closed-loop matrices of subsystem dynamics are not listed here.

Since the diagonal elements of A_{cij}, $i, j = 1, 2, 3$, are dominant, in terms of Gerschgorin theorem [27, 28], all eigenvalues of A_{cij} are real, resulting in the aperiodic dynamics, that is,

$$\tilde{n}(\mathbf{A}_{c11}) = \{-0.6751, \quad -1.0816, \quad -5.4598\}, \quad \tilde{n}(\mathbf{A}_{c21}) = \{-0.6756, \quad -1.0842, \quad -5.4620\},$$
$$\tilde{n}(\mathbf{A}_{c31}) = \{-0.6757, \quad -1.0861, \quad -5.4613\}, \quad \tilde{n}(\mathbf{A}_{c12}) = \{-0.6745, \quad -1.0805, \quad -5.4593\},$$
$$\tilde{n}(\mathbf{A}_{c22}) = \{-0.6750, \quad -1.0831, \quad -5.4615\}, \quad \tilde{n}(\mathbf{A}_{c32}) = \{-0.6751, \quad -1.0851, \quad -5.4609\},$$
$$\tilde{n}(\mathbf{A}_{c13}) = \{-0.6742, \quad -1.0795, \quad -5.4588\}, \quad \tilde{n}(\mathbf{A}_{c23}) = \{-0.6748, \quad -1.0820, \quad -5.4610\},$$
$$\tilde{n}(\mathbf{A}_{c33}) = \{-0.6748, \quad -1.0840, \quad -5.4604\}.$$

Figure 1 gives the associated TS fuzzy static output control structure in a forced mode.

For Eqs. (88)–(91), it can find the following feasible solutions by using the given design procedure

$$\boldsymbol{S}^{\circ} = \begin{bmatrix} 0.6194 & -0.0614 & 0.0000 & 0.0000 & 0.0000 & 0.0000 \\ -0.0614 & 0.1305 & 0.0000 & 0.0000 & 0.0000 & 0.0000 \\ 0.0000 & 0.0000 & 0.8724 & 0.0000 & 0.0000 & 0.0000 \\ 0.0000 & 0.0000 & 0.0000 & 0.7066 & 0.0000 & 0.0000 \\ 0.0000 & 0.0000 & 0.0000 & 0.0000 & 0.7066 & 0.0000 \\ 0.0000 & 0.0000 & 0.0000 & 0.0000 & 0.0000 & 0.7066 \end{bmatrix},$$

$$\boldsymbol{H}^{\circ} = \begin{bmatrix} 0.7066 & 0.0000 & 0.0000 & 0.0000 & 0.0000 \\ 0.0000 & 0.7066 & 0.0000 & 0.0000 & 0.0000 \\ 0.0000 & 0.0000 & 0.7066 & 0.0000 & 0.0000 \\ 0.0000 & 0.0000 & 0.0000 & 0.6808 & -0.0614 \\ 0.0000 & 0.0000 & 0.0000 & 0.4889 & 0.0691 \end{bmatrix},$$

$$\boldsymbol{Y}_1 = \begin{bmatrix} -0.5668 & 0.0000 & 0.0000 & 0.0000 & 0.0000 \\ 0.0000 & -0.5668 & 0.0000 & -0.0001 & 0.0000 \\ 0.0000 & 0.0000 & -0.5667 & 0.0000 & 0.0000 \\ 0.0000 & 0.0000 & 0.0000 & 0.3612 & -0.2783 \\ 0.0000 & 0.0000 & 0.0000 & 0.7396 & -1.1397 \end{bmatrix},$$

$$\boldsymbol{Y}_2 = \begin{bmatrix} -0.5668 & 0.0000 & 0.0000 & 0.0000 & 0.0000 \\ 0.0000 & -0.5668 & 0.0000 & -0.0001 & 0.0000 \\ 0.0000 & 0.0000 & -0.5667 & 0.0000 & 0.0000 \\ 0.0000 & 0.0000 & 0.0000 & 0.3615 & -0.2784 \\ 0.0000 & 0.0000 & 0.0000 & 0.7397 & -1.1397 \end{bmatrix},$$

$$\boldsymbol{Y}_3 = \begin{bmatrix} -0.5667 & -0.0001 & 0.0000 & 0.0000 & 0.0000 \\ -0.0001 & -0.5667 & 0.0000 & -0.0001 & 0.0000 \\ 0.0000 & 0.0000 & -0.5668 & 0.0000 & 0.0000 \\ 0.0000 & 0.0000 & 0.0000 & 0.3519 & -0.2859 \\ -0.0001 & 0.0000 & -0.0001 & 0.7486 & -1.1421 \end{bmatrix}$$

and, computing the biproper dynamic output controller parameters, then

$$J_1 = \begin{bmatrix} -0.8022 & 0.0000 & 0.0000 \\ 0.0000 & -0.8021 & 0.0000 \\ 0.0000 & 0.0000 & -0.8021 \end{bmatrix}, \quad L_1 = 10^{-3}\begin{bmatrix} 0.0394 & 0.0048 \\ -0.3143 & 0.3318 \\ 0.0221 & -0.0508 \end{bmatrix},$$

$$M_1 = 10^{-4}\begin{bmatrix} -0.2041 & 0.1504 & -0.0600 \\ -0.5318 & 0.0915 & -0.3275 \end{bmatrix}, \quad N_1 = \begin{bmatrix} 2.0889 & -2.1701 \\ 7.8914 & -9.4765 \end{bmatrix},$$

$$J_2 = \begin{bmatrix} -0.8022 & 0.0001 & 0.0000 \\ 0.0001 & -0.8022 & 0.0000 \\ 0.0000 & 0.0000 & -0.8021 \end{bmatrix}, \quad L_2 = 10^{-3}\begin{bmatrix} -0.2009 & 0.3531 \\ 0.0453 & -0.2017 \\ -0.1903 & 0.1765 \end{bmatrix},$$

$$M_2 = 10^{-4}\begin{bmatrix} -0.2022 & 0.1779 & 0.1796 \\ -0.4575 & 0.0413 & 0.2985 \end{bmatrix}, \quad N_2 = \begin{bmatrix} 2.0897 & -2.1707 \\ 7.8915 & -9.4766 \end{bmatrix},$$

$$J_3 = \begin{bmatrix} -0.8020 & -0.0001 & 0.0000 \\ -0.0001 & -0.8021 & 0.0000 \\ 0.0000 & 0.0000 & -0.8022 \end{bmatrix}, \quad L_3 = 10^{-3}\begin{bmatrix} -0.0641 & 0.0139 \\ -0.1516 & 0.2382 \\ -0.2116 & 0.2630 \end{bmatrix},$$

$$M_3 = 10^{-4}\begin{bmatrix} -0.0135 & 0.0020 & -0.0238 \\ -0.0917 & 0.0218 & -0.1102 \end{bmatrix}, \quad N_3 = \begin{bmatrix} 2.1286 & -2.2445 \\ 7.9148 & -9.4907 \end{bmatrix}.$$

It is evident that all matrices J_i, $i = 1.2.3$ are Hurwitz, which rise up a TS fuzzy stable dynamic output controller, and based on the solutions obtained, the TS fuzzy dynamic controller can be designed via the concept of PDC.

Verifying the closed-loop stability, it can compute the eigenvalue spectra as follows

$$\begin{aligned}
\tilde{n}(A_{c11}) &= \{-0.8022, \ -0.8021, \ -0.8021, \ -4.3774, \ -1.2919 \pm 0.2804i\}, \\
\tilde{n}(A_{c21}) &= \{-0.8022, \ -0.8021, \ -0.8021, \ -4.3774, \ -1.2919 \pm 0.2804i\}, \\
\tilde{n}(A_{c21}) &= \{-0.8022, \ -0.8021, \ -0.8021, \ -4.3774, \ -1.2919 \pm 0.2804i\}, \\
\tilde{n}(A_{c12}) &= \{-0.8022, \ -0.8021, \ -0.8021, \ -4.3774, \ -1.2919 \pm 0.2805i\}, \\
\tilde{n}(A_{c22}) &= \{-0.8022, \ -0.8021, \ -0.8021, \ -4.3774, \ -1.2919 \pm 0.2805i\}, \\
\tilde{n}(A_{c32}) &= \{-0.8022, \ -0.8021, \ -0.8021, \ -4.3788, \ -1.2946 \pm 0.2963i\}, \\
\tilde{n}(A_{c13}) &= \{-0.8020, \ -0.8023, \ -0.8023, \ -4.3713, \ -1.2919 \pm 0.2797i\}, \\
\tilde{n}(A_{c23}) &= \{-0.8020, \ -0.8023, \ -0.8023, \ -4.3713, \ -1.2919 \pm 0.2797i\}, \\
\tilde{n}(A_{c33}) &= \{-0.8020, \ -0.8023, \ -0.8023, \ -4.3728, \ -1.2945 \pm 0.2958i\}.
\end{aligned}$$

7. Concluding remarks

New approach for static and dynamic output feedback control design, taking into account the affine properties of the TS fuzzy model structure, is presented in the chapter. Applying the fuzzy output control schemes relating to the parallel-distributed output compensators, the method presented methods that significantly reduces the conservativeness in the control design conditions. Sufficient existence conditions of the both output controller realization, manipulating the global stability of the system, implies the parallel decentralized control framework which stabilizes the nonlinear system in the sense of Lyapunov, and the design of controller parameters, resulting directly from these conditions, is a feasible numerical problem. An additional benefit of the method is that controllers use minimum feedback information with respect to desired system output and the approach is flexible enough to allow the inclusion of additional design conditions. The validity and applicability of the approach is demonstrated through numerical design examples.

Acknowledgement

The work presented in this chapter was supported by VEGA, the Grant Agency of Ministry of Education and Academy of Science of Slovak Republic, under Grant No. 1/0608/17. This support is very gratefully acknowledged.

Author details

Dušan Krokavec* and Anna Filasová

*Address all correspondence to: dusan.krokavec@tuke.sk

Department of Cybernetics, Artificial Intelligence, Faculty of Electrical Engineering, Informatics, Technical University of Košice, Košice, Slovakia

References

[1] Takagi T, Sugeno M. Fuzzy identification of systems and its applications to modeling and control. IEEE Transactions on Systems, Man, and Cybernetics. 1985;**15**(1):116–132. DOI: 10.1109/TSMC.1985.6313399.

[2] Wang HO, Tanaka K, Griffin MF. An approach to fuzzy control of nonlinear systems: Stability and design issues. IEEE Transactions on Fuzzy Systems. 1996;**4**(1):14–23. DOI: 10.1109/91.481841.

[3] Johansson M, Rantzer A, Arzen KE. Piecewise quadratic stability of fuzzy systems. Transactions on Fuzzy Systems. 1999;**7**(6):713–722. DOI: 10.1109/91.811241

[4] Tanaka K, Wang HO. Fuzzy Control Systems Design and Analysis. A Linear Matrix Inequality Approach. New York: John Wiley & Sons; 2001:309 p.

[5] Lam HK. Polynomial Fuzzy Model-Based Control Systems. Stability Analysis and Control Synthesis Using Membership Function-Dependent Techniques. Cham: Springer-Verlag; 2016:307 p.

[6] Michels K, Klawonn F, Kruse R, Nürnberger N. Fuzzy Control. Fundamentals, Stability and Design of Fuzzy Controllers. Berlin: Springer-Verlag; 2006:416 p.

[7] Abdelmalek I, Golea N, Hadjili ML. A new fuzzy Lyapunov approach to non-quadratic stabilization of Takagi-Sugeno fuzzy models. International Journal of Applied Mathematics and Computer Science. 2007;**17**(1):39–51. DOI: 10.2478/v10006-007-0005-4.

[8] Krokavec D, Filasová A. Optimal fuzzy control for a class of nonlinear systems. Mathematical Problems in Engineering. 2012;**2012**. 29 p. DOI: 10.1155/2012/481942.

[9] Pan J, Fei S, Ni Y, Xue M. New approaches to relaxed stabilization conditions and H-infinity control designs for T-S fuzzy systems. Journal of Control Theory and Applications. 2012;**10**(1):82–91. DOI: 10.1007/s11768-012-0088-9.

[10] Boyd B, El Ghaoui L, Peron E, Balakrishnan V. Linear Matrix Inequalities in System and Control Theory. Philadelphia: SIAM;1994:205 p. DOI: http://dx.doi.org/10.1137/1.9781611970777.

[11] Liu X, Zhang Q. New approaches to H_∞ controller designs based on fuzzy observers for T-S fuzzy systems via LMI. Automatica. 2003;**39**(9):1571-1582. http://dx.doi.org/10.1016/S0005-1098(03)00172-9.

[12] Nguang SK, Shi P. H_∞ fuzzy output feedback control design for nonlinear systems: An LMI approach. IEEE Transactions on Fuzzy Systems. 2003;**11**(3):331–340. DOI: 10.1109/TFUZZ.2003.812691.

[13] Kau SW, Lee HJ, Yang CM, Lee CH, Hong L, Fang CH. Robust H_∞ fuzzy static output feedback control of T-S fuzzy systems with parametric uncertainties. Fuzzy Sets and Systems. 2007;**158**(2):135–146. DOI: 10.1016/j.fss.2006.09.010.

[14] Tognetti ES, Oliveira RCLF, Peres PLD. Improved stabilization conditions for Takagi-Sugeno fuzzy systems via fuzzy integral Lyapunov functions. In Proceedings of the 2011 American Control Conference; 29 Jun–01 Jul 2011; San Francisco, USA, pp. 4970–4975.

[15] Chen BS, Tseng CS, Uang HJ. Mixed H_2/H_∞ fuzzy output feedback control design for nonlinear dynamic systems: An LMI approach. IEEE Transactions on Fuzzy Systems. 2000;**8**(3):249–265. DOI: 10.1109/91.855915.

[16] Krokavec D, Filasová A. Stabilizing fuzzy output control for a class of nonlinear systems. Advances in Fuzzy Systems. 2013;**2013**:9 p. DOI:10.1155/2013/294971.

[17] Krokavec D, Filasová A, Liščinský P. Dynamic output control of nonlinear systems described by Takagi-Sugeno models. In: Proceedings of the 2014 IEEE Multi-conference on Systems and Control MSC 2014; 8-10 October 2014; Antibes, France, pp. 959–964. DOI: 10.1109/CCA.2014.6981460.

[18] Huang D, Nguang SK. Static output feedback controller design for fuzzy systems: An ILMI approach. Information Sciences. 2007;**177**(14):3005–3015. DOI: 10.1016/j.ins.2007.02.014.

[19] Kim E, Lee H. New approaches to relaxed quadratic stability condition of fuzzy control systems. IEEE Transactions on Fuzzy Systems. 2000;**8**(5):523–534. DOI: 10.1109/91.873576.

[20] Krokavec D, Filasová A. Relaxed design conditions for Takagi-Sugeno unknown input observers. In: Proceedings of the 15th IEEE International Symposium on Applied Machine Intelligence and Informatics SAMI 2017; 26-28 January 2017; Herlany, Slovakia: pp. 61–66.

[21] Passino KM, Yurkovich S. Fuzzy Control. Berkeley: Addison-Wesley Longman; 1998:522 p.

[22] Haddad WM, Chellaboina V. Nonlinear Dynamical Systems and Control. A Lyapunov-Based Approach. Princeton: Princeton University Press; 2008:976 p.

[23] Kailath T. Linear Systems. Englewood Cliffs: Prentice-Hall; 1980:704 p.

[24] Doyle JC, Glover K, Khargonekar PP, Francis BA. State-space solutions to standard H_2 and H_∞ control problems. IEEE Transactions on Automatic Control. 1989;**34**(8):831–847. DOI: 10.1109/9.29425.

[25] Krokavec D, Filasová A, Serbák V FTC structures with virtual actuators and dynamic output controllers. IFAC-PapersOnLine. 2015;**48**(21):511–516. DOI: 10.1016/j.ifacol.2015.09.577.

[26] Peaucelle D, Henrion D, Labit Y, Taitz K. User's Guide for SeDuMi Interface 1.04. Toulouse: LAAS-CNRS; 2002:36 p.

[27] Gerschgorin S. On the limitation of the eigenvalues of a matrix. Bulletin of the Academy of Sciences of the USSR, Section of Mathematical and Natural Sciences. 1931;**1931**(6):749–754. (in German).

[28] Feingold DG, Varga RS. Block diagonally dominant matrices and generalizations of the Gerschgorin circle theorem. Pacific Journal of Mathematics. 1962;**12**(4):1241–1250. DOI: 10.2140/pjm.1962.12.1241.

Chapter 2

Non-Fragile Guaranteed Cost Control of Nonlinear Systems with Different State and Input Delays Based on T-S Fuzzy Local Bilinear Models

Junmin Li, Jinsha Li and Ruirui Duan

Additional information is available at the end of the chapter

http://dx.doi.org/10.5772/intechopen.69777

Abstract

This paper focuses on the non-fragile guaranteed cost control problem for a class of Takagi-Sugeno (T-S) fuzzy time-varying delay systems with local bilinear models and different state and input delays. A non-fragile guaranteed cost state-feedback controller is designed such that the closed-loop T-S fuzzy local bilinear control system is delay-dependent asymptotically stable, and the closed-loop fuzzy system performance is constrained to a certain upper bound when the additive controller gain perturbations exist. By employing the linear matrix inequality (LMI) technique, sufficient conditions are established for the existence of desired non-fragile guaranteed cost controllers. The simulation examples show that the proposed approach is effective and feasible.

Keywords: fuzzy control, non-fragile guaranteed cost control, delay-dependent, linear matrix inequality (LMI), T-S fuzzy bilinear model

1. Introduction

In recent years, T-S (Takagi-Sugeno) model-based fuzzy control has attracted wide attention, essentially because the fuzzy model is an effective and flexible tool for the control of nonlinear systems [1–8]. Through the application of sector nonlinearity approach, local approximation in fuzzy partition spaces or other different approximation methods, T-S fuzzy models will be used to approximate or exactly represent a nonlinear system in a compact set of state variables. The merit of the model is that the consequent part of a fuzzy rule is a linear dynamic subsystem, which makes it possible to apply the classical and mature linear systems theory to nonlinear systems. Further, by using the fuzzy inference method, the overall fuzzy model will be obtained. A fuzzy controller is designed via the method titled 'parallel distributed compensation (PDC)' [3–6], the main idea of which is that for each linear subsystem, the corresponding linear controller is carried out. Finally, the overall nonlinear controller is obtained via fuzzy blending of each individual linear controller. Based on the above content, T-S fuzzy model has been widely studied, and many results have been obtained [1–8]. In practical applications, time delay often occurs in many dynamic systems such as biological systems, network systems, etc. It is shown that the existence of delays usually becomes the source of instability and deteriorating performance of systems [3–8]. In general, when delay-dependent results were calculated, the emergence of the inner product between two vectors often makes the process of calculation more complicated. In order to avoid it, some model transformations were utilized in many papers, unfortunately, which will arouse the generation of an inequality, resulting in possible conservatism. On the other hand, due to the influence of many factors such as finite word length, truncation errors in numerical computation and electronic component parameter change, the parameters of the controller in a certain degree will change, which lead to imprecision in controller implementation. In this case, some small perturbations of the controllers' coefficients will make the designed controllers sensitive, even worse, destabilize the closed-loop control system [9]. So the problem of non-fragile control has been important issues. Recently, the research of non-fragile control has been paid much attention, and a series of productions have been obtained [10–13].

As we know, bilinear models have been widely used in many physical systems, biotechnology, socioeconomics and dynamical processes in other engineering fields [14, 15]. Bilinear model is a special nonlinear model, the nonlinear part of which consists of the bilinear function of the state and input. Compared with a linear model, the bilinear models have two main advantages. One is that the bilinear model can better approximate a nonlinear system. Another is that because of nonlinearity of it, many real physical processes may be appropriately modeled as bilinear systems. A famous example of a bilinear system is the population of biological species, which can be showed by $\frac{d\theta}{dt} = \theta v$. In this equation, v is the birth rate minus death rate, and θ denotes the population. Obviously, the equation cannot be approximated by a linear model [14].

Most of the existing results focus on the stability analysis and synthesis based on T-S fuzzy model with linear local model. However, when a nonlinear system has of complex nonlinearities, the constructed T-S model will consist of a number of fuzzy local models. This will lead to very heavy computational burden. According to the advantages of bilinear systems and T-S fuzzy control, so many researchers paid their attentions to the T-S fuzzy models with bilinear rule consequence [16–18]. From these papers, it is evident that the T-S fuzzy bilinear model may be suitable for some classes of nonlinear plants. In Ref. [16], a nonlinear system was transformed into a bilinear model via Taylor's series expansion, and the stability of T-S fuzzy bilinear model was studied. Moreover, the result was stretched into the complex fuzzy system with state time delay [17]. Ref. [18] presented robust stabilization for a class of discrete-time fuzzy bilinear system. Very recently, a class of nonlinear systems is described by T-S fuzzy models with nonlinear local models in Ref. [19], and in this paper, the scholars put forward a new fuzzy control scheme with local nonlinear feedbacks, the advantage of which over the existing methods is that a fewer fuzzy rules and less computational burden. The non-fragile guaranteed cost controller was designed for a class of T-S discrete-time fuzzy bilinear systems in Ref. [20]. However, in Refs. [19, 20], the time-delay effects on the system is not considered. Ref. [17] is only considered the fuzzy system with the delay in the state and the derivatives of time-delay, $\dot{d}(t) < 1$ is required. Refs. [21–23] dealt with the uncertain fuzzy systems with time-delay in different ways. It should be pointed out that all the aforementioned works did not take into account the effect of the control input delays on the systems. The results therein are not applicable to systems with input delay. Recently, some controller design approaches have been presented for systems with input delay, see [2, 3, 4, 18, 24–32] for fuzzy T-S systems and [8, 15, 33, 34] for non-fuzzy systems and the references therein. All of these results are required to know the exact delay values in the implementation. T-S fuzzy stochastic systems with state time-vary or distributed delays were studied in Refs. [35–39]. The researches of fractional order T-S fuzzy systems on robust stability, stability analysis about "$0 < \alpha < 1$", and decentralized stabilization in multiple time delays were presented in Refs. [40–42], respectively. For different delay types, the corresponding adaptive fuzzy controls for nonlinear systems were proposed in Refs. [33, 43, 44]. In Refs. [45, 46], to achieve small control amplitude, a new T-S fuzzy hyperbolic model was developed, moreover, Ref. [46] considered the input delay of the novel model. In Ref. [25, 47], the problems of observer-based fuzzy control design for T-S fuzzy systems were concerned.

So far, the problem of non-fragile guaranteed cost control for fuzzy system with local bilinear model with different time-varying state and input delays has not been discussed.

In this paper, the problem of delay-dependent non-fragile guaranteed cost control is studied for the fuzzy time-varying delay systems with local bilinear model and different state and input delays. Based on the PDC scheme, new delay-dependent stabilization conditions for the closed-loop fuzzy systems are derived. No model transformation is involved in the derivation. The merit of the proposed conditions lies in its reduced conservatism, which is achieved by circumventing the utilization of some bounding inequalities for the cross-product between two vectors as in Ref. [17]. The three main contributions of this paper are the following: (1) a non-fragile guaranteed cost controller is presented for the fuzzy system with time-varying delay in both state and input; (2) some free-weighting matrices are introduced in the derivation process, where the constraint of the derivatives of time-delay, $\dot{d}(t) < 1$ and $\dot{h}(t) < 1$, is eliminated; and (3) the delay-dependent stability conditions for the fuzzy system are described by LMIs. Finally, simulation examples are given to illustrate the effectiveness of the obtained results.

The paper is organized as follows. Section 2 introduces the fuzzy delay system with local bilinear model, and non-fragile controller law for such system is designed based on the parallel distributed compensation approach in Section 3. Results of non-fragile guaranteed cost control are given in Section 4. Two simulation examples are used to illustrate the effectiveness of the proposed method in Section 5, which is followed by conclusions in Section 6.

Notation: Throughout this paper, the notation $P > 0 (P \geq 0)$ stands for P being real symmetric and positive definite (or positive semi-definite). In symmetric block matrices, the asterisk (*) refers to a term that is induced by symmetry, and diag{....} denotes a block-diagonal matrix.

The superscript T means matrix transposition. The notion $\sum_{i,j=1}^{s}$ is an abbreviation of $\sum_{i=1}^{s}\sum_{j=1}^{s}$. Matrices, if the dimensions are not explicitly stated, are assumed to have compatible dimensions for algebraic operations.

2. System description and assumptions

In this section, we introduce the T-S fuzzy time-delay system with local bilinear model. The ith rule of the fuzzy system is represented by the following form:

$$\begin{aligned} &\textit{Plant Rule } i:\\ &\textit{IF } \vartheta_1(t) \textit{ is } F_{i1} \textit{ and } \ldots \textit{ and } \vartheta_v(t) \textit{ is } F_{iv}, \textit{ THEN}\\ &\dot{x}(t) = A_i x(t) + A_{di}x(t-d(t)) + B_i u(t) + B_{hi}u(t-h(t)) + N_i x(t)u(t) + N_{di}x(t-d(t))u(t-h(t))\\ &\qquad x(t) = \phi(t), \quad t \in [-\tau_1, 0]\ , i = 1, 2, \ldots, s \end{aligned} \tag{1}$$

where F_{ij} is the fuzzy set, s is the number of fuzzy rules, $x(t) \in R^n$ is the state vector, and $u(t) \in R$ is the control input, $\vartheta_1(t)$, $\vartheta_2(t)$,…,$\vartheta_v(t)$ are the premise variables. It is assumed that the premise variables do not depend on the input $u(t)$. A_i, A_{di}, N_i, $N_{di} \in R^{n\times n}$, B_i, $B_{hi} \in R^{n\times 1}$ denote the system matrices with appropriate dimensions. $d(t)$ is a time-varying differentiable function that satisfies $0 \le d(t) \le \tau_1$, $0 \le h(t) \le \tau_2$, where τ_1, τ_2 are real positive constants as the upper bound of the time-varying delay. It is also assumed that $\dot{d}(t) \le \sigma_1$, $\dot{h}(t) \le \sigma_2$, and σ_1, σ_2 are known constants. The initial conditions $\phi(t)$, $\varphi(t)$ are continuous functions of t, $t \in [-\tau,\ 0]$, $\tau = \min(\tau_1, \tau_2)$.

Remark 1: The fuzzy system with time-varying state and input delays will be investigated in this paper, which is different from the system in Ref. [17]. In Ref. [17], only state time-varying delay is considered. And also, here, we assume that the derivative of time-varying delay is less than or equal to a known constant that may be greater than 1; the assumption on time-varying delay in Ref. [17] is relaxed.

By using singleton fuzzifier, product inferred and weighted defuzzifier, the fuzzy system can be expressed by the following globe model:

$$\begin{aligned} \dot{x}(t) = &\sum_{i=1}^{s} h_i(\vartheta(t))[A_i x(t) + A_{di}x(t-d(t)) + B_i u(t) + B_{hi}u(t-h(t)) + N_i x(t)u(t)\\ &+ N_{di}x(t-d(t))u(t-h(t))] \end{aligned} \tag{2}$$

where

$h_i(\vartheta(t)) = \omega_i(\vartheta(t))/\sum_{i=1}^{s}\omega_i(\vartheta(t)), \omega_i(\vartheta(t)) = \prod_{j=1}^{v}\mu_{ij}(\vartheta(t))$, μ_{ij} $(\vartheta(t))$ is the grade of membership of $\vartheta_i(t)$ in F_{ij}. In this paper, it is assumed that $\omega_i(\vartheta(t)) \ge 0$, $\sum_{i=1}^{s}\omega_i(\vartheta(t)) > 0$ for all t. Then, we have the following conditions $h_i(\vartheta(t)) \ge 0$, $\sum_{i=1}^{s} h_i(\vartheta(t)) = 1$ for all t. In the consequent, we use abbreviation h_i, h_{hi}, $x_d(t)$, $u_d(t)$, $x_h(t)$, $u_h(t)$, to replace $h_i(\vartheta(t))$, $h_i(\vartheta(t-h(t)))$, $x(t-d(t))$, $u(t-d(t))$, $x(t-h(t))$, $u(t-h(t))$, respectively, for convenience.

The objective of this paper is to design a state-feedback non-fragile guaranteed cost control law for the fuzzy system (2).

3. Non-fragile guaranteed cost controller design

Extending the design concept in Ref. [17], we give the following non-fragile fuzzy control law:

$$\begin{aligned}&IF\ \vartheta_1(t)\ is\ F_1^i\ and\ \ldots\ and\ \vartheta_v(t)\ is\ F_v^i\\&THEN\ u(t)=\frac{\rho(K_i+\Delta K_i)x(t)}{\sqrt{1+x^T(K_i+\Delta K_i)^T(K_i+\Delta K_i)x}}=\rho\sin\theta_i=\rho\cos\theta_i(K_i+\Delta K_i)x(t)\end{aligned} \tag{3}$$

where $\rho>0$ is a scalar to be assigned, and $K_i\in R^{l\times n}$ is a local controller gain to be determined. ΔK_i represents the additive controller gain perturbations of the form $\Delta K_i = H_iF_i(t)E_{ki}$ with H_i and E_{ki} being known constant matrices, and $F_i(t)$ the uncertain parameter matrix satisfying $F_i^T(t)F_i(t)\le I$. $\sin\theta_i=\frac{\overline{K}_ix(t)}{\sqrt{1+x^T\overline{K}_i^T\overline{K}_ix}}$, $\cos\theta_i=\frac{1}{\sqrt{1+x^T\overline{K}_i^T\overline{K}_ix}}$, $\theta_i\in[-\frac{\pi}{2},\frac{\pi}{2}]$, $\overline{K}_i=K_i+\Delta K_i(t)=K_i+H_iF_i(t)E_{ki}$.

The overall fuzzy control law can be represented by

$$u(t)=\sum_{i=1}^{s}h_i\frac{\rho\overline{K}_ix(t)}{\sqrt{1+x^T\overline{K}_i^T\overline{K}_ix}}=\sum_{i=1}^{s}h_i\rho\sin\theta_i=\sum_{i=1}^{s}h_i\rho\cos\theta_i\overline{K}_ix(t) \tag{4}$$

When there exists an input delay $h(t)$, we have that

$$u_h(t)=\sum_{l=1}^{s}h_{hl}\rho\sin\varphi_l=\sum_{l=1}^{s}h_{hl}\rho\cos\varphi_l\tilde{K}_lx_h(t) \tag{5}$$

where $\sin\varphi_l=\frac{\tilde{K}_lx_h(t)}{\sqrt{1+x_h^T\tilde{K}_l^T\tilde{K}_lx_h}}$, $\cos\varphi_l=\frac{1}{\sqrt{1+x_h^T\tilde{K}_l^T\tilde{K}_lx_h}}$, $\varphi_l\in[-\frac{\pi}{2},\frac{\pi}{2}]$, $\tilde{K}_l=K_l+\Delta K_l(t-h(t))=K_l+H_lF_l(t-h(t))E_{kl}$.

So, it is natural and necessary to make an assumption that the functions h_i are well defined all $t\in[-\tau_2,0]$, and satisfy the following properties:

$h_i(\vartheta(t-h(t)))\ge0$, for $i = 1,2,\ldots,s$, and $\sum_{i=1}^{s}h_i(\vartheta(t-h(t)))=1$.

By substituting Eq. (5) into Eq. (2), the closed-loop system can be given by

$$\dot{x}(t)=\sum_{i,j,l=1}^{s}h_ih_jh_{hl}(\Lambda_{ij}x(t)+\Lambda_{dij}x_d(t)+\Lambda_{hil}x_h(t)) \tag{6}$$

where

$\Lambda_{ij}=A_i+\rho\sin\theta_jN_i+\rho\cos\theta_jB_i\overline{K}_j$, $\Lambda_{dil}=A_{di}+\rho\sin\varphi_lN_{di}$, $\Lambda_{hil}=\rho\cos\varphi_lB_{hi}\tilde{K}_l$.

Given positive-definite symmetric matrices $S \in R^{n \times n}$ and $W \in R$, we take the cost function

$$J = \int_0^{\infty} [x^{\mathrm{T}}(t)Sx(t) + u^{\mathrm{T}}(t)Wu(t)]dt \tag{7}$$

Definition 1. The fuzzy non-fragile control law $u(t)$ is said to be non-fragile guaranteed cost if for the system (2), there exist control laws (4) and (5) and a scalar J_0 such that the closed-loop system (6) is asymptotically stable and the closed-loop value of the cost function (7) satisfies $J \leq J_0$.

4. Analysis of stability for the closed-loop system

Firstly, the following lemmas are presented which will be used in the paper.

Lemma 1 [20]: *Given any matrices M and N with appropriate dimensions such that* $\varepsilon > 0$, *we have* $M^T N + N^T M \leq \varepsilon M^T M + \varepsilon^{-1} N^T N$.

Lemma 2 [21]: *Given constant matrices G, E and a symmetric constant matrix S of appropriate dimensions. The inequality* $S + GFE + E^T F^T G^T < 0$ *holds, where* $F(t)$ *satisfies* $F^T(t)\, F(t) \leq I$ *if and only if, for some* $\varepsilon > 0$, $S + \varepsilon GG^T + \varepsilon^{-1} E^T E < 0$.

The following theorem gives the sufficient conditions for the existence of the non-fragile guaranteed cost controller for system (6) with additive controller gain perturbations.

Theorem 1. *Consider system (6) associated with cost function (7). For given scalars* $\rho > 0$, $\tau_1 > 0$, $\tau_2 > 0$, $\sigma 1 > 0$, $\sigma_2 > 0$, *if there exist matrices* $P > 0$, $Q_1 > 0$, $Q2 > 0$, $R_1 > 0$, $R_2 > 0$, K_i, $i = 1, 2, \ldots, s$, X_1, X_2, X_3, X_4, Y_1, Y_2, Y_3, Y_4, *and scalar* $\varepsilon > 0$ *satisfying the inequalities (8), the system (6) is asymptotically stable and the control law (5) is a fuzzy non-fragile guaranteed cost control law, moreover,*

$$J \leq x^{\mathrm{T}}(0)Px(0) + \int_{-d(0)}^{0} x^{\mathrm{T}}(s)Q_1x(s)ds + \int_{-\tau_1}^{0}\int_{\theta}^{0} \dot{x}^{\mathrm{T}}(s)R_1\dot{x}(s)dsd\theta$$
$$+ \int_{-h(0)}^{0} x^{\mathrm{T}}(s)Q_2x(s)ds + \int_{-\tau_2}^{0}\int_{\theta}^{0} \dot{x}^{\mathrm{T}}(s)R_2\dot{x}(s)dsd\theta = J_0$$

$$\begin{bmatrix} T_{ijl} & * & * \\ \tau_1 X^{\mathrm{T}} & -\tau_1 R_1 & * \\ \tau_2 Z^{\mathrm{T}} & 0 & -\tau_2 R_2 \end{bmatrix} < 0, \quad i, j, l = 1, 2, \ldots, s \tag{8}$$

$$\textit{where } T_{ijl} = \begin{bmatrix} T_{11,ij} & * & * & * \\ T_{21,i} & T_{22,i} & * & * \\ T_{31,i} & T_{32,ij} & T_{33,il} & * \\ T_{41,i} & T_{42,i} & T_{43} & T_{44} \end{bmatrix},$$

$$T_{11,ij} = Q_1 + Q_2 + X_1 + X_1^{\mathrm{T}} + Y_1 A_i + A_i^{\mathrm{T}} Y_1^{\mathrm{T}} + S + 2\varepsilon\rho^2 Y_1 Y_1{}^{\mathrm{T}} + 4\varepsilon^{-1}(N_i^{\mathrm{T}} N_i + (B_i \overline{K}_j)^{\mathrm{T}}(B_i \overline{K}_j))$$
$$+ Z_1^T + Z_1 + \rho^2 \overline{K}_i^{\mathrm{T}} W \overline{K}_i,$$
$$T_{21,i} = -X_1^{\mathrm{T}} + X_2 + Z_2 + Y_2 A_i + A_{di}^{\mathrm{T}} Y_1^{\mathrm{T}}, \quad T_{31,i} = Z_3 - Z_1 + X_3 + Y_3 A_i,$$
$$T_{22,ij} = -(1-\sigma_1)Q_1 - X_2 - X_2^{\mathrm{T}} + Y_2 A_{di} + A_{di}^{\mathrm{T}} Y_2^{\mathrm{T}} + 2\varepsilon\rho^2 Y_2 Y_2{}^{\mathrm{T}} + 4\varepsilon^{-1} N_{di}^{\mathrm{T}} N_{di},$$
$$T_{32,i} = -X_3 + Y_3 A_{di} - Z_2^{\mathrm{T}}, \quad T_{33,il} = -(1-\sigma_2)Q_2 - Z_3 - Z_3^{\mathrm{T}} + 2\varepsilon\rho^2 Y_3 Y_3{}^{\mathrm{T}} + 4\varepsilon^{-1}(B_{hi}\tilde{K}_l)^{\mathrm{T}} B_{hi}\tilde{K}_l$$
$$T_{41,i} = P + X_4 + Z_4 + Y_4 A_i - Y_1^{\mathrm{T}}, T_{42,i} = -X_4 + Y_4 A_i - Y_2^{\mathrm{T}},$$
$$T_{43} = -Z_4 - Y_3^{\mathrm{T}}, T_{44} = \tau_1 R_1 + \tau_2 R_2 - Y_4 - Y_4{}^{\mathrm{T}} + 2\varepsilon\rho^2 Y_4 Y_4^{\mathrm{T}}.$$

Proof: Take a Lyapunov function candidate as

$$\begin{aligned} V(x(t),t) &= x^{\mathrm{T}}(t)Px(t) + \int_{t-d(t)}^{t} x^{\mathrm{T}}(s)Q_1 x(s)ds + \int_{-\tau_1}^{0}\int_{t+\theta}^{t} \dot{x}^{\mathrm{T}}(s)R_1\dot{x}(s)dsd\theta \\ &+\int_{t-h(t)}^{t} x^{\mathrm{T}}(s)Q_2 x(s)ds + \int_{-\tau_2}^{0}\int_{t+\theta}^{t} \dot{x}^{\mathrm{T}}(s)R_2\dot{x}(s)dsd\theta \end{aligned} \tag{9}$$

The time derivatives of $V(x(t),t)$, along the trajectory of the system (6), are given by

$$\begin{aligned} &\dot{V}(x(t),t) = 2x^{\mathrm{T}}(t)P\dot{x}(t) + x^{\mathrm{T}}(t)(Q_1+Q_2)x(t) \\ &-(1-\dot{d}(t))x_d^{\mathrm{T}}(t)Q_1 x_d(t) + \dot{x}^{\mathrm{T}}(t)(\tau_1 R_1 + \tau_2 R_2)\dot{x}(t) \\ &-\int_{t-\tau_1}^{t} \dot{x}^{\mathrm{T}}(s)R_1\dot{x}(s)ds - (1-\dot{h}(t))x_h^{\mathrm{T}}(t)Q_2 x_h(t) - \int_{t-\tau_2}^{t} \dot{x}^{\mathrm{T}}(s)R_2\dot{x}(s)ds \end{aligned} \tag{10}$$

Define the free-weighting matrices as $X = [X_1^{\mathrm{T}} \ \ X_2^{\mathrm{T}} \ \ X_3^{\mathrm{T}} \ \ X_4^{\mathrm{T}}]^{\mathrm{T}}$, $Y = [Y_1^T \ \ Y_2^T \ \ Y_3^T \ \ Y_4^T]^{\mathrm{T}}$, $Z = [Z_1^{\mathrm{T}} \ \ Z_2^{\mathrm{T}} \ \ Z_3^{\mathrm{T}} \ \ Z_4^{\mathrm{T}}]^{\mathrm{T}}$, where $X_k \in R^{n\times n}$, $Y_k \in R^{n\times n}$, $Z_k \in R^{n\times n}$, k = 1, 2, 3, 4 will be determined later.

Using the Leibniz-Newton formula and system equation (6), we have the following identical equations:

$$\begin{aligned} &[x^T(t)X_1 + x_d^T(t)X_2 + x_h^T(t)X_3 + \dot{x}^T(t)X_4][x(t) - x_d(t) - \int_{t-d(t)}^{t} \dot{x}(s)ds] \equiv 0, \\ &[x^T(t)Z_1 + x_d^T(t)Z_2 + x_h^T(t)Z_3 + \dot{x}^T(t)Z_4][x(t) - x_h(t) - \int_{t-h(t)}^{t} \dot{x}(s)ds] \equiv 0, \\ &\sum_{i,j=1}^{s} h_i h_j h_l [x^T(t)Y_1 + x_d^T(t)Y_2 + x_h^T(t)Y_4 + \dot{x}^T(t)Y_4][\Lambda_{ij}x(t) + \Lambda_{dil}x_d(t) + \Lambda_{hil}x_h(t) - \dot{x}(t)] \equiv 0 \end{aligned} \tag{11}$$

Then, substituting Eq. (12) into Eq. (11) yields

$$\begin{aligned}
\dot{V}(x(t),t) &= 2x^{\mathrm{T}}(t)P\dot{x}(t) + x^{\mathrm{T}}(t)(Q_1+Q_2)x(t) + \dot{x}^{\mathrm{T}}(t)(\tau_1R_1+\tau_2R_2)\dot{x}(t)\\
&\quad -(1-\dot{d}(t))x_d^{\mathrm{T}}(t)Q_1x_d(t) - (1-\dot{h}(t))x_h^{\mathrm{T}}(t)Q_2x_h(t)\\
&\quad -\int_{t-\tau_1}^{t}\dot{x}^{\mathrm{T}}(s)R_1\dot{x}(s)ds + 2\eta^{\mathrm{T}}(t)X[x(t)-x_d(t)-\int_{t-d(t)}^{t}\dot{x}(s)ds]\\
&\quad -\int_{t-\tau_2}^{t}\dot{x}^{\mathrm{T}}(s)R_2\dot{x}(s)ds + 2\eta^{\mathrm{T}}(t)Z[x(t)-x_h(t)-\int_{t-h(t)}^{t}\dot{x}(s)ds]\\
&\quad + 2\eta^{\mathrm{T}}(t)Y\sum_{i,j,l=1}^{s}h_ih_jh_{hl}[\Lambda_{ij}x(t)+\Lambda_{dil}x_d(t)+\Lambda_{hil}x_h(t)-\dot{x}(t)]\\
&\le 2x^{\mathrm{T}}(t)P\dot{x}(t) + x^{\mathrm{T}}(t)(Q_1+Q_2)x(t) + \dot{x}^{\mathrm{T}}(t)(\tau_1R_1+\tau_2R_2)\dot{x}(t)\\
&\quad -(1-\sigma_1)x_d^{\mathrm{T}}(t)Q_1x_d(t) - (1-\sigma_2)x_h^{\mathrm{T}}(t)Q_2x_h(t)\\
&\quad -\int_{t-d(t)}^{t}\dot{x}^{\mathrm{T}}(s)R_1(s)\dot{x}(s)ds + 2\eta^{\mathrm{T}}(t)X[x(t)-x_d(t)-\int_{t-d(t)}^{t}\dot{x}(s)ds]\\
&\quad -\int_{t-h(t)}^{t}\dot{x}^{\mathrm{T}}(s)R_2(s)\dot{x}(s)ds + 2\eta^{\mathrm{T}}(t)Z[x(t)-x_h(t)-\int_{t-h(t)}^{t}\dot{x}(s)ds]\\
&\quad + 2\eta^{\mathrm{T}}(t)Y\sum_{i,j,l=1}^{s}h_ih_jh_{hl}[\Lambda_{ij}x(t)+\Lambda_{dil}x_d(t)+\Lambda_{hil}x_h(t)-\dot{x}(t)] + x^{\mathrm{T}}(t)Sx(t)\\
&\quad + \sum_{i,j=1}^{s}h_ih_j\rho^2x^{\mathrm{T}}(t)\overline{K}_i^{\mathrm{T}}\cos\theta_iW\overline{K}_j\cos\theta_jx(t) - [x^{\mathrm{T}}(t)Sx(t)+u^{\mathrm{T}}(t)Wu(t)]
\end{aligned} \tag{12}$$

where $\eta(t) = [x^{\mathrm{T}}(t),\ x_d^{\mathrm{T}}(t),\ x_h^{\mathrm{T}}(t),\ \dot{x}^{\mathrm{T}}(t)]^{\mathrm{T}}$.

Applying Lemma 1, we have the following inequalities:

$$\begin{aligned}
&2x^{\mathrm{T}}(t)Y_1\Lambda_{ij}x(t) \le 2x^{\mathrm{T}}(t)Y_1A_ix(t) + \varepsilon\rho^2x^{\mathrm{T}}(t)Y_1Y_1^{\mathrm{T}}x(t) + \varepsilon^{-1}x^{\mathrm{T}}(t)(N_i^{\mathrm{T}}N_i + (B_i\overline{K}_j)^{\mathrm{T}}(B_i\overline{K}_j))x(t),\\
&2x^{\mathrm{T}}(t)Y_1\Lambda_{dil}x_d(t) \le 2x^{\mathrm{T}}(t)Y_1A_{di}x_d(t) + \varepsilon\rho^2\sin^2\phi_lx^{\mathrm{T}}(t)Y_1Y_1^{\mathrm{T}}x(t) + \varepsilon^{-1}x_d^{\mathrm{T}}(t)N_{di}^{\mathrm{T}}N_{di}x_d(t),\\
&2x^{\mathrm{T}}(t)Y_1\Lambda_{hil}x_h(t) \le \varepsilon\rho^2\cos^2\phi_lx^{\mathrm{T}}(t)Y_1Y_1^{\mathrm{T}}x(t) + \varepsilon^{-1}x_h^{\mathrm{T}}(t)(B_{hi}\tilde{K}_l)^{\mathrm{T}}(B_{hi}\tilde{K}_l)x_h(t),\\
&2x_d^{\mathrm{T}}(t)Y_2\Lambda_{ij}x(t) \le 2x_d^{\mathrm{T}}(t)Y_2A_ix(t) + \varepsilon\rho^2x_d^{\mathrm{T}}(t)Y_2Y_2^{\mathrm{T}}x_d(t) + \varepsilon^{-1}x^{\mathrm{T}}(t)(N_i^{\mathrm{T}}N_i + (B_i\overline{K}_j)^{\mathrm{T}}(B_i\overline{K}_j))x(t),\\
&2x^{\mathrm{T}}(t)Y_1\Lambda_{hil}x_h(t) \le \varepsilon\rho^2\cos^2\phi_lx^{\mathrm{T}}(t)Y_1Y_1^{\mathrm{T}}x(t) + \varepsilon^{-1}x_h^{\mathrm{T}}(t)(B_{hi}\tilde{K}_l)^{\mathrm{T}}(B_{hi}\tilde{K}_l)x_h(t),\\
&2x_d^{\mathrm{T}}(t)Y_2\Lambda_{ij}x(t) \le 2x_d^{\mathrm{T}}(t)Y_2A_ix(t) + \varepsilon\rho^2x_d^{\mathrm{T}}(t)Y_2Y_2^{\mathrm{T}}x_d(t) + \varepsilon^{-1}x^{\mathrm{T}}(t)(N_i^{\mathrm{T}}N_i + (B_i\overline{K}_j)^{\mathrm{T}}(B_i\overline{K}_j))x(t),\\
&2x_d^{\mathrm{T}}(t)Y_2\Lambda_{dil}x_d(t) \le 2x_d^{\mathrm{T}}(t)Y_2A_{di}x_d(t) + \varepsilon\rho^2\sin^2\phi_lx_d^{\mathrm{T}}(t)Y_2Y_2^{\mathrm{T}}x_d(t) + \varepsilon^{-1}x_d^{\mathrm{T}}(t)N_{di}^{\mathrm{T}}N_{di}x_d(t),\\
&2x_d^{\mathrm{T}}(t)Y_2\Lambda_{hil}x_h(t) \le \varepsilon\rho^2\cos^2\phi_lx_d^{\mathrm{T}}(t)Y_2Y_2^{\mathrm{T}}x_d(t) + \varepsilon^{-1}x_h^{\mathrm{T}}(t)(B_{hi}\tilde{K}_l)^{\mathrm{T}}(B_{hi}\tilde{K}_l)x_h(t),\\
&2x_h^{\mathrm{T}}(t)Y_3\Lambda_{ij}x(t) \le 2x_h^{\mathrm{T}}(t)Y_3A_ix(t) + \varepsilon\rho^2x_h^{\mathrm{T}}(t)Y_3Y_3^{\mathrm{T}}x_h(t) + \varepsilon^{-1}x^{\mathrm{T}}(t)(N_i^{\mathrm{T}}N_i + (B_i\overline{K}_j)^{\mathrm{T}}(B_i\overline{K}_j))x(t),\\
&2x_h^{T}(t)Y_3\Lambda_{dil}x_d(t) \le 2x_d^{\mathrm{T}}(t)Y_3A_{di}x_d(t) + \varepsilon\rho^2\sin^2\phi_lx_h^{T}(t)Y_3Y_3^{\mathrm{T}}x_h(t) + \varepsilon^{-1}x_d^{T}(t)N_{di}^{\mathrm{T}}N_{di}x_d(t),\\
&2x_h^{\mathrm{T}}(t)Y_3\Lambda_{hil}x_h(t) \le \varepsilon\rho^2\cos^2\phi_lx_h^{\mathrm{T}}(t)Y_3Y_3^{\mathrm{T}}x_h(t) + \varepsilon^{-1}x_h^{\mathrm{T}}(t)(B_{hi}\tilde{K}_l)^{\mathrm{T}}(B_{hi}\tilde{K}_l)x_h(t),\\
&2\dot{x}^{\mathrm{T}}(t)Y_4\Lambda_{ij}x(t) \le 2\dot{x}^{\mathrm{T}}(t)Y_4A_ix(t) + \varepsilon\rho^2\dot{x}^{\mathrm{T}}(t)Y_4Y_4^{\mathrm{T}}\dot{x}(t) + \varepsilon^{-1}x^{\mathrm{T}}(t)(N_i^{\mathrm{T}}N_i + (B_i\overline{K}_j)^{\mathrm{T}}(B_i\overline{K}_j))x(t),\\
&2\dot{x}^{\mathrm{T}}(t)Y_4\Lambda_{dil}x_d(t) \le 2\dot{x}^{\mathrm{T}}(t)Y_4A_{di}x_d(t) + \varepsilon\rho^2\dot{x}^{\mathrm{T}}(t)Y_4Y_4^{\mathrm{T}}\dot{x}(t) + \varepsilon^{-1}x_d^{\mathrm{T}}(t)N_{di}^{\mathrm{T}}N_{di}x_d(t),\\
&2\dot{x}^{\mathrm{T}}(t)Y_4\Lambda_{hil}x_h(t) \le \varepsilon\rho^2\cos^2\phi_l\dot{x}^{\mathrm{T}}(t)Y_4Y_4^{\mathrm{T}}\dot{x}(t) + \varepsilon^{-1}x_h^{\mathrm{T}}(t)(B_{hi}\tilde{K}_l)^{\mathrm{T}}(B_{hi}\tilde{K}_l)x_h(t)
\end{aligned} \tag{13}$$

Substituting Eq. (13) into Eq. (12) results in

$$
\begin{aligned}
\dot{V}(x(t),t) \le & \sum_{i,j,l=1}^{s} h_i h_j h_{hl} \eta^{\mathrm{T}}(t) T_{ij} \eta(t) - \int_{t-d(t)}^{t} \dot{x}^{\mathrm{T}}(s) R_1 \dot{x}(s) ds - \int_{t-h(t)}^{t} \dot{x}^{\mathrm{T}}(s) R_2 \dot{x}(s) ds \\
& -2\eta^{\mathrm{T}}(t) X \int_{t-d(t)}^{t} \dot{x}(s) ds - 2\eta^{\mathrm{T}}(t) Z \int_{t-h(t)}^{t} \dot{x}(s) ds - [x^{\mathrm{T}}(t) S x(t) + u^{\mathrm{T}}(t) W u(t)] \\
& \le \sum_{i,j,l=1}^{s} h_i h_j h_{hl} \eta^{\mathrm{T}}(t) (T_{ijl} + \tau_1 X R_1^{-1} X^{\mathrm{T}} + \tau_2 Z R_2^{-1} Z^{\mathrm{T}}) \eta(t) \\
& - \int_{t-d(t)}^{t} \left(\eta^{\mathrm{T}}(t) X + \dot{x}^{\mathrm{T}}(s) R_1\right) R_1^{-1} \left(\eta^{\mathrm{T}}(t) X + \dot{x}^{\mathrm{T}}(s) R_1\right)^{\mathrm{T}} ds \\
& - \int_{t-h(t)}^{t} \left(\eta^{\mathrm{T}}(t) Z + \dot{x}^{\mathrm{T}}(s) R_2\right) R_2^{-1} \left(\eta^{\mathrm{T}}(t) X + \dot{x}^{\mathrm{T}}(s) R_2\right)^{\mathrm{T}} ds - [x^{\mathrm{T}}(t) S x(t) + u^{\mathrm{T}}(t) W u(t)] \\
& \le \sum_{i,j,l=1}^{s} h_i h_j h_{hl} \eta^{\mathrm{T}}(t) (\tilde{T}_{ijl} + \tau_1 X R_1^{-1} X^{\mathrm{T}} + \tau_2 Z R_2^{-1} Z^{\mathrm{T}}) \eta(t) - [x^{\mathrm{T}}(t) S x(t) + u^{\mathrm{T}}(t) W u(t)]
\end{aligned} \tag{14}
$$

where

$$
\tilde{T}_{ijl} = \begin{bmatrix} \tilde{T}_{11,ij} & * & * & * \\ T_{21,i} & T_{22,i} & * & * \\ T_{31,i} & T_{32,ij} & T_{33,il} & * \\ T_{41,i} & T_{42,i} & T_{43} & T_{44} \end{bmatrix}, \quad \tilde{T}_{11,ij} = T_{11,ij} + \rho^2 \overline{K}_i^{\mathrm{T}} \cos\theta_i W \overline{K}_j \cos\theta_j - \rho^2 \overline{K}_i^{\mathrm{T}} W \overline{K}_j.
$$

In light of the inequality $\overline{K}_i^{\mathrm{T}} W \overline{K}_j + \overline{K}_j^{\mathrm{T}} W \overline{K}_i \le \overline{K}_i^{\mathrm{T}} W \overline{K}_i + \overline{K}_j^{\mathrm{T}} W \overline{K}_j$, we have

$$
\dot{V}(x(t),t) \le \sum_{i,j,l=1}^{s} h_i h_j h_{hl} \eta^{\mathrm{T}}(t) (T_{ijl} + \tau_1 X R_1^{-1} X^{\mathrm{T}} + \tau_2 Z R_2^{-1} Z^{\mathrm{T}}) \eta(t) - [x^{\mathrm{T}}(t) S x(t) + u^{\mathrm{T}}(t) W u(t)] \tag{15}
$$

Applying the Schur complement to Eq. (8) yields

$$
T_{ii} + +\tau_1 X R_1^{-1} X^{\mathrm{T}} + \tau_2 Z R_2^{-1} Z^{\mathrm{T}} < 0, T_{ij} + T_{ji} + 2\tau_1 X R_1^{-1} X^{\mathrm{T}} + 2\tau_2 Z R_2^{-1} Z^{\mathrm{T}} < 0.
$$

Therefore, it follows from Eq. (15) that

$$
\dot{V}(x(t),t) \le -[x^{\mathrm{T}}(t) S x(t) + u^{\mathrm{T}}(t) W u(t)] < 0 \tag{16}
$$

which implies that the system (6) is asymptotically stable.

Integrating Eq. (16) from 0 to T produces

$$
\int_0^T [x^{\mathrm{T}}(t) S x(t) + u^{\mathrm{T}}(t) W u(t)] dt \le -V(x(T),T) + V(x(0),0) < V(x(0),0)
$$

Because of $V(x(t),t) \geq 0$ and $\dot{V}(x(t),t) < 0$, thus $\lim_{T\to\infty} V(x(T),T) = c$, where c is a nonnegative constant. Therefore, the following inequality can be obtained:

$$J \leq x^{\mathrm{T}}(0)Px(0) + \int_{-d(0)}^{0} x^{\mathrm{T}}(s)Q_1x(s)ds + \int_{-\tau_1}^{0}\int_{\theta}^{0} \dot{x}^{\mathrm{T}}(s)R_1\dot{x}(s)dsd\theta + \int_{-h(0)}^{0} x^{\mathrm{T}}(s)Q_2x(s)ds + \int_{-\tau_2}^{0}\int_{\theta}^{0} \dot{x}^{\mathrm{T}}(s)R_2\dot{x}(s)dsd\theta = J_0 \tag{17}$$

This completes the proof.

Remark 2: In the derivation of Theorem 1, the free-weighting matrices $X_k \in R^{n\times n}$, $Y_k \in R^{n\times n}$, $k = 1, 2, 3, 4$ are introduced, the purpose of which is to reduce conservatism in the existing delay-dependent stabilization conditions, see Ref. [17].

In the following section, we shall turn the conditions given in Theorem 1 into linear matrix inequalities (LMIs). Under the assumptions that Y_1, Y_2, Y_3, Y_4 are non-singular, we can define the matrix $Y_i^{-\mathrm{T}} = \lambda Z$, $i = 1, 2, 3, 4$, $Z = P^{-1}$, $\lambda > 0$.

Pre- and post-multiply (8) and (9) with $\Theta = diag\{Y_1^{-1}, Y_2^{-1}, Y_3^{-1}, Y_4^{-1}, Y_4^{-1}, Y_4^{-1}\}$ and $\Theta^T = diag\{Y_1^{-T}, Y_2^{-T}, Y_3^{-T}, Y_4^{-T}, Y_4^{-T}, Y_4^{-T}\}$, respectively, and letting $\overline{Q}_1 = Y_1{}^{-1}Q_1Y_1{}^{-T}$, $\overline{Q}_2 = Y_1{}^{-1}Q_2Y_1{}^{-T}$, $\overline{R}_k = Y_4{}^{-1}R_kY_4{}^{-\mathrm{T}}, k = 1, 2$, $\overline{X}_i = Y_i{}^{-1}X_iY_i{}^{-\mathrm{T}}$, $\overline{Z}_i = Y_i{}^{-1}Z_iY_i{}^{-\mathrm{T}}$, i = 1, 2, 3, 4, we obtain the following inequality (18), which is equivalent to (8):

$$\begin{bmatrix} \overline{T}_{11,ij} & * & * & * & * & * \\ \overline{T}_{21,i} & \overline{T}_{22,i} & * & * & * & * \\ \overline{T}_{31,i} & \overline{T}_{32,i} & \overline{T}_{33,il} & * & * & * \\ \overline{T}_{41,i} & \overline{T}_{42,i} & \overline{T}_{43} & \overline{T}_{44} & * & * \\ \tau_1 X_1 & \tau_1 X_2 & \tau_1 X_3 & \tau_1 X_4 & -\tau_1\overline{R}_1 & * \\ \tau_2 Z_1 & \tau_2 Z_2 & \tau_2 Z_3 & \tau_2 Z_4 & 0 & -\tau_2\overline{R}_2 \end{bmatrix} < 0, \quad i,j,l = 1,2,\ldots,s \tag{18}$$

where

$$\begin{aligned}
\overline{T}_{11,ij} &= \overline{Q}_1 + \overline{Q}_2 + \overline{X}_1 + \overline{X}_1^{\mathrm{T}} + \lambda A_i Z + \lambda Z A_i^{\mathrm{T}} + \lambda^2 ZSZ + 2\varepsilon\rho^2 I + 4\varepsilon^{-1}\lambda^2 ZN_i^{\mathrm{T}}N_i Z \\
&\quad + Z_1 + Z_1^{\mathrm{T}} + 4\varepsilon^{-1}\lambda^2(B_i\overline{K}_j Z)^{\mathrm{T}}(B_i\overline{K}_j Z) + \rho^2\lambda^2 Z\overline{K}_i^{\mathrm{T}} W\overline{K}_i Z, \\
\overline{T}_{21,i} &= -\overline{X}_1^{\mathrm{T}} + \overline{X}_2 + \overline{Z}_2 + \lambda A_i Z + \lambda Z A_{di}^{\mathrm{T}}, \quad \overline{T}_{31,i} = \overline{Z}_3 - \overline{Z}_1 + \overline{X}_3 + \lambda A_i Z, \\
\overline{T}_{41,i} &= \lambda^2 Z + \lambda A_i Z - \lambda Z + \overline{X}_4 + \overline{Z}_4, \\
\overline{T}_{22,i} &= -(1-\sigma_1)\overline{Q}_1 - \overline{X}_2 - \overline{X}_2^{\mathrm{T}} + \lambda A_{di} Z + \lambda Z A_{di}^{\mathrm{T}} + 2\varepsilon\rho^2 I + 4\varepsilon^{-1}\lambda^2 Z N_{di}^{\mathrm{T}} N_{di} Z, \\
\overline{T}_{32,i} &= -\overline{X}_3 - \overline{Z}_2 + \lambda A_{di} Z - \lambda Z A_{di}^{\mathrm{T}}, \quad \overline{T}_{42,i} = -\overline{X}_4 + \lambda A_i Z - \lambda Z, \\
\overline{T}_{33,il} &= -(1-\sigma_2)\overline{Q}_2 - \overline{Z}_3 - \overline{Z}_3^{\mathrm{T}} + 4\varepsilon^{-1}\lambda^2(B_{hi}\tilde{K}_l Z)^{\mathrm{T}} B_{hi}\tilde{K}_l Z + 2\varepsilon\rho^2 I, \\
\overline{T}_{43} &= -\overline{Z}_4 - \lambda Z, \quad \overline{T}_{44} = \tau_1 R_1 + \tau_2 R_2 - \lambda Z - \lambda Z^{\mathrm{T}} + 2\varepsilon\rho^2 I.
\end{aligned}$$

Applying the Schur complement to Eq. (18) results in

$$\Gamma_{ijl} = \begin{bmatrix} \Phi_{11,i} & * & * \\ \Phi_{21,ij} & \Phi_{22} & * \\ \Phi_{31,il} & 0 & \Phi_{33} \end{bmatrix} < 0, \quad i, j, l = 1, 2, \ldots, s \tag{19}$$

where

$$\Phi_{11,i} = \begin{bmatrix} \overline{\overline{T}}_{11,i} & * & * & * & * & * \\ \overline{T}_{21,i} & \overline{\overline{T}}_{22,i} & * & * & * & * \\ \overline{T}_{31,i} & \overline{T}_{32,i} & \overline{\overline{T}}_{33,il} & * & * & * \\ \overline{T}_{41,i} & \overline{T}_{42,i} & \overline{T}_{43} & \overline{T}_{44} & * & * \\ \tau_1 X_1 & \tau_1 X_2 & \tau_1 X_3 & \tau_1 X_4 & -\tau_1 \overline{R}_1 & * \\ \tau_2 \overline{Z}_1 & \tau_2 \overline{Z}_2 & \tau_2 \overline{Z}_3 & \tau_2 \overline{Z}_4 & 0 & -\tau_2 \overline{R}_2 \end{bmatrix}$$

With

$$\begin{aligned} \overline{\overline{T}}_{11,i} &= \overline{Q}_1 + \overline{Q}_2 + \overline{X}_1 + \overline{X}_1^T + \lambda A_i Z + \lambda Z A_i^T + Z_1 + Z_1^T + 2\varepsilon\rho^2 I, \\ \overline{\overline{T}}_{22,i} &= -(1-\sigma_1)\overline{Q}_1 - \overline{X}_2 - \overline{X}_2^T + \lambda A_{di} Z + \lambda Z A_{di}^T + 2\varepsilon\rho^2 I, \\ \overline{\overline{T}}_{33} &= -(1-\sigma_2)\overline{Q}_2 - \overline{Z}_3 - \overline{Z}_3^T + 2\varepsilon\rho^2 I. \end{aligned}$$

$$\Phi_{21,ij} = \begin{bmatrix} \lambda Z & 0 & 0 & 0 \\ \lambda N_i Z & 0 & 0 & 0 \\ \lambda B_i \overline{K}_j Z & 0 & 0 & 0 \\ \rho\lambda \overline{K}_j Z & 0 & 0 & 0 \end{bmatrix}$$

$$\Phi_{31,il} = \begin{bmatrix} 0 & \lambda B_{hi} \tilde{K}_l Z & 0 & 0 \\ 0 & 0 & \lambda N_{di} Z & 0 \end{bmatrix}$$

$$\Phi_{22} = \begin{bmatrix} -S^{-1} & 0 & 0 & 0 \\ 0 & -\frac{\varepsilon}{4} I & 0 & 0 \\ 0 & 0 & -\frac{\varepsilon}{4} I & 0 \\ 0 & 0 & 0 & -W^{-1} \end{bmatrix}$$

$$\Phi_{33} = \begin{bmatrix} -\frac{\varepsilon}{4} I & 0 \\ 0 & -\frac{\varepsilon}{4} I \end{bmatrix}$$

Obviously, the closed-loop fuzzy system (6) is asymptotically stable, if for some scalars $\lambda > 0$, there exist matrices $Z > 0, \overline{Q} > 0, \overline{R} > 0$ and $\overline{X}_1, \overline{X}_2, \overline{X}_3, \overline{K}_i, i = 1, 2, .., s$ satisfying the inequality (19).

Theorem 2. *Consider the system (6) associated with cost function (7). For given scalars $\rho > 0$, $\tau_1 > 0$, $\tau_2 > 0$, $\sigma_1 > 0$, $\sigma_2 > 0$ and $\lambda > 0$, $\delta > 0$, if there exist matrices $Z > 0, \overline{Q}_1 > 0, \overline{R}_1 > 0, \overline{Q}_2 > 0, \overline{R}_2 > 0$ and $\overline{X}_1, \overline{X}_2, \overline{X}_3, \overline{X}_4$, M_i, $i = 1,2,\ldots,s$ and scalar $\varepsilon > 0$ satisfying the following LMI (20), the system (6) is asymptotically stable and the control law (5) is a fuzzy non-fragile guaranteed cost control law*

$$\begin{bmatrix} \Theta_{1,ijl} & * \\ \Theta_{2,ijl} & \Theta_3 \end{bmatrix} < 0, \quad i, j, l = 1, 2, \ldots, s \tag{20}$$

Moreover, the feedback gains are given by

$$K_i = M_i Z^{-1}, i = 1, 2, \ldots, s$$

and

$$J \leq x^{\mathrm{T}}(0)Px(0) + \int_{-d(0)}^{0} x^{\mathrm{T}}(s)Q_1 x(s)ds + \int_{-\tau_1}^{0}\int_{\theta}^{0} \dot{x}^{\mathrm{T}}(s)R_1\dot{x}(s)dsd\theta + \int_{-h(0)}^{0} x^{\mathrm{T}}(s)Q_2 x(s)ds$$
$$+\int_{-\tau_2}^{0}\int_{\theta}^{0} \dot{x}^{\mathrm{T}}(s)R_2\dot{x}(s)dsd\theta = J_0$$

where

$$\Theta_{2,ijl} = \begin{bmatrix} \lambda E_{kj}Z & 0 & 0 & 0 & 0 & 0 & 0 & 0 & 0 & 0 & 0 & 0 \\ \lambda E_{ki}Z & 0 & 0 & 0 & 0 & 0 & 0 & 0 & 0 & 0 & 0 & 0 \\ 0 & 0 & \lambda E_{kl}Z & 0 & 0 & 0 & 0 & 0 & 0 & 0 & 0 & 0 \\ 0 & 0 & 0 & 0 & 0 & 0 & 0 & 0 & (B_iH_j)^{\mathrm{T}} & 0 & 0 & 0 \\ 0 & 0 & 0 & 0 & 0 & 0 & 0 & 0 & 0 & \rho H_i^{\mathrm{T}} & 0 & 0 \\ 0 & 0 & 0 & 0 & 0 & 0 & 0 & 0 & 0 & 0 & 0 & (B_{hi}H_l)^{\mathrm{T}} \end{bmatrix},$$

$$\Theta_3 = \mathrm{diag}\{-\delta I, -\delta I, -\delta I, -\delta^{-1}I, -\delta^{-1}I, -\delta^{-1}I\},$$

$$\Theta_{1,ijl} = \begin{bmatrix} \overline{\overline{T}}_{11,i} & * & * & * & * & * & * & * & * & * & * & * \\ \overline{T}_{21,i} & \overline{\overline{T}}_{22,i} & * & * & * & * & * & * & * & * & * & * \\ \overline{T}_{31,i} & \overline{T}_{32,i} & \overline{T}_{33} & * & * & * & * & * & * & * & * & * \\ \overline{T}_{41,i} & \overline{T}_{42,i} & \overline{T}_{43,i} & \overline{T}_{44} & * & * & * & * & * & * & * & * \\ \tau_1\overline{X}_1 & \tau_1\overline{X}_2 & \tau_1\overline{X}_3 & \tau_1\overline{X}_4 & -\tau_1\overline{R}_1 & * & * & * & * & * & * & * \\ \tau_2 Z_1 & \tau_2 Z_2 & \tau_2 Z_3 & \tau_2 Z_4 & 0 & -\tau_2\overline{R}_{21} & * & * & * & * & * & * \\ \lambda Z & 0 & 0 & 0 & 0 & 0 & -S^{-1} & * & * & * & * & * \\ \lambda N_i Z & 0 & 0 & 0 & 0 & 0 & 0 & -\frac{\varepsilon}{4}I & * & * & * & * \\ \lambda B_i M_j & 0 & 0 & 0 & 0 & 0 & 0 & 0 & -\frac{\varepsilon}{4}I & * & * & * \\ \rho\lambda M_i & 0 & 0 & 0 & 0 & 0 & 0 & 0 & 0 & -W^{-1} & * & * \\ 0 & \lambda N_{di}Z & 0 & 0 & 0 & 0 & 0 & 0 & 0 & 0 & -\frac{\varepsilon}{4}I & * \\ 0 & 0 & \lambda B_{hi}M_l & 0 & 0 & 0 & 0 & 0 & 0 & 0 & 0 & -\frac{\varepsilon}{4}I \end{bmatrix}$$

Proof: At first, we prove that the inequality (20) implies the inequality (19). Applying the Schur complement to Eq. (20) results in

$$\Phi_{1,ijl} + \delta \begin{bmatrix} 0 & 0 & 0 \\ 0 & 0 & 0 \\ 0 & 0 & 0 \\ 0 & 0 & 0 \\ 0 & 0 & 0 \\ 0 & 0 & 0 \\ 0 & 0 & 0 \\ 0 & 0 & 0 \\ B_iH_j & 0 & 0 \\ 0 & \rho H_i & 0 \\ 0 & 0 & 0 \\ 0 & 0 & B_{hi}H_l \end{bmatrix} \begin{bmatrix} 0 & 0 & 0 & 0 & 0 & 0 & 0 & 0 & (B_iH_j)^{\mathrm{T}} & 0 & 0 & 0 \\ 0 & 0 & 0 & 0 & 0 & 0 & 0 & 0 & 0 & \rho H_i^{\mathrm{T}} & 0 & 0 \\ 0 & 0 & 0 & 0 & 0 & 0 & 0 & 0 & 0 & 0 & 0 & (B_{hi}H_i)^{\mathrm{T}} \end{bmatrix}$$

$$+\delta^{-1}\begin{bmatrix} (\lambda E_{kj}Z)^{\mathrm{T}} & (\lambda E_{ki}Z)^{\mathrm{T}} & 0 \\ 0 & 0 & 0 \\ 0 & 0 & (\lambda E_{kl}Z)^{\mathrm{T}} \\ 0 & 0 & 0 \\ 0 & 0 & 0 \\ 0 & 0 & 0 \\ 0 & 0 & 0 \\ 0 & 0 & 0 \\ 0 & 0 & 0 \\ 0 & 0 & 0 \\ 0 & 0 & 0 \\ 0 & 0 & 0 \end{bmatrix}\begin{bmatrix} \lambda E_{kj}Z & 0 & 0 & 0 & 0 & 0 & 0 & 0 & 0 & 0 & 0 & 0 \\ \lambda E_{ki}Z & 0 & 0 & 0 & 0 & 0 & 0 & 0 & 0 & 0 & 0 & 0 \\ 0 & 0 & \lambda E_{kl}Z & 0 & 0 & 0 & 0 & 0 & 0 & 0 & 0 & 0 \end{bmatrix} < 0 \tag{21}$$

Using Lemma 2 and noting $M_i = K_iZ$, by the condition (21), the following inequality holds:

$$\Phi_{1,ijl} + \begin{bmatrix} 0 & * & * & * & * & * & * & * & * & * & * & * \\ 0 & 0 & * & * & * & * & * & * & * & * & * & * \\ 0 & 0 & 0 & * & * & * & * & * & * & * & * & * \\ 0 & 0 & 0 & 0 & * & * & * & * & * & * & * & * \\ 0 & 0 & 0 & 0 & 0 & * & * & * & * & * & * & * \\ 0 & 0 & 0 & 0 & 0 & 0 & * & * & * & * & * & * \\ 0 & 0 & 0 & 0 & 0 & 0 & 0 & * & * & * & * & * \\ 0 & 0 & 0 & 0 & 0 & 0 & 0 & 0 & * & * & * & * \\ \lambda B_i\Delta\overline{K}_jZ & 0 & 0 & 0 & 0 & 0 & 0 & 0 & 0 & * & * & * \\ \rho\lambda\Delta\overline{K}_iZ & 0 & 0 & 0 & 0 & 0 & 0 & 0 & 0 & 0 & * & * \\ 0 & 0 & 0 & 0 & 0 & 0 & 0 & 0 & 0 & 0 & 0 & * \\ 0 & 0 & \lambda B_{hi}\Delta\tilde{K}_lZ & 0 & 0 & 0 & 0 & 0 & 0 & 0 & 0 & 0 \end{bmatrix} < 0 \tag{22}$$

where $\Delta\tilde{K}_i = \Delta K_i(t - d(t))$.

Therefore, it follows from Theorem 1 that the system (6) is asymptotically stable and the control law (5) is a fuzzy non-fragile guaranteed cost control law. Thus, we complete the proof.

Now consider the cost bound of

$$J \le x^{\mathrm{T}}(0)Px(0) + \int_{-d(0)}^{0} x^{\mathrm{T}}(s)Q_1x(s)ds + \int_{-\tau_1}^{0}\int_{\theta}^{0} \dot{x}^{\mathrm{T}}(s)R_1\dot{x}(s)dsd\theta + \int_{-h(0)}^{0} x^{\mathrm{T}}(s)Q_2x(s)ds$$
$$+ \int_{-\tau_2}^{0}\int_{\theta}^{0} \dot{x}^{\mathrm{T}}(s)R_2\dot{x}(s)dsd\theta = J_0$$

Similar to Ref. [23], we supposed that there exist positive scalars $\alpha_1, \alpha_2, \alpha_3, \alpha_4, \alpha_5$, such that $Z^{-1} \le \alpha_1 I$, $\frac{1}{\lambda^2}P\overline{Q}_1P \le \alpha_2 I$, $\frac{1}{\lambda^2}P\overline{Q}_2P \le \alpha_3 I$, $\frac{1}{\lambda^2}P\overline{R}_1P \le \alpha_4 I$, $\frac{1}{\lambda^2}P\overline{R}_2P \le \alpha_5 I$.

Then, define $S_{Q1}=\overline{Q}_1^{-1}$, $S_{Q2}=\overline{Q}_2^{-1}$, $S_{R1}=\overline{R}_1^{-1}$, $S_{R2}=\overline{R}_2^{-1}$, by Schur complement lemma, we have the following inequalities:

$$\begin{bmatrix} -\alpha_1 I & I \\ I & -Z \end{bmatrix} \le 0, \begin{bmatrix} -\alpha_2 I & \frac{1}{\lambda}P \\ \frac{1}{\lambda}P & S_{Q1} \end{bmatrix} \le 0, \begin{bmatrix} -\alpha_3 I & \frac{1}{\lambda}P \\ \frac{1}{\lambda}P & S_{Q2} \end{bmatrix} \le 0, \begin{bmatrix} -\alpha_4 I & \frac{1}{\lambda}P \\ \frac{1}{\lambda}P & S_{R1} \end{bmatrix} \le 0,$$
$$\begin{bmatrix} -\alpha_5 I & \frac{1}{\lambda}P \\ \frac{1}{\lambda}P & S_{R2} \end{bmatrix} \le 0, \begin{bmatrix} Z & I \\ I & Z \end{bmatrix} \ge 0, \begin{bmatrix} S_{Q1} & I \\ I & \overline{Q}_1 \end{bmatrix} \ge 0, \begin{bmatrix} S_{Q2} & I \\ I & \overline{Q}_2 \end{bmatrix} \ge 0, \tag{23}$$
$$\begin{bmatrix} S_{R1} & I \\ I & \overline{R}_1 \end{bmatrix} \ge 0, \begin{bmatrix} S_{R2} & I \\ I & \overline{R}_2 \end{bmatrix} \ge 0,$$

Using the idea of the cone complement linear algorithm in Ref. [24], we can obtain the solution of the minimization problem of upper bound of the value of the cost function as follows:

$$\text{minimize}\{trace(PZ + S_{Q1}\overline{Q}_1 + S_{Q2}\overline{Q}_2 + S_{R1}\overline{R}_1 + S_{R2}\overline{R}_2 + \alpha_1 x^{\mathrm{T}}(0)x(0) + \alpha_2\int_{-d(0)}^{0} x^{\mathrm{T}}(s)x(s)ds$$
$$+\alpha_4\int_{-\tau_1}^{0}\int_{\theta}^{0}\dot{x}^{\mathrm{T}}(s)\dot{x}(s)dsd\theta + \alpha_3\int_{-h(0)}^{0} x^{\mathrm{T}}(s)x(s)ds + \alpha_5\int_{-\tau_2}^{0}\int_{\theta}^{0}\dot{x}^{\mathrm{T}}(s)\dot{x}(s)dsd\theta\} \tag{24}$$
$$\text{subject to } (20), (23),\ \varepsilon > 0, \overline{Q}_1 > 0, \overline{Q}_2 > 0, \overline{R}_1 > 0, \overline{R}_2 > 0, Z > 0, \alpha_i > 0, i = 1, \ldots, 5$$

Using the following cone complement linearization (CCL) algorithm [24] can iteratively solve the minimization problem (24). □

5. Simulation examples

In this section, the proposed approach is applied to the Van de Vusse system to verify its effectiveness.

Example: Consider the dynamics of an isothermal continuous stirred tank reactor for the Van de Vusse

$$\begin{aligned} \dot{x}_1 &= -50x_1 - 10x_1^3 + u(10 - x_1) + u(t-h) + u(t-h)(0.5x_1(t-d) + 0.2x_2(t-d)) + 5x_2(t-d) \\ \dot{x}_2 &= 50x_1 - 100x_2 - u(t-h) + u(t-h)(0.3x_1(t-d) - 0.2x_2(t-d)) + 10x_2(t-d) - 5x_1(t-d) \end{aligned} \tag{25}$$

From the system equation (25), some equilibrium points are tabulated in **Table 1**. According to these equilibrium points, $[x_e\ u_e]$, which are also chosen as the desired operating points, $[x'_e\ u'_e]$, we can use the similar modeling method that is described in Ref. [16].

x_e^T	x_{de}^T	u_e	u_{de}
[2.0422 1.2178]	[2.0422 1.2178]	20.3077	20.3077
[3.6626 2.5443]	[3.6626 2.5443]	77.7272	77.7272
[5.9543 5.5403]	[5.9543 5.5403]	296.2414	296.2414

Table 1. Data for equilibrium points.

Thus, the system (25) can be represented by

R^1 : *if* x_1 *is about* 2.0422 *then*

$$\dot{x}_\delta(t) = A_1 x_\delta(t) + A_{d1} x_{d\delta}(t) + B_1 u_\delta(t) + B_{h1} u_{d\delta}(t) + N_1 x_\delta(t) u_\delta(t) + N_{d1} x_{d\delta}(t) u_{h\delta}(t)$$

$$\begin{array}{l} R^2 : \text{ if } x_1 \text{ is about } 3.6626, \text{ then} \\ \quad \dot{x}_\delta(t) = A_2 x_\delta(t) + A_{d2} x_{d\delta}(t) + B_2 u_\delta(t) + B_{d2} u_{h\delta}(t) + N_2 x_\delta(t) u_\delta(t) + N_{d2} x_{d\delta}(t) u_{h\delta}(t) \\ R^3 : \text{ if } x_1 \text{ is about } 5.9543, \text{ then} \\ \quad \dot{x}_\delta(t) = A_3 x_\delta(t) + A_{d3} x_{d\delta}(t) + B_3 u_\delta(t) + B_{d3} u_{h\delta}(t) + N_3 x_\delta(t) u_\delta(t) + N_{d3} x_{d\delta}(t) u_{h\delta}(t) \end{array} \tag{26}$$

where

$$\begin{array}{l} A_1 = \begin{bmatrix} -75.2383 & 7.7946 \\ 50 & -100 \end{bmatrix}, A_2 = \begin{bmatrix} -98.3005 & 11.7315 \\ 50 & -100 \end{bmatrix}, A_3 = \begin{bmatrix} -122.1228 & 8.8577 \\ 50 & -100 \end{bmatrix}, \\ N_1 = N_2 = N_3 = \begin{bmatrix} -1 & 0 \\ 0 & -1 \end{bmatrix}; B_1 = B_2 = B_3 = \begin{bmatrix} 10 \\ 0 \end{bmatrix}; A_{d1} = A_{d2} = A_{d3} = \begin{bmatrix} 0 & 5 \\ 10 & -5 \end{bmatrix}, \\ N_{d1} = N_{d2} = N_{d3} = \begin{bmatrix} 0.5 & 0.2 \\ 0.3 & -0.2 \end{bmatrix}, B_{h1} = B_{h2} = B_{h3} = \begin{bmatrix} 1 \\ 0 \end{bmatrix}, x_\delta = x(t) - x_e^{'}, \\ u_\delta = u(t) - u_e^{'}, x_{d\delta} = x(t-d) - x_{de}^{'}, u_{h\delta} = u(t-d) - u_{he}^{'}. \end{array}$$

The cost function associated with this system is given with $S = \begin{bmatrix} 1 & 0 \\ 0 & 1 \end{bmatrix}$, $W = 1$. The controller gain perturbation ΔK of the additive form is give with $H_1 = H_2 = H_3 = 0.1$, $E_{k1} = [0.05 \quad -0.01]$, $E_{k2} = [0.02 \quad 0.01]$, $E_{k3} = [-0.01 \quad 0]$.

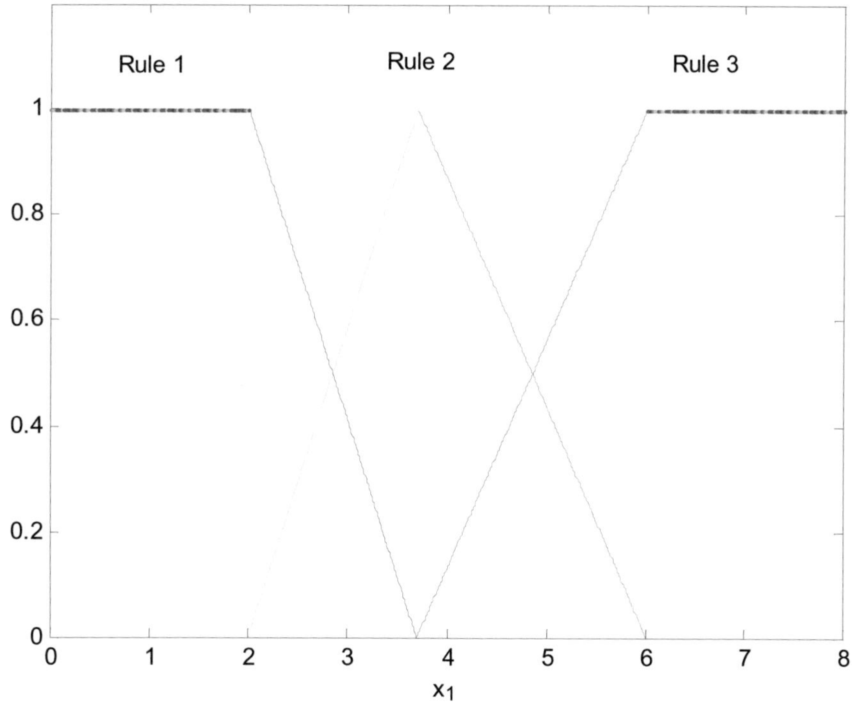

Figure 1. Membership functions.

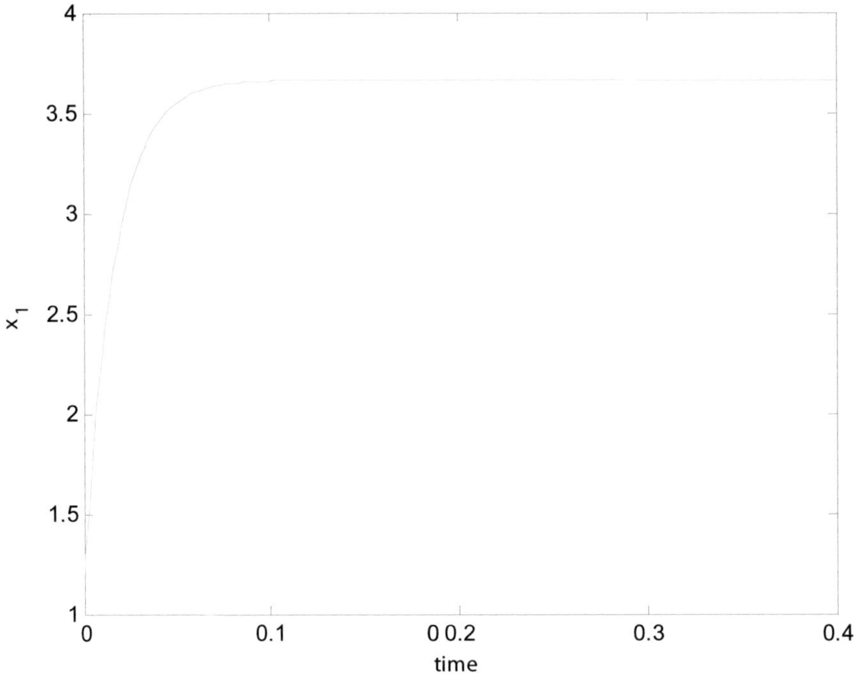

Figure 2. State responses of $x_1(t)$.

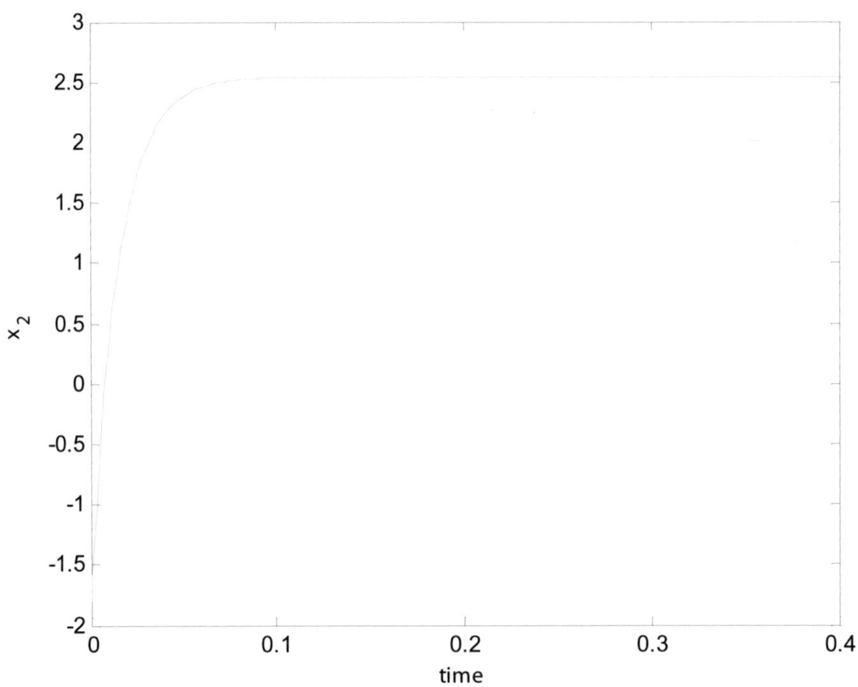

Figure 3. State responses of $x_2(t)$.

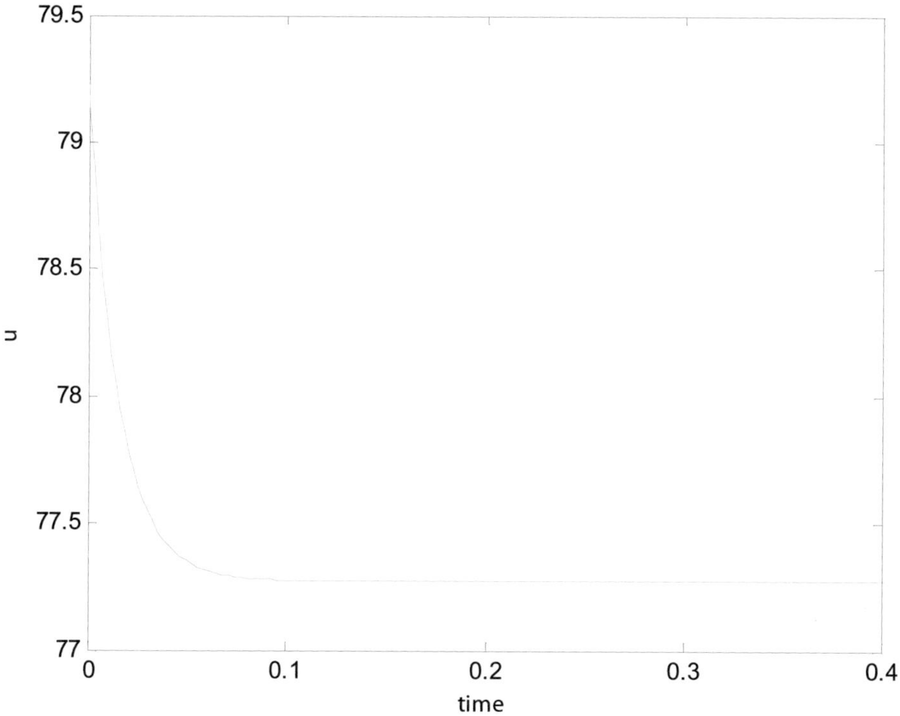

Figure 4. Control trajectory of system.

The membership functions of state x_1 are shown in **Figure 1**.

Then, solving LMIs (23) and (24) for $\rho = 0.45$, $\lambda = 1.02$ and $\delta = 0.11, \tau_1 = \tau_2 = 2, \sigma_1 = 0, \sigma_2 = 0$ gives the following feasible solution:

$$P = \begin{bmatrix} 4.2727 & -1.3007 \\ -1.3007 & 6.4906 \end{bmatrix}, Q_1 = \begin{bmatrix} 14.1872 & -1.9381 \\ -1.9381 & 13.0104 \end{bmatrix}, Q_2 = \begin{bmatrix} 3.1029 & 1.2838 \\ 1.2838 & 2.0181 \end{bmatrix},$$

$$R_1 = \begin{bmatrix} 8.3691 & -1.3053 \\ -1.3053 & 7.0523 \end{bmatrix}, R_1 = \begin{bmatrix} 5.2020 & 2.2730 \\ 2.2730 & 1.0238 \end{bmatrix}, \varepsilon = 1.8043,$$

$$K_1 = [-0.4233 \ -0.5031], K_2 = [-0.5961 \ -0.7049], K_1 = [-0.4593 \ -0.3874].$$

Figures 2–4 illustrate the simulation results of applying the non-fragile fuzzy controller to the system (25) with $x'_e = [3.6626 \quad 2.5443]^{\mathrm{T}}$ and $u'_e = 77.7272$ under initial condition $\varphi(t)$ = [1.2 −1.8]^{T}, $t \in [-2\ 0]$. It can be seen that with the fuzzy control law, the closed-loop system is asymptotically stable and an upper bound of the guaranteed cost is $J_0 = 292.0399$. The simulation results show that the fuzzy non-fragile guaranteed controller proposed in this paper is effective.

6. Conclusions

In this paper, the problem of non-fragile guaranteed cost control for a class of fuzzy time-varying delay systems with local bilinear models has been explored. By utilizing the Lyapunov stability theory and LMI technique, sufficient conditions for the delay-dependent asymptotically stability of the closed-loop T-S fuzzy local bilinear system have been obtained. Moreover, the designed fuzzy controller has guaranteed the cost function-bound constraint. Finally, the effectiveness of the developed approach has been demonstrated by the simulation example. The robust non-fragile guaranteed cost control and robust non-fragile H-infinite control based on fuzzy bilinear model will be further investigated in the future work.

Acknowledgements

This work is supported by NSFC Nos. 60974139 and 61573013.

Author details

Junmin Li*, Jinsha Li and Ruirui Duan

*Address all correspondence to: jmli@mail.xidian.edu.cn

School of Mathematics and Statistics, Xidian University, Xi'an, PR China

References

[1] Pang CT, Lur YY. On the stability of Takagi-Sugeno fuzzy systems with time-varying uncertainties. IEEE Transactions on Fuzzy Systems. 2008;**16**:162–170

[2] Zhou SS, Lam J, Zheng WX. Control design for fuzzy systems based on relaxed non-quadratic stability and H∞ performance conditions. IEEE Transactions on Fuzzy Systems. 2007;**15**:188–198

[3] Zhou SS, Li T. Robust stabilization for delayed discrete-time fuzzy systems via basis-dependent Lyapunov-Krasovskii function. Fuzzy Sets and Systems. 2005;**151**:139–153

[4] Gao HJ, Liu X, Lam J. Stability analysis and stabilization for discrete-time fuzzy systems with time-varying delay. IEEE Transactions on Systems, Man, Cybernetics, Part B. 2009;**39**:306–316

[5] Wu HN, Li HX. New approach to delay-dependent stability analysis and stabilization for continuous-time fuzzy systems with time-varying delay. IEEE Transactions on Fuzzy Systems. 2007;**15**:482–493

[6] Chen B, Liu XP. Delay-dependent robust H∞ control for T-S fuzzy systems with time delay. IEEE Transactions on Fuzzy Systems. 2005;**13**:238–249

[7] Chen M, Feng G, Ma H, Chen G. Delay-dependent H∞ filter design for discrete-time fuzzy systems with time-varying delays. IEEE Transactions on Fuzzy Systems. 2009;**17**:604–616

[8] Zhang J, Xia Y, Tao R. New results on H∞ filtering for fuzzy time-delay systems. IEEE Transactions on Fuzzy Systems. 2009;**17**:128–137

[9] Keel LH, Bhattacharryya SP. Robust, fragile, or optimal?. IEEE Transactions on Automatic Control. 1997;**42**:1098–1105

[10] Yang GH, Wang JL, Lin C. H∞ control for linear systems with additive controller gain variations. International Journal of Control. 2000;**73**:1500–1506

[11] Yang GH, Wang JL. Non-fragile H∞ control for linear systems with multiplicative controller gain variations. Automatica. 2001;**37**:727–737

[12] Zhang BY, Zhou SS, Li T. A new approach to robust and non-fragile H∞ control for uncertain fuzzy systems. Information Sciences. 2007;**177**:5118–5133

[13] Yee JS, Yang GH, Wang JL. Non-fragile guaranteed cost control for discrete-time uncertain linear systems. International Journal of Systems Science. 2001;**32**:845–853

[14] Mohler RR. Bilinear Control Processes. New York, NY: Academic; 1973

[15] Elliott DL. Bilinear Systems in Encyclopedia of Electrical Engineering. New York, NY: Wiley; 1999

[16] Li THS, Tsai SH. T-S fuzzy bilinear model and fuzzy controller design for a class of nonlinear systems. IEEE Transactions on Fuzzy Systems. 2007;**15**:494–505

[17] Tsai SH, Li THS. Robust fuzzy control of a class of fuzzy bilinear systems with time-delay. Chaos, Solitons and Fractals. 2009;**39**:2028–2040

[18] Li THS, Tsai SH, et al. Robust H∞ fuzzy control for a class of uncertain discrete fuzzy bilinear systems. IEEE Transactions on Systems, Man, Cybernetics, Part B. 2008;**38**:510–526

[19] Dong J, Wang Y, Yang G. Control synthesis of continuous-time T-S fuzzy systems with local nonlinear models. IEEE Transactions on Systems, Man, Cybernetics, Part B. 2009;**39**:1245–1258

[20] Zhang G, Li JM. Non-fragile guaranteed cost control of discrete-time fuzzy bilinear system. Journal of Systems Engineering and Electronics. 2010;**21**:629–634

[21] Ho DWC, Niu Y. Robust fuzzy design for nonlinear uncertain stochastic systems via sliding-mode control. IEEE Transactions on Fuzzy Systems. 2007;**15**:350–358

[22] Yang DD, Cai KY. Reliable guaranteed cost sampling control for nonlinear time-delay systems. Mathematics and Computers in Simulation. 2010;**80**:2005–2018

[23] Chen WH, Guan ZH, Lu XM. Delay-dependent output feedback guaranteed cost control for uncertain time-delay systems. Automatica. 2004;**44**:1263–1268

[24] Chen B, Lin C, Liu XP, Tong SC. Guaranteed cost control of T–S fuzzy systems with input delay. International Journal of Robust Nonlinear Control. 2008;**18**:1230–1256

[25] Chen B, Liu XP, Tong SC, Lin C. Observer-based stabilization of T–S fuzzy systems with input delay. IEEE Transactions on Fuzzy Systems. 2008;**16**:652–663

[26] Chen B, Liu X, Tong S, Lin C. guaranteed cost control of T-S fuzzy systems with state and input delay. Fuzzy Sets and Systems. 2007;**158**:2251–2267

[27] Du BZ, Lam J, Shu Z. Stabilization for state/input delay systems via static and integral output feedback. Automatica. 2010;**46**:2000–2007

[28] Kim JH. Delay-dependent robust and non-fragile guaranteed cost control for uncertain singular systems with time-varying state and input delays. International Journal of Control, Automation, and Systems. 2009;**7**:357–364

[29] Li L, Liu XD. New approach on robust stability for uncertain T–S fuzzy systems with state and input delays. Chaos, Solitons and Fractals. *2009*;**40**:2329–2339

[30] Yu KW, Lien CH. Robust H-infinite control for uncertain T–S fuzzy systems with state and input delays. Chaos, Solitons and Fractals. 2008;**37**:150–156

[31] Yue D, Lam J. Non-fragile guaranteed cost control for uncertain descriptor systems with time-varying state and input delays. Optimal Control Applications and Methods. 2005;**26**:85–105

[32] Zhang G, Li JM. Non-fragile guaranteed cost control of discrete-time fuzzy bilinear system with time-delay. Journal of Dynamic Systems, Measurement and Control. 2014;**136**:044502–044504

[33] Yue HY, Li JM. Output-feedback adaptive fuzzy control for a class of nonlinear systems with input delay and unknown control directions. Journal of the Franklin Institute. 2013;**350**:129–154

[34] Yue HY, Li JM. Adaptive fuzzy tracking control for a class of perturbed nonlinear time-varying delays systems with unknown control direction. International Journal of Uncertainty, Fuzziness and Knowledge-based Systems. 2013;**21**:497–531

[35] Wang RJ, Lin WW, Wang WJ. Stabilizability of linear quadratic state feedback for uncertain fuzzy time-delay systems. IEEE Transactions on Systems, Man, Cybernetics, Part B. 2004;**34**:1288–1292

[36] Xia ZL, Li JM, Li JR. Delay-dependent fuzzy static output feedback control for discrete-time fuzzy stochastic systems with distributed time-varying delays. ISA Transaction. 2012;**51**:702–712

[37] Xia ZL, Li JM. Switching fuzzy filtering for nonlinear stochastic delay systems using piecewise Lyapunov-Krasovskii function. International Journal of Fuzzy Systems. 2012;**14**:530–539

[38] Li JR, Li JM, Xia ZL. Delay-dependent generalized H2 fuzzy static-output-feedback control for discrete T-S fuzzy bilinear stochastic systems with mixed delays. Journal of Intelligent and Fuzzy Systems. 2013;**25**:863–880

[39] Xia ZL, Li JM, Li JR. Passivity-based resilient adaptive control for fuzzy stochastic delay systems with Markovian switching. Journal of the Franklin Institute-Engineering and Applied Mathematics. 2014;**351**:3818–3836

[40] Li JM, Li YT. Robust stability and stabilization of fractional order systems based on uncertain T-S fuzzy model with the fractional order. Journal of Computational and Nonlinear Dynamics. 2013;**8**:041005

[41] Li YT, Li JM. Stability analysis of fractional order systems based on T-S fuzzy model with the fractional order α: $0<\alpha<1$. Nonlinear Dynamics. 2014;**78**:2909–2919

[42] Li YT, Li JM. Decentralized stabilization of fractional order T-S fuzzy interconnected systems with multiple time delays. Journal of Intelligent and Fuzzy Systems. 2016;**30**:319–331

[43] Li JM, Yue HY. Adaptive fuzzy tracking control for stochastic nonlinear systems with unknown time-varying delays. Applied Mathematics and Computation. 2015;**256**:514–528

[44] Yue HY, Yu SQ. Adaptive fuzzy tracking control for a class of stochastic nonlinearly parameterized systems with distributed input delay. Journal of the Franklin Institute-Engineering and Applied Mathematics. 2016;**353**:713–734

[45] Duan RR, Li JM, Zhang YN, Yang Y, Chen GP. Stability analysis and H-inf control of discrete T-S fuzzy hyperbolic systems. International Journal of Applied Mathematics and Computer Science. 2016;**26**:133–145

[46] Wang JX, Li JM. Stability analysis and feedback control of T-S fuzzy hyperbolic delay model for a class of nonlinear systems with time-varying delay. Iranian Journal of Fuzzy Systems. 2016;**13**:111–134

[47] Li JR, Li JM, Xia ZL. Observer-based fuzzy control design for discrete time T-S fuzzy bilinear systems. International Journal of Uncertainty, Fuzziness and Knowledge-Based Systems. 2013;**21**:435–454

Chapter 3

Fuzzy Interpolation Systems and Applications

Longzhi Yang, Zheming Zuo, Fei Chao and
Yanpeng Qu

Additional information is available at the end of the chapter

http://dx.doi.org/10.5772/68051

Abstract

Fuzzy inference systems provide a simple yet effective solution to complex non-linear problems, which have been applied to numerous real-world applications with great success. However, conventional fuzzy inference systems may suffer from either too sparse, too complex or imbalanced rule bases, given that the data may be unevenly distributed in the problem space regardless of its volume. Fuzzy interpolation addresses this. It enables fuzzy inferences with sparse rule bases when the sparse rule base does not cover a given input, and it simplifies very dense rule bases by approximating certain rules with their neighbouring ones. This chapter systematically reviews different types of fuzzy interpolation approaches and their variations, in terms of both the interpolation mechanism (inference engine) and sparse rule base generation. Representative applications of fuzzy interpolation in the field of control are also revisited in this chapter, which not only validate fuzzy interpolation approaches but also demonstrate its efficacy and potential for wider applications.

Keywords: fuzzy inference systems, fuzzy interpolation, adaptive fuzzy interpolation, sparse rule bases, fuzzy control

1. Introduction

Fuzzy logic and fuzzy sets have been used successfully as tools to manage the uncertainty of fuzziness since their introduction in the 1960s, which have been applied to many fields, including [1–6]. The most widely used fuzzy systems are fuzzy rule-based inference systems, each comprising of a rule base and an inference engine. Different inference engines were invented to support different situations, such as the Mamdani inference engine [7] and the TSK inference engine [8]. The rule bases are usually extracted from expert knowledge or learned from data. The TSK model produces crisp outputs due to its polynomial rule consequences in TSK-style rule bases, while the Mamdani model is more appealing in handling inferences based on human natural language due to its fuzzy rule consequences. Despite of the wide applications, these conventional fuzzy inference mechanisms are only workable with dense rule bases which fully cover the entire input domain.

Fuzzy interpolation systems (FISs) were proposed to address the above issue [9], and they also help in complexity reduction for fuzzy models with too complex (dense) rule bases. If there is only a spare rule base available and a given input does not overlap with any rule antecedent, conventional fuzzy inference systems will not be applicable. However, FISs are still able to generate a conclusion by means of fuzzy interpolation in such situations, thus enhancing the applicability of conventional fuzzy inference systems. FISs can also improve the efficiency of complex fuzzy inference systems by excluding those rules that can be accurately interpolated or extrapolated using other rules in a complex rule base. Various fuzzy interpolation methods based on Mamdani-style rule bases have been proposed in the literature such as Refs. [9–20], with successful applications in the fields of decision-making support, prediction and control, amongst others.

FISs have also been developed to support TSK-style sparse fuzzy rule bases by extending the traditional TSK fuzzy inference system [21]. This approach was developed based on a modified similarity degree measure that enables the effective utilisation of all rules during inference process to generate a global result. In particular, the modified similarity measure guarantees that the similarity degree between any given input and any rule antecedent is greater than 0 even when they do not overlap at all. Therefore, all the rules in the rule base can be fired to certain degrees such that they all contribute to the final result to some extents and consequently a conclusion still can be generated even when no rule antecedent is overlapped with the given observation. The extended TSK fuzzy model enjoys the advantages of both TSK model and fuzzy interpolation, which is able to obtain crisp inference results from either sparse, dense or unevenly distributed (including dense parts and spare parts) TSK-style fuzzy rule bases.

FISs have been successfully applied to real-world problems. In some real world scenarios, neither complete expert knowledge nor complete data set is available or readily obtainable to generate evenly distributed dense rule bases. FISs therefore have been applied in such situations. For instance, a FIS has been applied to building evaluation in the work of Molnárka et al. [22] in an effort to help estate agencies making decisions for residential building maintenance, when some necessary relevant data have been lost. In Ref. [23], a FIS system was applied successfully to reduce the complexity and improve the efficiency of a fuzzy home heating control system. The work of Bai et al. [24] applied a FIS to calibrate parallel machine tools for industry use. A behaviour-based fuzzy control system is introduced in Ref. [25], which applied a FIS to make decisions when only incomplete knowledge base has been provided or available. Most recently, FISs have also been used to support network quality of service [26] and network intrusion detection [27].

The remainder of this chapter is organised as follows. Section 2 reviews the theoretical underpinnings of conventional fuzzy inference systems, that is, the Mamdani inference system and the TSK inference system. Section 3 discusses different fuzzy interpolation approaches to support sparse Mamdani-style rule bases. Section 4 presents the extension of the conventional TSK inference system in supporting sparse TSK-style rule bases. Section 5 reports two representative examples of fuzzy interpolation systems in the field of system control. Section 6 concludes the chapter and points out the directions for future work.

2. Fuzzy inference systems

The process of fuzzy inference is basically an iteration of computer paradigm based on fuzzy set theory, fuzzy-if-then-rules and fuzzy reasoning. Each iteration takes an input which can be an observation or a previously inferred result, crisp or fuzzy. Then, these inputs are used to fire the rules in a given rule base, and the output is the aggregation of the inferred results from all the fired rules. There are generally two primary ways to construct a rule base for a given problem. The first way is directly translating expert knowledge to rules, and the fuzzy inference systems with such rule bases are usually called fuzzy expert systems or fuzzy controllers [28]. In this case, rules are fuzzy representations of expert knowledge, and the resultant rule base offers a high semantic level and a good generalisation capability. The difficulty of building rule bases for complex problems has resulted in the development of another approach of rule base construction, which is driven by data, that is, fuzzy rules are obtained from data by employing machine learning techniques rather than expert knowledge [29, 30]. In contrast, the rule bases built in this way lack comprehensibility and transparency. There are two types of rule bases depending on the expression of the consequences of the fuzzy rules composing the rule base. Mamdani-style fuzzy rules consider fuzzy terms or linguistic values in the consequence, while TSK-style fuzzy rules represent the consequences as polynomial functions of crisp inputs.

2.1. Inference with Mamdani-style rule bases

There are a number of fuzzy inference mechanisms that can be utilised to derive a consequence from a given observation using a Mamdani rule base. The two most significant modes are the compositional rule of inference (CRI) [31] and analogy-based reasoning [24, 33], which are introduced below.

2.1.1. *Compositional rule of inference*

The introduction of CRI marks the era of fuzzy inference [31]. Given a rule 'IF x is A, THEN y is B'' and an observation 'x is A^*', the conclusion B^* can be generated through CRI as:

$$\mu_{B^*}(v) = sup_{u \in U_x} T\left(\mu_{A^*}(u), \mu_R(u, v)\right), \tag{1}$$

where T is a triangular norm, sup represents supremum, and R is the relationship between variables x and y. Essentially, CRI is a fuzzy extension of classical modus ponens which can be viewed from two perspectives. Firstly, classical modus ponens only supports predicates concerning singleton elements, but CRI is able to deal with predicates which concern a set of elements in the variable domain. This is achieved by representing a fuzzy rule as a fuzzy relation over the Cartesian product of the domains of the antecedent and consequent variables. Various fuzzy implication relations have been proposed [7, 32–34], each of which may have its own properties and therefore is suitable for a certain group of applications. Secondly, classical modus ponens only supports Boolean logic, but CRI supports multi-value logic. That is, CRI is able to deal with predicates with partial truth values, which are implemented by a compositional operator sup T, where T represents a t-norm [35].

A number of existing fuzzy reasoning methods based on CRI have been developed [36, 37], including the first successful practical approach, that is, the Mamdani inference [28]. This approach is also the most commonly seen fuzzy methodology in physical control systems thus far. It was originally proposed as an attempt to control a steam engine and boiler combination by synthesising a set of linguistic control rules obtained from experienced human operators. Mamdani inference implements CRI using *minimum* as the t-norm operator due to its simplicity. In particular, the inferred result from each fired rule is a fuzzy set which is transformed from the rule consequence by restricting the membership of those elements whose memberships are greater than the firing strength. The firing strength is also sometimes termed the satisfaction degree, which is the supremum within the variable domain of the minimum of the rule antecedent and the given observation. A defuzzification process is needed when crisp outputs are required.

2.1.2. Analogy-based fuzzy inference

Despite the success of CRI in various fuzzy system applications, it suffers various criticisms including its complexity and vague underlying semantics [34, 38]. This has led to another group of fuzzy reasoning approaches which are based on similarity degree, usually called analogy-based fuzzy reasoning [38–41]. Similarity considerations play a major role in human cognitive processes [42], so do they in approximate reasoning. It is intuitive that if a given observation is similar to the antecedent of a rule, the conclusion from the observation should also be similar to the consequence of the rule. Different to CRI-based fuzzy reasoning, analogy-based fuzzy reasoning does not require the construction of a fuzzy relation. Instead, it is based on the degree of similarity (given a certain similarity metric) between the given observation and the antecedent of a rule. Utilising the computed similarity degree, the consequence of the fired rule can be modified to the consequence of the given observation.

Approximate analogical reasoning schema is a typical analogy-based fuzzy inference approach [34, 38]. In this method, rules are fired according to the similarity degrees between a given observation and the antecedents of rules. If the degree of similarity between the given observation and the antecedent of a rule is greater than a predefined threshold value, the rule will be fired and the consequence of the observation is deduced from the rule consequence by a given modification procedure. Another analogy-based fuzzy inference approach was proposed in Refs. [39, 40], which particularly targets medical diagnostic problems. This approach is based on the cosine angle between the two vectors that represent the actual and the user's specified values of the antecedent variable. Several modification procedures can be found in Refs. [43, 44]. Particularly, a fuzzy reasoning method which employs similarity measures based on the degree of subsethood between the propositions in the antecedent and a given observation is proposed in Ref. [45]. This method has also been extended to consider the weights of the propositions in the antecedent [46]. Analogy-based fuzzy inference approaches usually arrive at solutions with more natural appeal than those introduced in the last section.

2.2. Inference with TSK-style rule bases

The TSK fuzzy inference system was proposed for the direct generation of crisp outputs [8]. In difference with the Mamdani-style fuzzy rule bases, TSK-style rule bases are usually generated from data using a clustering algorithm such as K-Means and an algorithm to determine the number of clusters such as Ref. [47]. Also, the consequence of a TSK fuzzy rule is a polynomial function rather than a fuzzy set. A typical TSK fuzzy rule can be defined as:

$$\text{IF } x_1 \text{ is } A_1 \wedge \ldots \wedge x_m \text{ is } A_m \text{ THEN } z = f(x_1, \ldots, x_m), \tag{2}$$

where $A_1,\ldots A_m$ are fuzzy values with regard to antecedent variables $x_1, \ldots, x_m$ respectively, and $f(x_1,\ldots,x_m)$ is a crisp polynomial function of crisp inputs determining the crisp output value. The rule consequent polynomial functions $f(x_{1,\ldots,}x_m)$ are usually zero order or first order. For simplicity, suppose that a TSK-style rule base is formed by two-antecedent rules as follows:

$$\begin{aligned} R_i &: \text{IF } x \text{ is } A_i \wedge y \text{ is } B_i \text{ THEN } z = f_i(x, y) \\ R_j &: \text{IF } x \text{ is } A_j \wedge y \text{ is } B_j \text{ THEN } z = f_j(x, y). \end{aligned} \tag{3}$$

Suppose that (x_0,y_0) is the crisp input pair, then the inference process can be shown in **Figure 1**. As the input values overlap with both rule antecedents, both rules are fired. Using rules R_i and R_j, the given input then leads to system outputs $f_i(x_0,x_0)$ and $f_j(x_0,x_0)$, respectively. The consequences from both rules are then integrated using weighted average function, where

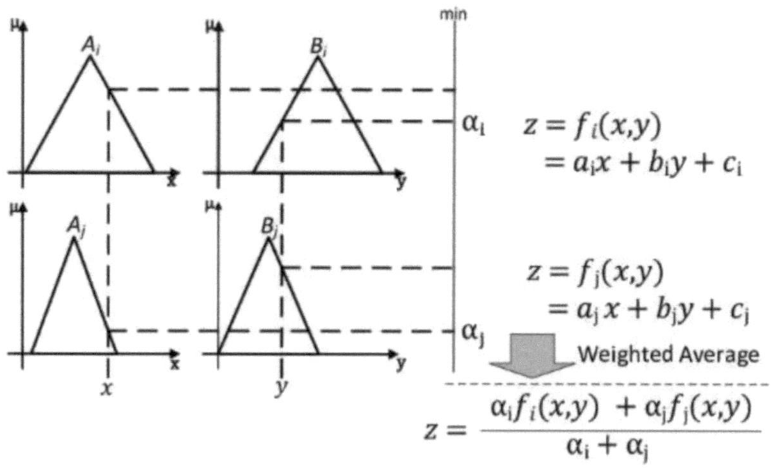

Figure 1. TSK fuzzy inference [21].

the values of weights represent the matching degrees between the given input and the rule antecedents (often referred to as firing strengths). Assume that $\mu_{A_i}(x_0)$ and $\mu_{B_i}(y_0)$ are the matching degree between inputs (x_0 and y_0) and rule antecedents (A_i and B_i), respectively. The firing strength of rule R_i, denoted as α_i, is calculated as:

$$\alpha_i = \mu_{A_i}(x_0) \wedge \mu_{B_i}(y_0), \tag{4}$$

where ∧ stands for a t-norm operator. Different implementations can be used for the t-norm operator, with the minimum operator being used most widely. Of course, if another system input (x_1, y_1) is presented and it is not covered by the rule base, the matching degrees between this new input and rule antecedents of R_i and R_j are equal to 0. In this case, no rule will be fired, and thus traditional TSK is not applicable. In this case, fuzzy interpolation is required, which is introduced in Section 4.

3. Fuzzy interpolation with sparse Mamdani-style rule bases

FISs based on Mamdani-style rule bases can be categorised into two classes. One group of approaches were developed based on the decomposition and resolution principle, termed as 'resolution principle-base interpolation'. In particular, the approach represents each fuzzy set as a series of α-cuts ($\alpha \in (0,1]$), and the α-cut of the conclusion is computed from the α-cuts of the observation and the α-cuts of rules. The final fuzzy set is assembled from all the α-cut consequences using the resolution principle [48–50]. The other group of fuzzy interpolation approaches were developed using the analogy reasoning system, thus termed as 'analogy-based fuzzy interpolation'. This group of approaches firstly generates an intermediate rule whose antecedent maximally overlaps with the given observation, then the system output is produced from the observation using the intermediate rule. Two representative approaches of the two classes, the KH approach [10] and the scale and move transformation-based approach [9, 51, 52], are discussed in this section based on simple rule bases with two antecedent rules. Despite of the simple examples used herein, both of these approaches have been extended to work with multiple multi-antecedent rules.

3.1. Resolution principle-based interpolation

Single step interpolation approaches are computationally efficient, such as the KH approach proposed in Refs. [9, 10, 53]. Following these approaches, all variables involved in the reasoning process must satisfy a partial ordering, denoted as $\prec$ [31]. According to the decomposition principle, a normal and convex fuzzy set A can be represented by a series of α-cut intervals, each denoted as A_α, $\alpha \in (0,1)$. Given fuzzy sets A_i and A_j which are associated with the same variable, the partial ordering $A_i \prec A_j$ is defined as:

$$\inf\{A_{i\alpha}\} < \inf\{A_{j\alpha}\} \text{ and } \sup\{A_{i\alpha}\} < \sup\{A_{j\alpha}\}, \quad \forall \alpha \in (0,1], \tag{5}$$

where $\inf\{A_{i\alpha}\}$ and $\sup\{A_{i\alpha}\}$ denote the infimum and supremum of $A_{i\alpha}$, respectively.

Take the KH approach as an example here. For simplicity, suppose there are two fuzzy rules: If x is A_i, then y is B_i, and If x is A_j then y is B_j, shorten as $A_i \Rightarrow B_i$ and $A_j \Rightarrow B_j$, respectively. Also, suppose that these two rules are adjacent, in other words, there is no rule $A \Rightarrow B$ existing such that $A_i \prec A \prec A_j$ or $A_j \prec A \prec A_i$. Given an observation A^* which satisfies $A_i \prec A^* \prec A_j$ or $A_j \prec A^* \prec A_i$, a conclusion B^* can be computed as:

$$\frac{D(A_{i\alpha}, A^*_\alpha)}{D(A^*_\alpha, A_{j\alpha})} = \frac{D(B_{i\alpha}, B^*_\alpha)}{D(B^*_\alpha, B_{j\alpha})}, \tag{6}$$

where given any $0 < \alpha \leq 1$, the distance $D(A_{i\alpha}, A_{j\alpha})$ between the α-cuts $A_{i\alpha}$ and $A_{j\alpha}$ is defined by the interval $[D^L(A_{i\alpha}, A_{j\alpha}), D^U(A_{i\alpha}, A_{j\alpha})]$ with:

$$D^L(A_{i\alpha}, A_{j\alpha}) = inf\{A_{j\alpha}\} - inf\{A_{i\alpha}\}, D^U(A_{i\alpha}, A_{j\alpha}) = sup\{A_{j\alpha}\} - sup\{A_{i\alpha}\}. \tag{7}$$

Following Eqs. (4) and (5), the following is resulted:

$$\begin{cases} min\{B^*_\alpha\} = \dfrac{\dfrac{inf(B_{i\alpha})}{D^L(A_{i\alpha}, A^*_\alpha)} + \dfrac{inf(B_{j\alpha})}{D^L(A^*_\alpha, A_{j\alpha})}}{\dfrac{1}{D^L(A_{i\alpha}, A^*_\alpha)} + \dfrac{1}{D^L(A^*_\alpha, A_{j\alpha})}} \\ max\{B^*_\alpha\} = \dfrac{\dfrac{sup(B_{i\alpha})}{D^U(A_{i\alpha}, A^*_\alpha)} + \dfrac{sup(B_{j\alpha})}{D^U(A^*_\alpha, A_{j\alpha})}}{\dfrac{1}{D^U(A_{i\alpha}, A^*_\alpha)} + \dfrac{1}{D^U(A^*_\alpha, A_{j\alpha})}} \end{cases} \tag{8}$$

For simplicity, let

$$\begin{cases} \Lambda^L_\alpha = \dfrac{\inf\{A^*_\alpha\} - \inf\{A_{i\alpha}\}}{\inf\{A_{j\alpha}\} - \inf\{A_{i\alpha}\}} \\ \Lambda^U_\alpha = \dfrac{\sup\{A^*_\alpha\} - \sup\{A_{i\alpha}\}}{\sup\{A_{j\alpha}\} - \sup\{A_{i\alpha}\}} \end{cases} \tag{9}$$

Also, denote $\Lambda = [\Lambda^L_\alpha, \Lambda^U_\alpha]$ hereafter. From this, Eq. (8) can be re-written as:

$$\begin{cases} \min\{B^*_\alpha\} = (1 - \Lambda^L_\alpha)\inf\{B_{i\alpha}\} + \Lambda^L_\alpha \inf\{B_{j\alpha}\} \\ \max\{B^*_\alpha\} = (1 - \Lambda^U_\alpha)\sup\{B_{i\alpha}\} + \Lambda^U_\alpha \sup\{B_{j\alpha}\} \end{cases} \tag{10}$$

This means $B^*_\alpha = [\min\{B^*_\alpha\}, \max\{B^*_\alpha\}]$ is generated. The final consequence B^* is then reassembled as:

$$B^* = U_{\alpha \in (0,1]} \alpha B^*_\alpha. \tag{11}$$

The KH approach may generate invalid interpolated results [54], which is usually called 'the abnormal problem'. To eliminate this deficiency, a number of modifications or improvements have been proposed, including Refs. [9, 10, 13, 14, 18, 53, 55–60]. Approaches such as Refs. [15, 16, 61–63] also belong to this group.

3.2. Analogy-based interpolation

The scale and move transformation-based fuzzy interpolation [51, 52, 64] is a representative approach in the analogy-based interpolation group. For simplicity, following the same assumption of a simple rule base containing two rules with two antecedents, the transformation-based approach is shown in **Figure 2** and outlined as follows.

Given neighbouring rules If x is A_i, then y is B_i, and If x is A_j then y is B_j and observation A^*, this method first maps fuzzy sets A_i, A_j and A^* to real numbers a_i, a_j and a^* (named as *representative values*) respectively, using real function f_1. Then, the location relationship between A^* and rule antecedents (A_i and A_j) is computed. This is achieved by another mapping function f_2, which results in the *relative placement factor* λ. In contrast to the resolution-based interpolation approaches, the generated *relative placement factor* in analogy-based fuzzy interpolation approach is a crisp real number. Finally, linear interpolation is implemented using mapping function f_3 of λ, which leads to the intermediate rule $A^{*'} \Rightarrow B^{*'}$.

Note that the *representative value* of intermediate rule antecedent $A^{*'}$ equals to that of A^* (the given observation), although $A^{*'}$ and A^* are not identical for most of the situations. In the scale and move transformation-based fuzzy interpolation approach, the similarity degree between two fuzzy sets A^* and $A^{*\prime}$ with the same representative value is expressed as the scale rate s, scale ratio $\mathbb{S}$ and move rate $\mathbb{M}$, which is obtained by real function f_4. From this, the consequence B^* is calculated from $B^{*'}$ using a transformation function f_5 which imposes the similarity degree between A^* and $A^{*\prime}$. Different approaches have been developed for intermediate rule generation and final conclusion production from the intermediate rule [17, 55, 63, 65].

3.3. Adaptive fuzzy interpolation

Fuzzy interpolation strengthens the power of fuzzy inference by enhancing the robustness of fuzzy systems and reducing the systems' complexity. Common to both classes of fuzzy

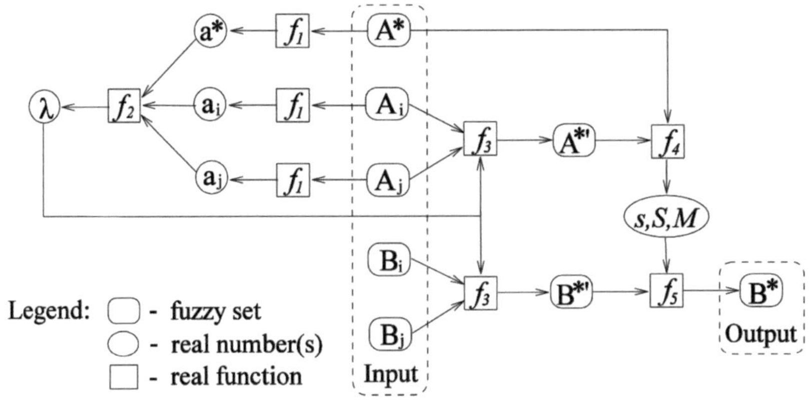

Figure 2. Transformation-based fuzzy interpolation [12].

interpolation approaches discussed above is the fact that interpolation is carried out in a linear manner. This may conflict with the nature of some realistic problems and consequently this may lead to inconsistencies during rule interpolation processes. Adaptive fuzzy interpolation was proposed to address this [12, 66–68]. It was developed upon FIS approaches, which detects inconsistencies, locates possible fault candidates and modifies the candidates in order to remove all the inconsistencies.

Each pair of neighbouring rules is defined as a *fuzzy reasoning component* in adaptive fuzzy interpolation. Each fuzzy reasoning component takes a fuzzy value as input and produces another as output. The process of adaptive interpolation is summarised in **Figure 3**. Firstly, the interpolator carries out interpolation and passes the interpolated results to the truth maintenance system (ATMS) [69, 70], which records the dependencies between an interpolated value (including any contradiction) and its proceeding interpolation components. Then, the ATMS relays any β_0-contradictions (i.e. inconsistency between two different values for a common variable at least to the degree of a given threshold β_0 ($0 \leq \beta_0 \leq 1$)) as well as their dependent fuzzy reasoning components to the general diagnostic engine (GDE) [71] which diagnoses the problem and generates all possible component candidates. After that, a modification process takes place to correct a certain candidate to restore consistency by modifying the original linear interpolation to become first-order piecewise linear.

The adaptive approach has been further generalised [11, 72, 73], which allows the identification and modification of observations and rules, in addition to that of interpolation procedures that were addressed in Ref. [12]. This is supported by introducing extra information of certainty degrees associated with such basic elements of FIS. The work also allows for all candidates for modification to be prioritised, based on the extent to which a candidate is likely to lead to all the detected contradictions, by extending the classic ATMS and GDE. This study has significantly improved the efficiency of the work in Ref. [12] by exploiting more information during both the diagnosis and modification processes. Another alternative implementation of the adaptive approach has also been reported in Ref. [74].

3.4. Sparse rule base generation

A Mamdani-style fuzzy rule base is usually implemented through either a data-driven approach [75] or a knowledge-driven approach [76]. The data-driven approach using artificial intelligence approach extracts rules from data sets, while the knowledge-driven approach generates rules by human expert. Due to the limited availability of expert knowledge, data-driven approaches have been increasingly widely applied. However, the application of such

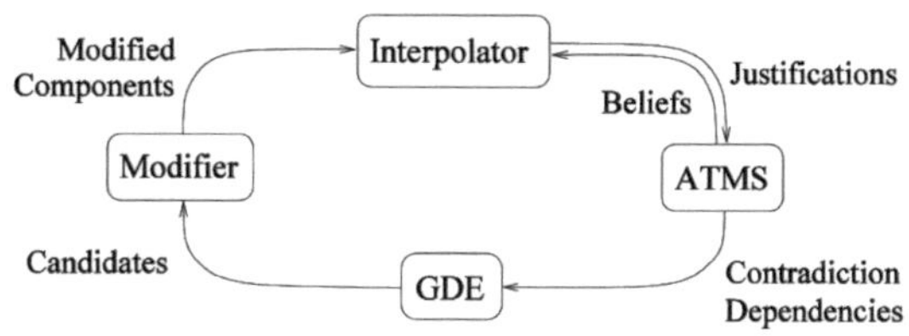

Figure 3. Adaptive fuzzy interpolation [12].

approaches usually requires a large amount of training data, and it often leads to dense rule bases to support conventional fuzzy inference systems, despite of the availability of rule simplification approaches such as Refs. [77, 78].

A recent development or rule base generation has been reported with compact sparse rule bases targeted [79]. This approach firstly partitions the problem domain into a number of sub-regions and each sub-region is expressed as a fuzzy rule. Then, the importance of each sub-region is analysed using curvature value by artificially treating the problem space as a geography object (and high-dimensional problem space is represented as a collection of sub-three-dimensional spaces). Briefly, the profile curvature of a surface expresses the extent to which the geometric object deviates from being 'flat' or 'straight', the curvature values of the sub-regions are then calculated to represent how important they are in terms of linear interpolation. Given a predefined threshold, important sub-regions can be identified, and their corresponding rules are selected to generate a raw sparse rule base. The generated raw rule base can then be optimised by fine-tuning the membership functions using an optimisation algorithm. Generic algorithm has been widely used for various optimisation problems, such as Ref. [80], which has also been used in the work of Ref. [79].

Compared to most of the existing rule base generation approaches, the above approach differs in its utilisation of the curvature value in rule selection. Mathematically, curvature is the second derivate of a surface or the slope of slope. The profile curvature [81] is traditionally used in geography to represent the rate at which a surface slope changes whilst moving in the direction, which represents the steepest downward gradient for the given direction. Given a sub-region $f(x, y)$ and a certain direction, the curvature value is calculated as the directional derivative which refers to the rate at which any given scalar field is changing. The overall linearity of a sub-region can thus be accurately represented as the maximum profile curvature value on all directions. From this, those rules corresponding to sub-regions with higher profile curvature values (with respect to a given threshold) are selected, which jointly form the sparse rule base to support fuzzy rule interpolation.

FISs relax the requirement of complete expert knowledge or large data sets covering the entire input domain from the conventional fuzzy inference systems. However, it is still difficult for some real-world applications to obtain sufficient data or expert knowledge for rule base generation to support FISs. In addition, the generated rule resulted from most of the existing rule base generation approaches are fixed and cannot support changing situations. An experience-based rule base generation and adaptation approach for FISs has therefore been proposed for control problems [82]. Briefly, the approach initialises the rule base with very limited rules first. Then, the initialised rule base is revised by adding accurate interpolated rules and removing out-of-date rules guided by the performance index from a feedback mechanism and the performance experiences of rules.

4. Fuzzy interpolation with sparse TSK-style rule base

The traditional TSK inference system has been extended to work with sparse TSK fuzzy rule base [21]. This approach, in the same time, also enjoys the benefit from its original version, which directly generates crisp outputs. The extended TSK inference approach is built upon a modified similarity measure which always generates greater than zero similarity degrees between observations and rule antecedents even when they do not overlap at all. Thanks to this property, a global consequence can always be generated by integrating the results from all rules in the rule base.

4.1. Rule firing strength

The modified similarity measure is developed from the work described in Ref. [83]. Suppose there are two fuzzy sets A and A' in a normalised variable domain. Without loss generality, a fuzzy set with any membership can be approximated by a polygonal fuzzy membership function with n odd points. Therefore, A and A' can be represented as $A = (a_1, a_2, \ldots a_n)$ and $A' = (a'_1, a'_2, \ldots a'_{n'})$, as shown in **Figure 4**. The similarity degree $S(A,A')$ between A and A' is computed as:

$$S(A, A') = \left(1 - \frac{\sum_{i=1}^{n} |a_i - a'_i|}{n}\right)(DF)^{\tilde{B}(supp_{A}, supp_{A'})} \frac{\min\left(\mu(C_A), \mu(C_{A'})\right)}{\max\left(\mu(C_A), \mu(C_{A'})\right)}, \tag{12}$$

where c_A is the centre of gravity of fuzzy sets A, and $\mu(c_A)$ is the membership of the centre of gravity of fuzzy set A; DF represents a distance factor which is a function of the distance between two concerned fuzzy sets, and $B(supp_{A}, supp_{A'})$ is defined as follows:

$$B(supp_{A}, supp_{A'}) = \begin{cases} 1, & if supp_A + supp_{A'} \neq 0, \\ 0, & if supp_A + supp_{A'} = 0, \end{cases} \tag{13}$$

where $supp_A$ and $supp_{A'}$ are the supports of A and A', respectively.

In Eq. (13), $B(supp_{A}, supp_{A'})$ is used to determine whether distance factor is considered. That is, if both A and A' are of crisp values, the distance factor DF will not take into consideration during the calculation of the similarity degree; otherwise, DF will be considered. The centre of gravity of a fuzzy set is commonly approximated as the average of its odd points. That is:

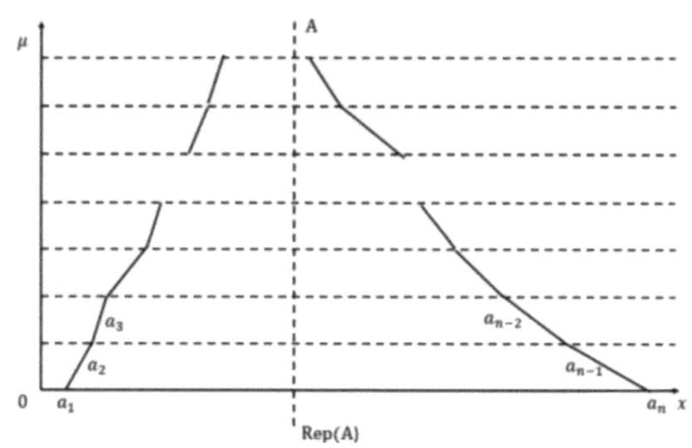

Figure 4. An arbitrary fuzzy set with n odd points.

$$c_A = \frac{a_1 + a_2 + \ldots + a_n}{n}, \tag{14}$$

$$\mu(c_A) = \frac{\mu(a_1) + \mu(a_2) + \ldots + \mu(a_n)}{n}. \tag{15}$$

The distance factor *DF* is represented as:

$$DF = 1 - \frac{1}{1 + e^{-hd+5}} \tag{16}$$

where d is the distance between the two fuzzy sets, and $h(h > 0)$ is a sensitivity factor. The smaller the value of h is, the more sensitive the similarity degree to their distance is. The value of h is usually within the range of (20, 60), but the exact value is problem specific.

4.2. Fuzzy interpolation

Using the modified similarity measure as traduced above, the similarity between any given observation and a rule antecedent is always greater than zero. This means that all the rules in the rule base are fired for inference. Therefore, if only a sparse rule base is available and a given observation is not covered by the sparse rule base, a consequence still can be generated by firing all the rules in the rule base. The inference process is summarised as below:

1. Calculate the matching degree $S(A^*, A_i)$ and $S(B^*, B_i)$ between each pair of rule antecedent (A_i, B_i) and the input values (A^*, B^*) based on Eq. (12).

2. Determine the firing strength of each rule by integrating the matching degrees between the input items and rule antecedents as calculated in Step 1:

$$\alpha_i = S(A^*, A_i) \wedge S(B^*, B_i). \tag{17}$$

3. Compute the consequence of each rule in line with the given input and the polynomial function in rule consequent:

$$f_i(A^*, A_i) = a_i \cdot c_{A^*} + b_i \cdot c_{B^*} + c_i. \tag{18}$$

4. Obtain the final result z by integrating the sub-consequences from all m rules in the rule base:

$$z = \frac{\sum_{i=1}^{n} \alpha_i f_i(A^*, B^*)}{\sum_{i=1}^{n} \alpha_i}. \tag{19}$$

5. Applications of fuzzy interpolation

Fuzzy interpolation systems have been successfully applied to a number of real-world problems including Refs. [23, 22, 25, 52, 57], two of which are reviewed in the section below.

5.1. Truck backer-upper control

Backing a trailer truck to a loading dock is a challenging task for all yet the most skilled truck drivers. Due to the difficulties, this challenge has been used as a control benchmark problem with various solutions proposed [75, 84, 85]. For instance, an artificial neural network has been applied to this problem, but a large amount of training data is required [84]. An adaptive fuzzy control system was also proposed for this problem, but the generation of the rule base is computationally expensive. Another solution combines empirical knowledge and data [85]. That is, a combined fuzzy rule base is generated by joining the previously generated rules (data-driven) and linguistic rules (expert knowledge-driven). More recently, a supervisory control system was proposed with fewer number of state variables required due to its capability to the decomposition of the control task, thus relieving the curse of dimensionality [86].

Fuzzy interpolation system has also been applied to the trailer truck backer-upper problem [52] to further reduce the system complexity. The problem can be formally formulated as $\theta = f(x, y, \emptyset)$. Variables x and y represent the coordinate values corresponding to horizontal and vertical axes; $\emptyset$ refers to the azimuth angle between the truck's onward direction and the horizontal axis; and θ is the steering angle of the truck. Given that enough clearance is present between the truck and loading lock in most cases, variable y can be safely omitted and hence results in a simplified formula to $\theta = f(x, \emptyset)$. By evenly partitioning each variable domain into three fuzzy sets, nine (i.e. 3*3) fuzzy rules were generated using FISMAT [87] and each of which is denoted as IF x is A AND $\emptyset$ is B THEN θ is C, where A, B and C are three linguistic values. Noting that domain partitions appear to be symmetrical in some sense, the three rules which are flanked by other rule pairs were removed from the rule base resulting a more compact rule base with only six fuzzy rules.

If the traditional fuzzy inference system were applied, the sparse rule base would cause a sudden break of the truck for some situations as no rule would be fired when the truck is in the position that can be represented by the omitted rules. In this case, fuzzy interpolation is naturally applied and the sudden break problem can be avoided. In addition, thanks to the great generalisation ability of the fuzzy interpolation systems, smooth performance is also demonstrated compared to the conventional fuzzy inference approaches. This study clearly demonstrates that fuzzy interpolation systems are able to simplify rule bases and support inferences with sparse rule bases.

5.2. Heating system control

The demotic energy waste contributes a large part of CO^2 emissions in the UK, and about 60% of the household energy has been used for space heating. Various heating controllers have been developed to reduce the waste of energy on heating unoccupied properties, which are usually programmable and developed using a number of sensors. These systems are able to successfully switch off heating systems when a property is unoccupied [88–92], but they cannot intelligently preheat the properties by warming the property before users return home without manual inputs or leaving the heating systems on unnecessarily for longer time. A smart home heating controller has been developed using a FIS, which allows efficient home heating by accurately predicting the users' home time using users' historic and current location data obtained from portable devices [23].

The overall flow chart of the smart home heating system is shown in **Figure 5**. The controller first extracts the resident's location and moving information. There are four types of residents' location and moving information that need to be considered: At Home, Way Back Home, Leaving Home and Static (i.e. at Special Location). The user's current location and moving states are obtained effectively using the GPS information provided by user's portable devices. From this, if the resident's current state is At Home, the algorithm terminates; and if the residents' current state is Leaving Home, that is the residents are moving away from home,

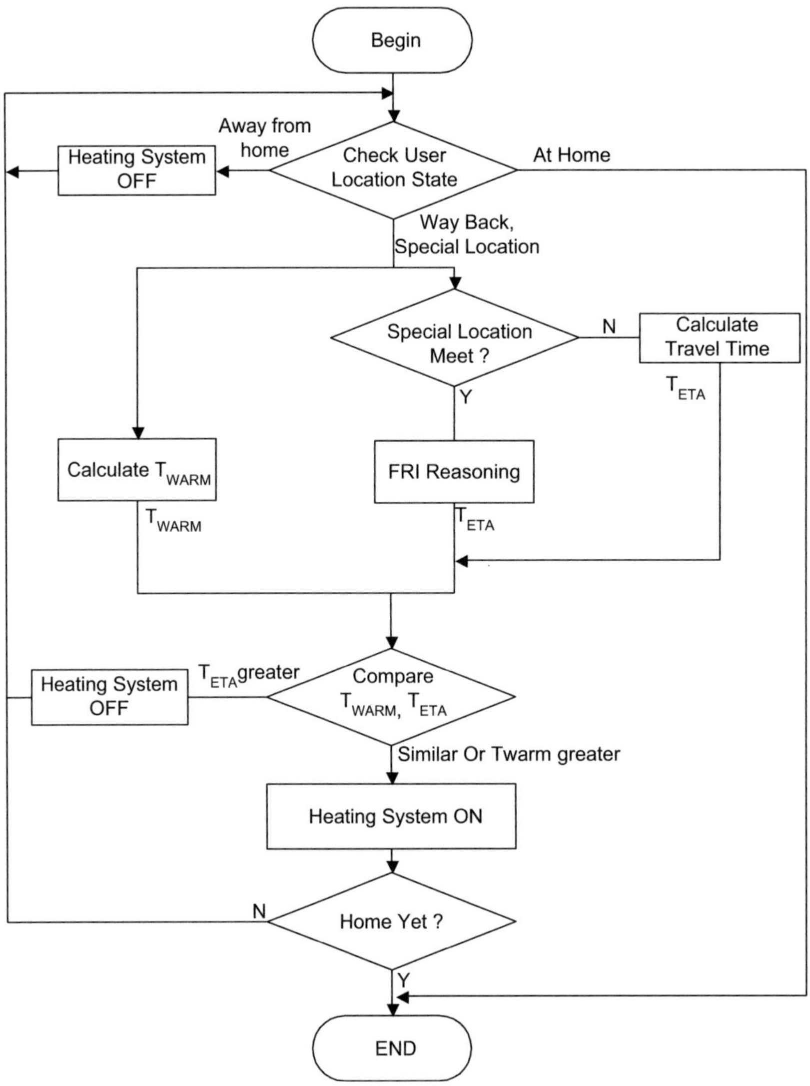

Figure 5. The flow chart of the heating controller [23].

the boiler is off and the system will check the resident's location and moving information again in a certain period of time. Otherwise, the time to arriving home (denoted as TAH) is predicted and the time to preheat the home to a comfortable temperature (denoted as TPH) is also calculated, based on the resident's current situation and the current environment around home.

The user's current travel modes (i.e. driving, walking or bicycling) can be detected by employing a naïve Bayes classifier [93] using the GPS information. Then the travel distance and time between the current location and home can be estimated using Google Distance Matrix API. Note that the time spent on different locations may vary significantly, and also different residents usually spend different amount of times at the same special location as people have their own living styles. The time that the residents spent at the current location is therefore estimated using fuzzy interpolation systems, thanks to the complexity of the problem. In particular, the fuzzy interpolation engine takes five fuzzy inputs and produces one fuzzy output which is the estimate of the time to getting home. The five inputs are the current location, the day of the week, the time of the day, the time already spent at the current location and the estimated travel between the current location and home.

If each input domain is fuzzy partitioned by 5 to 13 fuzzy, tens of thousands of rules will be resulted which requires significant resources during inferences. The proposed system, however, has selected the most important 72 rules forming a sparse rule base to support fuzzy rule interpolation, which significantly improve the system performance. Once the home time is calculated, the home can then be accurately preheated based on a heating gain table developed based on the particular situation and environment of a concerned property [91]. This work has been applied to a four-bedroom detached house with a total hearing space of $100\ m^2(floor\ area) \times 2.4\ m(hight)$. The house is heated by a 15 kW heating boiler. The study has shown that the controller developed using fuzzy inference has successfully reduced the burning time of the boiler for heating and more accurately preheat the home.

Despite of the success of the applications introduced above, there is a potential for FISs to be applied to more and larger scales real-world problems, especially in the field of system control. Note that robotics has taken the centre in the control field to perform tasks from basic robot calligraphy system [94] to complex tasks which require hand-eye (camera) coordination [95]. FISs can also be applied to such advanced areas in the field of robotics, which require further investigation.

6. Conclusions

This chapter reviewed fuzzy interpolation systems and their applications in the field of control. There are basically two groups of fuzzy interpolation approaches using the two most common types of fuzzy rule bases (i.e. Mamdani-style rule bases and TSK-style rule bases) to supplement the two groups of widely used fuzzy inference approaches (i.e. the Mamdani inference and the TSK inference). The applications of fuzzy interpolation systems have also been discussed in the chapter which demonstrate the power of the approaches. FISs can be further improved despite of its promising performance. Firstly, type-2 FISs have already been proposed in the literature, but how type-2 FISs can be applied in real-world applications requires further investigation. Also, more theoretical analysis for FISs is needed to mathematically prove the convergence property of the approaches. In addition, most of the existing fuzzy interpolation approaches are proposed as a supplementary of the existing fuzzy inference models. It is interesting to investigate the development of a united platform which integrates both the existing fuzzy models and fuzzy inference systems such that the new system can benefit from both approaches.

Acknowledgements

This work was jointly supported by the National Natural Science Foundation of China (No. 61502068) and the China Postdoctoral Science Foundation (Nos. 2013M541213 and 2015T80239).

Author details

Longzhi Yang[1*], Zheming Zuo[1], Fei Chao[2] and Yanpeng Qu[3]

*Address all correspondence to: longzhi.yang@northumbria.ac.uk

1 Department of Computer and Information Sciences, Northumbria University, Newcastle, UK

2 School of Information Science and Engineering, Xiamen University, Xiamen, PR China

3 Information Science and Technology College, Dalian Maritime University, Dalian, PR China

References

[1] Fu X, Zeng XJ, Wang D, Xu D, Yang L. Fuzzy system approaches to negotiation pricing decision support. Journal of Intelligent & Fuzzy Systems. 2015;**29**(2):685–699

[2] Yang L, Neagu D. Integration strategies for toxicity data from an empirical perspective. In: 2014 14th UK Workshop on Computational Intelligence (UKCI). 2014. pp. 1–8

[3] Yang L, Neagu D. Toxicity risk assessment from heterogeneous uncertain data with possibility-probability distribution. In: 2013 IEEE International Conference on Fuzzy Systems (FUZZ-IEEE). 2013. pp. 1–8

[4] Yang, Neagu D, Cronin MTD, Hewitt M, Enoch SJ, Madden JC, Przybylak K. Towards a fuzzy expert system on toxicological data quality assessment. Molecular Informatics. 2012;**32**(1):65–78

[5] Yang L, Neagu D. Towards the integration of heterogeneous uncertain data. In: 2012 IEEE 13th International Conference on Information Reuse & Integration (IRI). 2012. pp. 295–302

[6] Guo Q, Qu Y, Deng A, Yang L. A new fuzzy-rough feature selection algorithm for mammographic risk analysis. In: 2016 12th International Conference on Natural Computation, Fuzzy Systems and Knowledge Discovery (ICNC-FSKD). 2016. pp. 934–939

[7] Mamdani EH. Application of fuzzy logic to approximate reasoning using linguistic synthesis. IEEE Transactions on Computers. 1976;**26**(12):1182–1191

[8] Takagi T, Sugeno M. Fuzzy identification of systems and its applications to modeling and control. IEEE Transactions on Systems, Man, and Cybernetics. 1985;**1**:116–132

[9] Kóczy L, Hirota K. Approximate reasoning by linear rule interpolation and general approximation. International Journal of Approximate Reasoning. 1993;**9**(3): 197–225

[10] Kóczy L, Hirota K. Interpolative reasoning with insufficient evidence in sparse fuzzy rule bases. Information Sciences. 1993;**71**(1–2):169–201

[11] Yang L, Chao F, Shen Q. Generalised adaptive fuzzy rule interpolation. IEEE Transactions on Fuzzy Systems. 2016. DOI: 10.1109/TFUZZ.2016.2582526

[12] Yang L, Shen Q. Adaptive Fuzzy Interpolation. IEEE Transactions on Fuzzy Systems. 2011;**19**(6):1107–1126

[13] Yang L, Shen Q. Closed form fuzzy interpolation. Fuzzy Sets and Systems. 2013;**225**:1–22

[14] Tikk D, Baranyi P. Comprehensive analysis of a new fuzzy rule interpolation method. IEEE Transactions on Fuzzy Systems. 2000;**8**(3):281–296

[15] Kovács S. Extending the fuzzy rule interpolation "FIVE" by fuzzy observation. In: Computational Intelligence, Theory and Applications. Springer; 2006. pp. 485–497

[16] Johanyák ZC, Kovács S. Fuzzy rule interpolation based on polar cuts. In: Computational Intelligence, Theory and Applications. Springer; 2006. pp. 499–511

[17] Baranyi P, Kóczy LT, Gedeon TD. A generalized concept for fuzzy rule interpolation. IEEE Transactions on Fuzzy Systems. 2004;**12**(6):820–837

[18] Dubois D, Prade H. On fuzzy interpolation. International Journal of General System. 1999;**28**(2–3):103–114

[19] Jin S, Diao R, Quek C, Shen Q. Backward fuzzy rule interpolation. IEEE Transactions on Fuzzy Systems. 2014;**22**(6):1682–1698

[20] Chen C, Parthaláin NM, Li Y, Price C, Quek C, Shen Q. Rough-fuzzy rule interpolation. Information Sciences. 2016;**351**:1–17

[21] Li J, Qu Y, Shum HP, Yang L. TSK inference with sparse rule bases. In: Advances in Computational Intelligence Systems. 2017. pp. 107–123

[22] Molnárka GI, Kovács S, Kóczy LT. Fuzzy rule interpolation based fuzzy signature structure in building condition evaluation. In: 2014 IEEE International Conference on Fuzzy Systems (FUZZ-IEEE). 2014. pp. 2214–2221

[23] Li J, Yang L, Shum HP, Sexton G, Tan Y. Intelligent home heating controller using fuzzy rule interpolation. In: 2015 15th UK Workshop on Computational Intelligence (UKCI). 2015

[24] Bai Y, Zhuang H, Wang D. Apply fuzzy interpolation method to calibrate parallel machine tools. The International Journal of Advanced Manufacturing Technology. 2012;**60**(5–8):553–560

[25] Kovács S, Koczy LT. Application of interpolation-based fuzzy logic reasoning in behaviour-based control structures. In: 2004 IEEE International Conference on Fuzzy Systems. 2004

[26] Li J, Yang L, Fu X, Chao F, Qu Y. Dynamic QoS solution for enterprise networks using TSK fuzzy interpolation. In: 2017 IEEE International Conference on Fuzzy Systems (FUZZ-IEEE). 2017

[27] Yang L, Li J, Fehringer G, Barraclough P, Sexton G, Cao Y. Intrusion detection system by fuzzy interpolation. In: 2017 IEEE International Conference on Fuzzy Systems (FUZZ-IEEE). 2017

[28] Mamdani EH, Assilian S. An experiment in linguistic synthesis with a fuzzy logic controller. International Journal of Man-Machine Studies. 1975;**7**(1):1–13

[29] Castro JL, Zurita JM. An inductive learning algorithm in fuzzy systems. Fuzzy Sets and Systems. 1997;**89**(2):193–203

[30] Chen SM, Lee SH, and Lee CH. A new method for generating fuzzy rules from numerical data for handling classification problems. Applied Artificial Intelligence. 2001;**15**(7):645–664

[31] Zedeh LA. Outline of a new approach to the analysis of complex systems and decision processes. IEEE Transactions on Systems, Man, and Cybernetic. 1973;**3**:28–44

[32] Bandler W, Kohout L. Fuzzy power sets and fuzzy implication operators. Fuzzy Sets and Systems. 1980;**4**(1):13–30

[33] Turksen IB. Four methods of approximate reasoning with interval-valued fuzzy sets. International Journal of Approximate Reasoning. 1989;**3**(2):121–142

[34] Turksen IB, Zhong Z. An approximate analogical reasoning approach based on similarity measures. IEEE Transactions on Systems, Man, and Cybernetics. 1988;**18**(6):1049–1056

[35] Klement EP, Mesiar R, Pap E. Triangular norms. Position paper I: Basic analytical and algebraic properties. Fuzzy Sets and Systems. 2004;**143**(1):5–26

[36] Mizumoto M, Zimmermann HJ. Comparison of fuzzy reasoning methods. Fuzzy Sets and Systems. 1982;**8**(3):253–283

[37] Nakanishi H, Turksen IB, Sugeno M. A review and comparison of six reasoning methods. Fuzzy Sets and Systems. 1993;**57**(3):257–294

[38] Turksen IB, Zhong Z. An approximate analogical reasoning schema based on similarity measures and interval-valued fuzzy sets. Fuzzy Sets and Systems. 1990;**34**(3): 323–346

[39] Chen SM. A new approach to handling fuzzy decision-making problems. IEEE Transactions on Systems, Man, and Cybernetics. 1988;**18**(6):1012–1016

[40] Chen SM. A Weighted fuzzy reasoning algorithm for medical diagnosis. Decision Support Systems. 1994;**11**(1):37–43

[41] Yeung DS, Tsang ECC. A comparative study on similarity-based fuzzy reasoning methods. IEEE Transactions on Systems, Man, and Cybernetics, Part B (Cybernetics). 1997;**27**(2):216–227

[42] Tversky A. Features of similarity. Psychological Review. 1977;**84**(4):327

[43] Yeung DS, Tsang ECC. Fuzzy knowledge representation and reasoning using petri nets. Expert Systems with Applications. 1994;**7**(2):281–289

[44] Yeung DS, Tsang ECC. Improved Fuzzy Knowledge Representation and Rule Evaluation Using Fuzzy Petri Nets and Degree of Subsethood. International Journal of Intelligent Systems. 1994;**9**(12):1083–1100

[45] Kosko B. Neural Networks and Fuzzy Systems: A Dynamical Systems Approach to Machine Intelligence. New Jersey: Prentice-Hall; 1992

[46] Yeung DS, Ysang ECC. A multilevel weighted fuzzy reasoning algorithm for expert systems. IEEE Transactions on Systems, Man, and Cybernetics-Part A: Systems and Humans. 1998;**28**(2):149–158

[47] Thorndike RL. Who belongs in the family? Psychometrika. 1953;**18**(4):267–276

[48] Klir GJ, Yuan B. Fuzzy Sets and Fuzzy Logic: Theory and Applications. Prentice Hall PTR; 1995

[49] Robinson JA. A machine-oriented logic based on the resolution principle. Journal of the ACM (JACM). 1965;**12**(1):23–41

[50] Zadeh LA. Quantitative fuzzy semantics. Information Sciences. 1971;**3**(2):159–176

[51] Huang Z, Shen Q. Fuzzy interpolative reasoning via scale and move transformations. IEEE Transactions on Fuzzy Systems. 2006;**14**(2):340–359

[52] Huang Z, Shen Q. Fuzzy interpolation and extrapolation: A practical approach. IEEE Transactions on Fuzzy Systems. 2008;**16**(1):13–28

[53] Kóczy LT, Hirota K. Size reduction by interpolation in fuzzy rule bases. IEEE Transactions on Systems, Man, and Cybernetics, Part B (Cybernetics). 1997;**27**(1):14–25

[54] Yan S, Mizumoto M, Qiao WZ. Reasoning conditions on Koczy's interpolative reasoning method in sparse fuzzy rule bases. Fuzzy Sets and Systems. 1995;**75**(1):63–71

[55] Chen SM, Ko YK. Fuzzy interpolative reasoning for sparse fuzzy rule-based systems based on ?-cuts and transformations techniques. IEEE Transactions on Fuzzy Systems. 2008;**16**(6):1626–1648

[56] Hsiao WH, Chen SM, Lee CH. A new interpolative reasoning method in sparse rule-based systems. Fuzzy Sets and Systems. 1998;**93**(1):17–22

[57] Wong KW, Tikk D, Dedeon TD, Kóczy LT. Fuzzy rule interpolation for multidimensional input spaces with applications: A case study. IEEE Transactions on Fuzzy Systems. 2005;**13**(6):809–819

[58] Yam Y, Kóczy LT. Representing membership functions as points in high-dimensional spaces for fuzzy interpolation and extrapolation. IEEE Transactions on Fuzzy Systems. 2000;**8**(6):761–772

[59] Yeung Y, Wong ML, Baranyi P. Interpolation with function space representation of membership functions. IEEE Transactions on Fuzzy Systems. 2006;**14**(3):398–411

[60] Yang L, Chen C, Jin N, Fu X, Shen Q. Closed form fuzzy interpolation with interval type-2 fuzzy sets. In: 2014 IEEE International Conference on Fuzzy Systems (FUZZ-IEEE). 2014. pp. 2184–2191

[61] Chang YC, Chen SM, Liau CJ. Fuzzy interpolative reasoning for sparse fuzzy-rule-based systems based on the areas of fuzzy sets. IEEE Transactions on Fuzzy Systems. 2008;**16** (5):1285–1301

[62] Tikk D, Joó I, Kóczy L, Várlaki P, Moser B, Gedeon TD. Stability of interpolative fuzzy KH controllers. Fuzzy Sets and Systems. 2002;**125**(1):105–119

[63] Ughetto L, Dubois D, Prade H. Fuzzy interpolation by convex completion of sparse rule bases. In: 2000 IEEE International Conference on Fuzzy Systems (FUZZ-IEEE). 2000. pp. 465–470

[64] Shen Q, Yang L. Generalisation of scale and move transformation-based fuzzy interpolation. Journal of Advanced Computational Intelligence and Intelligent Informatics. 2011;**15**(3):288–298

[65] Jenei S, Klement EP, Konzel R. Interpolation and extrapolation of fuzzy quantities-The multiple-dimensional case. Soft Computing. 2002;**6**(3–4):258–270

[66] Yang L, Shen Q. Towards adaptive interpolative reasoning. In: 2009 IEEE International Conference on Fuzzy Systems (FUZZ-IEEE). 2009. pp. 542–549

[67] Yang L, Shen Q. Extending adaptive interpolation: From triangular to trapezoidal. In: 2009 9th UK Workshop on Computational Intelligence (UKCI). 2009. pp. 25–30

[68] Yang L, Shen Q. Adaptive fuzzy interpolation and extrapolation with multiple-antecedent rules. In: 2010 IEEE International Conference on Fuzzy Systems (FUZZ-IEEE). 2010. pp. 1–8

[69] Kleer JD. An assumption-based TMS. Artificial Intelligence. 1986;**28**(2):127–162

[70] Kleer JD. Extending the ATMS. Artificial Intelligence. 1986;**28**(2):163–196

[71] Kleer JD, Williams BC. Diagnosing multiple faults. Artificial Intelligence. 1987;**32**(1):97–130

[72] Yang L, Shen Q. Adaptive fuzzy interpolation with prioritized component candidates. In: 2011 IEEE International Conference on Fuzzy Systems (FUZZ-IEEE). 2011. pp. 428–435

[73] Yang L, Shen Q. Adaptive fuzzy interpolation with uncertain observations and rule base. In 2011 IEEE International Conference on Fuzzy Systems (FUZZ-IEEE). 2011. pp. 471–478

[74] Cheng SH, Chen SM, Chen CL. Adaptive fuzzy interpolation based on ranking values of polygonal fuzzy sets and similarity measures between polygonal fuzzy sets. Information Sciences. 2016;**342**:176–190

[75] Wang LX, Mendel JM. Generating fuzzy rules by learning from examples. IEEE Transactions on Systems, Man, and Cybernetics. 1992;**22**(6):1414–1427

[76] Johanyák ZC, Kovács S. Sparse fuzzy system generation by rule base extension. In: 2007 11th IEEE International Conference on Intelligent Engineering Systems. 2007. pp. 99–104

[77] Bellaaj H, Ketata R, Chtourou M. A new method for fuzzy rule base reduction. Journal of Intelligent & Fuzzy Systems. 2013;**25**(3):605–613

[78] Tao CW. A reduction approach for fuzzy rule bases of fuzzy controllers. IEEE Transactions on Systems, Man, and Cybernetics, Part B (Cybernetics). 2002;**32**(5):668–675

[79] Tan Y, Li J, Wonders M, Chao F, Shum HP, Yang L. Towards sparse rule base generation for fuzzy rule interpolation. In: 2016 IEEE International Conference on Fuzzy Systems (FUZZ-IEEE). 2016. pp. 110–117

[80] Cowton J, Yang L. A smart calendar system using multiple search techniques. In: 2015 15th UK Workshop on Computational Intelligence (UKCI). 2015

[81] Peckham SD. Profile, plan and streamline curvature: A simple derivation and applications. In: Proceedings of Geomorphometry. 2011. pp. 27–30

[82] Li J, Shum PH, Fu X, Sexton G, Yang L. Experience-based rule base generation and adaptation for fuzzy interpolation. In: 2016 IEEE International Conference on Fuzzy Systems (FUZZ-IEEE). 2016. pp. 102–109

[83] Chen SJ, Chen SM. Fuzzy risk analysis based on similarity measures of generalized fuzzy numbers. IEEE Transactions on Fuzzy Systems. 2003;**11**(1):45–56

[84] Nguyen D, Widrow B. The truck backer-upper: An example of self-learning in neural networks. In: Proceedings of the International Joint Conference on Neural Networks (IJCNN). 1989;**2**:357–363

[85] Kong SG, Kosko B. Adaptive fuzzy systems for backing up a truck-and-trailer. IEEE Transactions on Neural Networks. 1992;**3**(2):211–223

[86] Riid A, Rustern E. Fuzzy logic in control: Truck backer-upper problem revisited. In: 2001 IEEE International Conference on Fuzzy Systems (FUZZ-IEEE). 2001;**1**:513–516

[87] Lotfi A. Fuzzy Inference Systems Toolbox for Matlab (FISMAT). 2000

[88] Haissig C. Adaptive fuzzy temperature control for hydronic heating systems. IEEE Control Systems. 2000;**20**(2):39–48

[89] Lu J, Sookoor T, Srinivasan V, Gao G, Holben B, Stankovic J, Field E, Whitehouse K. The Smart Thermostat: Using Occupancy Sensors to Save Energy in Homes. In: Proceedings of the 8th ACM Conference on Embedded Networked Sensor Systems. 2010. pp. 211–224

[90] Nevius M, Pigg S. Programmable thermostats that go berserk: Taking a social perspective on space heating in Wisconsin. In: Proceedings of the 2000 ACEEE Summer Study on Energy Efficiency in Buildings. 2000;8:e44

[91] Scott, J, Bernheim Brush AJ, Krumm J, Meyers B, Hazas M, Hodges S, Villar N. PreHeat: Controlling home heating using occupancy prediction. In: Proceedings of the 13th International conference on Ubiquitous Computing. 2011 pp. 281–290

[92] Von Altrock, C, Arend HO, Krausse B, Steffens C. Customer-adaptive fuzzy logic control of home heating system. In: 1994 IEEE International Conference on Fuzzy Systems (FUZZ-IEEE). 1994;**3**:1713–1718

[93] Friedman N, Geiger D, Goldszmidt M. Bayesian network classifiers. Machine Learning. 1997;**29**(2–3):131–163

[94] Chao F, Huang Y, Zhang X, Shang C, Yang L, Zhou C, Hu H, Lin CM. A robot calligraphy system: From simple to complex writing by human gestures. Engineering Applications of Artificial Intelligence. 2017;**59**:1–14

[95] Chao F, Zhu Z, Lin CM, Hu H, Yang L, Shang C, Zhou C. Enhanced robotic hand-eye coordination inspired from human-like behavioral patterns. IEEE Transactions on Cognitive and Developmental Systems. 2017. DOI: 10.1109/TCDS.2016.2620156

Chapter 4

Implementing Complex Fuzzy Analysis for Business Planning Systems

Danil Dintsis

Additional information is available at the end of the chapter

http://dx.doi.org/10.5772/67974

Abstract

The chapter deals with implementing fuzzy logic for transition of descriptions in natural language to formal fuzzy and stochastic models and their further optimization in terms of effectiveness and efficiency of information modeling and prediction systems. The theoretical methods are implemented in lifelong learning business for development-specific virtual trainings for adult students.

Keywords: fuzzy, fuzzy set, fuzzy risk, lifelong learning, virtual learning, training method prediction, fuzzy analysis, complex definition area

1. Introduction

In the first part of the chapter the author examines challenges in transforming subject description in a natural language to a formal model. Establishing collaboration between specialists in different knowledge areas (technology, information, business, etc.) is usually a very complicated problem. In the process of common work there is a strong need to estimate model clearness, effectiveness, and efficiency for specialists in different knowledge areas. We consider effectiveness as correct interpretation of a subject in a real world by a model. And efficiency is calculated based on the share of the service states of the model.

In the second part, the author shows the implementation of the approach in virtual training design. Developing specific training methods for adult lifelong students is a very complicated task. We implemented fuzzy-based analysis to determine the best learning methods for different student groups and course types.

The special section is devoted to adapting and implementing virtual training for hard of hearing people.

2. Extension for fuzzy model definition area

2.1. Introduction

Information systems are widely spread in our daily life: offices, industry, and home. Each system represents or controls an extra object, such as household appliances or industrial control devices. Efficiency of those information and control systems depends on a wide range of factors. Wide-known methods for control systems' synthesis demand the formal model represented in analytical format [1]. The main problem of the discrete automate models is the strong necessity of formalization at the beginning of the process. It is a rather difficult task for complicated technological and information processes. Moreover, it is more appropriate to set a task of the adaptive control system with parametric adaptation features synthesis.

Let us consider the control system for an informational object as a "black box" with two sets of inputs and outputs [3]. Two main inputs are a math (or analytical) model, and data. One of the important features is a number and percentage of internal states of an information system/ control device. Internal states are always necessary to control stability of work, prevent user's mistakes, etc. These internal states and their properties require additional computing resources. The author examines an approach aimed at balancing advantages and resource usage of internal states [2].

Meanwhile these both inputs are (or can be) cross-dependent. Creating an analytical model as the first step of the system synthesis is based on intrinsic and semantic analysis of input data. Establishment of collaboration between specialists in different knowledge areas (technology, information, business, or others) is usually one of the main problems and challenges almost in all practice-based projects [3]. Factor analysis or lingual models may provide partial decision of the problem. Lingual models combine a system description in both formal and descriptive terms. A restriction of linguistic models is in their insufficient formalization and a high level of dependence on subjective expert appraisals.

Fuzzy models implementation ensures an analysis of processes in technical and information-based systems with nonlinear and/or multifactor-based behavior [1, 5]. The author presents the approach that combines model analysis based on natural language with strict formal systems by implementing fuzzy logic approaches.

2.2. General approach based on lingual models

We consider a process of analytical model synthesis as a first phase of a control model synthesis. The first important step at this phase is transition of a description based on natural language to a formal model. We should consider an interaction between specialists in different knowledge areas while creating such models as they may use different terms. Another important task is establishing back coupling between input data and information/control system. It is quite necessary for creating information and control systems with adaptive features. We implement fuzzy logic to solve the task of information system synthesis for complicated technological objects.

Implementing fuzzy logic ensures the opportunity of a semantic analysis of the object description made on natural-based language due to partial entrance of an element in one or several sets. The conceptual structure of the research area is represented as a set of abstract entities relying on concepts and terms of both a natural language and fuzzy sets. Next, we need to extend the existing fuzzy sets' model as an analytical model contains features, appropriate to elements of analyzable object, and own properties of the model, that provide its integrity.

Second step should be to deal with input data analysis to adapt the model and control system to environment changes. We do not need to change our model but make it adaptive. We need to establish clear borders for research object and its analytical model. To resolve this task, we need also to extend current fuzzy sets and models' methods and features. In this work, we present an extension range of definition of fuzzy sets and its elements' logical division to two sets: corresponding to object properties and internal properties of a system itself.

2.3. Model analysis in a partially defined environment based on fuzzy sets

Any analytical model has several internal features that ensure its internal integrity and reliability of a control system to controlled object. One may implement internal model states for internal data integrity control, exchange, additional logic control, etc.

Let's name a set of objects to be synthesized as $S = \{a[i]\}$. Each noun in a lingual model compares to an object $a[i]$ with value (weight) $a(i)$ and functionality $H(a[i])$. Lets' define an expansion of a definition range values for fuzzy attributes to imaginary area:

$$\begin{cases} j{*}a\,(i), & \text{if an object exists only in a formal model} \\ a\,(i) = a\,(i) & \text{if an object compares to a modelling subject} \end{cases} \tag{1}$$

Based on the above, it is obvious that model S always consists of objects with both rational and imaginary features and functionality weights. $H+$ and $H-$[4, 6, 7]. In this case, total number of object $a[i]$ features is h(n).

1. Number of features compared to lingual model is h+.

2. Total number of object features in fuzzy model– S.

3. Weights of each feature is $s(i)$.

4. A functional as a summary (vector) weight of features divided into rational and imaginary parts $a[i]$ -> H.

A model functional parameter may be represented as the following complex number:

$$H = H + +jH - \tag{2}$$

2.4. Extending fuzzy logic procedures for analysis and synthesis of information systems

Below, we show the extension is correct and does not break the fuzzy logic postulates. The definition area would be the following:

$$H\ (A|B) \in\ [0;\infty[+j([0,\infty[) \tag{3}$$

Lemmas:

If A is the element of a real object and B is the internal element:

$$H\ (A|B) \in\ [0;\infty[\tag{4}$$

If both A, B are internal elements:

$$H\ (A|B) \in j([0;\infty[) \tag{5}$$

$$H(U) = \infty \tag{6}$$

U is a universal set.

We provided a transformation of the traditional fuzzy sets axioms and features to a specific fuzzy model for using complex-based definition area to specify features of models for synthesis of control automation systems.

2.4.1. Fuzzy model synthesis based on lingual model analysis for technological and information objects

First, we need to transform a lingual model based on subject area terms and definitions to a metalanguage-based model. This model still may include some terms from its predecessor—the certain lingual model. Let us mark sets of keywords, and consider them as fuzzy model objects. At the first step, we may consider all nouns as fuzzy model objects [1–3].

Next let's define a transition between fuzzy objects:

$$\begin{cases} \quad\ \text{fi} \\ a[i] \rightarrow\ a[i{+}1], \text{fi is an } a[i] \text{ object's method} \end{cases} \tag{7}$$

If weight $a(i)$ is an imaginary one, then the object $a[i]$ does not match any object in a lingual model, and only meta model contains it.

Then, based on δ-operation, as it is defined in common fuzzy logic, we provide the following transformation between logically tied objects:

$$\begin{cases} a[i]\ \ \delta ia[i{+}1]\ \ \Rightarrow\ a[i{+}1] \\ \qquad\qquad \text{fi}+1 \\ a[i]\ \delta i\ {+}1\ [i{+}1]\ \Rightarrow\ a[i]\quad \rightarrow\quad a[i{+}2] \end{cases} \tag{8}$$

So, we can formalize an object's method based on a 2-step δ-operation.

Let's introduce a weight of a fuzzy object $H\ (M)=A$. We consider $\sum H(a[i])>=H(P))$, as the model contains both objects and their methods $f(i)$, that define a consequence of transitions between objects in a model. $f(i)$ methods are described by γ and δ operations. δ -operation defines a consequence of operations, events, and nodes of a model, and γ-operation defines weights (for

http://dx.doi.org/10.5772/67974

example, possibility, availability, resilience, etc.) for nodes, events, operations. If we take into consideration both types and an optimization factor $H(M)=A\rightarrow$max, then the following equality is true:

$$H(P) = H(a[1]\gamma H(a[2]\gamma\ \gamma H(a[i]\gamma\ \gamma H(a[n] \tag{9}$$

The following set of conditions provides model and based on it control system integrity:

$$\begin{cases} \mathrm{H(M)} \quad \rightarrow \max; \\ \mathrm{H(M)} = \Gamma a[i]\ \Delta a[i]; \\ \nabla a[i], \mathrm{H}(a[i] \neq 0; \\ \nabla a[i], \mathrm{H}(a[i]) \rightarrow \max; \\ \gamma \mathrm{i} \in a[i],\ \delta \mathrm{i} \in\ a[i]; \\ \mathrm{f(i)} = \gamma \mathrm{i} \cup \delta \mathrm{i} \end{cases} \tag{10}$$

The following operations provide structuration of fuzzy sets in A:

1. Establish relations between sets:

1.1. Relation of entrance Ai ∈ Aj

1.2. Relation of inheritance. Ai inherits Aj if a(i) ∈ Ai has the fixed set of values, and methods no less than a(j) ∈ A(j).

2. Let's introduce meta-set M, which describes a set of sets:

$$\mathrm{Ai}:\ \mathrm{M} = \cup(\mathrm{n(i)}, \cup \mathrm{h(i,j)}), \tag{11}$$

And a system of their internal relationships, where *n*(*i*), Ai's unique identifier (usually string name); *h*(*i,j*), a weight of relation between Ai and Aj sets.

An optimization of model parameters provides by consequent iterations. A rate, which weights reliability between fuzzy meta-model and preceding lingual model, is one of the most important optimization criteria. It is based on a feature $\nabla a[i] \in M$, that indicates a belongingness of a set element to a real or imaginary areas.

2.4.2. Fuzzy model synthesis algorithm based on a preceding lingual model

1. Allocate described essences based on the linguistic analysis. Create a set S of subessences.

2. Allocate a set of described attributes of essences. Based on it create a set T of fuzzy model attributes as mentioned in [2, 3].

3. Create a set K of basic knowledge sources, including experts, knowledgebase, and experimental data.

4. Create set C as a united set of weight criteria of each element. Primary weight is proportional to number of its occurrences into subsets *C*(*i*).

5. Create subsets $S(i)$ as based on the S set. $S(i)$ elements are united into a certain subset according to $C(i)$ criteria.

6. Create sets H(j) containing experts', knowledgebase articles and other data sources' weights: $Ci \to Hj(Ci)$.

7. Create sets $A(i)$ including attributes $a(i)$, as Ai ∇ S,t(i)A1, An, where

$$\mathrm{n} >= 1, \nabla \mathrm{h(j)}(Ai) = \mathrm{f(h(j), A(i), Hj,} [Ai] \backslash \mathrm{Ci}).$$

8. Establish fuzzy model M (Ai, H[a(j)].

Synthesis of a control system for technical and/or information object is based on the defined fuzzy set M.

2.5. Scope definition for fuzzy sets usage in control and simulation systems for technical and information systems

While designing formal model (including fuzzy one), it is necessary to estimate the following items:

- Possibility of implementing a certain formalization method to a certain object;

- Degree of model compliance to an object. Consider fuzzy set A, containing $a[i]$ elements. Each $a[i]$ is compliant with an object feature or internal feature of a model. We consider a method of compliance between model and object based on research of mutual consistency of elements. The method is based on a ∪-operation, and indicates subsets, which can describe an object behavior if used in common [2, 3]:

$$\mathrm{M} = \mathrm{A} \cup \mathrm{B} \cup \mathrm{C} \cup \ \cup \mathrm{N} \tag{12}$$

2.5.1. Internal consistency criteria

1. Redundancy property:

$$\forall \mathrm{H}(a[i]\beta a[j] > 0$$

2. Compliance property:

$$\forall \mathrm{H}(a[i]|a[j] \neq \infty$$

3. Efficiency and resilience balance:

$$\begin{cases} \sum a[i](\mathrm{h}+) > 0 \\ \sum a[i](\mathrm{h}-) > 0 \\ \sum a[i](\mathrm{h}) > 0 \end{cases} \tag{13}$$

2.5.2. Compliance to modeling object criteria

(1) Compliance between M - model transformations results, and an object fact features: implementing of any possible track of operations of M model cannot lead an object to a prohibited state.

$$\forall \mathrm{H}(Xi(Ai)) > 0, \quad (14)$$

or

Implementing of any possible track in M, model can transform the controlled system or object R into a possible state or common null state

$$\forall H(Xi(Ai)) >= 0 \quad (15)$$

Thus, we consider a fuzzy model is applicable in case that its real functional is positive and there is at least one set of allowed methods that transform a control system from its initial to final state.

2.6. Conclusion

Finally, we found that the synthesis of control systems based on descriptive models of natural language may be adequately implemented based on fuzzy sets. Logical separation of elements of fuzzy sets, in which the real domain includes the attributes and functional elements that describe the state of an object, but to the imaginary one—own internal state of the model and the management system that are required to make it operational. Based on this logical separation, we may estimate effectiveness and resilience of control system.

Finally, the authors resume that automated systems' synthesis is appropriately presented and formalized by fuzzy sets' models. Fuzzy logic definition area has been extended to an imaginary area. We established the logical division of model components to real and imaginary areas per their role. Internal objects of a model are presented in the imaginary area, and objects that describe the modeling system to the real area. We introduced necessary functional extensions for fuzzy logic to operate with logical extension.

Transformation algorithm is developed, and we recommend the certain implementation area for it.

3. Defining appropriate training methods for lifelong learning organization

3.1. Introduction

The author has developed a method and algorithms of fuzzy analysis for lingual models with complex digits' implementation. The author used the approach that differentiates native data, attributes of an object and internal model data, and attributes [2, 3]. Dividing these classes into real and imaginary leads to the decrease of dimensions in a model, and in this way to the decrease of computing capacity. This fact allows to decrease the risk of incorrect interpretation of results, and it provides also an opportunity to estimate costs of efficiency.

This article is devoted to implementing fuzzy analysis to define and implement various virtual training methods in a lifelong learning educational organization and reaches the highest possible satisfaction level by different categories of adult students as defined in Ref. [4].

3.2. Big challenges in lifelong learning

The lifelong professional learning training center offers short-term trainings and postdiploma programs to upgrade professional skills or gain a new specialization for adult professionals. A lot of students take multiple courses as bundles or periodically in accordance to new versions of software, technological equipment, or professional standards. It's of great importance for the training organization to analyze big data interdependencies to find out trends, develop new courses, make targeted offers for students, and create specific training methods for certain client groups. Since 2009, the author has been deeply involved into developing and implementing various virtual—online, and blended—training methods. During this work, the author carried out a regular analysis of data from different sources to determine customer requirements, demands for courses, and ways of their representation, technical, and mythological opportunities [4, 5]. The goal of those continual research efforts was the development of strongly targeted training options for certain student groups and courses. The fuzzy-based modeling is considered as the most appropriate approach to the task, because students', customers', trainers', and other staff's feedback, requirements, as well as demand estimations, are mostly represented as a nonformal or mixed way. For example, rating A in a feedback means "more than I can expect." It is obvious that the level of expectations differs among students, and customer representatives.

3.3. Opportunities and threats in lifelong learning

Based on M_o_R™ and Total Risk Management® concepts fundamental characteristics of any risk define organization behavior for it, for example: tolerance level, impact, mitigation and contingency strategies, management level, as well as level of financial reserves. While examining risk nature we often consider that a single risk belongs to different characteristic sets. For example, a risk of incorrect professional behavior can belong to human and organizational, and technical sets simultaneously. Therefore, we can create a fuzzy description of a risk:

$$r(i) \in O, r(i) \in T, \mathrm{r(i)}\ H \tag{16}$$

where ***O, T, H,*** fuzzy sets (organizational, technical, human features).

Implementing risk analysis in fuzzy terms ensures complex analysis for risk source, impact, mitigation, and contingency. The author examined complex risk analysis for portfolio (both projects and operational activities) of virtual learning methods in a lifelong adult training center. As a service-based and private user (a student) oriented business, its success depends dramatically on a subjective personalized opinion of students and partially of corporate HR managers. Their feedbacks are represented both in partially formalized manner, and comments in a natural language.

Another challenge concerns representing risk dependencies, or so-called domino effect. As it's a rather complicated task to formalize risks interdependencies, we can implement an approach, starting with an informal description in a natural language with further formalizing it by means of fuzzy-based algorithms.

The fuzzy analysis is the very appropriate tool to transfer statements in a natural language into a formal model, and explore threats and opportunities. The fuzzy analysis is implemented as described by the author in Ref. [5]. Identifying and analyzing risks, and their interdependencies, we include both negative (threat) and positive (opportunity) parts of risk analysis with the primary aim for finding new opportunities for development and quality improvement.

Main threats for an adult professional training organization are in customer dissatisfaction, and on the opposite main opportunities are based on reaching continuous education of students personally, and corporate customers. Let's examine a simple example about modern technology-based virtual learning implementation, and consider online learning process. There are several main opportunities of online learning for an educational organization, which are:

1. Improve organization innovative brand.
2. Attracting more students from distant regions.
3. More students in a class in a certain group.

On the other side, there are threats:

1. Changing in teaching methods.
2. Decrease of teaching quality.
3. Student's and corporate customers' rejection of new training method.
4. Technical issues.

Those risks—both positive and negative—are well-known when we talk about them in a natural language, but training organization's decision-making process requires qualitative and quantitative estimates. As shown in reference [1] we can implement fuzzy analysis to transform natural language to a weak formalized fuzzy model, by placing model-internal risks into an imaginary area, and objective risks into a real area of the model.

3.4. Building current data analysis with fuzzy logic

We investigated our students, trainers, corporate clients, and internal administrative staff feedbacks to discover additional training opportunities.

In 2009 we started online webinar trainings, which are held as simultaneous trainings in groups consisting of online (webinar), and class-based students (named as "webinar-in-class™"). The example of feedbacks is given in a **Table 1**. Total number of feedbacks: 10,000+ student feedbacks, 700+ by trainers, 1000+ by training center administrative staff, and 500+ by corporate customers' representatives.

We compared and analyzed feedbacks of webinars with the excerpt feedbacks of traditional class-based trainings. As the "specialist computer training center (CCT)" has been operated since 1991, we extracted feedbacks for class trainings for previous 5 years, e.g., we included 25,000 students', 5000 trainers', 5000 administrative staff's, and 2000 customer representatives' feedbacks into comparative analysis against "webinar-in-class" feedbacks, which are presented in **Table 2**.

The "trapeze" form of a fuzzy interpretation, as shown at **Figure 1**, is used to represent fuzzy component, because rating values are subjective and personal oriented. For example, we use ratings from "1" or "E" (minimum value, means that a client is completely dissatisfied) to "5" or "A" (maximum value, indicates that a service exceeds customer's expectations). Rate "3" or "C" indicates customer's general satisfaction. These estimations are personally based and depend on a lot of factors, such as professional specialization, job function and rank, individual specialties. Only ratings "E" and "A" strongly indicate satisfactory level. For example, rating "3" or "C" at design or HR trainings is mainly considered as more dissatisfied than satisfied. On the other side a "C" rating at business or project management trainings is mainly considered as satisfied' and rating A" is very rare, because business managers are mostly not as emotional.

Rate	Parameter (%)			
	Trainer	Technical facilities	Course	Willing for further training
A	32	27	45	42
B	48	51	42	47
C	12	14	10	7
D	5	6	2	2
E	3	2	1	1

Table 1. An excerpt from webinar-N-class studies.

Rate	Parameter (%)			
	Trainer	Technical facilities	Course	Willing for further training
A	39	42	41	42
B	46	41	44	47
C	11	13	13	7
D	3	4	2	2
E	1	2	1	1

Table 2. An excerpt from traditional class-based studies.

Figure 1. Trapeze interpretation of satisfaction level.

In fact, the total number of analyzed attributes is more than 100, and it changes regularly to follow customer, market demands, technical, and methodological facilities.

Below is a partial list of main attributes in the information system, which contain basic data and we consider them as a real area attributes in an analytical model:

- Student, Client/Customer, Learning format, Country/region, Course, Year/season/month, Vendor, Product Trainer, Trainer rating, Course rating, Training method rating, Number of courses taken by a student afterwards

The total number of real area attributes is more than 50.

Next I show an excerpt from a list of more than 25 additional (information model based) attributes. We consider these attributes in an imaginary area of our fuzzy model:

- Total rating of a training method, comparative rating to class-based training, comparative rating of webinars against class and self-learning combined method, views and filters across client types, regions, time, season, etc.

To build an integral customer satisfaction rating we use multidimensional fuzzy analysis of different partial (single-parameter) rankings as shown at **Figure 2**. Also, different filters and constraints are implemented to localize problems, challenges and find grow-points.

3.5. Investigating students' satisfaction against educational organization efficiency

We can investigate an integral satisfaction/dissatisfaction level of a training group based on the following attributes: trainer ratings, course ratings, willingness to continue training at a next course, and ratings of technical facilities. Each set is fuzzy and contains ranking values (ratings) by each student.

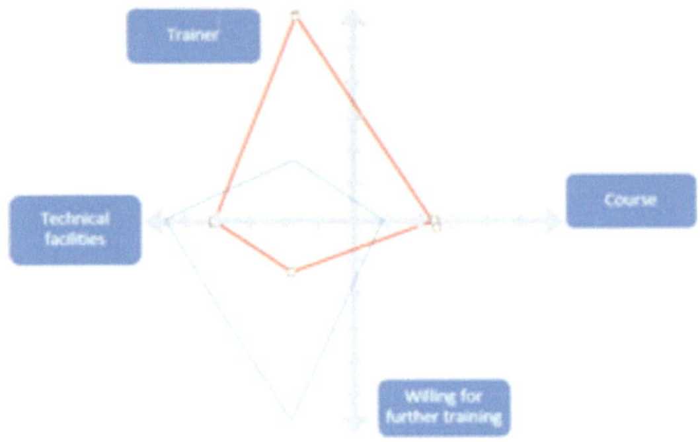

Figure 2. Determining the satisfaction area on an example of four parameter analysis.

In each case (for a student/group/trainer/course, etc.) we form a fuzzy area, in which we may consider that a course/study/trainer/class, etc. are satisfactory. According to company goals and statistical data we can also implement additional weights for attributes. For example, a trainer rank has coefficient "1, 5" and class ranking—"0, 8". Attribute "willing for further training" will have the maximum weight coefficient "2", as it's the most important factor for commercial adult learning company. This model shows a subjective satisfaction level as a set (family) of polygons. Each polygon draws by connecting points, which reflect partial ratings. For example, an area of subjective satisfaction is defined by the following partial ratings:

Trainer rating = 2 AND

Course rating = 4 AND

Technical facilities = 3 AND

Willing to continue education = 5.

In this case, we define that a group is partially satisfied/partially dissatisfied with a trainer, but if most of students of a certain group are ready to continue education at next courses, we can mark this group as "satisfied."

For complex estimation of job effectiveness of an adult life-long training organization we developed more complicated approaches, which include statistical data based on more than 70 attributes (both in real and imaginary areas) and collected them in multidimensional databases—OLAP cubes. This cube has the appropriate number (more than 70) of dimensions, and we need to build a set of fuzzy models, and optimize them for daily calculations and analysis. A real part of a model includes basic facts, and on the other side an imaginary part of a model includes filters, views, and additional states of a model or database. Due to this approach, we decrease number of dimensions to 50 in total, which leads to decrease in computing capacity requirements. Practical result: we have an opportunity to process analytical reports in a real-time mode, and postpone few complicated reports for nonworking hours (night time and Sundays).

Let us examine a comparatively simple set of fuzzy sets, which describes an integral satisfactory factor and training organization efficiency:

- By course, a certain trainer and/or trainer group, a company—customer, learning location, a certain period, a training method, a training branch (e.g., Management, ITSM, software development, HR, etc.).

The developed model contains both real area attributes, which reflect basic states, and imaginary area attributes, which reflect temporary, service states, filter conditions, identifiers, etc.

Below, I show an excerpt from a model. Attributes named in a lingual model terms to simplify understanding

$$\begin{cases} \mathrm{M1} = (\mathrm{CN} + \mathrm{St} + \mathrm{Tr} + \mathrm{TM}) + \mathrm{j}(\mathrm{V} + \mathrm{TP} + \mathrm{ST}), \\ \mathrm{M2} = (\mathrm{CN} + \mathrm{Tr} + \mathrm{CCF} + \mathrm{TCF} + \mathrm{WFT}) + \mathrm{j}(\mathrm{TP(i)} + \mathrm{ACR(i)} + \mathrm{ATR(i)} + \mathrm{V}), \\ \mathrm{M3} = (\mathrm{CN} + \mathrm{TL} + \mathrm{C} + \mathrm{TM}) + \mathrm{j}(\mathrm{TP(i)} + \mathrm{ACR(i)} + \mathrm{AMR(i)} + \mathrm{V}), \end{cases} \tag{17}$$

where CN, Course name; St, Student identifier; Tr, trainer identifier; V, view name; TP(i), selected time period; St, threshold level of students' satisfaction for the certain model; TM, training method; CCF, course cash flow; TCF, cash flow on courses by a certain trainer; WFT, student's willing for continuous education; ACR, average course ranking for selected period; ATR, average trainer ranking for selected period; TL, training location; C, corporate customer name; and AMR, average ranking of a certain training method for selected period.

Set M1 reflects mean level of students' satisfaction for a certain course and a certain trainer for selected time period.

Set M2 reflects current level of economic efficiency of a certain trainer, based on dynamic trend of students' satisfaction across a number of time periods (for example, month to month or quarter to quarter).

Set M3 reflects dynamics of corporate customers' satisfaction for a certain training location, course, and a training method. This set gives a control how a certain training location provides quality for a certain course and a training method, for example, webinar, or blended, or self-paced, etc.

3.6. Defining training methods and models

Adult learning training organization should offer various training opportunities for its students, such as long- and short-term trainings, class-based, virtual, blended, synchronous, and asynchronous, etc. Based on the analysis model at our Specialist CCT we develop balanced cost-effective vs. "student satisfactory" training methods for precisely defined customer audience and course bundles.

While analyzing results of modeling, we find several maximums. Each of the maximums is characterized by a certain set of parameters, as shown in **Figure 3**.

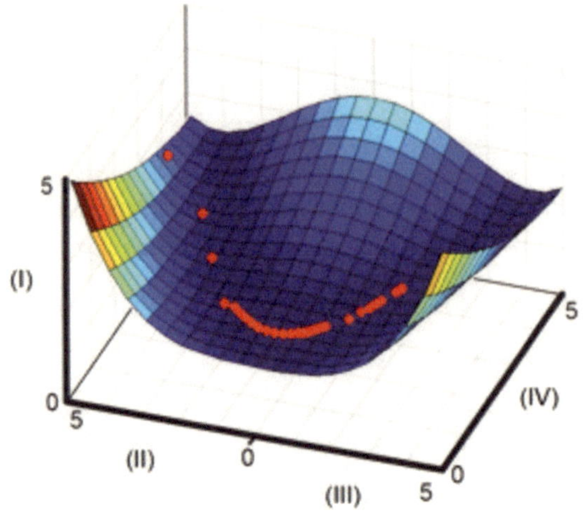

Figure 3. Multidimensional estimation, where I – Individual approach level; II – Trainer level; III – Technical facilities; IV – Training center economical effectiveness.

For example, the maximum satisfaction level which is found in an analytical cube is defined by the following attributes:

- student total expenses (minimal);

- high quality skills received;

- high qualified trainers;

- number of students in a class (maximal);

- collaboration facilities in the group (maximal);

- technical expenses are minimal.

The "webinar-in-class" training option is the best for mentioned attributes, because a webinar-based student has an opportunity to study anywhere, and has no travel or accommodation expenses. Simultaneously, he/she has an access to the best trainers, and can collaborate with classmates in a class and other webinar students, using a very simple software. The webinar students have an access to the same technical facilities such as labs. Thus, the webinar-in-class training method becomes very popular solution for students studying technical (Microsoft, Oracle, CISCO, etc.) courses, as well as project, and IT service management courses.

We worked further to analyze maximums, and another point is based on the following attributes:

- appropriate (frequent) course start date;

- individual approach;

- introvert students;

- students with high level of self-organization capabilities;

- full classes;

- low expenses on training organization by a training center.

In this example, we see more attributes, and it was a bit more difficult to create an appropriate training method. The result was a kind of blended learning, which we named an "open learning."

- blended "open learning" training method is the best solution for very specialized trainings, and self-organized students;

- unlimited webinar subscription is for students, who are unemployed or wish to change profession. Both categories need cheap training at a large number of courses.

To resume I want to stress that the estimating process is everlasting, as well as optimization of the research model. While preparing this article, one more training method was developed, which proves the efficiency of the described approach. The described methods won several professional awards [6, 7].

4. Identifying special tools for virtual training of hard of hearing people

4.1. Introduction

More than 10% of people on the Earth suffers from different hearing impairs as the World Health Organization data shows. Many of them are young people, or employees, which are involved into lifelong learning. They have difficulties with taking both class and virtual trainings, if they do not have or use special hearing aids. Meanwhile, many young people do not use special devices due to medical recommendations or having scruples.

Based on our research of students' with hearing impairs demands our research team deliver a special computer-based technology—named as Petralex©. It is implemented in mobile Apps and Windows driver, which works as a personal hearing assistant [6]. A student passes an "*in cito*" hearing test and the application creates a personal hearing profile for each place or environment (for example, public transport, café, car, room at home, classroom, and workplace). Mobile App acts as a hearing aid in a smartphone, so a student can easily attend classes. The Windows-based driver creates a Virtual audio device (VAD), which adapts streaming audio for both online and asynchronous virtual learning to an appropriate user's hearing profile.

Different virtual training methods—synchronous, asynchronous, blended—which are defined in Ref. [4], include online and/or off-line listening in videos, podcasts, as well as online training delivery, including real-time discussion with a trainer and classmates. So, students with hearing disabilities should have opportunities to be involved into the entire training process.

4.2. Synchronous learning methods

A student with partial hearing losses can feel uncomfortable while studying in class, or at online webinar. If a student studies in a class, he or she can implement the Petralex® mobile app as a hearing assistant, and involve in-depth into a learning process. In a synchronous learning training content is delivered in an online mode as shown in **Figure 4**. A student accesses it using the special audio driver, which ensures audio stream adjustment to personal hearing profile. As a result, a student can attend studies for a long time—up to 8 hours per day—due to improved hearing tolerance, reducing fatigue for long listening sessions, and attenuation of excessive sound pressure [6, 7].

4.2.1. Typical learning cases for online trainings

1. A trainer explains learning materials: a student studies at home. S/he connects to an online tool (for example, Skype®, Citrix GoTo®, WebEx, Adobe® Connect®, or any other), activates "My room" profile. Next, our driver transforms audio stream in a real-time mode with only 10–50 ms delay, so a student can hear a teacher clearly, have concentration on studying, and ask or answer questions; present his/her work, and discuss with other students in a real-time mode.

2. A business game: business games and other forms of group studies are very popular according to our model (Eq. (16)). An online student usually plays a role of a virtual team member or help desk agent. Implementing audio-driver provides both parties an opportunity to collaborate in a clear mode without delays.

4.3. Asynchronous learning methods

In this section let us consider different scenarios for asynchronous learning of students with partial hearing losses. The most popular tools for asynchronous learning are: learning management systems (LMS), stream and offline video, and audio services.

As shown in **Figure 5**, an external audio signal from a learning tool goes through the virtual audio driver, which transforms it according to an activated user hearing profile. Thanks to it, a student can study anywhere. At our training center, and with our partners, the following scenarios, as shown in **Figure 5**, were tested:

In a special class,

At home,

At a workplace, and

On a beach.

Figure 4. The learning schema for synchronous virtual learning.

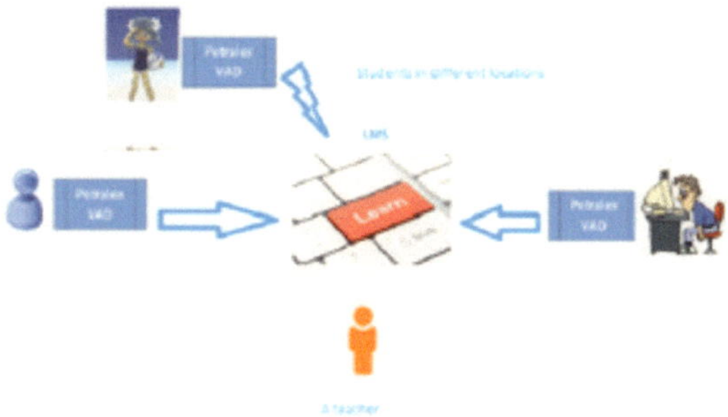

Figure 5. Asynchronous learning scenarios.

One of the most impressive cases in our training practices is short-term learning for adult busy people. We defined lifelong business students as a separate category in our model. Lifelong learners often study during their vacations or weekends. Usually they are strongly motivated, so they can easily combine their rest and studies. As an example, a student creates a "beach" profile and can listen learning records on a comfortable manner for his/her hearing.

5. Resume

Implementing extended definition area for fuzzy set analysis provides vast opportunities for representation of control objects by information systems, their analysis and optimization. Based on implementation of fuzzy analysis the author succeeded in creating and launching various virtual training models for lifelong learning, including people suffering with partial hearing losses.

Author details

Danil Dintsis

Address all correspondence to: consult@dintsis.org

Educational Private organization "Specialist", Moscow, Russian Federation

References

[1] Zadeh L. A. Decision Analysis and Fuzzy logic. In: Conference on Fuzzy Sets and Soft Computing in Economics and International Finance (FSSCEF 2004). Saint-Petersburg, Russia; 2004.

[2] Dintsis D. A modification of discrete-event models algebra for industrial technological systems. Devices and systems. Control, Monitoring, And Diagnostics Journal # 1, 2008.

[3] Dintsis D, Simankov V. The control automation system method based on combined implementation of fuzzy logic and discrete automation systems. In: European Modelling Symposium EMS2013 - 03-C Methodologies, Tools and Operations Research; November 2013; Manchester. Print ISBN:978-1-4799-2577-3 INSPEC Accession Number:14199480. DOI: 10.1109/EMS.2013.15. IEEE. p. 87–90

[4] Ferdinand D.S. Flexible Learning Environments-Theories-Trends-Issues. Monography. University of West-Indies. West Indies. St.-Augustine. DOI: 10.13140/RG.2.1.3958.2488. 2016. 32p

[5] Dintsis D. Implementing Fuzzy Sets For "Big Data" Analysis Based on Large Training Centre Feedbacks. In: IEEE AEIT Annual 2015 Conference. Naples, Italy. Publisher: IEEE. October, 2015.

[6] Dintsis D, Bredikhin A. Virtual Learning for People with Hearing Impairs. In: IEEE AEIT Annual 2015 Conference. Naples, Italy. Publisher: IEEE. October, 2015.

[7] LERN International Award Winners 2016. https://www.flipsnack.com/LearningResourcesNetwork/2016-lern-international-award-winners.html. 2016. LERN. Chicago. [Accessed: 2017-01-11]

A New Methodology for Tuning PID-Type Fuzzy Logic Controllers Scaling Factors Using Genetic Algorithm of a Discrete-Time System

Wafa Gritli, Hajer Gharsallaoui and
Mohamed Benrejeb

Additional information is available at the end of the chapter

http://dx.doi.org/10.5772/intechopen.69604

Abstract

In this chapter, a proportional-integral derivative (PID)-type fuzzy logic controller (FLC) is proposed for a discrete-time system in order to track a desired trajectory generated using the flatness property. In order to improve the performance of the proposed controller, genetic algorithm (GA) based on minimizing the integral of the squared error (ISE) is used for tuning the input and output PID-type FLC scaling factors online. The considered controller is applied to an electronic throttle valve (ETV). GA tuning shows a better and robust performance compared to Simulink design optimization (SDO) algorithm in terms of tracking a desired trajectory with disturbances rejection.

Keywords: PID-type FLC, scaling factors, genetic algorithm, integral of the squared error, Simulink design optimization technique, flatness, electronic throttle valve

1. Introduction

Fuzzy logic control (FLC) has been widely used in many successful industrial applications. The first FLC algorithm was implemented by Mamdani in 1974. Unlike conventional control, which is based on mathematical model of a plant, an FLC usually embeds the intuition and experience of a human operator and may provide a nonlinear relationship induced by membership functions, rules and defuzzification. In that respect, FLC has been reported to be successfully used for a number of complex and nonlinear systems and are proved to be more robust and their performances are less sensitive to parametric variations than conventional controllers.

In the literature, various types such as proportional integral (PI), proportional derivative (PD) and proportional-integral derivative (PID) of FLCs have been proposed. For example, PI-type FLCs have been successfully implemented in many physical applications such the control of the temperature and pressure of a steam engine and control the steering and speed of an automobile. However, performance of PI-type FLCs for higher order systems and nonlinear systems may be poor due to the large overshoot and the excessive oscillation. PD-type FLCs are suitable for a limited class of systems and they are not recommendable in the presence of measurement noise and sudden load disturbances. Theoretically, PID-type FLCs provide a good performance. However, there are difficulties associated with the generation of an efficient rule base and the tuning of parameters.

In the proposed PID-type FLC, the design of parameters within two groups: structural parameters and tuning parameters. Basically, structural parameters include input/output (I/O) variables to fuzzy inference, fuzzy linguistic sets, membership functions, fuzzy rules, inference mechanism and defuzzification mechanism, which are usually determined during offline design. Tuning parameters include I/O scaling factors (SF) and parameters of membership functions (MF), which can be calculated during online adjustments of the controller in order to enhance the process performance [1].

The appropriate selection of input and output scaling factors is very important because they have significant effects on the dynamic of fuzzy controller. This leads researchers to explore the best method in searching optimum PID-type FLC parameters. Various strategies or methods have been used up to now. In Ref. [2], Qiao and Mizumoto proposed a peak observer mechanism-based method to adjust the PID-type FLC parameters. This self-tuning mechanism decreases the equivalent integral control component of the fuzzy controller gradually with the system response process time. Furthermore, Woo et al. [3] developed a method based on two empirical functions evolved with the system's error information. In Ref. [1], Guzelkaya et al. proposed a technique that adjusts the scaling factors, corresponding to the derivative and integral components, using a fuzzy inference mechanism. However, the major disadvantages of all these PID-type FLC tuning method are the difficult choice of their relative parameters and mechanisms. To overcome these difficulties, differential search algorithm (DSA) meta-heuristic technique is proposed for systematically tuning the scaling factors of the PID-type FLC in Ref. [4]. The fuzzy control design is formulated as a constrained optimization problem, which is efficiently solved based on an improved DSA. In this proposed technique, different optimization criteria such as integral square error (ISE) and maximum overshoot are considered in order to guarantee more robustness and performance control objectives.

In this chapter, a genetic algorithm (GA)-based heuristic optimization technique has been implemented to obtain better performance compared to the Simulink design optimization (SDO) technique. GA is based upon minimizing the error between the output system and the desired trajectory starting from a flat output variable generated using the flatness property. Various performance indices can be used. In this study, the integral of squared error (ISE) index has been used in order to minimize the error between the output and the desired flat trajectory. The methods are applied in a discrete-time framework to an electronic throttle valve as a case of study.

2. PID-type fuzzy logic controller description

In this study, we will deal with fuzzy PID-type controllers formed using one PD-type FLC with an integrator at the output.

The PID-type fuzzy logic controller structure is shown in **Figure 1** [5], where K_e and K_d ($K_e, K_d \in \mathbb{R}^+$) are the input scaling factors are α and β ($\alpha, \beta \in \mathbb{R}^+$) the output scaling factors.

The inputs variables, well known as the error e_k between the desired trajectory y_k^d and the measure y_k, as well as the error variation Δe_k given by Eqs. (1) and (2) where T_e is the sampling period.

$$e_k = y_k^d - y_k \tag{1}$$

$$\Delta e_k = \frac{e_k - e_{k-1}}{T_e} \tag{2}$$

The output variable Δu_k of such a controller is the variation of the control signal u_k which can be defined as Eq. (3).

$$\Delta u_k = \frac{u_k - u_{k-1}}{T_e} \tag{3}$$

The output of the PID-type fuzzy is given by Eq. (4) [5]

$$\begin{aligned} u_k &= \alpha \Delta u_k + \beta \int \Delta u_k dt \\ &= \alpha (A + PK_e e_k + DK_d \Delta e_k) \\ &+ \beta \int (A + PK_e e_k + DK_d \Delta e_k) dt \\ &= \alpha A + \beta A t + (\alpha K_e P + \beta K_d D) e_k \\ &+ \beta K_e P \int e_k dt + \alpha K_d D \Delta e_k \end{aligned} \tag{4}$$

Thus, the equivalent control components of the PID-type FLC are such that

Proportional gain: $\alpha K_e P + \beta K_d D$

Integral gain: $\beta K_e P$

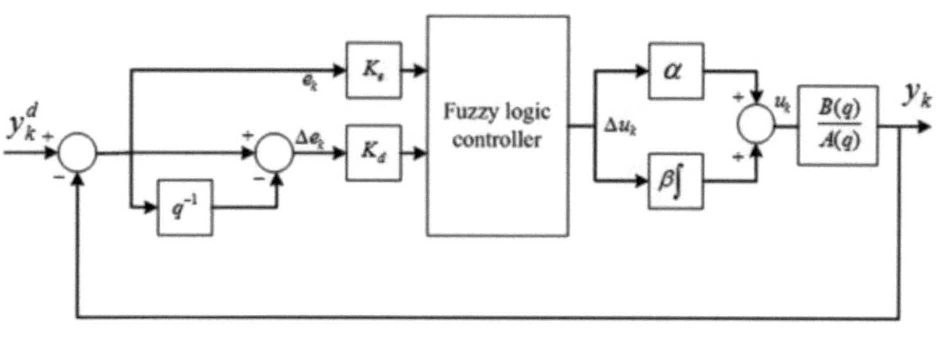

Figure 1. PID-type FLC.

$e_k \backslash \Delta e_k$	N	ZR	P
N	NL	N	ZR
ZR	N	ZR	P
P	ZR	P	PL

Table 1. Fuzzy rules-base.

Derivative gain: $\alpha K_d D$

where the terms P and D are given by Eqs. (5) and (6) [2].

$$P = \frac{\Delta u_{(i+1)j} - \Delta u_{ij}}{e_{i+1} - e_i} \tag{5}$$

$$D = \frac{\Delta u_{i(j+1)} - \Delta u_{ij}}{\Delta e_{j+1} - \Delta e_j} \tag{6}$$

The fuzzy controllers with a product-sum inference method, centroid defuzzification method and triangular uniformly distributed membership functions for the inputs and a crisp output proposed in Refs. [2, 6] are used in our case of study.

Table 1 gives the linguistic levels, assigned to the variables e_k, Δe_k and Δu_k, as follows: *NL*: negative large; *N*: negative; *ZR*: zero; *P*: positive; *PL*: positive large.

3. Scaling factors tuning using genetic algorithm

In this work, a new method is proposed for tuning the coefficients of PID-type FLCs. This method adjusts the input scaling factor corresponding to the derivative coefficient and the output scaling factor corresponding to the integral coefficient of the PID-type FLC using genetic algorithm, as shown in **Figure 2**. The integral of squared error (ISE) index has been used in order to minimize the error between the output and the desired flat trajectory.

3.1. Genetic algorithm

Genetic algorithm was first proposed by Holland [7]. It is a heuristic optimization technique inspired by the mechanism of natural selection. It is used in order to solve highly complex problems. GA starts with an initial population containing a number of parameters, where each one is regarded as the genes of a chromosome and can be structured by a string of concatenated values. Each chromosome represents a solution of the problem and its performance is evaluated based on fitness function.

In the beginning, an initial chromosome population is randomly generated. The chromosomes are candidate solutions to the problem. Then, the fitness values of all chromosomes are evaluated by calculating the objective function. So, a group of the best chromosomes is selected based on the fitness of each individual. In this 'surviving' population, the genetic operators of

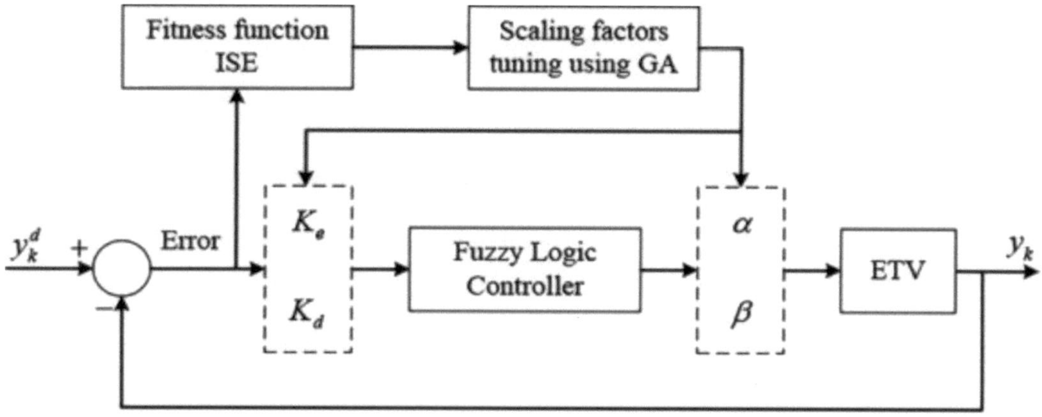

Figure 2. PID-type FLC scaling factors tuning.

crossover and mutation are applied in order to create the next population solution. The above steps are repeated until a specific termination criterion is found.

- **Reproduction**: Create a part of the new population by simply copying without changing the selected individuals from the present population. Also, new population has the possibility of selection by already developed solutions [8].
- **Crossover**: Create new individuals as offspring of two parents. It is a recombination operator that combines selected subparts called crossover points of two parent chromosomes. The individuals resulting in this way are the offspring [8].
- **Mutation**: Create a new individual for the new population by randomly mutating a selected individual. The modifications can consist of changing one or more values in the representation or adding/deleting parts of the representation [8].

To compute the fitness of each chromosome, the objective functions are used. Many authors use integral of time multiplied by absolute error (ITAE), mean of the squared error (MSE), integral of absolute error (IAE) and integral of the squared error (ISE) as performance index [9, 10].

In this chapter, the method of tuning PID-type FLC parameters using GA consists in finding the optimal I/O scaling factors, which minimize the defined objective function, chosen as the ISE in order to specify more performance in terms of tracking a desired trajectory.

If $y^d(t)$ is the desired trajectory and y is the output trajectory, then error $e(t)$ is

$$e(t) = y^d(t) - y(t) \tag{7}$$

and the ISE can be defined by

$$ISE = \int_0^{\tau} e(t)^2 dt \tag{8}$$

Fitness function is taken as inverse of error, i.e., performance index.

$$Fitness\ value = \frac{1}{Performance\ index} \tag{9}$$

3.2. Tuning procedure

The overall flowchart for optimization using GA is shown in **Figure 3**. Initially, a number of populations N have been generated for the scaling factors K_e, K_d, α and β. Each individual of these N sets in the current population is evaluated using the objective function ISE. Based on the values of the objective function, out of these N possible solutions, the good solutions are retained and the others are eliminated. A new population is formed by applying the genetic operators (reproduction, crossover and mutation) to these selected individuals. This process of

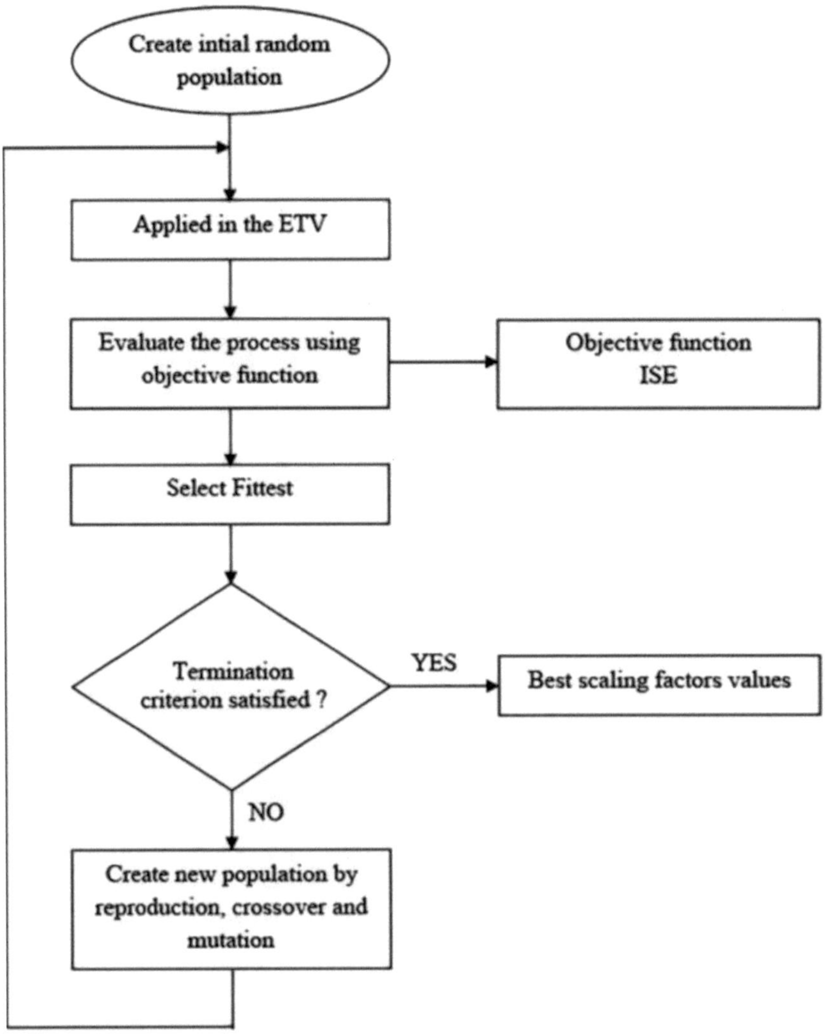

Figure 3. Flowchart of the GA optimization algorithm.

production of a new generation and its evaluation is performed repetitively. The algorithm continues until the population converges to the stop criterion.

4. Flatness and trajectory planning

The flat property has been introduced in Ref. [11] for continuous-time nonlinear systems. It can be stated in a discrete-time version which leads for the design of a control which ensures a tracking of a desired trajectory. One major property of differential flatness is that the state and the input variables can be directly expressed, without integrating any differential equation, in terms of the flat output and a finite number of its derivatives. The flatness approach will be used in this chapter in a discrete-time framework.

The studied dynamic linear discrete system is described by Eq. (10).

$$A(q)y_k = B(q)u_k \tag{10}$$

where q is the forward operator, u_k and y_k are the input and the output, respectively, and $A(q)$ and $B(q)$ are polynomials defined by

$$A(q) = q^n + a_{n-1}q^{n-1} + \ldots + a_1 q + a_0 \tag{11}$$

$$B(q) = b_{n-1}q^{n-1} + \ldots + b_1 q + b_0 \tag{12}$$

where the parameters a_i and b_i are constants, i=0,1,...,$n-1$. The partial state of such a dynamic system can be considered as a discrete flat output z_k which can be expressed as a function of input and output signals as following

$$A(q)z_k = u_k \tag{13}$$

$$B(q)z_k = y_k \tag{14}$$

Often, the real output signal y_k to be controlled is not a flat output. Then, it is necessary to plan a desired trajectory for the flat output [11] and to consider thereafter the relation (14).

The open loop control law can be determined by the following relations [12].

$$u^d(t) = f(z^d(t), \ldots, z^{d^{(r+1)}}(t)) \tag{15}$$

$$y^d(t) = g(z^d(t), \ldots, z^{d^{(r)}}(t)) \tag{16}$$

where f and g are the vectorial functions. Then, it is sufficient to find a desired continuous flat trajectory $t \mapsto z^d(t)$ that must to be differentiable at the $(r+1)$ order.

The polynomial interpolation technique is used in order to plan the desired flat trajectory $z^d(t)$. Let consider the state vector $Z^d(t) = (z^d(t) \;\; \dot{z}^d(t) \;\; \ldots \;\; z^{d^{(r+1)}}(t))^T$ containing the desired continuous flat output and its successive derivatives. The expression of $Z^d(t)$ can be given in Eq. (17); t_0 and t_f are the two moments known in advance.

$$Z^d(t) = M_1(t-t_0)c_1(t_0) + M_2(t-t_0)c_2(t_0, t_f) \tag{17}$$

where M_1 and M_2 are such as [12]

$$M_1 = \begin{pmatrix} 1 & t & \cdots & \frac{t^{n-1}}{(n-1)!} \\ 0 & 1 & \cdots & \frac{t^{n-2}}{(n-2)!} \\ \vdots & \ddots & \ddots & \vdots \\ 0 & \cdots & 0 & 1 \end{pmatrix} \tag{18}$$

$$M_2 = \begin{pmatrix} \frac{t^n}{n!} & \frac{t^{n+1}}{(n+1)!} & \cdots & \frac{t^{2n-1}}{(2n-1)!} \\ \frac{t^{n-1}}{(n-1)!} & \frac{t^n}{n!} & \cdots & \frac{t^{(n-2)}}{(n-2)!} \\ \vdots & \ddots & \ddots & \vdots \\ t & \cdots & \frac{t^{n-1}}{(n-1)!} & \frac{t^n}{n!} \end{pmatrix} \tag{19}$$

and the vectors c_1 and c_2 defined by

$$c_1 = Z^d(t_0) \tag{20}$$

$$c_2 = M_2^{-1}(t_f - t_0)(Z^d(t_f) - M_1(t_f - t_0)Z^d(t_0)) \tag{21}$$

Then, the output desired trajectory y_k^d is defined. In the discrete-time framework, the real output y_k has asymptotically to track this such as Eq. (22).

$$y_k^d = B(q)z_k^d \tag{22}$$

In the following section of this chapter, the efficiency of the proposed methodology for tuning PID-type FLC scaling factors has been validated on a discrete-time system: an electronic throttle valve for a defined desired trajectory generated using the flatness property and compared to the Simulink design optimization (SDO) technique.

5. Case of study: electronic throttle valve (ETV)

Throttle valve is one of the most important devices in the engine management system. In conventional engine, the amount of airflow into the combustion system has been adjusted by the throttle valve, which is connected mechanically to an accelerator pedal [13]. The electronic throttle body (ETB) regulates air inflow into the car engine. Compared to the mechanical throttle, a well-controlled ETB can reduce fuel consumption.

5.1. System modelling

The case of the ETV is described in **Figure 4**.

The electrical part is modelled by Eq. (23)

$$u(t) = L\frac{d}{dt}i(t) + Ri(t) + k_v\omega_m(t) \tag{23}$$

where L is the inductance R is the resistance $u(t)$ and $i(t)$ are the voltage and the armature current, respectively, k_v is an electromotive force constant and $\omega_m(t)$ is the motor rotational speed.

The mechanical part of the throttle is modelled by a gear reducer characterized by its reduction ratio γ such as Eq. (24)

$$\gamma = \frac{C_g}{C_L} \tag{24}$$

where C_L is the load torque and C_g is the gear torque. The mechanical part is modelled according to Eq. (25), such that [14, 15].

$$J\frac{d}{dt}\omega_m(t) = C_e - C_f - C_r - C_a \tag{25}$$

and

$$\frac{d}{dt}\theta(t) = (180/\pi/\gamma)\omega_m(t) \tag{26}$$

where $\theta(t)$ is the throttle plate angle, J is the overall moment of inertia, $C_e = k_e i(t)$ is the electrical torque where K_e is a constant, C_f is the torque caused by mechanical friction, C_r is the spring resistive torque and C_a is the torque generated by the airflow. The electronic throttle valve involves two complex nonlinearities due to the nonlinear spring torque Cr and the friction

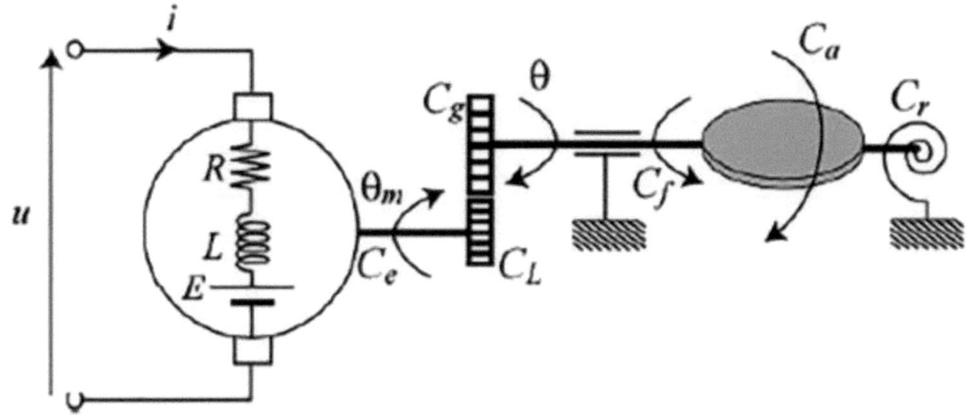

Figure 4. Electronic throttle valve system.

torque C_f. They are given by their static characteristics [16]. The static characteristic of the nonlinear spring torque C_r is defined by

$$C_r = \frac{k_r}{\gamma}(\theta - \theta_0) + D\,\mathrm{sgn}(\theta - \theta_0) \tag{27}$$

For $\theta_{min} \le \theta \le \theta_{max}$, k_r, is the spring constant, D is a constant, θ_0 is the default position and sgn(.) is the following signum function

$$\mathrm{sgn}(\theta - \theta_0) = \begin{cases} 1, & \text{if } \theta \ge \theta_0 \\ -1, & \text{else} \end{cases} \tag{28}$$

The friction torque function C_f of the angular velocity of the throttle plate can be expressed as

$$C_f = f_v\omega + f_c\,\mathrm{sgn}(\omega) \tag{29}$$

where f_v and f_c are two constants. By substituting in Eq. (25) the expressions of C_g, C_f and C_r and by neglecting the torque generated by the airflow C_a, the two nonlinearities $\mathrm{sgn}(\theta - \theta_0)$ and $\mathrm{sgn}(\omega)$ and the two constants $\frac{k_r}{\gamma}\theta_0$ and f_v the transfer function of the linear model becomes (30) [15].

$$H(q) = \frac{(180/\pi/\gamma)k_e}{JLq^3 + JRq^2 + (k_e k_v + Lk_s)q + Rk_s} \tag{30}$$

with $k_s = (180/\pi/\gamma^2)k_r$ and q as the Laplace operators.

5.2. Simulation results

The identified parameters of $H(q)$ are given in **Table 2** at 25°C temperature [15].

The corresponding discrete-time transfer function is given by Eq. (31) for the sample time $T_e = 0.002$s.

$$H(q^{-1}) = \frac{0.007833q^{-1} + 0.01396q^{-2} + 0.0007724q^{-3}}{1 - 1.948q^{-1} + 0.954q^{-2} - 0.006152q^{-3}} \tag{31}$$

Parameters	Values
R(Ω)	2.8
L(H)	0.0011
ke (N.m/A)	0.0183
kv (v/rad/s)	0.0183
J (kg.m^2)	4×10^{-6}
γ	16.95

Table 2. Model's parameters.

The desired continuous time flat trajectory $z^d(t)$ can be computed according to the following polynomial form

$$z^d(k) = \begin{cases} \dfrac{cst1}{B(1)}, & \text{if } 0 \le k \le k_0 \\ Poly_1(k), & \text{if } k_0 \le k \le k_1 \\ \dfrac{cst2}{B(1)}, & \text{if } k_1 \le k \le k_2 \\ Poly_2(k), & \text{if } k_2 \le k \le k_3 \\ \dfrac{cst1}{B(1)}, & \text{if } k \ge k_3 \end{cases} \quad (32)$$

where $cst1$ and $cst2$ are constants, $k_0 = 3s$, $k_1 = 6s$, $k_2 = 10s$ and $k_3 = 15s$ are the instants of transitions, $B(1)$ is the static gain between the flat output z_k and the output signal y_k for each operating mode and $Poly_1(k)$ and $Poly_2(k)$ are polynomials calculated using the technique of polynomial interpolation.

The desired trajectory is then given in **Figure 5**.

The obtained optimal I/O scaling factors $(Ke,\ Kd,\ \alpha,\ \beta)$ for Simulink design optimization technique and GA are summarized in **Table 3**.

The obtained results are given in **Figures 6–9**.

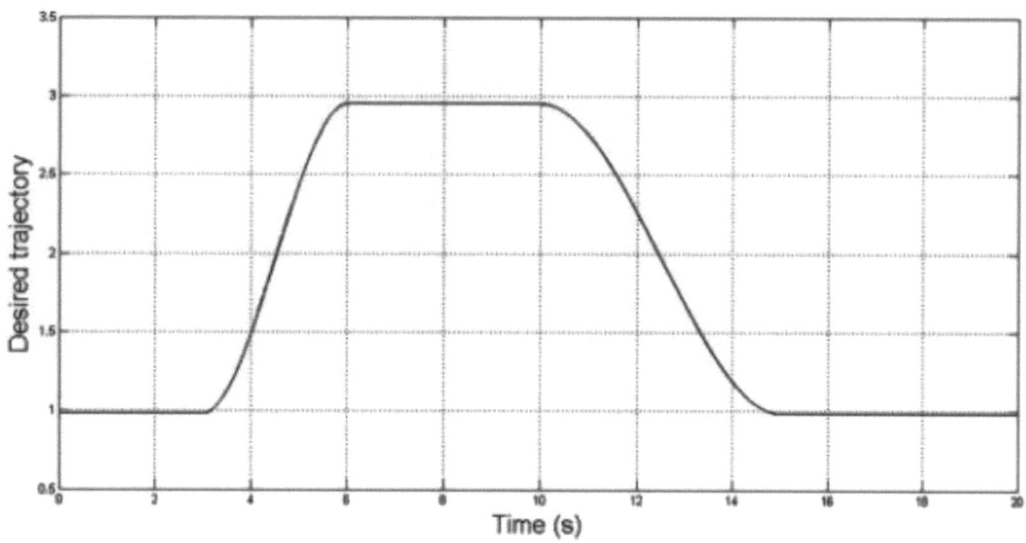

Figure 5. Desired trajectory.

SF\Method	SDO	GA
Ke	0.1144	0.0086
Kd	1.5997	0.4612
α	2.6239	0.1108
β	0.0001	0.1934

Table 3. PID-type fuzzy scaling factors values.

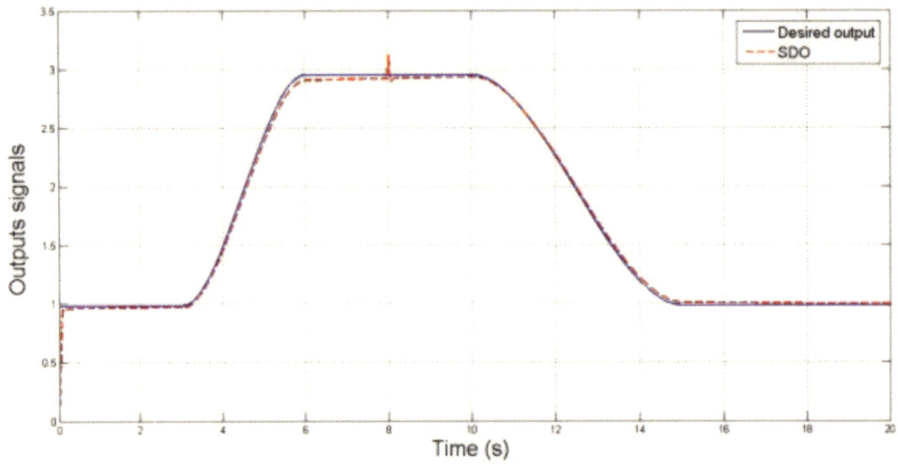

Figure 6. System outputs using Simulink design optimization.

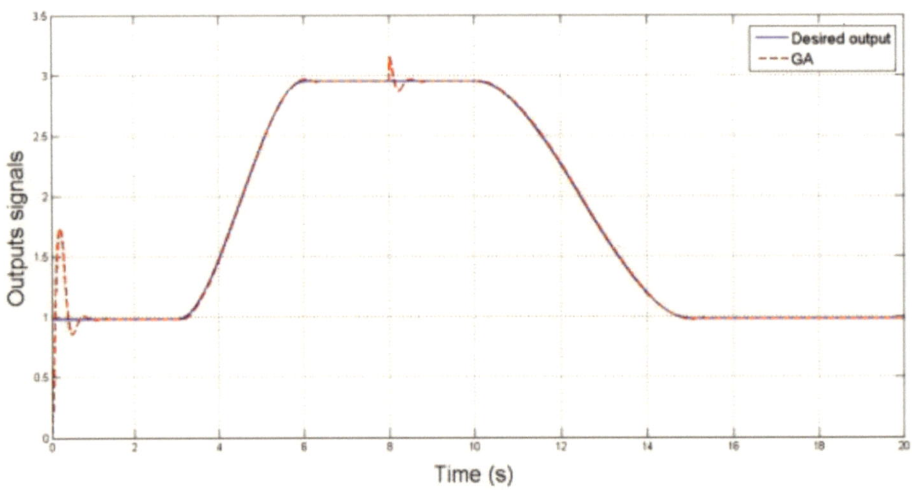

Figure 7. System outputs using genetic algorithm.

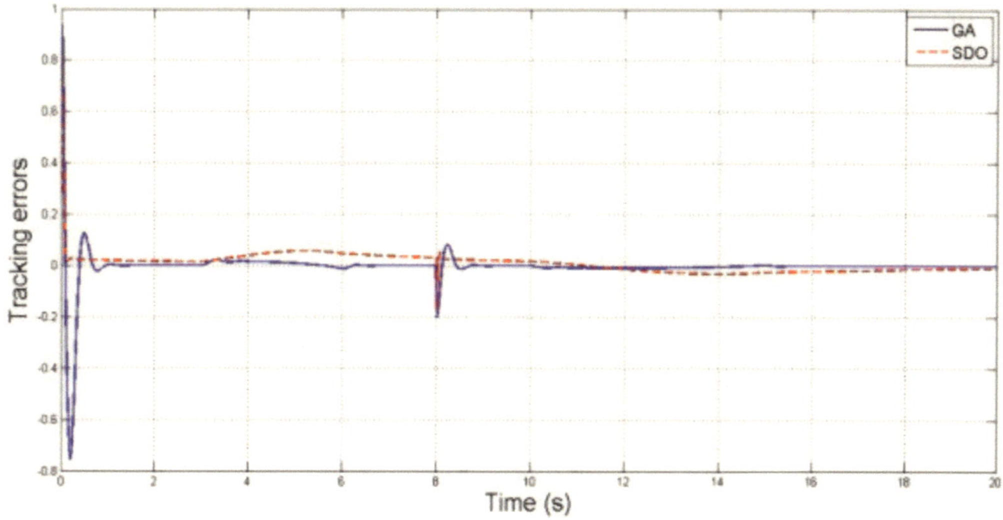

Figure 8. Tracking errors.

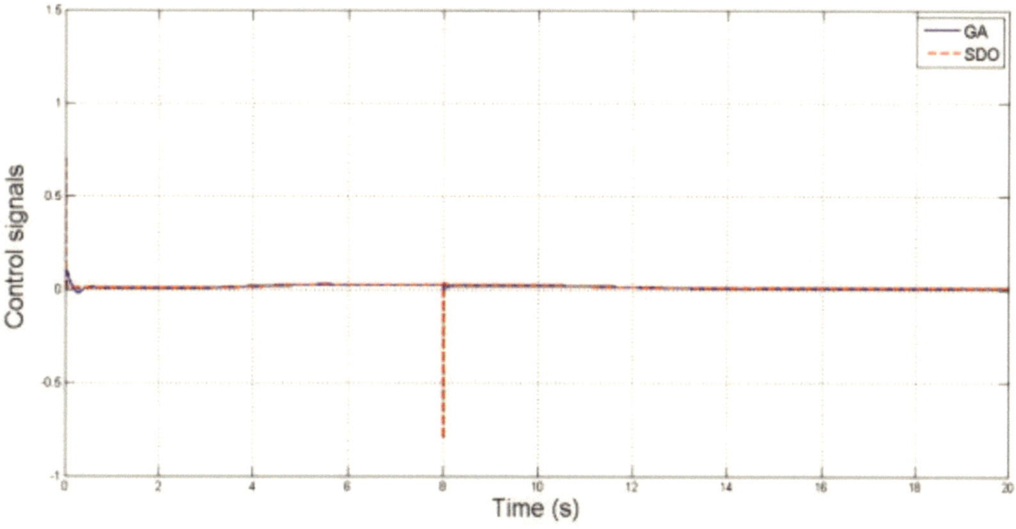

Figure 9. Control signals.

Figures 8 and **9** show the responses with Simulink design optimization technique and GA tuning using ISE criterion. Based on a comparative analysis, better results were there obtained with the GA tuning method.

All results, for obtained scaling factors values, are acceptable and show the effectiveness of the proposed GA tuning method in terms of the tracking desired trajectory with disturbances rejection in comparison with the SDO technique.

6. Conclusion

In this chapter, an optimization technique was introduced to tune the parameters of PID-type fuzzy logic controller (FLC). The idea is to use the genetic algorithm (GA)-based heuristic optimization technique in order to solve highly complex problems. In order to specify more robustness and performance of the proposed GA-tuned PID-type FLC, optimization criteria such as integral square error (ISE) is considered.

The proposed controller is applied to an electronic throttle valve (ETV) in the discrete-time framework in order to track a desired trajectory starting from a flat output generated using flatness property. The performance comparison with the Simulink design optimization (SDO) technique shows the efficiency of the proposed GA-tuned approach in terms of tracking a desired trajectory with disturbances rejection.

Author details

Wafa Gritli*, Hajer Gharsallaoui and Mohamed Benrejeb

*Address all correspondence to: wafa_gritli@yahoo.fr

National Engineering School of Tunis, Tunis, Tunisia

References

[1] Guzelkaya M, Eksin I, Yesil E. Self-tuning of PID-type fuzzy logic controller coefficients via relative rate observer. Engineering Applications of Artificial Intelligence. 2003;**16** (3):227–236

[2] Qiao WZ, Mizumoto M. PID type fuzzy controller and parameters adaptive method. Fuzzy Sets and Systems. 1996;**78**:23–35

[3] Woo ZW, Chung HY, Lin JJ. A PID type fuzzy controller with self-tuning scaling factors. Fuzzy Sets and Systems. 2000;**115**:321–326

[4] Toumi F, Bouallègue S, Haggège J, Siarry P. Differential search algorithm-based approach for PID-type fuzzy controller tuning. In: International Conference on Control, Engineering & Information Technology (CEIT'14); 2014. pp. 329–334

[5] Gritli W, Gharsallaoui H, Benrejeb M. PID-type fuzzy scaling factors tuning using genetic algorithm and Simulink design optimization for electronic throttle valve. In: International Conference on Control, Decision and Information Technologies (CoDIT'16); 6-8 April; Malta. 2016. pp. 216–221

[6] Galichet S, Foulloy L. Fuzzy controllers: Synthesis and equivalences. IEEE Transactions on Fuzzy Systems. 1995;**3**:140–148

[7] Holland JJ. Adaptation in Natural and Artificial Systems, University of Michigan Press; 1975

[8] Kim JS, Kim JH, Park JM, Park SM, Choe WY, Heo H. Auto tuning PID controller based on improved genetic algorithm for reverse osmosis plant. International Journal of Computer, Electrical, Automation, Control and Information Engineering. 2008;**2**(11):3707–3712

[9] Gaing ZL. Particle swarm optimization approach for optimum design of PID controller in AVR system. IEEE Transactions on Energy Conversion. 2004;**19**:384–391

[10] Mahony TO, Downing CJ, Fatla K. Genetic algorithm for PID parameter optimization: minimizing error criteria. In: Process Control and Instrumentation; 26–28 July; 2000. pp. 148–153

[11] Fliess M, Levine J, Martin P, Rouchon P. On differentially flat non linear systems. In: Proc IFAC Symposium on Nonlinear Control Systems Design (NOLCOS), IFAC, Laxenburg, Austria; 1992, pp. 408–412

[12] Gharsallaoui H, Ayadi M, Benrejeb M, Borne P. Robust flatness-based multi-controllers approach. Studies in Informatics and Control. 2010;**19**(4):357–368

[13] Costin M, Schaller R, Maiorana M, Purcell J. An architecture for electronic throttle control systems. In SAE Technical Paper, SAE International; 2003.

[14] Lebbal MI, Chafouk H, Hoblos G, Lefebvre D. Modelling and identification of non-linear systems by a multimodel approach: Application to a throttle valve. International Journal Information and Systems Science. 2007;**3**:67–87

[15] Yang C. Model-based analysis and tuning of electronic throttle controllers. In: Visteon Corporation, SAE Paper; March 8-11; Detroit, Michigan. 2004.

[16] Aidi I, Ayadi M, Benrejeb M, Borne P. Flatness-based control of throttle valve using neural observer. International Journal of Research and Surveys. 2012;**12**:333–344

Section 2

Applications of Fuzzy in Energy and Power Systems

Applications of the Fuzzy Logic to the Energy Conversion Systems on Board of UAVs

Dinca Liviu and Corcau Jenica Ileana

Additional information is available at the end of the chapter

http://dx.doi.org/10.5772/67992

Abstract

This chapter intends to present some applications of the fuzzy controllers to the automatic control of the DC-to-DC converters type boost, buck and buck-boost. For the mathematical modelling of these converters, one used averaged models; taking into account that controlled parameter is the average output voltage. One considered only the continuous conduction mode in the mathematical model. For each converter, one makes a short description of the principal scheme and the functioning mode and presents the averaged model. One obtains the transfer functions and finally builds fuzzy controllers in order to stabilize the output voltage with respect to the input voltage variations. The control is realized by modifying the duty cycle of the PWM command pulses. Obtained systems, both in closed loop and in open loop, are implemented in MATLAB/Simulink and simulations results are also presented.

Keywords: DC-to-DC converters, fuzzy logic control, averaged models, UAV, power systems

1. Introduction

Type UAV (Unmanned Air Vehicle) platforms have realised in the last period a special development due to their large area of applicability. UAVs can perform a variety of missions like military observation, combat, crops observations, disaster areas surveillance, in conditions of very high econommic and energy efficiency against classical aviation. UAVs offers many advantages like reduced manufacturing and operating costs, low energy consumption, possibility to access safely in hazardous areas for manned aircrafts, ability to penetrate spaces more restricted, inaccessible to classic aircraft, and due to technological developments obtained recently, the ability to achieve continuous flight missions with durations of the order of several days to several weeks. In this sense, the development of UAVs is expected to play an important role in High Altitude Pseudo-Satellites (HAPS) with continuous flight duration of the order of 5 years.

Achieving such high flight duration was possible due to the use of hybrid power sources that use at least two types of energy sources. In this sense, the most used energy sources are presently high capacity batteries or super capacitors that are recharged from renewable energy sources, especially high-efficiency photovoltaic cells developed in the last period. Fuel cells present a high efficiency, so it represents an alternative for the electrically powered manned airplanes. This solution is extensively studied and planned to be used in the near future. Airbus Company has already developed an electric airplane for two flights that last approximately for 1 hour. Currently, it is working to develop a plane for four persons with hybrid power source (high capacity battery plus two-stroke combustion engine that acts as an electric generator). Flight duration is expected to be extended to two hours. Research department of Airbus is considering replacing the thermal engine with the fuel cell to further expand the duration of flight.

Titan Aerospace was tested in August 2013, a concept vehicle for future high-altitude SOLAR 50 and 60 UAVs. SUN SOLAR 50 was scheduled to fly in 2014. SOLAR 50 was conceived as a UAV which uses high-capacity batteries and 3000 solar cells as power sources. The maximum flight duration of 50 SOLAR is expected to be 5 years at last. It is designed to be equipped with telecommunications and recognition systems, atmospheric sensors, etc. SUN 60 is planned for a payload of 125 kg [1–3].

The High Altitude Long Endurance Boeing (HALE) UAS 2012 aims at developing high altitude UAS for missions to ensure transmissions for the disposal of existing infrastructure and prolonged surveillance missions to the areas of interest.

Phantom Eye Project uses hydrogen internal combustion engine with hydrogen stored cryogenically. Solar Eagle Project, of Boeing as well, expects the use of high efficiency solar cells and SOFC type fuel cells. Wingspan aircraft is expected to be 120 m and propulsion will use six electric motors with permanent magnets [1–3].

Project Zephyr 7 High Altitude Pseudo-satellites from Airbus Defence & Space is to be used for surveillance, communications and monitoring services on surfaces in the order of tens of thousands km^2. This UAV holds the record for the longest flight—336 hours. Solar energy has been used for the propulsion and storage in lithium-ion battery with 3 kWh capacities to drive at night [1–3].

Since the parameters of power sources such as solar cells or fuel cells vary greatly in relation to the operating conditions, and loads useful on board UAVs (cameras, sensors, communications systems, etc.) usually require stabilized voltage, power buses of UAVs need to be maintained at a constant voltage.

This can be achieved by using power converters to adapt output sources to the requirements of bus power sources. Most sources on board UAVs are DC sources and loads are mostly (except some propulsion motors) of DC types. Following are required DC-to-DC converter types which have a higher efficiency and weight as low as possible aboard UAVs. The most convenient in this sense are the converters that have not transformed, part of this class are buck, boost and buck-boost converters. They are suitable for use on UAVs because they adapt the power source voltage to the power bus voltage by simply varying the duty cycle of the command pulses.

For the converters control, there are many techniques in literature, more or less sophisticated, among these are fuzzy techniques. These techniques offer a robust control and have an advantage which is related to human logic that these techniques do not require to know a detailed mathematical model of the system and can be tuned relatively easily granted by numerical simulations.

In this chapter, development of models used in the study of electrical systems of UAVs follows the modelling of three types of DC-DC converters, used on board UAVs, and development of fuzzy controllers is used to stabilize the output voltage.

Mathematical models developed for these converters follow the behaviour in terms of the averaged parameters (voltage and current) and neglect their ripples.

Although all models are being averaged, technique for obtaining these models is the introduction of the circuit voltage or current sources to make up in terms of DC parameters switch, coil or diodes present in converters. We consider this method more viable in terms of logic than writing equations split by two periods of operation of the converter and then their average as shown in Ref. [4].

2. Boost converter

The principal scheme of the boost converter is shown in **Figure 1a** and **b** equivalent scheme for continuous mode is presented. Functioning of this converter contains two steps. In the first step when the switch is in position 1, the inductor current rises and the inductor accumulate energy in its electromagnetic field. In the second step when the switch is in position 2, the inductor is now in series with the input source and pushes its energy on the capacitor. Inductor current drops, so the inductor behaves as voltage source in series with the input source and the capacitor will be charged to a voltage higher than the input voltage. For this reason, the boost converter raises the input voltage [5, 6].

Diagram in **Figure 1a** is equivalent in terms of DC components with the diagram in **Figure 1b**. Given that, coil behaviour was modelled in terms of averaged voltage with a voltage source dependent on the duty cycle of the control pulses $U_L(d)$, and the current through the switch was modelled by a current source dependent on the duty cycle of control pulses $I_T(d)$.

Coil voltage was found to be $U_L = \frac{d \cdot U_{in}}{1-d}$ so as to respect the converter transfer characteristic in steady state. Regarding the currents, it was considered that I_T = I_D in terms of the average

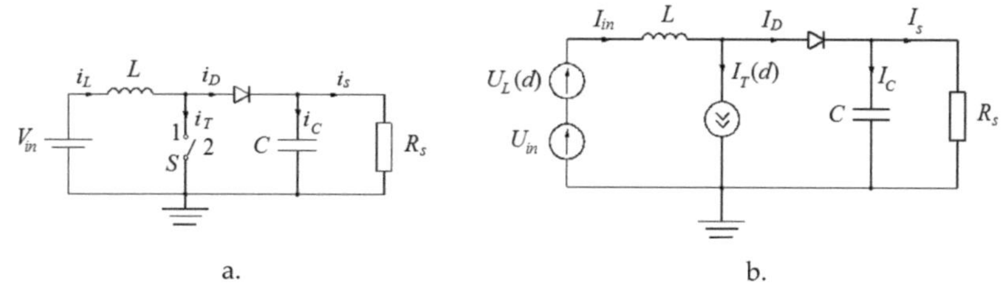

Figure 1. Boost converter. (a) Principal scheme, (b) Equivalent scheme for continuous mode.

values. In stationary regime, coil L behaves like a voltage source $U_L(d)$. At a duty cycle variation, in transitory regime. In these conditions, one can write relations:

$$\begin{cases} \dfrac{dI_{\text{in}}}{dt} = -\dfrac{1}{L}U_C + \dfrac{U_{\text{in}}}{L(1-d)} \\ \dfrac{dU_C}{dt} = \dfrac{1}{2C}I_{\text{in}} - \dfrac{U_C}{CR_S} \end{cases} \tag{1}$$

For linearization, notations, $I_{\text{in}} = I_{\text{in}0} + \Delta I_{\text{in}}$, $U_C = U_{C0} + \Delta U_C$, $U_{\text{in}} = U_{\text{in}0} + \Delta U_{\text{in}}$, $d = d_0 + \Delta d$ have been used. Relation obtained after linearization is (2)

$$\begin{cases} \begin{bmatrix} \Delta I_{\text{in}} \\ \Delta U_C \end{bmatrix} = \begin{bmatrix} 0 & -\dfrac{1}{L} \\ \dfrac{1}{2C} & -\dfrac{1}{CR_S} \end{bmatrix} \begin{bmatrix} \Delta I_{\text{in}} \\ \Delta U_C \end{bmatrix} + \begin{bmatrix} \dfrac{1}{L(1-d)} & \dfrac{U_{\text{in}}}{L(1-d)^2} \\ 0 & 0 \end{bmatrix} \begin{bmatrix} \Delta U_{\text{in}} \\ \Delta d \end{bmatrix} \\ [\Delta U_{\text{out}}] = [0 \quad 1] \begin{bmatrix} \Delta I_{\text{in}} \\ \Delta U_C \end{bmatrix} + [0 \quad 0] \begin{bmatrix} \Delta I_{\text{in}} \\ \Delta d \end{bmatrix} \end{cases} \tag{2}$$

Using this system, one can deduce the transfer functions given in Eqs. (3) and (4)

$$H_1(s) = \frac{\Delta U_{\text{out}}(s)}{\Delta U_{\text{in}}(s)} = \frac{\frac{1}{2LC(1-d)}}{s^2 + s\frac{1}{CR_S} + \frac{1}{2LC}} = \frac{k_1}{(s-s_1)(s-s_2)} \tag{3}$$

$$H_2(s) = \frac{\Delta U_{\text{out}}(s)}{\Delta d(s)} = \frac{\frac{U_{\text{in}}}{2LC(1-d)^2}}{s^2 + s\frac{1}{CR_S} + \frac{1}{2LC}} = \frac{k_2}{(s-s_1)(s-s_2)} \tag{4}$$

In **Figure 2**, simulation scheme of boost converter in open loop is shown. They were used ΔU_{in} and Δd as input, the output voltage was considered as ΔU_{out}. Simulated boost converter parameters were as follows $L = 47$ μF, $C = 2.2$ mF, $R_S = 1\ \Omega$, $d = 0.5$, $U_{\text{in}} = 24$ V. In the simulations using transfer functions, variations of output voltage about output voltage reference,

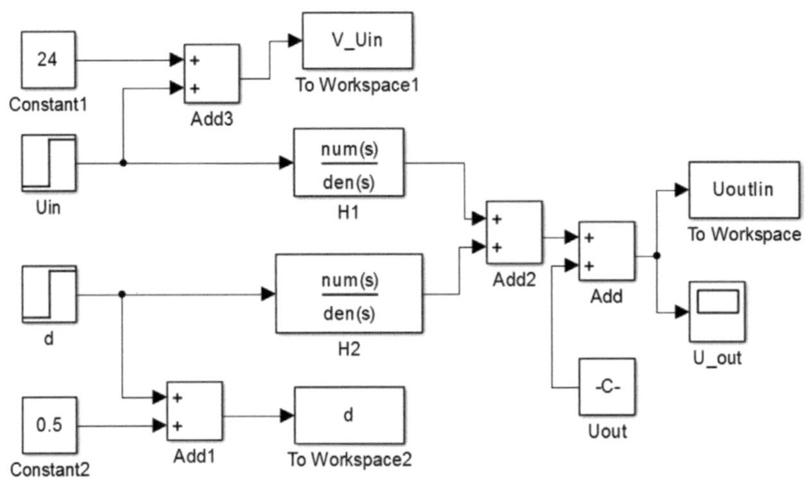

Figure 2. Open loop simulation scheme of boost converter.

when there are some variations in input voltage and duty cycle compared to reference values, are obtained. To capture the absolute values, some sum blocks are introduced in **Figure 2** whose variations overlap reference values achieved through the transfer or the absolute values of the input parameters. Simulation results are presented in **Figure 3a–c**.

Fuzzy control technique for DC-DC converters is applied in some studies in literature [7–10]. Other control techniques for these converters are studied in Refs. [11–17].

Closed loop scheme of the boost converter, implemented in MATLAB/Simulink is shown in **Figure 4**. One used a fuzzy controller with two inputs, output voltage error with respect to the

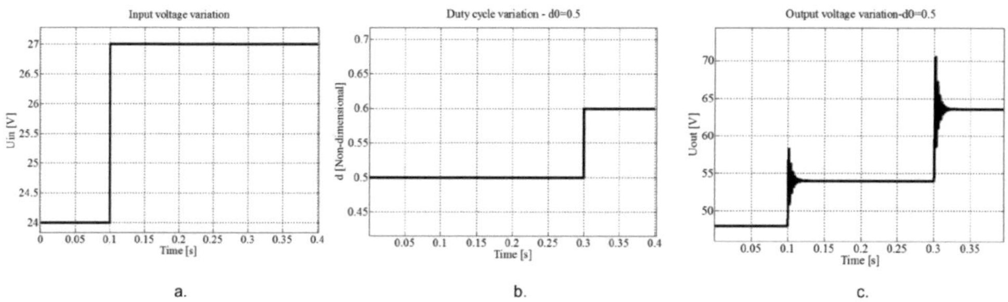

Figure 3. Boost converter response: (a) input voltage variation; (b) duty cycle variation; (c) output voltage variation.

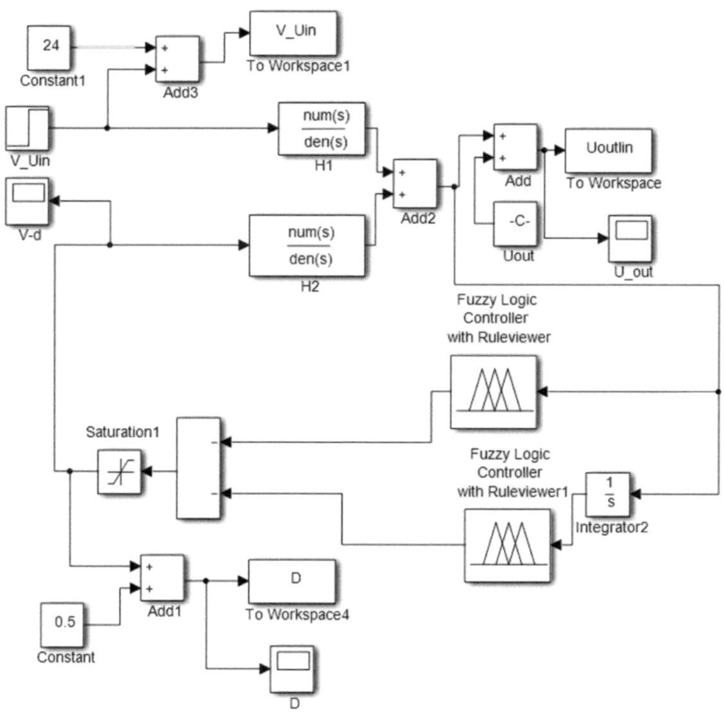

Figure 4. Closed loop simulation scheme for boost converter.

nominal output voltage and its integral, so one can say that this is a PI fuzzy controller. The advantage of the fuzzy controllers is that one can modify conveniently the control characteristics slopes in order to obtain the desired system behaviour. Usually one wishes a higher slope for points far from the origin and smaller slope for points near the origin. In this way, when the system is far from the nominal functioning point, one can accelerate its return to the nominal point. When it comes near the nominal point, parameters variations are sluggish in order to obtain a good stability in the nominal point. Fuzzy system in **Figure 4** has a particularity for an error and its integral are defined by two separate fuzzy controllers with four membership functions on input and output. In this way, one can simplify the controller tuning. There are necessarily eight inference rules (four rules for each controller) in contrast with the classical fuzzy controller with two inputs and one output having 16 inference rules. This difference increases once the membership function for each input increases. In **Figure 5**, the membership functions for fuzzy proportional controller P and its control surface are presented.

In **Figure 6**, the membership functions for fuzzy integrator controller I and its control surface are presented. Converter operation with these fuzzy controllers has been tested to a step type signal and ramp type signal. In hybrid power systems of UAVs containing fuel cell or solar cells, parameters variations are not very sudden, so it is exciting to study the converters behaviour at sluggish signals, such as the ramp type which is taken into consideration here. In **Figure 7**, the boost converter behaviour at step signal is shown and in **Figure 8**, its behaviour to a ramp signal with a slope of 3 V/s is shown.

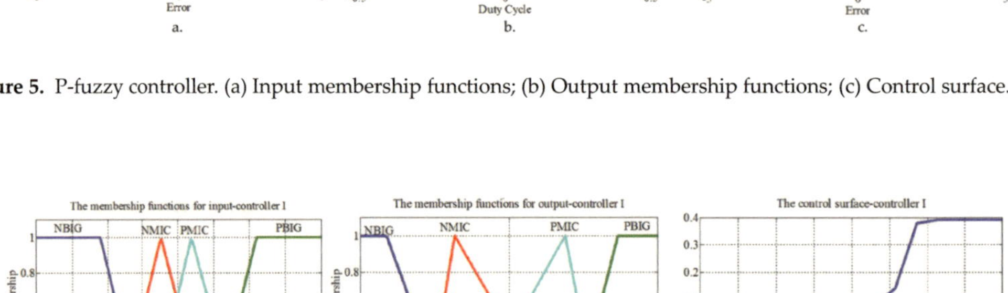

Figure 5. P-fuzzy controller. (a) Input membership functions; (b) Output membership functions; (c) Control surface.

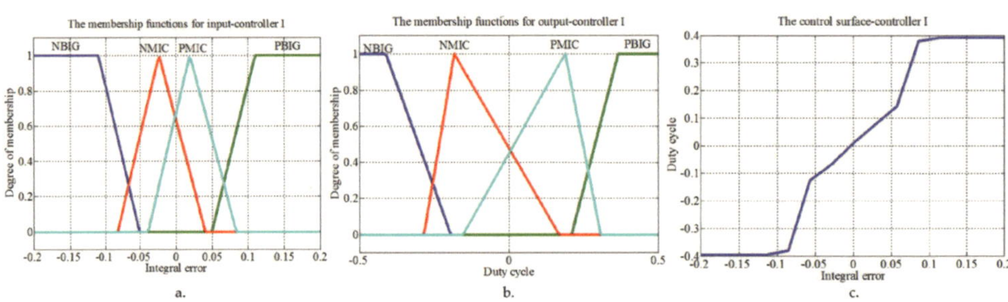

Figure 6. I-fuzzy controller. (a) Input membership functions; (b) Output membership functions; (c). Control surface.

http://dx.doi.org/10.5772/67992

Linguistic terms for both the controllers and for both inputs and outputs are NBIG (large negative), NMC (small negative), PMC (small positive) and PBIG (big positive). For both controllers, inference rules are: If input is NBIG, then output is NBIG; If input is NMC, then output is NMC; If input is PMC, then output is PMC; If input is PBIG, then output is PBIG.

For defining the controllers in this case, it is intended to simplify them as much so as to reduce the computation time required. In this purpose, one eliminated the membership function corresponding to linquistic term ZERO, which is used in the usual manner in fuzzy controllers. By conveniently change of the membership function one obtained control surfaces in concordance with the followed strategy - lower slope around the origin and greater slope farther origin.

The behaviour of the converter closed loop is one adequate operation of power systems on UAVs. Peak voltages at step input are attenuated and voltage returns quickly enough to the prescribed value. For the ramp input, deviation from set point is only 0.025 V, which is a very good performance.

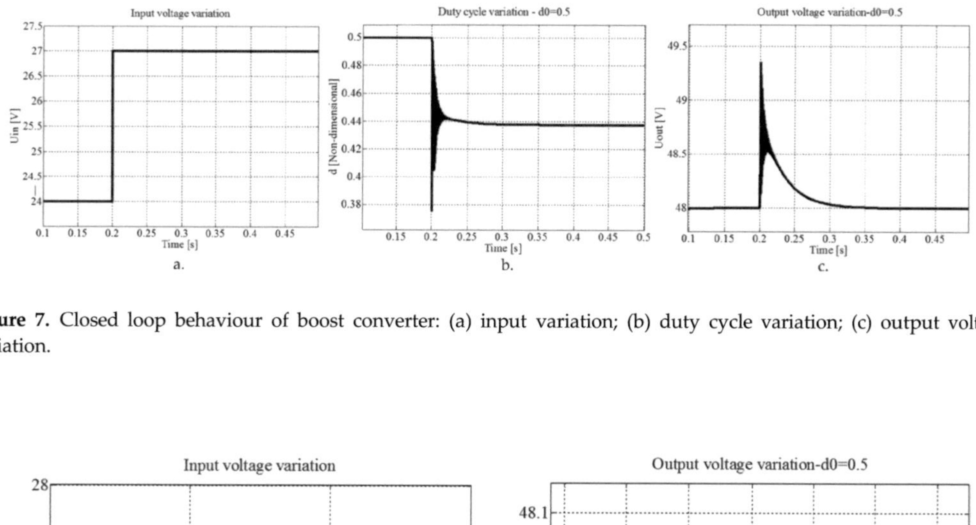

Figure 7. Closed loop behaviour of boost converter: (a) input variation; (b) duty cycle variation; (c) output voltage variation.

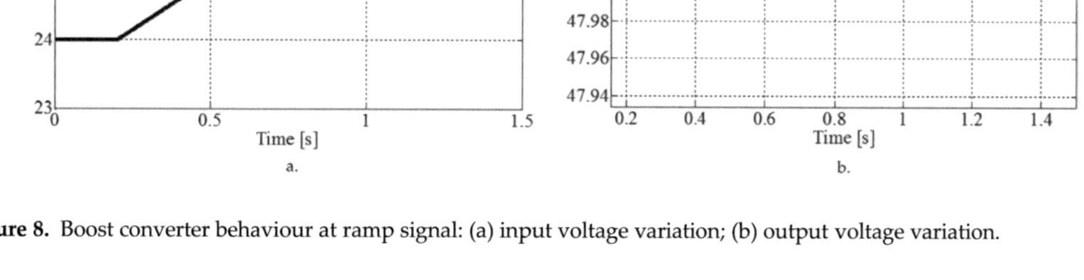

Figure 8. Boost converter behaviour at ramp signal: (a) input voltage variation; (b) output voltage variation.

3. Buck converter

Principal scheme of the buck converter is shown in **Figure 9**. This converter has also two steps in its functioning. In the second step when the switch is in position 1, the capacitor loads through the inductor and inductor current increases, so the inductor behaves as a voltage source with inverse polarity with respect to the input voltage. In this manner on the capacitor, one can obtain a voltage smaller than the input voltage. In the second step when the switch is in position 2, energy accumulated in the inductor is pushed on the capacitor and thus Inductor current decreases, so it behaves as a voltage source with the same polarity as the input voltage but with a smaller value. So, the buck converter decreases the input voltage. In **Figure 10**, equivalent scheme for continuous mode is presented.

Following **Figure 10**, one can write the following relations: $U_{\text{in}} - U_L = U_{\text{out}} = d{\cdot}U_{\text{in}},\ i_T + i_D = i_L = i_S,\ U_{in}{\cdot}i_T = U_{out}{\cdot}i_S$. After several transformations, following system of equations give:

$$\begin{cases} U_{\text{out}} = U_{\text{in}}d - L\dfrac{d}{dt}\left(\dfrac{I_T}{d}\right) \\ \dfrac{U_{\text{out}}}{R_S} = I_T\dfrac{1}{d} - C\dfrac{dU_{\text{out}}}{dt} \end{cases} \tag{5}$$

By linearization, like the case of the boost converter, resulted linearized model is:

$$\begin{cases} \Delta\dot{I}_T = -\dfrac{d_0}{L}\Delta U_{\text{out}} + \dfrac{d_0^2}{L}\Delta U_{\text{in}} + \dfrac{V_{\text{in}0}\cdot d_0}{L}\Delta d + \dfrac{I_{T0}}{d_0}\Delta\dot{d} \\ \Delta\dot{U}_{\text{out}} = \dfrac{1}{C\cdot d_0}\Delta I_T - \dfrac{1}{R_S\cdot C}\Delta U_{\text{out}} - \dfrac{I_{T0}}{C\cdot d_0^2}\Delta d \end{cases} \tag{6}$$

Applying Laplace in system zero initial conditions, Eq. (6) results in transfer functions of the form

$$\frac{\Delta U_{\text{out}}}{\Delta U_{\text{in}}} = \frac{R_S\cdot d_0}{s^2(R_S\cdot L\cdot C) + L\cdot s + R_S} \tag{7}$$

$$\frac{\Delta U_{\text{out}}}{\Delta d} = \frac{R_S\cdot U_{\text{in}0}}{s^2(R_S\cdot L\cdot C) + L\cdot s + R_S} \tag{8}$$

Buck converter parameters considered in this chapter are $L = 56.5\ \mu\text{F},\ C = 166.7\ \mu\text{F},\ R_S = 2\ \Omega$, $d = 0.5,\ U_{\text{in}0} = 24$ V, $U_{\text{out}0} = 12$ V. Simulation scheme for the buck converter with transfer functions, open loop is shown in **Figure 11**. Simulation results are presented in **Figure 12**. Utilizing a closed loop scheme and the same fuzzy controllers, like in the case of boost converter shown in **Figure 4**, closed-loop simulation results are obtained shown in **Figure 13**. For the buck converter control, a classical PI fuzzy controller has been designed with two inputs, the error between the output voltage and the voltage prescribed and integral of this error. Simulation scheme in this case is the one in **Figure 14**. The membership functions and control surface of this

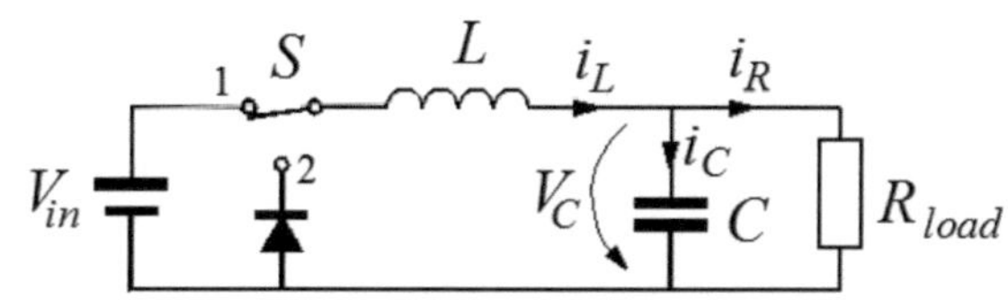

Figure 9. Principal scheme of buck converter.

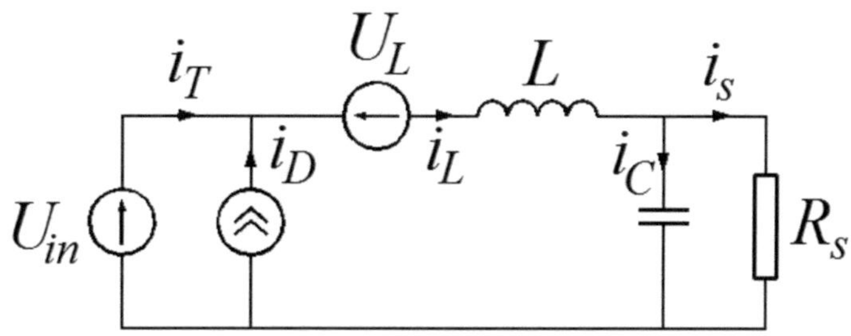

Figure 10. Equivalent scheme for continuous mode of buck converter.

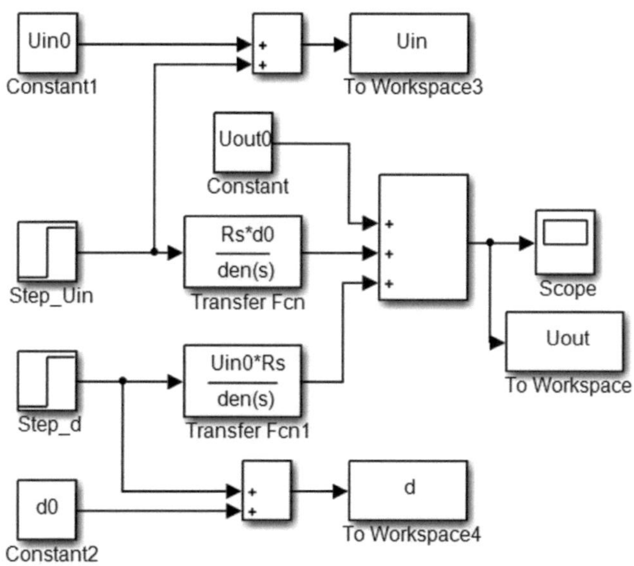

Figure 11. Open loop simulation scheme for buck converter.

controller are presented in **Figure 15**. The inference rules of fuzzy PI controller here are defined in table of this **Figure 16**. Buck converter behaviour with this control is shown in **Figure 17**. Note that the peak voltage obtained from step input is reduced to one third of the peak voltage obtained with the controller in the previous case (fuzzy controllers P and I are put in parallel), and the time to restore the prescribed value is also diminished.

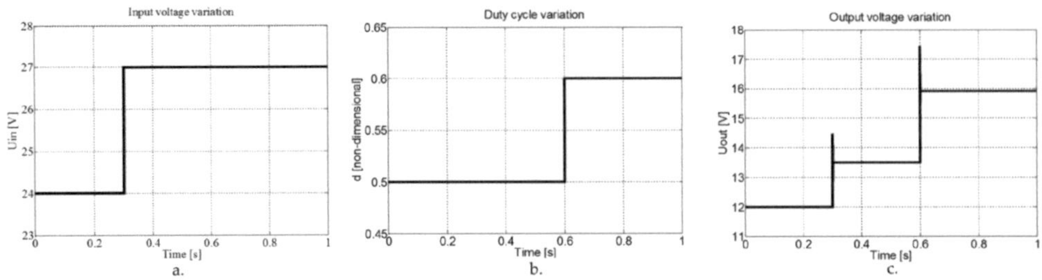

Figure 12. Open loop behaviour of buck converter: (a) input voltage variation; (b) duty cycle variation; (c) output voltage variation.

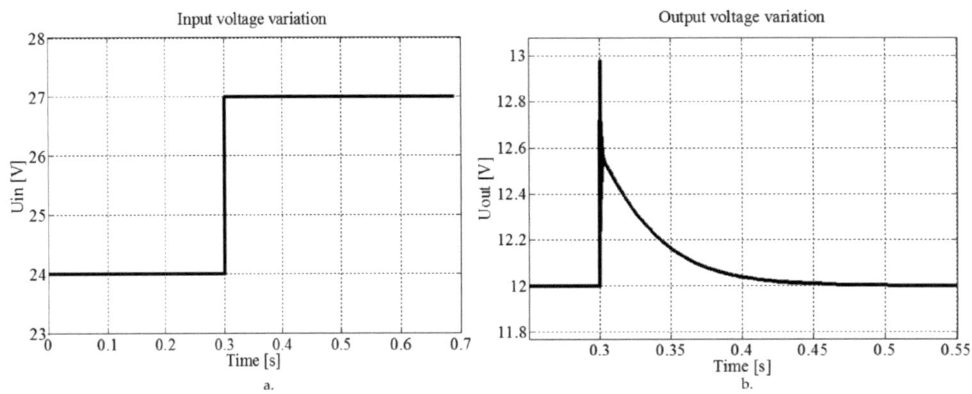

Figure 13. Closed loop behaviour of buck converter with two fuzzy controllers in parallel P+I: (a) input voltage variation; (b). output voltage variation.

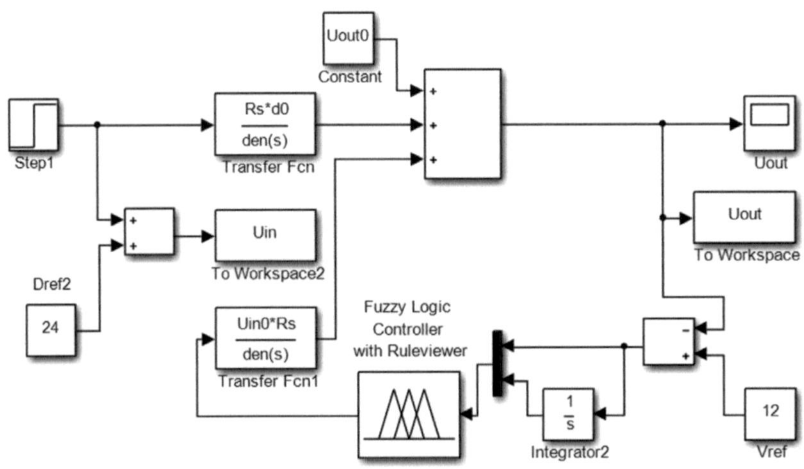

Figure 14. Closed loop simulation scheme for buck converter with one PI fuzzy controller with two inputs.

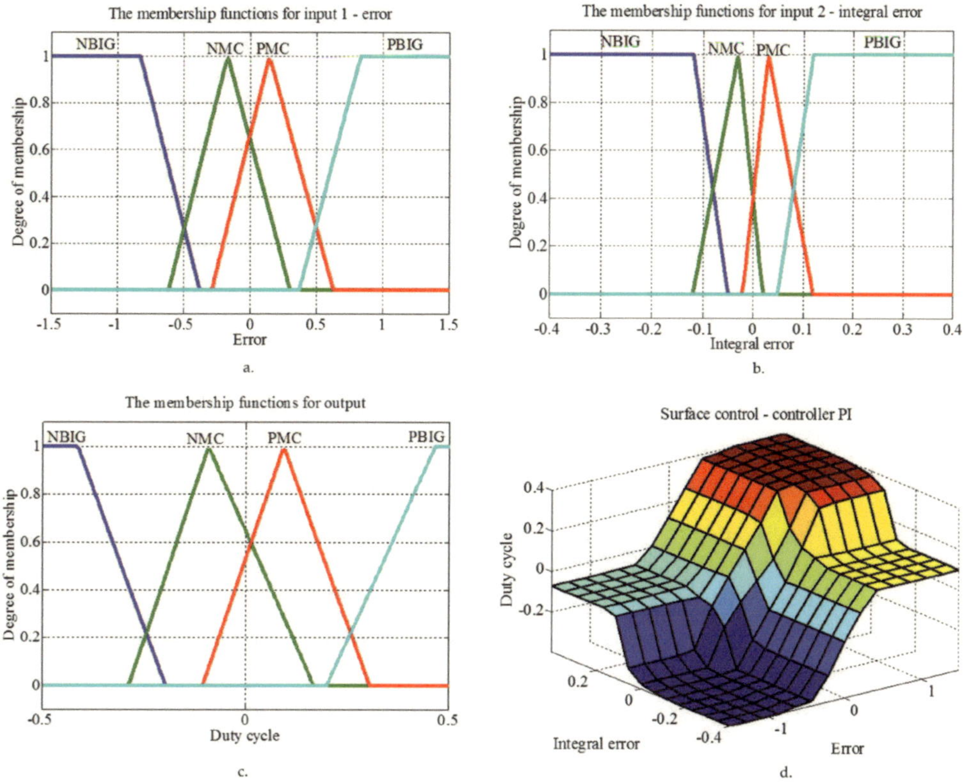

Figure 15. PI fuzzy controller with two inputs: (a) membership functions for error input; (b) membership functions for integral error input; (c) membership functions for output; (d) control surface.

		ERROR			
		NBIG	NMC	PMC	PBIG
ERROR INTEGRAL	NBIG	NBIG	NBIG	PMC	PMC
	NMC	NBIG	NBIG	PMC	PBIG
	PMC	NBIG	NMC	PBIG	PBIG
	PBIG	NMC	NMC	PBIG	PBIG

Figure 16. Inference rules for PI fuzzy controller with two inputs.

So, we can say that this second fuzzy PI controller with two inputs provides better behaviour, but at the expense of greater computation time, so microcontroller that is implemented should have better performance.

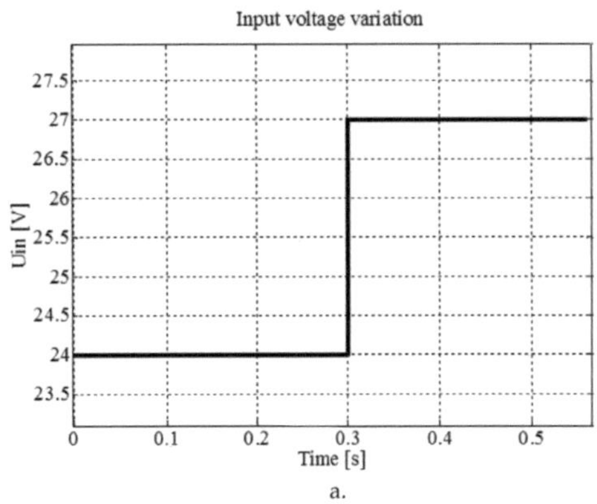

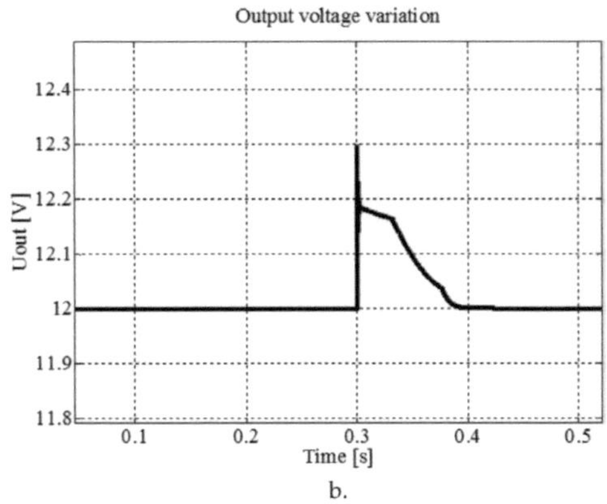

Figure 17. Closed loop behaviour of buck converter with PI fuzzy controller with two inputs: (a) input voltage variation; (b) output voltage variation.

4. Buck-boost converter

Principle scheme of the buck-boost converter is in **Figure 18**. This converter outputs a higher or a lower voltage than the input voltage but with inverse polarity, so it is known as controller-inverter. It also functions in two steps. In step 1, when the switch is in position 1, the diode is inversely polarized and the current flows from the source through the inductor and charges its electromagnetic field. In the second step, when the switch is in position 2, the energy stored in the inductor is released on the capacitor and the output load with inverse polarity of the voltage.

From the point of view of the average values, it was considered a voltage drop occurs on the switch. Also one considered the coil average voltage is equal with continuous output voltage. Following **Figure 19**, one can write the following relations: $i_S + i_C = i_D$, $i_D + i_T = i_L$, $U_L = U_{\text{in}} \cdot \frac{d}{d-1}$, $i_T(d-1) = i_D \cdot d$. It has considered among others stationary input-output characteristics of buck-boost converter. In addition, one can write the following equations:

$$U_{\text{in}} + U_T - U_L = -L\frac{di_L}{dt} \tag{9}$$

$$U_{\text{in}} = -L\frac{di_L}{dt} + R_S i_S \tag{10}$$

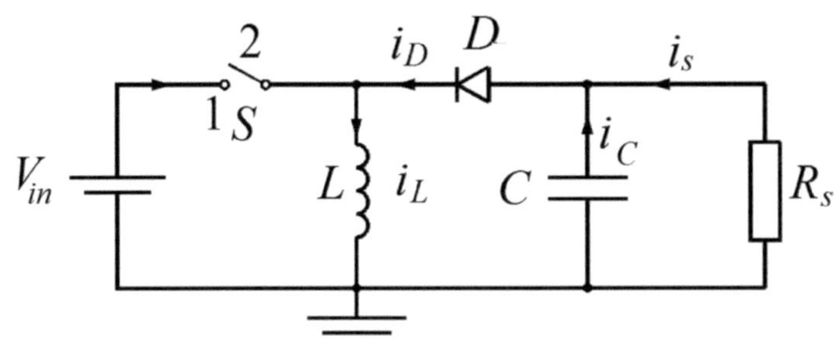

Figure 18. Principle scheme of buck-boost converter.

http://dx.doi.org/10.5772/67992

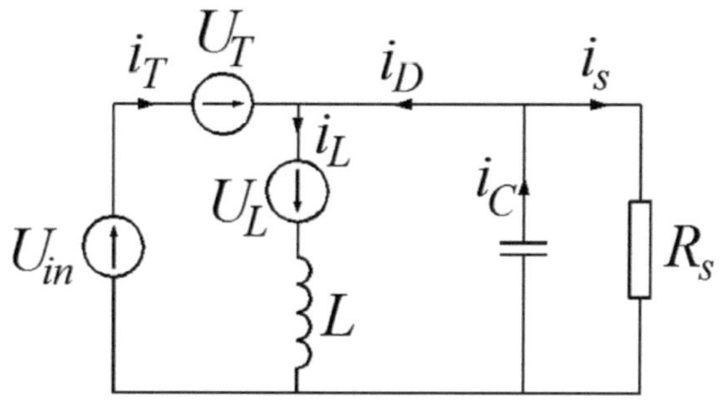

Figure 19. Equivalent scheme for continuous mode of buck-boost converter.

$$C\frac{dU_{\text{out}}}{dt}+\frac{U_{\text{out}}}{R_S}+i_T=i_L \tag{11}$$

After several transformations, we obtain:

$$U_{\text{out}}-U_{\text{in}}\frac{d}{d-1}=-L\cdot C\frac{d^2U_{\text{out}}}{dt^2}\cdot\frac{1}{d-1}+\frac{L}{R_S}\frac{dU_{\text{out}}}{dt}\cdot\frac{1}{d-1} \tag{12}$$

By linearization and applying Laplace transform in null initial conditions, transfer functions result as follows:

$$\frac{\Delta U_{\text{out}}(s)}{\Delta U_{\text{in}}(s)}=\frac{\frac{d_0}{d_0-1}}{-L\cdot C\frac{d_0}{d_0-1}s^2-\frac{L}{R_S\cdot}\cdot\frac{d_0}{d_0-1}s+1} \tag{13}$$

$$\frac{\Delta U_{\text{out}}(s)}{\Delta d(s)}=\frac{-U_{\text{in}0}\frac{1}{(d_0-1)^2}}{-L\cdot C\frac{d_0}{d_0-1}s^2-\frac{L}{R_S\cdot}\cdot\frac{d_0}{d_0-1}s+1} \tag{14}$$

Parameters used in the simulation buck-boost converter are: $L=56.5\ \mu\text{F},\ \ C=166.7\ \mu\text{F}$, $R_S=2\ \Omega,\ d=0.5,\ U_{\text{in}0}=15\ \text{V},\ U_{\text{out}0}=-15\ \text{V}$. Scheme simulation for buck-boost converter in open circuit is similar to that for buck converter in **Figure 11**. Responses to step signals applied to both voltage input, and on the duty cycle are shown in **Figure 20**. Although for this converter two versions of fuzzy controllers were made, one version with two fuzzy controllers disposed in parallel, one for the proportional component and one for the integrative component, and a second version with a single fuzzy controller with two inputs (proportional and integrative) and one output (duty cycle variation). For the first variant, fuzzy controllers are shown in **Figures 21** and **22**. Buck-boost converter case has chosen five membership functions, one of which corresponds to the term linguistic ZERO (ZE in **Figures 21** and **22**), the other membership functions corresponding to the same linguistic terms as with previous fuzzy regulators. The inference rules in this case were the same for both controllers: If the input is NBIG, then output is NBIG; If the input is NMIC, then output is NMIC; If the input is ZE, then output is ZE; If the input is PMIC, then output is PMIC; If the input is PBIG, then output is PBIG.

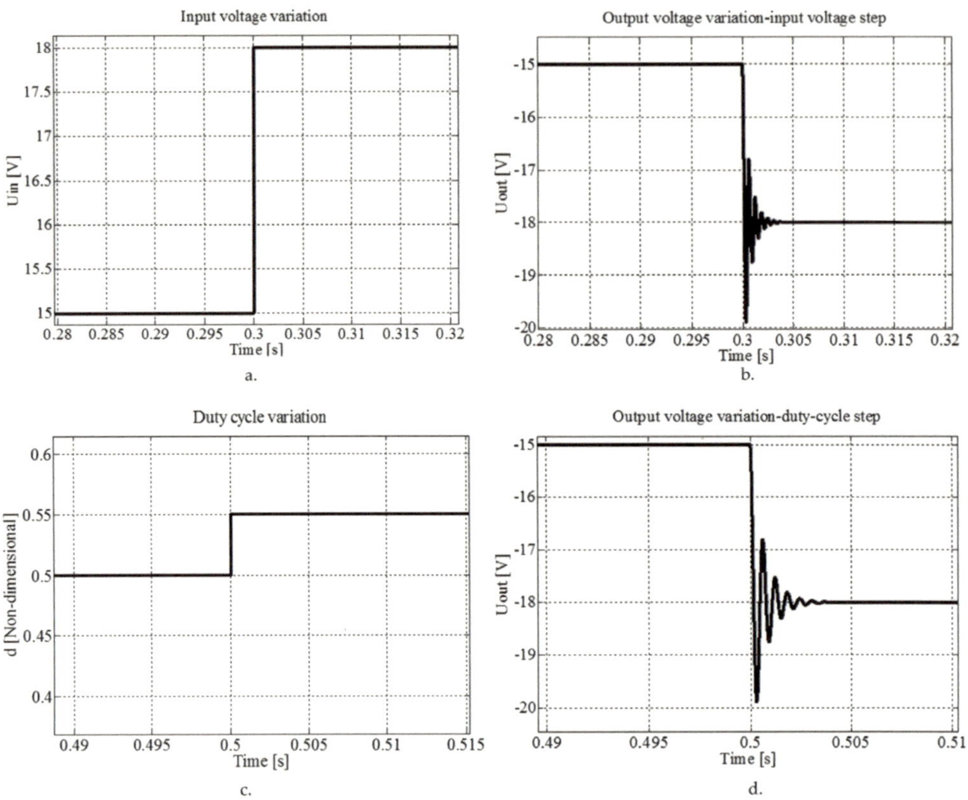

Figure 20. Buck-boost converter behaviour in open loop: (a) input voltage step; (b) output voltage at input voltage step; (c) duty cycle step; (d) output voltage at duty cycle step.

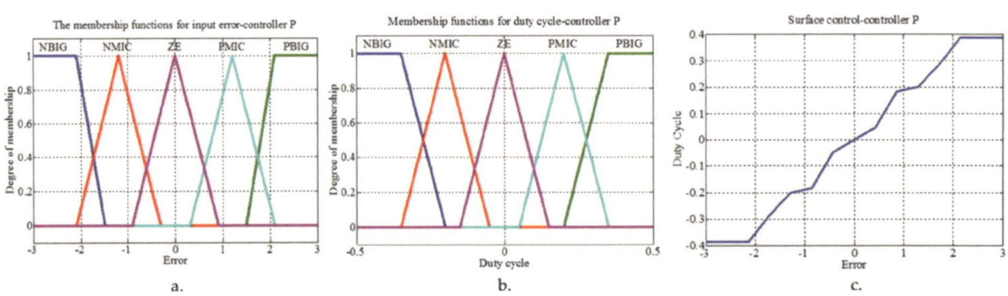

Figure 21. P fuzzy controller for buck-boost converter: (a) membership functions for input; (b) membership functions for output; (c) control surface.

Closed-loop simulation scheme for this case is similar to that for boost converter in **Figure 4** but with transfer functions and controllers shown in buck-boost converter. Buck-boost converter's response in a closed loop and step input are shown in **Figure 23**.

The behaviour of the converter for sinusoidal input voltage variation has been studied in this case which is shown in **Figure 24**.

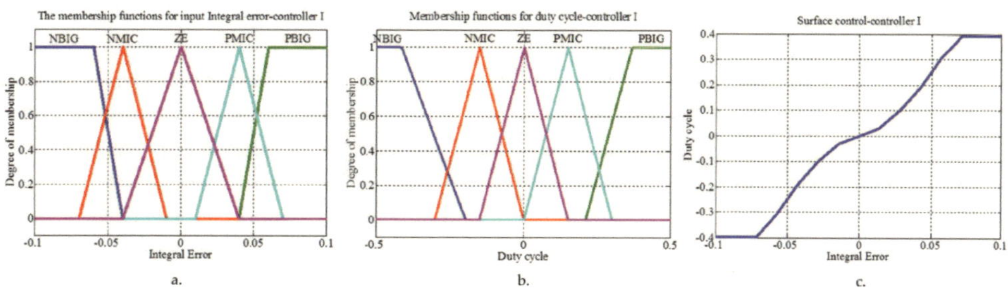

Figure 22. I fuzzy controller for buck-boost converter: (a) membership functions for input; (b) membership functions for output; (c) control surface.

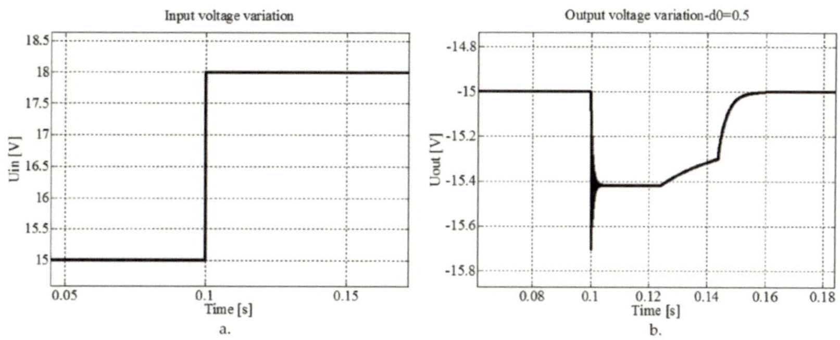

Figure 23. Closed loop behaviour of buck-boost converter with two fuzzy controllers in parallel (P+I): (a) input voltage variation; (b) output voltage variation.

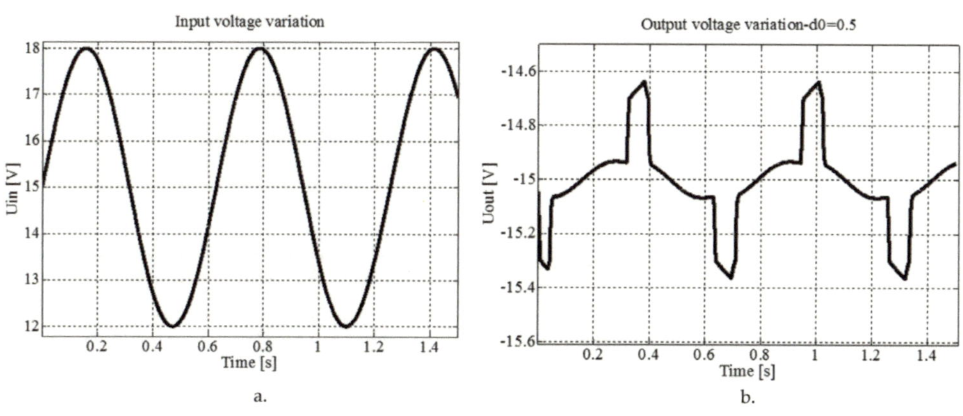

Figure 24. Closed loop behaviour of buck-boost converter with two fuzzy controllers in at sinusoidal input: (a) input voltage variation; (b) output voltage variation.

We notice here a better behaviour of fuzzy controller at step signal, the restore time to prescribed value is lower than the other controller presented converters but notice a worsened behaviour for sinusoidal signal response. At zero-crossing of the input signal, one obtained jumps in output voltage. The amplitude is not large, about 0.3 V, but these jumps means higher

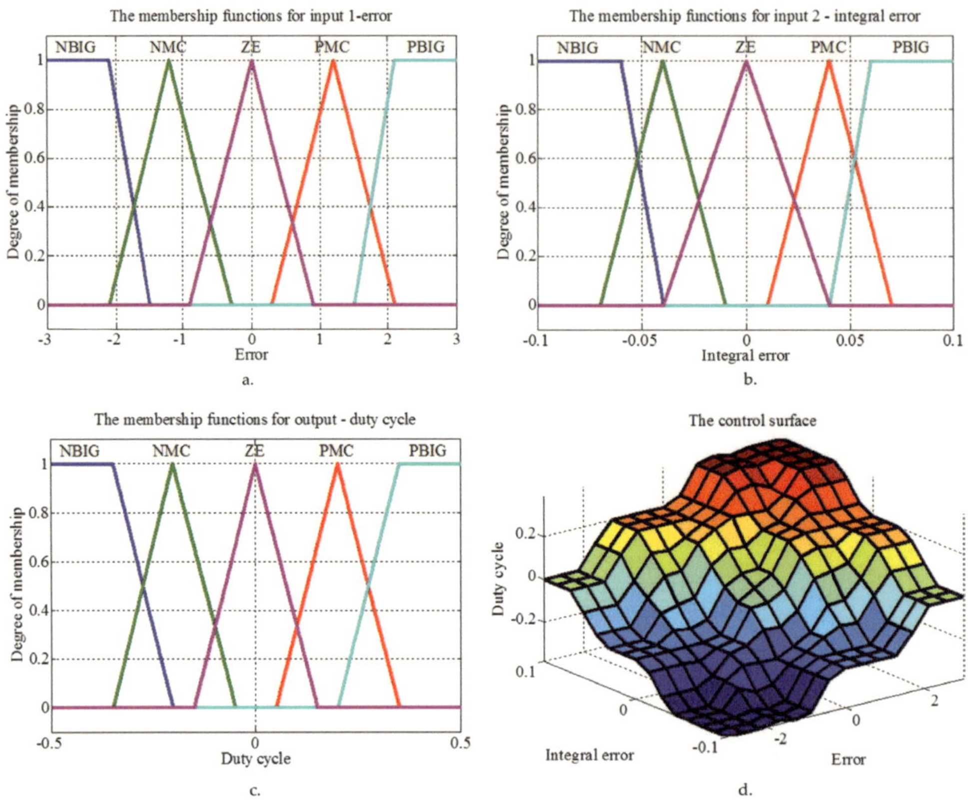

Figure 25. PI fuzzy controller with two inputs for buck-boost converter: (a) membership functions for proportional input; (b) membership functions for integral input; (c) membership functions for output; (d) control surface.

		ERROR				
		NBIG	NMC	ZE	PMC	PBIG
ERROR INTEGRAL	NBIG	NBIG	NBIG	NMC	NMC	ZE
	NMC	NBIG	NMC	NMC	ZE	PMC
	ZE	NMC	NMC	ZE	PMC	PMC
	PMC	NMC	ZE	PMC	PMC	PBIG
	PBIG	ZE	PMC	PMC	PBIG	PBIG

Figure 26. Inference rules for PI fuzzy controller.

http://dx.doi.org/10.5772/67992

harmonics induced on power bus and thus worsening power quality on board of the UAVs. A second version of fuzzy controller designed for buck-boost converter is that of a controller with two inputs and one output, with five membership functions on each input and the output. This fuzzy controller is shown in **Figure 25**. Inference rules for this fuzzy controller are presented in table in **Figure 26**.

One observes a good response at step input in this case; but at sinusoidal input, voltage peaks appear again at input zero crossing. That means again worse energy quality on board.

5. Conclusions

This paper presents some applications of the fuzzy control technique to DC-DC converters in usual configurations (boost, buck, buck-boost). For each converter, first averaged mathematical models are presented, obtained upon some equivalent schemes for continuous regime. Upon these averaged models, transfer functions for each converter are deduced. These transfer functions are applicable for the transitory regimes in the small perturbation hypothesis. One has to take into account that the averaged models and transfer functions are available for the continuous conduction regime for each converter. In order to obtain the continuous conduction regime, each of the studied converters needs a big enough switching frequency. This frequency does not appear explicitly in the transfer functions, but is determined by the circuit components.

Fuzzy controllers developed here for the studied converters are less usual. First, for the boost and buck converters, one used fuzzy controllers with four membership functions two on each input and output. In order to diminish the calculus volume for each functioning step as much as possible, one used triangular membership functions at the middle of the interval and trapezoidal membership functions at extremities. Although, one followed to reduce the number of membership functions.

For a more stable behaviour of the converters, one tried to reduce the control surface slope around the origin and to keep a bigger slope far from the origin, in order to accelerate converter return to the desired functioning point. By a convenient choose of the membership functions, one could reach this purpose in conditions of a small number of membership functions, as one can see in **Figures 5c**, **6c** and less in **Figure 15d**.

To reduce further the calculus volume at each functioning step and to simplify inference rules definition, one tested the possibility to decompose a PI fuzzy controller with two inputs in two simpler fuzzy controllers, each of them with one input. These new fuzzy controllers are disposed in parallel and one has as input the error between the real output voltage and the desired output voltage, and the other has as input the integral of this error. In this way, one can reduce the inference rules from $n \times n$ to $2 \times n$ where n is the number of the membership functions. Fuzzy controllers defined for the boost converter were used successfully to control the buck converter.

For buck converter, one designed a classical PI fuzzy controller with two inputs and one output, with four membership functions, two on each input an on the output (see **Figure 15**). Performances obtained with this controller are improved. The peaks on the output voltage are reduced, and time to return to the prescribed voltage is also reduced (see **Figure 17**). In **Figure 17**, one observes response slope changes when the controller passes from a control interval to another and the return acceleration when the integrative component reaches the limit of membership function PMC. One observes further response of slope diminishing when the error decreases and smooth approach of the real output voltage to the prescribed one.

For the buck-boost converter, one tested classical fuzzy converters with odd number of membership functions on each input (one of them corresponding to the linguistic term ZERO). One aimed to obtain for this controller a small slope of the control surface near the origin and a bigger slope far from it. Designed controllers can be seen in **Figures 21, 22** and **25**. Buck-boost converter behaviour is good when step input is used, as one can see in **Figures 23** and **27**. But when sinusoidal input is used, problems appear. In **Figures 24** and **27** one can see the output voltage peaks when the input passes through zero. That means worse quality of energy on board.

As an important conclusion, one can say that control with fuzzy technique offers a special flexibility to the controller design. One can reach very good performances even for simplified configurations of the controller. Possibility to obtain a convenient control surface for each

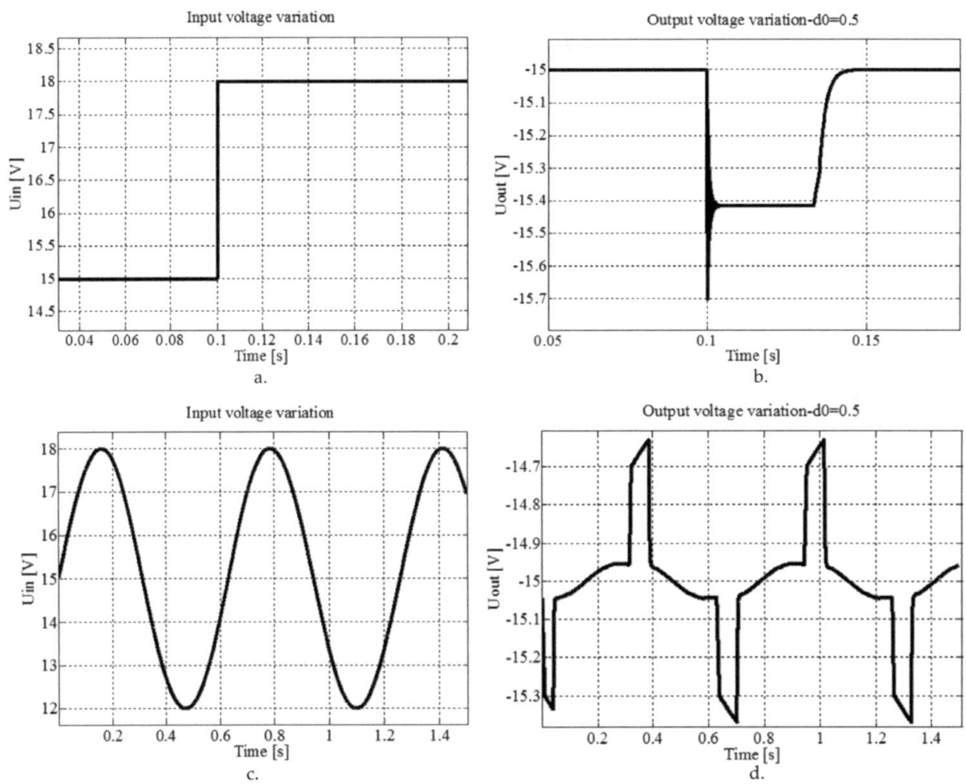

Figure 27. Buck-boost converter behaviour with PI fuzzy controller: (a) voltage step input; (b) response to step input; (c) sinusoidal voltage input; (d) response to sinusoidal input.

http://dx.doi.org/10.5772/67992

application is an advantage that cannot be reached with other control techniques. Although, in some cases, design of a good fuzzy controller needs some experience and testing of many variants, in order to select the better one. Simplistic definition of the membership functions and inference rules can lead either to inconvenient control surfaces or to linear ones. In the second case, the fuzzy controller could be replaced very well with a classical P or PI controller, with same performances.

Author details

Dinca Liviu* and Corcau Jenica Ileana

*Address all correspondence to: ldinca@elth.ucv.ro

Department of Electrical, Energetic and Aerospace Engineering, Faculty of Electrical Engineering, University of Craiova, Craiova, Romania

References

[1] P. O'Neil. Boeing High Altitude Long Endurance (HALE) UAS, available at https://higherlogicdownload.s3.amazonaws.com/AUVSI/656942e4-4448-41c3-877d-0c5f3ea40e63/UploadedImages/KnowledgeVault/HALE.pdf

[2] N. Owano. Titan Aerospace readies solar-powered, long-endurance UAVs; 22 August 2013. www.boeing.com/defense/phantom-eye [Internet].

[3] H. Ross. www.solarimpulse.com [Internet]; May 2009.

[4] F. Misoc. A comparative study of dc–dc converters' effects on the output characteristics of direct ethanol fuel cells and NiCd batteries [dissertation]. Thesis of Dissertation Kansas State University, Manhattan, Kansas; 2007.

[5] M. H. Rashid. Power Electronics Circuits, Devices, and Applications. 3rd edition. University of West Florida, Pearson Prentice Hall; 2003.

[6] R. W. Erickson, D. Maksimovic. Fundamentals of Power Electronics. 2nd edition. 912 p. Kluwer Academic Publishers 2001, Sixth Printing 2004, DOI: ISBN 0-7923-7270-0.

[7] G. Feng, W. Zhang, Y.-F. Liu. An Adaptive Current Mode Fuzzy Logic Controller for Dc-to-Dc Converters. Eighteenth Annual IEEE Applied Power Electronics Conference and Exposition; Miami Beach, Florida, 2003; pp. 983–989.

[8] P. Mattavelli, L. Rossetto, G. Spiazzi, P. Tenti. General-Purpose Fuzzy Controller for Dc/Dc Converters. In: Applied Power Electronics Conference, Dallas, Texas 1995.

[9] A. G. Perry, G. Feng, Y.-F. Liu, P.C. Sen. A New Design Method for PI-like Fuzzy Logic Controllers for DC-to-DC Converters. In: 35th Annual IEEE Power Electronics Specialists Conference; 2004; pp. 3751–3757.

[10] A. Gad, M. Farooq. Application of Fuzzy Logic in Engineering Problems. Proceedings of the 27th annual conference of the IEEE Industrial Electronics Society (IECON 01), Denver, CO, 29 November-2 December 2001; p. 2044–2049.

[11] B. Johansson. DC-DC converters-dynamic model design and experimental verification [thesis]. Lund University: Doctoral Dissertation in Industrial Automation Department of Industrial Electrical Engineering and Automation; 2004.

[12] M. Salimi, A. Zakipour. Direct voltage regulation of DC-DC buck converter in a wide range of operation using adaptive input-output linearization. In: IEEE Transactions on Electrical and Electronic Engineering, January 2015; pp. 85–91. DOI: ISSN 1931–4973.

[13] J. Chen, D. Maksimovic, R. Erickson. Buck-boost PWM converters having two independently controlled switches. In: Proc. IEEE Power Electronics Specialists Conference; Vancouver Canada, June 17–21, 2001; pp. 736–741.

[14] L. Guo. Design and implementation of digital controllers for buck and boost converters using linear and nonlinear control methods [dissertation]. Auburn University, Alabama; 2006.

[15] H. Fadali. Fuel cell distributed generation, power conditioning, control and energy management [dissertation]. Master of Applied Science in Electrical and Computer Engineering, Ontario, Canada; 2008.

[16] P. Hemachander, A. D. VimalRaj, M. Sudhakaran. Analysis design and implementation f soft single switched boost converter. In: International Journal of Computer Applications; 26(11), July 2011; pp. 38–46.

[17] P. S. Priambodo, D. Sukoco, W. Purnomo, H. Sudibyo, D. Hartanto., Electric Energy Management and Engineering in Solar Cell System. Chapter 12 in book "Solar cells – Research and Application Perspectives" editor Arturo Morales-Acevedo, published by INTECH on 6th March 2013; pp. 327–351.

Chapter 7

Fuzzy Logic Energy Management for a Residential Power System Using Renewable Energy Sources

Stefan Breban, Ioana Gros, Calin Marginean and
Petre Teodosescu

Additional information is available at the end of the chapter

http://dx.doi.org/10.5772/intechopen.69486

Abstract

A fuzzy logic energy management algorithm is proposed for a hybrid wind/photovoltaic (PV) power generation unit, an electric vehicle battery, and a heat pump for household applications. The proposed concept refers to two independent power systems—a light electric vehicle and a household that interact through light, interchangeable batteries; moreover, they are powered from a renewable energy system comprising PV panels, wind generator, and appropriate MPPT-based converters. The main features of the concept are the heat pump load that produces thermal energy, as the main electric load of the system, and the storage element that is alternately used by the vehicle, which can be recharged from renewable sources. The presented algorithm allows the implementation, by means of fuzzy tools, of an appropriate energy management control system in order to obtain maximum utilization of the renewable energy. The results show that most of the energy required to charge the battery and to feed the heat pump can be covered from renewable sources.

Keywords: fuzzy logic energy management, PVs, wind turbine, heat pump, electric vehicle, battery

1. Introduction

Over the last decades, renewable energy sources have witnessed major (annual) growth rates, mainly the solar energy ones, which offer competitive, environmental friendly, low-cost solutions, accessible at a mass production level. The renewable energy producing units located close to energy loads are advantageous as the transportation energy losses are practically eliminated. The energy needed for a typical residential home is relatively small and can be covered mainly from renewable sources.

Considering the two major renewable energy sources, such as the wind and sun, the efficient energy harvesting from these sources is one major target of the energy industry. Although these energies seem to be free, with low negative impact, only the way these energies are extracted and used can state if the system is a sustainable system or not. In view of this, one can state that a system is efficiently harvesting the energy from these sources in a sustainable way if the overall efficiency of the system is high and if the energy produced is managed in such a way that these systems have minimum negative effect on the power grid. In other words, it is more efficient to use or store locally the energy produced from renewable sources instead of injecting the energy into the grid. In this way, the power grid is not perturbed by a small production facility with high dynamic behavior. Moreover, if the system is designed in an overall cost-efficient way, the number of PV panels, the wind turbine power, and battery storage capacity can all be minimized in order to obtain an optimized solution in which the power grid can be used only as an energy buffer supply.

The renewable systems of the future will have to integrate all the energy-dependent applications into a system that can be centrally controlled to obtain the best cost-performance balance. In view of this, one interesting way is to combine the electric car and the home electric systems. Concepts like vehicle-to-grid are recently being introduced [1], imposing that the energy storage of the electric vehicle can be integrated into the home electric system to maximize the overall system performance. An overview on solar heat pump systems is presented in Ref. [2]. Garcia et al. [3] present an optimal energy management system for standalone wind turbine/photovoltaic/hydrogen/battery hybrid based on fuzzy logic. Also, Ben Salah et al. [4] and Athari and Ardehali [5] are using fuzzy logic to facilitate the integration of renewable energy into residential or decentralized small power grid applications. Several papers, written by the same research group, present optimal control of different combinations of power sources that feed a heat pump and other loads. A wind/PV + power grid combination is used in Ref. [6], whereas a PV/diesel/battery + power grid combination and a fuel cell/wind/PV + power grid combination are used in Refs. [7] and [8], respectively. These three papers present cost savings derived from optimal control strategies.

The following sections present the power system under study, the presentation of the fuzzy logic energy management, the simulation results, and a conclusion.

2. Power system under study

This chapter proposes a concept in which two power systems interact by means of two interchangeable small batteries: a lightweight electric vehicle and a household partially powered from renewable energy sources. The overall schematic of the studied system is represented in **Figure 1**. The power generation system is composed of renewable energy conversion equipment: PV panels and a wind generator, both components being connected at the outputs on a DC-Link by means of MPPT-based converters. This DC-Link can be considered the main power line of the system on which also all the DC house loads are connected. This power system is also connected to the grid with a bidirectional DC-AC converter. The

http://dx.doi.org/10.5772/intechopen.69486

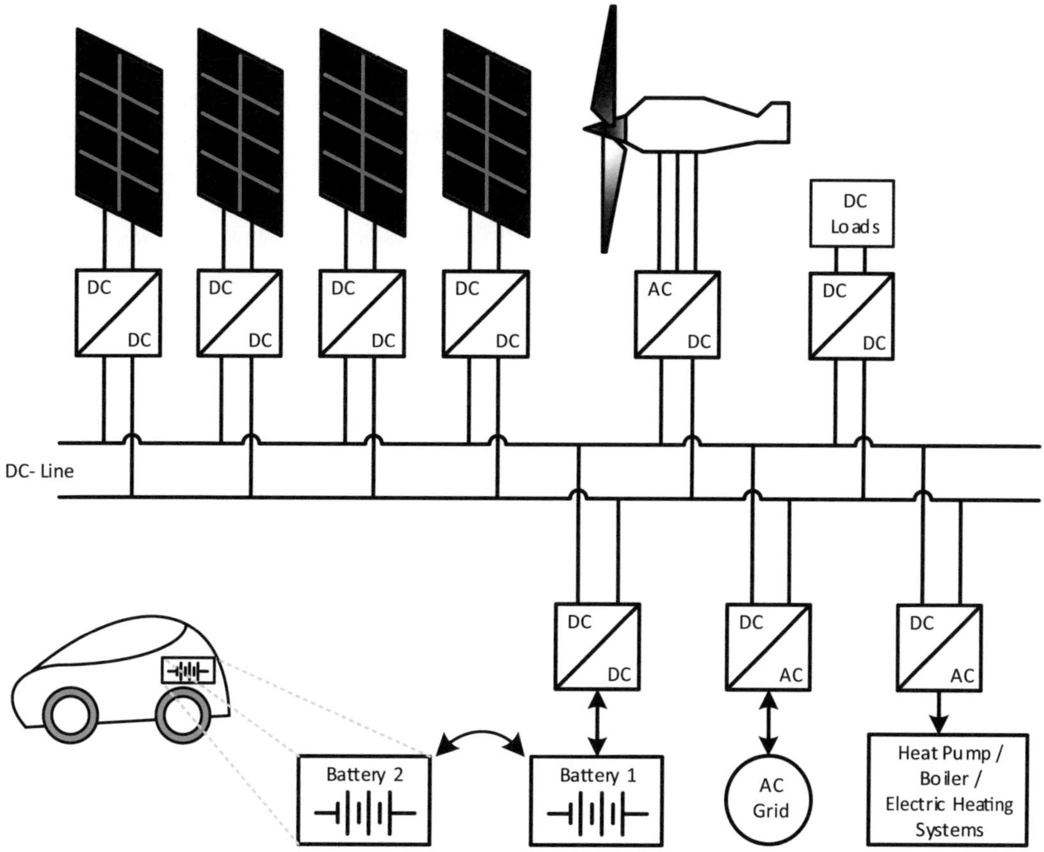

Figure 1. Schematic of the power system.

storage element is a battery from a lightweight electric vehicle. This small capacity battery is exchangeable with a second battery that is installed in the vehicle. The heat pump load is a key element in the proposed system, mainly because it will be the main electric load in balancing the energy from the renewable sources. Also, a key element of the proposed system is that the battery storage capacity is kept to a minimum; the two batteries are alternately used by the vehicle and are recharged mainly from the renewable sources. The battery can be used by the home electric system only when the power grid is unavailable. It must be stated that, to obtain a feasible approach, the battery has to be as light as possible in order to be easily exchanged.

The main objective of this study is to develop an energy management control system, based on fuzzy logic, which performs in such way that all the energy from the renewable sources is used, with minimum or no energy injection into the power grid and with a small energy storage capacity. The electrical energy can also be obtained from the power grid. In this way, the best cost-performance tradeoff with minimum storage capacity and maximum utilization of the renewable energy can be obtained.

3. Fuzzy logic energy management

The fuzzy logic energy management is developed with the aim of splitting the renewable energy power to be either stored into the battery or transformed into thermal energy, with a multiplication factor by a heat pump. The fuzzy algorithm is using as inputs the battery state of charge (SoC) and the required heat energy (hot water and heating during cold periods). The output is the battery power (P_{batt}), the heat pump power (P_{hp}) being calculated with the simple Eq. (1):

$$P_{hp} = 1 - P_{batt} \tag{1}$$

Figure 2 presents the two inputs and one output of the fuzzy logic supervisor.

Figures 3–5 present the membership functions (MFs) of the two inputs and output. The implementation of the fuzzy logic supervisor is done using Fuzzy logic toolbox and Matlab/Simulink® software from Mathworks®. It should be noted that those membership functions are built considering that the sum of them is on the entire interval 1 and the variation on both axes is expressed in per unit.

The rule base of the fuzzy algorithm is presented in **Table 1**. It contains nine rules according to the fact that each input has three MFs, all of them being considered in the rule-editing phase.

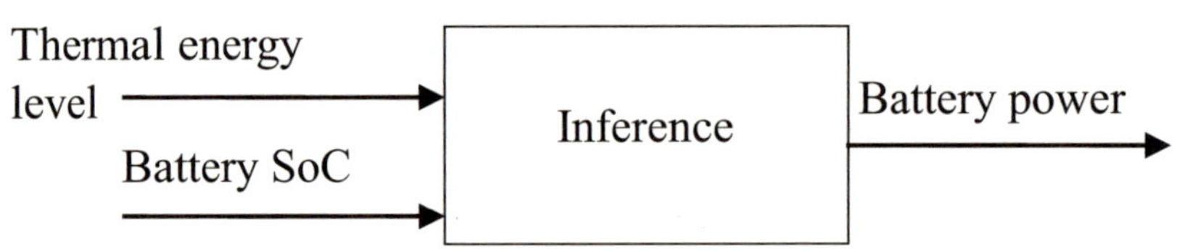

Figure 2. Fuzzy logic energy management inputs and output.

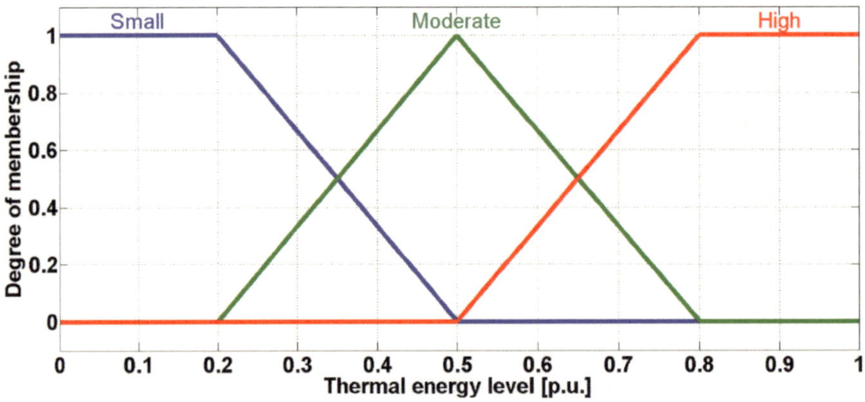

Figure 3. MFs for the first input of the fuzzy logic management system.

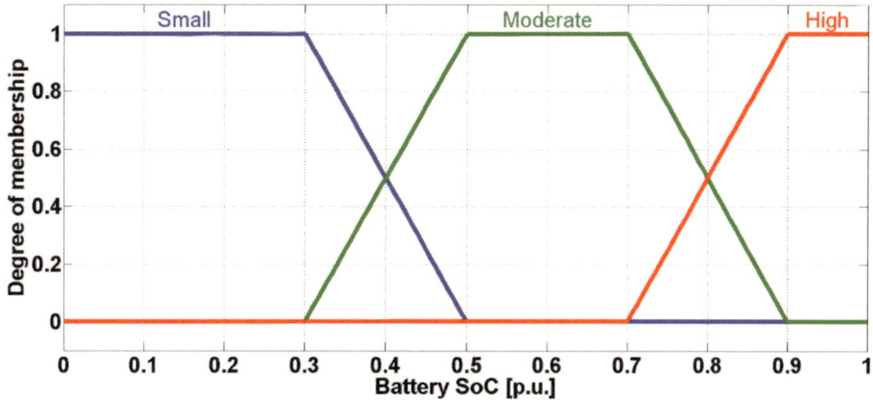

Figure 4. MFs for the second input of the fuzzy logic management system.

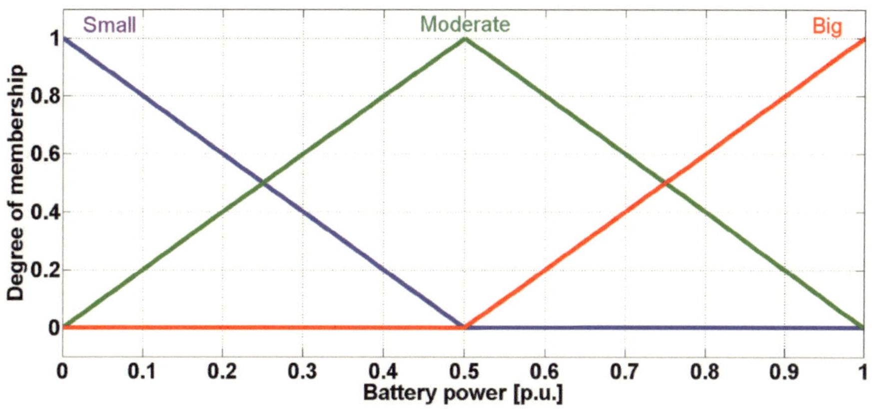

Figure 5. The output of the fuzzy logic management system.

Battery power		Battery SoC		
		Small	**Moderate**	**High**
Thermal energy level	Small	Moderate	Small	Small
	Moderate	Big	Moderate	Small
	High	Big	Big	Small

Table 1. Rule base of the fuzzy logic power management.

These rules were established considering the next principle: the thermal energy level and the battery SoC should increase similarly if they have a similar level (in per unit), but if one is lower than the other, it will be favored in order to reach a similar level (in per unit).

The defuzzification method used is centroid, which returns the center of area under the response surface. **Figure 6** presents the response surface generated for this fuzzy logic supervisor. As it can be seen from this figure, the output variation does not reach the full range between 0 and 1 (its range is between 0.163 and 0.837). Thus, in order to extend the output range, the output result is subtracted with 0.163 and then multiplied with 1.485.

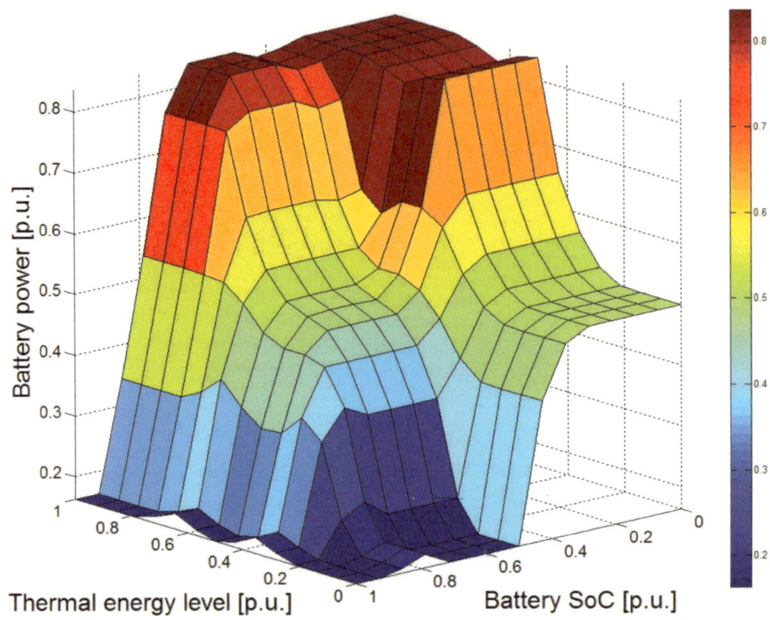

Figure 6. Response surface for the fuzzy logic supervisor.

4. Simulation results

The following figures present the simulation results for a power system with the following assumptions:

- Wind turbine and PV panels have an installed power of 1 kW each.
- The electric vehicle battery has a capacity of 1 kWh with an SoC of 40% when the charge process starts.
- It is considered that the vehicle is an ultralight version having two interchangeable batteries: one that equips the vehicle and the other left at home for charging.
- The heat pump has a coefficient of performance (COP) of 3, thus it consumes one-third electrical energy and generates three times more heat energy.

- The necessary heat energy considered is 30 kWh, thus a 10 kWh of electrical energy is required for the heat pump.
- The heat pump has the possibility to adjust power according to renewable energy production.
- Simulations are made for 1 day (24 h).

Figures 7 and **8** present the power and energy produced by the wind/PV hybrid system. The PV power curve has similar variations with some classic production curves for a sunny day, and the wind power has approximate variations according to a wind measurement from Brasov area (Romania).

The necessary electrical and thermal powers required are presented in **Figure 9**. The electrical power consumption is taken from a figure presented in Ref. [9] and the thermal power is estimated considering 30 kWh of needed energy (for a cold period) and no energy consumption during workhours.

Figure 10 presents the power grid failures. Even if it is unlikely to have two grid voltage drops during 1 day, it was considered in the simulation for demonstration purposes. Each voltage drop has a period of 15 min, first, from 14h30 to 14h45 and second from 21h30 to 21h45.

Figures 11 and **12** present the thermal energy level and the battery SoC, respectively, of the heat pump and battery powers, for the considered application. Both the thermal energy level and the SoC have an initial level. Due to the fact that at 18h00, the thermal power consumption restarts, and even if the heat pump is still working, the thermal energy level decreases (**Figure 11**). At the same time (18h00), the batteries are switched, the charged one is placed on the electric vehicle and the discharged one is put to charge. In order to use mainly the renewable energy to charge the battery and considering that the thermal energy demand is high, during the remaining hours of the day (18h00–24h00), the charging of the battery is stopped.

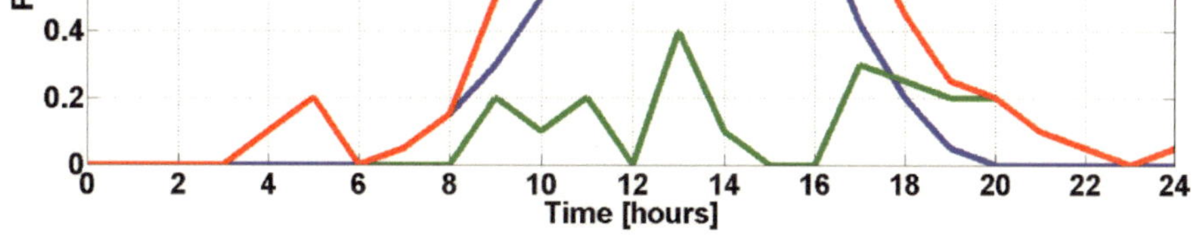

Figure 7. Renewable sources power curves.

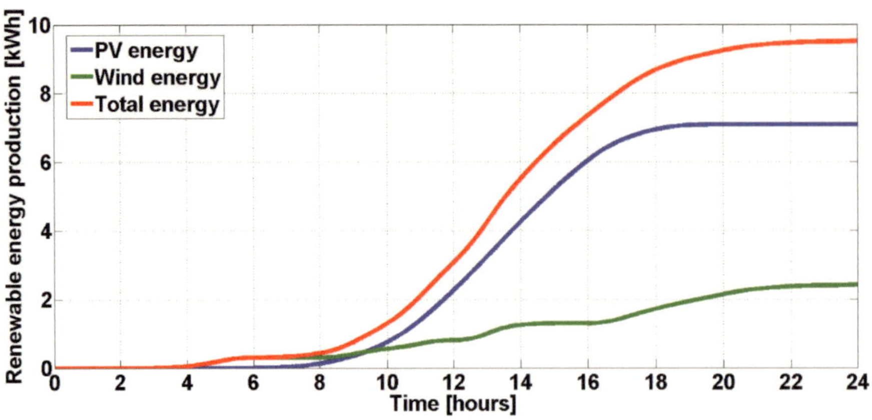

Figure 8. Renewable energy production.

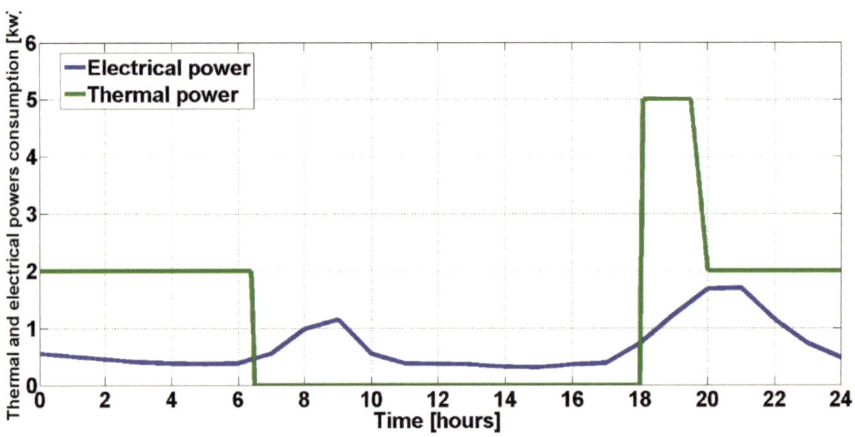

Figure 9. Estimated consumption of thermal and electrical powers.

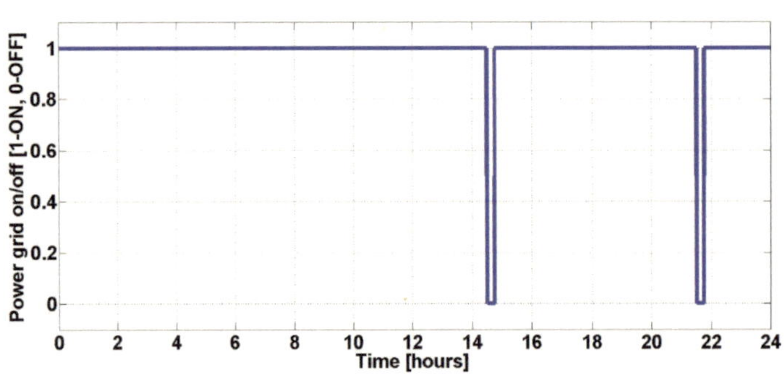

Figure 10. Power grid failure.

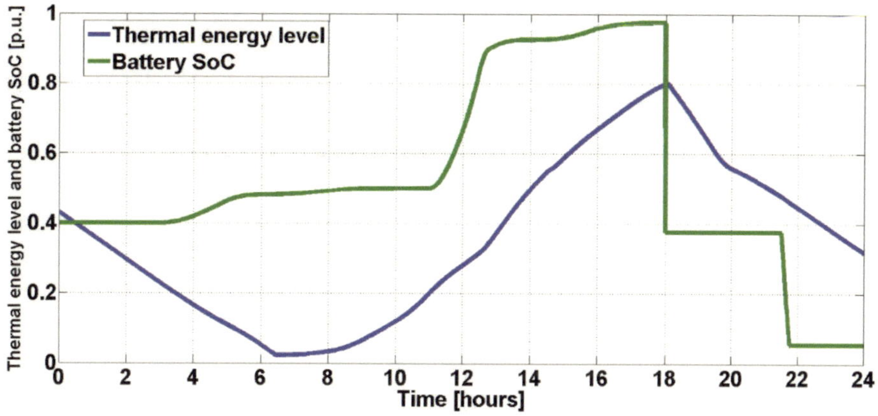

Figure 11. Thermal energy level and the battery SoC.

By the end of the day, the thermal energy level and the battery SoC are lower compared with the situation at the beginning of the day. It is clear that not all required energy can be obtained from the renewable sources, thus, some has to come from the power grid to arrive to the necessary level. From **Figure 12** one can observe that the first power grid drop is easily covered from renewable energy sources, just a sudden drop in the heat pump power can be observed. But the second one takes a lot of energy from the battery, which arrives to an SoC of about 5%. If the power grid loss would be longer, the feeding of all the loads would not be possible, and if considered appropriate, only the mandatory loads (pumps, lighting, etc.) should be fed. Or, if charging the discharged one is started at 18h00, the SoC would not decrease that much (**Figure 13**). Of course, in this case, the thermal energy level will decrease. For this situation, the heat pump and battery working powers are changed (**Figure 14**).

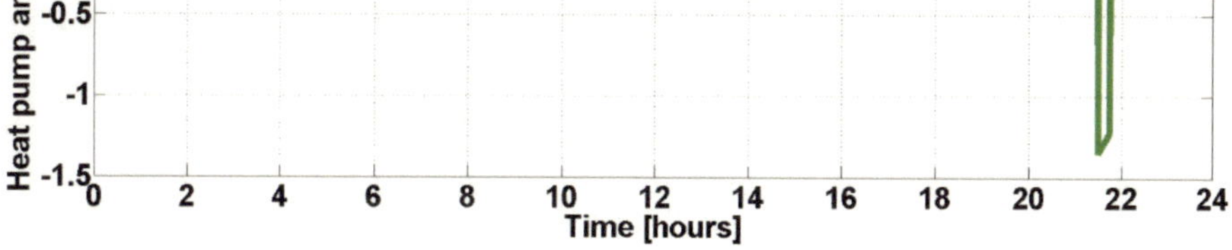

Figure 12. Heat pump power and battery power.

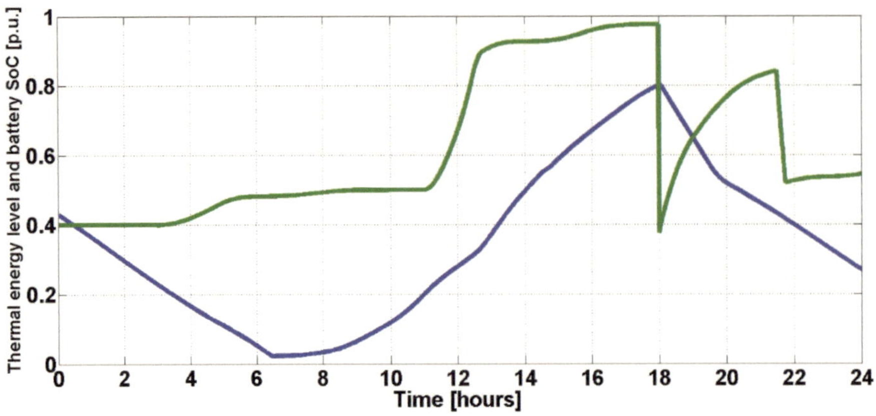

Figure 13. Thermal energy level and the battery SoC for the immediate charging of the discharged battery.

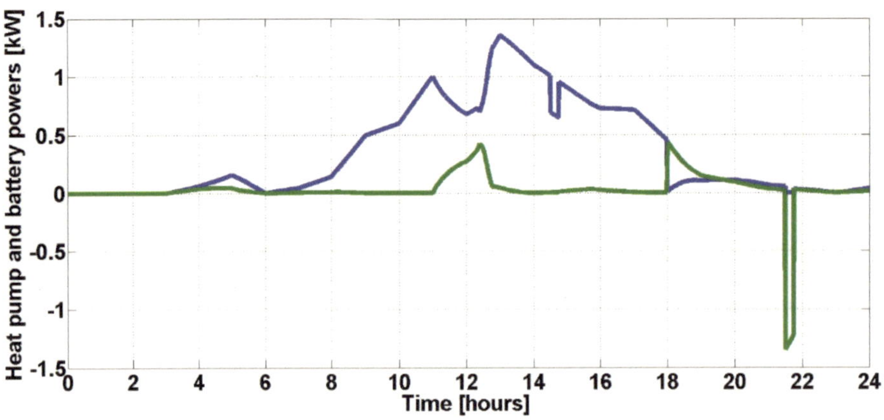

Figure 14. Heat pump and battery powers for the immediate charging of the discharged battery.

5. Conclusion

A fuzzy logic energy management algorithm has been proposed and validated by simulations, for a household application. This algorithm allows the distribution of the renewable energy to charge a battery and also to feed a heat pump that produces thermal energy. The results show that the battery charges to around 97%, and the thermal energy level from renewable sources is around 88% for the first case (discharged battery is not charged during the evening) and around 83% for the second (discharged battery is charged during the evening). The rest of the needed energy should be covered from the power grid.

Author details

Stefan Breban*, Ioana Gros, Calin Marginean and Petre Teodosescu

*Address all correspondence to: stefan.breban@emd.utcluj.ro

Technical University of Cluj-Napoca, Cluj-Napoca, Romania

References

[1] Guille C, Gross G. A conceptual framework for the vehicle-to-grid (V2G) implementation. Energy Policy. 2009;**37**(11):4379-4390. ISSN: 0301-4215

[2] Sparber W, Vajen K, Herkel S, Ruschenburg J, Thür A, Fedrizzi R, D'Antoni M. Overview on solar heat pump systems and review of monitoring results. In: ISES World Solar Congress, 28 August -2 September 2011, Kessel, Germany; Published by International Solar Energy Society, Freiburg Germany:1-12

[3] Garcia P, Torreglosa JP, Fernandez LM, Jurado F. Optimal energy management system for standalone wind turbine/photovoltaic/hydrogen/battery hybrid system with supervisory control based on fuzzy logic. International Journal of Hydrogen Energy. 2013;**23**:14146-14158

[4] Ben Salah C, Chaabene M, Ben Ammar M. Multi-criteria fuzzy algorithm for energy management of a domestic photovoltaic panel. Renewable Energy. 2008;**33**(5):993-1001. ISSN: 0960-1481

[5] Athari MH, Ardehali MM. Operational performance of energy storage as function of electricity prices for on-grid hybrid renewable energy system by optimized fuzzy logic controller. Renewable Energy. 2016;**85**:890-902

[6] Sichilalu S, Mathaba T, Xia X. Optimal control of a wind–PV-hybrid powered heat pump water heater. Applied Energy. 2017;**185**:1173-1184

[7] Sichilalu S, Xia X. Optimal energy control of grid tied PV–diesel–battery hybrid system powering heat pump water heater. Solar Energy. 2015;**115**:243-254

[8] Sichilalu S, Tazvinga H, Xia X. Optimal control of a fuel cell/wind/PV/grid hybrid system with thermal heat pump load. Solar Energy. 2016;**135**:59-69

[9] Arif MT, Oo AMT, Ali ABMS. Estimation of energy storage and its feasibility analysis. In: Zobaa AF, editor. Energy Storage—Technologies and Applications. InTech, Rijeka, Croatia; 2013: 41-78. ISBN: 978-953-51-0951-8

Chapter 8

Robust Adaptive Fuzzy Control for a Class of Switching Power Converters

Cheng-Lun Chen

Additional information is available at the end of the chapter

http://dx.doi.org/10.5772/67895

Abstract

This chapter provides the reader with a control-centric modeling and analysis approach along with a nonlinear control design for a class of switching power converters. A comprehensive model combining the respective state variable models of the interval subsystems is established. Comparison with PSpice simulation justifies the credibility of the model. Based on this model, internal/BIBO stability can be studied for each interval subsystem. Moreover, controllability and observability can also be analyzed to help determine subsequent control configuration. The established model is further investigated for advanced control design, i.e., robust adaptive fuzzy control.

Keywords: adaptive fuzzy control, dc-dc power converter, modeling, switching power converter

1. Introduction

Switching power converters are increasingly taking over conventional linear power converters due to their being compact, lightweight, high efficiency, and larger input voltage range. With the rapid advancement and popularity of personal computers, mobile communication devices, and automotive electronics, the need for stability and efficiency of converters is rising. Among the switching power converters, the phase-shifted pulse-width modulated (PSPWM) full-bridge soft switched power converter [1, 2] and corresponding alteration [3–8] have become a widely used circuit topology due to various beneficial characteristics, e.g., reduction of switching losses and stresses, and elimination of primary snubbers. The circuit is capable of high switching frequency operation with improved power density and conversion efficiency.

Feedback control has been incorporated into switching power converters to not only stabilize, but also improve the performance robustness of the output voltage. In spite of its advantageous features, feedback control for soft switched PSPWM full-bridge converters is still confined to simple linear time-invariant design, e.g., proportional-integral (PI) or lead-lag compensators based on a linearized model [9–12]. As pointed out by [13, 14], due to the increased number of topological stages and the PWM duty cycle being affected by input voltage, output voltage, and load current, the dynamics of a PSPWM full-bridge converter is much more sophisticated than that of a simple buck converter. A trade-off needs to be made regarding whether a simple model (e.g., linearized model) or a complex one (e.g., switched model) is to be established for the purpose of control design.

For model-based control design, a mathematical model of appropriate sophistication and capture of desired dynamics is normally the initial step. Such models can usually be obtained by simplifying a complex model, i.e., model reduction, or linearizing a nonlinear around specified operation point. For design of linear controllers, transfer function is a matured modeling tool. For more advanced control design, state variable model is usually a prerequisite. Various modeling approaches for switching power converters of complex topology have been proposed in the literature [15–20]. Most of them have been successful in terms of modeling the "local" behavior (i.e., small signal model) or analysis of the fundamental characteristics. However, few of the results can carry over to the next stage of control design. To be specific, a variety of crucial information for control design cannot be retrieved from those "not control centric" modeling approaches. That essential information includes stability, controllability, and observability of the open-loop system.

This chapter will provide the reader with a control-centric modeling and analysis approach along with robust adaptive fuzzy control design for switching power converters of complex topology and resort to PSPWM full-bridge power converters as a design example. The outline of the chapter is as follows:

Sections 2 and 3 demonstrate how to establish a control-centric mathematical model for a PSPWM full-bridge soft switched power converter system. The set of differential equations and the corresponding state variable model are established for each operation interval. The subsystem models for all intervals are integrated to form a comprehensive model. Numerical simulation of the model is performed and compared to that of the corresponding PSpice model to verify its validity. Section 4 will perform stability analysis for the system. Specifically, stability analysis is performed for each interval subsystem (of the established model) to determine whether the subsystem is internally/BIBO stable. Section 5 will conduct controllability/observability analysis for the system. Controllability and observability of the subsystems are analyzed to determine which signals/variables can actually be manipulated by control effort and which can be estimated using output feedback control structure. The established comprehensive model is further exploited for advanced control design. For example, by getting rid of uncontrollable and unobservable variables and dynamics, an LPV gain-scheduling control design may be conducted as in Ref. [21]. Model reduction and robust adaptive fuzzy control design are presented in Sections 6 and 7. Conclusion and future work are given in Section 8.

http://dx.doi.org/10.5772/67895

2. Control-centric mathematical model

In this section, operation of a PSPWM full-bridge dc-dc power converter will be briefly described. Note that there are eight operation intervals. Due to switching, operation of adjacent intervals is discontinuous. This implies that the parameters and initial conditions change when the converter switches. It will be demonstrated how a comprehensive control-oriented state variable model for each operation interval can be established for subsequent analysis and numerical simulation. The circuit diagram of the converter is shown in **Figure 1**. **Figure 2** is the waveform timing for various signals in the converter, where $i_{L_{lK}}$ is the primary current, v_{ab} is the voltage between a and b, i_L is the secondary current, v_s is the secondary voltage, Q_A,Q_B,Q_C, and Q_D are the four switches, ΔD is the duty cycle loss, and ZVS delay is the dead time.

2.1. Positive half cycle: trailing-leg (passive-to-active) transition (t_0~t_1)

During this operation interval, only Q_D is conducting. **Figure 3** shows the equivalent circuit of state 1. Initial conditions are $v_{C_A}(t_0) = v_i$, $v_{C_B}(t_0) = 0$, $v_{C_C}(t_0) = v_i$, and $v_{C_D}(t_0) = 0$. During this interval, C_A is discharging and C_B is charging until v_{C_A}equals zero. $Q_A$is turned on at zero voltage. Utilizing Kirchhoff's voltage law (KVL) and Kirchhoff's current law (KCL), we arrive at the following set of differential equations:

$$\frac{dv_{C_A}(t)}{dt} = \frac{i_{L_{lk}}}{C_A + C_B}, \frac{dv_{C_B}(t)}{dt} = -\frac{i_{L_{lk}}}{C_A + C_B}, \frac{dv_{C_C}(t)}{dt} = 0, \frac{dv_{C_D}(t)}{dt} = 0 \quad (1)$$

$$-\frac{di_{L_{lk}}}{dt} = \frac{di_L}{dt} = \frac{-1}{n^2 L_{lk} + L} v_C + \frac{n}{n^2 L_{lk} + L} v_{C_A} + \frac{-n}{n^2 L_{lk} + L} v_i, \frac{dv_C(t)}{dt} = \frac{1}{C} i_L - \frac{1}{RC} v_C \quad (2)$$

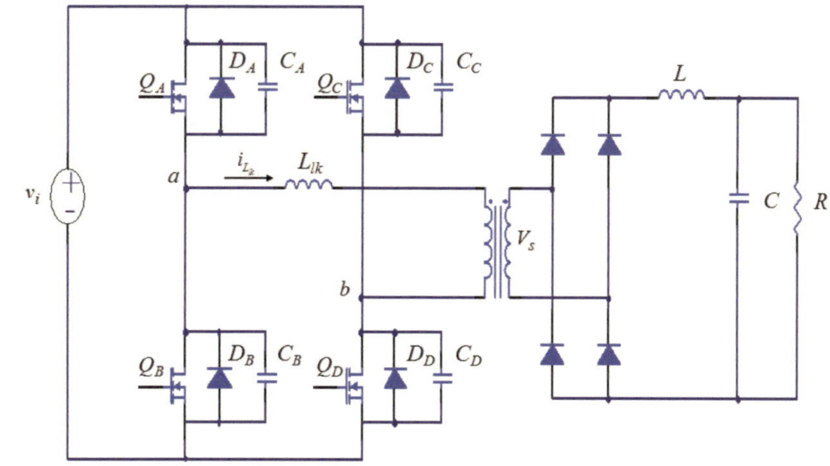

Figure 1. Circuit topology of a PSPWM full-bridge converter.

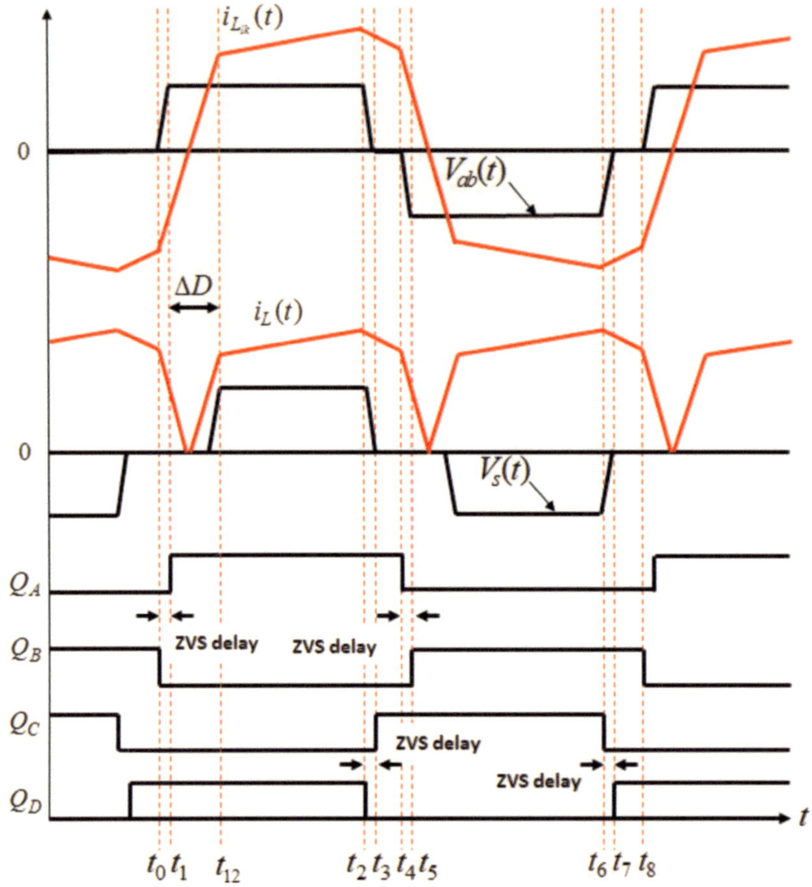

Figure 2. Waveform timing for a PSPWM full-bridge converter.

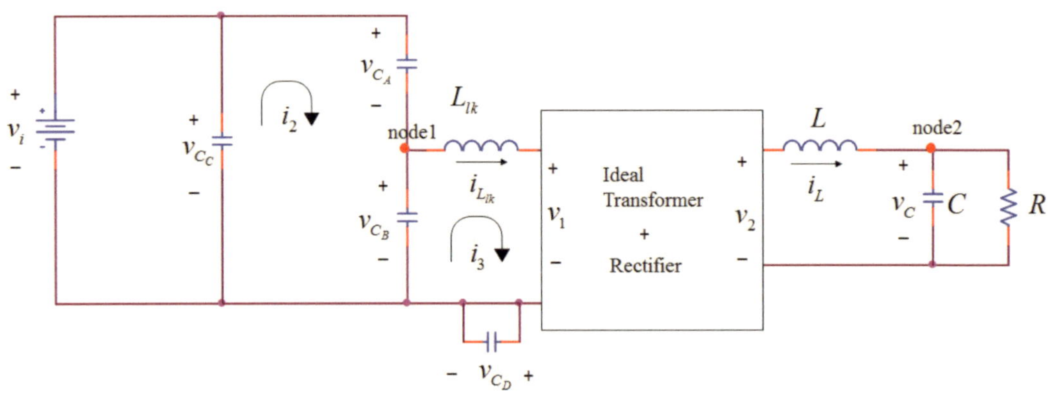

Figure 3. The equivalent circuit of operation interval 1 (t_0~t_1).

Define $\vec{x}(t) = [i_{L_{lK}}(t) \quad i_L(t) \quad v_C(t) \quad v_{C_A}(t) \quad v_{C_B}(t) \quad v_{C_C}(t) \quad v_{C_D}(t)]^T$, where $i_{L_{lK}}$ is leakage inductance current, i_L is inductance current, v_C is output voltage, v_{C_A} is the voltage across C_A, v_{C_B} is the voltage across C_B, v_{C_C} is the voltage across C_C, and v_{C_D} is the voltage across C_D. Therefore, a state variable model can be obtained as follows:

$$\dot{\vec{x}}(t) = \begin{bmatrix} 0 & 0 & \frac{n}{n^2L_{lk}+L} & \frac{-n^2}{n^2L_{lk}+L} & 0 & 0 & 0 \\ 0 & 0 & \frac{-1}{n^2L_{lk}+L} & \frac{n}{n^2L_{lk}+L} & 0 & 0 & 0 \\ 0 & \frac{1}{C} & -\frac{1}{RC} & 0 & 0 & 0 & 0 \\ \frac{1}{C_A+C_B} & 0 & 0 & 0 & 0 & 0 & 0 \\ -\frac{1}{C_A+C_B} & 0 & 0 & 0 & 0 & 0 & 0 \\ 0 & 0 & 0 & 0 & 0 & 0 & 0 \\ 0 & 0 & 0 & 0 & 0 & 0 & 0 \end{bmatrix} \vec{x}(t) + \begin{bmatrix} \frac{n^2}{n^2L_{lk}+L} \\ \frac{-n}{n^2L_{lk}+L} \\ 0 \\ 0 \\ 0 \\ 0 \\ 0 \end{bmatrix} v_i(t) \quad (3)$$

where v_i is input voltage, $N_2/N_1 = n$ is the transformer turns ratio, L is inductance, C is capacitance, and R is resistance.

2.2. Positive half cycle: active region (t_1~t_2)

During this operation interval, Q_A and Q_D are conducting. Initially, due to leakage inductance, the secondary side will experience a short period of "no energy" passing through, called duty cycle loss. **Figure 4** shows the equivalent circuit for this period. Initial conditions are $v_{C_A}(t_1) = 0$, $v_{C_B}(t_1) = v_i$, $v_{C_C}(t_1) = v_i$, and $v_{C_D}(t_1) = 0$. Similarly, we may derive a set of differential equations and the corresponding state variable model for this short period is

$$\dot{\vec{x}}(t) = \begin{bmatrix} 0 & 0 & 0 & 0 & 0 & 0 & 0 \\ 0 & 0 & -\frac{1}{L} & 0 & 0 & 0 & 0 \\ 0 & \frac{1}{C} & -\frac{1}{RC} & 0 & 0 & 0 & 0 \\ 0 & 0 & 0 & 0 & 0 & 0 & 0 \\ 0 & 0 & 0 & 0 & 0 & 0 & 0 \\ 0 & 0 & 0 & 0 & 0 & 0 & 0 \\ 0 & 0 & 0 & 0 & 0 & 0 & 0 \end{bmatrix} \vec{x}(t) + \begin{bmatrix} \frac{1}{L_{lk}} \\ 0 \\ 0 \\ 0 \\ 0 \\ 0 \\ 0 \end{bmatrix} v_i(t) \quad (4)$$

After the short period of duty cycle loss, energy passes through the transformer again. **Figure 5** shows the equivalent circuit. Initial conditions are $v_{C_A}(t_{12}) = 0$, $v_{C_B}(t_{12}) = v_i$, $v_{C_C}(t_{12}) = v_i$, and $v_{C_D}(t_{12}) = 0$. We may derive a set of differential equations and the corresponding set of differential equations and state variable model are

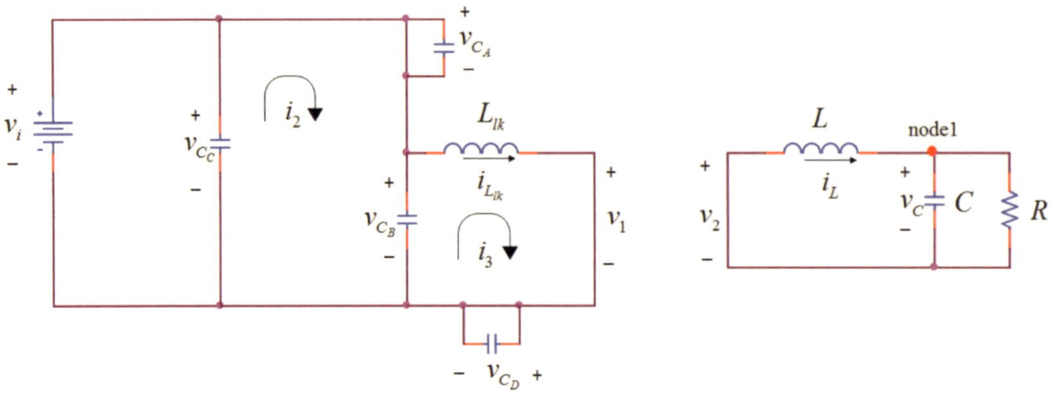

Figure 4. The equivalent circuit of the period of duty cycle loss (t_1~t_2).

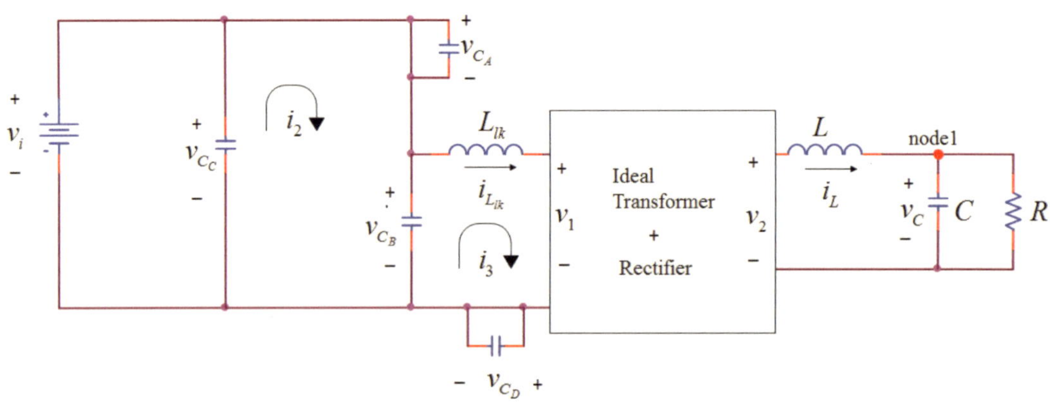

Figure 5. The equivalent circuit of operation interval 2 (t_1~t_2).

$$\dot{\vec{x}}(t)=\begin{bmatrix} 0 & 0 & -\frac{n}{n^2L_{lk}+L} & 0 & 0 & 0 & 0 \\ 0 & 0 & -\frac{1}{n^2L_{lk}+L} & 0 & 0 & 0 & 0 \\ 0 & \frac{1}{C} & -\frac{1}{RC} & 0 & 0 & 0 & 0 \\ 0 & 0 & 0 & 0 & 0 & 0 & 0 \\ 0 & 0 & 0 & 0 & 0 & 0 & 0 \\ 0 & 0 & 0 & 0 & 0 & 0 & 0 \\ 0 & 0 & 0 & 0 & 0 & 0 & 0 \end{bmatrix}\vec{x}(t)+\begin{bmatrix} \frac{n^2}{n^2L_{lk}+L} \\ \frac{n}{n^2L_{lk}+L} \\ 0 \\ 0 \\ 0 \\ 0 \\ 0 \end{bmatrix}v_i(t) \tag{5}$$

The duration of duty cycle loss may be derived based on falling range of i_L is equal to rising range of i_L, i.e.,

$$\frac{t_{passive}}{n^2L_{lk}+L}v_C+\frac{\Delta D}{L}v_C=\left(\frac{-1}{n^2L_{lk}+L}v_C+\frac{n}{n^2L_{lk}+L}v_i\right)\times(t_{active}-\Delta D)\phi=\frac{t_{passive}}{t_{passive}+t_{active}}\times 180^\circ \quad (6)$$

where ΔD is duty cycle loss time, $t_{passive}$ is the time of passive region, t_{active} is the time of active region, and ϕ is the difference of phase between Q_A and Q_C.

2.3. Positive half cycle: leading-leg (active-to-passive) transition $(t_2{\sim}t_3)$

During this operation interval, only Q_A is conducting. Initial conditions are $v_{C_A}(t_2)=0$, $v_{C_B}(t_2)=v_i$, $v_{C_C}(t_2)=v_i$, and $v_{C_D}(t_2)=0$. In this interval, C_C is discharging and C_D is charging until v_{C_C}equals zero. Q_C is turned on at zero voltage. Applying KVL/KCL in a similar way, we obtain

$$\dot{\vec{x}}(t)=\begin{bmatrix} 0 & 0 & -\frac{n}{n^2L_{lk}+L} & 0 & 0 & 0 & -\frac{n^2}{n^2L_{lk}+L} \\ 0 & 0 & -\frac{1}{n^2L_{lk}+L} & 0 & 0 & 0 & -\frac{n}{n^2L_{lk}+L} \\ 0 & \frac{1}{C} & -\frac{1}{RC} & 0 & 0 & 0 & 0 \\ 0 & 0 & 0 & 0 & 0 & 0 & 0 \\ 0 & 0 & 0 & 0 & 0 & 0 & 0 \\ -\frac{1}{C_C+C_D} & 0 & 0 & 0 & 0 & 0 & 0 \\ \frac{1}{C_C+C_D} & 0 & 0 & 0 & 0 & 0 & 0 \end{bmatrix}\vec{x}(t)+\begin{bmatrix} \frac{n^2}{n^2L_{lk}+L} \\ \frac{n}{n^2L_{lk}+L} \\ 0 \\ 0 \\ 0 \\ 0 \\ 0 \end{bmatrix}v_i(t) \quad (7)$$

2.4. Positive half cycle: Passive region $(t_3{\sim}t_4)$

During this interval, Q_A and Q_C are conducting. Initial conditions are $v_{C_A}(t_3)=0$, $v_{C_B}(t_3)=v_i$, $v_{C_C}(t_3)=0$, and $v_{C_D}(t_3)=v_i$. Applying KVL/KCL in a similar way, we obtain

$$\dot{\vec{x}}(t)=\begin{bmatrix} 0 & 0 & -\frac{n}{n^2L_{lk}+L} & 0 & 0 & 0 & 0 \\ 0 & 0 & -\frac{1}{n^2L_{lk}+L} & 0 & 0 & 0 & 0 \\ 0 & \frac{1}{C} & -\frac{1}{RC} & 0 & 0 & 0 & 0 \\ 0 & 0 & 0 & 0 & 0 & 0 & 0 \\ 0 & 0 & 0 & 0 & 0 & 0 & 0 \\ 0 & 0 & 0 & 0 & 0 & 0 & 0 \\ 0 & 0 & 0 & 0 & 0 & 0 & 0 \end{bmatrix}\vec{x}(t) \quad (8)$$

2.5. Negative half cycle

The subsequent four operation intervals basically "mirror" those in positive cycle. Therefore, the derivations are omitted for brevity.

3. Solution and numerical simulation

Numerical simulation based on a typical PSPWM full-bridge power converter circuit (with parameters: transformer turns ratio $n = 0.5$, $V_i = 160$ volt, $V_o = 50$ volt, $R = 6\ \Omega$, C = 940 μF, L = 300 μH, $L_{lk} = 20\mu H$, $C_A = C_B = C_C = C_D = 5nF$, $f_s = 50$ kHz) in our laboratory is performed. A "realistic" model of the circuit is built using PSpice, and the developed mathematical model is realized using MATLAB/Simulink. Comparison of the simulation results validates the correctness and effectiveness of the established model.

4. Stability analysis

4.1. Zero-state response

A SISO system (A, B, C) with proper rational transfer function $G(s) = C(sI - A)^{-1}B$ is BIBO stable if and only if every pole of $G(s)$ has a negative real part or, equivalently, lies inside the left-half s-plane. For both positive and negative half cycles, we can obtain the following transfer function for each operation interval:

- Trailing-leg (passive-to-active) transition:

$$G_1 = \frac{-nR(C_A + C_B)s}{RC(C_A + C_B)(n^2L_{lk} + L)s^3 + (C_A + C_B)(n^2L_{lk} + L)s^2 + (R(C_A + C_B) + n^2RC)s + n^2} \quad (9)$$
$$pole = p11, p12, p13 \text{ (in complicated form)}$$

- Active region (duty cycle loss): $G_{2\,loss} = 0$
- Active region:

$$G_2 = \frac{nR}{RC(n^2L_{lk} + L)s^2 + (n^2L_{lk} + L)s + R}$$
$$pole = \frac{-(n^2L_{lk} + L) \pm \sqrt{(n^2L_{lk} + L)^2 - 4R^2C(n^2L_{lk} + L)}}{2RC(n^2L_{lk} + L)} \quad (10)$$

- Leading-leg (active-to-passive) transition:

$$G_3 = \frac{nR(C_C + C_D)s}{RC(C_C + C_D)(n^2L_{lk} + L)s^3 + (C_C + C_D)(n^2L_{lk} + L)s^2 + (R(C_C + C_D) + n^2RC)s + n^2} \quad (11)$$
$$pole = p31, p32, p33 \text{(in complicated form)}$$

- Passive region: no input:

Due to "mirroring" operation, corresponding intervals in positive and negative half cycles will have the same transfer function. The pole location for the operation interval of trailing-leg and leading-leg transition depends further on the values of the circuit elements. The poles for the operation interval of active region have negative real parts due to

$$(n^2L_{lk}+L) > \sqrt{(n^2L_{lk}+L)^2-4R^2C(n^2L_{lk}+L)} \tag{12}$$

Hence, the system is BIBO stable within this interval. Note that there is pole/zero cancelation for all intervals, which implies that each interval subsystem is either uncontrollable or unobservable.

4.2. Zero-input response

The equation $\dot{x}=Ax$ is marginally stable if and only if all eigenvalues of A have zero or negative real parts and those with zero real parts are simple root of the minimal polynomial of A. The equation $\dot{x}=Ax$ is asymptotically stable if and only if all eigenvalues of A have negative real parts. For both positive and negative half cycles, we can obtain the following set of eigenvalues for each operation interval:

- Trailing-leg (passive-to-active) transitions:

$$A_1=\begin{bmatrix} 0 & 0 & \frac{n}{n^2L_{lk}+L} & \frac{-n^2}{n^2L_{lk}+L} & 0 & 0 & 0\\ 0 & 0 & \frac{-1}{n^2L_{lk}+L} & \frac{n}{n^2L_{lk}+L} & 0 & 0 & 0\\ 0 & \frac{1}{C} & \frac{-1}{RC} & 0 & 0 & 0 & 0\\ \frac{1}{C_A+C_B} & 0 & 0 & 0 & 0 & 0 & 0\\ \frac{-1}{C_A+C_B} & 0 & 0 & 0 & 0 & 0 & 0\\ 0 & 0 & 0 & 0 & 0 & 0 & 0\\ 0 & 0 & 0 & 0 & 0 & 0 & 0 \end{bmatrix}, \lambda=0,0,0,0,\lambda 11,\lambda 12,\lambda 13 \tag{13}$$

- Active region (duty cycle loss):

$$\lambda=0,0,0,0,0,\frac{-1\pm\sqrt{(L-4R^2C)/L}}{2RC} \tag{14}$$

- Active region:

$$\lambda=0,0,0,0,0,\frac{-(n^2L_{lk}+L)\pm\sqrt{(n^2L_{lk}+L)^2-4R^2C(n^2L_{lk}+L)}}{2RC(n^2L_{lk}+L)} \tag{15}$$

- Leading-leg (active-to-passive) transitions:

$$\lambda = 0, 0, 0, 0, \lambda 31, \lambda 32, \lambda 33(\text{in complicated form}) \tag{16}$$

- Passive region:

$$\lambda = 0, 0, 0, 0, 0, \frac{-(n^2 L_{lk} + L) \pm \sqrt{(n^2 L_{lk} + L)^2 - 4R^2 C(n^2 L_{lk} + L)}}{2RC(n^2 L_{lk} + L)} \tag{17}$$

Since all intervals have zero eigenvalue, we need to determine whether zero is a simple root of the minimal polynomial of A. The minimal polynomial (in x) for each operation interval (positive or negative half cycle) is summarized as follows:

- Trailing-leg (passive-to-active) transitions:

$$x^4 + \frac{1}{RC}x^3 + \frac{n^2 C + C_A + C_B}{C(C_A + C_B)(n^2 L_{lk} + L)}x^2 + \frac{n^2}{RC(C_A + C_B)(n^2 L_{lk} + L)}x \tag{18}$$

- Active region (duty cycle loss):

$$x^3 + \frac{1}{RC}x^2 + \frac{1}{LC}x \tag{19}$$

- Active region:

$$x^3 + \frac{1}{RC}x^2 + \frac{1}{C(n^2 L_{lk} + L)}x \tag{20}$$

- Leading-leg (active-to-passive) transitions:

$$x^4 + \frac{1}{RC}x^3 + \frac{n^2 C + C_C + C_D}{C(C_C + C_D)(n^2 L_{lk} + L)}x^2 + \frac{n^2}{RC(C_C + C_D)(n^2 L_{lk} + L)}x \tag{21}$$

- Passive region:

$$x^3 + \frac{1}{RC}x^2 + \frac{1}{C(n^2 L_{lk} + L)}x \tag{22}$$

Although all operation intervals have different state matrix (A), corresponding intervals in positive and negative half cycles actually possess the same set of eigenvalues. The eigenvalues for the operation interval of trailing-leg and leading-leg transition depend further on the values of the circuit elements. Both intervals of active (including duty cycle loss) and passive region are marginally stable due to

$$n^2L_{lk} + L > \sqrt{(n^2L_{lk} + L)^2 - 4R^2C(n^2L_{lk} + L)}, \quad 1 > \sqrt{\frac{L - 4R^2C}{L}} \tag{23}$$

and zero is the simple root of the minimal polynomial.

5. Controllability/observability analysis

The stability analysis indicates that all interval subsystems have uncontrollable or unobservable modes. We may decompose the state variable model of each subsystem into controllable and uncontrollable parts, and follow by decomposing each part into observable and unobservable parts to obtain

$$\begin{bmatrix} \dot{\overline{x}}_{co} \\ \dot{\overline{x}}_{c\overline{o}} \\ \dot{\overline{x}}_{\overline{c}o} \\ \dot{\overline{x}}_{\overline{c}\overline{o}} \end{bmatrix} = \left[\begin{array}{c|ccc} \overline{A}_{co} & 0 & \overline{A}_{13} & 0 \\ \hline \overline{A}_{21} & \overline{A}_{c\overline{o}} & \overline{A}_{23} & \overline{A}_{24} \\ 0 & 0 & \overline{A}_{\overline{c}o} & 0 \\ 0 & 0 & \overline{A}_{43} & \overline{A}_{\overline{c}\overline{o}} \end{array}\right] \begin{bmatrix} \overline{x}_{co} \\ \overline{x}_{c\overline{o}} \\ \overline{x}_{\overline{c}o} \\ \overline{x}_{\overline{c}\overline{o}} \end{bmatrix} + \begin{bmatrix} \overline{B}_{co} \\ \overline{B}_{c\overline{o}} \\ 0 \\ 0 \end{bmatrix} u, y = \left[\begin{array}{c|ccc} \overline{C}_{co} & 0 & \overline{C}_{\overline{c}o} & 0 \end{array}\right] \begin{bmatrix} \overline{x}_{co} \\ \overline{x}_{c\overline{o}} \\ \overline{x}_{\overline{c}o} \\ \overline{x}_{\overline{c}\overline{o}} \end{bmatrix} \tag{24}$$

The observability matrices of the controllable part for each operation interval (positive or negative half cycle) are summarized as follows:

- Trailing-leg (passive-to-active) transition:

$$O_{c1} = \begin{bmatrix} 0 & \frac{-n}{C(n^2L_{lk} + L)} & \frac{n}{RC^2(n^2L_{lk} + L)} \\ \frac{-n}{C(n^2L_{lk} + L)} & \frac{n}{RC^2(n^2L_{lk} + L)} & \frac{n(n^2C + C_A + C_B)}{C^2(C_A + C_B)(n^2L_{lk} + L)^2} - \frac{n}{R^2C^3(n^2L_{lk} + L)} \\ \frac{n}{RC^2(n^2L_{lk} + L)} & oc1.1 & oc1.2 \end{bmatrix} \tag{25}$$

- Active region (duty cycle loss):

$$O_{c2loss} = [0] \tag{26}$$

- Active region:

$$O_{c2} = \begin{bmatrix} 0 & \frac{n}{C(n^2L_{lk} + L)} \\ \frac{n}{C(n^2L_{lk} + L)} & \frac{-n}{RC^2(n^2L_{lk} + L)} \end{bmatrix} \tag{27}$$

- Leading-leg (active-to-passive) transitions:

$$O_{c3} = \begin{bmatrix} 0 & \frac{n}{C(n^2L_{lk}+L)} & \frac{-n}{RC^2(n^2L_{lk}+L)} \\ \frac{n}{C(n^2L_{lk}+L)} & \frac{-n}{RC^2(n^2L_{lk}+L)} & \frac{n}{R^2C^3(n^2L_{lk}+L)} - \frac{n(n^2C+C_C+C_D)}{C^2(C_C+C_D)(n^2L_{lk}+L)^2} \\ \frac{-n}{RC^2(n^2L_{lk}+L)} & oc3.1 & \text{oc3.2} \end{bmatrix} \quad (28)$$

Table 1 summarizes the rank of the observability matrix. The state variables of both controllable and observable are listed in **Table 2**. Since equivalent transformation does not affect the eigenvalues, Eq. (24) has the same set of eigenvalues as in stability analysis. For the operation intervals of trailing/leading leg and active region, uncontrollable or unobservable states (v_{C_A},v_{C_B},v_{C_C}, and v_{C_D}) are marginally stable corresponding to zero eigenvalue. Therefore, those states will stay constant within those intervals, which matches what is observed in numerical simulation. For the intervals of duty cycle loss and passive region, uncontrollable (i_L, v_C) states are asymptotically stable, which also matches what is observed during simulation.

	Interval	**Rank**
Positive half cycle	Trailing-leg (passive-to-active) transitions	3
	Active region (duty cycle loss)	0
	Active region	2
	Leading-leg (active-to-passive) transitions	3
Negative half cycle	Trailing-leg (passive-to-active) transitions	3
	Active region (duty cycle loss)	0
	Active region	2
	Leading-leg (active-to-passive) transitions	3

Table 1. Rank of the observability matrix for the controllable part.

	Interval	**Controllable and observable**
Positive half cycle	Trailing-leg (passive-to-active) transitions	i_L, v_C, v_{C_B}
	Active region	i_L, v_C
	Leading-leg (active-to-passive) transitions	i_L, v_C, v_{C_D}
Negative half cycle	Trailing-leg (passive-to-active) transitions	i_L, v_C, v_{C_B}
	Active region	i_L, v_C
	Leading-leg (active-to-passive) transitions	i_L, v_C, v_{C_D}

Table 2. State variables of both controllable and observable.

6. Model reduction

The goal is to obtain a low dimensional model that encompasses the imperative response characteristics of the comprehensive model. The reduced model is then utilized for subsequent control design. For control of the "steady-state" response, we neglect the transition intervals and take only the active region and passive region into consideration. Define d is duty cycle (ON) of the converter and $d' = 1 - d$ (OFF). Assuming that $L_{lk} \ll L$, we may derive the following differential inclusion model

$$\begin{bmatrix} \frac{di_L(t)}{dt} \\ \frac{dv_o(t)}{dt} \end{bmatrix} = \begin{bmatrix} 0 & -\frac{1}{L} \\ \frac{1}{C} & -\frac{1}{RC} \end{bmatrix} \begin{bmatrix} i_L(t) \\ v_o(t) \end{bmatrix} + \begin{bmatrix} \frac{nd}{L} \\ 0 \end{bmatrix} v_i(t) \tag{29}$$

Let $x_1 = i_L, x_2 = v_o, u = d, y = v_o$

$$\begin{bmatrix} \dot{x}_1 \\ \dot{x}_2 \end{bmatrix} = \begin{bmatrix} 0 & \frac{-1}{L} \\ \frac{1}{C} & \frac{-1}{RC} \end{bmatrix} \begin{bmatrix} x_1 \\ x_2 \end{bmatrix} + \begin{bmatrix} \frac{nv_i}{L} \\ 0 \end{bmatrix} u, \ddot{y} = \frac{-1}{RC^2} x_1 + \left(\frac{1}{R^2C^2} - \frac{1}{LC} \right) x_2 + \frac{nv_i}{LC} u \tag{30}$$

7. Indirect adaptive fuzzy control for uncertain switching power converters subject to external disturbances

In the following, we propose a robust adaptive fuzzy tracking controller for the PSPWM full-bridge soft switched power converter. Although the controller is designed based on the reduced model, its effectiveness and performance are subsequently verified with the comprehensive model.

Indirect adaptive fuzzy control with sliding model

Based on the input-output linearization concept, Eq. (30) can be represented by

$$y^{(2)} = f(\boldsymbol{x}) + g(\boldsymbol{x})u, f(\boldsymbol{x}) = \frac{-1}{RC^2} x_1 + \left(\frac{1}{R^2C^2} - \frac{1}{LC} \right) x_2,\ g(\boldsymbol{x}) = \frac{nv_i}{LC} \tag{31}$$

The control objective is to force y to follow a given bounded reference signal y_m, under the constraint that all signals involved must be bounded. Let $e = y_m - y$, $\boldsymbol{e} = (e, \dot{e})^T$ and $\boldsymbol{k} = (k_2, k_1)^T$ be such that all roots of the polynomial $s^2 + k_1 s + k_2$ are in the open left half-plane. If the functions f and g are known, then the control law

$$u = \frac{1}{g(\boldsymbol{x})} [-f(\boldsymbol{x}) + {y_m}^{(2)} + \boldsymbol{k}^T \boldsymbol{e}] \tag{32}$$

applied to Eq. (31) results in

$$e^{(2)} + k_1\dot{e} + k_2 e = 0 \tag{33}$$

which implies that $\lim_{t\to\infty} e(t) = 0$ a main objective of control.

However, f and g are unknown. We replace f and g in Eq. (32) by the fuzzy logic systems $\hat{f}(\boldsymbol{x}|\boldsymbol{\theta}_f)$ and $\hat{g}(\boldsymbol{x}|\boldsymbol{\theta}_g)$. The resulting control law

$$u_c = \frac{1}{\hat{g}(\boldsymbol{x}|\boldsymbol{\theta}_g)}\left[-\hat{f}(\boldsymbol{x}|\boldsymbol{\theta}_f) + {y_m}^{(2)} + \boldsymbol{k}^T\boldsymbol{e}\right] \tag{34}$$

is the so-called certainty equivalent controller. We use

$$u = u_c + u_s \tag{35}$$

where the additional control term u_s is called a supervisory control for stability. Substituting Eq. (35) into Eq. (31), we obtain the error equation

$$\dot{\boldsymbol{e}} = \Lambda_c\boldsymbol{e} + \boldsymbol{b}_c\left[\hat{f}(\boldsymbol{x}|\boldsymbol{\theta}_f) - f(\boldsymbol{x}) + (\hat{g}(\boldsymbol{x}|\boldsymbol{\theta}_g) - g(\boldsymbol{x}))u_c - g(\boldsymbol{x})u_s\right] \tag{36}$$

where

$$\Lambda_c = \begin{bmatrix} 0 & 1 \\ -k_2 & -k_1 \end{bmatrix}, \quad b_c = \begin{bmatrix} 0 \\ 1 \end{bmatrix} \tag{37}$$

Since Λ_c is a stable matrix ($|sI - \Lambda_c| = s^n + k_1 s^{(n-1)} + \cdots + k_n$ which is stable), we know that there exists a unique positive definite symmetric $n \times n$ matrix P which satisfies the Lyapunov equation:

$$\Lambda_c{}^T P + P\Lambda_c = -Q \tag{38}$$

where Q is an arbitrary 2×2 positive definite matrix. Let $V_e = \frac{1}{2}\boldsymbol{e}^T P\boldsymbol{e}$, in order for $x_i = y_m^{(i-1)} - e^{(i-1)}$ to be bounded, we require that V_e must be bounded, which means we require that $\dot{V}_e \leq 0$ when V_e is greater than a large constant $\overline{V}$. Using Eq. (35) and Eq. (38), we have

$$\begin{aligned}\dot{V}_e &= -\frac{1}{2}\boldsymbol{e}^T Q\boldsymbol{e} + \boldsymbol{e}^T P\boldsymbol{b}_c[(\hat{f}(\boldsymbol{x}|\boldsymbol{\theta}_f) - f(\boldsymbol{x})) + (\hat{g}(\boldsymbol{x}|\boldsymbol{\theta}_g) - g(\boldsymbol{x}))u_c - g(\boldsymbol{x})u_s] \\ &\leq -\frac{1}{2}\boldsymbol{e}^T Q\boldsymbol{e} + |\boldsymbol{e}^T P\boldsymbol{b}_c|[|(\hat{f}(\boldsymbol{x}|\boldsymbol{\theta}_f)| + |f(\boldsymbol{x})| + |\hat{g}(\boldsymbol{x}|\boldsymbol{\theta}_g)u_c| + |g(\boldsymbol{x})u_c|] - \boldsymbol{e}^T P\boldsymbol{b}_c g(\boldsymbol{x})u_s\end{aligned} \tag{39}$$

In order to design the u_s such that the right-hand side of Eq. (39) is not positive, we need to know the bounds of f and g. That is, we have to make the following assumption.

Assumption: We can determine functions $f^U(\boldsymbol{x})$, $g^U(\boldsymbol{x})$ and $g_L(\boldsymbol{x})$such that $|f(\boldsymbol{x})| \leq f^U(\boldsymbol{x})$ and $g_L(\boldsymbol{x}) \leq g(\boldsymbol{x}) \leq g^U(\boldsymbol{x})$ for $\boldsymbol{x} \in R^2$, where $f^U(\boldsymbol{x}) < \infty$, $g^U(\boldsymbol{x}) < \infty$, and $g_L(\boldsymbol{x}) > 0$ for $\boldsymbol{x} \in R^2$.

Based on $f^U(\underline{x})$, $g^U(\underline{x})$ and $g_L(\underline{x})$, and by observing Eq. (39), we choose the supervisory control u_s as

$$u_s = \begin{cases} \mathrm{sgn}(\boldsymbol{e}^T P \boldsymbol{b}_c) \dfrac{1}{g_L(\boldsymbol{x})} [|\hat{f}(\boldsymbol{x}|\boldsymbol{\theta}_f)| + |f^U(\boldsymbol{x})| + |\hat{g}(\boldsymbol{x}|\boldsymbol{\theta}_g)u_c| + |g^U(\boldsymbol{x})u_c|], & V_e \geq \overline{V} \\ 0, & V_e \leq \overline{V} \end{cases} \tag{40}$$

Substituting Eq. (40) into Eq. (39) and considering the case $V_e > \overline{V}$, we have

$$\begin{aligned} \dot{V}_e &\leq -\frac{1}{2}\boldsymbol{e}^T Q \boldsymbol{e} + |\boldsymbol{e}^T P \boldsymbol{b}_c| [|\hat{f}| + |f| + |\hat{g}u_c| + |gu_c| - \frac{g}{g_L}(|\hat{f}| + f^U + |\hat{g}u_c| + |g^U u_c|)] \\ &\leq -\frac{1}{2}\boldsymbol{e}^T Q\boldsymbol{e} < 0 \end{aligned} \tag{41}$$

In summary, using the control Eq. (35), we can guarantee that $V_e \leq \overline{V} < \infty$. Since P is positive definite, the boundedness of V_e implies the boundedness of $\boldsymbol{e}$, which in turn implies the boundedness of $\boldsymbol{x}$.

We employ the following fuzzy logic system:

$$\hat{f}(\boldsymbol{x}|\boldsymbol{\theta}_f) = \sum_{l=1}^{M} \theta_l \xi_l(\boldsymbol{x}) = \boldsymbol{\theta}^T \boldsymbol{\xi}(\boldsymbol{x}) \tag{42}$$

where $\boldsymbol{\theta} = (\theta_1, \ldots, \theta_M)^T$, $\xi(\boldsymbol{x}) = (\xi_1(\boldsymbol{x}), \ldots, \xi_M(\boldsymbol{x}))^T$, $\xi_l(\boldsymbol{x})$ is the fuzzy basis function defined by

$$\xi_l(\boldsymbol{x}) = \prod_{i=1}^{2} \mu_{F_i^l}(x_i) \Big/ \sum_{l=1}^{M} \prod_{i=1}^{2} \mu_{F_i^l}(x_i) \tag{43}$$

θ_l are adjustable parameters, and $\mu_{F_i^l}$ are given membership functions.

We present the detailed design steps of the adaptive fuzzy controller.

- Step 1 Define p_i fuzzy sets $A_i^{l_i}(l_i = 1, 2, \ldots, p_i)$ for variable $x_i (i = 1, 2)$
- Step 2 There are $\prod_{i=1}^{2} p_i$ rules to construct fuzzy systems $\hat{f}(\boldsymbol{x}|\boldsymbol{\theta}_f)$:

$$\textit{if } x_1 \textit{ is } A_1^{l_1} \textit{ and } x_2 \textit{ is } A_2^{l_2}, \quad \textit{then } \hat{f} \textit{ is } E^{l_1 \ldots l_2}$$

There are $\prod_{i=1}^{2} q_i$ rules to construct fuzzy systems $\hat{g}(\boldsymbol{x}|\boldsymbol{\theta}_g)$:

$$\textit{if } x_1 \textit{ is } B_1^{l_1} \textit{ and } x_2 \textit{ is } B_2^{l_2}, \quad \textit{then } \hat{g} \textit{ is } H^{l_1 \ldots l_2}$$

Using product-inference rule, singleton fuzzifier, and center average defuzzifier, we obtain

$$\hat{f}(\boldsymbol{x}|\boldsymbol{\theta}_f) = \boldsymbol{\theta}_f^T\xi_f(\boldsymbol{x}), \hat{g}(\boldsymbol{x}|\boldsymbol{\theta}_g) = \boldsymbol{\theta}_g^T\xi_g(\boldsymbol{x}) \tag{44}$$

where

$$\xi_f^l(\boldsymbol{x}) = \frac{\prod_{i=1}^{2}\mu_{A_i^l}(x_i)}{\sum_{l=1}^{p_1\times p_2}\prod_{i=1}^{2}\mu_{A_i^l}(x_i)}, \xi_g^l(\boldsymbol{x}) = \frac{\prod_{i=1}^{2}\mu_{B_i^l}(x_i)}{\sum_{l=1}^{p_1\times p_2}\prod_{i=1}^{2}\mu_{B_i^l}(x_i)} \tag{45}$$

Our next task is to develop an adaptive law to adjust the parameters in the fuzzy logic systems for the purpose of forcing the tracking error to converge to zero.

Define

$$\begin{aligned} \boldsymbol{\theta}_f^* &= \arg\min_{\boldsymbol{\theta}_f}[sup_{\boldsymbol{x}\in R^2}|\hat{f}(\boldsymbol{x}|\boldsymbol{\theta}_f) - f(\boldsymbol{x})|] \\ \boldsymbol{\theta}_g^* &= \arg\min_{\boldsymbol{\theta}_g}[sup_{\boldsymbol{x}\in R^2}|\hat{g}(\boldsymbol{x}|\boldsymbol{\theta}_g) - g(\boldsymbol{x})|] \end{aligned} \tag{46}$$

Define the minimum approximation error

$$\omega = (\hat{f}(\boldsymbol{x}|\boldsymbol{\theta}_f^*) - f(\boldsymbol{x})) + (\hat{g}(\boldsymbol{x}|\boldsymbol{\theta}_g^*) - g(\boldsymbol{x}))u_c \tag{47}$$

Then the error equation can be rewritten as

$$\dot{\boldsymbol{e}} = \Lambda_c\boldsymbol{e} + \boldsymbol{b}_c[(\hat{f}(\boldsymbol{x}|\boldsymbol{\theta}_f)) - \hat{f}(\boldsymbol{x}|\boldsymbol{\theta}_f^*)) + (\hat{g}(\boldsymbol{x}|\boldsymbol{\theta}_g) - \hat{g}(\boldsymbol{x}|\boldsymbol{\theta}_g^*))u_c + \omega] \tag{48}$$

Substituting Eq. (43) into Eq. (47), we have

$$\dot{\boldsymbol{e}} = \Lambda_c\boldsymbol{e} + \boldsymbol{b}_c\omega + \boldsymbol{b}_c[(\boldsymbol{\theta}_f - \boldsymbol{\theta}_f^*)^T\xi_f(\boldsymbol{x}) + (\boldsymbol{\theta}_g - \boldsymbol{\theta}_g^*)^T\xi_g(\boldsymbol{x})u_c] \tag{49}$$

Consider the Lyapunov function candidate

$$V = \frac{1}{2}\boldsymbol{e}^T P\boldsymbol{e} + \frac{1}{2\gamma_1}(\boldsymbol{\theta}_f - \boldsymbol{\theta}_f^*)^T(\boldsymbol{\theta}_f - \boldsymbol{\theta}_f^*) + \frac{1}{2\gamma_2}(\boldsymbol{\theta}_g - \boldsymbol{\theta}_g^*)^T(\boldsymbol{\theta}_g - \boldsymbol{\theta}_g^*) \tag{50}$$

where γ_1 and $\gamma_2$are positive constants. The time derivative of V along the trajectory of Eq. (48) is

$$\begin{aligned} \dot{V} = &-\frac{1}{2}\boldsymbol{e}^T Q\boldsymbol{e} + \boldsymbol{e}^T P\boldsymbol{b}_c\omega + \frac{1}{\gamma_1}(\boldsymbol{\theta}_f - \boldsymbol{\theta}_f^*)^T[\dot{\boldsymbol{\theta}}_f + \gamma_1\boldsymbol{e}^T P\boldsymbol{b}_c\boldsymbol{\xi}_f(x)] \\ &+\frac{1}{\gamma_2}(\boldsymbol{\theta}_g - \boldsymbol{\theta}_g^*)^T[\dot{\boldsymbol{\theta}}_g + \gamma_2 e^T Pb_c\boldsymbol{\xi}_g(\boldsymbol{x})u_c] \end{aligned} \tag{51}$$

If we choose the adaptive law

$$\dot{\boldsymbol{\theta}}_f = -\gamma_1\boldsymbol{e}^T P\boldsymbol{b}_c\xi_f(\boldsymbol{x}), \dot{\boldsymbol{\theta}}_g = -\gamma_2\boldsymbol{e}^T P\boldsymbol{b}_c\xi_g(\boldsymbol{x})u_c \tag{52}$$

then from Eq. (50) we have

$$\dot{V} = -\frac{1}{2}\boldsymbol{e}^T Q\boldsymbol{e} + e^T P\boldsymbol{b}_c\omega \tag{53}$$

This is the best we can hope to get because the term $\boldsymbol{e}^T P \boldsymbol{b}_c \omega$ is of the order of the minimum approximation error. If $\omega = 0$, that is, the searching spaces for $\hat{f}$ and $\hat{g}$ are so big that f and g are included in them, then we have $\dot{V} \leq 0$. Eq. (51) cannot guarantee $\boldsymbol{\theta}_f$ and $\boldsymbol{\theta}_g$ are bounded, so we use projection algorithm. If the parameter vectors $\boldsymbol{\theta}_f$ and $\boldsymbol{\theta}_g$ are within the constraint sets or on the boundaries of the constraint sets but moving toward the inside of the constraint sets, then use the simple adaptive law Eq. (51). Otherwise, if the parameter vectors are on the boundaries of the constraint sets but moving toward the outside of the constraint sets, then use the projection algorithm to modify the adaptive law Eq. (51). such that the parameter vectors will remain inside the constraint sets.

$$\Omega_f = \left\{ \boldsymbol{\theta}_f \in R^{\prod_{i=1}^{2} p_i} \,|\, \|\boldsymbol{\theta}_f\| \leq M_f \right\} \tag{54}$$

$$\Omega_g = \left\{ \boldsymbol{\theta}_g \in R^{\prod_{i=1}^{2} q_i} \,|\, 0 < \varepsilon \leq \|\boldsymbol{\theta}_g\| \leq M_g \right\} \tag{55}$$

where Ω_f and Ω_g are constraint sets for $\boldsymbol{\theta}_f$ and $\boldsymbol{\theta}_g$, M_f, M_g, ε are constants

$$\dot{\boldsymbol{\theta}}_f = \begin{cases} -\gamma_1 \boldsymbol{e}^T P \boldsymbol{b}_c \xi_f(x) & \text{if } (\|\boldsymbol{\theta}_f\| < M_f) \text{ or } (\|\boldsymbol{\theta}_f\| = M_f \text{ and } \boldsymbol{e}^T P \boldsymbol{b}_c \xi_f(\boldsymbol{x}) \geq 0) \\ P\{-\gamma_1 \boldsymbol{e}^T P \boldsymbol{b}_c \xi_f(x)\} & \text{if } (\|\boldsymbol{\theta}_f\| = M_f \text{ and } \boldsymbol{e}^T P \boldsymbol{b}_{cc} \xi_f(\boldsymbol{x}) < 0) \end{cases} \tag{56}$$

where the projection operator $P\{*\}$ is defined as

$$P\{-\gamma_1 \boldsymbol{e}^T P \boldsymbol{b}_c \xi_f(\boldsymbol{x})\} = -\gamma_1 \boldsymbol{e}^T P \boldsymbol{b}_c \xi_f(\boldsymbol{x}) + \gamma_1 \boldsymbol{e}^T P \boldsymbol{b}_c \frac{\boldsymbol{\theta}_f \boldsymbol{\theta}_f^T \xi_f(\boldsymbol{x})}{\|\boldsymbol{\theta}_f\|^2} \tag{57}$$

Whenever an element $\boldsymbol{\theta}_{gi}$ of $\boldsymbol{\theta}_g = \varepsilon$, use

$$\dot{\theta}_{gi} = \begin{cases} -\gamma_2 \boldsymbol{e}^T P \boldsymbol{b}_c \xi_{gi}(\boldsymbol{x}) u_c & \text{if } \boldsymbol{e}^T P \boldsymbol{b}_c \xi_{gi}(\boldsymbol{x}) u_c < 0 \\ 0 & \text{if } \boldsymbol{e}^T P \boldsymbol{b}_c \xi_{gi}(\boldsymbol{x}) u_c \geq 0 \end{cases} \tag{58}$$

where $\xi_{gi}(\boldsymbol{x})$ is the ith component of $\xi_g(\boldsymbol{x})$.

Otherwise, use

$$\dot{\boldsymbol{\theta}}_g = \begin{cases} -\gamma_2 \boldsymbol{e}^T P \boldsymbol{b}_c \xi_g(\boldsymbol{x}) u_c & \text{if } (\|\boldsymbol{\theta}_g\| < M_g) \text{ or } (\|\boldsymbol{\theta}_g\| = M_g \text{ and } \boldsymbol{e}^T P \boldsymbol{b}_c \xi_g(\boldsymbol{x}) u_c \geq 0) \\ P\{-\gamma_2 \boldsymbol{e}^T P \boldsymbol{b}_c \xi_g(\boldsymbol{x})\} & \text{if } (\|\boldsymbol{\theta}_g\| = M_g \text{ and } \boldsymbol{e}^T P \boldsymbol{b}_c \xi_g(\boldsymbol{x}) u_c < 0) \end{cases} \tag{59}$$

where the projection operator $P\{*\}$ is defined as

$$P\{-\gamma_2 \boldsymbol{e}^T P \boldsymbol{b}_c \xi_g(\boldsymbol{x}) u_c\} = -\gamma_2 \boldsymbol{e}^T P \boldsymbol{b}_c \xi_g(\boldsymbol{x}) u_c + \gamma_2 \boldsymbol{e}^T P \boldsymbol{b}_c \frac{\boldsymbol{\theta}_g \boldsymbol{\theta}_g^T \xi_g(\boldsymbol{x}) u_c}{\|\boldsymbol{\theta}_g\|^2} \tag{60}$$

Theorem:

1. $\|\boldsymbol{\theta}_f(t)\| \leq M_f, \|\boldsymbol{\theta}_g(t)\| \leq M_g$, all elements of $\boldsymbol{\theta}_g \geq \varepsilon$,

$$\|\boldsymbol{x}(t)\| \leq \|\boldsymbol{y_m}\| + \left(\frac{2\overline{V}}{\lambda_{P\min}}\right)^{\frac{1}{2}} \tag{61}$$

$$|u(t)| \leq \frac{1}{\varepsilon}\left(M_f + |y_m^2| + \|\boldsymbol{k}\|\left(\frac{2\overline{V}}{\lambda_{P\min}}\right)^{\frac{1}{2}}\right) + \frac{1}{g_L(\boldsymbol{x})}$$
$$\left[M_f + |f^U(\boldsymbol{x})| + \frac{1}{\varepsilon}(M_g + g^U)\left(M_f + |y_m^2| + \|\boldsymbol{k}\|\left(\frac{2\overline{V}}{\lambda_{P\min}}\right)^{\frac{1}{2}}\right)\right] \tag{62}$$

for all $t \geq 0$, where $\lambda_{P\min}$ is the minimum eigenvalue of P, and $\boldsymbol{y_m} = (y_m, \dot{y}_m, \ldots, y_m^{(n-1)})^T$.

2.

$$\int_0^t \|\boldsymbol{e}(\tau)\|^2 d\tau \leq a + b\int_0^t |\omega(\tau)|^2 d\tau \tag{63}$$

for all $t \geq 0$, where a and b are constants, and ω is the minimum approximation error defined by Eq. (46)

3. If ω is squared integrable, that is, $\int_0^\infty |\omega(t)|^2 dt < \infty$, then $\lim_{t\to\infty} \|\boldsymbol{e}(t)\| = 0$

Design example:

The parameters of the converter are listed in **Figure 6**. Consider the following system:

Figure 6. Photo and parameters of a PSPWM full-bridge power converter circuit. Transformer turns ratio $n = 0.5$. $V_i = 160$ volt, $V_0 = 50$ volt, $R = 6\Omega$, $C = 940\mu F$ $L = 300\mu H$, $L_{lk} = 20\mu H$ $C_A = C_B = C_C = C_D = 5nF$, $f_s = 50$ kHz.

$$y^{(2)} = f(\boldsymbol{x}) + g(\boldsymbol{x})u \tag{64}$$

$$f(\boldsymbol{x}) = \frac{-1}{RC^2}x_1 + \left(\frac{1}{R^2C^2} - \frac{1}{LC}\right)x_2,\ g(\boldsymbol{x}) = \frac{nv_i}{LC} \tag{65}$$

The design steps of the adaptive fuzzy controller are provided in the following:

$$\begin{aligned} u &= u_c + u_s,\ u_c = \frac{1}{\hat{g}(\boldsymbol{x}|\boldsymbol{\theta}_g)}\left[-\hat{f}(\boldsymbol{x}|\boldsymbol{\theta}_f) + y_m{}^{(2)} + \boldsymbol{k}^T\boldsymbol{e}\right] \\ u_s &= \begin{cases} \text{sgn}(\boldsymbol{e}^T P \boldsymbol{b}_c)\dfrac{1}{g_L(\boldsymbol{x})}\left[|\hat{f}(\boldsymbol{x}|\boldsymbol{\theta}_f)| + |f^U(\boldsymbol{x})| + |\hat{g}(\boldsymbol{x}|\boldsymbol{\theta}_g)u_c| + |g^U(\boldsymbol{x})u_c|\right], & V_e \geq \overline{V} \\ 0, & V_e \leq \overline{V} \end{cases} \end{aligned} \tag{66}$$

Step 1:

Let

$$\Lambda_c = \begin{bmatrix} 0 & 1 \\ -100000 & -1000 \end{bmatrix}, Q = \begin{bmatrix} 200000 & 0 \\ 0 & 1 \end{bmatrix} \tag{67}$$

Use $\Lambda_c{}^T P + P\Lambda_c = -Q$ to find P

$$P = \begin{bmatrix} 1150 & 1 \\ 1 & 0.0015 \end{bmatrix}, \lambda_{P\min} = 0.00063 \tag{68}$$

Determine the range of $\boldsymbol{x}$

$$0 < x_1 < 20, 0 < x_2 < 60, 0 < \|\boldsymbol{x}\| < 63.24 \tag{69}$$

Determine the range of input and output

$$10 < u \leq 90, y_m = 50 \tag{70}$$

Obtain $\overline{V}$ by

$$\left(\frac{2\overline{V}}{\lambda_{P\min}}\right)^{\frac{1}{2}} \leq 13.24, \overline{V} = \frac{\lambda_{P\min}}{2}(13.24)^2 = 0.055 \tag{71}$$

Find $f^U(\boldsymbol{x})$, $g^U(\boldsymbol{x})$, $g_L(\boldsymbol{x})$ according to

$$\begin{aligned} |f(\boldsymbol{x})| &= \left|\frac{-1}{RC^2}x_1 + \left(\frac{1}{R^2C^2} - \frac{1}{LC}\right)x_2\right| \\ &= \left|\frac{-1}{6\times(940\times10^{-6})^2}x_1 + \left(\frac{1}{6^2(940\times10^{-6})^2} - \frac{1}{(300\times10^{-6})(940\times10^{-6})}\right)x_2\right| \\ &= 188,622.3027x_1 + 3,514,662.2403x_2 = f^U(\boldsymbol{x}) \\ |g(\boldsymbol{x})| &= \left|\frac{nv_i}{LC}\right| = \frac{0.5\times160}{(300\times10^{-6})(940\times10^{-6})} = 283,687,943.2624 = g^U(\boldsymbol{x}) = g_L(\boldsymbol{x}) \end{aligned} \tag{72}$$

Set the other parameters as

$$M_f = 1,000,000,000,\ M_g = 1,000,000,000,\ \varepsilon = 2,\ \gamma_1 = 10,000,000,000,\ \gamma_2 = 500,000,000 \quad (73)$$

Step 2:

Establish the following fuzzy rules

$$\begin{aligned}&\mu_{F_1^1}(x_1) = \exp\left(-\left(\tfrac{x_1-0}{2}\right)^2\right), \mu_{F_1^2}(x_1) = \exp\left(-\left(\tfrac{x_1-4}{2}\right)^2\right), \mu_{F_1^3}(x_1) = \exp\left(-\left(\tfrac{x_1-8}{2}\right)^2\right),\\ &\mu_{F_1^4}(x_1) = \exp\left(-\left(\tfrac{x_1-12}{2}\right)^2\right), \mu_{F_1^5}(x_1) = \exp\left(-\left(\tfrac{x_1-16}{2}\right)^2\right), \mu_{F_1^6}(x_1) = \exp\left(-\left(\tfrac{x_1-20}{2}\right)^2\right)\end{aligned} \quad (74)$$

$$\begin{aligned}&\mu_{F_2^1}(x_2) = \exp\left(-\left(\tfrac{x_1-0}{6}\right)^2\right), \mu_{F_2^2}(x_2) = \exp\left(-\left(\tfrac{x_1-12}{6}\right)^2\right), \mu_{F_2^3}(x_2) = \exp\left(-\left(\tfrac{x_1-24}{6}\right)^2\right),\\ &\mu_{F_2^4}(x_2) = \exp\left(-\left(\tfrac{x_1-36}{6}\right)^2\right), \mu_{F_2^5}(x_2) = \exp\left(-\left(\tfrac{x_1-48}{6}\right)^2\right), \mu_{F_2^6}(x_2) = \exp\left(-\left(\tfrac{x_1-60}{6}\right)^2\right)\end{aligned} \quad (75)$$

such that we have 36 rules

$$\hat{f}(\boldsymbol{x}|\boldsymbol{\theta}_f) = \boldsymbol{\theta}_f^T \xi_f(x), \hat{g}(\boldsymbol{x}|\boldsymbol{\theta}_g) = \boldsymbol{\theta}_g^T \xi_g(\boldsymbol{x}) \quad (76)$$

where

$$\xi_l(\boldsymbol{x}) = \prod_{i=1}^{2} \mu_{F_i^l}(x_i) \Big/ \sum_{l=1}^{36} \prod_{i=1}^{2} \mu_{F_i^l}(x_i), l = 1, 2 \ldots 36 \quad (77)$$

Step 3:

Use the adaptive law as described in Eq. 51 to Eq. 59

Numerical simulation is performed by augmenting the controller and parametric adaptive law with the comprehensive open-loop model. **Figure 7** is the input, output current, output

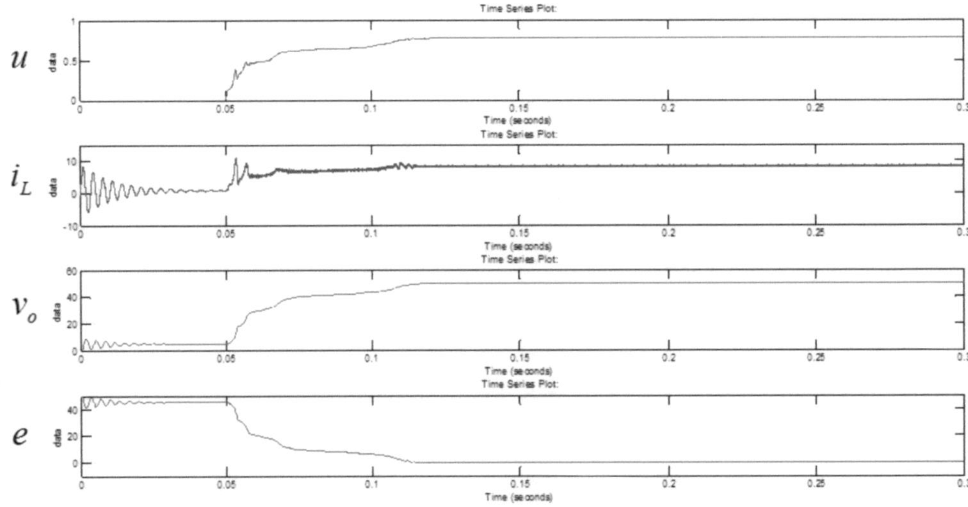

Figure 7. t = 0~0.3s.

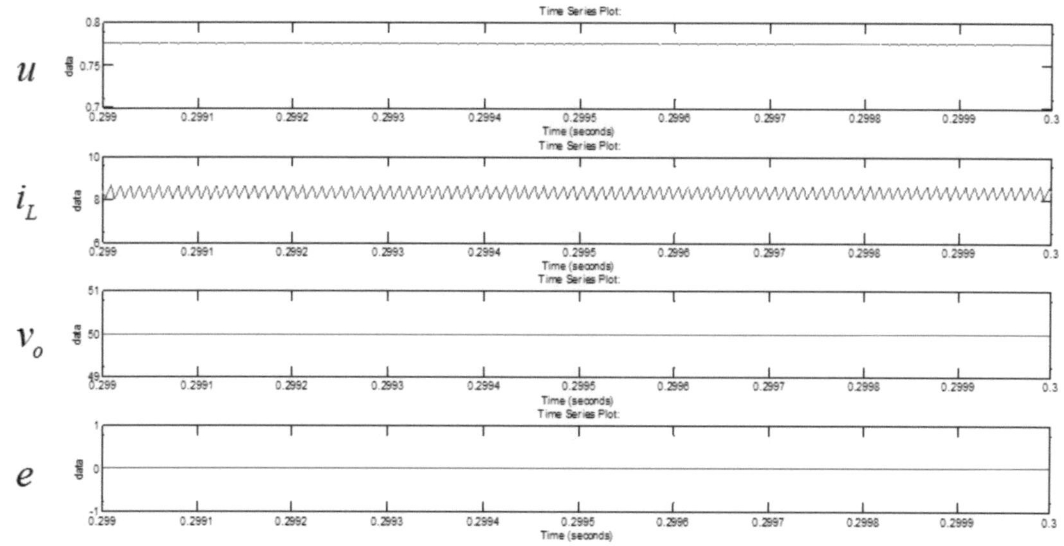

Figure 8. $t = 0.299$~0.3.

voltage, and tracking error within 0–0.3s. The input is not supplied until 0.05s to allow some transient response. Note that the tracking error converges around 0.12s. **Figure 8** is the zoomed input, output current, output voltage, and tracking error within 0.299–0.3s. We see that the output voltage converges to 50V and the tracking error converges to 0.

8. Conclusion

This chapter presents a control-oriented modeling and analysis approach for a class of PWM full-bridge power converters. The results can be extended to other categories of switching power converters with complex topology. The proposed modeling and analysis approach provides an assortment of essential information for subsequent control design, including selection of the values of circuit elements, stability characteristics of the open-loop system, controllable and observable signals/variables, and so on. Current research on feedback control of dc-dc power converters mostly focuses on systems with simple circuit topology (buck, boost, or buck/boost). In particular, control for soft switched PSPWM full-bridge converters is still limited to linearized design with PI or lead-lag compensators. The conventional linearized design approaches may overlook critical dynamics due to bilinear terms being neglected. For systems possessing nonlinearities and uncertainties of which accurate mathematical description is difficult to obtain, fuzzy control is definitely a sensible option. Moreover, in this study, we see that desirable properties are achieved (e.g., tracking, robustness) by integrating fuzzy control with parametric adaptation and sliding mode control. For future work, the experimental verification of the proposed control system is currently under progress. It is also a future plan to build a power factor correction (PFC) circuit to shape the input current of off-line power supplies for maximizing the actual power available from the mains. Another motivation to employ PFC is to comply with regulatory requirements.

Acknowledgements

The author gratefully acknowledges the support from the Ministry of Science and Technology, R.O.C., under grant MOST105-2221-E-005-047.

Author details

Cheng-Lun Chen

Address all correspondence to: chenc@dragon.nchu.edu.tw

National Chung Hsing University, Taiwan, Republic of China

References

[1] Mweene LH, Wright CA, Schlecht MF. A 1 kW 500 kHz front-end converter for a distributed power supply system. IEEE Transactions on Power Electronics. 1991;6(3):398–407.

[2] Redl R, Sokal NO, Balogh L. A novel soft-switching full-bridge DC/DC converter: analysis, design considerations, and experimental results at 1.5 kW, 100 kHz. IEEE Transactions on Power Electronics. 1991;6(3):408–418.

[3] Brunoro M, Vieira JLF. A high-performance ZVS full-bridge DC-DC 0-50-V/0-10-A power supply with phase-shift control. IEEE Transactions on Power Electronics. 1999;14(3):495–505.

[4] Cho J-G, Sabate JA, Hua G, Lee FC. Zero-voltage and zero-current-switching full bridge PWM converter for high-power applications. IEEE Transactions on Power Electronics. 1996;11(4):622–628.

[5] Hua G, Lee FC, Jovanovic MM. An improved full-bridge zero-voltage-switched PWM converter using a saturable inductor. IEEE Transactions on Power Electronics. 1993;8 (4):530–534.

[6] Jang Y, Jovanovic MM, editors. A new family of full-bridge ZVS converters. Applied Power Electronics Conference and Exposition, 2003 APEC'03 Eighteenth Annual IEEE; 2003: IEEE.

[7] Jang Y, Jovanovic MM, Chang Y-M. A new ZVS-PWM full-bridge converter. IEEE Transactions on Power Electronics. 2003;18(5):1122–1129.

[8] Redl R, Balogh L, Edwards DW, editors. Optimum ZVS full-bridge dc/dc converter with PWM phase-shift control: analysis, design considerations, and experimental results. Applied Power Electronics Conference and Exposition, 1994 APEC'94 Conference Proceedings 1994, Ninth Annual; 1994: IEEE Orlando, Florida (USA).

[9] Cho J-H, Seong H-W, Jung S-M, Park J-S, Moon G-W, Youn M-J, editors. Implementation of digitally controlled phase shift full bridge converter for server power supply. Energy Conversion Congress and Exposition (ECCE), 2010 IEEE; 2010: IEEE Atlanta, Georgia (USA).

[10] Lim J-G, Lim S-H, Chung S-K, editors. Digital control of phase-shifted full bridge PWM converter. 2007 ICPE'07 7th International Conference on Power Electronics; 2007: IEEE Daegu, Korea.

[11] Liu Y-F, Meyer E, Liu X. Recent developments in digital control strategies for DC/DC switching power converters. IEEE Transactions on Power Electronics. 2009;24(11):2567–2577.

[12] Tseng K-H, Chen C-L, editors. Design and hardware implementation for a full-bridge phase-shift PWM DC/DC converter system with FPGA-based PI gain-scheduling control. 2011 6th IEEE Conference on Industrial Electronics and Applications (ICIEA); 2011: IEEE Beijing, China.

[13] Schutten MJ, Torrey DA. Improved small-signal analysis for the phase-shifted PWM power converter. IEEE Transactions on Power Electronics. 2003;18(2):659–669.

[14] Vlatkovic V, Sabate JA, Ridley RB, Lee FC, Cho BH. Small-signal analysis of the phase-shifted PWM converter. IEEE Transactions on Power Electronics. 1992;7(1):128–135.

[15] Chen W, Lee FC, Jovanovic M, Sabate JA, editors. A comparative study of a class of full bridge zero-voltage-switched PWM converters. Applied Power Electronics Conference and Exposition, 1995 APEC'95 Conference Proceedings 1995, Tenth Annual; 1995: IEEE Dallas, Texas (USA).

[16] Alonge F, D'Ippolito F, Raimondi FM, Tumminaro S. Nonlinear modeling of DC/DC converters using the Hammerstein's approach. IEEE Transactions on Power Electronics. 2007;22(4):1210–1221.

[17] Ortiz CEC. On the circuit oriented average large-signal modeling of switching power converters and its applications. Virginia Polytechnic Institute and State University; 2003 Blacksburg, Virginia (this is a doctoral dissertation).

[18] Iannello C, Luo S, Batarseh I. Small-signal and transient analysis of a full-bridge, zero-current-switched PWM converter using an average model. IEEE Transactions on Power Electronics. 2003;18(3):793–801.

[19] Tsai F-S. Small-signal and transient analysis of a zero-voltage-switched, phase-controlled PWM converter using averaged switch model. IEEE Transactions on Industry Applications. 1993;29(3):493–499.

[20] Van Dijk E, Spruijt J, O'sullivan DM, Klaassens JB. PWM-switch modeling of DC-DC converters. IEEE Transactions on Power Electronics. 1995;10(6):659–665.

[21] Chen C-C, Chen C-L, Chang J-X, Yang C-F. LPV Gain-scheduling control for a phase-shifted PWM full-bridge soft switched converter. IFAC Proceedings Volumes. 2014;47(3):6135–6140.

Chapter 9

Fuzzy Optimization Control: From Crisp Optimization

Makoto Katoh

Additional information is available at the end of the chapter

http://dx.doi.org/10.5772/67969

Abstract

This section shows interesting contents from the development results of author's past crisp optimization combustion control concerning real boilers of fossil power plants to the upper and lower separation new fuzzy optimization control system plan. The fuzzy decision-type optimization is for elevators and the fuzzy table-like control with zero is for a single-element level control of one tank model. In addition, other researchers' recent researches concerning other applications are introduced to maintain fairness and balance.

Keywords: crisp optimization control, fuzzy optimization control

1. Introduction

In section 1, the section hierarchy and abstract of this chapter is introduced. Moreover, sometimes, abstract of many studies on many kinds of fuzzy optimization control systems are also introduced and discussed. Current interests [1] are mathematically steady and will become active in the future, whereas Mandani type fuzzy control is comprehensible and has been installed already to a marketing personal computer control system for a coupled tank level [2].

In section 2, the composition of the optimizing control system, which the author and the colleagues of an enterprise had developed [3] and the recent evolution of optimization part are introduced. The first real process application version consisted of the optimal search part with the restriction by an upper computer, a usual cascade control by decentralization digital control system and various input parts and an interface part for them. The optimization search method was developed from some general nonlinear programming to an integer programming that combined a local search and a boundary search using a simple pattern. Afterward, the optimization part has been enhanced to methods using the double patterns with more evaluation cells [4] and to methods in many books for optimization technique [5–7] except for boundary pattern search. In addition, the concept of cellular automata [8] and the technique of Q learning [9] were taken and it has been enhanced in the three universities and an institute of technology.

Zhang, Maeda, and Kawachi [10] presented an optimization model in order to allocate irrigation of water, which is withdrawn from a river, to paddy field blocks in irrigation system. A fuzzy linear programming is employed in order to solve the fuzzy decision-type optimization in the model formulation for dealing with uncertainties due to randomness of hydrologic and hydraulic parameters and fuzziness in management goals.

Multi various patterns and the other concept for optimization are introduced in a literature by the author [11].

Then, recent literature by Fujita, Tani, and Kawamura et al. [12] on fuzzy optimum control theory based on fuzzy maximization decision method of built structure are introduced.

In Section 3, a single input single output (SISO) feed-forward and feedback control system with a Table Base Controller with Zero (TBCZ) as one of Table Base System (TBS) is proposed using crisp number, which is expected to provide some advantages.

There, a feed-forward controller refers to an inverse transfer function of the controlled object.

It is proposed that a control table for three inputs of PID and three types of membership functions are not in fuzzy sets, but rather in crisp sets in expectation of the some advantages:

Some simulation results and evaluations are also shown there.

In Section 4, a fuzzy decision method based on crisp numbers with not only 1D-2D fuzzy evaluation membership functions and 1D-2D fuzzy restriction membership functions but also 3D description of their membership functions with overview plan, and a search method on the overview plan is proposed as a kind of fuzzy optimization part.

2. Examples of crisp and fuzzy optimization control systems

2.1. A crisp example in real plants

The optimizing control is divided into three parts, the optimization part in the upper system, the digital control part in the lower system, and the interface part where they are connected to former two parts. The method of the search for the combination pattern of the local search part and the boundary search part [4] was used as an upper system.

Figure 1 shows a system block diagram of the optimum combustion control system (MHI Operation Support System), which is a real example of crisp optimization controls [3]. The output chart of optimization part is shown in **Figure 2** [3] and control logic part is shown in **Figure 3** [11]. The patent [11] has more information on the optimization control.

http://dx.doi.org/10.5772/67969

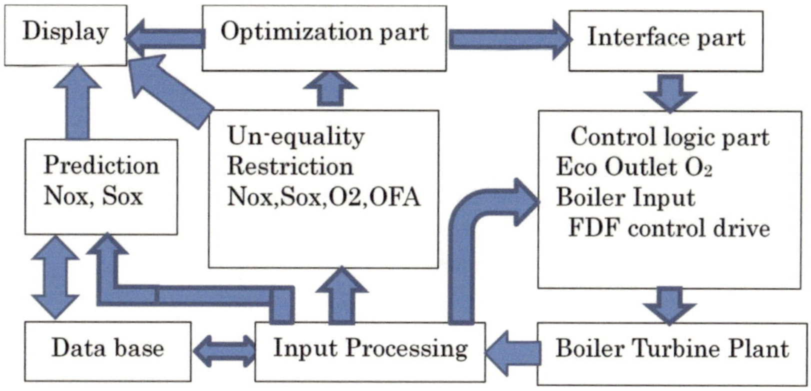

Figure 1. System block diagram of an optimization control.

This example is very important, for horizontally developing the optimizing control with the three-layer structures in the future, and as an example of applying a real machine of the optimizing control in a thermal power generation process that is a multivariable system with large changes though it has not arrived at "fuzzy" optimizing control yet.

However, the initially developed artificial intelligent language and the detail search algorithm of the local search agent and the boundary search agent could not open for secret of know-how. Then, the author has developed a new evolutionary cellular automata algorithm using

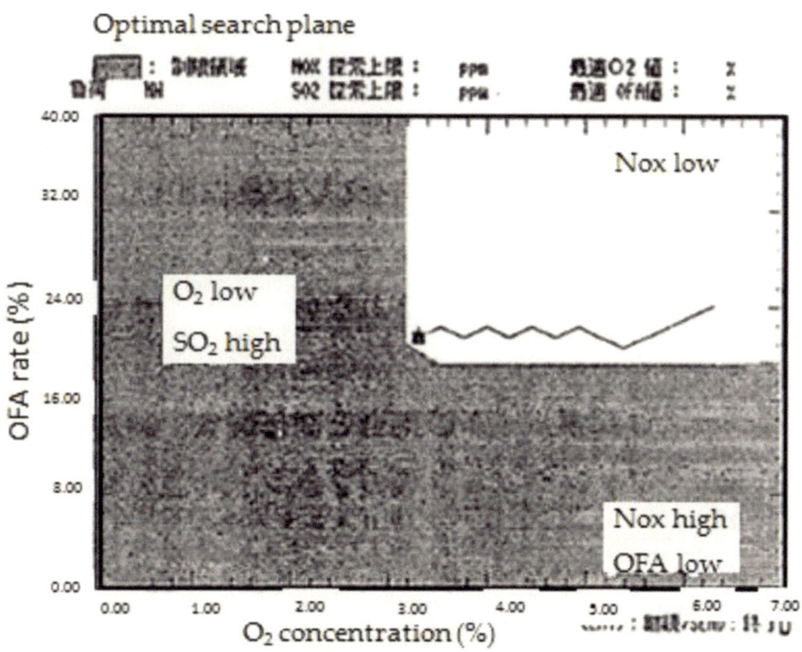

Figure 2. An example of optimization search for maximizing a boiler efficiency in OFA rate and O_2 concentration plane.

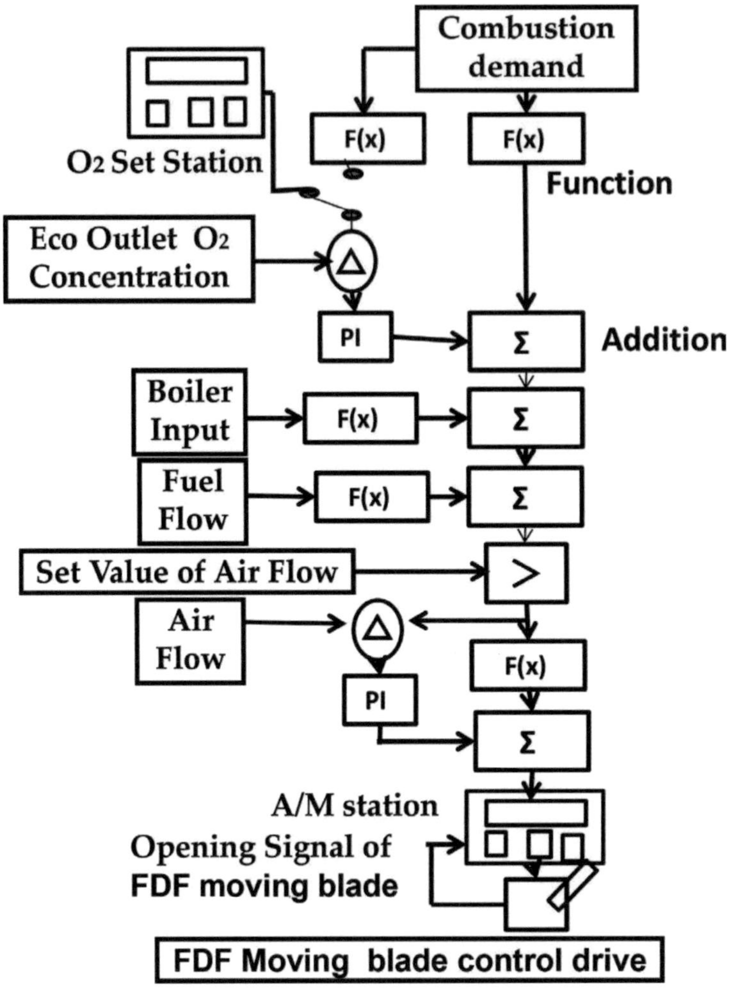

Figure 3. Eco outlet O_2 feedback control logic with optimization signal.

different patterns to low cost squares in the upper optimization control system though they are cost up. They are introduced in the next paragraph.

2.2. A crisp evolutionary algorithm for hybrid optimization control

In this paragraph, a crisp evolutionary algorithm for hybrid optimization control by multiagents is shown, because reliability and adaptability of the above hybrid optimization must be needed to increase and open.

Figures 4 and **5** show the cellular (remaining to use fuzzy number) structure of local search agents to search for the peak in the boundary line and boundary search agents to search for the peak out of the boundary line. Moreover, increasing reliability and adaptability of these search methods are shown [8].

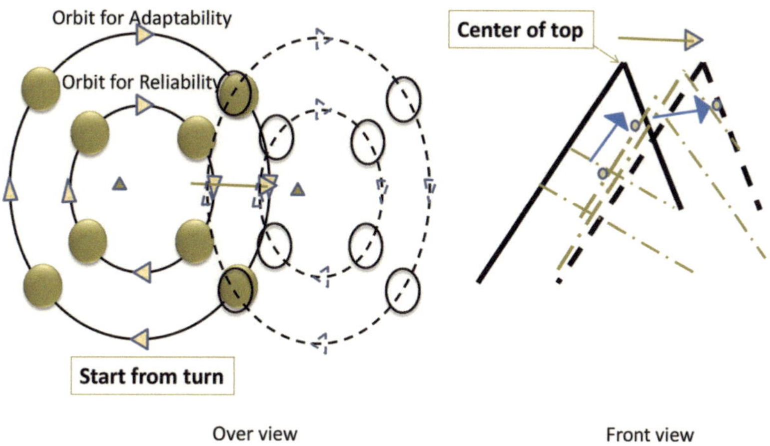

Figure 4. Cellular structure of local search agents for reliability and adaptability.

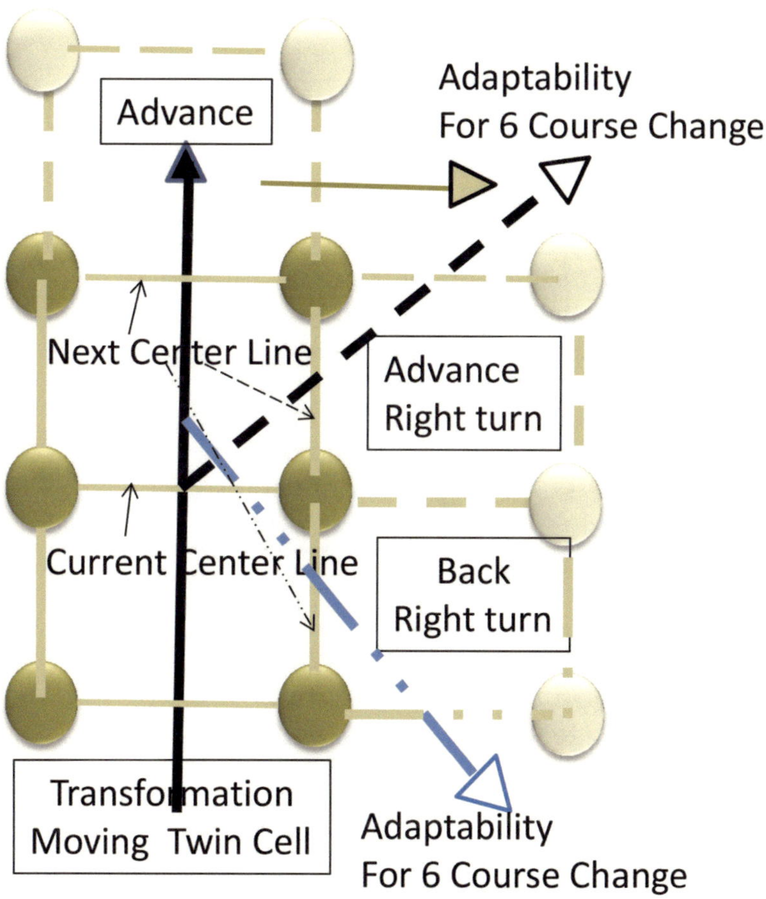

Figure 5. Cellular structure of boundary search agents for reliability and adaptability.

The following features are obtained by using eight cells group in an agent for evaluation of double circles as rotated machines as shown in **Figure 4**.

1. Adaptability to movement of mountains
2. Evaluation to continuous values
3. Application to multipeak

The following features are obtained by using six cells group in an agent for evaluation of double squares as moving cars with six wheels as shown in **Figure 5**.

1. Decrease of derailment probability
2. Evaluation to discrete values
3. Application to bifurcated boundary by using agents with different priority evaluation order

Oppositely, there are the following features in the quadrangle (square) agents adopted with an initial real system.

1. It is easy to develop because algorithm is easy.
2. There is a room to enlarge fuzzy circle.

Figure 6 shows local and boundary search algorithm using cellular automata.

The cellular automata algorithm has the following features.

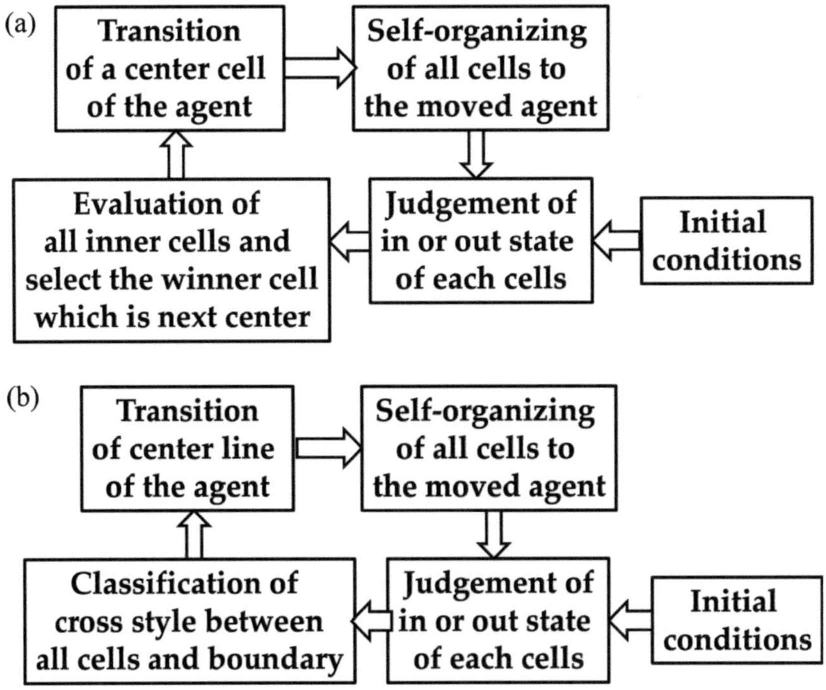

Figure 6. Cellular automata algorithm for local and boundary search algorithm. (a) Local search. (b) Boundary search.

1. Multistart points can set to any positions.
2. Four steps of center transition, self-organization of all cells to the moved agent, in or out state judgment, are repeated.

In the case of four-cell agents like four wheel cars in **Figure 2**, center line must be replaced to front line in automata (b) of **Figure 6**.

2.3. A fuzzy example of built structure

On the other hand, Fujita, Tani, and Kawamura et al. at Kobe University [12] execute fuzzy optimum control theory based on fuzzy maximization decision method (target conditions and restriction condition are expressed by fuzzy membership functions) to built structure, and it is called intelligent active control. Not only target response displacement and target control power but also structure identification values obtained from the responses of the structure and earthquake vibration forecast obtained from earthquake input measurement were used for the structural response forecast.

They decided the parameters using the fuzzy maximization decision method. Then, the effect of controls were examined by experiment and simulation about two methods for addition of control power added the feedback only of the ground vibration input acceleration power and the relative acceleration power's from base of structure.

They reported that the feedback control with optimal coefficients can imorove the rate of improvement by 30% higher or more.

This report feels the difficulty for execution of fuzzy optimization control though it can encourage the execution.

In the following section, a table base controller with zero (TBCZ) is introduced for easiness of tuning of membership functions more than conventional fuzzy control.

3. A fuzzy-like table base PID controller with zero

The purpose of this section is to design and evaluate a Table Base Controller with Zero (TBCZ) as one of Table Base System (TBS) [13].

3.1. Table base controller with zero

3.1.1. TBCZ configuration

In this paragraph, we propose the following SISO feed-forward and feedback control system with a TBCZ in **Figure 7**, which is expected to provide the following advantages:

1. Fast start up
2. Small overshoot
3. Easy maintenance using the tuning knowledge of the conventional PID control

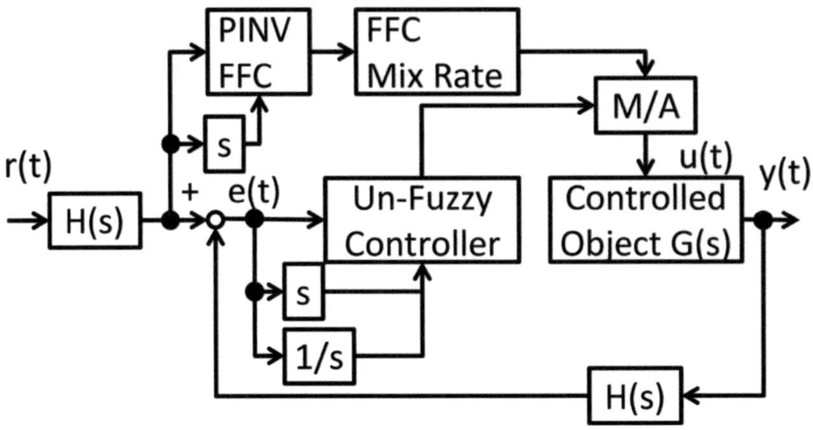

Figure 7. Configuration of a TBCZ.

Moreover, M/A means a manual/auto switch station, *s* means a differential operator, and *1/s* means an integral operator in **Figure 7**.

3.1.2. *Table of TBC*

Figure 8 shows a table of TBCZ and rectangular membership functions instead of usual triangle membership functions. Here, SUMi contains not only the proportional scaling factor SF_p and the differentiation scaling factor SF_d but also the integrator scaling factor SF_i. Finally, the scaling factor of the input SF_u must not be overlooked.

The rectangular membership functions are easy on computation because they are crisp.

e \ Δe	N	ZERO	P
N	SUMi	SUMi	ZERO
ZERO	SUMi	ZERO	SUMi
P	ZERO	SUMi	SUMi

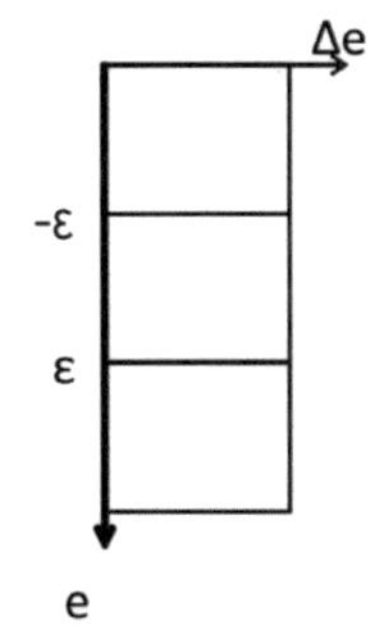

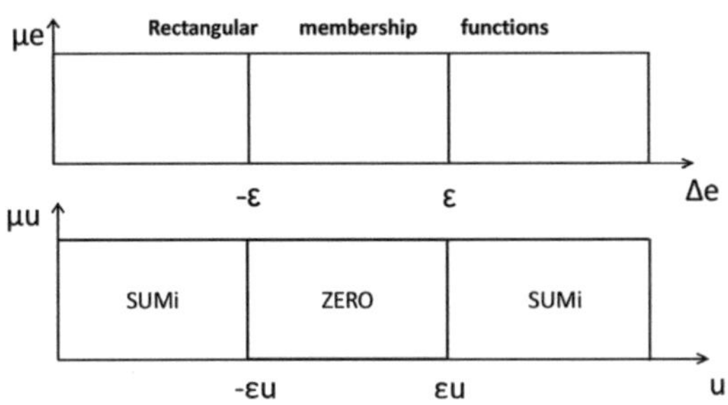

Figure 8. A table of TBCZ and membership functions.

3.2. Modeling, parameter tuning, and evaluation

Here, assumed characteristic of up-down symmetry without hysteresis, and neglected pump with long nose tube reached in water of tank and PWM control dynamics, which is sufficiently fast.

Then, we modeled the control using the following simple linear transfer functions:

$$G(s) = \frac{Ke^{-Ls}}{Ts+1}; K = 30, T = 20, L = 2 \tag{1}$$

$$H(s) = \frac{K_s}{T_s s+1}; K_s = 1, T_s = 0.1 \tag{2}$$

3.2.1. Parameter tuning

3.2.1.1. Scaling factor in the TBCZ

The followings are examples of scaling factors in the TBCZ.

$$\begin{aligned} &\mathrm{P} = \frac{1}{SF_p}, \quad \mathrm{D} = \frac{1}{SF_d}, \quad \mathrm{I} = \frac{\frac{1}{s}}{SF_i}, \quad \varepsilon_0 = 0 \\ &\frac{1}{SF_p} = 0.15, \quad \frac{1}{SF_i} = 0.0091, \quad \frac{1}{SF_d} = 0.1 \end{aligned} \tag{3}$$

3.2.2. Performances of 3 × 3 and 2 × 2 tables of TBCZ

An evaluation of the mean integral square error and input (MISEI) was compared for various values of the terminator ε in **Figure 9**.

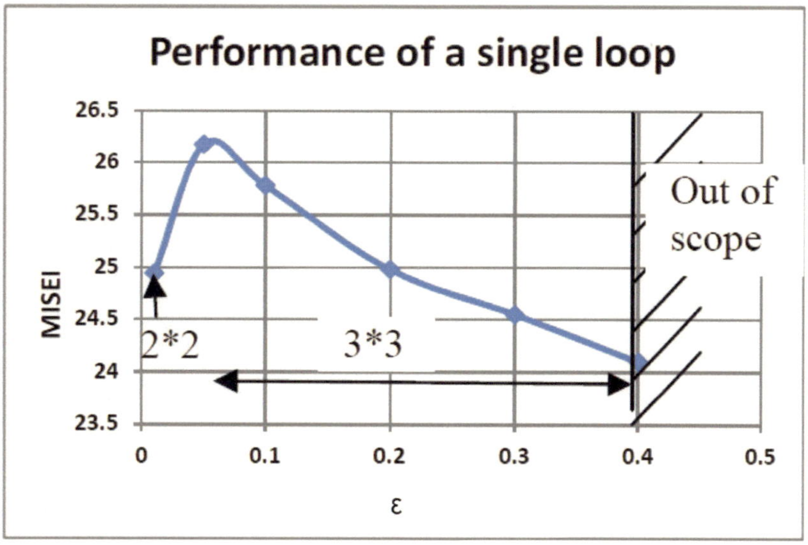

Figure 9. MISEI vs. ε (in a table of TBCZ).

The performance of ε = 0.01 is considered to be equivalent to that of the 2 × 2 table of TBCZ, it is one point except for line from 0.1 to 0.4 in **Figure 9**. Thus, the 3 × 3 table of TBCZ is superior than the 2 × 2 table of TBCZ since this table allows superior performance tuning. If performance is valued, then a larger table is better, although this results in high-cost and increased complexity. Performance of MISEI is superior in the case of smaller Center ZERO. There is a minimum point of MISEI on the edge of out of scope.

3.3. Other considerations to TBCZ

The robust modification of the experimental PID tuning method of Ziegler Nichols could be used here. **Figure 10** shows performance MISEI of a double loop.

The FFC mix rate in the double loop case must be less than half that in the single loop case.

The decoupling [14] study is omitted in this paragraph, then refer to in reference [13].

3.4. Subconclusion

A new concept of a TBCZ with rectangular membership functions based on crisp sets, which was featured by ZERO's in the rule table like the fuzzy-like control table as one of TBC and a feed-forward control line were proposed, and simulation results are presented for a tank level control as an example. Then, superior evaluation based on MISEI and performance was obtained.

The membership functions of the proposed TBCZ were able to easily tune only terminators, which mean the size of ZERO's through the evaluation of MISEI.

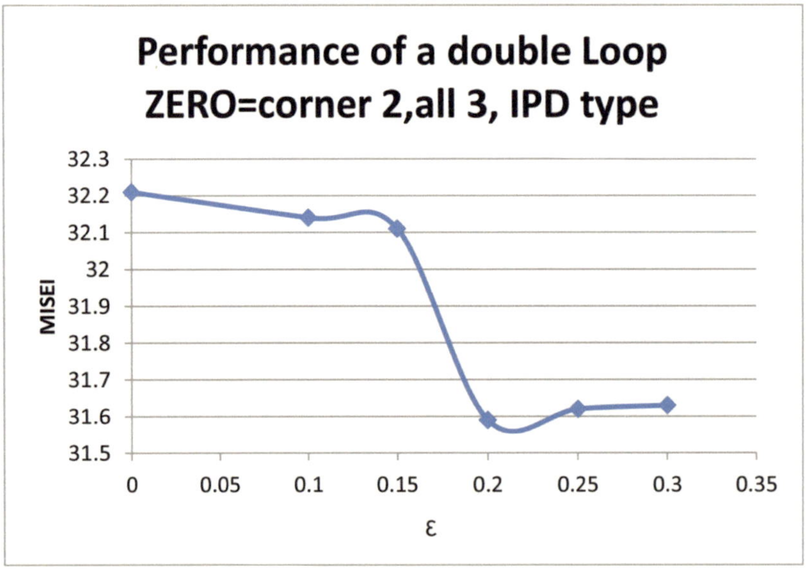

Figure 10. MISEI vs. ε (in a table of TBCZ).

Making and tuning of the controller are easier than in conventional fuzzy control because the membership functions are rectangular without common parts in crisp sets, and the only control rules are ZERO and SUMi in a 3 × 3 control rule table.

The development of a robust compensator of integrator for no-overshoot property is a future theme. The readers can find the literature [15] on conventional system.

4. Fuzzy decision-type optimization

It proposes a fuzzy decision-making type optimization technique in this paragraph as an example of the problem on elevators.

Wada and Kato propose an optimization technique of fuzzy rules according to the situation by using behavior acquisition based on emotional memory that uses fuzzy sets on pleasantness and unpleasantness of a robot. And the robot is made to acknowledge pleasantness and unpleasantness by using a source of light and an experiment toward a comfortable goal evading the obstacle is done [16].

The fuzzy decision-type optimization for an elevator is proposed firstly in ordinary 2D membership description [17]. In this paragraph, it is enhanced to 3D description.

The other fuzzy optimization studies [18–22] are interesting for this study.

4.1. Maximizing decision probability methods

In this paragraph, firstly, the problem of fuzzy decision-type optimization (maximization) with subjects is defined generally.

Assuming that x_1 and x_2 are fuzzy numbers, membership functions μ_1 and μ_2 are introduced according to x_1 and x_2 and λ is a scalar for λ-cut.

$$\begin{array}{lll} \underset{x \in X}{\text{maximize}} & & \min\{\mu_1(x_1),\ \mu_2(-x_1)\} \Leftrightarrow \\ \text{maximize} & & \lambda \\ \textit{subject} & \textit{to} & \alpha_1 x_1 + \beta_1 \geq \lambda,\ -\alpha_2 x_2 + \beta_2 \geq \lambda,\ x \in X \end{array} \tag{4}$$

This technique can be used for fuzzy deciding. For example, it is decided using this technique whether an "almost crowded" elevator should pass over a certain floor with "long queuing length". These two fuzzy sets "almost crowded" and "long queuing length" are described by using fuzzy numbers of passengers in the elevators and queues in the floor.

x_1 is defined to the number of passengers ride on an elevator, and x_2 is defined to the number of queuing in an elevator lobby of a floor. An example of detail equations and measurement method of person numbers can be referred to the literature [23].

The fuzzy decision method is called to maximizing (min-max) decision because it is optimize when the product (minimum) of two membership functions is max as in **Figure 11**.

where

$$\lambda_{max} = \max_{\{z_1^0, z_2^0\}} \min_{\{z_1^0, z_2^0\}} \{\mu(x_1), \mu(-x_2)\}$$
$$\lambda_c = \frac{1}{3}\lambda_{max}, \quad \lambda_0 = 0 \tag{5}$$

The notation of the trapezoid membership function is described by the three terms set ⟦*left_terminator, center, right_terminator*⟧ as same as the triangle membership function (center means the position of grade 1, and terminator means the position of grade 0) and is inserted the end point of right and left terminator by infinity mark ∞, −∞ as the following G and C. Then, the product (minimum calculation) D of G and C is described, if you devise it as multiplying the scalar corresponding the max value $\mu_D(x^*)$; x^* means center of D, then you can write as follows.

$$G = [g_1, g, \infty], \qquad C = [-\infty, c, c_2] \tag{6}$$

$$D = \mu_D(x^*)[g_1, x^*, c_2] \tag{7}$$

where x^*, $\mu_D(x^*)$ can be obtained easily because they are coordinate values of a cross point of two lines in the case of triangle or trapezoidal membership functions.

In multiobjective decision-making, generally it is common to narrow down to the only optimum solution by using preference function from among plural noninferior Pareto solutions, and depending on the shape of this preference function, decision maker's preference is divided into whether it is risk avoidance type, risk-oriented type, or risk neutral type.

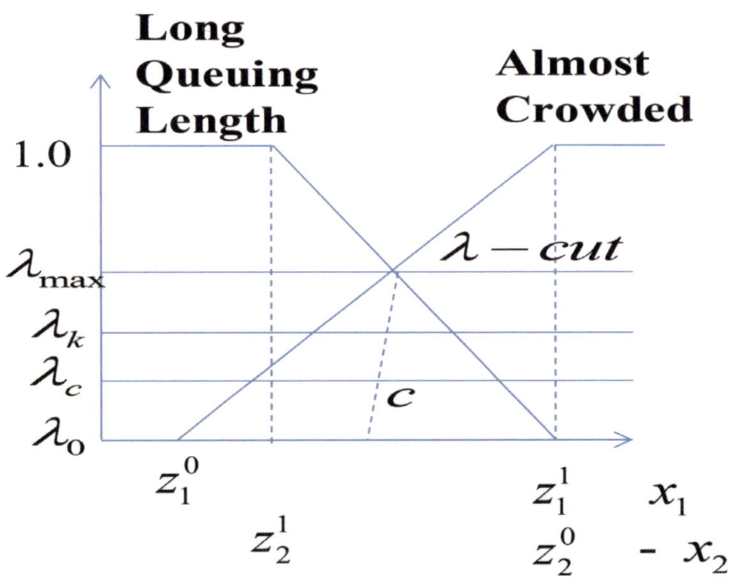

Figure 11. Membership functions of fuzzy sets "Almost crowded" and "Long Queuing Length" with a variable λ-level on fuzzy passage decision of "Maybe Pass a Waiting Floor".

When the preference function is convex, the case is the risk avoidance type. When it is concave, the case is the risk-oriented type.

When the case has multipurpose G_i (i = 1,…,n) and numerous restrictions C_j (j = 1,…,m), their common set D is defined equally as follows.

$$D = G_1 \cap G_2 \cap \cdots \cap G_n \cap C_1 \cap C_2 \cap \cdots \cap C_m \tag{8}$$

When getting a minimum of the respective membership function μ_{Gi}, μ_{Cj}, the membership function of D is found as μ_D.

$$\mu_D = \mu_{G_1} \wedge \mu_{G_2} \wedge \cdots \wedge \mu_{G_n} \wedge \mu_{C_1} \wedge \mu_{C_2} \wedge \cdots \wedge \mu_{C_n} \tag{9}$$

When adopting notation like Eq. (7), it can be written as follows.

$$D = \mu_D(x^*)\left[\{minx, s.t., \mu_D(x) = 0\}, x^*, \{max\ x, s.t., \mu_D(x) = 0\}\right] \tag{10}$$

Here, about (x^*, $\mu_D(x^*)$), because they are the coordinates of an intersection point of two straight lines, they can be found easily when finding the set of which D is composed. About both end points, the finding method is like the same.

4.2. Expendability of the proposed method

4.2.1. 3D description

Figure 11 with only a front view becomes **Figure 12** with a front view, a side view and an over view when elevator passenger number x_1 and queuing line number x_2 is selected as independent logically double axis if three view description is adopted in this fuzzy optimization. This change may be increase possibility of fuzzy optimization approach.

The following **Figure 12** comes next **Figure 13** not making the independent double axis of regular axis and reverse axis like **Figure 11** but making psychological one axis by describing variable X_3 (the underside is unpleasant and the upper side is pleasant) in addition to physical two axis by describing variables X_1 and X_2.

Here, "Almost Crowded" and "Long Queuing Length" are unequality conditions and "Passenger pleasure" is an objective function. It is the fuzzy decision-making method that "pass of an elevator to a floor" is done by maximizing decision method using min-max of three member ship functions (a blue point by minimax method). Whereas it is the crisp decision-making that the decision-making is done at the grey point of which pleasant degree is higher slightly in grey area which grade of all membership functions are one. It's a problem that the comfort level of the waiting line isn't considered against the comfort level of the passenger in this fuzzy decision-making area. Though the comfort level of the passenger is sacrificed, it can be said that the comfort level of the waiting line is considered in the crisp decision-making area.

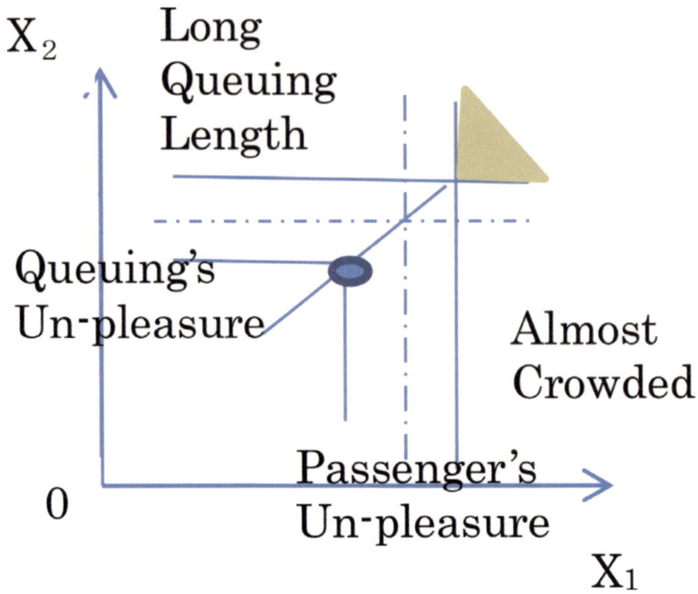

Figure 12. Over view plan on passengers' un-pleasure caused by almost crowded and queuing's unpleasure caused by long queuing length for a fuzzy optimization problem.

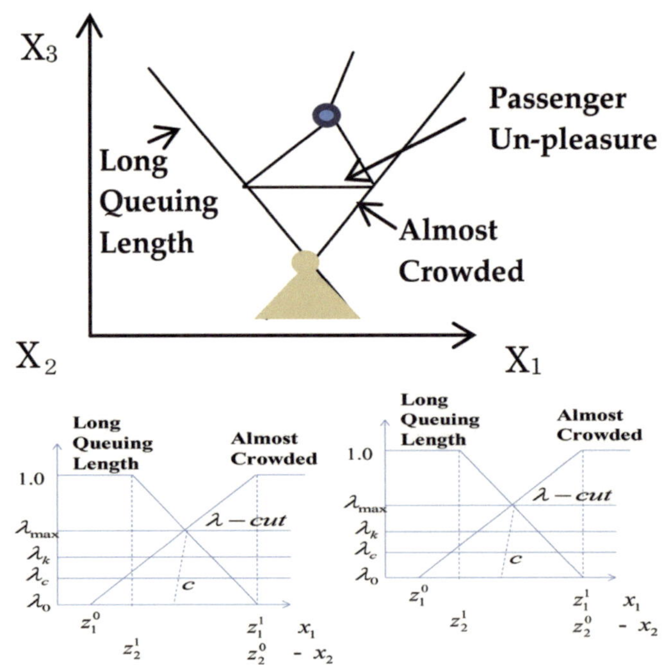

Figure 13. Three side plan in physics and psychological crossing at right angle axis of two dimensions on a fuzzy optimization problem.

The three side of plans (over view, front view and side view) are described like **Figure 13** instead of the above plan like **Figure 12** when X_2 and X_3 out of the three axes are changed like **Figure 13**. Usually, only front view plan is indicated as **Figure 11** and others are omitted for economy and maximum decision policy. Here, overview plan is also indicated because minimum decision policy in overview plan is better than the maximum decision policy as understand if you see.

These plans permit two kinds of optimization methods, minimizing in overview plan and maximizing in front plan.

4.2.2. Search on overview plan

For searching minimum point on overview plan less than grade 1 such as **Figure 14**, mobile method of freely writing lines and circles like dance on the plane in order to evaluate grade values using the following equations with the rotation matrix is proposed here. This is a discrete mobile model which generates left and right double velocities v_L and v_R of interval d.

$$v_G(k) = v_L(k) + v_R(k); r = 0.5d \tag{11}$$

$$\omega_G(k) = \frac{1}{r}(|v_R(k)| - |v_L(k)|) \tag{12}$$

$$\Delta\theta_G(k) = h\omega_G(k); \theta_G(0) = \theta_{G0} \tag{13}$$

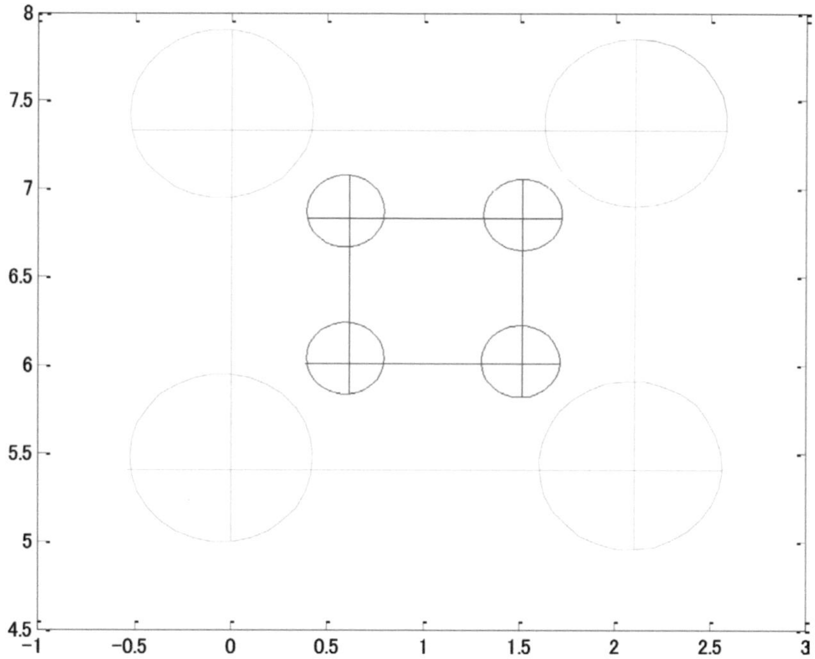

Figure 14. An example pattern of searched course for reduced repeat by a discrete mobile agent.

$$\theta_G(k+1) = \theta_G(k) + \Delta\theta_G(k) \tag{14}$$

$$v_G(k+1) = \begin{bmatrix} \cos\Delta\theta_G(k) & -\sin\Delta\theta_G(k) \\ \sin\Delta\theta_G(k) & \cos\Delta\theta_G(k) \end{bmatrix} v_G(k) \tag{15}$$

$$p_G(k+1) = p_G(k) + hv_G(k+1); p_G(0) = p_{G0} \tag{16}$$

where suffix G means center of the mobile object, suffix L means left, suffix R means right, p means position, ω means angle velocity, θ means angle from x-axis, and h means sample time.

Figure 14 shows an example pattern by a mobile agent and the reduced inner copy for design of searched course in the plane.

Addition of inverse kinematics will make easier the process of drawing figures filled in the canvas.

4.3. Action mode, effective, influence, yield analysis (AMEIYA)

A table on action mode effect, influence, yield analysis (AMEIYA) may be obtained by imitating reasoning methods from left-column action mode to right-column effect, influence and yield things.

A Q&A table used to obtain the above table on AMEIYA may also be made by readers referred to Ref. [23].

In the future, new strategic systems for fuzzy optimization control on elevators will expect to born from these tables.

5. Conclusions

In this chapter, a fuzzy optimization control system by combining fuzzy decision-type optimization parts and a fuzzy-like table base PID controller parts was proposed separately instead of the conventional united crisp optimization control system.

Double future themes on fuzzy optimization, which may be used in fuzzy optimization control, were also presented. One is three side plans with an overview plan, which omitted conventional studies. The other is a new search method on a plane using a new simple discrete mobile model, which generates left and right double vector velocities.

These two separate parts are expected to be united and to be evaluated. Moreover, optimization based on fuzzy numbers and calculations may be used for absorbing uncertainty of sensors output and digitalizing of input to computer.

Moreover, some idea of strategic systems on fuzzy optimization control was provided for readers in the future.

http://dx.doi.org/10.5772/67969

Acknowledgements

The author presents cordial acknowledgments to the cooperation of his students Ms. Natsuki Imura, Ms. Xue Li, Mr. Keibun Wang, Mr. Toru Ueno, Mr. Junnichi Sawaki, Mr. Takayuki Ozeki, Mr. Takuma Nishikawa, and Mr. Koichi Wada for their cooperation and presentations in the references on this study. Moreover, he thanks and apologies for Dr. Shuichi Isomura and many colleagues of MHI Co. Ltd, Mr. Mizuno and Mr. Manabe of Hokkaido Power Co. Ltd., many teachers and students of Osaka University, Toin Yokohama University, Osaka Institute of Technology, Hosei University, and family, who cooperated some works, lectures, and a life.

Author details

Makoto Katoh

Address all correspondence to: makoto.kato@oit.ac.jp

Osaka Institute of Technology, Hosei University, Osaka, Japan

References

[1] N. Furukawa, "Mathematical Theory of Fuzzy Optimization", Morikita Publisher, Tokyo, 1999 (in Japanese).

[2] Manual of "A Coupled Tank", PID Co. Ltd, Tokyo, 2011.

[3] M. Mizuno, K. Katsuki, M. Katoh, K. Mizuochi, K. Takahashi and K. Ryotaro, "Operation Assistance System for Fossil Thermal Power Plant for Sunagawa Power Station No.3 and 4 Units of The Hokkaido Electric Power Co. Inc ", MHI Technical Reports, 27, 33–36.

[4] M. Katoh and K. Hasegawa, A Cooperative Heuristic Search Approach By Multi Agents, Proceedings of SMC 2000, IEEE, New York, pp. 277–279, 2000.

[5] S. S. Rao. "Engineering Optimization Theory and Practice Third Edition", Wiley Interscience Publication, John Wiley & Sons, Inc, New York, Chichester, Brisbane, Toronto, Singapore, 1995.

[6] Edgar, Himmelblau, Lasdon, "Optimization of Chemical Processes", Mcgrowhill, Boston, Burr Ridge, Dubuque, Madison, New York, San Francisco, St. Louice, Bangkok, Bogota, Caracas, Kuala Lumpur, Lisbon, London, Madrid, Mexico City, Milan, Montreal, New Delhi, Santiago, Seoul, Singapore, Sydney, Taipei, Toronto, 2001.

[7] N. S. Rau. "Optimization Principles", Wiley Interscience, 2003.

[8] M. Katoh, K. Wada and T. Nishikawa. "A Cooperative Heuristic Search Approach By Multi Agent Cellular Automata", Proceedings of JSME No.1–63, Cellular Automata Symposium, pp. 171–173, 2001.

[9] M. Katoh, R. Simotani and K. Tokushige "Integrated Multiagent Course Search to Goal by ε-Greedy Learning Strategy—Dual Probability Approximation Searching-", Proceedings of SMC 2015, IEEE, New York, pp. 382–392, 2015.

[10] Qin Zhang, Shigeya Maeda and Toshihiko Kawachi, "Fuzzy Optimization Model for Allocating Irrigation Water to Paddy Fields", Transactions of JSIDRE, 2007, No.249, pp.55–62, 2007.

[11] M. Katoh, J P 3 7 0 8 5 8 1, B 2 2 0 0 5 . 1 0 . 1 9, (in Japanese)

[12] H. Fujita, A. Tani, H. Kawamura and A. Takizawa, "Study on Activation Methods of Control Forces", Experimental Research on Intelligent Fuzzy Optimal Active Control Forces, 279/280, 2001 (in Japanese).

[13] M. Katoh and T. Ueno, "Design of an Un-fuzzy Controller for a Tank Level Control", Proceedings of 2013 International Conference on Fuzzy Theory and Its Application, pp. 340–345, 2013.

[14] M. Matsuzawa and M. Saeki, "Entrainment in interpersonal motor coordination of oscillating lower legs", Annual Bulletin of Showa Woman's University, Vol. 11, 9–15, 2008.

[15] M. Katoh, No Overshoot Step Response of an Integral Controller with a Simple Robust Lead Inverse Compensator to Improve Safety of Vehicles, IJACE, Vol. 5, 1/11, 2016.

[16] K. Wada and M. Katoh, "Behavior Acquisition of Mobile Robots with Emotional Memory System", Proceedings of SICE System Integration Conference, 69/70, 2002.

[17] M. Katoh and N. Imura, "A Plan for Fuzzy Passage Decision of Elevators Using Equivalent Person Number and AMEA", IJACE, Vol. 5, Issue (1), pp. 47–52, 2012.

[18] H. Suito and H. Kawarada, "An algorithm of Fuzzy Optimization", Department Bulletin Paper of Kyoto University, Kyoto. Research Information Repository. 1129, 95/107, 2000.

[19] M. Sakawa and K. Katoh, "Large Scale of Multi-Objective Fuzzy Programming, Operation Research", Science of Business, 47(5), 2002, pp. 295–301 (in Japanese).

[20] R.E. Bellman and Zadeh, "Decision making in a fuzzy environment", Management Science, Vol. 17, 1970, pp. 141–164.

[21] Y. Gao, G. Zhang, J. Ma and J. Lu, "A λ–Cut and Goal-Programming-Based Algorithm for Fuzzy-Linear Multiple-Objective Bi-level Optimization", IEEE Transactions On Fuzzy Systems, Vol. 18, No. 1, 2010, pp. 1–13.

[22] P. T. Chang and K. C. Hung, "α- Cut Fuzzy Arithmetic: Simplifying Rules and a Fuzzy Function Optimization with a Decision Variable", IEEE Transactions On Fuzzy Systems, Vol. 14, No. 4, 2006, pp. 496–510.

[23] M. Katoh, T. Ozeki, and J. Sawaki, "A Strategic Question Inquiry System for Fuzzy Decision of Elevator Street", 2012 9th International Conference on Fuzzy Systems and Knowledge Discovery, 2012.

Section 3

Applications of Fuzzy in Navigation Systems

Chapter 10

An Approach of Fuzzy Logic H_∞ Filter in Mobile Robot Navigation Considering Non-Gaussian Noise

Hamzah Ahmad, Nur Aqilah Othman and
Saifudin Razali

Additional information is available at the end of the chapter

http://dx.doi.org/10.5772/intechopen.68866

Abstract

This chapter has presented an analysis of H_∞ filter-based mobile robot navigation with fuzzy logic to tolerate in non-Gaussian noise conditions. The technique exploits the information obtained through H_∞ filter measurement innovation to reduce the noises or the uncertainties during mobile robot observations. The simulation results depicted that the proposed technique has improved the mobile robot estimation as well as any landmark being observed. Different aspects such as γ values, noise parameters, intermittent measurement data lost and finite escape time issues are also analysed to investigate their effects in estimation. Different fuzzy logic design configurations were also studied to achieve better estimation results. As demonstrated in this work, fuzzy logic offers reliable estimation results compared to the conventional technique.

Keywords: navigation, mobile robot, H_∞ filter, estimation, fuzzy logic

1. Introduction

Working in a hazardous area has always been an issue to human safety and health. This is a situation where robotics offers an alternate solution to perform any given task. In realization of this problem, an autonomous robot is providing a suitable approach with less human monitoring system. Since the term of 'robot' defined by Karel Capek in 1920s, robotics has experienced a lot of interesting advancements and evolutions.

In general, robotics is classified into several categories stated by the Robotics Institute of America such as the variable-sequence robot, playback robot, numerical control robot and intelligent robot. There are also a number of available robotic configurations such as manipulator robot, SCARA robot, spherical type robot, cartesian robot and cylindrical robot.

The robotic research, development and technology have been immense recently considering various fields and applications. In fact, the technology has spread and is widely used in home-based appliances such as the lawn mower robot, vacuum cleaner robot and floor-washing robot. These domestic robots are sometimes designed to work automatically and independently with their own pre-described algorithm or system. Considering these conditions, two main features must be made available for the robot to perform their task, which are navigation and autonomous capabilities.

Mobile robot navigation is the main concern being focussed in this chapter. In modelling the mobile robot navigation, three approaches are available [1, 2]. First, the mathematical modelling which develops the mathematical model of the robot and the environment through the robot dimensions, configurations and the environment conditions. Therefore, designer must understand clearly the mobile robot configurations as well as the environment conditions.

Sensor information is one of the influential elements during information processing in mobile robot navigation. The information enables better mobile robot confidence in the navigation processes. This is the second type of navigation modelling which uses the information extracted from sensors. Several types of sensors can be equipped in a mobile robot to assist the navigation. Compared to the mathematical model, this technique is less complex as sensors provide relative measurements between mobile robot and any observed landmarks.

The third type is focussing on how to tolerate the noises when the mobile robot navigates. Based on probabilistic rules, particularly the Bayes rule, the mobile robot attempts to identify its location and the landmarks observed with some uncertainties. This method consumed lower mathematical computation and does not rely heavily on the sensors information. Owing to these advantages, the third type has been famously applied nowadays. This chapter focusses on the probabilistic technique for mobile robot navigation.

On the aspect of making an autonomous mobile robot, a number of approaches have been proposed especially by using an artificial intelligence technique, for example [3–5]. Jaradat and Abdel-Hafez [4], for example, have successfully applied the neural network to process information obtained through the sensors, IMU and GPS, to navigate the vehicle. Vehicle-measured force and angular velocity are measured via IMU and for the vehicle location, GPS is employed to do the work. Two hidden layer with at least six neurons are designed to calculate and analyse the vehicle movement. Their results seem to provide a good accuracy of vehicle estimation.

Inspired by the human nature of thinking and based on preceding literatures, the mobile robot can be guaranteed to move with less human monitoring and relies on the artificial intelligence to observe any environment. In this context, artificial intelligence is recognized as one of the possible techniques to provide efficient solutions. This research applies the fuzzy logic control to improve a probabilistic technique known as H_∞ filter by analysing several issues such as finite escape time [6], operating in non-Gaussian Noise, intermittently on lost measurement data and parameter effects to the estimation.

http://dx.doi.org/10.5772/intechopen.68866

2. Mobile robot navigation

The celebrated Kalman filter is the famously used approach for mobile robot navigation. The technique which is based on the minimum mean square error utilizes the prior information obtained by the sensor to update its current location. Even though Kalman filter offers reliable estimations, it is still incompetent in an environment that holds non-Gaussian noise characteristics. The noise can be considered as noise when it is not zero mean and holds a characteristic of not uniformly distributed noise. Therefore, a number of new methods are proposed such as the unscented Kalman filter, particle filter and ensemble Kalman filter, which generally use a lot of particles to infer the mobile robot position [2]. These techniques exhibit higher computational cost if the number of particles is increased. In fact, they require a fast processor to calculate and store the information during mobile robot observation. Thus, a simple and robust filter than the above-mentioned techniques is welcome to solve the navigation issues. Some of the Kalman filter limitations can be listed as follows:

- The mean and correlation of the process and measurement noise need to be known at each time instant.
- The covariances of the noises must be determined beforehand, since the Kalman filter uses these covariances as design parameters.
- The Kalman filter is the minimum variance estimator if the noise is Gaussian. However, if a different cost function (such as the worst-case estimation error) should be applied to the system, then the Kalman filter may not be suitable to accomplish the objectives.

H_∞ filter is one of Kalman filter families. The filter is proven to work in non-Gaussian noise environment and it is assumed that the noises are bounded. By defining the tuning parameters known as γ, the filter can provide a better solution than Kalman filter as well as other techniques. In this chapter, the non-Gaussian noise is assumed to be bounded and acts as a random noise when mobile robot does its observations. It is also important to note that H_∞ filter is facing one issue, that is the finite escape time where the estimation can be unbounded if the system is not satisfying the designed criteria [7]. Despite these shortcomings, the H_∞ filter performance analysis is the main objective in this chapter to propose an alternative solution for mobile robot navigation. To ensure the finite escape time is not perceived during observation, fuzzy logic technique is applied for the proposed system. The system design consisting of these approaches is presented in the next section.

3. H_∞ filter and fuzzy logic technique modelling

One of the main parts in navigation is the simultaneous localization and mapping (SLAM) problem. It states a situation where the mobile robot builds up an environment based on its sensor readings while at the same time localizes itself on the map [8, 9]. Typically, two models that simulate the system are referred to describe how the SLAM problem is being solved. The first model is known as the kinematic model and the latter is defined as measurement model. Both models are essential to define the behaviour of the system with reference to the environmental conditions.

The kinematic model simply illustrates how the mobile robot movements are recorded. The following equation demonstrates the model.

$$X_{k+1} = F_k X_k + B_k u_k + w_k \tag{1}$$

where X is the augmented states of the mobile robot (x, y) and landmark i (x_i, y_i). F and B are the state transition matrix and the control input matrix, respectively. u and w are the control input and the associated noise during mobile robot movements.

Mobile robot measures relative distance and angle to any observed landmarks location when moving throughout the environment. The measurement model holds the following equation.

$$y_{k+1} = H_k X_k + v_k \tag{2}$$

	The extended Kalman filter	The H_∞ filter
The system	$X_{k+1} = f(X_k, u_k, w_k, k)$	$X_{k+1} = f(X_k, u_k, w_k, k)$
	$z_k = h(X_k, v_k, k)$	$z_k = h(X_k, v_k, k)$
		$y_k = D_k X_k$
The prediction step	$\hat{X}^-_{k+1} = f(\hat{X}_k, u_k, 0, k)$	
	$P^-_{k+1} = \nabla F_X P_k \nabla F_X^T + \nabla F_w Q_k \nabla F_w^T$	$\hat{X}_{k+1} = \nabla F_k \hat{X}_k + \nabla F_k K_k (z_k - \nabla H_k \hat{X}_k) + u_k$
The update step	$\hat{X}^+_{k+1} = \hat{X}^-_{k+1} + K_{k+1}\ \mu_{k+1}$	$P_{k+1} = \nabla F_k P_k \left(I - \gamma^{-2} P_k + \nabla H_k^T R_k^{-1} \nabla H_k P_k\right)^{-1} \tilde{N} F_k^T + Q_k$
	$P^+_{k+1} = (I_n - K_{k+1} \nabla H_i) P^-_{k+1}$	
	$\mu_{k+1} = z_{k+1} - h(\hat{X}^-_{k+1}, 0, k)$	$K_k = P_k \left(I - \gamma^{-2} P_k + \nabla H_k^T R_k^{-1} \nabla H_k P_k\right)^{-1} \nabla H_k^T R_k^{-1}$
	$S_{(\mu\mu)k+1} = \nabla H_i P^-_{k+1} \nabla H_i^T + R_{k+1} K_{k+1} = P^-_{k+1} \nabla H_i^T \left(S_{(\mu\mu)k+1}\right)^{-1}$	

Table 1. Kalman filter and H_∞ filter comparison [10].

Here, y contains the information of relative distance and angle measurements and H is showing the measurement matrix. v is the associated noise occurred during measurements.

H_∞ filter algorithm is almost similar to the Kalman filter. The only differences are notated on the state covariance and state update. For convenience, the comparison between H_∞ filter and Kalman filter is presented in **Table 1**.

3.1. Fuzzy Logic-based navigation

In this chapter, the measurement innovation is referred as the main reference in designing the fuzzy logic. Different to Ref. [3] which utilizes the heading angle and relative distance range as their inputs to the system, fuzzy logic is designed in this research to process the angle and distance errors as its inputs. Our objective is to decrease those errors by configuring the fuzzy sets in producing smaller errors. By choosing the output appropriately, the effects or measurement error due to sensor inaccuracies can be minimized further.

There were also some researchers attempts to applied fuzzy Logic in navigation such as in Refs. [11–17]. Each of them demonstrated that fuzzy logic or artificial intelligence technique is capable to fuse the information obtained from the sensors for mobile robot estimation. Some of the researches are referred to evaluate the performance of this work.

Literature has stated that mobile robot has some confidence on its estimation especially when Kalman filter is applied for estimation [3]. For non-Gaussian noise characteristic, the sensors reading might be interfered and exhibits bigger error and hence results in bigger measurement noise covariance, R. If the gain K is small at all times during observations, then it is possible to have smaller measurement errors. Inspired by this fact, fuzzy logic is proposed to find the best value of measurement innovation to pursue lower error. Kobayashi et al. [16] have selected the P, Q and R from fuzzy logic to gain smaller uncertainties. Work by Wang et al. [18] has recognized this as one of the ways to realize smaller measurement noise even when it was being applied to the other H_∞ filter family, the Kalman filter. Mamdani technique is used for analysis purposes in determining the output of the system. The technique is proposed as it calculates the output by considering and utilizing the maximum information gained from measurement.

The general design is illustrated in **Figures 1–3**, that consists of the input and output and their respective fuzzy sets. The fuzzy sets are changed whenever the mobile robot moves in different motion and therefore, the values are not included. The following describes the rules of fuzzy logic that are used to define the output of the measurement innovation.

- IF angle error is negative and distance error is negative, THEN angle is negative.
- IF angle error is negative and distance error is normal, THEN angle is normal.
- IF angle error is negative and distance error is positive, THEN angle is negative and distance is normal.

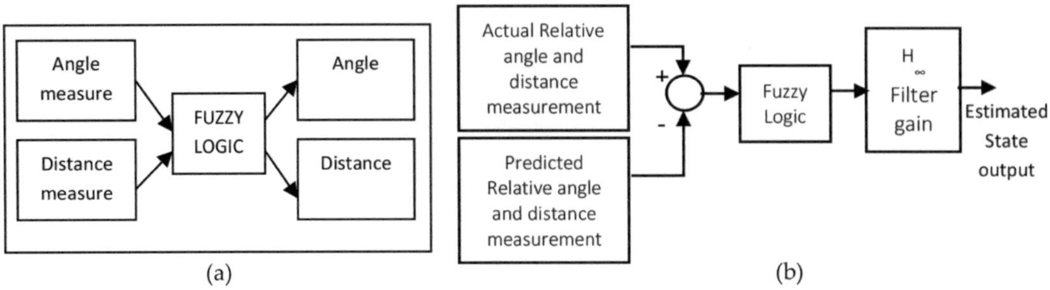

Figure 1. Fuzzy logic with inputs and outputs (a) and its associated block diagram to the proposed technique (b).

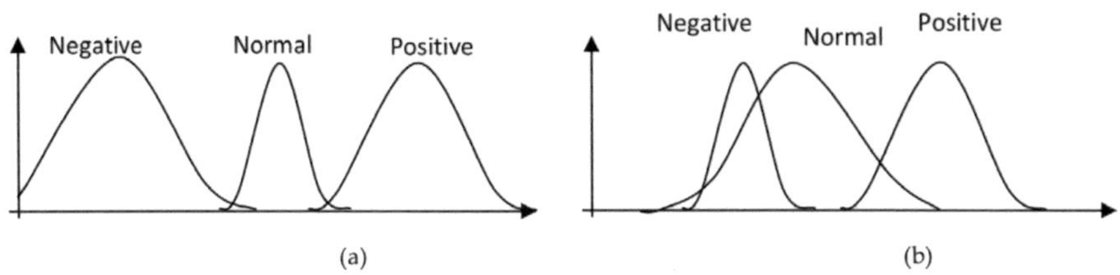

Figure 2. (a) Angle measurement and (b) distance measurement.

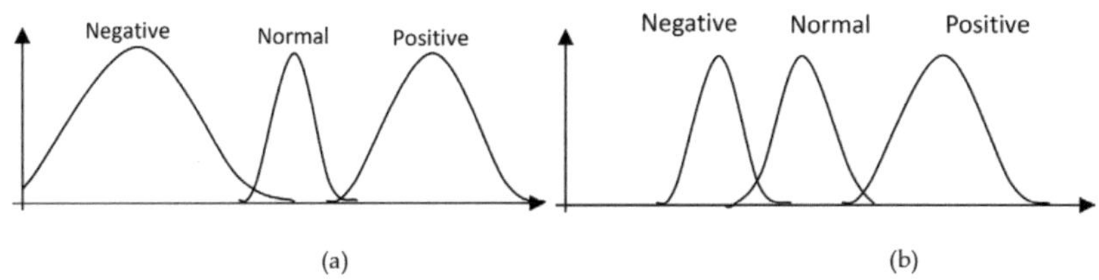

Figure 3. (a) Fuzzified angle and (b) fuzzified distance measurement.

- IF angle error is positive and distance error is normal, THEN angle is negative.
- IF angle error is positive and distance error is negative, THEN distance is normal.
- IF angle error is positive and distance error is positive, THEN angle is negative and distance is normal.

Generally, Gaussian and triangular membership functions are considered in this chapter for evaluation purposes. Only three fuzzy sets are defined which are divided into three different categories: the negative, normal and positive regions. The scale of each of the fuzzy sets is selected based on the normal condition that has high errors. The value differs with each of the fuzzy sets, and it has been tuned several times to obtain the best estimation results. The tuning is taking into account the uncertainties behaviour throughout the simulation. Other than that, both angle and distance measurement characteristics are observed prior to the tuning process. Wang et al. [18] designed the membership function of the angle error to be positive at all times. However, random mobile robot's movements may also show a negative angle especially when a global coordinate system is being considered. This aspect is one of the major differences between our approach and what they have investigated.

http://dx.doi.org/10.5772/intechopen.68866

4. Results and discussions

This section describes the performance of the fuzzy logic-based H_∞ filter by considering several factors such as the finite escape time problem, localization and mapping problem and the effects of noise parameters. These three issues are important to be solved concurrently during the navigation. If not, then the estimation results will not be as expected. Analysis is based on the design parameters included in **Table 2**.

Figures 4 and **5** demonstrate the estimation results when $\gamma = 0.7$, which determines that fuzzy logic-based H_∞ filter outperforms the normal H_∞ filter about the mobile robot and landmarks estimations. Erroneous estimations are perceived for the mobile robot location estimations as well as the landmarks positions. The measurement details are presented in **Figure 2** for each of the estimation error. **Figure 6** illustrates the state covariance update performance. It can also be seen from this figure that the proposed method attempts to avoid the finite escape time from happening.

4.1. Effect of changing γ

The γ effect to the proposed technique has also been analysed to ensure that our proposed technique has consistent and reliable results. **Figures 7** and **8** describe the performance of the state estimations for both system of normal H_∞ filter and fuzzy logic-based H_∞ filter, respectively, when $\gamma = 0.23$, which is smaller than the previous case. It is plotted clearly that normal H_∞ filter exhibits erroneous results compared to our proposed technique.

4.2. Effect of initial state covariance

Generally, in SLAM, mobile robot does not have any prior information about its location or the environment. Therefore, the initial state covariance is designed to pose high uncertainties. Owing to these conditions, the mobile robot can probably have lost its way and is unable to navigate effectively on the environment. The analysis covers this issue by simulating the results using both normal H_∞ filter and fuzzy logic-based H_∞ filter in the next few figures.

Variables	Parameter values
Process noise:	
Q_{min}, Q_{max}	−0.002, 0.001
Measurement noise:	
$R_{\theta min}$, $R_{\theta max}$	−0.04, 0.01
$R_{dist\text{-}min}$, $R_{dist\text{-}max}$	−0.15, 0.3
Initial covariance:	
P_{robot}, $P_{landmark}$	0.001, 100
Simulation time	1000 (s)

Table 2. Simulation parameters.

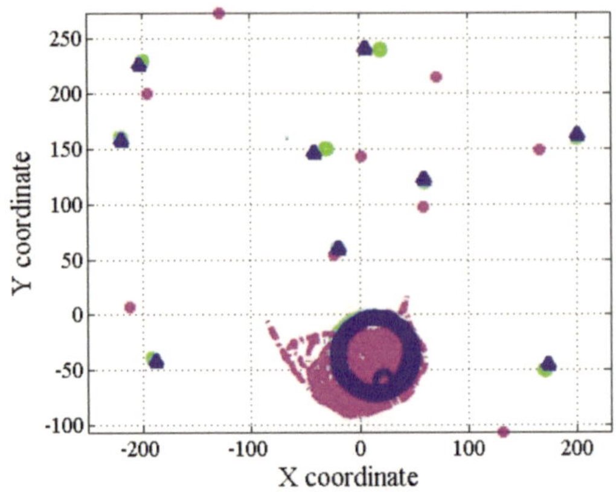

Figure 4. The mobile robot movements through the environment. Lighter colour of round shape defines the actual positions while the triangle shape and darker colour round shape presents the H_∞ Filter with Fuzzy Logic(FHF) and normal H_∞ Filter (HF) estimation performance, respectively.

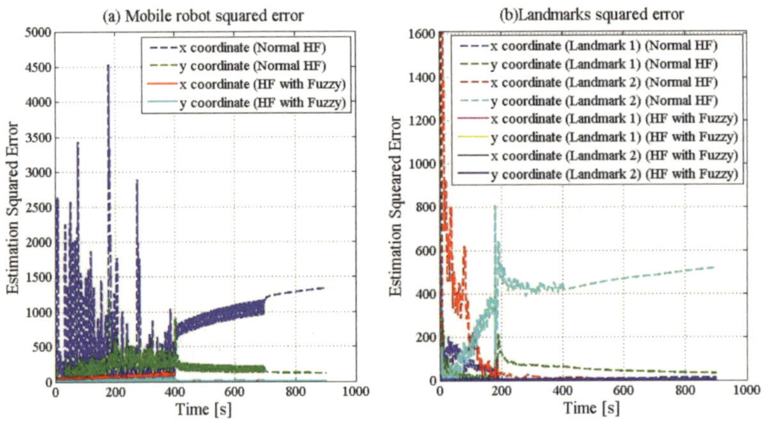

Figure 5. A performance comparison between FHF and normal HF estimations for both mobile robot (a) and landmarks (b) estimations about the errors.

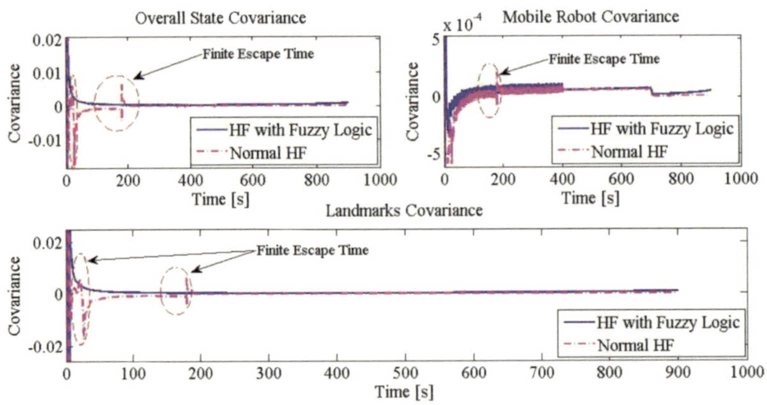

Figure 6. The state covariance conditions between H_∞ Filter (HF) and H_∞ Filter with Fuzzy Logic(FHF). Normal HF exhibits frequent Finite Escape Time(FET) compared to the H_∞ Filter with Fuzzy Logic(FHF).

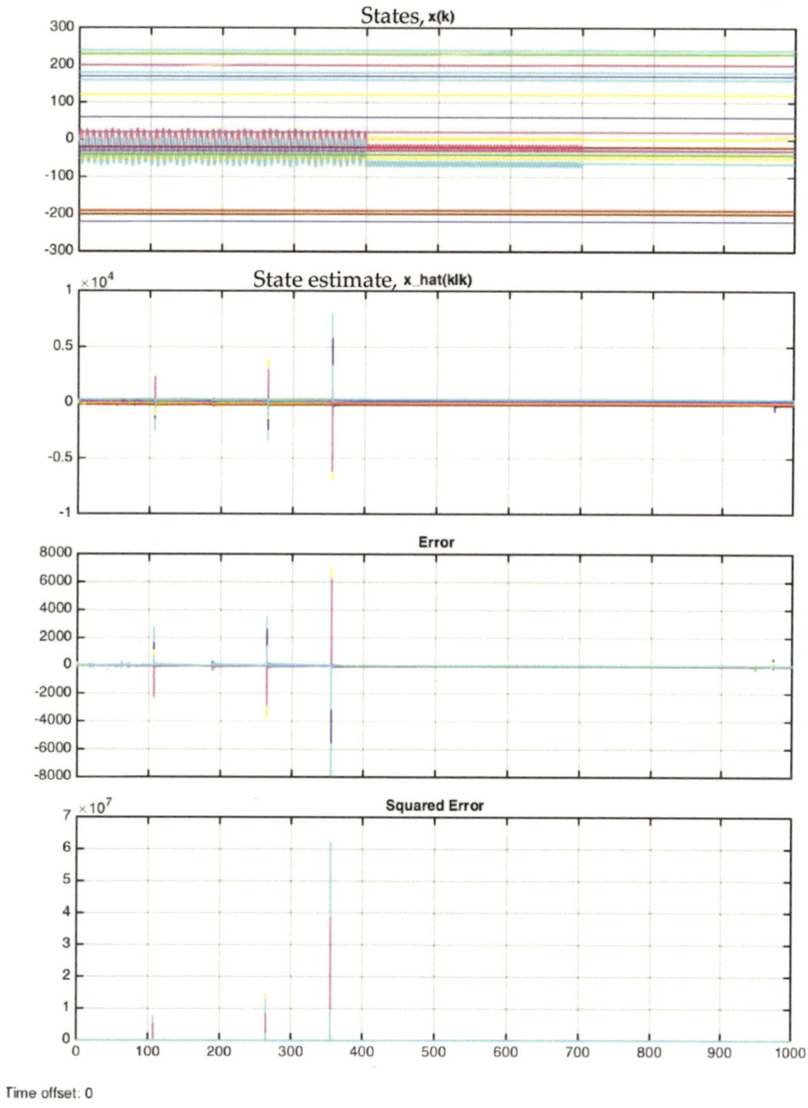

Figure 7. The state update conditions of normal HF generate a lot of estimation errors.

Figure 9 demonstrates the results of estimation when mobile robot is in the above-mentioned condition. As expected, the normal H_∞ filter did not deliver good estimation results. On the other hand, the fuzzy logic-based H_∞ filter guarantees a good estimation that can still be preserved with a considerable estimation. In this analysis, the Gaussian membership is used for decision making.

Further inspection is done by using different fuzzy membership with the similar range of fuzzy sets in the case when the mobile robot attempts to localize itself in a given environment. Other associated and related simulation parameters remain unchanged to observe any significant improvement that fuzzy logic can offer. The initial state covariance for the landmarks is now known and mobile robot does not have any information of its location. Even other types

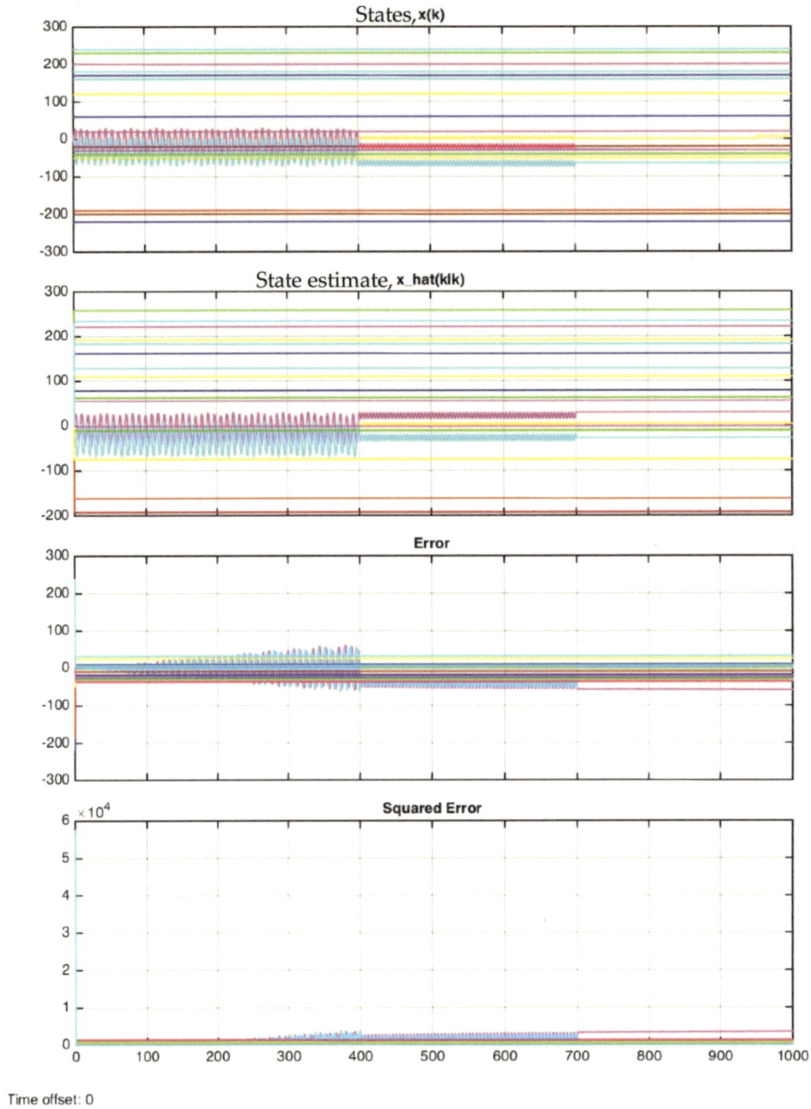

Figure 8. The state update conditions of FHF do not encounter high uncertainties or errors during mobile robot navigation.

of membership functions are applied for decisions; the fuzzy logic-based H_∞ filter surpassed the normal filter performance as depicted in **Figure 10**.

Assessment on different mobile robot movements during its observations with the same simulation parameters is also conducted to evaluate the consistency of estimation. The results are not disappointing and show a reliable estimation when comparing it to the normal H_∞ filter as illustrated in **Figure 11**.

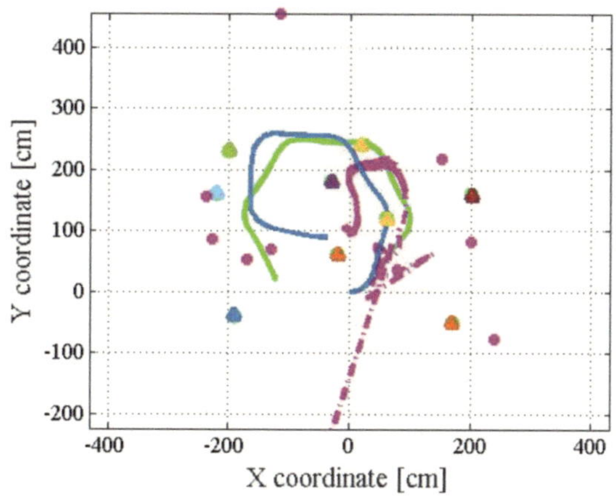

Figure 9. The normal H_∞ Filter(dotted line) shows erroneous results compared to the true mobile robot path(lighter line) and Fuzzy Logic based H_∞ Filter(darker line). Landmarks estimation of normal H_∞ Filter(round shape) is also results in erroneous estimation.

4.3. Effect of non-Gaussian measurement noise

H_∞ filter is known to be more robust to the extended Kalman filter (EKF) especially whenever non-Gaussian noise is available. The measurement noise is increased to observe whether the H_∞ filter with fuzzy logic still able to preserve a good estimation. **Figure 12** presents the results that still define that the H_∞ filter with fuzzy logic (triangular memberships) maintains better performance than the normal H_∞ filter. Similar performance is also observed with different fuzzy memberships even though the results are not included in this chapter.

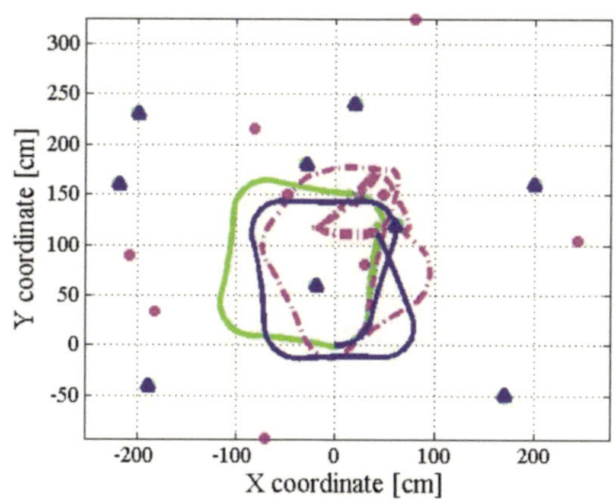

Figure 10. The normal H_∞ Filter (dotted line) cannot localize itself as well as the available landmarks even though the landmarks information is given compared to the Fuzzy Logic based H_∞ Filter (darker line) with reference to the truth locations (lighter line). Triangle shows the Fuzzy Logic based H_∞ Filter and round shape showing the normal H_∞ Filter for landmarks estimation.

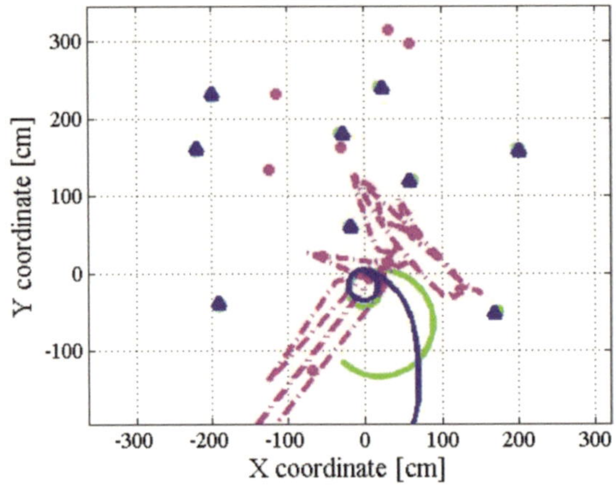

Figure 11. Normal H_∞ Filter (dotted line) produces erroneous results compared to the proposed technique (FHF in darker line) for different mobile robot movements. Triangle shows the Fuzzy Logic based H_∞ Filter and round shape showing the normal H_∞ Filter for landmarks estimation.

The performance analysis between fuzzy logic-based H_∞ filter and extended Kalman filter is also considered. A condition where mobile robot loses its measurement data at random time is referred in this case. Interesting results are obtained in **Figure 13** showing that the H_∞ filter with fuzzy logic inference is still better compared to the normal EKF estimation in non-Gaussian noise. The landmarks estimation for H_∞ filter with fuzzy logic outperforms the EKF. Therefore, based on the figure, H_∞ filter with fuzzy logic technique offers better solutions when non-Gaussian noise as well as when measurement data is lost unexpectedly during mobile robot observations.

There are few remarks to be considered in designing the fuzzy logic control for mobile robot navigation. The measurement innovation characteristics must be first examined prior to the estimation to ensure the results achieve the desired conditions. Besides measurement

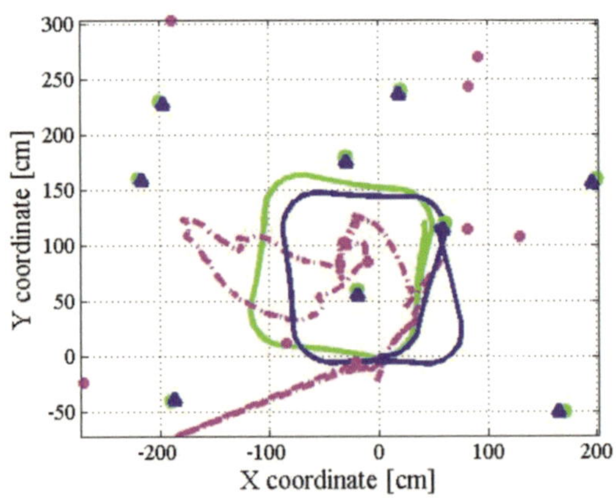

Figure 12. Normal H_∞ Filter (dotted line) performance is still low compared to the Fuzzy Logic based H_∞ Filter (darker line) with reference to the truth positions(lighter line) for bigger measurement noise. Triangle shows the Fuzzy Logic based H_∞ Filter and round shape showing the normal H_∞ Filter for landmarks estimation.

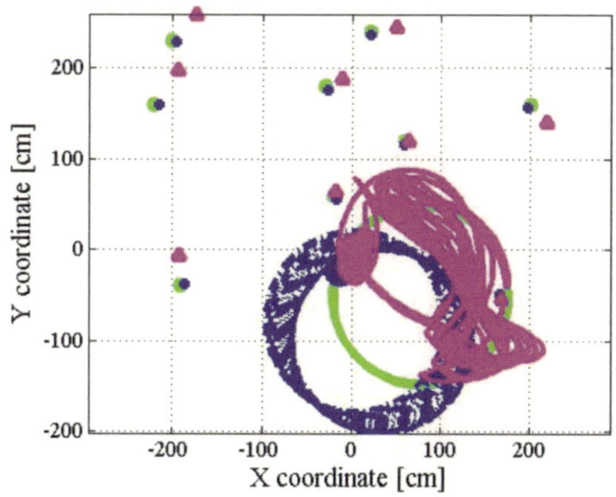

Figure 13. An analysis when measurement data is randomly unavailable. Fuzzy Logic based H∞ Filter(darker line) performance is surpassing the EKF(triangle) especially on the landmarks estimations with reference to the truth positions(lighter line).

innovation, the noise characteristics can also influence the estimation performance. Therefore, the designer should carefully understand and model the noise according to the environment to be observed. H_∞ filter is also sensitive to some parameters as stated by Bolzern and Maroni [7] and those parameters must be studied before conducting further analysis on the proposed technique.

5. Concluding remarks

This research has presented the analysis and study of H_∞ filter for mobile robot navigation using fuzzy logic control. The investigation was mainly focussing on the development of the fuzzy logic control to analyse the relative angle and distance measurement as its input to produce smaller error of navigation. Besides, fuzzy logic control was also found to be a possible technique to avoid finite escape time problem in H_∞ filter. A number of tests have been conducted for the proposed technique which includes the effects of having different γ value, different noise parameters and intermittently data lost. Preliminary results describe that the fuzzy logic-based H_∞ filter is able to tolerate the problem by using only few number of rules and fuzzy sets. Thus, the technique can be one of the alternative solutions for navigation.

Author details

Hamzah Ahmad*, Nur Aqilah Othman and Saifudin Razali

*Address all correspondence to: hamzah.ahmd@gmail.com

Faculty of Electrical & Electronics, Universiti Malaysia Pahang, Pekan, Pahang, Malaysia

References

[1] Thrun, S. Burgard, W. Fox, D., 2000. A Real Time Algorithm for Mobile Robot Mapping with Applications to Multi-Robot and 3d Mapping, IEEE International Conference on Robotics and Automation, Vol.1, San Francisco, CA, USA, USA , pp. 321-328

[2] Thrun, S. Burgard, W. Fox, D. 2005, Probabilistic Robotics, MIT Press, pp. 39-79.

[3] Valavanis KP, Doitsidis L, Long M, Murphy RR. A case study of fuzzy-logic-based robot navigation. IEEE Robotics & Automation Magazine. 2006;**13**(3):93-107

[4] Jaradat MAK, Abdel-Hafez MF. Non-linear autoregressive delay-dependent INS/GPS navigation system using neural networks. IEEE Sensors Journal. 2017;**17**(4):1105-1115

[5] Átila V. F. M. de Oliveira; Marcelo A. C. Fernandes, 2014. Dynamic Planning Navigation Technique Optimized with Genetic Algorithm, 2014 Joint Conference on Robotics: SBR-LARS Robotics Symposium and Robot Control, Sao Carlos, Brazil, pp. 91-96

[6] Ahmad H, Namerikawa T. Extended Kalman filter based mobile robot localization with intermittent measurement. System Science & Control Engineering: An Open Access Journal. 2013;**1**:113-126

[7] Bolzern P, Maroni M. New condition for the convergence of H_∞ filters and predictors. IEEE Transactions of Automatic Control. 1999; **44**:1564-1568

[8] Huang S, Dissayanake G. Convergence and consistency analysis for extended Kalman filter based SLAM. IEEE Transaction on Robotics. 2007;**23**(5):1036-1049

[9] Ahmad H, Othman N. The impact of cross-correlation on mobile robot localization. International Journal of Control, Automation and Systems. 2015;**13**(5):1251-1261

[10] Lewis, F. L., Xie, L., & Popa, D. 2008.Optimal and Robust Estimation With an Introduction to Stochastic Control Theory. USA: CRC Press, pp. 353-385

[11] Gualda, D, Urena, J. Garcia J.C, Garcia E, Ruiz D, Lindo A, 2014, Fusion of Data from Ultrasonic LPS and Isolated Beacons for Improving MR Navigation, 2014 IEEE International Conference on Instrumentation and Measurement Technology Conference, Montevideo, Uruguay, pp.1552-1555

[12] Abdelnour, G. Chand, S. Chiu, S. and Kido T., 1993, On-line detection and correction of Kalman filter divergence by fuzzy logic, in Proc. Amer. Control Conf., 1993, San Francisco, CA, USA, USA, pp. 1835-1839

[13] Asadian A., Moshiri, B., Sedigh A.K, 2005, A Novel Data Fusion Approach in an Integrated GPS/INS System Using Adaptive Fuzzy Particle Filters, In Proc. of the 5th IEEE Conf. on Technology and Automation (ICTA), Greece, pp. 125-130

[14] Choomang R, Afzulpurkar N. Hybrid Kalman filter/fuzzy logic-based position control of autonomous mobile robot. International Journal of Advanced Robotic System. 2005;**2**(3):197-208

[15] Ip YL, Rad AB, Wong YK, Liu Y, Ren X. A localization algorithm for autonomous mobile robots via a fuzzy tuned extended Kalman filter. Advnced Robotics. 2010;**24**179-206

[16] Kobayashi K, Cheok KC, Watanabe K, Munekata F. Accurate differential global positioning system via fuzzy logic Kalman filter sensor fusion technique. IEEE Transactions on Industrial Electronics. 1998;**45**(3):510-518

[17] Raimondi, F.M, Melusso, M, 2006. Fuzzy EKF Control for Wheeled nonholonomic Vehicle, 32th Anuual Conference on IEEE Indusrial Electronics, Paris, France, pp. 43-48

[18] Wang JH, Song CL, Chen JB. Sigma point H infinity filter for initial alignment in marine strapdown inertial navigation system. In: 2nd International Conference on Signal Processing Systems. 2010; pp. 580-584

Indoor Mobile Positioning Using Neural Networks and Fuzzy Logic Control

Anatoly D. Khomonenko, Sergey E. Adadurov,
Alexandr V. Krasnovidow and Pavel A. Novikov

Additional information is available at the end of the chapter

http://dx.doi.org/10.5772/68009

Abstract

Indoor mobile navigation systems are becoming more prevalent in many areas (transport, public institutions, logistics, etc.). The interior navigation based on the access points, arranged according to the radio fingerprints, is becoming increasingly popular. The model of artificial neural networks (ANN) is often used as a mechanism for storing and processing radio fingerprints. The task of selection of the access point in WLAN network in the case of high user density is quite topical. Such selection must take into account not only the level of the signal received by the mobile device, but also a width in the dedicated channel bandwidth. The main issues related to the creation of program complex for the mobile indoors navigation using neural networks is discussed in the chapter as well as the method of access point selection based on analysis not only the signal level but also the other parameters. To solve this task, fuzzy logic is used.

Keywords: Wi-Fi radio network, neural network, mobile navigation indoor, navigation systems, learning algorithms, mobile devices 802.11k standard, the mobile subscribers, fuzzy logic, MATLAB, frame transmission time

1. Introduction

The modern world cannot exist without precise navigation systems. Satellite navigation systems such as Global Positioning System (GPS) and Global Navigation Satellite System (GLONASS) [1] are widely used in a variety of areas of activity, such as navigation transport, engineering, surveying, and other cellular communication. GPS receivers' consumer level is set in almost all modern phones. In open areas, such receivers allow for positional accuracy in the region of 1–5 m.

At the same time, these receivers do not meet the existing demand to navigate indoors. In such circumstances, GPS does not work or provide location data with a very high error of about 100–150 m.

Navigation systems indoors can be applied in many fields. Including navigation inside large shopping malls, warehouses, or different systems, "smart home," in which different home systems (heating, lighting, air conditioning, and so on) can be centrally managed in automatic mode.

In particular, obvious argument for the need for indoor navigation can serve as tasks of navigation in transportation systems, such as airports and railway stations [2]:

(A) For visitors: positioning; search check-in desks/offices/storage rooms and a cafe, parking, taxi, etc.; installation of the route, taking into account number of storeys of buildings; search colleagues inside the airport/train station, social activity; service based on knowledge of the position of the visitor (location-based service).

(B) For airports, railway stations, and tenants: an additional service to visitors; analyst extensive movements of visitors and staff; advertising opportunities geo contextual advertising (location-based advertising) airlines/shops/cafes, etc., as well as products/services; promotions based on the location of the visitor.

There are now possibilities to improve the methods and the practical use of mobile navigation indoors. Let us briefly consider the modern approaches to the solution of this task and proposals for justifying the choice of learning algorithms and neural networks of the individual parameters of the same interests.

Mobile navigation systems indoors are becoming more widespread in many areas (transport, public institutions, logistics, and others.). It is becoming popular navigation based on fingerprint radio access points Wi-Fi. As a mechanism for storing and processing the radio fingerprint is often considered a model of artificial neural networks (ANN). The chapter examines the main issues related to the creation of complex programs for mobile navigation indoors using neural networks, which are one of the parts of fuzzy logic. We justify the choice of learning algorithms ANN navigation mobile devices indoors.

The data for the neural network training in MATLAB taken from the file "train.txt" is used for functional testing of software navigation. To verify the performance of ANN training, the original file should be divided into two sets (training and testing).

Wireless local area networks 802.11k as precise navigation systems have become very popular for several reasons. They operate in the unlicensed frequency bounds and they need not in large time and cost for deployment. The appearance of vast number of mobile devices supporting Wi-Fi technology gives possibilities of free choice and cost saving for the various kinds of users. WLAN networks have a number of advantages over traditional wired networks.

- Much easier and cheaper to deploy a local wireless network in the new location than a traditional network.

http://dx.doi.org/10.5772/68009

- The subscriber does not have to be next to his desk or local network socket-outlet. He can move freely inside the area coverage.

However, the number of users working in the unlicensed frequency bounds increases day by day. In this context, a very important problem is the way by which the wireless device uses to select an access point to connect to the network. Now, the device selects the access point according to signal power. This way allows determining the nearest access point. In other words, currently used 802.11k standard is aimed at the implementation of load balancing of radio Wi-Fi networks. Nevertheless, the high level of the signal does not always mean high network bandwidth. Suppose, for example, that most of the notebooks operating in the certain conference halls connect to WLAN by using the access point which is above the entrance door. In such a case, the number of subscribers connected to it would be up to tens if not hundreds, while other access points would not be fully loaded. As a result, the network bandwidth per subscriber reduces to a significant degree that leads to network reduction of productivity of the network in whole. Hence, the task of selection of the access point in WLAN network in the case of high user density is quite topical. Such selection must take into account not only the level of the signal received by the mobile device, but also a width in the dedicated channel bandwidth that depends on the number of connected subscribers to the access point. In this chapter, we examined the method based on analysis of not only signal level but also other parameters. To solve this task, fuzzy logic is used in the chapter. Constructed membership functions and linguistic rules are examined. Structure of the developed model and simulation results are presented.

2. Indoor mobile positioning using neural networks

2.1. Characteristics of modern approach to navigation mobile

First, we note that the modern mobile devices contain a variety of different sensors and receivers. The main of them include:

1. GPS.
2. Wi-Fi, a radio signal operates at frequencies of 2.4 and 5 GHz.
3. Bluetooth operates on the same frequencies.
4. The accelerometer and gyroscope—inertial sensors measuring linear and angular acceleration.
5. Magnetometer—a sensor that measures the intensity of magnetic fields.
6. Other sensors—humidity, light, proximity, barometer, etc.

Sensors 1–5 are well represented in many of today's mobile devices that are running operating systems iOS and Android. These systems provide programmatic access to the sensors through its own application program interfaces (API), and any application can receive data from the sensors (with certain insignificant limitations).

Currently, there are several ways to use the mobile navigation device.

1. GPS (Global Positioning System) and GLONASS (Global Navigation Satellite System) [1]: these are navigation using satellites. Well suited for positioning in the open spaces. Because of the need to be in the field of view of at least three satellites, these are poorly suited for closed premises, as they greatly impair the satellite signal. The initial search for satellites might take few minutes.

2. AGPS (Assisted GPS). Navigation using radio signals from cell towers. Typically it used in conjunction with GPS. It accelerates the initial determination of the coordinates by the fact that the use of data on the mobile device serving cell towers.

3. Navigate using radio beacons operating on technology iBeacons (Bluetooth Low Energy) [3]. This is a relatively young technology, the impetus for the development of which will serve standard Bluetooth 4.0 (+) with low power consumption. Tracker is a chip with a radio module, which is a predetermined frequency radio, sends packets with information about themselves. The receiver, knowing the map of the location beacons and signal strength to the nearest of them, calculates their relative location.

4. Navigation on the basis of fingerprints of radio access points Wi-Fi [4]. Such navigation systems are more exploratory in nature. Due to the relative novelty of this approach, ways to implement proven accurate navigation have not yet developed.

5. Inertial navigation system [5]. Navigation is based on data from the inertial sensor device, which is constructed on the basis of the model orientation in space.

Comparative characteristics of the main approaches of navigation using mobile devices [2] are given in **Table 1**.

In addition to these approaches, following approaches are also included. In Ref. [6], the integration of Wi-Fi and an inertial navigation system is considered. In Refs. [7, 8], mobile navigation services and the use of technology OpenCellID to determine the location of mobile devices are studied.

System	Dignity	Disadvantages
GPS	Average precision (5 m) Ease of use Good compatibility	Inability To work Indoor
GSM	Ease of use Good compatibility	Ease of use Good compatibility
Wi-Fi	Average Accuracy (5 m) Work indoors	The need for network deployment Limited compatibility High power consumption
iBeacon	High accuracy (1–2 m) Ease of use Good compatibility Low power consumption	The need to deploy BLE-network

Table 1. Comparative characteristics of approaches.

The article [9] considered a relatively new approach to the positioning of mobile devices in the premises on the basis of two-dimensional barcodes. In Ref. [10], the model of context-aware computing (context-dependent browser) based on network proximity is considered. This mobile phone is considered as a proximity sensor and geo replaced positional information network proximity. An algorithm for calculating the trajectories of mobile networks on the basis of information about network proximity is also considered.

2.2. Rationale for navigation of mobile devices using neural networks

Employment of navigation system via radio fingerprint from the access points of Wi-Fi consists of two parts:

1. Preparation of a database of known fingerprints radio with known coordinates associated with them.
2. Getting new coordinates from the database on the new print of radio signals.

As a mechanism for storing and processing the radio fingerprint has considered a model of artificial neural networks (ANNs).

This approach is interesting for the following reasons:

- The problem of determining the location based on a previously defined radio fingerprint can be viewed as a problem of classification of multidimensional data.
- Main computational load of this method accounts for the learning process, which is performed only once and can be executed on an independent computer system.
- Data on the surrounding Wi-Fi points can be supplemented by the data from other sources, such as Bluetooth LE beacons.

The possibilities of modern ANN are the subject of active research. The approach to the use of ANN to solve the above problem is considered in several publications, for example, Refs. [10–12].

2.3. Characteristics of software navigation

Here examined a complex program [13, 14] to navigate through the Wi-Fi signals using a mobile phone. It consists of two components:

1. Mobile application that
   a. collects data on the surrounding Wi-Fi access points;
   b. uses trained neural network to determine the current location of the mobile device based on the new data on Wi-Fi signals.
2. Desktop application that
   a. prepares data radio prints for further training of the neural network;
   b. training and testing neural network.

As an implementation of a neural network, we used multilayer ANN provided free library FANN (Fast Artificial Neural Network Library). To train the ANN learning algorithm, Resilient Propagation (*RProp*) has been used [15].

Unlike standard algorithm *Backprop, RProp* uses only partial signs for adjusting the weighting coefficients. The algorithm uses the so-called "training periods" when the correction occurs after the presentation of the balance of the network of examples from the training set.

To determine the amount of correction using the following rule [15]:

$$\Delta_{ij}^{(t)} = \left\{ \begin{array}{ll} \eta^{+}\Delta_{ij}^{(t)}, & \frac{\partial E^{(t)}}{\partial \omega_{ij}} \frac{\partial E^{(t-1)}}{\partial \omega_{ij}} > 0 \\ \eta^{-}\Delta_{ij}^{(t)}, & \frac{\partial E^{(t)}}{\partial \omega_{ij}} \frac{\partial E^{(t-1)}}{\partial \omega_{ij}} < 0 \end{array} \right\}, \tag{1}$$

$$0 < \eta^{-} < 1 < \eta^{+} \tag{2}$$

If the partial derivative of the corresponding weight $\partial\omega_{ij}$ had changed its sign at the current step, it means that the latest update was great, and the algorithm passed a local minimum and therefore the value of change must be reduced by η and return the previous weight value: in other words, you must make 'roll back'.

$$\partial\omega_{ij}(t) = \partial\omega_{ij}(t) - \Delta_{ij}^{(t-1)}. \tag{3}$$

If the sign of the partial derivative is not changed, it is necessary to increase the amount of compensation η^{+} to achieve a more rapid convergence. Fixing factors η^{-} and η^{+} can be dispensed with global settings of the neural network, which can also be seen as an advantage of the algorithm to the standard algorithm *Backprop*.

Recommended values are $\eta^{-} = 0.5$, $\eta^{+} = 1.2$, but there are no restrictions on the use of other values for these parameters.

To prevent too large or small weight values, the correction value limit from above the maximum Δ_{max} and below the minimum Δ_{min} value of the correction value, which default, respectively, shall equal 50 and 1.0E–6.

The initial values for all Δ_{ij} are set to 0.1. Again, this should be seen only as a recommendation, and in the practical implementation it can specify a different value for the initialization.

The current implementation of the navigation software package is a simple one-dimensional classifier, which, by new radio-prints, is able to determine the room in which the mobile device is located.

2.4. Working with software navigation

Working with a program complex navigation is implemented using desktop and mobile software, and includes the following steps:

http://dx.doi.org/10.5772/68009

1. Collection of data on radio-prints in the studied areas. The output of this stage is a set of files, each of which contains a set of vectors (matrix) measuring Wi-Fi signals points (for example, **Table 2**).

   In the first row, there are names of all the access points that were visible in the data collection process. For clarity, we have been selected to work with network names, rather than their mac address. Each line represents the measured signal strength to a point. A value 0 means that the point was not available at this time.

2. Combine all the source files into one.

   The result is a common image with the measurements in all the studied areas which is shown in **Table 3**.

   The first column is stored name space in which the measurement was performed. Null values (0) are replaced by −100 (very weak signal).

   In addition, a file is created "names.txt," which lists the names of all the networks in the same order in which they are placed in the file of training (step 4).

3. Represented by map matching names with their premises of formal numerical representation:

   Cabinet [0] Room [0.5] Kitchen [1]

4. On the basis of the contents of the files of the previous steps 2 and 3, create a file with the data for network training:

   60 15 1

   −86.0 −100.0 −54.0 −69.0 −100.0 −100.0 −88.0 −100.0 −100.0 −100.0 −100.0 −100.0 −100.0 −100.0 −100.0

Grimas	DSL-2640U	NETGEAR	STERH76	InterZet-at-home	InetBezP	ASUS 26
−86	0	−54	−69	0	0	−88
−89	−83	−56	−66	0	0	−91
…	…	…	…	…	…	…

Table 2. Set of vectors (matrix) measuring signals Wi-Fi points.

	Grimas	DSL-2640U	NETGEAR	STERH76	InterZet-at-home	InetBezP	ASUS 26	…
Study	−86	−100	−54	−69	−100	−100	−88	…
Study	−89	−83	−56	−66	−100	−100	−91	…
…	…	…	…	…	…	…	…	…

Table 3. Common image with the measurements in all the studied areas.

0.0

−89.0 −83.0 −56.0 −66.0 −100.0 −100.0 −91.0 −100.0 −100.0 −100.0 −100.0 −100.0 −100.0 −100.0 −100.0

0.0

−90.0 −71.0 −51.0 −67.0 −100.0 −100.0 −88.0 −100.0 −100.0 −100.0 −100.0 −100.0 −100.0 −100.0 −100.0

0.0

The first line shall include: the number of hidden layer neurons, the number of neurons in the input layer and output layer.

The following are a couple of lines:

A. The input vector.

B. The value that the National Assembly should be trained for this vector.

The result of this step is to file "network," which is a trained neural network in the library "FANN." The network name is chosen at random.

The following steps describe the workflow of software application that is running on a mobile device.

5. A mobile application launches library "FANN" and passes the file to the trained neural network.

6. Mobile application starts scanning the surrounding Wi-Fi hotspots. After the completion of the next iteration of the scanning application, an array of the currently available access points is received to which the device may try to connect. Each object of the array, among other things, contains in its structure the following two fields:

A. Title.

B. The strength of the signal to a predetermined point.

With using the array of file names "names.txt" obtained in the second stage and the current array of visible access points built radio fingerprint, which, in essence, is a simple array of numbers. Create a radio footprint that includes the following steps:

A. Create an array of numbers (float []) size which determine the number of elements in the array of names.

B. Go through the array of names in the cycle:

a. if the array access point is a point with the current name, the add-in radio signals strength to mark this point of the current iteration of the index;

b. otherwise, add the current iteration index of the default value for the missing point that was made in step 2 (−100).

C. At the output of iteration scanning radio fingerprint is obtained as a vector of values (−100, −50, −75, −100, ...).

7. The resulting vector is passed to the neural network, which returns the result vector (float []). Since the ANN has been trained on the basis of dimensional data, we are only interested in the first element of the resulting vector.

8. The resulting value is compared with the data from the map space. In the version of the software complex in the map, closest value to the current value obtained from the network is searched. After finding, it displays the name of the current premises that corresponds to the measured radio-imprint.

2.5. Training a neural network in MATLAB

The data for the neural network training in MATLAB are taken from the file "train.txt" and used for functional testing of software navigation. To verify the performance of ANN training, the original file should be divided into two sets (training and testing).

By analogy with [16], we make a comparative analysis of the main learning algorithms of the neural network used for the navigation of mobile applications for indoors, with position accuracy and complexity ratios depending on the number of neurons in the hidden layer. The corresponding results are shown in **Table 4**.

Table 4 uses the following symbols: *trainlm*—learning algorithm Levenberg-Markarth, *trainscg* —related gradient method scaled association, and *trainbr*—Bayesian regularization method. MSE—mean square error, epochs (s)—the number of training cycles, and in brackets the number of seconds.

Based on the analysis of **Table 1**, we can conclude that the optimal number of hidden layer neurons for a given set of source data is in the range 15–25. It showed the highest accuracy Bayesian regularization algorithm, but it is considerably more time-consuming compared with

Number of neurons	Trainlm		Trainscg		Trainbr	
	MSE	Epochs(s)	MSE	Epochs(s)	MSE	Epochs(s)
5	0.11	12(0.1)	0.08	15(0.1)	0.03	592(25)
10	0.06	16(0.1)	0.09	20(0.2)	0.02	700(20)
15	0.07	9(0.1)	0.05	13(0.2)	0.013	800(23)
20	0.05	11(0.2)	0.13	14(0.2)	0.015	269(48)
30	0.13	6(0.2)	0.19	23(0.3)	0.018	889(120)
50	0.23	13(0.2)	0.21	30(0.5)	0.02	483(187)

Table 4. Characteristics of the complexity and accuracy of learning algorithms NA.

other algorithms. The optimal ratio of accuracy and convergence provides the Levenberg-Marquardt algorithm.

Note that low accuracy is common to all algorithms, which is associated with the original data. Probably, it reduces the dimension of the input vector (data used by Wi-Fi 15 points, most of which are available), and increases the total number of measurements. At the same time in the original data set of vector measurements corresponded to only three different rooms, so get enough accuracy.

3. Using fuzzy logic control for selection of the access point

3.1. Introduction

Wireless local area networks of 802.11k standards become very popular due to some reasons. They operate in the unlicensed frequency bounds and they need not require large time and cost for deployment. The appearance of vast number of mobile devices supporting Wi-Fi technology gives possibilities for free choice and cost saving for the various kinds of users. WLAN networks have a number of advantages over traditional wired networks.

- Much easier and cheaper to deploy a local wireless network in the new location than with a traditional network.
- The subscriber does not need to be next to his desk or local network socket. He can move freely inside the area coverage.

However, number of users working in the unlicensed frequency bounds increases day by day. In this context, a very important problem is the way by which the wireless devices are used to select an access point to connect to the network.. Now device selects the access point according with signal power [17]. This way allows determining the nearest access point. In other words, currently used 802.11k standard is aimed at the implementation of load balancing of radio Wi-Fi networks. Nevertheless, the high level of the signal does not always mean high network bandwidth. Suppose, for example, that most notebooks operating in the certain conference hall connects to WLAN using access point which is above the entrance door. In such a case, the amount of subscribers connected to it would be to tens if not hundreds, while other access points would not be fully loaded. As a result the network bandwidth per subscriber reduces to a significant degree that leads to the network reduction of productivity of the network in whole. Hence, the task of selection of the access point in WLAN network in the case of high user density is quite topical. Such selection must take into account not only the level of the signal received by the mobile device, but also a width in the dedicated channel bandwidth that depends on the number of connected subscribers to the access point.

Further, examined the method based on analysis not only signal level but also other parameters. To solve this task, fuzzy logic is used (See Ref. [18, 19]). Constructing membership functions and linguistic rules are examined. Structure of the developed model and simulation results are presented.

3.2. Choice of access points for connecting a mobile device

Signal level and the bandwidth of a system are linked by the well-known Shannon formula, which allows determining the capacity of the data transmission system:

$$C = \Delta F \times \log_2(1 + S/N). \quad (4)$$

The total bandwidth allocated in the range, is divided equally among all active subscribers [17]. If the length of the transmitted packet is L bits, one can determine the time required for transmission of a packet:

$$T(M, S/N) = \frac{L \times M}{F \times \log_2(1 + S/N)}, \quad (5)$$

where $\Delta F = F/M$ is the network bandwidth, and S/N is signal/noise ratio at the receiver input and M is the number of subscribers already connected. Thus, the functional dependence between bandwidth and the number of connected subscribers is linear. Then the task of selecting the best access point to connect can be formulated as follows:

Find the value of function (5), the T_{max} of which does not exceed a predetermined time with the following restrictions: $S/N > P_0$ and $\Delta F \geq \Delta F_{min}$, where P_0 is some ratio threshold signal/noise at which the operation of mobile subscriber receiver is possible and ΔF_{min} is a minimum possible bandwidth width. In other words, the problem is reduced to finding such a pair of values (ΔF, S/N) for which the transmission time has the minimum possible value.

This problem can be solved by various methods:

- Analytical methods that use differential and variational calculus.
- Numerical methods which use prior information in order to find improved solutions using iterative algorithms.
- Mathematical methods (linear and nonlinear programming).

In the case of choice of any method, the finding of a point of extremum of the considered function is usually required. The type of function (5) is shown in **Figure 1**.

Figure 1 shows that the function (5) has minimum values for various combinations of parameters and the number of connected subscribers M. Function (5) has no maximum that determines the above-mentioned formulation of the problem of choosing the access point. **Table 5** shows the values of the function (5) calculated for various combinations of parameters and M.

Analysis of **Table 5** confirms that the high signal level does not always provide an acceptable transmission time. It's known for solve the problem of the selection the access point to connect mobile subscribers on the basis of function (5) using the above methods it is necessary and sufficient that the following conditions are met (as it's shown for instance in Ref. [20, 21]):

1. Function (5) should be continuous and differentiable at the point of extremum.
2. The Hessian matrix of function (5) must be negative definite (for a point of minimum).

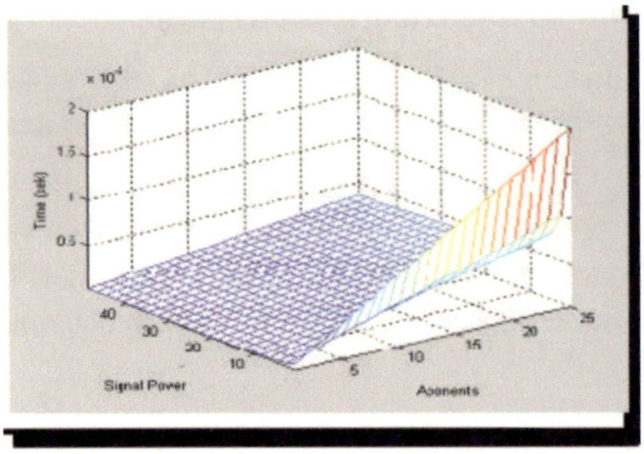

Figure 1. Dependence of the packet transmission time from the signal level and the number of subscribers.

Signal power, db	The bandwidth width, hz	The transmission time, s
30	4E + 6	2.514E − 4
30	8E + 5	1.257E − 3
30	4.444E + 5	2.263E − 3

Table 5. Values of the time of packet transmission for different ratios of parameters.

Clearly, the condition 1 for the function (5) is performed. To simplify the analysis of the condition (2), it is convenient to consider function 1, taking into account that low values of transmission time are achieved with large bandwidth. Then, for the function (4), condition 2 changes to opposite. Condition 2 for the function (5) is not satisfied due to the fact that $\partial^2 C/\partial(\Delta F)^2 = 0$, which implies that:

$$\begin{vmatrix} \dfrac{\partial^2 C}{(\Delta F)^2} & \dfrac{\partial^2 C}{\partial(\Delta F)\partial\left(\dfrac{S}{N}\right)} \\ \dfrac{\partial^2 C}{\partial\left(\dfrac{S}{F}\right)\partial(\Delta F)} & \dfrac{\partial^2 C}{\partial(\frac{S}{N})^2} \end{vmatrix} = \begin{vmatrix} 0 & \dfrac{1}{\log_2 e\left(1+\dfrac{S}{N}\right)} \\ \dfrac{1}{\log_2 e\left(1+\dfrac{S}{N}\right)} & \left(\dfrac{1}{\log_2 e\left(1+\dfrac{S}{N}\right)}\right)^2 \end{vmatrix} \leq 0. \quad (6)$$

It follows that the function in question is not concave. However, the requirement of convexity or concavity of the function is a serious restriction that is far from always performed in practical problems. This is why the concept of such functions is generalized by the introduction of pseudo-convex unimodal functions [21]. The f: $X \to R$ function is called the pseudo-unimodal in the interim $[a, b] \subset X$, if $\exists$ an arbitrary interval $I^* \subset [a, b]$ such that the f function:

- Strictly increasing in the interval $[a, b]$;
- Equals to some constant, $\leq \min\{f(c), f(d)\}$ on the some interval I^*;
- Strictly decreasing in the interval $[d, b]$.

http://dx.doi.org/10.5772/68009

The points c and d determined by the following way:

$$c = \inf_{x \subset I^*} x; \quad d = \sup_{x \subset I^*} x. \tag{7}$$

In this case, the interval I^* is the solution of the problem max $\{f(x) : x \subset [a, b]\}$. In the particular case if $c = d$, then $I^* = \{x^0\}$ and $x = c = d$, the f function is called unimodal. The example of unimodal function is shown in **Figure 2**.

Then, if we set

$$a = [\Delta F_{\min}, P_0]; \quad b = [\Delta F_{\min}, P_{\max}]. \tag{8}$$

Here $P_{\max}$ is some maximum possible value of S/N ratio, then unimodality of the function (4) and, consequently, function (5) derived from their definitions. Consequently, the optimization problem has a solution in the following formulation: find the minimum of function (4) in the interval (8). This solution can be found by any of the methods listed above. However, their use is associated with a large number of calculations (solution of the corresponding equations), or with a large amount of stored data, requiring constant modification (various search methods).

In Ref. [18], to solve above-mentioned problem is proposed to use the apparatus of fuzzy logic, free from above disadvantages. Fuzzy logic operators are very similar to conventional Boolean operators and allow simplified algorithms for solving this problem. Complicated mathematical modeling can be replaced by evaluation of membership functions and rules of fuzzy logic [22]. Various approaches to solving this type of problems are considered in Refs. [23, 24]. In Ref. [25] there is an example of controlling the operation of the charger using intelligent controller which applies various algorithms of fuzzy inference. One of the most powerful tools for solving such problems is the MATLAB system, which provides users a various species of software, including the visual ones. With the help of visual programming, the necessary model can be built, and then can run the simulation in program mode.

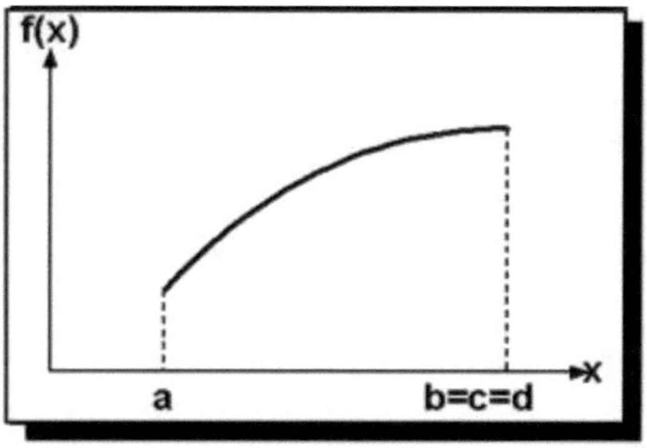

Figure 2. The example of unimodal function.

3.3. Model of algorithm of determination of the best access point

In accordance with the 802.11 standard, the mobile station scans the individual channels for the detection of the best signal from the access point. APs periodically send a Beacon signal in a broadcast mode. Mobile network station accepts these beacon signals and takes note of the relevant signal level. Thus, the received signal level actually characterizes the relative location of the subscriber and the access point. During this process, a mobile subscriber searches such access point. To determine whether channel is free, the well-known algorithm, which is named clear channel assessment (CCA), is used. Its essence lies in the measurement of the signal power at the antenna and determining the received signal strength (RSSI). If the received signal strength is below a certain threshold, then the channel is declared free, and the MAC layer receives the CTS status. If the power is above the threshold, the data transfer is delayed in accordance with the protocol rules. The standard provides another opportunity to determine the idle channel, which can be used either separately or together with the RSSI measurement—method of checking of the carrier. The most appropriate method to use depends on the level of interference in the work area. Various 802.11 standards use one of the five possible CCA modes:

- solution that channel is free and is based on the detection of channel power which exceeds a certain threshold value;
- the decision that the channel is free based on the detection of a carrier signal corresponding to the 802.11 standard;
- carrier signal detection and discovering of power (combination of modes 1 and 2);
- carrier signal detection with message that the ether is free, if neither signal is detected during 3.5 ms timeslot;
- detection of the power corresponding to the increased transmission rate in a physical layer and carrier detection at mode 3, but with reference to the ERP.

802.11 MAC layer is responsible for the way in which the subscriber is connected to the access point. When a subscriber enters into the coverage area of one or more access points, it selects the access point based on signal power values and observed number of errors, selects one of them and connects thereto. Once the subscriber receives confirmation that it is accepted by the access point, it tunes to a radio channel in which it operates. From time to time, it checks all the channels to see if there is another point which provides access services of higher quality. If such an access point is found, then the subscriber connects to it and readjusts to its frequency.

In accordance with this principle, the diagrams of signal levels and levels of channel diagrams of signal levels and levels of channel congestion can be constructed for each access point. These diagrams look like shown in **Figure 3**.

Analysis of these diagrams and subject area makes it possible to apply fuzzy logic to make a decision on selecting the best access point to connect. As shown above, a signal power can be characterized through signal/noise ratio, and a congestion level of AP can be described with the help of number of connected subscribers. Then it becomes obvious that proposed model

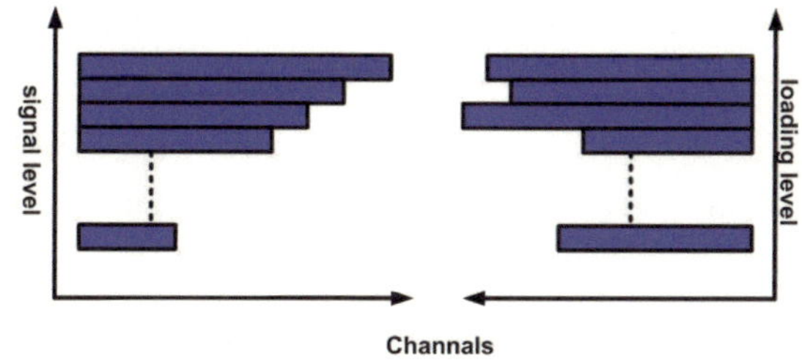

Figure 3. Diagrams of signal levels and levels of channel congestion.

must have two input linguistic variables (*S/N* ratio and the degree of loading of the access point). As the set of terms of the first linguistic variable "signal/noise" the following set is used:

$$S/N = \{\text{'weak'},\ \text{'medium'},\ \text{'high'}\}. \tag{9}$$

As the set of terms of the second linguistic variable "degree of loading" the following set is used:

$$\text{NUMUSERS} = \{\text{'few'},\ \text{'medium'},\ \text{'many'}\}. \tag{10}$$

As the set of terms of the output linguistic variable the following set is used:

$$\text{RESULT} = \{\text{'is a candidate for connection'}, \text{'is not a candidate for connection'}\} \tag{11}$$

Since the algorithm Mamdami is used for fuzzy inference, the following methods of execution had been chosen for the stages of composition:

1. To perform a logical conjunction in conditions of the fuzzy rules (And method), the minimum value (min) method was chosen;
2. To perform a logical disjunction in conditions of the fuzzy rules (Or method), the maximum value (max) method was chosen;
3. To perform a logical conclusion in the every of fuzzy rules (Implication), the minimum value (min) method was chosen;
4. For aggregation of membership function values of each input variable in conclusions of fuzzy rules (Aggregation), the maximum (max) method was chosen;
5. To perform defuzzification of output variables (Defuzzification), the method of center of gravity for a discrete set of values of the membership functions (centroid) was chosen.

In the process of fuzzy inference, it is necessary to select such an access point for which the convolution of the functions of the accessory of the signal/noise ratio and the degree of congestion of the access point give the best result. To implement the fuzzy inference, the creation of the rule base of fuzzy inference system is required. To implement the fuzzy inference, the creation of the rule base of fuzzy inference system is required. Such base of rules for the solution of discussed task looks like:

- If the *S/N* ratio is weak and access point is lightly loaded (the number of users *M* is few), then the access point is not considered as a candidate for connection.
- If the *S/N* ratio is weak and access point is moderately loaded (the number of users *M* is medium), then the access point is not considered as a candidate for connection.
- If the *S/N* ratio is weak and access point is heavily loaded (the number of users *M* is many), then the access point is not considered as a candidate for connection.
- If the *S/N* ratio is medium and access point is lightly loaded (the number of users *M* is few), then the access point is considered as a candidate for connection.
- If the *S/N* ratio is medium and access point is moderately loaded (the number of users *M* is medium), then the access point is not considered as a candidate for connection.
- If the *S/N* ratio is medium and access point is heavily loaded (the number of users *M* is many), then the access point is not considered as a candidate for connection.
- If the *S/N* ratio is high and access point is lightly loaded (the number of users *M* is few), then the access point is considered as a candidate for connection.
- If the *S/N* ratio is high and access point is moderately loaded (the number of users *M* is medium), then the access point is considered as a candidate for connection.
- If the *S/N* ratio is high and access point is heavily loaded (the number of users *M* is many), then the access point is not considered as a candidate for connection.

The following notations are in use here to simplify the formalization of fuzzy productions:

1. if (*S/N* is A1) and (Numusers is B1) then (Result is E1) (1);
2. if (*S/N* is A1) and (Numusers is B2) then (Result is E1) (1);
3. if (*S/N* is A1) and (Numusers is B3) then (Result is E1) (1);
4. if (*S/N* is A2) and (Numusers is B1) then (Result is E2) (1);
5. if (*S/N* is A2) and (Numusers is B2) then (Result is E1) (1);
6. if (*S/N* is A2) and (Numusers is B3) then (Result is E1) (1);
7. if (*S/N* is A3) and (Numusers is B1) then (Result is E2) (1);
8. if (*S/N* is A3) and (Numusers is B2) then (Result is E2) (1);
9. if (*S/N* is A3) and (Numusers is B3) then (Result is E1) (1).

A1 means the *S/N* ratio is weak, A2 means the *S/N* ratio is medium, A3 means the *S/N* ratio is high, B1 means the number of users is few, B2 means the number of users is medium, B3 means the number of users is many, E1 means that the access point is not a candidate for connection, E2 means that the access point is a candidate for connection. Thus, the connection to the access point occurs only when the access point is not overloaded and has an acceptable signal level. The numbers in round brackets mean the weight of rule (0 or 1). The next step in the construction of the model is to determine the membership functions of input and output variables. Kind of membership functions for the signal/noise ratio, the number of subscribers, and the output variable are shown in **Figure 4**.

For analyzing the adequacy of the developed fuzzy model, the surface fuzzy inference can be quite useful. It allows assessing the impact of changes in the values of the input fuzzy variables on the values of the output fuzzy variables. This surface is shown in **Figure 5**. A comparison of this figure with **Figure 1** allows concluding that the nature of the surface of the fuzzy inference coincides in general with the frame transmission time dependent on the signal level and the number of subscribers. This confirms the adequacy of the proposed model.

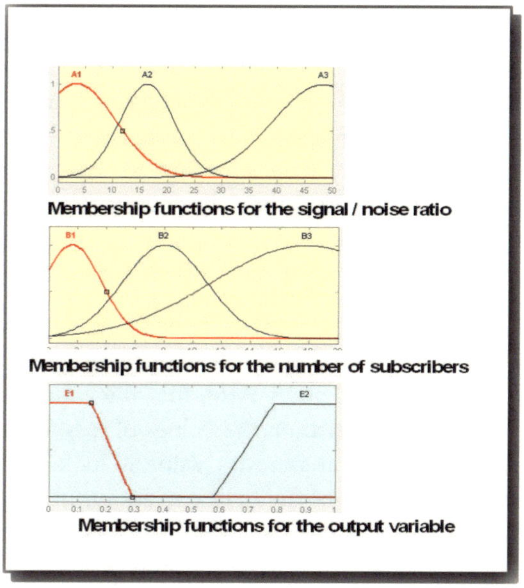

Figure 4. Membership functions for input and output variables.

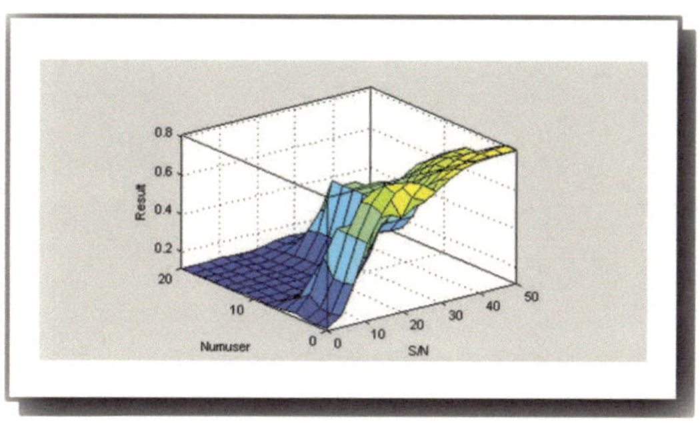

Figure 5. The surface fuzzy inference.

3.4. Simulation of the algorithm for determining the best access point

For checking the correctness and efficiency of the proposed algorithm, its styling is first carried out. MATLAB provides the ability to download a model developed with the help of visual aids into the MATLAB—program and perform simulation in program mode. The data structure of a software model is shown in **Figure 6**.

Each access point is represented by the data structure containing the following data fields:

- Index of the access point (its number in a vector of access points).
- The current value of the signal level at that point.
- The current number of connected subscribers at this point.
- The current time of the frame transmission for this point.

At initialization of this model, these structures combine into a vector. Thus, the sequential number of structure in the vector imitates the Beacon signal. All other fields of structures are set to zero. Also, the vector of signal levels is forming. This vector simulates a signal power at which every access point radiates, and thereby, the location of subscriber is relative to access points in the moment of its connection. The signal levels are forming according with normal distribution and exponential distribution. This adequately reflects the real situation when connecting subscribers grouped in a small area, for example, when entering a room or in some corner of the room, as it is shown in **Figure 7**.

Simulation process starts with the loading of model into the program. Further, the current values of signal level out of the vector of signal levels are moving into the fields of current values of the signal level in every access point. After this, the values of membership functions are calculated for every access point. Further, the values of membership functions are calculated for every access point and their maximum value is looked for. The access point for which this value is maximal, the value of the field current number of connected subscribers to this point is increased by 1. This simulates the RSSI level correction for the next simulation step. Furthermore, the current time of the frame transmission for this point is calculated according with function (5). This process is repeated for all the elements of the vector of

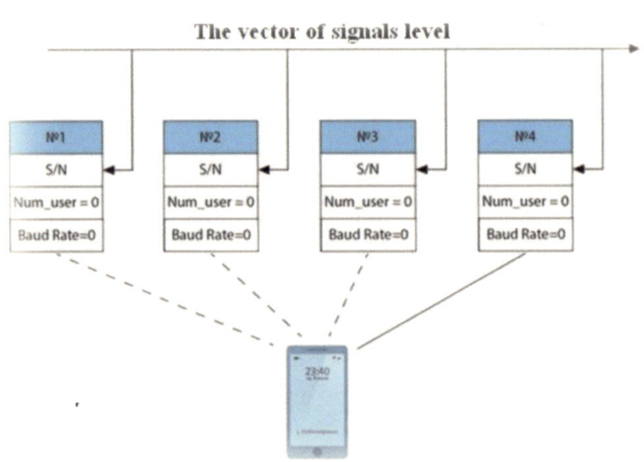

Figure 6. The data structure of model.

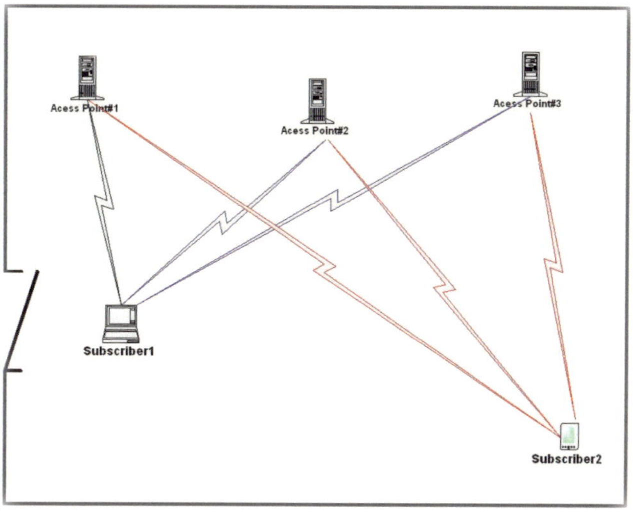

Figure 7. The example of the subscriber's location indoors.

signal levels. Thus, after the completion of simulation process, total loading and the frame transmission time in every point would be determined.

The modeling process is based on a method that was considered in Ref. [26]. In this case the sufficient characteristics are current values of congestion of access point and frame transmission time on every simulation step; the final characteristics are the distributions of load of access points, transmission time, mean value, and dispersion of the transmission time. For the confirmation of efficiency of the proposed algorithm, its result was compared with the result of simulation of classic, which is based on the measuring of maximal signal power. The results of the comparison of the analyzed algorithms are shown in **Figure 8**. In this picture, "Sample Model" means the model is based on the measuring of maximal signal power. This figure shows that the algorithm that uses fuzzy logic (the right-hand side of **Figure 8**), taking into account the current load on each access point, ensures a more even loading of the entire LAN as a whole. The frame transmission time for the second algorithm is approximately same for all access points (the dispersion is equal to 0.0799). For the first algorithm, the transmission time

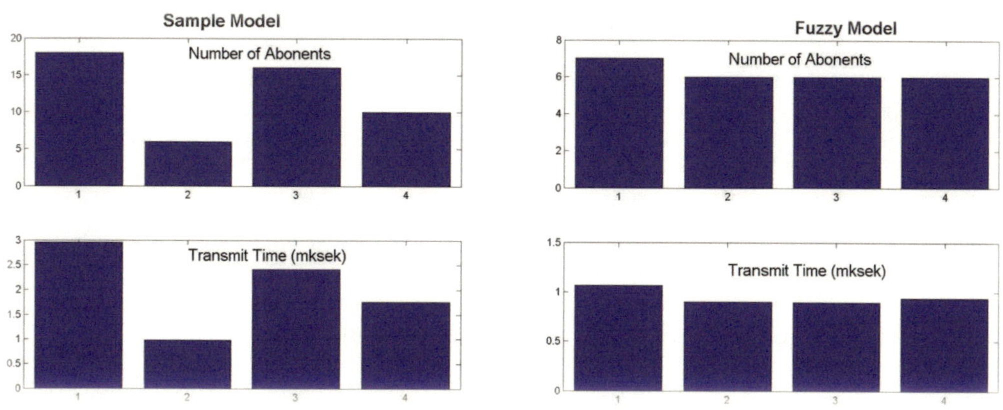

Figure 8. The results of the comparison of the analyzed algorithms.

fluctuates strongly (the dispersion is equal to 0.3129). Hence, it follows that accounting the degree of congestion of access points increases the efficiency of the wireless LAN.

Developed in the framework of this chapter, MATLAB—the program cannot be used directly in some hardware for the following reasons:

- Developed program is a model.
- MATLAB—program runs in interpreted mode and cannot be loaded directly into the controller.

However, there are various possibilities to convert MatLab-program into equivalent program on one of the high-level programming languages. In particular, there is the possibility of constructing an equivalent program on high level language such as C# or C++. Such program can be installed on the controller or other device that will automatically determine the best access point to connect to the local network.

4. Conclusions

We have reviewed the advanced solutions for indoor navigation and briefly reviewed the program complex to navigate through radio fingerprints. Navigation software package uses the model ANN. In addition, by using the MATLAB performed rationale for the selection of ANN learning algorithm and determine the recommended number of neurons of the hidden layer for the considered number of Wi-Fi networks.

As a recommendation, to the use of this software system, one should take into account the importance of using the mac-address of an access point instead of its name because it is not unique. Emerged collision may disrupt the entire complex.

The general advantages of this approach include the fact that the software package can run on the existing infrastructure of Wi-Fi networks that are deployed in a variety of areas, such as residential buildings and shopping malls.

The disadvantages are a feature of this approach lie in the fact that before using it you must manually map the radio fingerprint.

Further studies, in our opinion, are advisable to continue in the following areas.

- The implementation multivariate classifier for converting radio prints multidimensional coordinates (x, y, height).
- Conducting research on the influence of other learning algorithms ANN (Quickprop, Batch, and Incremental) the speed and accuracy of navigation training.
- Development of a method for reducing the dimensionality of the input vector radio fingerprint.
- Improving the quality of navigation when combined with inertial navigation system.
- The addition of processing signals from Bluetooth LE beacons to signals from Wi-Fi access points.

The task of selection of the access point in WLAN network in the case of high user density is quite topical. Such selection must take into account not only the level of the signal received by the mobile device, but also a width in the dedicated channel bandwidth that depends on the number of connected subscribers to the access point.

- Uneven loading of access points in a wireless LAN tends to reduce the overall effectiveness of its work. This is manifested in the fact that the frame transmission time for the overloaded access point may increase significantly.
- Connecting the mobile subscriber to the access point in accordance with the maximum received signal level leads to that some access points may be overloaded.
- Accounting of degree of current workload of access points when a subscriber connects to them provides better balancing of total workload in the network.
- To find the best access point for connecting a subscriber with account of its workload is necessary to solve the optimization task.
- The most effective method for solving this task is to use the apparatus of fuzzy logic.
- The developed model can be used in controllers or in embedded devices by converting into an executable program on high-level programming language.

Author details

Anatoly D. Khomonenko, Sergey E. Adadurov, Alexandr V. Krasnovidow*, Pavel A. Novikov

*Address all correspondence to: alexkrasnovidow@mail.ru

Petersburg State Transport University, St. Petersburg, Russia

References

[1] Yuri S. Satellite Navigation Systems. Moscow: ECO Trendz; 2000. p. 267. http://www.indoorsnavi.com/

[2] Hatami A, Pahlavan K. A comparative performance evaluation of RSS-based positioning algorithms used in WLAN networks. In: Proceedings of the IEEE Wireless Communications and Networking Conference (WCNC '05); Vol. 4. March; New Orleans, LA, USA; 2005. pp. 2331-2337

[3] Kukolj D, Vuckovic M, Pletl S. Indoor location fingerprinting based on data reduction. International Conference on Broadband and Wireless Computing, Communication and Applications (BWCCA); 26-28 October; 2011. pp. 327-332

[4] Butakov NA. Applicability inertial navigation in mobile devices. International Journal of Open Information Technologies. 2014;**2**(5):24-32. ISSN: 2307-8162.

[5] Evennou F, Marx F. Advanced integration of wi-fi and inertial navigation systems for indoor mobile positioning. EURASIP Journal on Applied Signal Processing. 2006;**2006:**1-11. Article ID 86706.

[6] Wang S, Min J, Yi BK. Location based services for mobiles: Technologies and standards. LG Electronics Mobile Research, USA. 2008. 122 pp.

[7] Dworkina NB, Namiot DE, Dvorkin BA. Mobile navigation services and the use of technology to determine the location OpenCellID. Geomatics. 2010:**2**;80-87. http://sovzond.ru/upload/iblock/fbf/2010_02_014.pdf.

[8] Abdrakhmanova AM, Namiot DE. Using two-dimensional bar code to create a system of positioning and navigation inside. Applied Informatics. 2013; **1**(43):31-39

[9] Namiot DE, Shneps-Shneppe MA. Analysis of the trajectories in mobile networks on the basis of information about network proximity. Automation and Computer Science. 2013;**3**:S48-S60

[10] Mok E, Cheung Bernard KS. An improved neural network training algorithm for wi-fi fingerprinting positioning. ISPRS International Journal of Geo-Information. 2013;**2**:854-868

[11] Mehmood H, Tripathi NK, Tipdecho T. Indoor positioning system using artificial neural network. Journal of Computer Science. 2010;**6**(10):1219-1225

[12] Fang SH, Lin TN. Indoor location system based on discriminant-adaptive neural network in IEEE 802.11 environments. IEEE Transactions on Neural Networks. 2008;**19**(11):1973-1978

[13] Novikov PA, Khomonenko AD, Yakovlev EL. Justification of the choice of neural networks learning algorithms for indoor mobile positioning. Proceeding CEE-SECR '15 Proceedings of the 11th Central & Eastern European Software Engineering Conference in Russia. Moscow, Russian Federation; 22-24 October 2015; ACM New York, NY, USA©; 2015. Article No. 9. DOI: 10.1145/2855667.2855677

[14] Novikov PA, Khomonenko AD, Yakovlev EL. Software for mobile indoor navigation using neural networks. Informatsionno-Upravliaiushchie Sistemy, [Information and Control Systems]. 2016;**1**:32-39. (In Russian). DOI: 10.15217/issn1684-8853.2016.1.32.

[15] Akobir S. The learning algorithm RProp – mathematical apparatus. http://basegroup.ru/community/articles/rprop.

[16] Khomonenko AD, Yakovlev EL. Neural network approximation of characteristics of multi-channel non-Markovian queuing systems. SPIIRAS Proceedings. 2015;**4**(41):81-93. (In Russian). DOI: http://dx.doi.org/10.15622/sp.41.4.

[17] IEEE. Wireless LAN medium access control (MAC) and physical layer (PHY) specifications. The Institute of Electrical and Electronics Engineers, IEEE 802.11. 2006.

[18] Krasnovidow AV. Model of algorithm to determine a best access point to connect a mobile device to the LAN. Intellectual Technologies on Transport. 2016;**2**(6):36-42 (In Russian).

[19] Larsson D, Merty R. Adaptivniy Podhod k optimizacii proizvoditelnosty besprovodnyh setey/ Dylan Larsson, Ravi Merty//[Adaptive approach to optimize the performance of wireless networks] Technology@Intel Magazine, - 2004, No 8, pp. 24–27.

[20] Himmelblaw D. Prikladnoe nelineinoe programmirovanie. [Applied Nonlinear Programming]. Moscow: Mir; 1976. p. 448

[21] Elter K-H, Reinhardt R, Schauble M, Donath G. Vvedenie v Nelineinoe Programmirovanie. [Introduction to the Nonlinear Programming]. Moscow: Nauka; 1985. p. 264

[22] Leonenko A. Nechetkoie modelirovanie v srede matlab i fuzzyTECH. [Fuzzy Modelling in MATLAB Environment and FuzzyTECH]. Saint-Petersburg: BHV – Petersburg; 2005. p. 736

[23] Terano T, Asai K, Sugeno M, editors, Prikladnye nechetkie sistemy. [Applied Fuzzy Systems]. Moscow: Mir; 1993. p. 386

[24] Ihmig M. Dynamic channel allocation for self-managing WLAN access points in chaotic wireless networks [Thesis]. Diplomarbeit, Master's Thesis, Technische Universität München, Mar 2006.

[25] Leith DJ, Clifford P. A self-managed distributed channel selection algorithm for WLANs. 4th International Symposium on Modelling and Optimization in Mobile, Ad Hoc and Wireless Networks. 2006; pp. 1-9. DOI: 10.1109/WIOPT.2006.1666484.

[26] Krasnovidow AV. An approach to the construction of algorithms for the statistical analysis of error flows in digital communications channels. Intellectual Technologies on Transport. 2015;**2**(2):20-25. (In Russian).

Chapter 12

A Fuzzy Logic Approach for Separation Assurance and Collision Avoidance for Unmanned Aerial Systems

Brandon Cook, Tim Arnett and Kelly Cohen

Additional information is available at the end of the chapter

http://dx.doi.org/10.5772/68126

Abstract

In the coming years, operations in low altitude airspace will vastly increase as the capabilities and applications of small unmanned aerial systems (sUAS) continue to multiply. Therefore, finding solutions to managing sUAS in highly congested airspace will facilitate sUAS operations. In this study, a fuzzy logic-based approach was used to help mitigate the risk of collisions between aircraft using separation assurance and collision avoidance techniques. The system was evaluated for its effectiveness at mitigating the risk of mid-air collisions between aircraft. This system utilizes only current state information and can resolve potential conflicts without knowledge of intruder intent. The avoidance logic was verified using formal methods and shown to select the correct action in all instances. Additionally, the fuzzy logic controllers were shown to always turn the vehicles in the correct direction. Numerical testing demonstrated that the avoidance system was able to prevent a mid-air collision between two sUAS in all tested cases. Simulations were also performed in a three-dimensional environment with a heterogeneous fleet of sUAS performing a variety of realistic missions. Simulations showed that the system was 99.98% effective at preventing mid-air collisions when separation assurance was disabled (unmitigated case) and 100% effective when enabled (mitigated case).

Keywords: fuzzy logic, UAS, collision avoidance, separation assurance, formal methods, satisfiability modulo theories

1. Introduction

In recent times, there have been substantial advances in the capability of mobile robots in several aerospace applications. These advances include autonomous intelligence, surveillance, and reconnaissance (ISR) efforts [1], aerial firefighting [2], and aerial delivery services [3]. However, despite the potential benefits, these advancements are currently being under-utilized due to several unresolved safety issues with integrating these platforms into the National Airspace System (NAS). As a result of these shortcomings, there is a need to develop algorithms that allow a heterogeneous team of small unmanned aerial systems (sUAS) to interact autonomously and perform time-critical tasks in complex environments. As the applications and capabilities of sUAS continue to proliferate, it is imperative to address the safe integration of these vehicles into the NAS.

Most of the work in this area deals with separation assurance, as it typically takes priority in NAS conflict resolution scenarios [4, 5]. However, most methods necessitate the communication of state information between the vehicles in order to properly select resolution actions [6]. For collision avoidance, several intelligent systems have been developed with promising results [7–12], but few have also shown behavioral verification using formal methods [13, 14].

To facilitate real-time control of a large number of sUAS, a fuzzy logic approach was implemented. This approach was utilized to mitigate the risk of losses of separation and ultimately collisions, between the sUAS. In order to generate scenarios to test the sUAS's ability to avoid collisions, a realistic simulation environment was created. This simulation environment was developed in a modular fashion, such that various algorithms could be implemented to coordinate sUAS maneuvers. This enables various vehicle platforms, sensor models, software packages, and traffic management methodologies to be tested and evaluated.

The main goal of this research is to develop a high-level concept of operations for a UAS traffic management (UTM) system. This system must address the challenges of collision avoidance and separation assurance. Each of these platforms will utilize fuzzy logic controllers (FLCs) to enable real-time decision-making and dynamic control. Additionally, the confidence in correct decision-making and avoidance control outputs needs to be extremely high. Therefore, formal methods were employed for behavioral verification.

The remainder of the paper is as follows. In Section 2, background material on some of the methods and tools used in this work is described. Section 3 details the proposed solution, which includes the separation assurance and collision avoidance methods. This includes detailed development procedures for the decision-making and fuzzy avoidance controllers. Section 4 presents the methodology for implementing and evaluating the decision-making and fuzzy avoidance controllers using formal methods. Section 5 then explains the test cases, their implementations, and an overview of the simulation environment and constraints. Section 6 presents results from the formal methods evaluations and simulation runs, and finally, Section 7 discusses conclusions and opportunities for future work.

2. Background

2.1. Hybrid fuzzy systems

Most fuzzy inference systems (FISs) involve multiple operations that associate inputs to outputs based on multiple if-then rules that are resolved to a singleton. The output space is typically nonlinear and difficult to describe as a function of the input variables. However, by constraining the FIS to have particular properties and association methods, an explicit expression can be more easily found.

Hybrid systems are systems that have regions of continuous behavior separated by discrete transitions [15]. This is analogous to a subset of FISs that contain membership functions that are constrained to a finite domain. Such FISs can be represented as hybrid systems after an explicit expression is found. This expression maps an input set to an output set using a set of mathematical functions. This is useful due to the number of low-level tools that have been developed for analyzing hybrid systems. Among these are formal methods tools [16], which are described in the following section.

2.2. Formal methods

In systems such as UTM collision avoidance algorithms, the level of confidence that they will always behave as intended needs to be extremely high. Typical methods for evaluating these algorithms usually involve simulation, but simulation and other numerical methods can miss critical cases that result in undesired behavior. To increase the confidence that the avoidance algorithms presented work as intended, formal methods were employed. Formal methods are defined by NASA as "mathematically rigorous techniques and tools for the specification, design, and verification of software and hardware systems." [17] There are numerous types of tools that fall under this definition, but in this work, satisfiability modulo theories (SMT) solvers and model checkers were used.

SMT solvers are tools that extend the Boolean Satisfiability (SAT) problem to first order logic (FOL) sentences and incorporate other theories for evaluating the truth assignments of variables (real values, bitvectors, etc.). If a behavior can be described in FOL, it can be encoded and evaluated by an SMT solver to find truth assignments that violate this behavior. If a behavior is found, it is returned as satisfying the behavioral specification. If there are no possible assignments to the variables that render the specification true, it is then said to be unsatisfiable. Therefore, "safety" properties, properties which should always hold true, can be evaluated by negating a specification that encapsulates the respective behavior. If a satisfying case is found for the negated specification, this means that there are conditions that violate the original specification. If no satisfying cases are found, then the original specification will hold for all possible conditions.

Model checkers are tools that exhaustively check the states of a system to search for combinations of variable assignments that violate behavioral specifications. In finite state systems, they use deductive proofs, and in infinite state systems, they can use inductive methods. These tools can also use SAT or SMT solvers in conjunction with their own search methods for finding counterexample cases. Encoding safety properties in model checkers is slightly different, however, as model checkers typically use some type of temporal operator in conjunction with logical sentences. However, there are methods for relating quantified FOL sentences for safety properties to temporal representations for use in model checkers [18].

In this work, an infinite-state model checker named JKind [19, 20] was used. JKind is a Java implementation of the Kind model checker which uses *k*-induction. To evaluate the truth assignments for variables within each state, JKind employs SMT solvers. The SMT solver used for this work is Z3 [21], a state-of-the-art SMT solver developed by Microsoft.

3. Proposed solution

To ensure that two or more sUAS do not collide with one another, an intruder avoidance system was developed. This avoidance problem is broken down into two sub-systems: strategic (separation assurance) and tactical (collision avoidance). The strategic separation assurance platform uses a centralized approach to coordinate trajectory modifications for sUAS to ensure that vehicles do not get too close to one another. This separation assurance technique is employed when two or more vehicles come within 0.4 nmi laterally of one another (separation alert threshold), 100 ft. vertically, and are predicted to have a loss of separation (LOS), defined by when vehicles come within 0.1 nmi laterally and 100 ft. vertically of one another. This LOS threshold was determined based on the characteristics of the vehicle platforms and feasible sensing abilities of sUAS. Based on the system constraints, the avoidance platform would have roughly 2 sec to resolve maximum closure rate encounters. If the separation assurance system fails to prevent an LOS, the vehicles will employ their onboard sense and avoid systems to prevent a mid-air collision. A mid-air collision occurs when two vehicles come within 60 m of one another. This collision threshold is intentionally conservative to introduce a notion of spatial uncertainty. Since the sensor models provide perfect state information, as described in Section 5, all vehicle locations are precisely known. For this study, the collision avoidance platform uses a de-centralized approach, that is, all vehicles attempt to avoid intruding vehicles independently. Thus, no communication between aircraft is available (i.e., uncoordinated maneuvers). In **Table 1**, the various distance thresholds used to describe the separation boundaries are shown.

Prior to presenting the details of each avoidance sub-system, Section 3.1 provides an overview of the avoidance system architecture. This overarching logic is used to determine when a vehicle should perform an avoidance maneuver. When deemed necessary, the system will activate the appropriate avoidance platform. In each avoidance platform, a set of heuristics are used to determine the appropriate action to resolve a conflict. These details are presented in Sections 3.2 and 3.3. Once the appropriate action has been decided, an FLC is used to control the vehicle's turn rate in the desired direction. The details of each FLC are shown in Section 3.4.

Finally, it is important to note that these two sub-systems use different approaches when trying to resolve conflicts between aircraft. This is primarily due to the overall purpose each

Threshold label	Lateral distance	Vertical distance
Separation alert	0.4 nmi	100 ft.
LOS	0.1 nmi	100 ft.
Collision	60 m	50 ft.

Table 1. Separation threshold values.

sub-system serves, and the information that is available to each. If vehicles are reporting their state information to a ground-based station, a centralized separation assurance platform could be used to coordinate a trajectory modification to one or more of the vehicles. Thus, coordinated maneuvers are possible. However, when two vehicles are within seconds away from a collision, minimizing the time between sensing the vehicle and performing an action is critical. Therefore, when the collision avoidance system is activated, the vehicles must independently choose the appropriate action using onboard processors. In these collision avoidance scenarios, there is no communication between aircraft. Thus, each sub-system requires a different set of rules to determine the appropriate action.

3.1. Overarching control logic

The overarching control logic determines whether to perform a separation assurance maneuver, activate the collision avoidance system, or allow vehicles to continue along their desired trajectories. This logic is shown in flow chart form in **Figure 1**. First, the system will find the distance separating all aircraft pairs. With this information, a calculation is made to see how much time can pass prior to two vehicles coming within 0.4 nmi. To calculate this value, the current separation, minus the 0.4 nmi threshold, is divided by the maximum closure rate of the aircraft pair. Therefore, if both vehicles moved directly toward one another at their maximum allowable speeds, this is the time it would take them to reach the 0.4 nmi separation threshold. This future time is known as the "time threshold", as shown in **Figure 1**. Using this time threshold, if two vehicles cannot possibly be within 0.4 nmi of one another, the system will not unnecessarily check if the two aircraft are in conflict. Rather, it remains idle between checks to improve the performance of the system.

Once enough time has passed and an aircraft pair reaches their assigned time threshold, the system will again check their separation. If the two aircraft are still more than 0.4 nmi apart, a new time threshold is calculated and set. However, if the aircraft pair has reached the 0.4 nmi threshold, it will next check to see if an LOS has occurred. If the vehicles have violated the 0.1 nmi LOS threshold, the collision avoidance system is enabled. Otherwise, the separation assurance system may be needed to ensure that two vehicles do not have an LOS. This decision is based on two criteria: if the predicted closest point of approach (pCPA) creates an LOS and if the time to LOS (tLOS) is within 2 min.

If both criteria are met, a final check is used to see if the aircraft pair has already been assigned a separation assurance maneuver. If neither vehicle has been assigned a maneuver to avoid an impending LOS, a resolution advisory is sent from the centralized system to one of the sUAS. However, if the sUAS was already assigned a maneuver and is currently in the middle of its resolution, a check is used to see if turning back toward its preferred heading will cause another predicted LOS. If resuming its originally intended mission will not cause an LOS, it will do so, otherwise, the sUAS will continue on its current bearing.

If neither criterion (pCPA and tLOS) is met, then no separation assurance command is given and the aircraft will continue toward its respective target using its navigation controller. After determining the appropriate separation assurance action, or deciding that no action is needed, the algorithm calculates another time threshold for each aircraft pair.

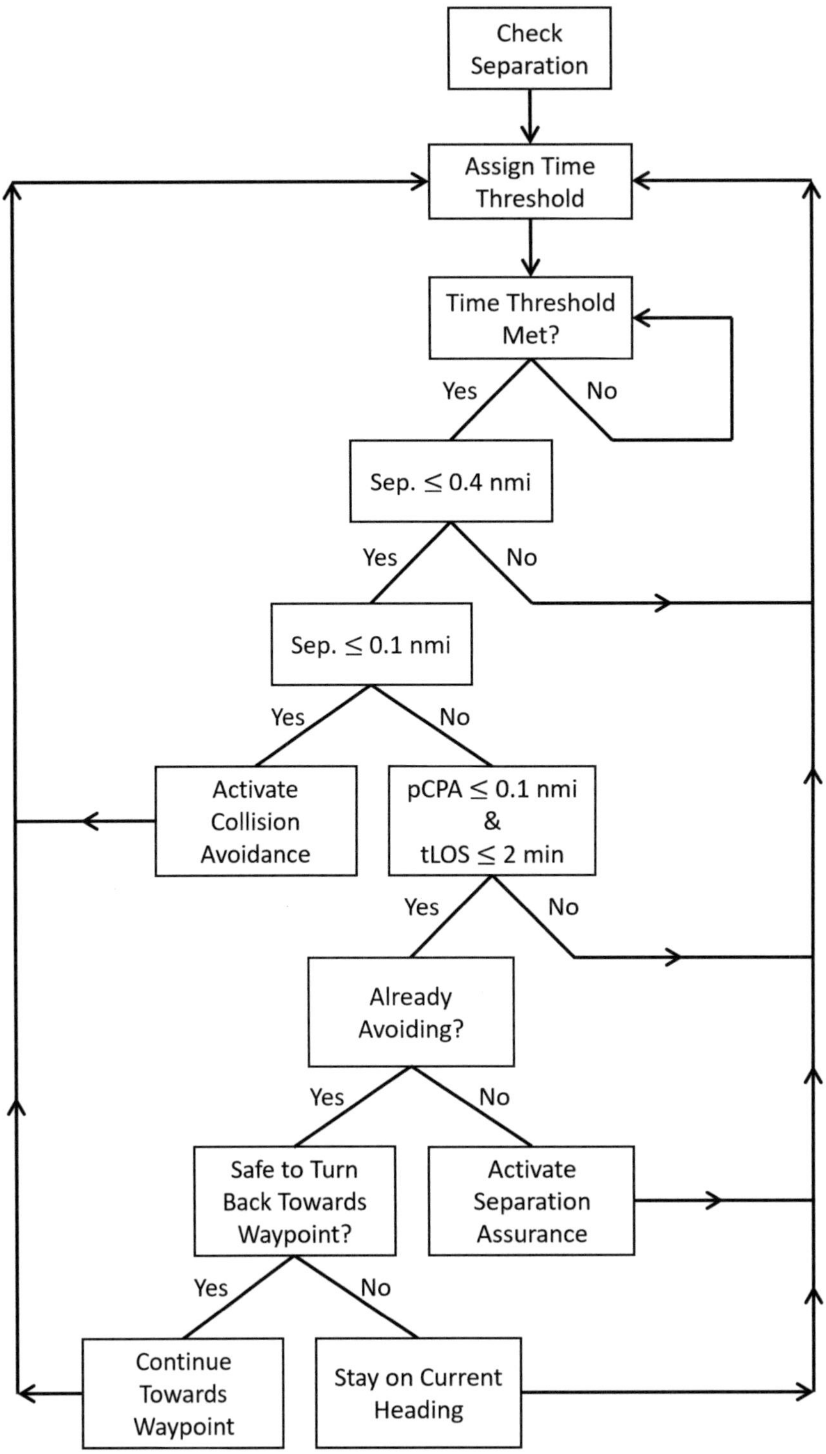

Figure 1. Control logic flow chart.

3.2. Separation assurance logic

The centralized separation assurance system observes only the current position, heading, and velocity of each vehicle. With this limited information, vehicle intent is unknown. Therefore, the system must be robust to dynamic scenarios and resolve conflicts without the knowledge of other vehicles' goals.

The separation assurance platform will be enabled if three criteria are met: separation less than 0.4 nmi, the pCPA is less than 0.1 nmi, and the tLOS is within 2 min. If two aircraft meet all three of these criteria, the separation assurance platform will activate and assign one of the vehicles a new trajectory in an effort to avoid the predicted LOS.

To determine what action the separation assurance platform should use to avoid a potential LOS, a series of conflict classification techniques are used. For this study, three pieces of information are used to classify all conflict scenarios: relative heading, relative angle, and crossing time. These parameters can be described using the more common parameters: location, speed, and heading. In **Figure 2**, a sample conflict scenario is shown. Here, the triangular objects each represent an sUAS, the arrows represent the velocities of each aircraft (both magnitude and direction), and the "x" represents the heading intersection point location. The heading intersection point is not to be confused with the pCPA. It is simply the point where the projected headings intersect with one another. For this example, the vehicle with the small circle represents aircraft 1, and the other represents aircraft 2.

The relationship used to describe the heading of vehicle 2 relative to vehicle 1's perspective is shown in Eq. (1).

$$R_{H_1} = H_2 - H_1 \tag{1}$$

where H_1 is the heading of vehicle 1, H_2 is the heading of vehicle 2, and R_{H_1} is the relative heading from vehicle 1's perspective. In this study, $0° \leq R_{h_i} < 360°$ for all vehicles, where i represents the index for each vehicle. Therefore, if $H_1 > H_2$, a 360° phase shift must be added to R_{H_1} to remain within the constrained range. Computing the relative heading for the example shown in **Figure 2**, R_{H_1} would be 140° (i.e., moving to the left with respect to vehicle 1), whereas, from vehicle 2's perspective, R_{H_2} would be roughly 220° after applying the 360° phase shift (i.e., moving to the right with respect to vehicle 2).

Similarly, the relative angle between two vehicles describes the relative position of one vehicle with respect to the other. This relationship has been shown in Eq. (2).

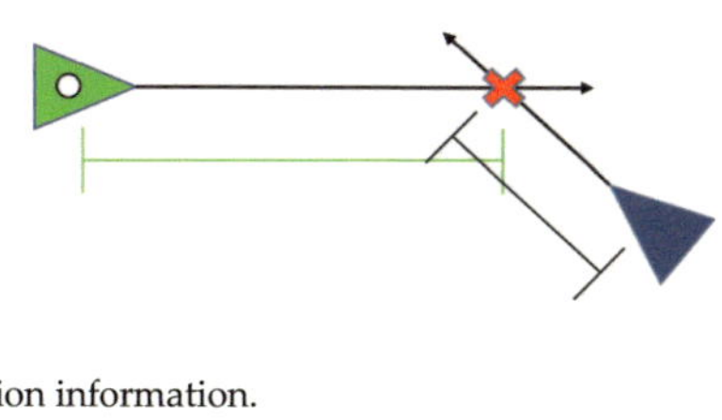

Figure 2. Conflict scenario classification information.

$$R_{A_1} = \tan^{-1}\left(\frac{y_2 - y_1}{x_2 - x_1}\right) - H_1 \tag{2}$$

where x_2 and y_2 are the two-dimensional coordinates of vehicle 2 in the global frame, x_1 and y_1 are the two-dimensional coordinates of vehicle 1 in the global frame, H_1 is the heading of vehicle 1, and R_{A_1} is the relative angle of vehicle 2 from vehicle 1's perspective. Like the relative heading, the relative angle is constrained. Here, we constrain the relative angle by the following relationship: $-180° \leq R_{A_1} \leq 180°$. Therefore, if the relative angle is less than $-180°$, a 360° phase shift is added to meet this constraint, or if the angle is greater than 180°, a 360° phase shift is subtracted. Again, using **Figure 2** as an example, R_{A_1} would be roughly $-15°$ after subtracting a 360° phase shift (i.e., vehicle 2 is to the right of vehicle 1 from vehicle 1's perspective), and R_{A_2} would be roughly 25° (i.e., vehicle 1 is on the left of vehicle 2 from vehicle 2's perspective).

Lastly, the crossing time, t_1, can be defined by the relationship shown in Eq. (3). This relationship defines how long it would take for vehicle 1 to reach the heading intersection point, as shown in **Figure 2**, if they remained on their current trajectory.

$$t_1 = sign\left(\vec{V_1} \cdot \vec{C_1}\right)\frac{c_1}{v_1} \tag{3}$$

where *sign* is a function used to determine the sign (positive or negative) of an expression, $\vec{V_1}$ is the velocity of vehicle 1, $\vec{C_1}$ is the vector used to describe the location of the heading intersection point relative to the vehicle's current position, v_1 is the speed of vehicle 1, and c_1 is the magnitude of the vector $\vec{C_1}$.

With these relationships, all possible encounter scenarios can be described. To aid in understanding what each of these parameters represent, **Table 2** describes, in linguistic terms, what each range of values represents in the physical conflict scenarios. Here, the term crossing point

Parameter	Range	Meaning
Relative heading	$0° < R_{H_1} < 180°$	Moving from right to left
	$180° < R_{H_1} < 360°$	Moving from left to right
	$R_{H_1} = 0°$	Same direction
	$R_{H_1} = 180°$	Head-on
Relative angle	$R_{A_1} > 0°$	On left
	$R_{A_1} < 0°$	On right
	$R_{A_1} = 0°$	Straight ahead
Time to crossing point	$t_1 > 0$	Crossing point in front
	$t_1 < 0$	Crossing point behind
	$t_1 > t_2$	Farther from crossing point
	$t_1 < t_2$	Closer to crossing point

Table 2. Linguistic descriptions of encounter scenarios.

is synonymous to the heading intersection point. Each of these descriptions has been listed to describe the motion, location, or crossing time of vehicle 2 from vehicle 1's perspective.

Although the primary goal of this system is to ensure safe separation of vehicles, it is also important to try and limit the number of unnecessary flight adjustments. This is particularly important when operating sUAS due to their typical limitations in power and endurance. Because vehicle intent is unknown in this study, a predicted LOS does not guarantee an LOS is imminent. Therefore, there is a tradeoff between using strict and relaxed criteria when determining if a trajectory modification is necessary. The criteria should be relaxed to ensure sUAS do not repetitively perform unnecessary adjustments but strict enough to ensure safe operation.

Aside from optimizing this time to predicted LOS threshold, a second way to limit the number of vehicles that divert from their desired flight paths is to assign vehicles priority. This priority assignment ranks all vehicles in conflict from highest priority to lowest priority. Therefore, the vehicle with the highest priority will continue on its preferred trajectory without modification. However, all vehicles with a lower priority must avoid all other vehicles with a higher priority.

To determine which aircraft has a lower priority, a series of evaluations are made. First, the system will be checked to see if the two aircraft are moving in a similar direction. If two vehicles have a heading within 5° of one another, that is, $355° \le R_{H_1} < 360°$ or $0° \le R_{H_1} \le 5°$, the trailing aircraft will have lower priority. This encounter scenario can be seen in **Figure 3**. If several aircraft have similar headings, the aircraft furthest behind will be assigned the lowest priority so must avoid all other aircraft. However, the vehicle in the front of the group will have the highest priority and will disregard all other aircraft.

If the vehicles in conflict do not have similar headings (i.e., more than a 5° difference), the vehicle closest to its next waypoint is given priority. Since the separation assurance system logic does not use the location of a vehicle's next waypoint (i.e., intent is unknown), this priority assignment was simply a means to an end. In practice, the priority of each vehicle in these scenarios would be randomly assigned.

To predict whether an LOS will occur, the separation assurance algorithm uses the current location and velocity of each aircraft to calculate a projected flight path for each. Using these projected trajectories, the pCPA between the aircraft is found. If the pCPA will result in an LOS within the next 2 min, a resolution is calculated and employed to prevent the predicted LOS.

Recalling the sample encounter scenario shown in **Figure 2**, the definitions described by Eqs. (1) through (3), and the constraints shown in **Table 2**, all encounter scenarios can be described. **Figure 4** depicts all the possible conflict scenarios when vehicle 2 is located to the right of vehicle

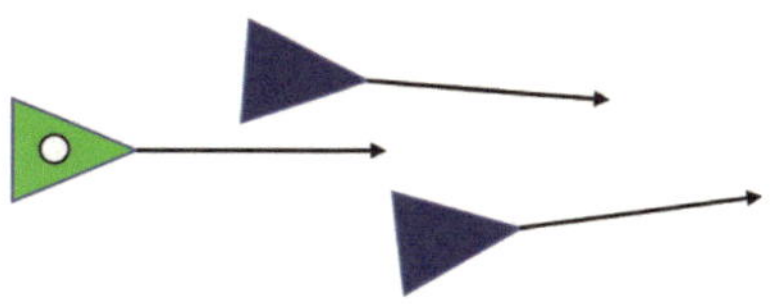

Figure 3. Encounter example with relative heading within 5°.

1. In each diagram, the aircraft with the small circle represents vehicle 1 and the other is vehicle 2. Thus, all parameters used to describe a particular conflict scenario are from the perspective of vehicle 1. Within the scope of the separation assurance system, vehicle 2 is classified as the vehicle that has been assigned the higher priority. Thus, vehicle 1 (lower priority) must perform a maneuver to prevent an LOS.

In **Figure 4(a)**, vehicle 2 is moving to the left, vehicle 1 is approaching from behind, and vehicle 1 is more than 45 sec closer to the heading intersection point. In this case, vehicle 1 would decide to go in front of vehicle 2. However, if vehicle 1 is not at least 45 sec closer, it will go behind.

In **Figure 4(b)**, vehicle 2 is still going to the left, but in this case, it is coming toward vehicle 1. In these scenarios, vehicle 1 must be more than 30 sec closer to the intersection point to go in front. In **Figure 4(c)**, vehicle 2 is to the right of vehicle 1 but is also going to the right. In these instances, the logic will determine vehicle 1 should turn left to avoid a potential LOS. This also holds for when vehicle 2 is located directly in front of vehicle 1. If, however, vehicle 2 is located on the left of vehicle 1 and going left, it will be instructed to turn right. (NOTE: The 30 and 45 sec buffers were selected after testing a handful of design iterations. Optimizing these buffer thresholds is left to future work.)

In **Figure 4(d)**, vehicle 2's heading is parallel and coming toward vehicle 1. If vehicle 2 is directly in front of, or to the left of, vehicle 1, the logic will instruct vehicle 1 to turn right. However, if vehicle 2 is located to the right of vehicle 1, it will be instructed to turn left.

To prevent an aircraft from prematurely exiting an avoidance maneuver, the system checks if reverting to the navigation controller generates another predicted LOS. Since the avoidance controller has only local sensor knowledge (i.e., $-90° \leq R_{A_1} \leq 90°$), switching back to the navigation controller can result in a turning action that generates another predicted LOS. If turning back to its desired target would create another predicted LOS, the avoiding aircraft will continue with its trajectory until authorized to resume its desired mission.

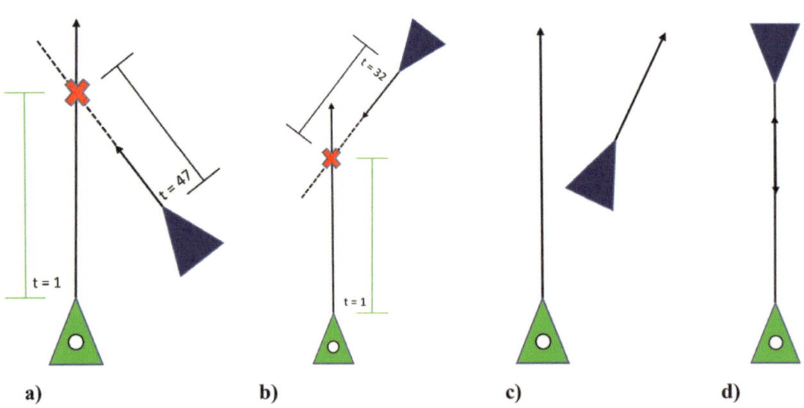

Figure 4. Separation assurance conflict scenario classifications: (a) from behind, (b) coming toward, (c) diverging, and (d) head-on.

Condition	Relative angle or time to cross	Connector	Relative heading	Action
IF	Any angle within sensor range $(-90° \leq R_{A_1} \leq 90°)$	AND	Similar direction $(R_{H_1} < 5°$ OR $R_{H_1} > 355°)$	Go behind
ElseIF	On right OR straight ahead $(R_{A_1} \leq 0°)$	AND	Going right $(180° < R_{H_1} < 360°)$	Go behind
ElseIF	On left $(R_{A_1} > 0°)$	AND	Going left $(0° < R_{H_1} < 180°)$	Go behind
ElseIF	On right $(R_{A_1} < 0°)$	AND	Head-on $(R_{H_1} = 180°)$	Turn left
ElseIF	On left OR straight ahead $(R_{A_1} \geq 0°)$	AND	Head-on $(R_{H_1} = 180°)$	Turn right
ElseIF	I'm more than 45+ seconds closer $(t_1 < (t_2 - 45))$	AND	Approaching from behind $(R_{H_1} < 90°$ OR $R_{H_1} > 270°)$	Go in front
ElseIF	I'm NOT 45+ seconds closer $(t_1 \geq (t_2 - 45))$	AND	Approaching from behind $(R_{H_1} < 90°$ OR $R_{H_1} > 270°)$	Go behind
ElseIF	I'm more than 30+ seconds closer $(t_1 < (t_2 - 30))$	AND	Coming towards $(90° \leq R_{H_1} \leq 270°)$	Go in front
ElseIF	I'm NOT 30+ seconds closer $(t_1 \geq (t_2 - 30))$	AND	Coming towards $(90° \leq R_{H_1} \leq 270°)$	Go behind

Table 3. Summary of separation assurance logic.

A summary of the separation assurance logic can be found in **Table 3**. For all cases, the crossing time is strictly positive. That is, the absolute value of the true crossing time found using Eq. (3) is used for all separation assurance logic.

3.3. Collision avoidance logic

When two vehicles have an LOS, are converging, and are within one another's sensor ranges, the collision avoidance system will be activated. In this study, each sUAS will attempt to avoid all intruders within its sensor range; therefore, no vehicle priority will be assigned. Like the approach used for the separation assurance platform to classify conflict scenarios, the collision avoidance system will use the same inputs to decide the appropriate action (i.e., relative angle, relative heading, and time to crossing point). Here, it is important to note that no two vehicles can communicate with one another. Therefore, the same logic is used onboard each system independently. This means that from each vehicle's perspective, we need to ensure that both vehicles will choose complementary actions, that is, the action will not force the vehicles to turn toward one another.

In **Figure 5**, all possible encounter scenarios are shown. Here, the black triangle and arrow represent the "ownship" (vehicle 1) location and heading respectively, the filled circle represents the "intruder" (vehicle 2) location, the dashed line connecting the two vehicles represents the relative position, and the other two dashed lines represent intruder headings that are either parallel or perpendicular to the ownship's heading. In this figure, the intruder can have any heading between 0° and 360°.

Using these dashed lines to divide the possible intruder heading into cases, the geometry of each encounter scenario can be broken down into twelve cases, provided that two of the three dashed lines do not coincide with one another. In **Figure 5(a)**, the twelve cases are depicted: two cases where the intruder has a parallel relative heading (vertical line), two cases where the headings are perpendicular (horizontal line), two cases where the intruder heading is directly toward or away from the ownship position (line connecting the two vehicles), and all headings that lie in between these angles each count as one case (i.e., six angle ranges in between the lines). If two of the three dashed lines coincide, this possible geometric space reduces to eight possible cases, as shown in **Figure 5(b)** and **c**).

Now that the geometric configurations have been defined, let us now introduce the third characteristic, time to heading intersection point. Unlike the separation assurance platform, the time to the heading intersection point, or "crossing time," can be either positive or negative. Therefore, if the crossing point lies behind the vehicle, the crossing time becomes negative. Using

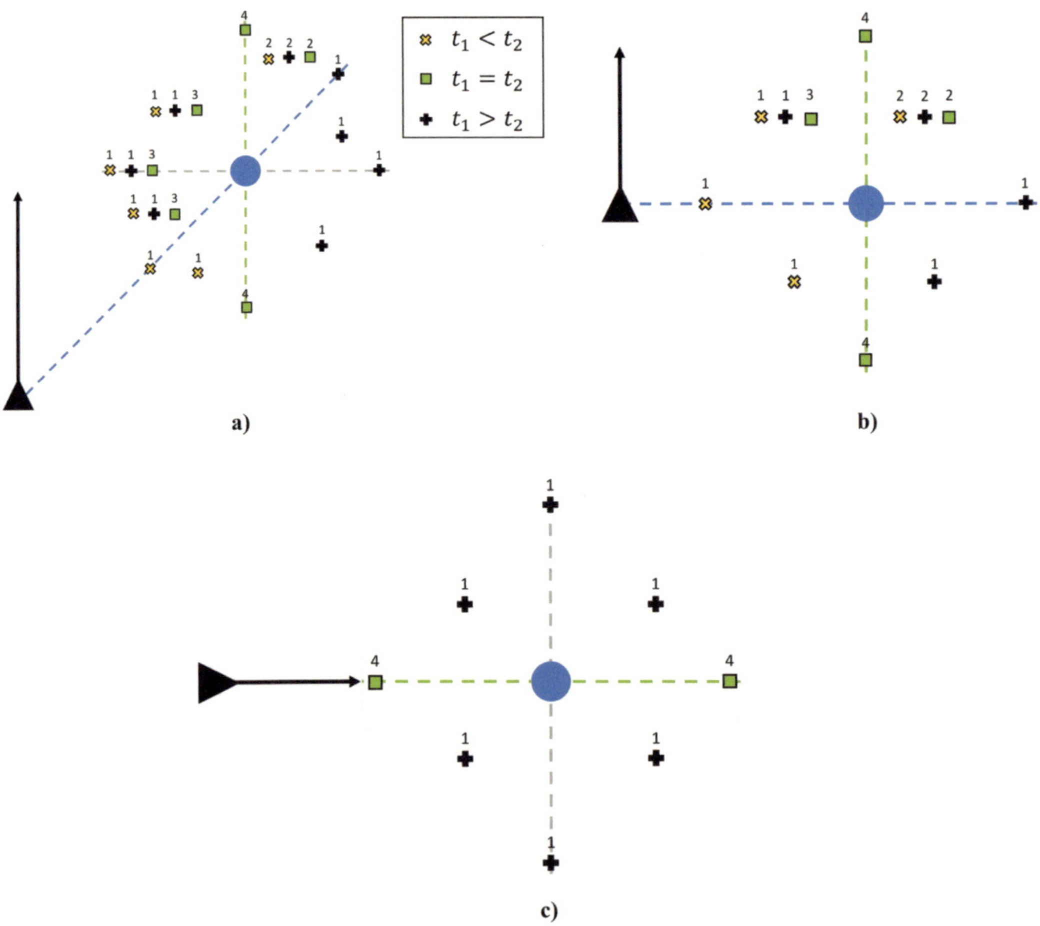

Figure 5. Classification of all possible encounter scenarios: (a) intruder not straight ahead or on side, (b) intruder on side, (c) intruder straight ahead.

the crossing time, there are up to three possible new situations for each of the twelve cases (or eight cases): the times are equal, the ownship crossing time is greater than that of the intruder, or the intruder crossing time is greater than that of the ownship. All possible crossing time scenarios based on the relative heading and angle have been shown. All instances where the ownship can have a crossing time less than the intruder crossing time have been marked by an "x." The black cross represents scenarios where the intruder crossing time can be less than the ownship crossing time. Finally, the square represents the situations where the two vehicles can have equal crossing times.

Using these three designations, all pairwise encounters where an intruder is not directly in front of or beside the ownship can be described by 20 possible cases, as shown in **Figure 5(a)**. In **Figure 5(b)**, cases where the intruder is directly beside the ownship are shown, resulting in 12 possible cases. Lastly, if the intruder is directly in front of the ownship, 8 additional cases can be attained, as shown in **Figure 5(c)**. Thus, a total of 40 cases can be attained using these three parameters to describe the pairwise encounter space.

Knowing that 40 possible cases have been shown, a total of four conflict classifications can be used to solve all possible encounter scenarios. In **Figure 5**, the number above each icon represents which of the four conflict classifications the scenario belongs. In **Figures 6** and **7**, the four conflict scenarios have been shown. In each figure, the ownship (vehicle 1) location is marked by a black triangle and its heading is designated by the black arrow. The small circle represents the location of the intruder (vehicle 2). This intruder can have any heading, but the

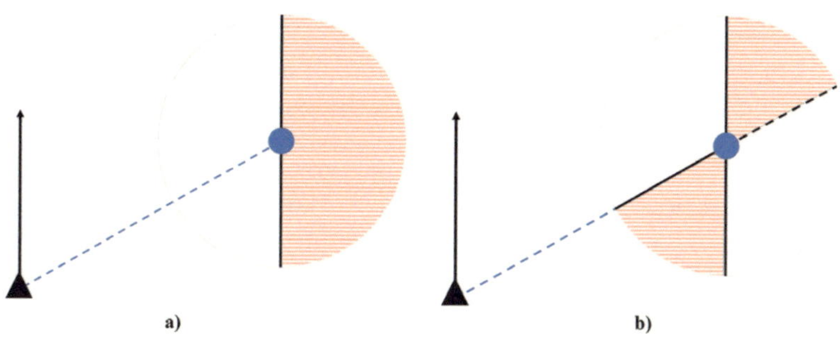

Figure 6. Conflict classification #1 where $(t_1 > 0) \vee (t_2 > 0)$ and: (a) $t_1 < t_2$, (b) $t_1 > t_2$.

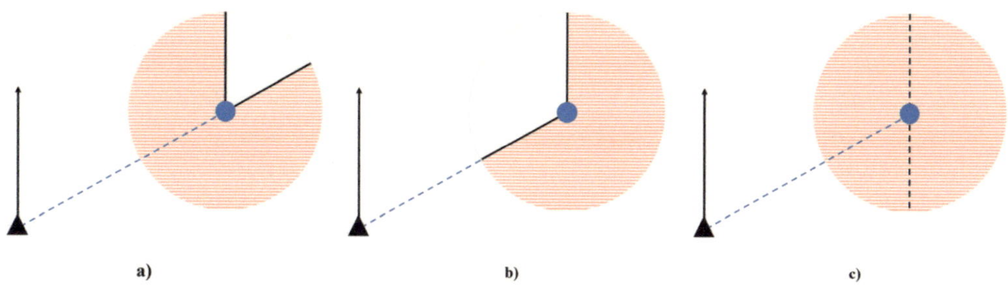

Figure 7. Conflict classifications #2, #3, and #4: (a) $(t_1 < 0) \wedge (t_2 < 0)$, (b) $(t_1 = t_2) \wedge (t_1 \neq \infty) \wedge (t_1 > 0)$, (c) $(t_1 = t_2) \wedge (t_1 = \infty)$.

different headings have been separated into different sections, as designated by the shaded regions. In each figure, the shaded regions represent the intruder headings that are excluded by that particular encounter scenario, whereas, the unshaded regions are the possible intruder headings allowed by that encounter scenario. In addition, like **Figure 5**, there is a dashed line connecting the two vehicles to represent the relative position. Lastly, if a dashed black line is on the boundary of the included and excluded regions, this represents an inclusive boundary (i.e., that heading is included in the possible headings allowed), whereas, the solid black line between the regions represents an exclusive boundary.

Figure 6 shows the first conflict classification scenario. Here, at least one of the vehicles must have a positive crossing time (t_i), and they are not equal to one another. As shown in **Figure 6(a)**, the ownship crossing time must be strictly less than that of the intruder. Thus, the intruder crossing time can be negative, or equal to zero (i.e., pointed directly at the ownship). If the intruder heading passes into the excluded region, either both values are negative, or the ownship value must be greater than the intruder value. For all the cases that satisfy this relationship, the ownship would determine to go in front of the intruder. Including the instances where the intruder is to the left of the ownship.

In **Figure 6(b)**, the ownship crossing time must be strictly greater than that of the intruder crossing time. As seen in the figure, the boundary along the relative position line is excluded when the intruder is pointing toward the ownship (i.e., the ownship crossing time is zero, but the intruder crossing time is positive), violating the relationship. However, when the intruder is pointed directly away from the ownship, the boundary is inclusive. In all possible instances, the ownship will determine to go behind the intruder. This includes the cases when the intruder is located to the left of the ownship or is directly in front of the ownship (i.e., $t_2 = 0$). For all cases represented by conflict classification #1, if one vehicle decides to go in front, the other will decide to go behind, given that they are both within one another's sensor field of view.

In the second conflict classification, as shown in **Figure 7(a)**, both the intruder and the ownship have crossing times less than zero. These negative crossing times, and the fact that the vehicles cannot have the same heading, result in the intruder never sensing the ownship. In these scenarios, if the intruder is on the left, the ownship will turn right. But, if the intruder is on the right, the ownship will turn left.

In **Figure 7(b)**, the scenarios where both the ownship and the intruder have the same time to the crossing point are shown. In this classification case, the intruder cannot have the same heading as the ownship, thus it must be crossing the ownship's path. In addition, this case is restricted to instances where both vehicles have strictly positive crossing times. For these instances, regardless of whether the vehicle is on the left, or on the right, both vehicles will decide to turn right. Since the crossing times are equivalent, and there is no coordination of intent with the other vehicle, both vehicles must choose the same action.

There is one remaining aircraft orientation in the encounter space. This occurs when an intruder is parallel to the ownship, either traveling in the same or opposite direction. This scenario has been shown in **Figure 7(c)**. As seen from this figure, only intruder headings that lie on the dashed line are included. If the intruder is to the right of the ownship, each vehicle will turn to the left (given that the intruder can see the ownship as well). However, if the intruder is directly in front of, or to the left of the ownship, both vehicles will be instructed to turn right.

Although the above classifications describe all possible encounters between moving aircraft, the quad-rotor vehicles can stop and hover. Therefore, a final classification must be described. If an intruder is stationary, the intruder crossing time is set to negative infinity. When this is the case, if the intruder is to the left or directly in front of the ownship, the controller will instruct the vehicle to turn right. However, if the intruder is to the right of the ownship, the logic will instruct the vehicle to turn left.

Table 4 describes the different encounter scenarios using both linguistic and mathematical descriptions, as well as the decided actions. Each of the above conflict classification numbers numerically matches the respective cases in this table. However, the encounter scenario where the intruder vehicle is stationary is referred to as case 0. Like **Table 3**, the linguistic descriptions in **Table 4** represent how the ownship perceives the intruder. Furthermore, all values with the subscript 1 represent the ownship, whereas, all values with the subscript 2 represent the intruder. (NOTE: the * designation indicates that the crossing time is negative infinity for that vehicle.)

3.4. Fuzzy inference systems

Using the methodology described in Sections 3.2 and 3.3, four possible actions can be used to avoid an intruding sUAS: go behind, go in front, turn right, and turn left. When the command

Case	Position	Direction	Time to cross	Crossing point location	Action
0	On left or straight ($0° \leq R_{A_1} \leq 90°$)	N/A	More ($t_1 > t_2{}^*$)	N/A	Turn right
	On right ($-90° \leq R_{A_1} < 0°$)	N/A	More ($t_1 > t_2{}^*$)	N/A	Turn left
1	Any ($-90° \leq R_{A_1} \leq 90°$)	Not parallel ($R_{H_1} \neq \{180, 0\}$)	Less ($t_1 < t_2$)	Not behind both ($t_1 \geq 0$ OR $t_2 \geq 0$)	Go in front
	Any ($-90° \leq R_{A_1} \leq 90°$)	Not parallel ($R_{H_1} \neq \{180, 0\}$)	More ($t_1 > t_2$)	Not behind both ($t_1 \geq 0$ OR $t_2 \geq 0$)	Go behind
2	On left ($R_{A_1} > 0°$)	Going left away ($R_{H_1} < 90°$)	Any	Behind both ($t_1 < 0$ AND $t_2 < 0$)	Turn right
	On right ($R_{A_1} < 0°$)	Going right away ($R_{H_1} > 270°$)	Any	Behind both ($t_1 < 0$ AND $t_2 < 0$)	Turn left
3	On left ($R_{A_1} > 0°$)	Going right ($180° < R_{H1} < 360°$)	Equal ($t_1 = t_2$)	In front of both ($t_1 > 0$ AND $t_2 > 0$)	Turn right
	On right ($R_{A_1} < 0°$)	Going left ($0° < R_{H_1} < 180°$)	Equal ($t_1 = t_2$)	In front of both ($t_1 > 0$ AND $t_2 > 0$)	Turn right
4	On left or straight ($R_{A_1} \geq 0°$)	Head-on or same ($R_{H_1} = \{180, 0\}$)	Equal ($t_1 = t_2$)	N/A	Turn right
	On right ($R_{A_1} < 0°$)	Head-on or same ($R_{H_1} = \{180, 0\}$)	Equal ($t_1 = t_2$)	N/A	Turn left

Table 4. Summary of collision avoidance logic.

action is either go in front or go behind, the corresponding FIS is activated to execute the maneuver. Therefore, two FISs were developed for this study: go in front and go behind. If a turn left or turn right command is selected, the turn rate of the vehicle will always be a constant value, either positive or negative, depending on which way it should turn. This constant value is one half of the maximum turn rate for collision avoidance maneuvers and one eighth of the maximum turn rate for separation assurance maneuvers.

Both the go in front and go behind fuzzy systems are of Mamdani-type and were constructed in such a way that the input-output relationship can be described using a simple mathematical representation. By using a hybrid representation, as described in Section 2.1, the fuzzy system can easily be expressed mathematically. In this study, the fuzzy systems have a common architecture: triangular membership functions, normalized inputs and outputs, membership function partitioning, product "and" method, minimum implication method, sum aggregation, and mean of maximum defuzzification. If more than one membership function exists for a particular input or output, membership functions are partitioned such that the endpoints of one membership function coincides with the center points of the neighboring membership functions.

Each FIS was developed using a three-input one-output structure. Each FIS uses the distance separating the two aircraft, their relative heading, and their closure rate as inputs to determine the appropriate turn rate output. Since a heterogeneous system is used in this study, the FIS must provide a sufficient turn rate output to avoid a collision for all vehicle type combinations. By considering the separation and closure rate, the conflict can be solved without expelling more energy than required.

In order to use the FISs for both the separation assurance and the collision avoidance platforms, all inputs and outputs must be normalized. Regardless of the avoidance platform being used, the relative heading and closure rate inputs are always normalized by the same values. The relative heading which falls between 0° and 360° is divided by 360°. Thus, a normalized relative heading of 0.5 would represent a head-on encounter. The closure rate is normalized by dividing the true closure rate by the maximum possible closure rate between two vehicles (i.e., 61.762 m/sec in this study). This maximum closure rate would be a result of two fixed wing vehicles approaching one another in a head-on scenario.

The third input, distance, is normalized by 0.4 nmi for separation assurance cases and 0.1 nmi for collision avoidance cases. Lastly, the turn rate output is always between −1 and 1. This output is then scaled by the respective vehicle platform's maximum turn rate. In the case of the collision avoidance system, the output is also multiplied by 1.58. This is to compensate for the fuzzy output not providing a sufficient turn rate command to avoid a collision, especially in head-on scenarios.

Figure 8 shows the structure of each avoidance FIS. Here, the number of membership functions and the corresponding classification can be seen for each input and output. Both the relative heading and the turn rate are partitioned such that the endpoints of the center membership functions coincide with the center points of the adjacent membership functions. As a result of this, and using a product "and" connector, for all possible inputs, at most two of the

http://dx.doi.org/10.5772/68126

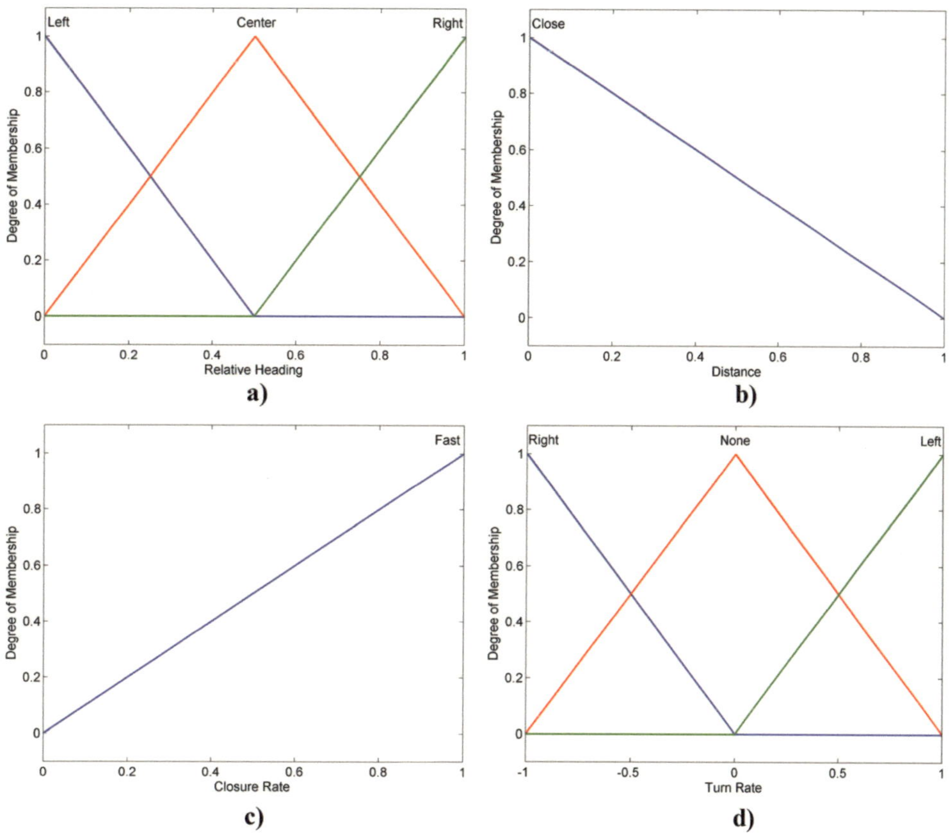

Figure 8. Avoidance FIS structure: (a) Input 1, (b) Input 2, (c) Input 3, (d) Output.

three rules will be active at one time. This drastically reduces the possible solution space, making it much easier to represent the system mathematically.

The input and output membership function sets for both the go behind and go in front FISs are identical. However, the rule bases associating the inputs and outputs are opposite, therefore, when one FIS outputs "turn right" (i.e., negative turn rate), the other FIS would output "turn left" (i.e., positive turn rate), and vice versa. In **Table 5**, the respective rule bases can be seen. Because the other two inputs only have one membership function, they have been excluded from the table.

Rule #	Input 1: relative heading	Go in front output	Go behind output
1	Left	Left	Right
2	Center	Center	Center
3	Right	Right	Left

Table 5. FIS logic.

Mode #	$\mathcal{D}_1$	$\mathcal{D}_2$	$\mathcal{D}_3$
1	[0,0.5)	[0,1]	[0,1]
2	[0.5]	[0,1]	[0,1]
3	(0.5,1]	[0,1]	[0,1]

Table 6. Input domains.

To map the fuzzy system to a set of nonlinear expressions, the following process was used. First, the system was discretized into three possible modes based on the domain of the ith input, D_i. These modes are described in **Table 6**. Here, let a square bracket represent an inclusive bound and a round bracket represent an exclusive bound. Based on the structure of the FIS, if input 1 is exactly 0.5 (Mode 2 in **Table 6**), the output will yield a turn rate of 0. It remains to find the explicit input-to-output mappings for modes 1 and 3.

When the FIS is in mode 1, only the "Left" and "Center" membership functions are active. Thus, the output will be independent of rule 3. Therefore, given that a product "and" method and a mean of maximum aggregation technique is used, the following relationship describes how the output is calculated.

$$\dot{\Psi} = \left[\left(\prod_{i=1}^{3} h_{i_1} \right) + \left(1 - \prod_{i=1}^{3} h_{i_2} \right) \right] / 2 \tag{4}$$

where, $\dot{\Psi}$, is the turn rate, h_{i_1} is the degree of membership for the ith input membership function when using rule 1 (i.e., left, close, fast membership functions), and h_{i_2} is the degree of membership for each input when using rule 2. When substituting the equation of each membership function into Eq. (4), the expression shown in Eq. (5) is found, which can be reduced to the polynomial shown in Eq. (6).

$$\dot{\Psi} = \{ [(-2\mu_1 + 1)(-\mu_2 + 1)(\mu_3)] + [1 - (2\mu_1)(-\mu_2 + 1)(\mu_3)] \} / 2 \tag{5}$$

$$\dot{\Psi} = (4\mu_1\mu_2\mu_3 - 4\mu_1\mu_3 - \mu_2\mu_3 + \mu_3 + 1)/2 \tag{6}$$

where μ_1, μ_2, and μ_3 are inputs 1, 2, and 3, respectively. The polynomial shown in Eq. (6) can be used to map any combination of inputs that belong to D_1 to an output for the go in front FIS.

Using the same methodology for mode 3, Eqs. (7) through (9) can be found.

$$\dot{\Psi} = \left[\left(-\prod_{i=1}^{3} h_{i_3} \right) + \left(\prod_{i=1}^{3} h_{i_2} - 1 \right) \right] / 2 \tag{7}$$

$$\dot{\Psi} = \{ [-(2\mu_1 + 1)(-\mu_2 + 1)(\mu_3)] + [(-2\mu_1 + 2)(-\mu_2 + 1)(\mu_3) - 1] \} / 2 \tag{8}$$

$$\dot{\Psi} = (4\mu_1\mu_2\mu_3 - 4\mu_1\mu_3 - 3\mu_2\mu_3 + 3\mu_3 - 1)/2 \tag{9}$$

The polynomial shown in Eq. (9) will map any combination of inputs belonging to D_3 to an output for the go in front FIS.

This same approach was used to map any combination of inputs for the go behind FIS to an output using a polynomial function. Since the rule bases are exactly opposite to one another, the output is the negation of Eqs. (6) and (9). Eqs. (10) and (11) describe the input-output relationships for modes 1 and 3, respectively, for the go behind FIS.

$$\dot{\Psi} = (-4\mu_1\mu_2\mu_3 + 4\mu_1\mu_3 + \mu_2\mu_3 - \mu_3 - 1)/2 \tag{10}$$

$$\dot{\Psi} = (-4\mu_1\mu_2\mu_3 + 4\mu_1\mu_3 + 3\mu_2\mu_3 - 3\mu_3 + 1)/2 \tag{11}$$

It is important to note that if three or more vehicles are in conflict, each pair of vehicles will be evaluated separately. Thus, for a single ownship, several turn rate outputs will be obtained. Once all respective turn rates are calculated for each intruder, the values are averaged into a single value. This mean turn rate serves as the final vehicle turn rate command.

4. Avoidance algorithm verification

To verify that the avoidance algorithm performed as intended, two levels of avoidance control were evaluated. The first was the high-level, decision-making logic that determined which action a vehicle would take during potential collision scenarios as shown in **Table 4**. The second that was evaluated was the low-level logic in the avoidance FLCs. These FLCs determine the actual vehicle turn rate output after being selected by the decision-making heuristics. For each of these levels, specifications about their behavior were developed and then translated into FOL sentences that could be evaluated by an SMT solver. For the first cases that deal with the avoidance decision-making logic, the specifications and system model were implemented in JKind, with Z3 being used as the SMT solver. The evaluation of the avoidance FISs was performed directly in Z3. The difference was purely a practical one, as the language JKind uses, Lustre [22], is more conducive to easily create a more detailed environment model.

4.1. Collision avoidance decision-making logic

As shown in **Table 4**, there are several conditions for which there are different output actions for avoidance. The specifications were first translated into FOL sentences. The sentences are shown in the following equations. Note that *RAL* is a predicate that acts on the relative angle limit to represent the vehicle's sensors detecting an intruder. Since they can only sense intruders between $\pm 90°$, the relative angle values were limited in the specifications such that being in that range implied that the output action would be the correct one. If the intruder is outside of this range, the specifications do not say what the output action should be. Note that the variables $out1$ and $out2$ are the output actions for the two vehicles. Since both vehicles pick actions based on the states and have no knowledge of the other vehicle's selected action, the specifications need to go both ways. Also, the actions are represented by integer values 1–4 such that {1,2,3,4} = {go behind, go in front, turn left, turn right}, respectively.

Although several avoidance actions could be taken by each vehicle, only cases 1, 3, and 4 required formal verification. In these cases, both vehicles can sense one another and will perform some type of avoidance maneuver. Thus, their actions must be such that they do not move closer to one another. In cases 0 and 2, however, there was no need to write requirements because only one vehicle performs an avoidance action. In case 0, vehicle 2 is stationary, so it does not try to avoid vehicle 1. In case 2, vehicle 2 cannot sense vehicle 1, thus, there is no need to check that the two vehicles' actions produce a converging result.

Eqs. (12) through (15) are the specifications for case 1 in **Table 4** and are intended to ensure that the vehicles do not both choose to go in front, or go behind.

$$S_{1_1} = \forall R_{A_1} \forall R_{A_2} \forall R_{H_1} \forall R_{H_2} (RAL \rightarrow (out1 = 1 \rightarrow out2 = 2)) \tag{12}$$

$$S_{1_2} = \forall R_{A_1} \forall R_{A_2} \forall R_{H_1} \forall R_{H_2} (RAL \rightarrow (out1 = 2 \rightarrow out2 = 1)) \tag{13}$$

$$S_{1_3} = \forall R_{A_1} \forall R_{A_2} \forall R_{H_1} \forall R_{H_2} (RAL \rightarrow (out2 = 1 \rightarrow out1 = 2)) \tag{14}$$

$$S_{1_4} = \forall R_{A_1} \forall R_{A_2} \forall R_{H_1} \forall R_{H_2} (RAL \rightarrow (out2 = 2 \rightarrow out1 = 1)) \tag{15}$$

Eqs. (16) and (17) are the specifications for case 3 and ensure that the vehicles turn the same way when resolving these particular conflicts. Recall that in this case, both vehicles are at the same time from the crossing point. This specification ensures that they will then be forced into new positions such that the crossing times are not equal and the FISs are then selected.

$$S_{3_1} = \forall R_{A_1} \forall R_{A_2} \forall R_{H_1} \forall R_{H_2} (RAL \rightarrow ((t_1 = t_2 \wedge t_1 > 0 \wedge t_2 > 0 \wedge ((R_{H_1} \geq 0 \wedge R_{H_1} < 180) \vee (R_{H_1} > 180 \wedge R_{H_1} < 360))) \rightarrow (out1 = 4 \wedge out2 = 4))) \tag{16}$$

$$S_{3_2} = \forall R_{A_1} \forall R_{A_2} \forall R_{H_1} \forall R_{H_2} (RAL \rightarrow ((t_1 = t_2 \wedge t_1 > 0 \wedge t_2 > 0 \wedge ((R_{H_2} \geq 0 \wedge R_{H_2} < 180) \vee (R_{H_2} > 180 \wedge R_{H_2} < 360))) \rightarrow (out2 = 4 \wedge out1 = 4))) \tag{17}$$

Finally, Eqs. (18) and (19) are for case 4. These two specifications are for cases where the vehicles are head-on, or traveling next to each other in the same direction, respectively. The first specification ensures that while in a head-on encounter, both vehicles turn the same direction (i.e., both left, or both right, forcing them to diverge). The second specification ensures that while traveling in the same direction, the vehicles turn in opposite directions.

$$S_{4_1} = \forall R_{A_1} \forall R_{A_2} \forall R_{H_1} \forall R_{H_2} (RAL \rightarrow ((t_1 = t_2 \wedge (R_{H_1} = 180 \wedge R_{H_2} = 180)) \rightarrow ((out1 = 3 \wedge out2 = 3) \vee (out1 = 4 \wedge out2 = 4)))) \tag{18}$$

$$S_{4_2} = \forall R_{A_1} \forall R_{A_2} \forall R_{H_1} \forall R_{H_2} (RAL \rightarrow ((t_1 = t_2 \wedge (R_{H_1} = 0 \wedge R_{H_2} = 0)) \rightarrow ((out1 = 3 \wedge out2 = 4) \vee (out1 = 4 \wedge out2 = 3)))) \tag{19}$$

where t_1 is the time until the crossing point, R_{H_1} is the relative heading from vehicle 1 to vehicle 2, and R_{A_1} is the relative angle from vehicle 1 to vehicle 2. Similarly, t_2, R_{H_2}, and R_{A_2} are for vehicle 2. In addition to these specifications, another specification was created to ensure that the vehicles always chose one of the valid actions. This specification is not shown but is similar to Eqs. (12) through (19) such that the output actions are one of the possible outcomes (1–4). These specifications were then translated to a temporal representation for evaluation in JKind using previously developed methods [18].

4.2. Avoidance fuzzy inference systems

The method for verifying the avoidance FISs is similar, but the specifications were left in FOL and then implemented directly into Z3. The main reason for this was that less of the system model was needed to check these outputs. The behavior that the specifications needed to encapsulate was that the turn rate output for each FIS needs to always be in the correct direction. The correct direction for each of these cases is shown in **Table 7**.

These can then be encoded in FOL sentences using the polynomial representation of the FISs shown in Section 3.4. These sentences are then negated to show that there are no possible variable values that make the negated sentences true. The negated FOL sentences are shown in Eqs. (20) through (23). Note that μ_1, μ_2, μ_3, $\dot{\Psi}$, and modes 1 and 3 are the same as detailed in Section 3.4.

$$S_{behind_1} = \exists\mu_1\exists\mu_2\exists\mu_3(\dot{\Psi}(\mu_1,\mu_2,\mu_3)\geq 0) \tag{20}$$

$$S_{behind_3} = \exists\mu_1\exists\mu_2\exists\mu_3(\dot{\Psi}(\mu_1,\mu_2,\mu_3)\leq 0) \tag{21}$$

$$S_{front_1} = \exists\mu_1\exists\mu_2\exists\mu_3(\dot{\Psi}(\mu_1,\mu_2,\mu_3)\leq 0) \tag{22}$$

$$S_{front_3} = \exists\mu_1\exists\mu_2\exists\mu_3(\dot{\Psi}(\mu_1,\mu_2,\mu_3)\geq 0) \tag{23}$$

These negated sentences were then implemented in Z3. If Z3 finds that these sentences were all unsatisfiable, there are no possible real-valued assignments to μ_1, μ_2, or μ_3 that allow the FISs to turn the vehicles in an undesired direction.

Mode	Action	Intruder position	Intruder direction	Turn direction
1	Go behind	On right	Going left	Right (negative)
3	Go behind	On left	Going right	Left (positive)
1	Go in front	On right	Going left	Left (positive)
3	Go in front	On left	Going right	Right (negative)

Table 7. Avoidance FIS outputs.

5. Simulation environment descriptions

5.1. Testing environment

Prior to integrating the controllers presented in Section 3 into the full simulation environment, each avoidance platform was tested using pairwise encounter scenarios. This component testing was used to ensure that each was operating as desired. In this section, the methods used to test each controller are described.

5.1.1. Separation assurance

A testing environment was created to evaluate a considerable amount of pairwise encounters between aircraft. This testing environment was used to identify potential controller failures. To accomplish this, various initial relative headings and relative angles were tested. In all cases, the initial location and heading of one aircraft was held constant. Then, by placing the intruding vehicle at a relative angle of −90° and just outside the vehicle's sensing radius, as shown in **Figure 9**, we evaluated the interactions for 720 initial intruder heading values. This was then repeated four more times by changing the initial relative angle between −90° (to the right) and 0° (straight ahead). A visualization of these scenarios is shown in **Figure 9**. Note that the radius of the sensing semi-circle for the separation assurance tests was set at 0.4 nmi.

Although intruders can approach a vehicle from the left, that is, a relative angle between 0° and 90°, symmetry allows us to limit the initial relative angles to lie between −90 and 0°. To verify, a sample scenario was tested for the full field of view. In all cases, the trajectories of the vehicles were symmetric to one another (i.e., when reflected across the vehicle's initial heading).

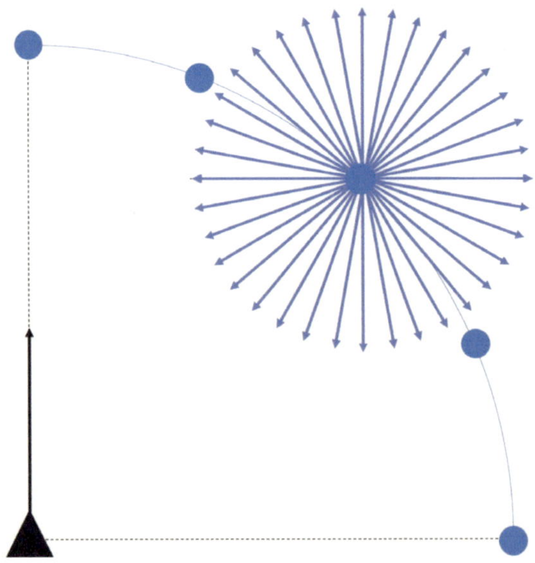

Figure 9. Separation and avoidance testing scenarios between ownship (triangle) and intruder UAS for multiple different initial positions (circles).

http://dx.doi.org/10.5772/68126

5.1.2. Collision avoidance

The collision avoidance platform was tested in the same manner as the separation assurance. The difference being that the sensing radius for the interaction tests as shown in **Figure 9** is 0.1 nmi. Thus, the initial position of each intruder scenario was just outside of this sensing radius.

5.2. Full simulation environment

5.2.1. Airspace description

The simulation environment created for this study models a portion of the US airspace over central Ohio. A depiction of the selected airspace can be seen in **Figure 10**. This airspace covers approximately 2500 square miles where sUAS can operate at a maximum altitude of 400 ft.

5.2.2. Mission types and objectives

Many sUAS will be active throughout this airspace, each with individual missions, such as, precision agriculture, forest monitoring, roadway surveillance, disaster management, and package delivery. During simulation runs, the UAS will travel to various waypoints to fulfill their assigned missions. After visiting all waypoints for a given mission, the aircraft will return to their respective starting locations. For more on the mission types, please refer to Refs. [11, 12].

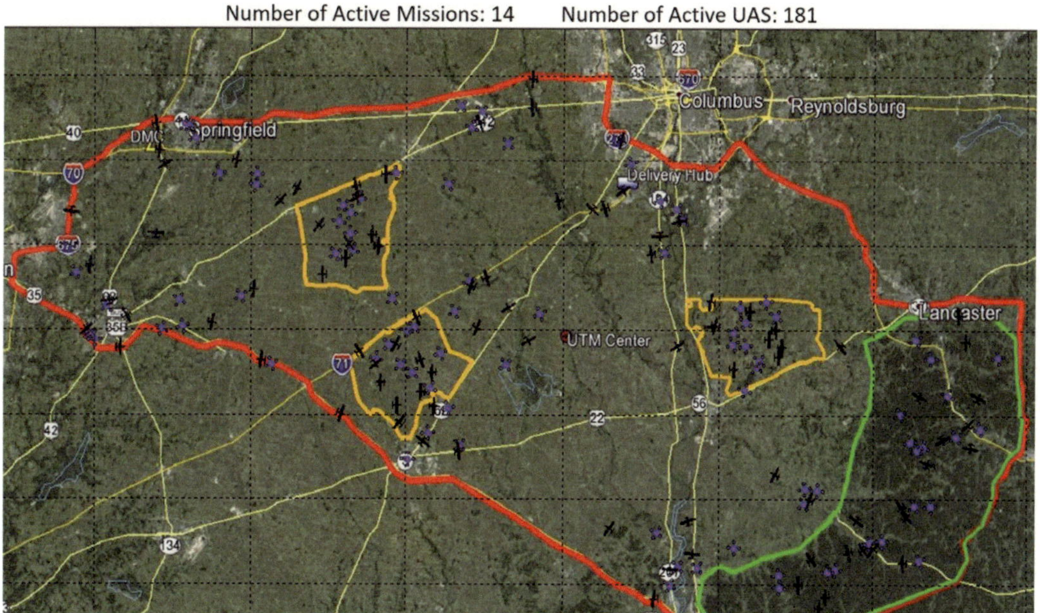

Figure 10. Simulation platform example.

5.2.3. *Scalability and operations*

The full simulation environment can accommodate any number of sUAS and missions. However, these numbers were constant throughout testing. A maximum of 184 sUAS can be airborne at any given time. This is a result of having centralized control of the aircraft. In a realistic environment, some of the computationally heavy components can be handled in parallel onboard each individual sUAS.

5.2.4. *Aircraft models*

Two vehicle platforms were used in all simulations: fixed wing and quad-rotor. Kinematic models were developed for each within the constraints of the environment. These constraints include maximum turn rate, maximum climb/decent rate, minimum/maximum speed, and maximum altitude.

For this study, each aircraft is assumed to climb at 4 ft./sec upon takeoff. The vehicles will continue to climb at this rate until an altitude of 400 ft. is reached. When vehicles are in conflict and are required to make trajectory modifications, all adjustments are constrained to lateral deviations in flight path (i.e., speed and altitude modifications are not used). Therefore, UAS are limited to level, two-dimensional flight during conflict resolution scenarios. Using this assumption, the maximum turn rate for each fixed wing vehicle is described in Eq. (24).

$$\dot{\Psi}_{max} = \frac{g\sqrt{n^2 - 1}}{V} \tag{24}$$

All fixed wing vehicles travel at 60 knots and have a maximum load factor of 3.5. Therefore, they are constrained to a maximum turn rate of 61.06 deg/sec. The multi-rotor systems, however, can travel at a maximum airspeed of 38 knots. Due to the nature of the quad-rotor sUAS, Eq. (24) does not accurately model the maximum turn rate for this vehicle type and it is assumed that the aircraft can yaw at a maximum rate of 45 deg/sec. Also, each aircraft is assumed to have the capability to detect an intruding aircraft at a distance of 0.1 nmi if within a 180° field of view in front of the aircraft (i.e., $-90^\circ \leq R_{A_1} \leq 90^\circ$).

As previously described in **Table 1**, the collision threshold is defined by aircraft coming within 60 m laterally and 50 ft. vertically of one other. To track the vehicles, ground-based sensors for detecting aircraft have been dispersed in the airspace. These sensors, along with telemetry data, provide continuous and reliable vehicle state information (i.e., position and velocity). This capability allows the implementation of global separation assurance practices. More details about the simulation environment and its properties can be found in Refs. [11, 12]. (NOTE: The collision buffer value was used to accommodate for position uncertainties while tracking the sUAS. In this study, the avoidance system has perfect knowledge of all vehicle state information acquired from ground based and onboard sensors.)

6. Results

6.1. Avoidance systems testing

For each pairwise encounter shown in **Figure 9**, the closest point of approach (CPA) was found. The CPA is defined as the minimum recorded distance between the two vehicles throughout the entire encounter. For both the collision avoidance and separation assurance platforms, a total of four different vehicle platform trials were conducted: fixed vs. fixed, quad vs. quad, quad vs. fixed, and fixed vs. quad. The first vehicle type designation represents the ownship vehicle's type in **Figure 9** (i.e., has the same starting position and heading for all tested cases), whereas, the second vehicle platform designation represents the intruder vehicle's type (i.e., initial conditions change for each tested scenario).

To measure the effectiveness of the avoidance logic, the minimum CPA was recorded for all initial relative angles tested. In addition, the total number of collisions for each case were tallied, and for the separation assurance case, the total number of LOSs. As a reminder, a collision is deemed by two vehicles coming within 60 m of one another, and an LOS is when two vehicles come with 0.1 nmi of one another.

The results for the separation assurance testing have been shown in **Table 8**. Here, the "angle case" refers to the various initial relative angles, from −90 to 0°, respectively. Thus, case 1 represents an intruder directly to the right, and case 5 represents an intruder directly in front of the ownship.

Although all vehicle platform combinations had more than one LOS, no LOS resulted in two vehicles colliding (i.e., a CPA less than 60 m). To try and understand where the separation assurance platform was breaking down to allow an LOS to occur, the CPA results were plotted for each initial intruder heading. When evaluating the results of each vehicle case, it was found that although several LOSs occurred, the failures tended to lie in groups near the same intruder angle depending on the initial position of the intruder. For example, in the fixed vs. quad case, it can be seen that no LOSs occurred in the first three trials. However, once the intruder was positioned such that it was nearly in front of or directly in front of the ownship,

Angle case	Fixed vs. Fixed		Quad vs. Quad		Fixed vs. Quad		Quad vs. Fixed	
	Min CPA (m)	LOS	Min CPA (m)	LOS	Min CPA (m)	LOS	Min CPA (m)	LOS
1	185.3	0	185.1	0	574.3	0	185.8	0
2	185.6	0	185.5	0	351.4	0	158.6	12
3	185.3	0	185.4	0	185.6	0	133.2	24
4	186.1	0	131.2	9	175.1	24	137.3	31
5	98.4	116	137.0	8	128.1	161	132.0	28

Table 8. Separation assurance testing results.

the separation assurance platform began to fail. This was caused by the fixed wing vehicle traveling at higher speeds than the quad-rotor vehicle. Therefore, when the intruder heading was away from the ownship, the ownship tended to approach the intruder from behind and begin to pass the vehicle. When approaching the vehicle from behind, it proved difficult for the avoidance platform to solve the conflict prior to an LOS in all vehicle configurations, as shown by angle case 5.

Table 9 shows the results of the collision avoidance testing. Each case had a minimum CPA greater than 60 m, implying no collisions were found throughout testing. Although not all possible encounter scenarios have been tested, this shows that the avoidance system logic is quite robust. As seen from the results, the homogeneous quad-rotor and the fixed vs. quad cases showed promising results. They consistently had a higher minimum CPA than the other two cases. The closure rates in the fixed vs. fixed cases were higher than cases involving quad-rotors. This resulted in consistently lower minimum CPA values. A noteworthy result is in the fixed vs. fixed scenario for angle case 5. The intruder being head-on and directly in front of the ownship resulted in a CPA of 60.8 m. Although this is close to the collision boundary, this shows that even in the highest closure rate scenario, the collision avoidance system was able to resolve the conflict.

6.2. Formal verification

6.2.1. Avoidance logic

After evaluating all the specifications outlined in Eqs. (12) through (19), JKind returned that they always held. This means that for all possible real-valued assignments to the variables (within the sensor domain limitations), the vehicles will always select the desired output action.

Although the final version of the avoidance logic adhered to all of the specifications, during development there were several cases where JKind found values that violated one or more of the specifications. These counterexamples are invaluable as they identify exact cases that result in undesired behavior. This gives way to corrections based on the counterexample conditions.

	Fixed vs. Fixed	**Quad vs. Quad**	**Fixed vs. Quad**	**Quad vs. Fixed**
Angle Case	**Min CPA (m)**	**Min CPA (m)**	**Min CPA (m)**	**Min CPA (m)**
1	132.1	138.9	154.7	125.3
2	111.9	125.2	133.3	121.0
3	101.9	113.5	122.9	111.1
4	82.9	104.4	103.5	97.8
5	60.8	96.7	98.9	99.0

Table 9. Collision avoidance testing results.

Specification	R_{H_1}	R_{A_1}	$out1$	R_{H_2}	R_{A_2}	$out2$
S_{3_2}	180	$-5 \cdot 10^{-15}$	3	180	$-5 \cdot 10^{-15}$	4

Table 10. Avoidance logic counter-example values.

As an example of this, one of the conditions that violated a specification found during development is shown in **Table 10**.

These conditions mean that one vehicle is heading in the exact opposite direction of the other and there is a slight position offset between them as shown in **Figure 11**. One vehicle selects the turn left action (3) while the other selects the turn right action (4). This implies they are not turning away from each other. The reason for this was that the range of angles that would force vehicle 1 into the correct action was not inclusive on one of its boundaries. This then meant that the conditions forced vehicle 1 into a different action and generated this counterexample.

6.2.2. *Avoidance FISs*

Similarly, after evaluating the specifications in Eqs. (20) through (23), Z3 showed that they were all unsatisfiable. This shows that the FLCs will always make the ownship turn away from an intruder.

6.3. Full simulation results

To test the algorithms in a dynamic environment, simulations were run both with the separation assurance mitigations and then without. The number of LOSs and number of collisions were recorded in order to directly compare the mitigated and unmitigated cases.

6.3.1. *Unmitigated study*

For this unmitigated study, the separation assurance system was disabled. The results of this study are shown in **Table 11**.

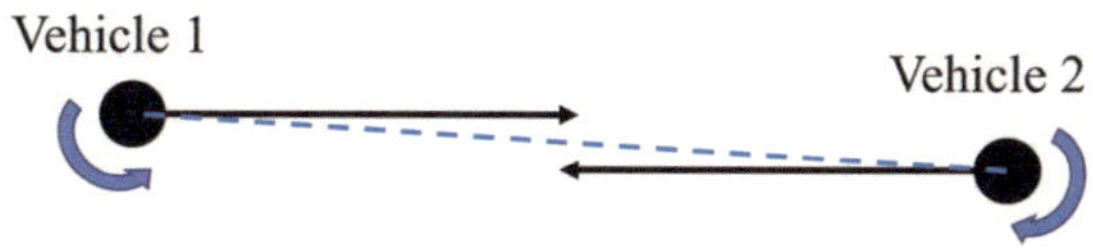

Figure 11. Counterexample showing head-on vehicles turning into each other.

Separation less than 0.4 nmi	LOSs	Collision avoidance maneuvers	Collisions	Collision avoidance success rate	Number of flight hours	LOSs per flight hour	Collisions per flight hour
26,576	8263	8252	2	99.98%	25,116	0.33	$7.96 \cdot 10^{-15}$

Table 11. Results for simulation without separation assurance.

Over the span of 25,116 flight hours, there were 26,576 recorded violations of the 0.4 nmi separation threshold. These resulted in 8263 LOSs. The collision avoidance algorithm was employed for all except for 11 LOS occurrences. In those cases, the vehicles were outside of one another's field of view, thus the collision avoidance system was not used. The collision avoidance system was 99.98% successful at resolving conflicts.

The only collisions that occurred throughout simulation can be attributed to the restriction on the detection sensor field of view and having no memory of state time histories. Therefore, if two vehicles were nearly parallel and directly beside one another, they would turn to resolve the conflict (i.e., turn away from one another). However, this turning again puts each intruder outside of the other vehicle's field of view. The lack of state memory combined with no sensor input caused a switch back to their navigation controllers. The navigation controller caused them to go back toward one another. Since the navigation controller had a higher turn rate output than the avoidance output, this cycle would continue (each vehicle turning away then toward) until they converged and were within 60 m of one another.

6.3.2. Mitigated study

For this study, the separation assurance features were enabled to help mitigate the risk of having an LOS. Although fewer LOSs were expected with the mitigations enabled, some LOSs were expected due to sub-optimal performance in head-on and trailing situations. The results of this mitigated study are shown in **Tables 12** and **13**.

Table 12 presents the results of the separation assurance platform. In this mitigated study, only 9277 flight hours were recorded. Thus, sUAS were able to complete their respective missions in a shorter period of time. Throughout the simulation aircraft came within 0.4 nmi on 33,550 occasions. However, of those instances, the separation assurance system predicted an LOS to occur within 2 min only 14,750 times. Of these resolution advisories, only 75.74% were successful, resulting in 3579 LOSs and 0.39 LOSs per flight hour. Although this number is slightly larger than the number of LOSs per flight hour in the unmitigated study, this additional layer of avoidance kept vehicles from entering any scenarios that resulted in collision. Thus, the overall safety of the UTM system has been improved.

Separation less than 0.4 nmi	Separation assurance maneuvers	LOSs	Separation assurance success rate	Number of flight hours	LOSs per flight hour
33,550	14,750	3579	75.74%	9277	0.39

Table 12. Separation assurance results.

Collision avoidance maneuvers	Collisions	Collision avoidance success rate	Number of flight hours	Collisions per flight hour
3568	0	100.00%	9277	0

Table 13. Collision avoidance results.

If an LOS occurred and the vehicles were within one another's sensor ranges, the sense and avoid software would activate. The results of the collision avoidance software can be seen in **Table 13**. Of the 3568 encounters, the collision avoidance software was 100.00% successful at resolving conflicts.

7. Conclusion

In this work, multiple fuzzy logic controllers and decision-making systems were used in conjunction to prevent potential losses of separation in a congested, three-dimensional airspace. This simulation environment allowed for extensive encounter scenarios between heterogeneous vehicles to test the two conflict resolution systems. First, a sense and avoid system was developed to prevent potential collisions using only current state information and without communication between vehicles. Next, a separation assurance platform was developed to further mitigate the risk of a potential collision. This platform uses global aircraft state information to predict if two aircraft will have an LOS within a given look-ahead time. If an LOS was predicted, the system would issue necessary resolution advisories to the proper aircraft to prevent an LOS.

Once the controllers were developed, numerical simulations and formal methods were used to verify the controllers performed as expected. Using a formal methods approach, we could show that the controller output was always in the correct direction (i.e., always performed as expected). In addition, we were able to verify that in all pairwise encounter scenarios between sUAS, the actions of each vehicle were such that they would never turn toward one another when avoiding a collision.

After a formal methods approach verified the control logic behavior and fuzzy logic controller outputs, numerical simulations were conducted. In all simulations, the avoidance system had perfect knowledge of all vehicle state information (i.e., speed, heading, and location). For the collision avoidance scenarios tested, the fuzzy system was successful at resolving all potential conflicts for both the homogeneous and heterogeneous cases. However, the separation assurance platform had trouble resolving certain types of encounter scenarios. Thus, it sometimes would not prevent an LOS between vehicles. However, when an LOS occurred, the collision avoidance system again prevented any mid-air collisions from occurring.

Several full simulation environment missions were also run to evaluate the effectiveness of the avoidance algorithms. These missions included cases where the separation assurance mitigations were both enabled and disabled. Overall, the results of this experiment were as expected. In the mitigated study, no collisions between aircraft occurred. However, when the mitigations were removed, vehicles encountered scenarios where the collision avoidance system could not prevent a collision. These collisions were not due to the collision avoidance logic or the fuzzy logic controllers, but were attributed to the limited vehicle sensor performance and lack of memory.

For future work, we aim to improve upon the separation assurance techniques to prevent an LOS in all encounter scenarios. Also, representing the system with a higher fidelity model of the environment in the formal methods tools would allow for more complete specifications (i.e., vehicles never lose separation) and then identify cases that violate them. In addition, since the avoidance system had perfect vehicle state information, we would like to introduce a level of uncertainty to the sensor models. Finally, we wish to implement the proposed avoidance software into hardware testing environments.

Appendix: Nomenclature

CPA	Closest Point of Approach
FIS	Fuzzy Inference System
FLC	Fuzzy Logic Controller
FOL	First Order Logic
LOS	Loss of Separation
NAS	National Airspace System
pCPA	Predicted Closest Point of Approach
SAT	Satisfiability (Boolean)
SMT	Satisfiability Modulo Theories
sUAS	Small Unmanned Aerial System
tLOS	Time to Loss of Separation
UAS	Unmanned Aerial System
UTM	UAS Traffic Management

Author details

Brandon Cook[1,2]*, Tim Arnett[2] and Kelly Cohen[2]

*Address all correspondence to: cookb9@mail.uc.edu

1 NASA Ames Research Center, Moffett Field, CA, USA

2 Department of Aerospace Engineering and Engineering Mechanics, University of Cincinnati, Cincinnati, OH, USA

References

[1] Cook K. The silent force multiplier: The history and role of UAVs in warfare. Proceedings of the IEEE Aerospace Conference; 3–10 March 2007; Big Sky, MT, USA. IEEE. 2007:1-7. DOI: 10.1109/AERO.2007.352737

[2] Ollero A, Merino L. Unmanned aerial vehicles as tools for forest-fire fighting. Forest Ecology and Management. 2006;**234**(1):S263

[3] Bamburry D. Drones: Designed for product delivery. Design Management Review. 2015;**26**(1):40-48. DOI: 10.1111/drev.10313

[4] Prevot T, Homola J, Mercer J. Human-in-the-loop evaluation of ground-based automated separation assurance for NEXTGEN. In: The 26th Congress of ICAS and 8th AIAA ATIO. 2008 Sep. 8885

[5] Erzberger H, Heere K. Algorithm and operational concept for resolving short-range conflicts. Proceedings of the Institution of Mechanical Engineers, Part G: Journal of Aerospace Engineering. 2010 Feb 1;**224**(2):225-243

[6] Lauderdale TA, Erzberger H. Automated separation assurance with weather and uncertainty. Air Traffic Management and Systems. 2014:35-47. Springer, Japan

[7] Temizer S, Kochenderfer M, Kaelbling L, Lozano-Pérez T, Kuchar J. Collision avoidance for unmanned aircraft using Markov decision processes. In: AIAA Guidance, Navigation, and Control Conference. 2010 Aug 2. 8040

[8] Durand N, Alliot JM, Noailles J. Collision avoidance using neural networks learned by genetic algorithms. In: IEA-AEI 1996, 9th International Conference on Industrial and Engineering Applications of Artificial Intelligence and Expert systems. 1996 Jun 1

[9] Hromatka M. A fuzzy logic approach to collision avoidance in smart UAVs. Honors Thesis. Collegeville, MN: College of Saint Benedict and Saint John's University; 2013

[10] Kuchar JE, Drumm AC. The traffic alert and collision avoidance system. Lincoln Laboratory Journal. 2007 Nov 2;**16**(2):277

[11] Cook B, Cohen K, Kivelevitch EH. A fuzzy logic approach for low altitude uas traffic management (UTM). In: AIAA Infotech@ Aerospace. 2016:1905. DOI: 10.2514/6.2016-1905

[12] Cook BM. Multi-Agent Control Using Fuzzy Logic. Electronic Thesis or Dissertation. University of Cincinnati. 2015. OhioLINK Electronic Theses and Dissertations Center. pp. 90-166

[13] Peled D. Software Reliability Methods. New York, NY: Springer Science & Business Media; 2013. 332

[14] Baier C, Katoen J, Larsen K. Principles of Model Checking. Cambridge, MA: MIT press; 2008. 984

[15] Clark M, Rattan K. Piecewise affine hybrid automata representation of a multistage fuzzy pid controller. In: Proceedings of the 2014 AAAI Spring Symposium Series: Formal Verification and Modeling in Human-Machine Systems; 24–26 March 2014

[16] Ross T. Fuzzy Logic with Engineering Applications. 3rd ed. Hoboken, NJ: John Wiley & Sons; 2009. 606

[17] Butler RW. "What is Formal Methods?" NASA LaRC Formal Methods Program [Internet]. 2001. Available from: https://shemesh.larc.nasa.gov/fm/fm-what.html. [Accessed: 03-02-2017]

[18] Wagner L, Fifarek A, DaCosta D, Gross K. SpeAR: Specification and Analysis of Requirements. InS5 Symposium. 2014

[19] Hagen GE. Verifying safety properties of Lustre programs: An SMT-based approach. Ann Arbor, MI: ProQuest; 2008

[20] Ghassabani E, Gacek A, Whalen MW. Efficient generation of inductive validity cores for safety properties. In: Proceedings of the 2016 24th ACM SIGSOFT International Symposium on Foundations of Software Engineering. 2016 Nov 1:314-325. ACM

[21] De Moura L, Bjørner N. Z3: An efficient SMT solver. In: International Conference on Tools and Algorithms for the Construction and Analysis of Systems. 2008 Mar 29. 337-340. Berlin Heidelberg: Springer

[22] Halbwachs N, Caspi P, Raymond P, Pilaud D. The synchronous data flow programming language LUSTRE. Proceedings of the IEEE. 1991 Sep;**79**(9):1305-1320

Chapter 13

Precision Improvement in Inertial Miniaturized Navigators Based on Fuzzy Logic Denoising of Sensors Signals

Teodor Lucian Grigorie and
Ruxandra Mihaela Botez

Additional information is available at the end of the chapter

http://dx.doi.org/10.5772/67966

Abstract

The chapter presents a new strategy to improve the precision of the inertial navigators processing in a fuzzy manner the signals provided by the miniaturized sensors equipping their inertial measurement units (IMU). To apply the developed technique, the hardware component of the inertial measurement units was specifically designed to include some redundant clusters of inertial sensors disposed in linear configurations on the measurement axes. The algorithm acts at the level of each detection cluster designed to measure an acceleration or angular speed along with an IMU axis by fusing the data obtained from the sensors in respective cluster. Based on the standard deviations of the sensors signals estimated for a data frame with a well-known size, the fuzzy logic mechanism provides a set of weights associated with each sensor in cluster, which are further used to fuse the data acquired from sensors at the current time. The algorithm has an adaptive character, the data frame used to estimate the standard deviations of the sensors signals being permanently updated with the new sensors measurements, and, in this way, the weights associated with each sensor are reestimated at each measurement step.

Keywords: inertial navigation, miniaturized sensors, data fusion, fuzzy logic, redundant configurations, noise reduction, precision improvement

1. Introduction

The permanent change at the level of the sensor fabrication technologies in the last two decades aiming especially toward miniaturization produced a significant change in the navigation and positioning systems field from the point of view of strap-down inertial navigation systems (SDINS). Therefore, more and more applications have been developed and are currently developed starting from this concept and based on miniaturized inertial measurement units and miniaturized data processing systems. These new technologies, MEMS (micro-electro-mechanical systems), NEMS (nano-electro-mechanical systems), MOEMS (micro-opto-electro-mechanical systems), or NOEMS (nano-opto-electro-mechanical systems), are currently successfully applied in the inertial sensors, dedicated to a large part of the SDINS applications. Actually, due to the miniaturization, the SDINS systems application field has been seriously extended beyond the classical navigation systems equipping the vehicles in the aerospace industry, at the level of land-vehicle applications, robotics, medicine, and assisted living [1–4]. The distribution of the inertial sensor fabrication technologies on future applications accuracy classes can be organized as in **Figure 1** [1]. It can be easily observed that the accuracy level for NEMS technologies (possibly NOEMS) are still at the commercial applications level; IOG—integrated optics gyro, IFOG—interferometric fiber optic gyro, HRG—hemispherical resonant gyro, PCF-FOG—photonic crystal fiber-fiber optic gyro.

Included in the dead-reckoning navigation systems category, the SDINS was succeeded on the market due to its availability to provide the vehicle speed and position with abundant dynamic information, due to the possibility to calculate at an extremely high rate comparatively with the GPS system, and due to its excellent short-term performance. Having

Very Precise Navigation	*Long-Range Guidance*	*1 nm/hr Navigator*	*Tactical Weapons*	*Comercial Consumer*
Accelerometer Technology				
Cold Atom	Cold Atom	MEMS	MEMS	MEMS / NEMS
	MEMS			
Mechanical	Mechanical			
Gyro Technology				
Cold Atom	Cold Atom	IOG	MEMS	MEMS / NEMS
	IFOG			
IFOG	HRG	PCF FOG		
	IOG			

Figure 1. Inertial sensor technologies on the application accuracy classes.

characteristics complementary to satellite positioning systems, in time, it became an important component of the modern navigation systems near the GPS [5–6]. If the two systems are used in an integrated configuration, for a short time, derived from the quality of the used inertial sensors, the inertial navigator can overcome the nonavailability of GPS signals and can maintain in this way the navigation solution at a high quality; a well-designed INS/GPS navigator provides improved performance in terms of accuracy, availability, and reliability in front of a simple GPS system.

From another point of view, the inertial navigators are subjected to various error sources, their main deficiency residing in the great accuracy degradation over time mainly due to the quality of the used inertial sensors [7–9]. New miniaturization technologies plays an important role in the reduction of the cost and the volume of the inertial sensors, which, apparently, offers an important advantage for the miniaturized inertial sensors, but, in fact, the fabrication processes of such sensors make these very sensitive to the changes of the environmental conditions: temperature, pressure, electric and magnetic fields, and vibrations. Therefore, the sensors output can vary quickly, widely, and sometimes randomly and is very hard to be modeled. In many cases, this sensitivity leads to the decreased sensor performance, adding more error types, and possibly, with higher values than those of classical sensors [3, 9].

In this context, the need to develop low-cost, small-size, and high-precision INS/GPS integrated navigators, which are able to be used also in GPS challenging environments, generated few research directions in the field; the main two being [4–6, 10–14]: (1) to develop standalone accurate SDINS structures with miniaturized IMUs; (2) to develop new INS/GPS data-fusion methods based on artificial intelligence algorithms, having objectives to overcome the limitations in terms of model dependency, prior knowledge dependency, and linearization dependency.

The increase of the SDINS accuracy in standalone configuration targets, in fact, the improvement of the quality for the signals provided by the IMUs to the navigation processor through the obtaining of high quality miniaturized sensors or through the development of various IMU hardware architectures accompanied by right numerical algorithms able to produce sensor errors estimation and compensation [9, 13, 14]. Two categories of errors parasitize the inertial sensors: deterministic errors and stochastic errors. For the deterministic errors, the specialists conceived various calibration procedures, which can easily estimate and eliminate it. Unfortunately, the stochastic error estimation necessitates a more complex process and they cannot be fully removed from the sensors data. For both categories of inertial sensors (accelerometers and gyros), the most important stochastic errors are caused by noise and by instability of bias and of scale factor [7].

The main difficulty with the noise is related to the impossibility to apply a direct filtering procedure by using classical methods, because, in terms of frequency spectrum, it is superimposed on the band 0–100 Hz of the navigation useful signal [7]. As a consequence, many researches have been initiated to find alternative solutions to limit its impact on the solution of navigation. On the other way, it must not forget that at the level of the noise there are present two components, long term noise (low frequency noise) and short term noise (high frequency noise), each one with its own spectral characteristics and with its own compensation mechanism: optimal low pass filtering or using algorithms that fusing the data from multiple sensors [13, 15–19].

Our identified way to improve the quality of the inertial sensor signals is based on the use of redundant detection clusters with the sensors disposed in linear configurations along each detection axis in IMU (**Figure 2**) [9, 13]. For each cluster, the measured data are subsequently fused to provide to the navigation processor a better measured signal of acceleration or angular speed for the respective axis in IMU. Simultaneously, this structure provides the advantage to have a redundant inertial navigator in terms of the detection unit components, i.e., when one or more sensors in a detection cluster break down they are removed from calculation process but the system still remains operational. The inertial sensors in the same cluster have the sensitivity axes parallel and oriented in the same sense with the detection axis.

The proposed fuzzy logic procedure was mainly designed to produce effects at the level of the sensor noise, but some benefits were also noted at the level of bias effects reduction.

The work exposed here is related to a research project developed in Romania and financed by the Executive Unit for the Financing of Higher Education and University of Scientific Research, which aimed at the strap-down inertial navigators' precision improvement by using redundant sensor networks for their IMUs and new adaptive numerical algorithms for data fusion.

In this chapter, the information is structured as follows: Section 1 is constituted by a short introduction; Section 2 exposes the algorithm basic elements and structure, including the fuzzy logic mechanism; Section 3 shows the numerical simulation results obtained with the developed algorithm for detection clusters with various sizes and sensors types; Section 4 provides the results obtained with some experimental data acquired with detection clusters with six MEMS sensors each (assisted by an integrated GPS/INS navigator as reference positioning system) and processed for a bidimensional INS navigator; and Section 5 highlights the benefits brought by the algorithm in the form of the conclusions.

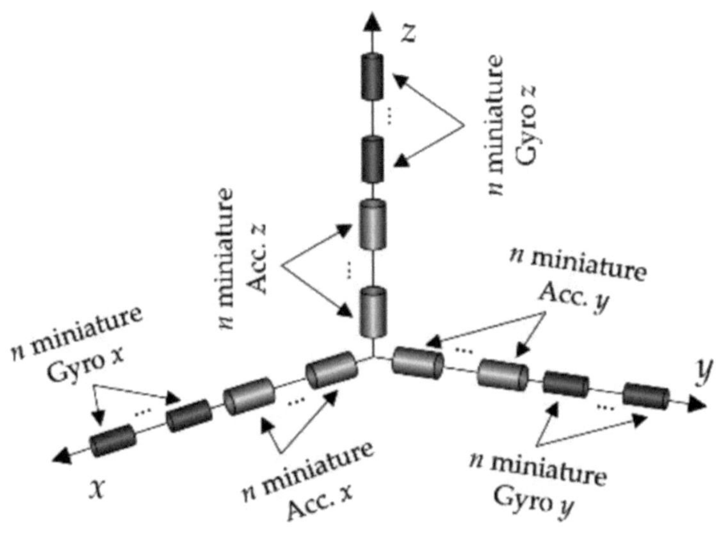

Figure 2. Inertial sensors disposed in linear configurations along each detection axis in IMU.

2. Data fusion algorithm basic elements and structure

To describe the generalized form of the proposed algorithm, it is denoted with n the number of collinear sensors included in a detection cluster, as in **Figure 2**. On the other way, if x_i $(i = 1 \div n)$ are the measurements of the x quantity provided by the n sensors, characterized by the standard deviations σ_i $(i = 1 \div n)$, then the weighted mean obtained by applying the data fusion algorithm conducts to the x_e estimate of the x quantity under the next form [14]:

$$x_e = w_1 \cdot x_1 + w_2 \cdot x_2 + \cdots + w_n \cdot x_n = \sum_{i=1}^{n} w_i \cdot x_i, \left(\sum_{i=1}^{n} w_i = 1\right); \tag{1}$$

w_i $(i = 1 \div n)$ are the weights of the x_i $(i = 1 \div n)$ measurements. For each sensor, the standard deviation is evaluated starting from the last m samples acquired from it. The algorithm has an adaptive character, the data frame used to estimate the standard deviations of the sensors signals being permanently updated with the new sensor measurements, and, in this way, the weights associated with each sensor are reestimated at each measurement step. The estimation of the new weight set is realized based on a fuzzy logic mechanism. The necessary data frame used in the estimation of sensors standard deviations is built using a FIFO (first in first out) buffer with n channels and able to memorize m successive samples on each channel. In this way, at each measurement time, the last column in the data frame goes out from the buffer and a new one comes in; two consecutive data frames are superposed with $(m - 1)$ samples [13]. The standard deviations of the n-independent channels result with the next equation [13, 14]:

$$\sigma_i(k) = \sqrt{\frac{1}{m} \sum_{j=1}^{m} [x_i(j) - \bar{x}_i(k)]^2}, \ (i = \overline{1, n}); \tag{2}$$

$x_i(j)$ is the jth measurement provided by the ith sensor in the kth data frame; $\sigma_i(k)$ is the standard deviation for the kth data frame, which characterize the ith sensor; and $\bar{x}_i x'_i(k)$ is the arithmetic mean of the ith sensor data for the kth data frame [14]. The value of $\sigma_i(k)$ is used to fuse the data at the next sensor reading.

According to the proposed algorithm, a fuzzy logic controller is used in each channel of a detection cluster, and, based on the standard deviation, $\sigma_i(k)$ of the m consecutive samples acquired from the sensor in the kth data frame provides a weight $w_i^{\text{fuzzy}}(k)(i = \overline{1, n}1, n)$ for the respective sensor. Based on the fact that in a weighted sum, the sum of all parts weight should be equal to 1, the sensors weights are recalculated and result with the equations [13, 14]):

$$w_1(k) = \frac{w_i^{\text{fuzzy}}(k)}{\sum_{i=1}^{n} w_i^{\text{fuzzy}}(k)}, \cdots, w_n(k) = \frac{w_n^{\text{fuzzy}}(k)}{\sum_{i=1}^{n} w_n^{\text{fuzzy}}(k)}. \tag{3}$$

As a consequence, by calculating the weighted sum from the data fusion equation, at the time $t_{k+1,}$ the x_e estimate of x is [13, 14]:

$$x_e(k+1) = w_1(k) \cdot x_1(k+1) + w_2(k) \cdot x_2(k+1) + \cdots + w_n(k) \cdot x_n(k+1), \tag{4}$$

i.e.,

$$x_e(k+1) = \sum_{i=1}^{n} w_i(k) \cdot x_i(k+1), \left(\sum_{i=1}^{n} w_i(k) = 1\right). \tag{5}$$

Based on the previous considerations, the architecture of the data fusion algorithm results as in **Figure 3** [14]. The chosen fuzzy logic controllers are based on rules set that created

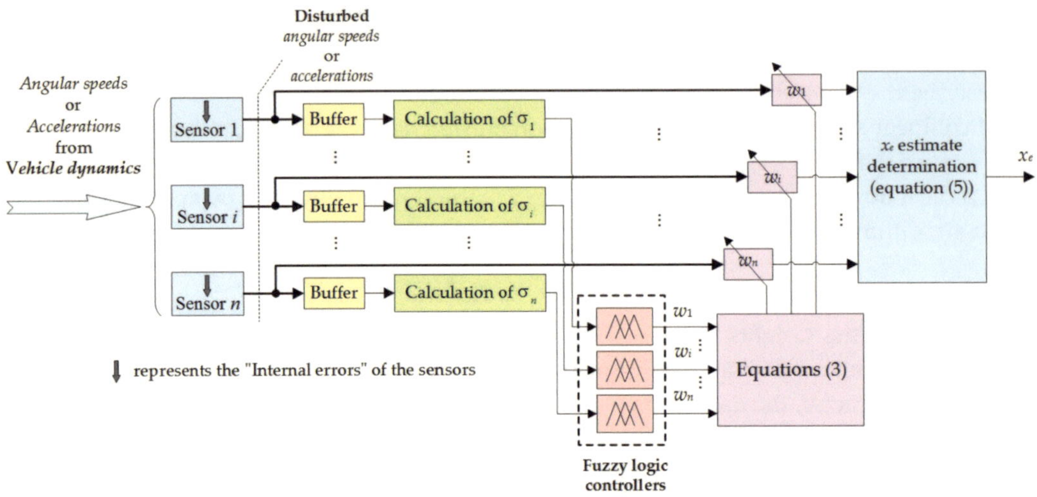

Figure 3. Data fusion algorithm.

a proportional dependence between the quality of the sensor signals and their associated weights, e.g., a "P" fuzzy logic controller. Therefore, the implemented fuzzy logic controllers established an inverse proportionality between the calculated standard deviations and sensor-associated weights.

The aim is to obtain a simpler fuzzy controller in order to reduce at minimum the computing time at each sample in order to implement the controller on a cheaper microcontroller. The way to reduce the computing time is to reduce on one hand the number of the membership functions and on the other hand to simplify these functions. The structure of fuzzy controller realized in Matlab/Simulink is presented in **Figure 4** [20, 21].

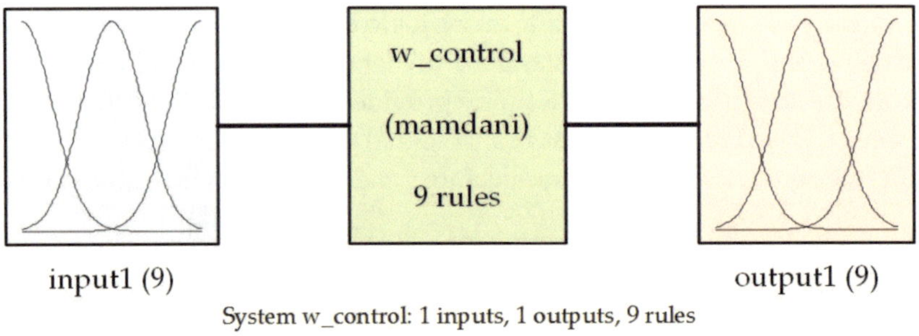

Figure 4. Fuzzy controller structure.

The universe of discourse for the controller input has considered a [0, 2] interval, where nine membership functions (*in1* to *in9*) with Gaussian shapes were uniformly distributed. In the same time, nine membership functions (*out1* to *out9*) with similar shapes were considered in the [0, 1] interval, taken as universe of discourse for the output. **Figure 5** presents the input/output membership functions.

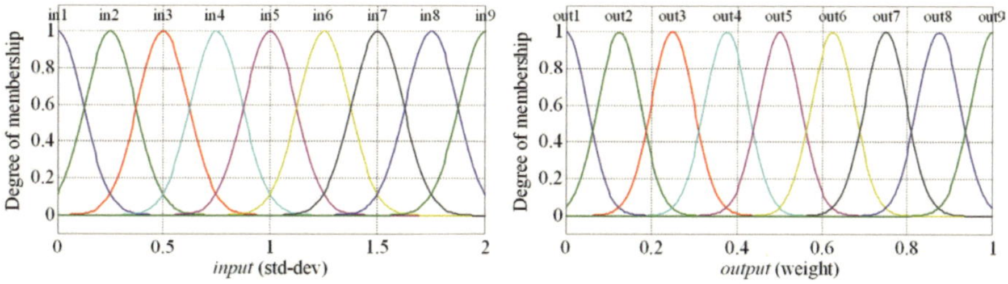

Figure 5. Membership functions (mfs) for input and output.

In the fuzzification process, the next nine rules were defined as: (1) If (*input* is *in1*) then (*output* is *out9*); (2) If (*input* is *in2*) then (*output* is *out8*); (3) If (*input* is *in3*) then (*output* is *out7*); (4) If (*input* is *in4*) then (*output* is *out6*); (5) If (*input* is *in5*) then (*output* is *out5*); (6) If (*input* is *in6*) then (*output* is *out4*); (7) If (*input* is *in7*) then (*output* is *out3*); (8) If (*input* is *in8*) then (*output* is *out2*); and (9) If (*input* is *in9*) then (*output* is *out1*). Widely accepted for capturing expert knowledge, a Mamdani controller type was used due to its simple structure of "min-max" operations [21]; and for defuzzification, the centroid method was applied.

3. Numerical simulation results

To simulate the algorithm, some models in Matlab/Simulink for detection clusters with various sizes and sensors types were realized. An example of such model is shown in **Figure 6**, being realized for a cluster with four sensors by the same type, with errors software modeled by the blocks "Model Acc" placed at the input of the simulation model. These models, realized by the authors, are based on the sensor data sheets and on the IEEE equivalent models for the inertial sensors [22, 23]. Accelerometers were modeled as in **Figure 7**, the obtained model having inputs, such as acceleration a_i, applied along the sensitive axis, and the cross-axis acceleration a_c, and as output the perturbed acceleration a [22, 23].

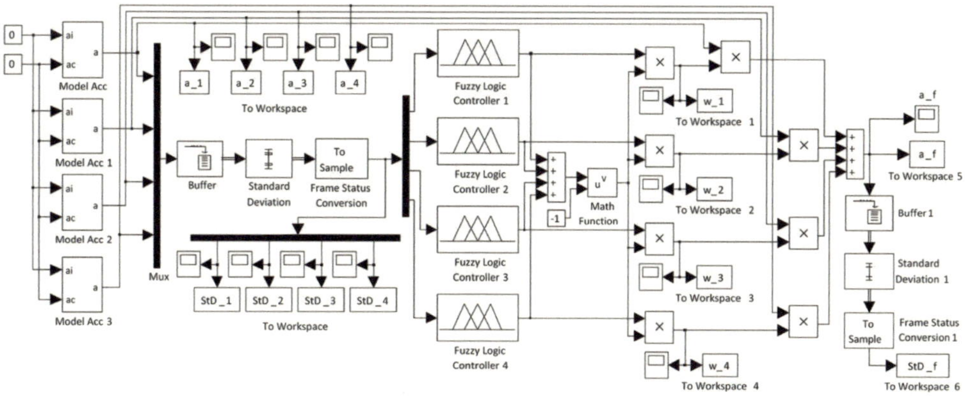

Figure 6. Simulation model with four sensors in cluster.

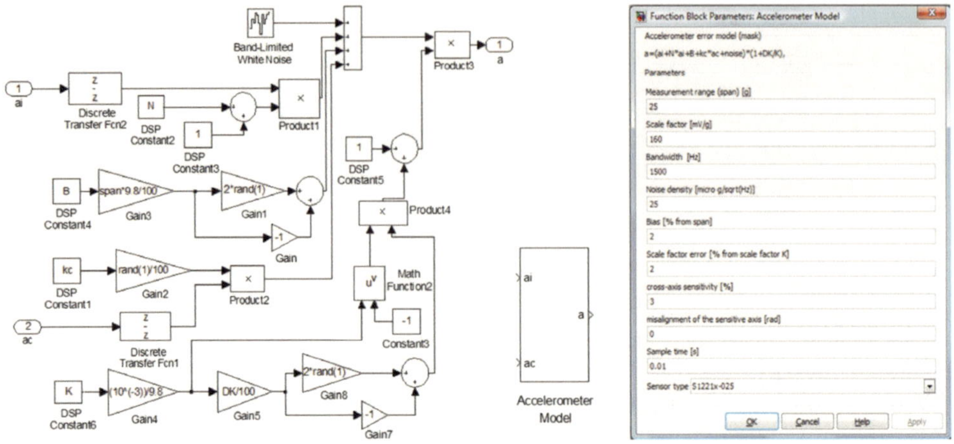

Figure 7. Accelerometers Matlab/Simulink model and its interface.

Studying the data sheets for various acceleration sensors, it was observed that a part of the included parameters is provided by using an interval inside which they can vary arbitrary. For example, for the bias it is provided a maximum absolute value (B) which is considered as percent from span, for the cross-axis sensitivity it is provided a maximal value (k_c) considered as a percent from the cross-axis acceleration (a_c), for the scale factor calibration error the producers give a maximum absolute value (ΔK) under the form of a percent from the scale factor (K), and for the noise is provided the maximum value of its density. Therefore, according to **Figure 7**, in the model, the "rand(1)" MATLAB function is used three times to generate a random value in the ($-B$, B) interval for the bias, a random value in the (0, k_c) interval for the cross-axis sensitivity, and a random value in the ($-\Delta K$, ΔK) interval for the scale factor calibration error. The noises are realized using the SIMULINK block "Band-Limited White Noise" and MATLAB function "RandSeed" through the generation of a random value of the density in the ($80\%\cdot\nu_d$, ν_d) interval [24].

In a simulation example at null accelerations as inputs, graphical characteristics resulted are shown in **Figures 8–10**. **Figure 8** shows the sensor measurements from "Acc-1" to "Acc-4" and also the results of the accelerometer data fusion "Acc-f". In the simulation, there was considered just the sensor noise overlapped on an ideal null input for four sensors with a span of 18 g, a bandwidth of 2500 Hz each, and with various noise densities taken aleatory between 280 and 380 μg/Hz$^{1/2}$. The noise pattern of the data fusion resulted a signal that proves an important reduction of the noise level, a similar observation resulting from **Figure 9** also depicting the evolution in time of the standard deviations for the sensors signals and for the estimate x_e. On the other way, **Figure 9** proves that the adaptive character of the algorithm is maintained, the standard deviations values being updated at each simulation time step. The best sensor in the cluster was sensor #3, the standard deviation of the fused signal being two times smaller (approx. $3\cdot10^{-3}$ m/s^2) than the raw values of the standard deviations of the sensors signals (approx. $6\cdot10^{-3}$ m/s^2). **Figure 10**, depicting the values of the sensor weights estimated at each calculation step, also proves that the best sensor in the cluster was the sensor #3, and it receives the biggest values from the four weights during the entire numerical simulation.

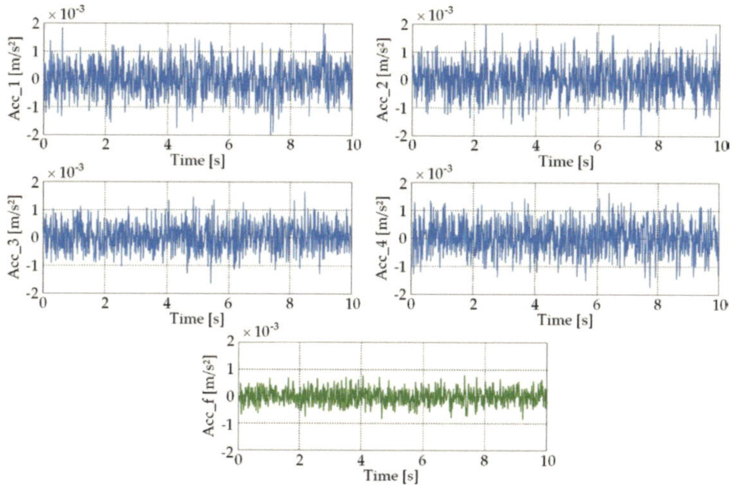

Figure 8. The sensor measurements and the data fusion results.

The sample time for the numerical simulation was 0.01 s, which means a data processing rate of 100 samples/s. Moreover, the data frame size for one detection channel was established to be m = 100, i.e., the buffers allow to be obtained data frames of 100 consecutive samples for each channel. In this way, any two consecutive data frames for one channel are overlapped with 99 samples. From **Figure 9**, the initialization phase of the algorithm with a duration of 1 s, time necessary for the first buffer to be full with 100 samples can be easily observed. All simulation results proved a good functioning of the algorithm which provided an important reduction of the sensor noise level, with a higher potential in the improvement of the solution of navigation accuracy when it is used in an inertial navigator.

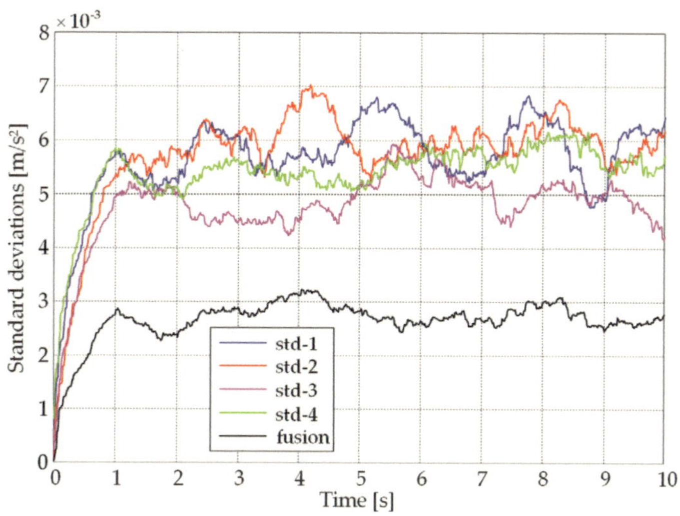

Figure 9. Standard deviations for the sensor signals and for the data fusion signals.

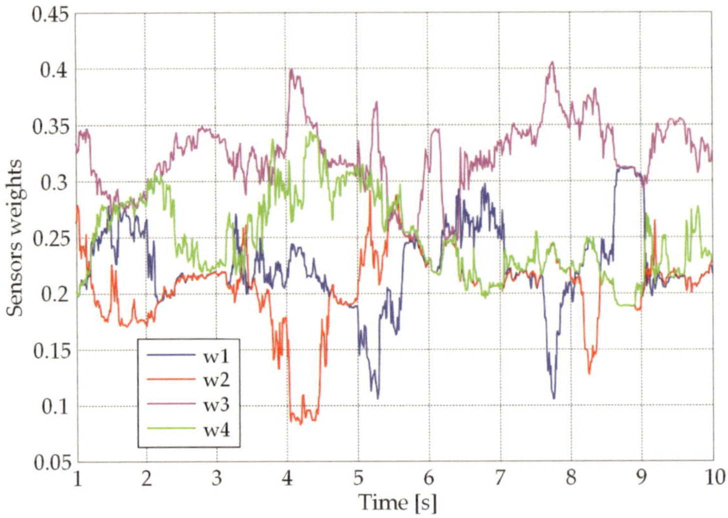

Figure 10. Sensor weights.

4. Results obtained with experimental data

The proposed algorithm was also tested using some experimental data acquired with an IMU equipped with detection clusters, which included six MEMS sensors each (assisted by an integrated GPS/INS navigator as reference positioning system). For the used IMU structure, with six sensors in each detection cluster, the Matlab/Simulink implementation of the data fusion algorithm resulted as shown in **Figure 11** [14]. Grouping the algorithm model resulted in "Fuzzy-logic data fusion" block from the right side of the figure. The data fusion block inputs are the measured data obtained from the sensors in the cluster ("Si" correspond to the

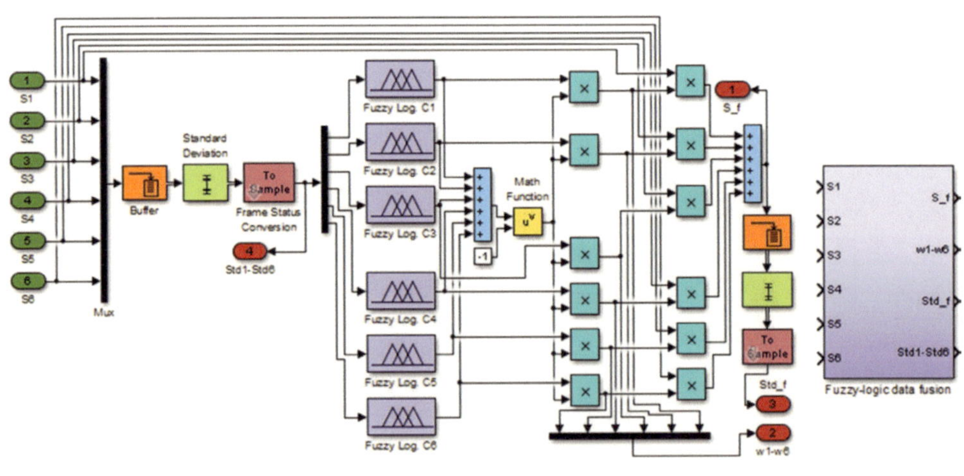

Figure 11. Matlab/Simulink implementation of the data fusion algorithm for experimental data.

measurements x_i, $i = 1 \div 6$), whereas its outputs are the fusion signal "S_f" (the estimate x_e), the sensor weights "w1–w6" (w_i, $i = 1 \div 6$), the standard deviation of the fusion signal "Std_f", and the standard deviations of the sensor data "Std1–Std6".

After the fusion, the obtained data were processed in a bidimensional INS navigator in horizontal plane, software implemented as in **Figure 12** [14]. Therefore, in this testing mechanism, three detection clusters were implied: two accelerometers clusters along the longitudinal and lateral axes of the vehicle and a gyro cluster along the vertical axis of the vehicle. The model included the "Flat Earth to LLA" Matlab/Simulink block in order to obtain the vehicle coordinates also in terms of latitude and longitude. The block "Horizontal plane navigator" in the right side of the figure has been obtained by grouping the navigator software model in the left side of the figure. The inputs of the navigator are the inertial measurements provided by the accelerometers and by the gyro: longitudinal (*f_xv*) and lateral (*f_yv*) accelerations in vehicle reference frame, and the angular speed (*w_zv*) along the vertical axis of the vehicle reference frame. As outputs, it provides information related to the angular orientation of the vehicle in horizontal plane (yaw angle—*psi*), to the vehicle speed and position components in local horizontal frame, *v_xl* (North speed), *v_yl* (East speed), *x_l* (position in North direction), and *y_l* (position in East direction), and to the latitude and longitude coordinates of the vehicle (*Lat* and *Lon*).

During tests, the IMU data were acquired simultaneously with the data from an integrated GPS/INS navigator used as reference positioning system; both equipments were boarded on a test car. The sensor acquired data were offline processed through fusion but also used together with the fusion signals in a Matlab/Simulink software model implementing seven bidimensional INS navigators ("Horizontal plane navigator" blocks) as in **Figure 13** [14]. It also includes three "Fuzzy-logic data fusion" blocks used to fuse the acquired data in each of the three detection clusters.

The sensors in IMU were organized in six groups, with three sensors each (two accelerometers and one gyro), one sensor from each of the three detection clusters. To avoid the complications in the data processing, the detection groups included sensors with the same number in

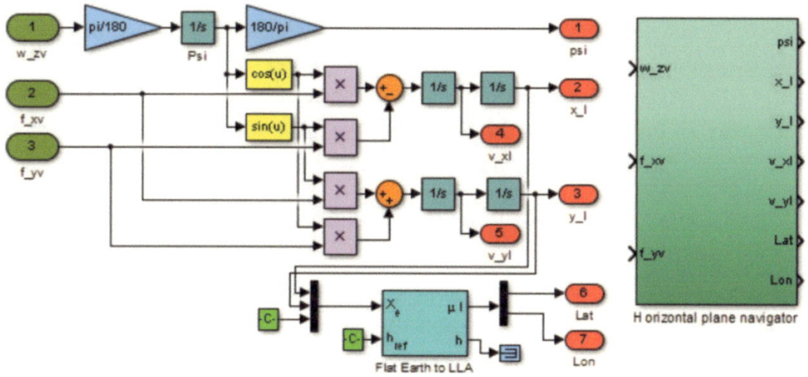

Figure 12. Software model of the bidimensional INS navigator in horizontal plane.

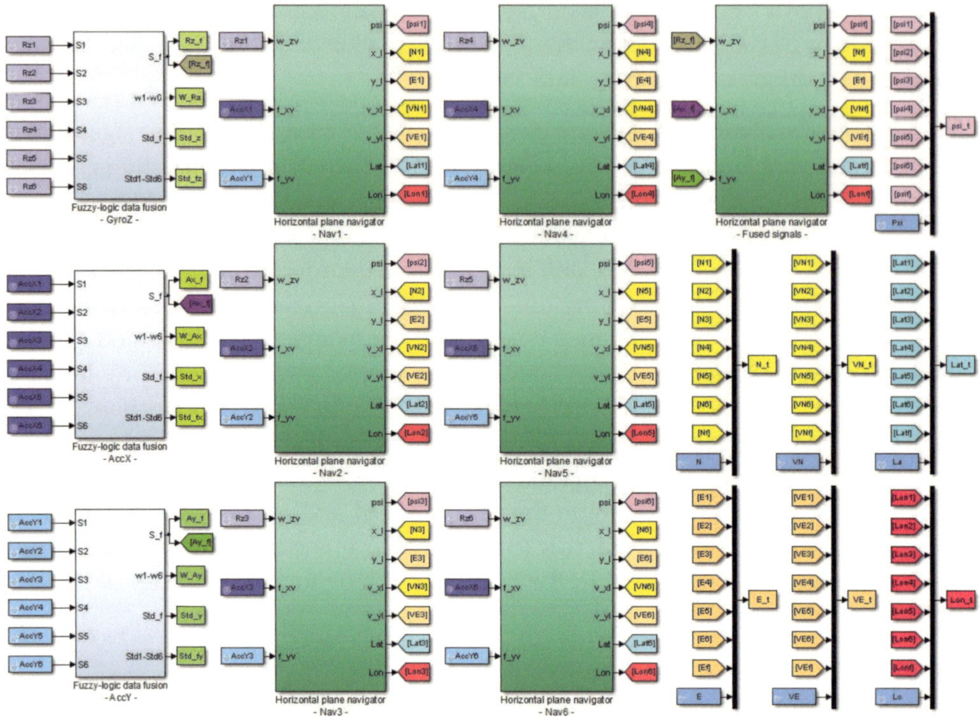

Figure 13. The evaluation Matlab/Simulink model.

the detection clusters. Six of the seven navigators in **Figure 13** ("Nav1" to "Nav6") processed data from the six detection groups, whereas the seventh navigator ("Horizontal plane navigator – Fused signals") processed data resulted from the fusion procedure applied on each of the three detection clusters. In this way, the evaluation model allows the obtaining of seven solutions of navigation, six solutions based on inertial sensor raw data (nonredundant INSs) and one solution based on inertial sensor fused data. All these solutions are further compared with the navigation solution provided by the integrated GPS/INS navigators which are used as reference positioning system.

For an evaluation situation, the experimental data acquired in each of the three detection clusters and the data fusion results for each cluster are shown in **Figure 14** for the accelerometers in the x-axis cluster, in **Figure 15** for the accelerometers in the y-axis cluster, and in **Figure 16** for the gyros in z-axis cluster. The data fusion signals are the last ones in each figure.

The processing of the acquired data with the model in **Figure 13** conducted at the navigation solutions components shown in the next figures. Together with the navigation solution, components have shown the associated errors, evaluated vis-à-vis of the reference navigation solution provided by the GPS/INS integrated navigator. **Table 1** centralizes the absolute maximal values of the positioning and speed errors for all six nonredundant INSs ("INS1" to "INS6" columns) and for the redundant INS ("Fusion" column), and also the mean values obtained for all these errors ("Mean value" column). In the same time, by calculating the mean of the absolute maximal values of the errors for the six nonredundant INSs in each line of **Table 1** and dividing it with the corresponding value in the "Fusion" column, the ratios in the "Mean/Fus" column have been obtained. Similarly, the "Max/Fus" and "Min/Fus" columns show the ratios between the maximal value of the tabled errors for the six nonredundant INSs in each line of **Table 1** and the corresponding value in the "Fusion" column, respectively, between the minimum value of the tabled errors for the six nonredundant INSs in each line of **Table 1** and the corresponding value in the "Fusion" column.

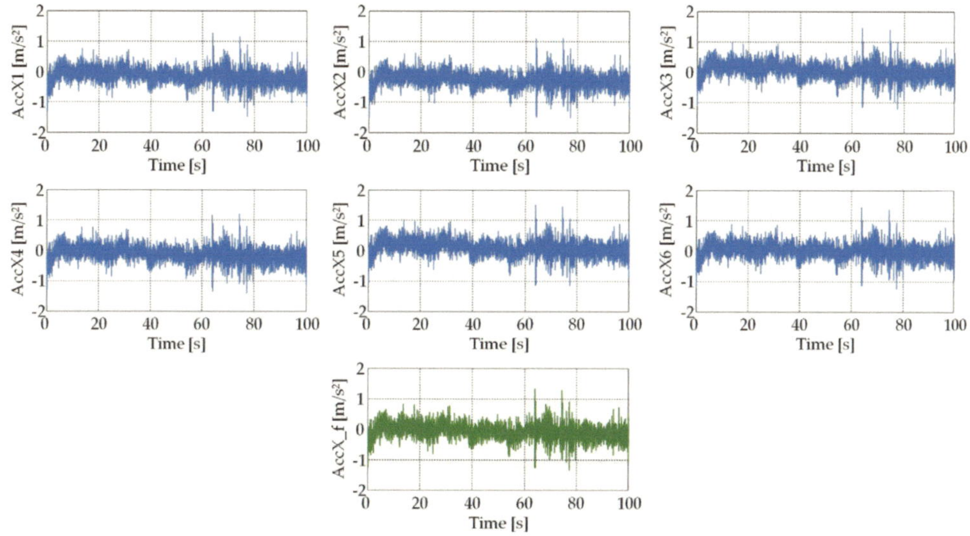

Figure 14. Experimental data and the data fusion results for accelerometers in the x-axis cluster.

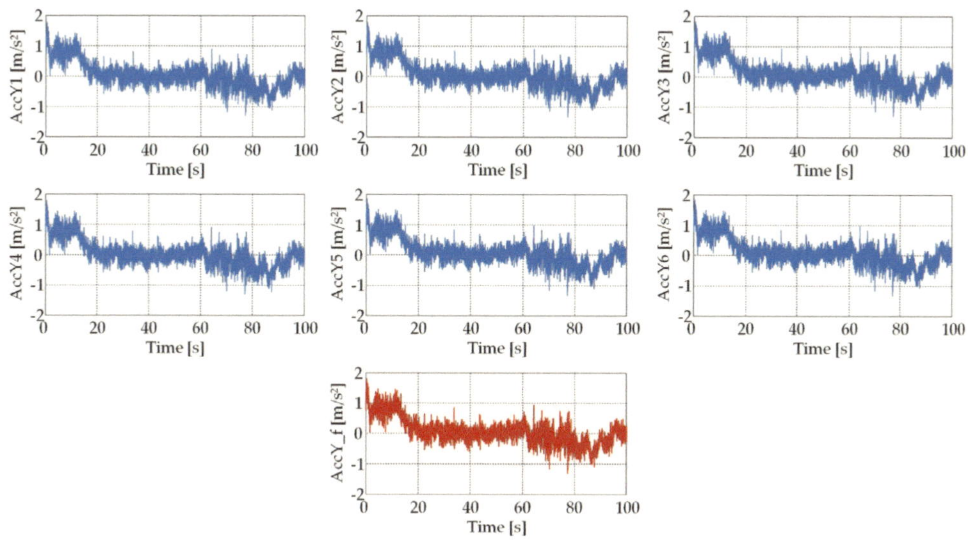

Figure 15. Experimental data and the data fusion results for gyros in the z axis cluster.

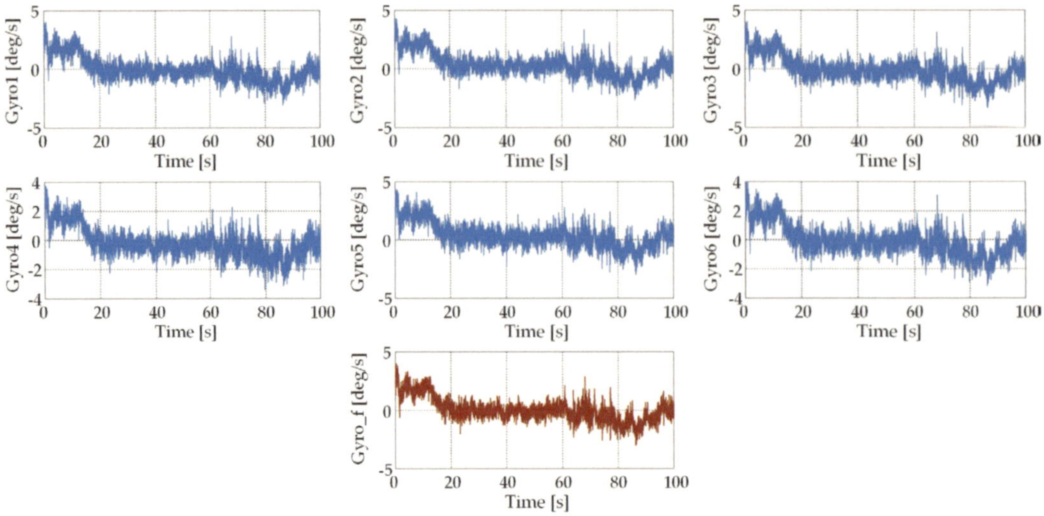

Figure 16. Experimental data and the data fusion results for gyros in the *z* axis cluster.

At the level of the vehicle attitude, the characteristics in **Figure 17** were obtained for the yaw angle. Also, from the point of view of the horizontal positioning, **Figure 18** presents the covered distances in the North direction, and **Figure 19** presents the covered distances in East direction. The North speed evolutions during time are depicted in **Figure 20**, whereas the speed evolutions during time in the East direction are shown in **Figure 21**. For each navigator, the incorporated "Flat Earth to LLA" block allows the calculus of the vehicle Latitude and Longitude as shown in **Figures 22** and **23**. Combining the positioning data in a horizontal plane, it resulted the vehicle trajectories (Latitude versus Longitude) and the deviations from the reference trajectory (Latitude errors versus Longitude errors), established by all seven navigators, as in **Figure 24**.

Figure 25 depicts the time evolution of the vehicle trajectories estimated with all seven navigators, whereas **Figure 26** presents their deviations from the reference trajectory.

Both graphical results and numerical values presented in **Table 1** prove an important positioning precision improvement by using the proposed data fusion mechanism. The errors of the navigation solution are substantially reduced comparatively with the nonredundant navigation solutions. The numerical data in the "Mean/Fus" column of **Table 1** highlighted a decrease of the mean of the absolute maximal values of the errors for the six nonredundant INSs of approximately 16.6 times for the positioning in the North direction, 5.1 times for the positioning in the East direction, 2.8 times for the speed component in North direction, 1.05 times for the speed component in the East direction, and 3.8 times in the angular positioning in the horizontal plane (yaw angle). It is very important to be mentioned that the inertial sensor outputs in the experimental model of the inertial measurement unit were not previously corrected with the biases values or with other errors values. The values in the "Fusion" column of **Table 1**, related to the positioning and speed errors for the redundant navigator after 100 s have shown us that the absolute maximal values of these errors were 32.67 m for the linear

http://dx.doi.org/10.5772/67966

Abs. max error	INS1	INS2	INS3	INS4	INS5	INS6	Fusion	Mean value	Mean/Fus	Max/Fus	Min/Fus
North [m]	411.3583	**990.5235**	599.5493	308.5221	692.1193	266.7808	**32.6736**	544.8089	16.6742	30.3156	8.1650
East [m]	73.0415	**318.1701**	262.5502	107.2303	310.4679	132.3827	**39.2776**	200.6405	5.1082	8.1005	1.8596
V-North [m/s]	9.0599	**19.4083**	11.1598	7.1313	14.0987	7.0160	**4.0333**	11.3123	2.8046	4.8119	1.7395
V-East [m/s]	10.9701	9.8672	3.4987	**13.6819**	**1.5752**	5.5067	**7.1445**	7.5166	1.0520	1.9150	0.2204
Latitude [degrees]	$37.019 \cdot 10^{-4}$	$\mathbf{89.141 \cdot 10^{-4}}$	$53.955 \cdot 10^{-4}$	$27.765 \cdot 10^{-4}$	$62.286 \cdot 10^{-4}$	$24.008 \cdot 10^{-4}$	$\mathbf{2.940 \cdot 10^{-4}}$	$49.029 \cdot 10^{-4}$	16.6742	30.3156	8.1659
Longitude [degrees]	$9.156 \cdot 10^{-4}$	$\mathbf{39.886 \cdot 10^{-4}}$	$32.914 \cdot 10^{-4}$	$13.442 \cdot 10^{-4}$	$38.921 \cdot 10^{-4}$	$16.595 \cdot 10^{-4}$	$\mathbf{4.923 \cdot 10^{-4}}$	$25.152 \cdot 10^{-4}$	5.1082	8.1005	1.8598
Yaw [degrees]	14.4621	25.2716	18.7894	**31.1004**	24.9227	19.4309	**5.7402**	22.3295	3.8899	5.4179	2.5194

Table 1. Error analysis for redundant INS and for nonredundant INSs.

positioning in North direction, 39.27 m for the linear positioning in East direction, 2.03 m/s for the speed component in North direction, 7.14 m/s for the speed component in East direction, $2.940 \cdot 10^{-4}$ degrees for the Latitude positioning, $4.923 \cdot 10^{-4}$ degrees for the Longitude positioning, and 5.74 degrees for the angular positioning in the horizontal plane (Yaw angle).

The nonredundant inertial navigation system's best configuration was switched between the detection groups associated with the first sensor in each detection cluster (East channel and Yaw angle channel) and with the sixth sensor in each detection cluster (North channel). The worst configurations for the nonredundant INSs are for the second sensor in each detection cluster, the absolute maximal values of errors in North position channel being 990.52 m, 318.17 m in East position channel, 19.40 m/s in North speed channel, $89.141 \cdot 10^{-4}$ degrees in the Latitude position channel, and $39.886 \cdot 10^{-4}$ degrees in the Longitude channel. From the point of view of East speed channel and yaw angle channel, the worst nonredundant INS is INS4, the East speed error being 13.68 m/s and the yaw angle error being 31.10 degrees.

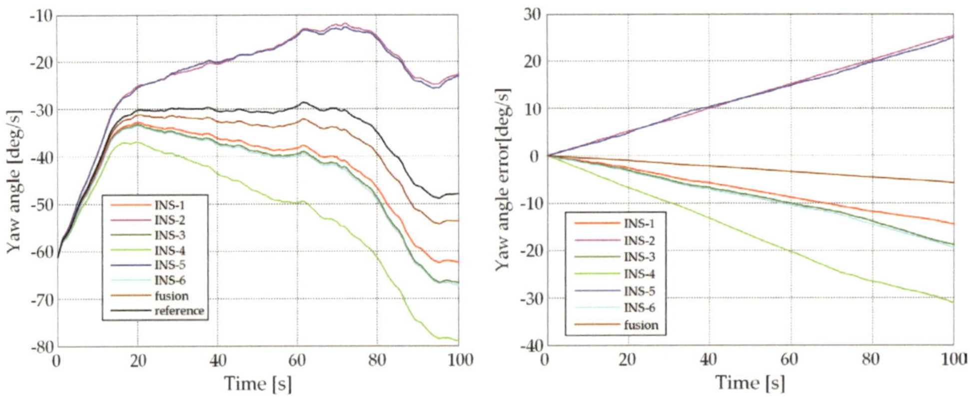

Figure 17. Yaw angle values and errors.

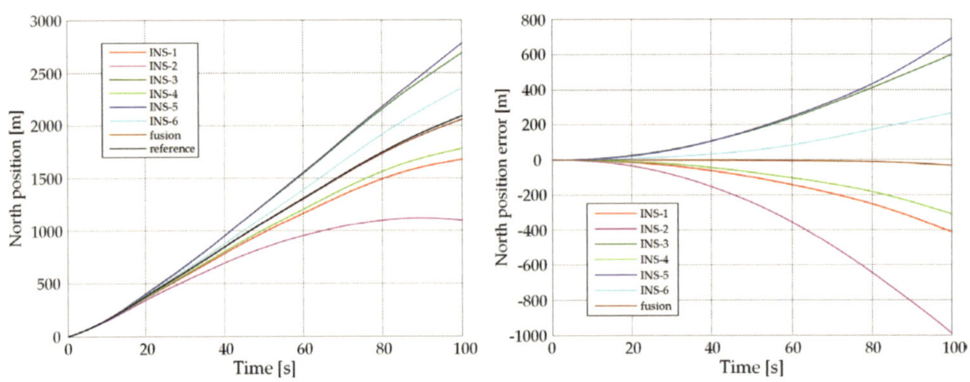

Figure 18. North positions and errors.

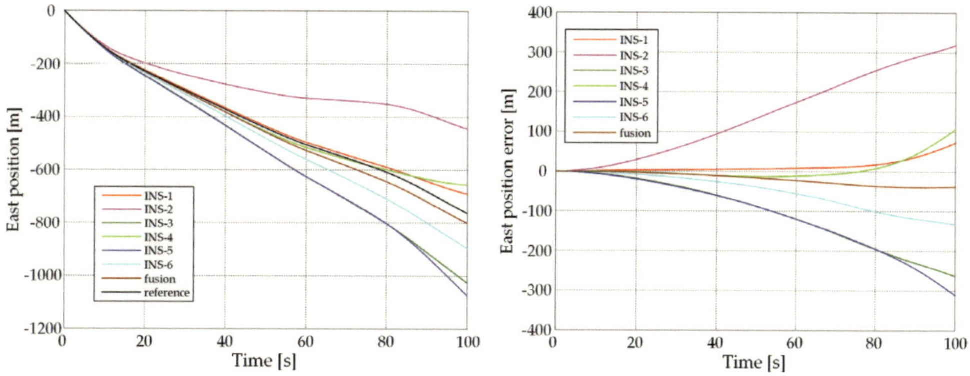

Figure 19. East positions and errors.

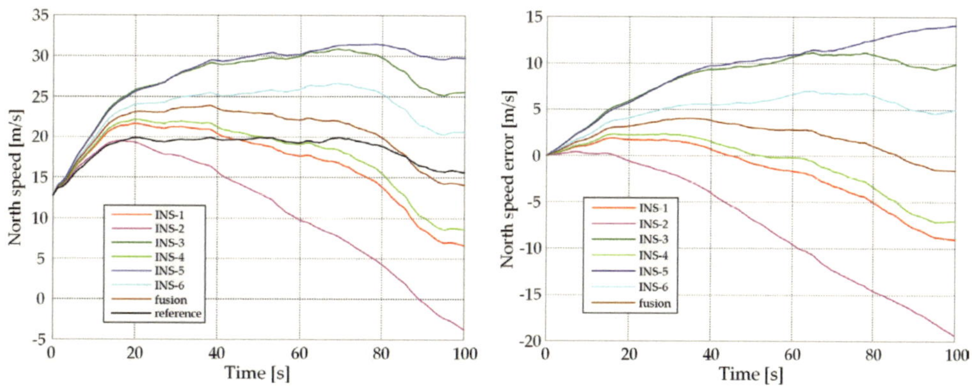

Figure 20. North speed evolutions during time.

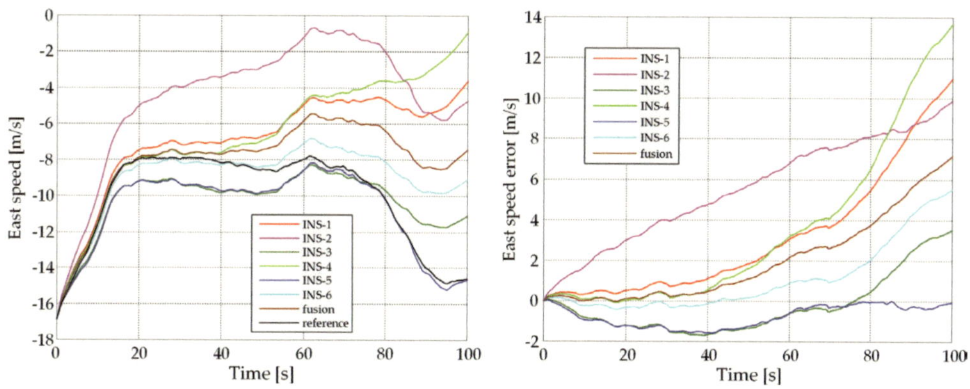

Figure 21. East speed evolutions during time.

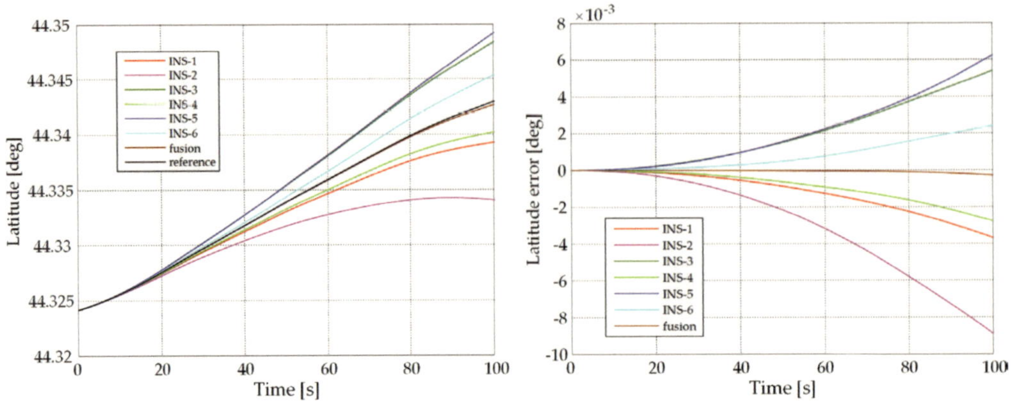

Figure 22. Latitude and errors.

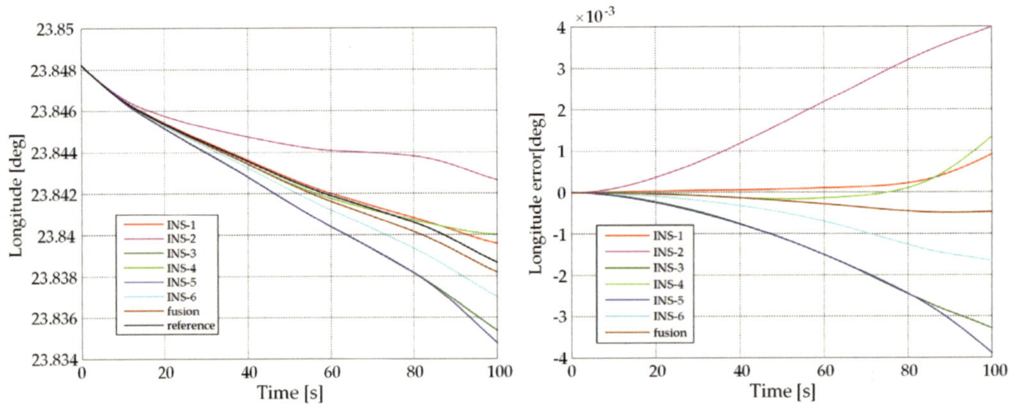

Figure 23. Longitude and errors.

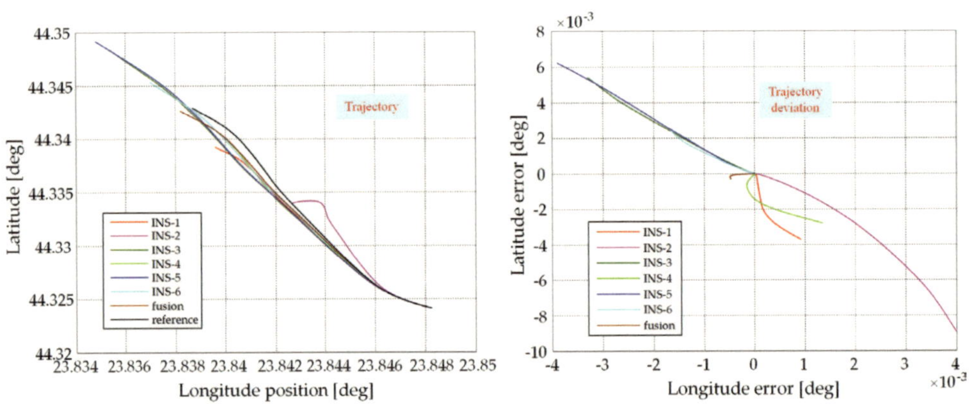

Figure 24. Vehicle trajectories and deviations from the reference trajectory in horizontal plane.

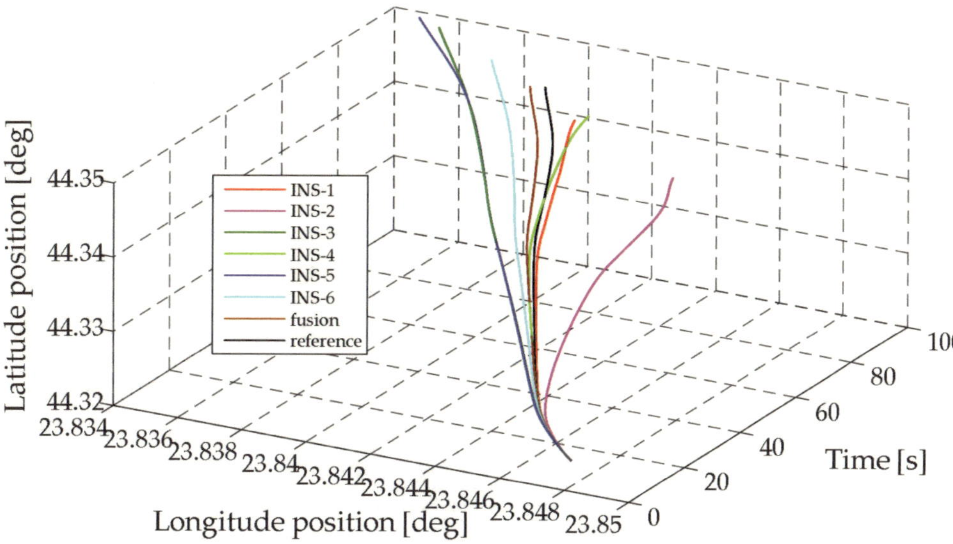

Figure 25. The time evolution of the vehicle trajectories for all seven INSs.

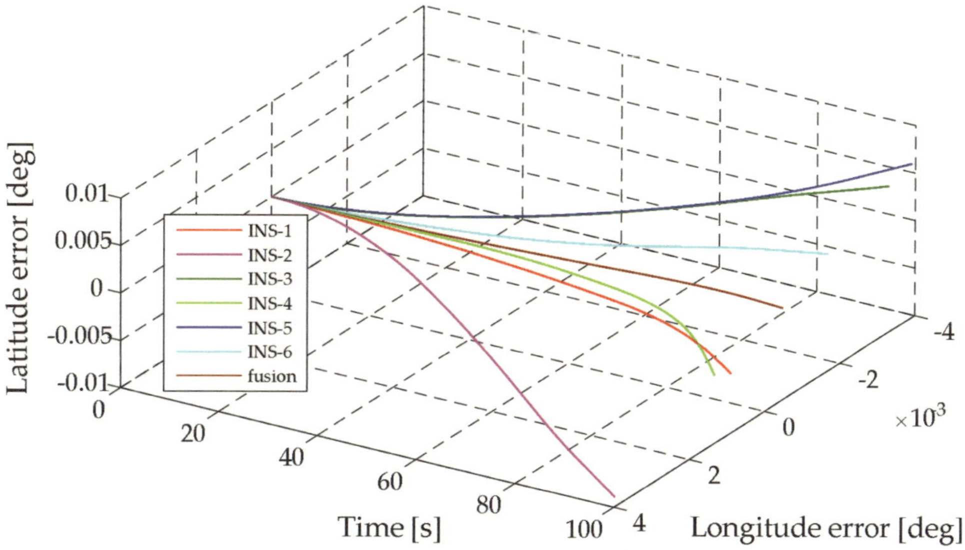

Figure 26. Time evolution of the deviations of the vehicle from the reference trajectory.

5. Conclusions

The chapter exposed a new way to reduce the noise in the inertial sensor data based on a fuzzy logic mechanism. As a special need of the developed mechanism is considered to be a redundant inertial measurement unit with inertial sensors disposed in detection clusters with linear configurations, and measuring the same quantity. The proposed algorithm was firstly numerically simulated based on some models in Matlab/Simulink realized for detection of clusters with various sizes and including various sensor types. All simulation results proved a good functioning of the algorithm which provided an important reduction of the sensor noise level, with a higher potential in the improvement of the solution of navigation accuracy when it is used in an inertial navigator. Second, the algorithm was integrated in an INS redundant structure, with an IMU with three detection clusters and six miniaturized sensors in each cluster. An integrated GPS/INS navigator was used as reference positioning system to estimate the errors of the redundant INS in each channel of the solution of navigation. The data acquired from the redundant IMU were used in a software model both in nonredundant configuration, to compute the solutions of navigations for six detection groups which included sensors with the same number in the detection clusters, but also for the INS redundant configuration when our data fusion methodology is preliminary applied. Both graphical results and tabled numerical values proved an important positioning precision improvement by using the proposed data fusion mechanism. The errors of the navigation solution were substantially reduced comparatively with the nonredundant navigation solutions. The numerical data highlighted a decrease of the mean of the absolute maximal values of the errors for the six nonredundant INSs of approximately 16.6 times for the positioning in the North direction, 5.1 times for the positioning in the East direction, 2.8 times for the speed component in North direction, 1.05 times for the speed component in the East direction, and 3.8 times in the angular positioning in the horizontal plane (yaw angle). It is very important to be mentioned that the inertial sensor outputs in the experimental model of the inertial measurement unit were not previously corrected with the biased values or with other errors values.

Acknowledgements

This work was supported by the CNCSIS-UEFISCSU, project PN II-RU, No. 1/28.07.2010, "High-precision strap-down inertial navigators, based on the connection and adaptive integration of the nano and micro inertial sensors in low cost networks, with a high degree of redundance," code TE-102/2010.

Author details

Teodor Lucian Grigorie[1]* and Ruxandra Mihaela Botez[2]

*Address all correspondence to: ltgrigorie@yahoo.com

1 University of Craiova, Craiova, Romania

2 École de Technologie Supérieure, Montréal, Canada

http://dx.doi.org/10.5772/67966

References

[1] Barbour, N.M. Inertial navigation sensors. In: NATO RTO-EN-SET-116; 2010.

[2] Barbour, N.M., Schmidt, G. Inertial sensor technology trends. IEEE Sensors Journal. 2001;1(4):332-339.

[3] Barbour, N.M. et al. Inertial MEMS Systems and Applications. In: NATO RTO-EN-SET-116-2011, Low-Cost Navigation Sensors and Integration Technology; Bagneux, France, 28-29 March, 2011.

[4] Schmidt, G.T. INS/GPS technology trends. In: NATO RTO-EN-SET-116-2011, Low-Cost Navigation Sensors and Integration Technology; Bagneux, France, 28-29 March, 2011.

[5] Farrell, J. Aided Navigation. GPS with High Rate Sensors. McGraw-Hill, New York; 2008.

[6] Mohinder,S., Lawrence, R., Angus, P. Global Positioning Systems, Inertial Navigation, and Integration. John Wiley & Sons, Inc., Hoboken, New Jersey, 2001.

[7] Titterton, D.H., Weston, J. Strapdown Inertial Navigation Technology. 2nd ed. IET, Stevenage Herts, United Kingdom; 2004.

[8] Grigorie, T.L., Sandu, D.G. Navigation Systems Synergic Architectures With Strap-Down Inertial Components. SITECH, Craiova, Romania; 2013.

[9] Grigorie, T.L., Edu, I.R. Inertial Navigation Applications With Miniaturized Sensors. SITECH, Craiova, Romania; 2013.

[10] He, X., Hu, X., Wu, M. Trends in GNSS/INS Integrated Navigation Technology. Coordinates. 2007; III(3):26-28.

[11] Al-Faiz, M.Z., Ismaeel, S.A. Design of Kalman filter for augmenting GPS to INS systems. In: International Conference on Advanced Remote Sensing for Earth Observation. 2005; 14-20.

[12] Lorinda, S., Aboelmagd, N. Bridging GPS outages using neural network estimates of ins position and velocity errors. Measurement of Science & Technology. 2006; 17:2783-2798.

[13] Grigorie, T.L., Edu, I.R., Corcau, J.I. Fuzzy logic denoising of the miniaturized inertial sensors in redundant configurations. In: 33rd International Conference on Information Technology Interfaces (ITI 2011); June 27-30; Cavtat, Croatia. 2011. p. 521-526.

[14] Grigorie, T.L., Botez, R.M. Positioning monitoring improvement in a horizontal plane ins by using fuzzy logic data fusion for denoising of inertial sensors in redundant clusters. International Journal of Fuzzy Systems And Advanced Applications. 2015; 2:33-40.

[15] Hasan, A.M., Samsudin, K., Ramli, A.R., Azmir, R.S. Wavelet-based pre-filtering for low cost inertial sensors. Journal of Applied Science. 2010;10:2217-2230.

[16] Mao, B., Wu, J.W., Wu, J.T., Zhou, X.M. MEMS gyro denoising based on second generation wavelet transform. In: First International Conference on Pervasive Computing; China. 2010. pp. 255-258.

[17] Sabat, S.L., Giribabu, N., Nayak, J., Krishnaprasad, K. Characterization of fiber optics gyro and noise compensation using discrete wavelet transform. In: Second International Conference on Emerging Trends in Engineering & Technology; Indian. 2009. pp. 909-913.

[18] Skaloud, B., Bruton, A.M., Schwartz, K. Detection and filtering of short-term (1/F) noise in inertial sensors. Journal of the Institute of Navigation. 1999; 46(2):97-107.

[19] Grigorie, T.L., Negrea, P., Edu, I.R., Adochiei, F.C. Assistive positioning system based inertial techniques and wavelet denoising. In: 8th International Conference on PErvasive Technologies Related to Assistive Environments (PETRA); July 1-3, 2015; Corfu, Greece. 2015

[20] Kovacic, Z., Bogdan, S. Fuzzy Controller Design – Theory and Applications. Taylor and Francis Group, USA; 2006.

[21] Castellano, G., Fanelli, A.M., Mencar, C. Design of transparent mamdani fuzzy inference systems. In: Design and Application of Hybrid Intelligent Systems. IOS, Amsterdam; 2003. pp. 468-476.

[22] Grigorie, T.L., Lungu, M., Edu, I.R., Obreja, R. Concepts for error modeling of miniature accelerometers used in inertial navigation systems. Annals of the University of Craiova, Electrical Engineering Series. 2010; 34:212-219.

[23] Grigorie, T.L., Botez, R.M. Modelling and simulation based Matlab/Simulink of a strapdown inertial navigation system' errors due to the inertial sensors. In: Kelly Bennett, editor. Matlab Applications for the Practical Engineer. InTech, Rijeka, Croatia; 2014. p. 305-338. DOI: 10.5772/57583

[24] Grigorie, T.L., Botez, R.M. Modeling and numerical simulation of an algorithm for the inertial sensors errors reduction and for the increase of the strap-down navigator redundancy degree in a low cost architecture. Transactions of the Canadian Society of the Mechanical Engineering (CSME). 2010; 34(1):1-16.

Section 4

Applications of Fuzzy in Imaging and Healthcare

Chapter 14

A Fuzzy Belief-Desire-Intention Model for Agent-Based Image Analysis

Peter Hofmann

Additional information is available at the end of the chapter

http://dx.doi.org/10.5772/67899

Abstract

Recent methods of image analysis in remote sensing lack a sufficient grade of robustness and transferability. Methods such as object-based image analysis (OBIA) achieve satisfying results on single images. However, the underlying rule sets for OBIA are usually too complex to be directly applied on a variety of image data without any adaptations or human interactions. Thus, recent research projects investigate the potential for integrating the agent-based paradigm with OBIA. Agent-based systems are highly adaptive and therefore robust, even under varying environmental conditions. In the context of image analysis, this means that even if the image data to be analyzed varies slightly (e.g., due to seasonal effects, different locations, atmospheric conditions, or even a slightly different sensor), agent-based methods allow to autonomously adapt existing analysis rules or segmentation results according to changing imaging situations. The basis for individual software agents' behavior is a so-called believe-desire-intention (BDI) model. Basically, the BDI describes for each individual agent its goal(s), its assumed current situation, and some action rules potentially supporting each agent to achieve its goals. The chapter introduces a believe-desire-intention (BDI) model based on fuzzy rules in the context of agent-based image analysis, which extends the classic OBIA paradigm by the agent-based paradigm.

Keywords: agent-based image analysis, fuzzy believe-desire-intention model, object-based image analysis, fuzzy control system, remote sensing

1. Introduction

Analyzing remote sensing data is strongly bound to methods of image processing and image analysis. In contrast to other imaging techniques, remote sensing as per definition is a method to acquire information about the earth's surface by detecting and analyzing its reflected or emitted electromagnetic radiation and without being in direct contact with it. Besides radiation of the visual spectrum also infrared (optical data) and microwave radiation (RADAR) is used to produce remote sensing images. The remote sensing instruments can be carried by space crafts (usually satellites) or airborne vehicles (airplanes, drones, etc.). In order to gather geo-information from remote sensing data, the produced images need to be analyzed, that is, preprocessed and classified. In this context, image classification means to assign pixels to meaningful object classes of the earth's surface, whereas the delineated and classified objects are finally stored in a geographic information system (GIS) as polygons, lines, or points (vector model). With the continuous increase of remote sensing images' spatial (and radiometric) resolution, image analysis in remote sensing became more and more complex. Until the late 1990s, the majority of remote sensing data was analyzed by means of classification methods taking into account the radiation stored in each single pixel. Meanwhile, rather sophisticated methods of pattern analysis, artificial intelligence, and computer vision are applied.

With the advent of very high resolution (VHR) satellite images, classic methods of image classification, as described above, failed since most of the objects of interest are represented in VHR data by numerous and spectrally inhomogeneous pixels. Moreover, properties such as shape, texture, and spatial context play a rather important role when identifying and delineating objects of interest in this kind of data [1–3]. Thus, more or less simultaneously with the advent of VHR satellite images, object-based image analysis (OBIA) has meanwhile evolved as a new and accepted paradigm for analyzing remote sensing data. In contrast to pixel-based analysis methods, OBIA deals with image objects as the building blocks for analysis. Image objects are initially generated by an arbitrary image segmentation followed by an initial classification of these image segments. The feature space for classification can be very high dimensional describing color, shape, texture, or the spatial context properties for the desired object classes. Numerous classifiers can be applied ranging from simple thresholding to Support Vector Machines (SVM), Bayesian Network Classifiers (BNC), and Artificial Neural Networks (ANN). Fuzzy set assignments are possible, too. For the latter the definition of fuzzy sets and the underlying fuzzy classification rules should reflect the ontology of the desired object classes [4–6]. Thus, the typical workflow of OBIA starts with an initial segmentation (and classification) as described above; followed by an iterative process of knowledge-based segmentation and classification improvement. The latter reflects the so-called task-ontology describing the necessary expert knowledge on image processing, which can be stored as an OBIA rule set and reapplied. However, the more precisely and reliebly remote sensing data has to be analyzed, the more complex are the methods and rule sets. The latter finally reduces the rule sets' transferability. In order to achieve acceptable results for different image data more manual interaction such as changing single rules or manually correcting object borders and/or class assignments is necessary [7, 8]. Consequently, in order to benefit from OBIA's advantages for numerous images or even whole image archives, intelligent, and flexible solutions are necessary, which are capable to autonomously adapt to image variability.

2. Agent-based and multiagent systems

Agent-based and multiagent systems (MAS) recently show a variety of applications: they range from simulation of complex systems such as social systems [9] and ecosystems [10, 11] to the automation and optimization of complex production systems such as industrial processes [12–14]. In software development, meanwhile the agent-oriented paradigm has evolved as a new paradigm that extends the classic object-oriented approach. Simply spoken, the general differences are: (1) objects behave rather passive than agents, that is, objects only change once they receive an appropriate signal while agents behave proactive and collaborative and (2) agents can be mobile while objects are static. Thus, agents have (individual) goals they intend to achieve; they have sensors and effectors that enable them to become aware about their current status and to interact with their environment. Agents can decide autonomously about their potential next action. The environment agents are interacting with can be of arbitrary complexity ranging from other (human) agents to factory plants, sports fields to traffic situations, etc. When embedded in (collaborative) MAS individual agents often have different roles but common goals. All these abilities allow software agents and MAS to react flexible but robust on unforeseeable changes of their environment.

Since each agent needs to have certain situation awareness, each agent must be capable to appraise its current situation. That is, evaluating the grade of goal achievement and the acting opportunities supporting the agent's goal achievement. This sort of situation awareness is commonly known as the belief-desire-intention (BDI) model [15–19]. Simply spoken, the BDI model allows an agent to analyze its current situation and to choose from a predefined list of plans the most promising one in order to achieve its goals. It is obvious that for the design of software agents and MAS ontologies are required, which are capable to formally describe an agent's or MAS' environment and which allow individual agents to infer their current situation. Further, ontologies are necessary to describe an agent's goals and to infer the most promising action for goal achievement [20]. Casali et al. [21] extend the classic BDI model to a graded BDI (g-BDI) model, which allows each agent to express its preferences among its acting opportunities, while Shen et al. [22] introduce an agent fuzzy decision making (AFDM) approach, which extends the classic BDI model "by making decisions based on quantified fuzzy judgment." Zarandi and Ahmadpour [23] present a fuzzy agent-based expert system for the steel making process that uses a fuzzy described knowledge base.

2.1. Agent-based image analysis

Although meanwhile matured, the agent-based paradigm and potential applications based on it are not very widespread in the image analysis community, yet. The most applications can be found in the field of image coregistration and image fusion [24–26]. In the field of object detection and delineation reported applications are still rare [27–31]—even more in the field of remote sensing image analysis [32–35].

2.2. Software agents and multiagent systems in OBIA

As already mentioned, in order to fully exploit the advantages of OBIA there is a strong need for more robustness and transferability of methods. The limiting factors are the rule sets' complexity and the unpredictable variations of the image objects' appearance in remote sensing data. At this background recently the integration of the agent-based paradigm with OBIA are investigated in order to improve its degree of automation, its robustness, and its transferability. MAS seem to have the potential to overcome OBIA's obstacles [33, 36–38]. Especially their ability to react flexibly on environmental perturbations, which is given in the remote sensing domain by varying illuminations, seasons, locations, sensors, and atmospheric conditions, are promising aspects to be investigated. Consequently, Hofmann et al. [38] developed a conceptual framework for agent-based image analysis (ABIA), which suggests two principle ways of integration: (1) adapting already existing OBIA rule sets(e.g., thresholds of single rules) by means of a MAS built by respective rule set adaptation agents (RSAAs) and (2) evolving OBIA image objects to image object agents (IOAs). In the first approach different RSAAs adapt a rule set's rules in order to improve its classification results. As constraint, adaptations must not violate the underlying ontology of the original rule set. The latter is controlled by one or more control agents (CAs), which also give feedback whether a to-be-defined minimum classification quality has been achieved after rule set adaptation (**Figure 1**).

In the second approach after initial segmentation and (fuzzy) classification IOAs build a hierarchical network of IOAs. Each IOA intends to become a best possible member of its assigned class (goal, desire). To achieve this goal, every IOA can change its shape by a number of predefined methods (effectors, intention). Further, every IOA is aware about its topological situation and can communicate within other IOAs in the hierarchical net. The underlying ontology for this approach is given by the (fuzzy) class descriptions (**Figure 2**).

The ABIA framework has been implemented in a typical environment for agent-based modeling (REPAST [39]) as well as in a typical OBIA environment (eCognition) [40], realizing

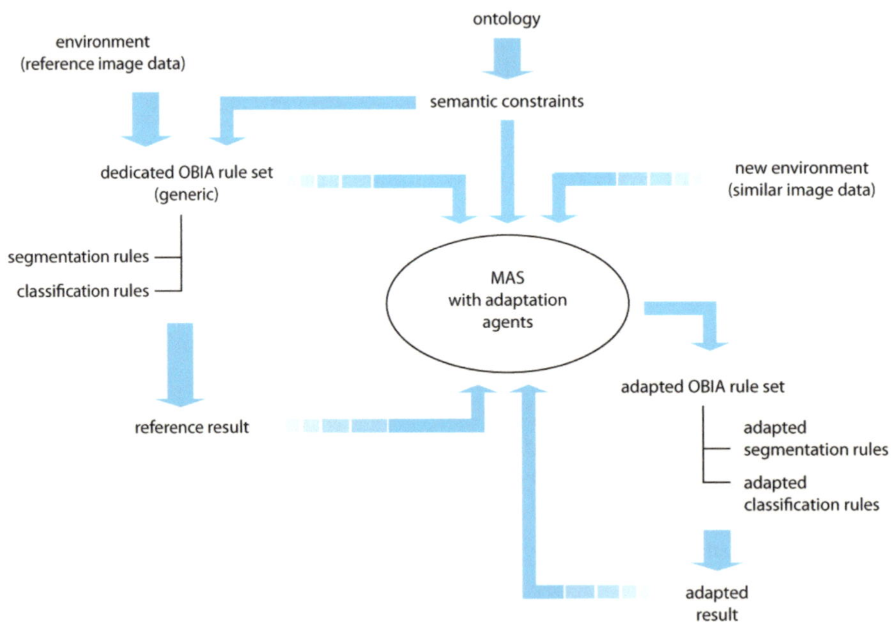

Figure 1. Principle workflow for OBIA rule set adaptation by means of a MAS built by RSAAs (Source: [38]).

the IOA approach. In both implementations the real world objects to be delineated and identified were described as fuzzy sets based on an appropriate ontology. As test scene a very high resolution digital orthophoto (0.08m) has been used together with an appropriate digital surface model (DSM) together with the calculated slope and curvature (slope of slope) per pixel (**Figure 3**).

The rule set was intentionally designed to delineate buildings in that particular scene following the ontology as outlined in **Figure 4**.

However, if the rule set is applied without any further adaptations it creates a rather over-segmented image—which would be a typical OBIA use case after reapplying a given rule set on similar images. The BDI model to solve this problem has been implemented as a hybrid model. That is, the class definitions were implemented as fuzzy sets, whereas the next action's decision rules were designed crisp. For the latter simply all three provided actions were virtually executed for each IOA. Every IOA then opted for that action that improved its class membership to "building" at best. In the example demonstrated the final result has been achieved after 17 iteration steps already (**Figure 5**).

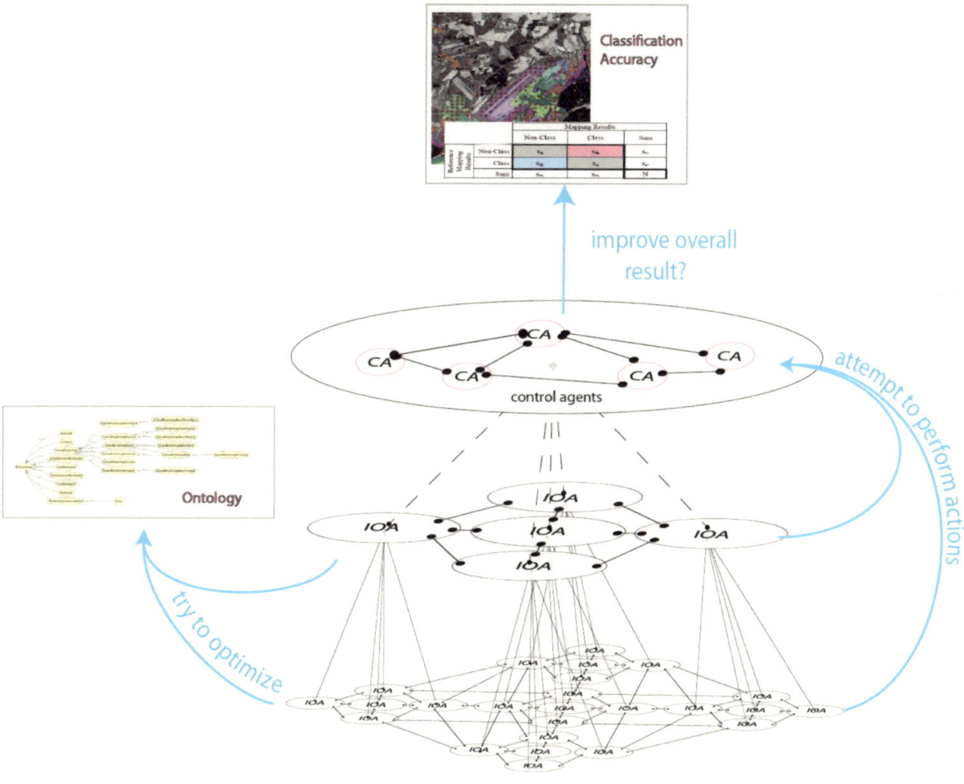

Figure 2. Scheme of a MAS built up by hierarchically organized IOAs and CAs (Source: [38]).

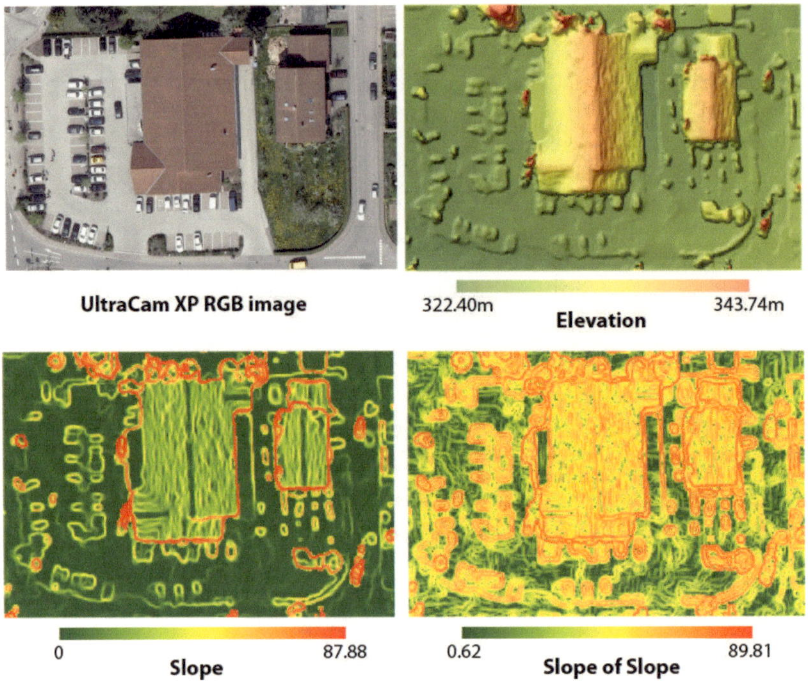

Figure 3. Used image and DSM data for first implementation of the ABIA framework (Source: [40]).

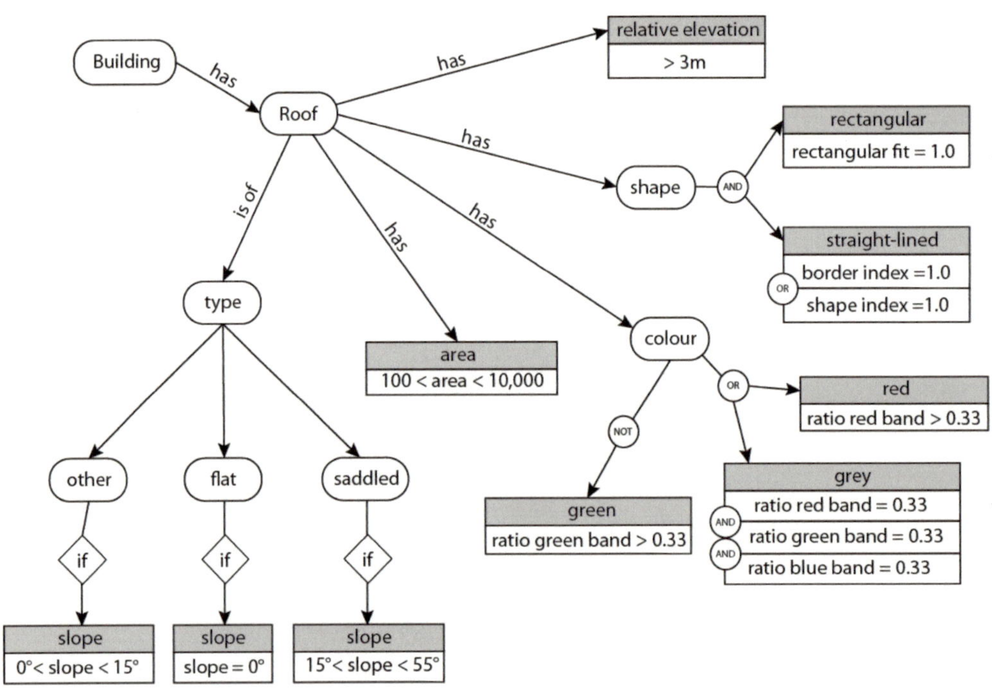

Figure 4. Ontology describing buildings and their appearance in the given data (Source: [38]).

http://dx.doi.org/10.5772/67899

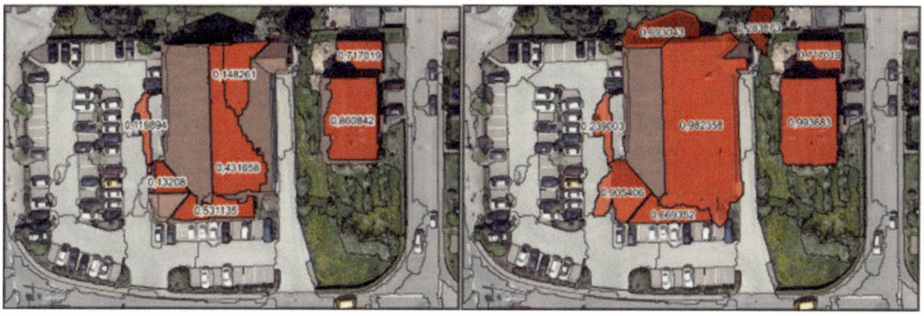

Figure 5. Segmentation and classification result before (left) and after (right) applying the agent-based optimization approach according to the ABIA framework. Numbers indicate the membership degree to "building" (Source: [38]).

3. A fuzzy believe-desire-intention model for agent-based image analysis

Based on the already achieved results with the relative simple crisp BDI model it has been investigated here whether the IOAs' intentions could also be expressed in fuzzy manner and whether this is of advantage compared to the above described BDI model. For this purpose the existing rule set, which has been developed in a software environment dedicated for OBIA (here: the commercial software eCognition), has been extended by the necessary components. In a standard fuzzy control-system control is given implicitly through the membership functions, e.g., "the colder room temperature the more open heater's valve." In this particular case the used software only allows fuzzy sets (alias classes) to be described in fuzzy manner. Consequently, the agents' acting opportunities had to be expressed as fuzzy sets whereas the membership degree to an "action-class" can be interpreted as the "intention-degree" or the willingness of an agent to perform that particular action. Another difference of the fuzzy BDI model developed here is that it only evaluates the current situation. That is, there is no virtual test for each potential action if and how it would improve an agent's situation (here: its class membership).

3.1. Components of the fuzzy BDI model

In order to fuzzify each IOA's believe its current status quo after segmentation and classification is analyzed. That is, each IOA had to be enabled not only to know its current class membership degree (degree of goal/desire achievement), but also for each classification rule the degree of its contribution to that particular classification result. The latter would allow each IOA to select an action that improves one of those conditions.

3.1.1. Fuzzy beliefs

For this purpose, the conditions that build a "building" as described in the ontology were separated into the four categories: color conditions, DSM conditions, and shape conditions. The class "roof"—which represents buildings—consequently was now described through these aggregated conditions whereas the property "Classification Value of ..." expresses the membership degree μ for this particular class or the degree of fulfillment (DOF) of that particular condition. Similarly, the condition classes were further deconstructed to classes describing the DOF's property conditions (feature-2-class) or operator conditions (operator-2-class). **Table 1** shows the cascaded classification scheme. With this decomposition every IOA now can determine the grade each of the fuzzy classification rules or rule groups (color conditions, shape conditions, or context conditions) it contributes to its final class assignment result (= grade of goal achievement). In the example displayed in **Figure 6** the **red** outlined IOA fulfills the criteria for "roof" only by 0.432, whereas the color conditions are fully fulfilled, the shape conditions are fulfilled by 0.555 and the DSM conditions are fulfilled by 0.432.The shape conditions are not fulfilled, because the IOA's rectangular fit value of 0.76 leads to a grade of fulfillment for that condition of 0.555 only.

Class	Properties	Operator	Membership function type	Function values	
				Lower bound	Upper bound
Roof	Classification value of 'Color conditions'	AND	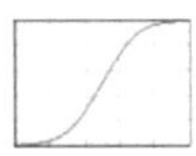	0.00	1.00
	Classification value of 'DSM conditions'			0.00	1.00
	Classification value of 'Shape conditions'		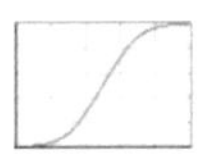	0.00	1.00
Color conditions	Classification value of 'feature-2-class "Ratio green"'	AND	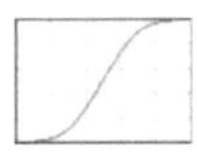	0.00	1.00
	Classification value of 'operator-2-class "Red-OR-Grey"'			0.00	1.00
DSM conditions	Classification value of 'feature-2-class "Mean difference to neighbors DSM"'	AND	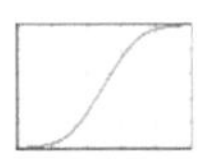	0.00	1.00
	Classification value of 'feature-2-class "Mean Slope-of-Slope"'		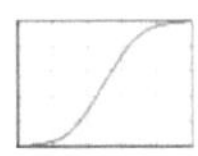	0.00	1.00
Shape conditions	Classification value of 'feature-2-class "Lower Bound of Area"'	AND		0.00	1.00
	Classification value of 'feature-2-class "Upper Bound of Area"'		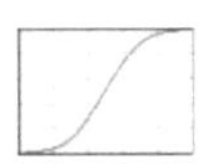	0.00	1.00
	Classification value of 'feature-2-class "Rectangular fit"'		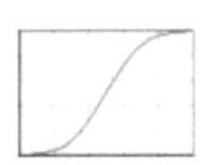	0.00	1.00
	Classification value of 'operator-2-class "Border Index-OR-Shape Index"'		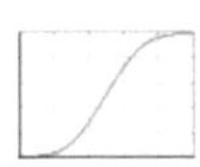	0.00	1.00

Class	Properties	Operator	Membership function type	Function values	
				Lower bound	Upper bound
Feature-2-class "Ratio green"	Ratio green	-	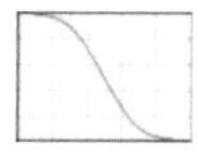	0.32	0.34
Feature-2-class "Ratio red"	Ratio red	-	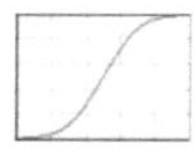	0.32	0.34
Feature-2-class "grey"	Ratio green	AND	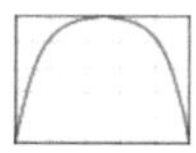	0.31	0.35
	Ratio red		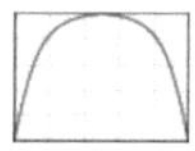	0.31	0.35
	Ratio blue		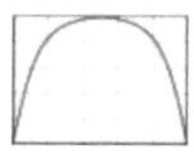	0.31	0.35
Operator-2-class "Red-OR-Grey"	Classification value of 'feature-2-class "grey"'	OR	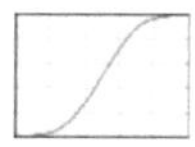	0.00	1.00
	Classification value of 'feature-2-class "Ratio red"'		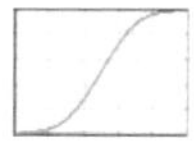	0.00	1.00
Feature-2-class "Mean difference to neighbours DSM"	Mean difference to neighbors DSM	-	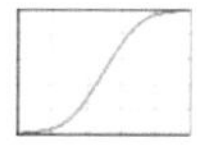	0.00	3.50
Feature-2-class "Mean Slope-of-Slope"	Mean Slope-of-Slope	-	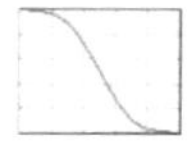	88.00	90.00
Feature-2-class "Lower Bound of Area"	Area	-	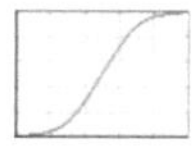	10.00	100.00
Feature-2-class "Upper Bound of Area"	Area	-	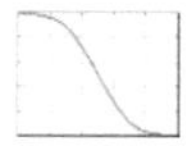	7391.60	7391.70
Feature-2-class "Border Index"	Border Index	-	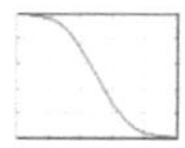	1.20	3.00
Feature-2-class "Rectangular fit"	Rectangular fit	-	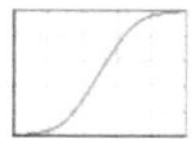	0.50	1.00
Feature-2-class "Shape Index"	Shape Index	-	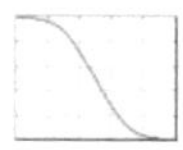	1.30	3.00
Operator-2-class "Border Index-OR-Shape Index"	Classification value of 'feature-2-class "Border Index"'	OR	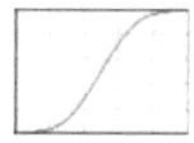	0.00	1.00
	Classification value of 'feature-2-class "Shape Index"'		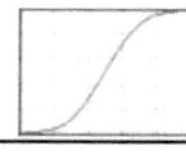	0.00	1.00

Table 1. Cascaded fuzzy classification scheme for the class "roof" as described in **Figure 4.**

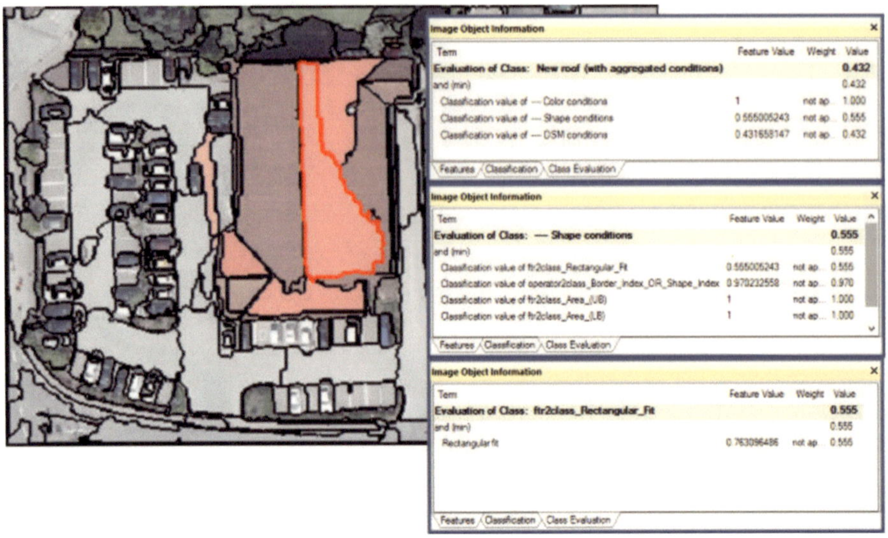

Figure 6. Evaluation of classification conditions of an IOA.

Similarly, the IOA's DSM conditions show a DOF of 0.432 because the mean difference to neighbors in the DSM is at 1.631 translating to a DOF for that feature of 0.432.

3.1.2. Fuzzy intentions

Based on these fuzzy beliefs an intended next action can be determined in fuzzy manner based on appropriate fuzzy decision rules—again expressed as fuzzy sets. In the example present the following possible actions were implemented: "grow," "shrink," "smooth," "merge," and "do nothing." While the latter action is obvious and applies only if an IOA has achieved its goal already, the former actions point to procedures eCognition offers and can be easily exchanged or adapted if necessary. In this particular example the actions translate to:

- "grow": grow the IOA of concern by one pixel into neighbor IOAs, which are unclassified.
- "shrink": shrink the IOA of concern by one pixel.
- "merge": merge the IOA of concern with its unclassified neighbors.
- "smooth": perform a grow-and-shrink sequence by one pixel each starting with shrink.

The appropriate intentions are defined as the fuzzy sets: "want grow," "want shrink," "want smooth," "want merge," or "do nothing." The degree or the intensity an agent wants to execute one of these actions can depend on the prior determined classification conditions or on the spatial situation in which IOA is embedded in, or a combination of both. Each of the action intensities is expressed gradually, that is, through an action-membership (**Table 2**). In the example given, an IOA wants to shrink if the situation is similar to that of "want grow;" it wants the more shrink the closer its size is at the upper bound of the area-rule for "building" ("Upper bound of Area"). An IOA wants to do nothing the more it fulfills already

http://dx.doi.org/10.5772/67899

Intention	Properties	Membership function	Operator	Function values	
				Lower bound	Upper bound
Do nothing	Classification value of "Color conditions"	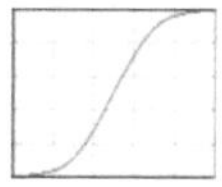	AND	0.00	1.00
	Classification value of "DSM conditions"	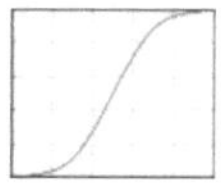		0.00	1.00
	Classification value of "Shape conditions"	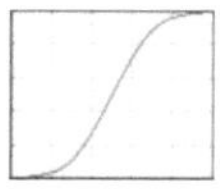		0.00	1.00
Want grow	Border contrast blue	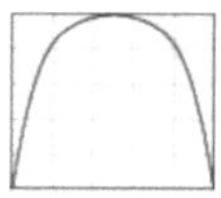	AND	Minimum/10	25.00
	Border contrast green	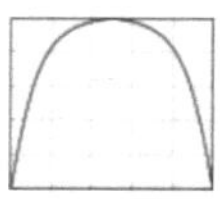		Minimum/10	25.00
	Border contrast red	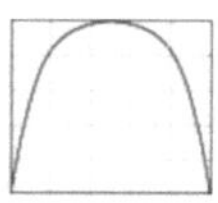		Minimum/10	25.00
	Border contrast DSM	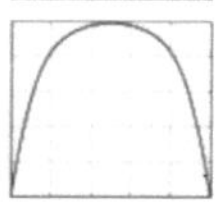		–2.00	2.00
Want merge	Area	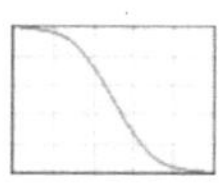	AND	10.00	2500.00
	Classification value of "Similarity to neighbors (DSM)"	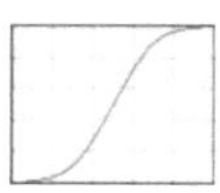		0.00	1.00
	Classification value of "Similarity to neighbors (Color)"	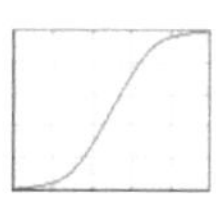		0.00	1.00
Want shrink	Want grow		AND		
	Feature-2-class "Upper bound of Area"				
Want smooth	NOT(feature-2-class "Rectangular Fit")		AND		
	NOT(operator-2-class "Border Index OR Shape Index"				

Table 2. Definition of intentions as fuzzy sets.

the "building" criteria. It wants the more grow, the lower its contrast and elevation difference is at its border. Its intention to merge increases, the lower its area is and simultaneously the more its DSM and color criteria are similar with those of its neighbors. Its intention to smooth its border increases the less the shape criteria for building are fulfilled.

As displayed in **Figure 7**, the **red** outlined IOA prefers to grow. However, similar to the ambiguity of class assignments of objects, each IOA can have ambiguous intentions in terms of a favorite, a second favorite, etc. action. Further, as with fuzzy class assignments [41], a minimum intention value should be defined (sensibly not less than 0.5) below which an intended action must be seen as not clearly enough wanted and thus not further considered.

In the example given in **Figure 7** although the IOA's most wanted action is to grow, this action is not the only one it intends to perform. Since the intention value for "want_grow" is not clearly around 1.0 the IOA seems to be not fully convinced about this action to achieve its goal. Merging (intention = 0.68) seems to be an option although it is second-best. In other words, the IOA could also be satisfied if it merges. The only thing that is clear, is, that the IOA does not want to shrink (intention = 0.0). And its willingness to smooth its border (intention = 0.0298) is even lower than that for doing nothing (intention = 0.356).

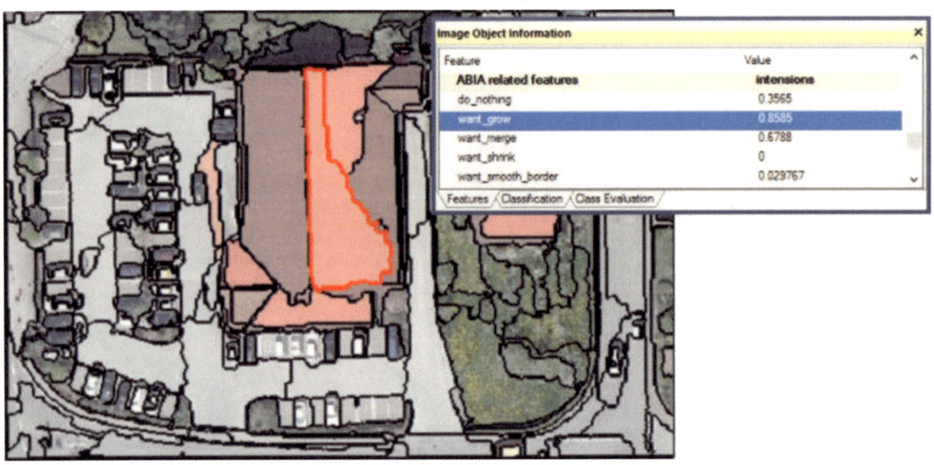

Figure 7. Grades of intentions for actions an IOA wants to perform for its improvement.

4. Conclusion

The amount of remote sensing data stored in archives is increasing continuously. At the background of an increasing demand for reliable, precise and timely geoinformation, searching these archives, and analyzing the image data stored in them urges for methods of reliable and automated image analysis. While OBIA is an accepted and highly accurate method for analyzing especially VHR remote sensing data, its robustness, and transferability, as well as degree of automation is still low. Major obstacles for automating OBIA and its methods are its sensibility against perturbations, that is, the images' and objects variability. ABIA as an extension of the OBIA paradigm has the potential to overcome these obstacles, since it has the ability to react more flexible and robust on unforeseeable perturbations. Nevertheless, research in this field is still in its beginning.

The example demonstrated in this chapter is just one aspect of this wide research field. It has been demonstrated how a fuzzy BDI model can be implemented in standard OBIA software, which is capable to allow individual IOAs to control their improvement actions within a MAS. Further research needs to be done in learning mechanisms for individual agents as well as in improved individual decision rules. Last but not the least, the performance issues are still another aspect that needs to be investigated in that field.

5. Outlook

The implemented fuzzy BDI model acts as the basis for a negotiation model that can be applied on ABIA agents: based on their individual action priorities, IOAs can negotiate their common next action(s) and therefore optimize the overall classification result of multiple but different images of the same kind as it is typical in remote sensing. While the fuzzy BDI model has been implemented for IOAs, it is also applicable in principle for rule set adaptation agents (RSAAs).

Author details

Peter Hofmann

Address all correspondence to: peter.hofmann@sbg.ac.at

Department of Geoinformatics – Z_GIS, University of Salzburg, Salzburg, Austria

References

[1] U. C. Benz, P. Hofmann, G. Willhauck, I. Lingenfelder, and M. Heynen, "Multi-resolution, object-oriented fuzzy analysis of remote sensing data for GIS-ready information," ISPRS Journal of Photogrammetry and Remote Sensing, vol. 58, pp. 239-258, 2004.

[2] T. Blaschke, M. Kelly, P. Hofmann, et al, "Geographic object-based image analysis – towards a new paradigm," ISPRS Journal of Photogrammetry and Remote Sensing, vol. 87, pp. 180-191, 2014.

[3] T. Blaschke and J. Strobl, "What's wrong with pixels? Some recent developments interfacing remote sensing and GIS," GeoBIT/GIS, vol. 6, pp. 12-17, 2001.

[4] P. Hofmann, "Detecting informal settlements from IKONOS image data using methods of object oriented image analysis-an example from Cape Town (South Africa)." In: Jürgens, C. (Ed.): Proceedings of the 2nd International Symposium on Remote Sensing of Urban Areas, Regensburg, Germany, June 22-23 2001, pp. 41-42, 2001.

[5] M. Belgiu, B. Hofer, and P. Hofmann, "Coupling formalized knowledge bases with object-based image analysis," Remote Sensing Letters, vol. 5, pp. 530-538, 2014.

[6] M. Belgiu and J. Thomas, "Ontology based interpretation of very high resolution imageries–grounding ontologies on visual interpretation keys," AGILE 2013—Leuven, pp. 14-17, 2013.

[7] P. Hofmann, "Übertragbarkeit von Methoden und Verfahren in der objektorientierten Bildanalyse-das Beispiel informelle Siedlungen," ed: Dissertation, University of Salzburg, 2005.

[8] P. Hofmann, T. Blaschke, and J. Strobl, "Quantifying the robustness of fuzzy rule sets in object-based image analysis," International Journal of Remote Sensing, vol. 32, pp. 7359-7381, 2011.

[9] A. Koch and D. Carson, "Spatial, temporal and social scaling in sparsely populated areas–geospatial mapping and simulation techniques to investigate social diversity," in Proceedings of the Geoinformatics Forum Salzburg, Offenbach, 2012, pp. 44-53.

[10] H. R. Gimblett, Integrating geographic information systems and agent-based modeling techniques for simulating social and ecological processes: Oxford University Press, Oxford, 2002.

[11] P. Göhner, Agentensysteme in der Automatisierungstechnik: Springer-Verlag, Heidelberg, 2013.

[12] W. Shen, D. H. Norrie, and J.-P. Barthès, Multi-agent systems for concurrent intelligent design and manufacturing: Taylor & Francis, London, New York, 2005.

[13] M. Wooldridge, An introduction to multiagent systems: 2nd Ed., John Wiley & Sons, Chichester, U.K., 2009.

[14] M. Wooldridge, An introduction to multiagent systems: John Wiley & Sons, 2009.

[15] M. Wooldridge and N. R. Jennings, "Intelligent agents: theory and practice," The knowledge Engineering Review, vol. 10, pp. 115-152, 1995.

[16] M. Wooldridge, "Agent-based software engineering," IEEE Proceedings-software, vol. 144, pp. 26-37, 1997.

[17] M. Georgeff, B. Pell, M. Pollack, M. Tambe, and M. Wooldridge, "The belief-desire-intention model of agency," in International Workshop on Agent Theories, Architectures, and Languages, 1998, pp. 1-10.

[18] A. S. Rao and M. P. Georgeff, "BDI agents: From theory to practice". In: Proceedings of the First International Conference on Multiagent Systems, June 12-14, 1995, San Francisco, California, USA. The MIT Press 1995 pp. 312-319.

[19] M. Bratman, Intention, plans, and practical reason, Center for the Study of Language and Information. 1987.

[20] K. Sycara and M. Paolucci, Ontologies in agent architectures, In: S. Staab, R. Studer (Eds.): Handbook on Ontologies, Springer, Berlin, Heidelberg, 2004, pp. 343-363.

[21] A. Casali, L. Godo, and C. Sierra, "A graded BDI agent model to represent and reason about preferences," Artificial Intelligence, vol. 175, pp. 1468-1478, 2011.

[22] S. Shen, G. M. O'Hare, and R. Collier, "Decision-making of BDI agents, a fuzzy approach," in Computer and Information Technology, 2004. CIT'04. The Fourth International Conference on, September 14-16, 2004, pp. 1022-1027.

[23] M. H. F. Zarandi and P. Ahmadpour, "Fuzzy agent-based expert system for steel making process," Expert Systems with Applications, vol. 36, pp. 9539-9547, 2009.

[24] J. Beyerer, M. Heizmann, J. Sander, and I. Gheta, "Bayesian methods for image fusion," In: T. Stathaki (Ed.): Image Fusion: Algorithms and Applications, Elsevier, Amsterdam, pp. 157-192, 2008.

[25] C. Elmas and Y. Sönmez, "A data fusion framework with novel hybrid algorithm for multi-agent Decision Support System for Forest Fire," Expert Systems with Applications, vol. 38, pp. 9225-9236, 2011.

[26] F. Castanedo, J. García, M. A. Patricio, and J. M. Molina, "A multi-agent architecture based on the BDI model for data fusion in visual sensor networks," Journal of Intelligent & Robotic Systems, vol. 62, pp. 299-328, 2011.

[27] E. G. P. Bovenkamp, J. Dijkstra, J. G. Bosch, and J. H. C. Reiber, "Multi-agent segmentation of IVUS images," Pattern Recognition, vol. 37, pp. 647-663, 2004.

[28] N. Richard, M. Dojat, and C. Garbay, "Automated segmentation of human brain MR images using a multi-agent approach," Artificial Intelligence in Medicine, vol. 30, pp. 153-176, 2004.

[29] V. Rodin, A. Benzinou, A. Guillaud, P. Ballet, F. Harrouet, J. Tisseau, et al., "An immune oriented multi-agent system for biological image processing," Pattern Recognition, vol. 37, pp. 631-645, 2004.

[30] K. E. Melkemi, M. Batouche, and S. Foufou, "A multiagent system approach for image segmentation using genetic algorithms and extremal optimization heuristics," Pattern Recognition Letters, vol. 27, pp. 1230-1238, 2006.

[31] J. Nolan, A. Sood, and R. Simon, "Agent-based, collaborative image processing in a distributed environment". In Proceedings of the 5th International Conference on Autonomous Agents (ACM Agents 2001), Montreal, QC, Canada, May 28 - June 01,2001, pp. 228-235.

[32] F. Samadzadegan, F. T. Mahmoudi, and T. Schenk, "An agent-based method for automatic building recognition from lidar data," Canadian Journal of Remote Sensing, vol. 36, pp. 211-223, 2010.

[33] K. Borna, A. Moore, and P. Sirguey, "A vector agent approach to extract the boundaries of real-world phenomena from satellite images," in Proceedings of Research at Locate'14 (R@Loc-2014), Canberra, Australia, 2014.

[34] K. Borna, A. Moore, and P. Sirguey, "Towards a vector agent modelling approach for remote sensing image classification," Journal of Spatial Science, vol. 59, pp. 283-296, 2014.

[35] F. Tabib Mahmoudi, F. Samadzadegan, and P. Reinartz, "Object oriented image analysis based on multi-agent recognition system," Computers & Geosciences, vol. 54, pp. 219-230, 2013.

[36] J. Liu and Y. Y. Tang, "Adaptive image segmentation with distributed behavior-based agents," IEEE Transactions on Pattern Analysis and Machine Intelligence, vol. 21, pp. 544-551, 1999.

[37] R. D. Labati, V. Piuri, and F. Scotti, "Agent-based image iris segmentation and multiple views boundary refining." In: Proceedings IEEE 3rd International Conference on Biometrics: Theory, Applications, and Systems 28-30 Sept., Washington DC, 2009, pp. 1-7.

[38] P. Hofmann, P. Lettmayer, T. Blaschke, M. Belgiu, S. Wegenkittl, R. Graf, et al., "Towards a framework for agent-based image analysis of remote-sensing data," International Journal of Image and Data Fusion, vol. 6, pp. 115-137, 2015.

[39] C. M. Macal and M. J. North, "Tutorial on agent-based modelling and simulation," Journal of Simulation, vol. 4, pp. 151-162, 2010.

[40] P. Hofmann, V. Andrejchenko, P. Lettmayer, M. Schmitzberger, M. Gruber, I. Ozan, et al., "Agent based image analysis (ABIA) - preliminary research results from an implemented framework." In: GEOBIA 2016 : Solutions and Synergies., 14 September 2016 - 16 September 2016, University of Twente Faculty of Geo-Information and Earth Observation (ITC), 2016.

[41] P. Hofmann, "Defuzzification strategies for fuzzy classifications of remote sensing data," Remote Sensing, vol. 8(6), doi:10.3390/rs8060467, 2016.

Chapter 15

ANFIS Definition of Focal Length for Zoom Lens via Fuzzy Logic Functions

Bahadır Ergün, Cumhur Sahin and Ugur Kaplan

Additional information is available at the end of the chapter

http://dx.doi.org/10.5772/67823

Abstract

The digital cameras have been effected from systematical errors which decreased metric quality of image. The digital cameras have been effected from systematical errors that decreases metric quality of image. The aim of this chapter is to explore usability of fuzzy logic on calibration of digital cameras. Therefore, a 145-pointed planar test field has been prepared in the laboratories of Department of Geodesy and Photogrammetric Engineering at the Gebze Technical University. The test field has been imaged from five points of view with the digital camera Nikon Coolpix-E8700 within maximum (71.2 mm) and minimum (8.9 mm) focal length. The input-output data have been determined from 10 calibration images obtained for fuzzy logic process. These data have also been used and formed for the space resection process. Adaptive neuro-fuzzy inference system (ANFIS) functions have been used for fuzzy process at MATLAB 7.0, and the results of these two distinct methods have been compared. Finally, the most convenient (least squares average error) or the most useful ANFIS *"Trimf, trapmf, gbellmf, gaussmf, gauss2mf, pimf, dsigmf and psigmf"* functions are determined and compared for space resection method for the conventional bundle adjustment process.

Keywords: ANFIS, zoom lens calibration, focal distance, MATLAB

1. Introduction

Fuzzy inference system (FIS) is a process of mapping from given inputs to outputs by using the theory of fuzzy sets [1]. FIS derives an output by using an inference engine, which is based on a form of **IF-THEN** rules. There are two well-established types of FIS [2–4]. While Mamdani FIS uses the technique of defuzzification of a fuzzy output, Sugeno FIS uses weighted average to compute the crisp output [4]. In fact, adaptive neuro-fuzzy inference system (ANFIS) structure is composed of a representation as a network structure, which has the neural learning ability of Sugeno-type fuzzy systems. This network is made of a combination of nodes, which are placed as layers in order to perform specific functions [5, 6].

Among the neuro-fuzzy models most used nowadays, the adaptive neuro-fuzzy inference system (ANFIS), which was proposed in 1992 by J.S Roger in his Ph.D. thesis, must be highlighted [7–9]. ANFIS adapts its parameters according to training data by using the hybrid learning algorithm. The algorithm consists of the gradient descent for tuning the non-linear antecedent parameters and the least square for tuning the linear consequent parameters [1]. Most of the success of ANFIS comes from its implementation in the MATLAB Fuzzy Logic Toolbox, with an excellent graphical interface personally developed by J.S. Roger in collaboration with N. Gulley, incorporating also diverse fuzzy logic pattern classification algorithms for the definition and dimensioning of the input membership functions (mf) [9, 10].

In the literature, there are several examples of the ANFIS, which enable it to achieve great success in a wide range of scientific applications. The advantageous features of an ANFIS include easy implementation, fast and accurate learning, strong generalization abilities, excellent explanation facilities through fuzzy rules, and easy to incorporate both linguistic and numeric knowledge for problem solving [11]. ANFIS models have recently gained much popularity not only for calibrating non-linear relationships because they offer more advantages over conventional modeling techniques, which include the ability to handle large amounts of noisy data from dynamic and non-linear systems, especially where the underlying physical relationships are not fully understood, but also for solving linear systems which include the interpolation modeling such as time series. Literature shows that lots of ANFIS and FIS methods were proposed for the determination of the uncertainty of pattern recognition, image matching, and three-dimensional (3D) position definition studies in computer vision applications. For instance, color recognition ANFIS model study for robust vision system has worked in Ref. [12]. Another application for ANFIS method is suggested for appropriate calibration method of stereo camera system used for non-intrusive distance measurement in [13]. Another example for ANFIS method in remote sensing study, functions for serving and prediction from satellite images of Prionace Glauca for pattern recognition in (Backpropagation network, RBF, functional separability network) and the neuro-diffuse networks (ANFIS) [14]. In Ref. [15], a new hybrid method of performing eye-to-hand coordination and manipulation to produce a working robot named COERSU. The method is an optimized combination of two neuro-fuzzy approaches developed by the authors: direct fuzzy servoing and fuzzy correction. The fuzzy methods are tuned by the adaptive neuro-fuzzy inference system (ANFIS). Human action recognition is an important research area in the field of computer vision having a great number of real-world applications. In Ref. [16], ANFIS controller method has been suggested for path tracking, a virtual field strategy for obstacle avoidance and path planning, and multiple sensors (an ultrasonic array, a thermal sensor, and a video streaming system) for obtaining information about the environment. In Ref. [17], development of a virtual robot tele-operation platform based on hand gesture recognition has been evolved from visual information by ANFIS and support vector machines (SVM). In Ref. [18], a multi-view action recognition framework that extracts human silhouette clues from different cameras was presented with fuzzy rule-based system for analyzing scene dynamics and interpreting human behaviors. The interaction tool has been located in 3D space then the 3D model has obtained by means of a structured light system which is calibrated using only the vanishing points extracted from a simple planar surface. Then, an immersive interaction technique was used to manipulate the 3D model using a fuzzy technique with the advantages of a low memory usage, real-time operation, and low positioning errors as compared to classical solutions in [19]. Quantitative analysis of the error in the reconstruction of a 3D scene, which has been captured with synthetic aperture integral imaging system, has been worked for two-dimensional (2D) images which was captured within unknown camera parameters with adaptive neuro-fuzzy inference system (ANFIS) in [20].

In this chapter, an ANFIS approach has been developed for the detection of effective camera focal distance parameters with additional parameters, which help the measurement of image coordinates that provide the basis for three-dimensional modeling obtained from two-dimensional images, which are similar to [19, 20] studies. Thus, the photogrammetric studies, which benefit from self-calibration conventional model, will be more usable with different cameras and different time points.

The main logic of this chapter is to process the selected radial distances via ANFIS functions and to obtain a focus distance value for each of the images. It is seen that ANFIS functions of optical distortion law can be validated by its effect on data sets. This also proves that ANFIS functions validate a physical reality and describe it in a highly reliable way. As a result of the research, it is seen that the approach of fuzzy logic can be used for the calibration process of digital cameras. Thus, it is concluded that it would be beneficial to use and study the fuzzy logic approach more in photogrammetric applications [21].

Information about fuzzy systems is described in the second part of this chapter. Third part of this chapter is the experimental application part. Test field has been photographed from five different angles with Nikon Coolpix-e8700 digital camera with maximum (71.2 mm) and minimum (8.9 mm) focal distances in the application. Input and output data are determined with the resulting 10 images; space resection approach is created and studied by using ANFIS functions on MATLAB 7.0 software. Trimf, trapmf, gbellmf, gaussmf, gauss2mf, pimf, dsigmf, and psigmf functions in Fuzzy Interface System (FIS), which are used for creating a relation between fuzzy sets, under anfisedit menu in fuzzy logic tools on MATLAB 7.0 software, are studied in terms of space resection approach, and the most appropriate functions are determined for the detection of focal distance. The fourth part presents the results and their interpretations.

2. Fuzzy logic and main principles

Fuzzy logic is based on the logic of clustering and determination of membership degrees depending of this clustering. Membership degrees generate rule-based work systematic which constitutes the rules of fuzzy systems. Fuzzy cluster sections which are placed in the inlets and outlets of the fuzzy rules express an approximation for each. In this respect, all expressions like "approximately 3," "nearly 9," "over 5 and approximate" always express a fuzzy number. Each of these approximations corresponds to a fuzzy cluster. It cannot be possible to solve mathematical operations with these fuzzy numbers. The operations are made with fuzzy numbers by defining some restrictions. In order for a fuzzy number to exist, fuzzy cluster of this needs to have an interval of normal, convex, limited support and at section of membership degree, closed and finite. For the fuzzy numbers to be normal, membership degree of at least one of the real numbers in the fuzzy expression must be 1.

Generally, two fuzzy numbers as a triangle and trapezoid are in question. Mathematical expression of a triangle fuzzy number showed with a fuzzy cluster is given as Eq. (1):

$$mf(x) = mf(x; a, b, c) = \begin{cases} (x-a)/(b-a) & a \le x \le b \\ (c-x)/(c-b) & b \le x \le c \\ 0 & x > c \quad \text{or} \quad x < a \end{cases} \tag{1}$$

In the expression of mf(x;a,b,c), a and c show the lower and upper limit values relatively and b shows the single number with full membership. Similarly, trapezoid fuzzy numbers are expressed with four full whole numbers as a, b, c, d. Here a and d show lower and upper limit values of trapezoid fuzzy numbers; b and c show the limits of the cluster of the trapezoid numbers whose membership degree is full between these two numbers. Mathematical indication of trapezoid fuzzy number is like Eq. (2):

$$mf(x) = mf(x; a, b, c, d) = \begin{cases} (x-a)/(b-a) & a \le x \le b \\ 1 & b \le x < c \\ (d-x)/(d-c) & c < x \le d \\ 0 & x > d \quad \text{or} \quad x < a \end{cases} \tag{2}$$

If we pay attention, when b = c, trapezoid fuzzy number is transformed into triangle fuzzy number. Graphical indication of these fuzzy numbers is as shown in **Figure 1**.

Generally, there are two reasons for the researchers to use fuzzy systems:

- Since the real-world incidents are very complicated, these incidents are not possible to be taken under control by being defined with specific equations. As a natural result of this,

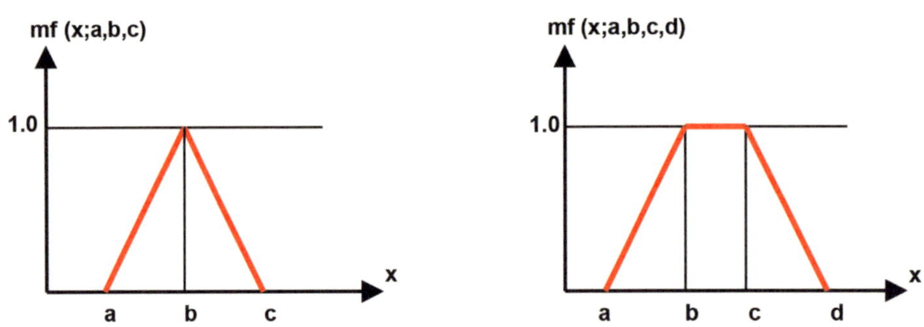

Figure 1. Fuzzy numbers: (a) triangle and (b) trapezoid.

the researcher always prefers applying to the methods that have approximate solubility even if not certain. As Einstein said, if it can be said that real-life incidents can be indicated with mathematical equations, either the accuracy of the equation cannot be mentioned or if the result that mathematical equations exactly depicts the reality is taken, then real-life incidents cannot be mentioned. So, the solutions are approximate to a certain extent in all studies conducted. Otherwise, many non-linear equations must be solved simultaneously which is known to lead to chaotic unspecified problems according to current information.

- All theories and equations in engineering express the real world approximately. Even though many real systems are not linear, every effort is given in order to accept the linearity in examination of these with classic methods. In order for the verbal data presented by people to be converted into numerical data and calculated by being perceived by computers and algorithms, fuzzy systems are needed.

Almost all of mathematical, stochastic, or conceptual systems so far are made up of three units given in **Figure 2**.

These are input unit, a transition function which converts this input to output and is called system behavior and output unit. In all of the units here, numerical data are processed. The difference of the fuzzy systems from the conventional systems is that the system behavior section is divided into two and there are four connected units between each other as shown in **Figure 3**.

Each of the units here has tasks, which are different but can be related to each other:

- **General information base unit:** It includes input variants to which the incident to be examined is subjected and all information about these. Because the it includes both numerical and/or textual data, the database is called as 'general database'.

Figure 2. Conventional system.

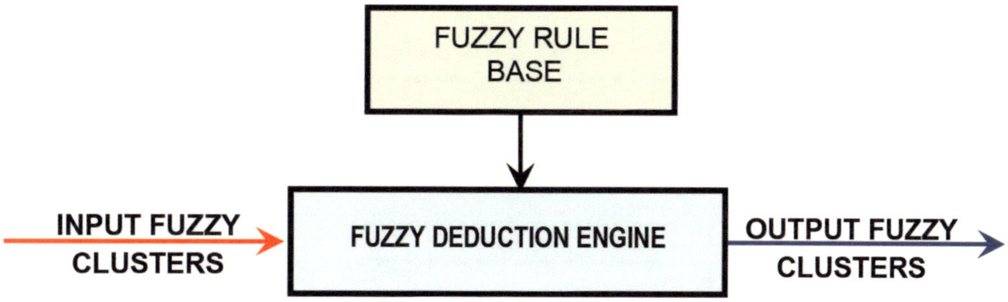

Figure 3. General fuzzy system.

- **Fuzzy rule base unit:** It includes all the rules which can be written in the type of logical **IF-THEN**, connecting the inputs in data base to output variants. In writing of these rules, the whole interval connections (fuzzy cluster) which can only be between input and output data are considered. Thus, each rule connects one piece of input space to the output space logically. All of these contexts form the base rule.
- **Fuzzy deduction engine unit:** It is a mechanism including the operations, which enable a system to behave as having one output by bringing the relations formed between input and output fuzzy clusters in a fuzzy rule base unit together. This engine serves for determination of what kind of output the whole system will give between the inputs by bringing deductions of each rule together.
- **Output unit:** It indicates the output values obtained as a result of the interaction of information and fuzzy rule bases with deduction engine.

Figure 3 represents a general fuzzy system. The point to be paid attention to here is information and outputs in the input; in other words, databases are fuzzy values. So each unit in **Figure 3** is made up of fuzzy clusters entirely. The most significant deficiency of basic fuzzy system is that numerical database cannot enter into such a fuzzy system and the outputs are not numerical, so they cannot be directly used in engineering designs.

In order to certainly eliminate the deficiencies of the general fuzzy system is extented to a new system which is proposed in Refs. [3, 22] This is called Takagi-Sugeno-Kank (TSK) fuzzy system. For each inputs and the outputs obtained as a result of operation of fuzzy rule and deduction engine are in the way of a function of inputs which are from input database. So like the input variants in rule base, it was thought that these variants are reflected to rule result section after THEN word as a linear function of these variants [21]. According to this, the rule is presented in the following part.

IF the speed of your car is high, **THEN** stepping on the gas force can be expressed as $\mathbf{y = ax}$. For example, in the case that three input variants exist (x_1, x_2, and x_3), in one of fuzzy system, y as output variant generally; **IF** x_1 is little and x_3 is wide, **THEN** it can be expressed as $\mathbf{y = a_0 + a_1x_1 + a_2x_2 + a_3x_3}$. Result sections of all rules are made up of a polynomial linear equation. Since in the fuzzy system which has such a structure output variants are not used in the deduction of fuzzy clustering, instead of fuzzy deduction unit in **Figure 3**, weighted deduction calculation unit comes as mainly the membership degrees calculated from the input section of each rule. This system is shown in **Figure 4**.

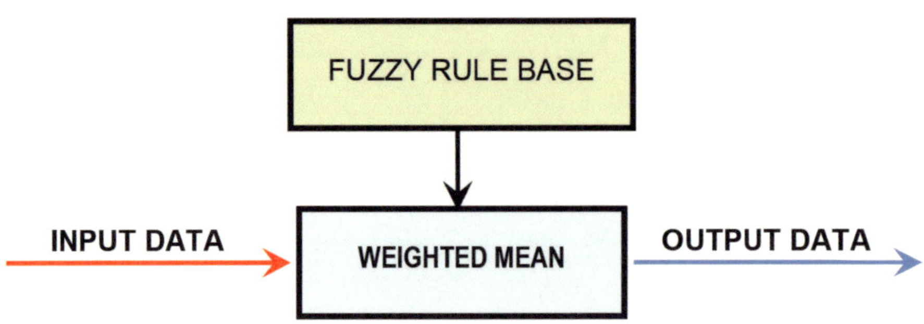

Figure 4. TSK fuzzy system.

http://dx.doi.org/10.5772/67823

Actually in such a fuzzy system, output space is represented as a function of the inputs for being a rule valid in each sub-space. Even in the case that the output surface is not linear, with TSK approach they are understood as modeled in the way of plain pieces of input variant kind on sub-spaces.

However, TSK fuzzy systems have deficiencies as there is a mathematical relation after **IF** part; output sections of the rules cannot model the verbal data given by human, and output sections of all rules, which are possible to be written between input and output variants cannot be written because they are fuzzy. In order to eliminate these deficiencies, fuzzy system, which is respectively blur, and clarification units are used in input and output units given in **Figure 5** [21].

Here, fuzzy rule base and deduction engine in a general fuzzy system remain as they are. In the case that the inputs are numerical, there are blurring agent unit, which is for blurring them with a process, and clarifying agent unit, which is for digitizing the fuzzy outputs. Blurring (fuzzification) and clarifying (defuzzification) mean, respectively, blurring the input numbers and digitizing the fuzzy numbers.

Among the main properties of fuzzy systems, multiple inputs, rule base, and conversion as single output by processing with deduction engine come as the most important issues. In some special situations, outputs may be more than one. Fuzzy system determines the behavior of the system by converting variants forming the inputs to output variant in a way of non-linear. By this way, it is possible to take the examined system under control in order to reach desired results by subjecting information base to non-linear conversions. Owing to the fuzzy systems, it becomes possible to process image, make guess based on time series, solve control problems, and perform applications on communication issues. Other than this, fuzzy systems can be used in many areas like engineering, medicine, sociology, psychology, business management, expert systems, artificial intelligence, signal processing, transportation, and signalization.

Exact numerical values are needed in practical applications particularly for sizing in engineering designs. In such cases, fuzzy information must be clarified in order for the required answers to be given by benefiting from information obtained or given as fuzzy. All processes made for conversion of fuzzy information to exact results are called clarification.

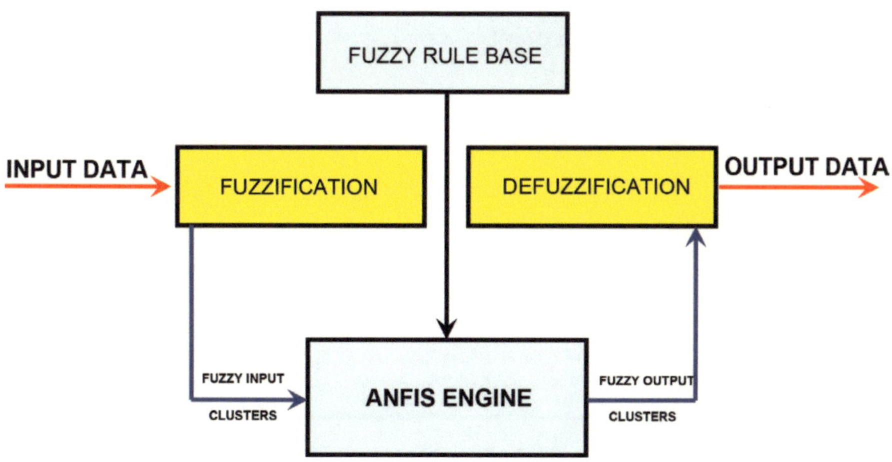

Figure 5. Fuzzy system with blurring-clarifying unit.

There are many clarification methods [23]. A few of them are presented below. In these processes, **z** indicates fuzzy deduction cluster, indicates components, and z* indicates clarified value.

- **The highest membership principle:** Another name of this process is height method. For this method to be used, fuzzy deduction clusters with peaks are needed. This method can be indicated with **Figure 6**.
- **Centroid method:** Another name of this method is center of gravity method. It could be the most common one of clarification methods. Mathematical expression is as shown in Eq. (3):

$$z^* = \frac{\int mf_C(z).zdz}{\int mf_C(z)dz} \tag{3}$$

Clarification is performed according to the center of gravity of fuzzy expression.

- **Weighted mean method:** In order for this method to be used, symmetric membership function (mf) is needed. Mathematical explanation is as shown in Eq. (4):

$$z^* = \frac{\sum mf_C(\overline{z}).\overline{z}}{\sum mf_C(\overline{z})} \tag{4}$$

Here Σ sign indicates algebraic sum. Graphical indication of this method is shown in **Figure 7**. Here, each membership function of fuzzy cluster constituting the output is multiplied with the highest membership degree value it has and their weighted mean is calculated. For instance, weighted mean (clarified value) of two fuzzy clusters in **Figure 7** is calculated as shown in Eq. (5).

$$z^* = \frac{a(0.5)+b(0.9)}{0.5+0.9} \tag{5}$$

Since this clarifying process is valid only for symmetric membership functions, a and b values are the mean of the figures they represent.

- **Averagely the highest membership degree method:** This method is also known as the mean of the highest ones. In this respect, it is very close to the first clarification principle. However, the position of the highest membership may not be singular. This means that the one having the highest membership degree in the membership function, mf (z) = 1, may exist in a plain section instead of a point.

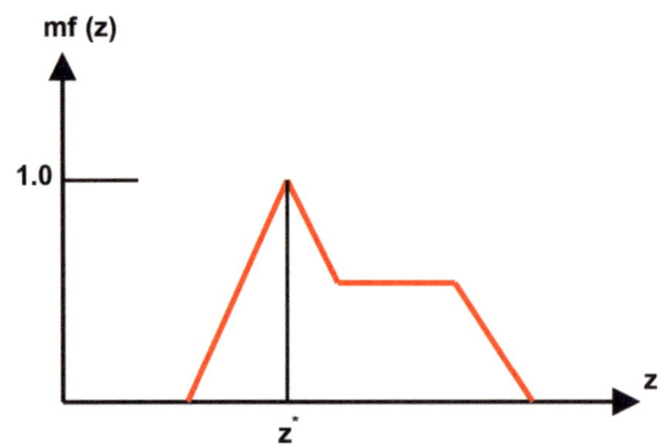

Figure 6. Clarification of the highest membership degree.

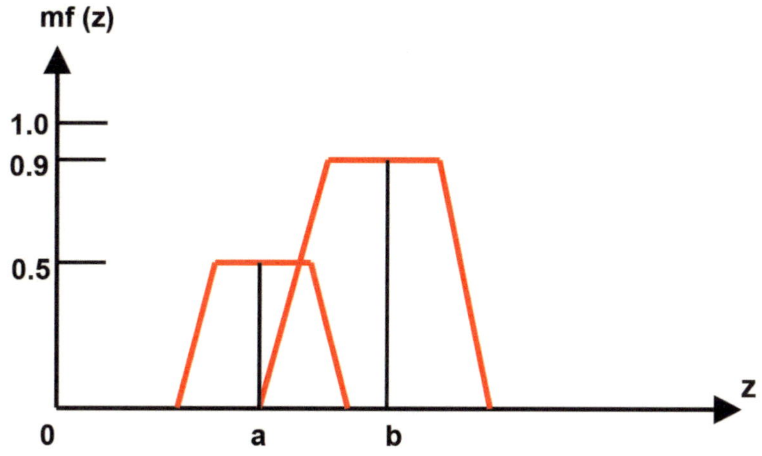

Figure 7. Clarification with weighted mean method.

Clarified value according to this method in graphical indication in **Figure 8** is as shown in Eq. (6):

$$z^* = \frac{a+b}{2} \tag{6}$$

- **Central management of the sums:** It is the fastest one among the clarification methods used. In this method, instead of combination of two fuzzy clusters, their algebraic sum is used. A deficiency of this is overlapping sections entering into the sum twice. Clarified value can be calculated as Eq. (7).

$$z^* = \frac{\int_z z \sum_{k=1}^{n} \mathrm{mf}_C(z)\, dz}{\int_z \sum_{k=1}^{n} \mathrm{mf}_C(z)\, dz} \tag{7}$$

In a way, this calculation form resembles weighted mean clarification. However, in the method of center of sums, weights are the areas of related membership functions. In the average weights method, this is a membership degree.

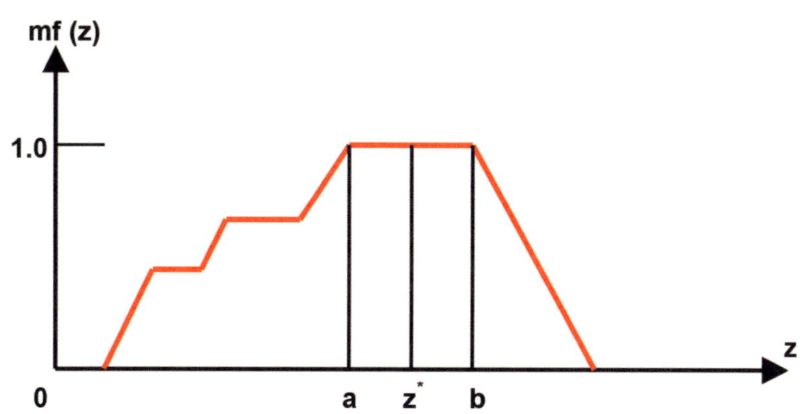

Figure 8. Averagely the highest membership clarification.

- **Center of the largest area:** If the outlet fuzzy cluster includes at least two convex fuzzy sub-clusters, center of gravity of the one with the biggest area of convex fuzzy clusters is used in clarification process. Mathematical indication of this method is as shown in Eq. (8):

$$z^* = \frac{\int mf_{Cm}(z)zdz}{\int mf_{Cm}(z)\,dz} \tag{8}$$

Here, $ü_{ebç}(z)$ indicates the sub-region where the convex fuzzy cluster with the biggest area dominates. This circumstance is used when all fuzzy deduction clusters are not convex; however, in the case that all deductions are convex, the result is the same with the one obtained with z^* centroid method.

- **The highest first and last membership degree method:** In this method, the lowest (or the highest) fuzzy cluster value with the highest membership degree in the fuzzy cluster that comes up as a combination of all outputs is selected. The equations below are valid for z^* value to be given by calculations. At first, the biggest height y_{eb} is determined in the combination of fuzzy cluster deduction, B.

$$y_{eb}(B) = \max\left[m\,f_B(z)\right] \tag{9}$$

Then, first highest value, z^* is found. Another way of this is to find the last highest fuzzy cluster value instead of the first one.

ANFIS name, which means fuzzy deduction system based on open adaptive networks or adaptive neural fuzzy detection system, is made of the first letters of adaptive network-based fuzzy interface system or adaptive neuro-fuzzy interface system. This network is the combination of the nodes, which are placed as layers for performing specific functions [5, 21]. ANFIS consists of six layers. This system is presented in **Figure 9**.

Node functions of each layers on ANFIS structure and the operations of layers are briefly explained [23, 24]:

- **First layer:** This is called the input layer. Input signals, which are taken from each of the nodes on this layer, are transferred into other layers.
- **Second layer:** This is called the fuzzification layer. Jang's ANFIS model uses the current Bell activation function as the membership function in order to divide input values into fuzzy sets. Here, output of each node is formed of input values, and membership degrees related to membership function and membership degrees, which are obtained from the second layer are shown as $\mu_{Aj}(x)$ and $\mu_{Bj}(y)$.
- **Third layer:** This is the rule layer. Each node in this layer expresses the rules and number of Sugeno fuzzy logic deduction system. Output of each rule node μ_i is the multiplication of membership degrees coming from second layer, and obtaining μ_i values are such as (*j*=1,2) and (*i*=1,…,n);

$$y_i^3 = \prod i = \mu_{A_j}(x)\, x\, \mu_{Bj}(y) = \mu_i \tag{10}$$

Here, y_i^3 represents the output values of the third layer and n represents the number of nodes in this layer.

http://dx.doi.org/10.5772/67823

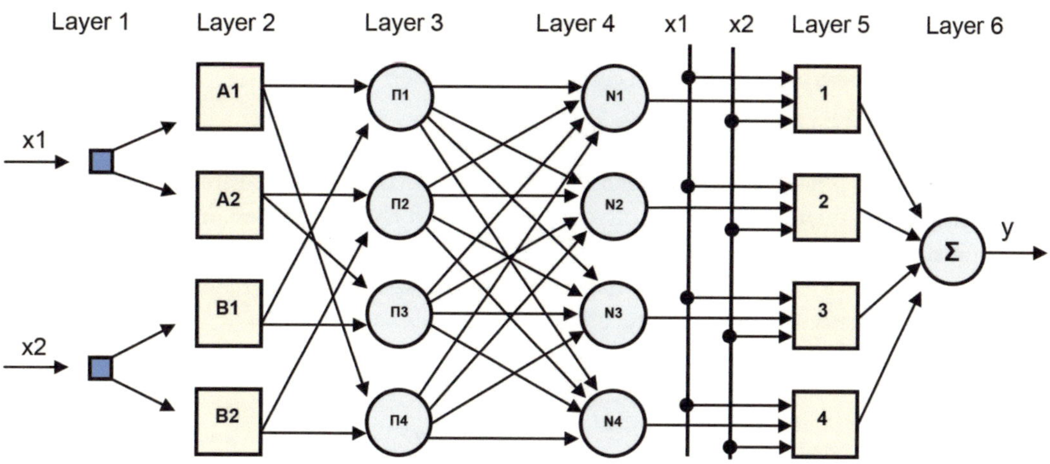

Figure 9. ANFIS network system.

- **Fourth layer:** This is the normalization layer. Each node in this layer assumes all the nodes coming from rule layer as the input value, and it calculates the normalized ignition level of each rule.

Calculation of normalized ignition level i μ is as shown in Eq. (11):

$$y_{i^4} = N_i = \frac{\mu_i}{\sum_{i=1}^{n} \mu_i} = \bar{\mu}, \; (i = 1, n) \tag{11}$$

- **Fifth layer:** This is the debugging layer. In this layer, weighted result values of a rule, which is given in each node, are calculated. Output value of i node in fifth layer is as shown in Eq. (12):

$$y_{i^5} = \bar{\mu}_i \left[p_i x_1 + q_i x_2 + r_i\right], \; (i = 1, n) \tag{12}$$

Here, (p_i, q_i, r_i) variables are result parameter set for i rule.

- **Sixth layer:** This is the sum layer. There is only one node in this layer, and it is tagged with $\sum$. Here, output values of each node in the fifth layer are added to each other, and a real value of ANFIS system is obtained. System output value y can be calculated with Eq. (13):

$$y = \sum_{i=1}^{n} {}^{-}\left[p_i x_1 + q_i x_2 + r_i\right] \tag{13}$$

3. Application

MATLAB is a software development instrument designed for technical calculations and the solutions of mathematical problems. MATLAB, which is the abbreviation of the words "MATrix LABoratory," works using matrixes as understood from its name or, in other words, using arrays. MATLAB, which is estimated to be used by over 500,000 academician, researcher, and students, is also described as the most advanced technical and scientific problem solving and application development instrument of the computer world with many interfaces it includes. MATLAB, particularly used in the analysis of the systems in engineering area, can perform the operations of data analysis and examination, visibility and image processing, generating algorithm prototype, modeling and simulation, programming and application development [25].

Application study consists of two basic steps. These are:

- Geometric camera calibration with conventional method
- Focal length calculation with ANFIS functions

In the application phase, geometric camera calibration study has been performed with bundle block adjustment method by using conventional method as the first step. In this study made by using zoom lens, camera calibration study has been preferred in both minimum and maximum focal lengths for Nikon Coolpix-E8700 camera whose focal length has been selected. The image of this camera is shown in **Figure 10**.

Figure 10. Nikon coolpix-E8700.

Images of test fields on which there are 145 points have been taken with Nikon Coolpix-E8700 digital camera from five angles in the situations of maximum (71.2 mm) and minimum (8.9 mm) focal length of the objective in the Photogrammetry Laboratory of Department of Geodesy and Photogrammetric Engineering of Gebze Technical University. This process is shown in **Figure 11**.

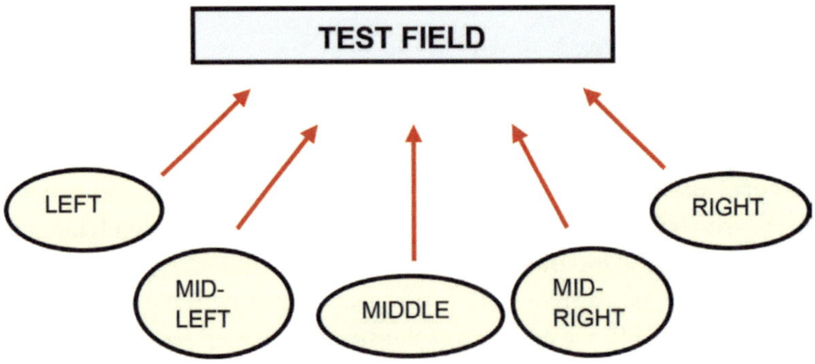

Figure 11. Test field shooting plan.

http://dx.doi.org/10.5772/67823

Coordinates of 145 points on test field were determined locally, and these values have been regarded as field coordinate (**ANNEX-1**). After that, 10 images obtained have been subjected to the evaluation in Topcon Pl-3000 software, and calibration parameters (inner orientation factors and distortion parameters) have been determined in situations of maximum focal length (71.2 mm) and minimum focal length (8.9 mm) of digital camera. These steps have been explained in **Figures 11** and **12** for both focal distances.

1	1.0000000e-002	1.0000000e-002	1.0000000e-002
2	9.5000000e-002	1.0000000e-002	1.0000000e-002
3	1.8000000e-001	1.0000000e-002	1.0000000e-002
4	2.6500000e-001	1.0000000e-002	1.0000000e-002
5	3.5000000e-001	1.0000000e-002	1.0000000e-002
6	4.3500000e-001	1.0000000e-002	1.0000000e-002
7	5.2000000e-001	1.0000000e-002	1.0000000e-002
8	6.0500000e-001	1.0000000e-002	1.0000000e-002
9	6.9000000e-001	1.0000000e-002	1.0000000e-002
10	7.7500000e-001	1.0000000e-002	1.0000000e-002
11	8.6000000e-001	1.0000000e-002	1.0000000e-002
12	9.4500000e-001	1.0000000e-002	1.0000000e-002
13	1.0300000e+000	1.0000000e-002	1.0000000e-002
14	1.1150000e+000	1.0000000e-002	1.0000000e-002
15	1.2000000e+000	1.0000000e-002	1.0000000e-002
16	9.5000000e-002	1.0000000e-002	1.0000000e-002
17	9.5000000e-002	9.5000000e-002	1.0000000e-002
18	9.5000000e-002	1.8000000e-001	1.0000000e-002
19	9.5000000e-002	2.6500000e-001	1.0000000e-002
20	9.5000000e-002	3.5000000e-001	1.0000000e-002
21	9.5000000e-002	4.3500000e-001	1.0000000e-002
22	9.5000000e-002	5.2000000e-001	1.0000000e-002
23	9.5000000e-002	6.0500000e-001	1.0000000e-002
24	9.5000000e-002	6.9000000e-001	1.0000000e-002
25	9.5000000e-002	7.7500000e-001	1.0000000e-002
26	9.5000000e-002	8.6000000e-001	1.0000000e-002
27	9.5000000e-002	9.4500000e-001	1.0000000e-002
28	9.5000000e-002	1.0300000e+000	1.0000000e-002
29	9.5000000e-002	1.1150000e+000	1.0000000e-002
30	9.5000000e-002	1.2000000e+000	1.0000000e-002
31	1.8000000e-001	1.0000000e-002	1.0000000e-002
32	1.8000000e-001	9.5000000e-002	1.0000000e-002
33	1.8000000e-001	1.8000000e-001	1.0000000e-002
34	1.8000000e-001	3.5000000e-001	1.0000000e-002
73	4.3500000e-001	6.0500000e-001	1.0000000e-002
74	4.3500000e-001	7.7500000e-001	1.0000000e-002
75	4.3500000e-001	8.6000000e-001	1.0000000e-002
76	4.3500000e-001	9.4500000e-001	1.0000000e-002
77	4.3500000e-001	1.0300000e+000	1.0000000e-002
78	4.3500000e-001	1.1150000e+000	1.0000000e-002
79	4.3500000e-001	1.2000000e+000	1.0000000e-002
80	5.2000000e-001	1.0000000e-002	1.0000000e-002
81	5.2000000e-001	9.5000000e-002	1.0000000e-002
82	5.2000000e-001	1.8000000e-001	1.0000000e-002
83	5.2000000e-001	3.5000000e-001	1.0000000e-002
84	5.2000000e-001	4.3500000e-001	1.0000000e-002
85	5.2000000e-001	5.2000000e-001	1.0000000e-002
86	5.2000000e-001	6.9000000e-001	1.0000000e-002
87	5.2000000e-001	7.7500000e-001	1.0000000e-002
88	5.2000000e-001	8.6000000e-001	1.0000000e-002
89	5.2000000e-001	1.0300000e+000	1.0000000e-002
90	5.2000000e-001	1.1150000e+000	1.0000000e-002
91	5.2000000e-001	1.2000000e+000	1.0000000e-002
92	6.0500000e-001	1.0000000e-002	1.0000000e-002
93	6.0500000e-001	9.5000000e-002	1.0000000e-002
94	6.0500000e-001	2.6500000e-001	1.0000000e-002
95	6.0500000e-001	4.3500000e-001	1.0000000e-002
96	6.0500000e-001	5.2000000e-001	1.0000000e-002
97	6.0500000e-001	6.0500000e-001	1.0000000e-002
98	6.0500000e-001	6.9000000e-001	1.0000000e-002
99	6.0500000e-001	7.7500000e-001	1.0000000e-002
100	6.0500000e-001	9.4500000e-001	1.0000000e-002
101	6.0500000e-001	1.1150000e+000	1.0000000e-002
102	6.0500000e-001	1.2000000e+000	1.0000000e-002
103	6.9000000e-001	1.0000000e-002	1.0000000e-002
104	6.9000000e-001	9.5000000e-002	1.0000000e-002
105	6.9000000e-001	1.8000000e-001	1.0000000e-002
106	6.9000000e-001	3.5000000e-001	1.0000000e-002

35 1.8000000e-001 4.3500000e-001 1.0000000e-002
36 1.8000000e-001 5.2000000e-001 1.0000000e-002
37 1.8000000e-001 6.0500000e-001 1.0000000e-002
38 1.8000000e-001 6.9000000e-001 1.0000000e-002
39 1.8000000e-001 7.7500000e-001 1.0000000e-002
40 1.8000000e-001 8.6000000e-001 1.0000000e-002
41 1.8000000e-001 1.0300000e+000 1.0000000e-002
42 1.8000000e-001 1.1150000e+000 1.0000000e-002
43 1.8000000e-001 1.2000000e+000 1.0000000e-002
44 2.6500000e-001 1.0000000e-002 1.0000000e-002
45 2.6500000e-001 9.5000000e-002 1.0000000e-002
46 2.6500000e-001 2.6500000e-001 1.0000000e-002
47 2.6500000e-001 4.3500000e-001 1.0000000e-002
48 2.6500000e-001 5.2000000e-001 1.0000000e-002
49 2.6500000e-001 6.0500000e-001 1.0000000e-002
50 2.6500000e-001 6.9000000e-001 1.0000000e-002
51 2.6500000e-001 7.7500000e-001 1.0000000e-002
52 2.6500000e-001 9.4500000e-001 1.0000000e-002
53 2.6500000e-001 1.1150000e+000 1.0000000e-002
54 2.6500000e-001 1.2000000e+000 1.0000000e-002
55 3.5000000e-001 1.0000000e-002 1.0000000e-002
56 3.5000000e-001 9.5000000e-002 1.0000000e-002
57 3.5000000e-001 1.8000000e-001 1.0000000e-002
58 3.5000000e-001 3.5000000e-001 1.0000000e-002
59 3.5000000e-001 4.3500000e-001 1.0000000e-002
60 3.5000000e-001 5.2000000e-001 1.0000000e-002
61 3.5000000e-001 6.9000000e-001 1.0000000e-002
62 3.5000000e-001 7.7500000e-001 1.0000000e-002
63 3.5000000e-001 8.6000000e-001 1.0000000e-002
64 3.5000000e-001 1.0300000e+000 1.0000000e-002
65 3.5000000e-001 1.1150000e+000 1.0000000e-002
66 3.5000000e-001 1.2000000e+000 1.0000000e-002
67 4.3500000e-001 1.0000000e-002 1.0000000e-002
68 4.3500000e-001 9.5000000e-002 1.0000000e-002
69 4.3500000e-001 1.8000000e-001 1.0000000e-002
70 4.3500000e-001 2.6500000e-001 1.0000000e-002
71 4.3500000e-001 3.5000000e-001 1.0000000e-002
72 4.3500000e-001 4.3500000e-001 1.0000000e-002

107 6.9000000e-001 4.3500000e-001 1.0000000e-002
108 6.9000000e-001 5.2000000e-001 1.0000000e-002
109 6.9000000e-001 6.0500000e-001 1.0000000e-002
110 6.9000000e-001 6.9000000e-001 1.0000000e-002
111 6.9000000e-001 7.7500000e-001 1.0000000e-002
112 6.9000000e-001 8.6000000e-001 1.0000000e-002
113 6.9000000e-001 1.0300000e+000 1.0000000e-002
114 6.9000000e-001 1.1150000e+000 1.0000000e-002
115 6.9000000e-001 1.2000000e+000 1.0000000e-002
116 7.7500000e-001 1.0000000e-002 1.0000000e-002
117 7.7500000e-001 9.5000000e-002 1.0000000e-002
118 7.7500000e-001 1.8000000e-001 1.0000000e-002
119 7.7500000e-001 2.6500000e-001 1.0000000e-002
120 7.7500000e-001 3.5000000e-001 1.0000000e-002
121 7.7500000e-001 4.3500000e-001 1.0000000e-002
122 7.7500000e-001 5.2000000e-001 1.0000000e-002
123 7.7500000e-001 6.0500000e-001 1.0000000e-002
124 7.7500000e-001 6.9000000e-001 1.0000000e-002
125 7.7500000e-001 7.7500000e-001 1.0000000e-002
126 7.7500000e-001 8.6000000e-001 1.0000000e-002
127 7.7500000e-001 9.4500000e-001 1.0000000e-002
128 7.7500000e-001 1.0300000e+000 1.0000000e-002
129 7.7500000e-001 1.1150000e+000 1.0000000e-002
130 7.7500000e-001 1.2000000e+000 1.0000000e-002
131 8.6000000e-001 1.0000000e-002 1.0000000e-002
132 8.6000000e-001 9.5000000e-002 1.0000000e-002
133 8.6000000e-001 1.8000000e-001 1.0000000e-002
134 8.6000000e-001 2.6500000e-001 1.0000000e-002
135 8.6000000e-001 3.5000000e-001 1.0000000e-002
136 8.6000000e-001 4.3500000e-001 1.0000000e-002
137 8.6000000e-001 5.2000000e-001 1.0000000e-002
138 8.6000000e-001 6.0500000e-001 1.0000000e-002
139 8.6000000e-001 6.9000000e-001 1.0000000e-002
140 8.6000000e-001 7.7500000e-001 1.0000000e-002
141 8.6000000e-001 8.6000000e-001 1.0000000e-002
142 8.6000000e-001 9.4500000e-001 1.0000000e-002
143 8.6000000e-001 1.0300000e+000 1.0000000e-002
144 8.6000000e-001 1.1150000e+000 1.0000000e-002
145 8.6000000e-001 1.2000000e+000 1.0000000e-002

ANNEX-1

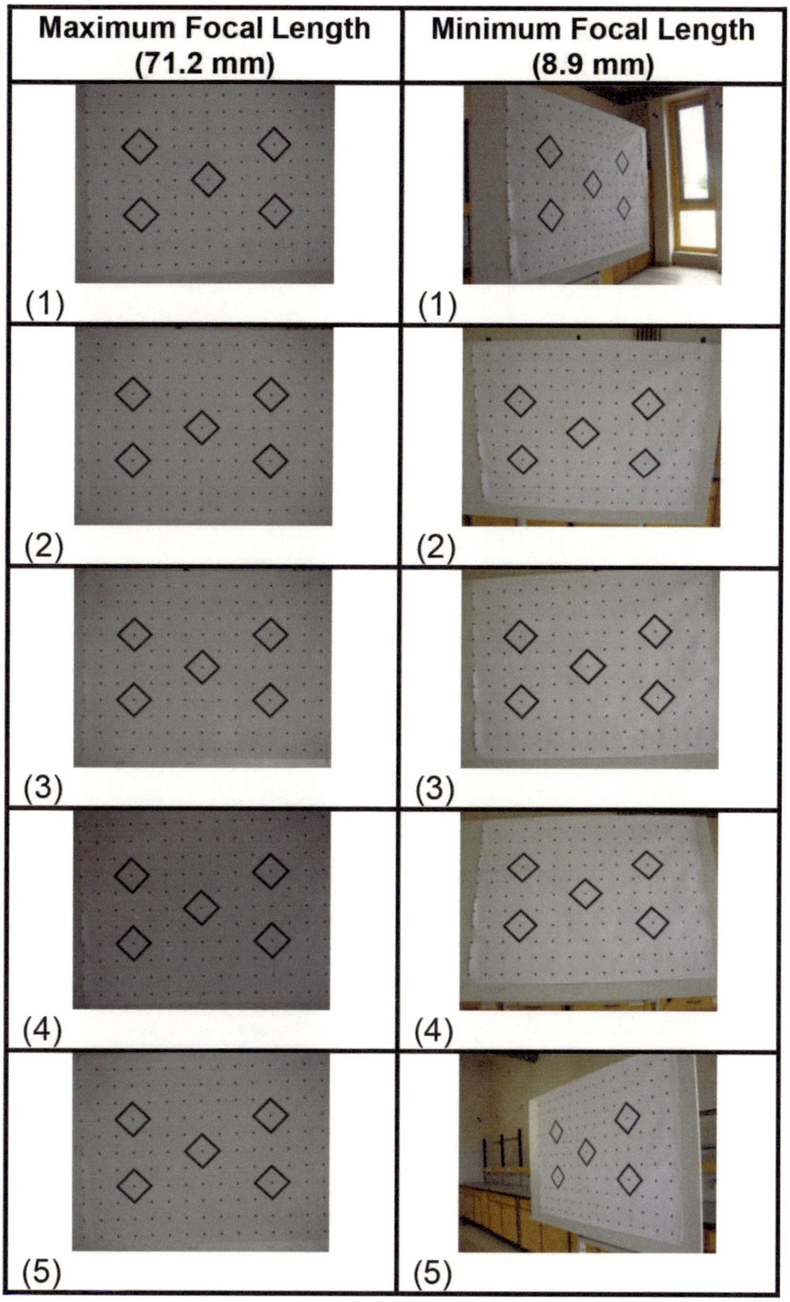

Figure 12. Camera calibration test field.

In **Figures 13** and **14**, focal length obtained with bundle block adjustment (conventional method), inner orientation parameters, and radial distortion parameters were summarized from Topcon PI-3000 software interface.

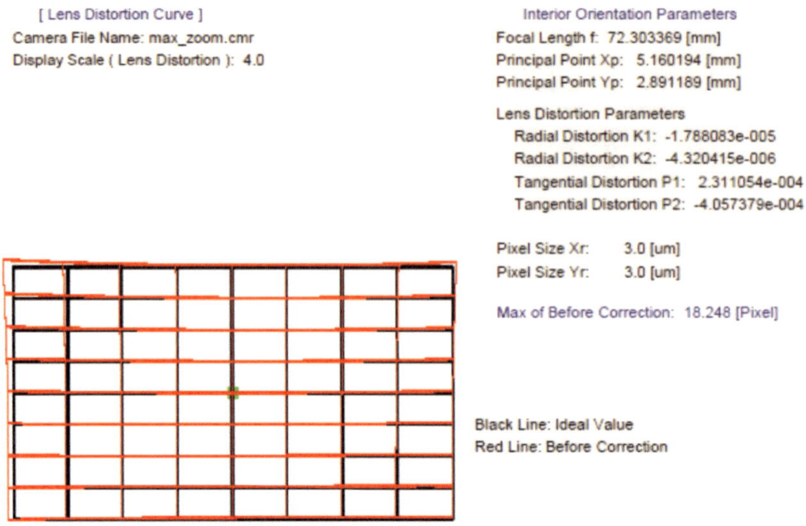

Figure 13. The one obtained by Topcon PI-3000 program in maximum focal length.

As the second step of the application, image coordinates measured over images have been indicated radially with Eq. (14) for maximum focal length and minimum focal length values of digital camera.

$$c_k = \frac{r^l}{\tan \tau} \tag{14}$$

c_k: calibrated focal length

r^1: radial distance

τ: distortion angle

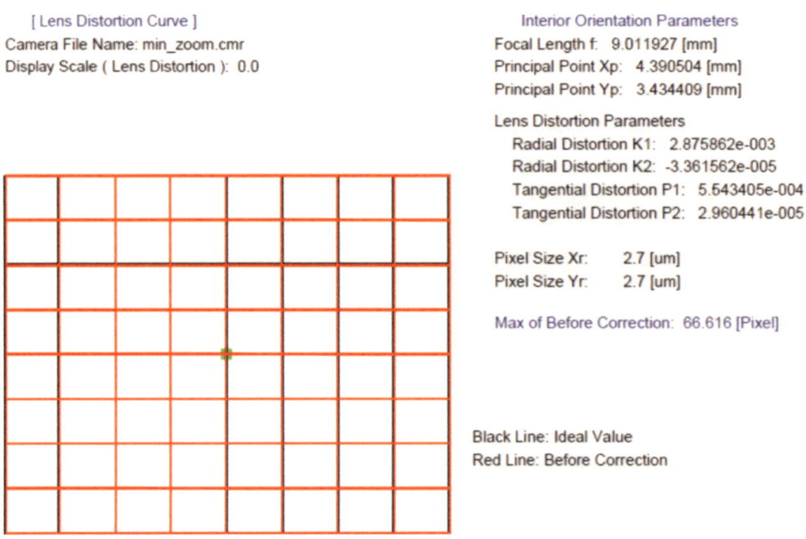

Figure 14. The one obtained by Topcon PI-3000 program in minimum focal length.

http://dx.doi.org/10.5772/67823

Twenty-three points have been selected and arranged in the format of training data (smalldata.dat) and remaining 122 points havc been arranged in the format of test data (bigdata.dat) as shown in **Figure 15**. ANFIS data set which are used in fuzzification and defuzzification have been composed from equation process with radial distance and angle. This process has been repeated for the images taken from 10 different angles for each zoom distance. ANFIS data set has been composed from equation process with radial distance and angles.

In **Figure 15**, first and fourth columns indicate x values of image coordinates, second and fifth columns indicate y values, and third and sixth columns indicate focal lengths calculated radially.

After ANFIS data set has been generated, an ANFIS application stated below has been performed by using MATLAB software. The basic logic in this application is radial distances selected as input is processed with the help of ANFIS functions, and as a result, a focal length value for each image has been obtained. Algorithm figure determined for this method is indicated below. Here, it is seen that training algorithm determined for generation of ANFIS artificial neural network operates iteratively. Artificial neural network structure shaped with training data described as training data and consisting of 23 points gives result with the last iteration where 122 points are also included for getting the exact result. Thus, a calculation, which includes again the same quantity of measurement (145 radial values) corresponding to focal length distance, is determined with conventional method. For this reason, a logical measurement number equality has been enabled in comparison of ANFIS results with conventional method. In brief, the measurement values in the first and second steps are in the same quantity. This process has been explained in **Figure 16**.

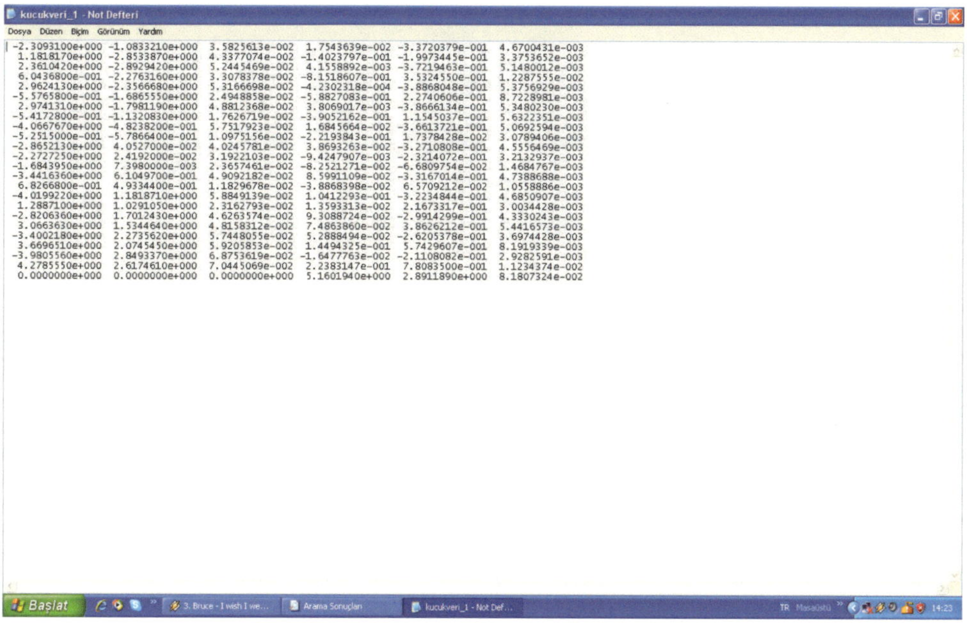
kucukveri_1 - Not Defteri
Dosya Düzen Biçim Görünüm Yardım

-2.3093100e+000	-1.0833210e+000	3.5825613e-002	1.7543639e-002	-3.3720379e-001	4.6700431e-003
1.1818170e+000	-2.8533870e+000	4.3377074e-002	-1.4023797e-001	-1.9973445e-001	3.3753652e-003
2.3610420e+000	-2.8929420e+000	5.2445469e-002	4.1558892e-003	-3.7219463e-001	5.1480012e-003
6.0436800e-001	-2.2763160e+000	3.3078378e-002	-8.1518607e-001	3.5324550e-001	1.2287555e-002
2.9624130e+000	-2.3566680e+000	5.3166698e-002	-4.2302318e-004	-3.8868048e-001	5.3756929e-003
-5.5765800e-001	-1.6865550e+000	2.4948858e-002	-5.8827083e-001	2.2740606e-001	8.7228981e-003
2.9741310e+000	-1.7981190e+000	4.8812368e-002	3.8069017e-003	-3.8666134e-001	5.3480230e-003
-5.4172800e-001	-1.1320830e+000	1.7626719e-002	-3.9052162e-001	1.1545037e-001	5.6322351e-003
-4.0667670e+000	-4.8238200e-001	5.7517923e-002	1.6845664e-002	-3.6613721e-001	5.0692594e-003
-5.2515000e-001	-5.7866400e-001	1.0975156e-002	-2.2193843e-001	1.7378428e-002	3.0789406e-003
-2.8652130e+000	4.0527000e-002	4.0245781e-002	3.8693263e-002	-3.2710808e-001	4.5556469e-003
-2.2727250e+000	2.4192000e-002	3.1922103e-002	-9.4247907e-003	-2.3214072e-001	3.2132937e-003
-1.6843950e+000	7.3980000e-003	2.3657461e-002	-8.2521271e-002	-6.6809754e-002	1.4684767e-003
-3.4416360e+000	6.1049700e-001	4.9092182e-002	8.5991109e-002	-3.3167014e-001	4.7388688e-003
6.8266800e-001	4.9334400e-001	1.1829678e-002	-3.8868398e-002	6.5709212e-002	1.0558886e-003
-4.0199220e+000	1.1818710e+000	5.8849139e-002	1.0412293e-001	-3.2234844e-001	4.6850907e-003
1.2887100e+000	1.0291050e+000	2.3162793e-002	1.3593313e-002	2.1673317e-001	3.0034428e-003
-2.8206360e+000	1.7012430e+000	4.6263574e-002	9.3088724e-002	-2.9914299e-001	4.3330243e-003
3.0663630e+000	1.5344640e+000	4.8158312e-002	7.4863860e-002	3.8626212e-001	5.4416573e-003
-3.4002180e+000	2.2735620e+000	5.7448055e-002	5.2888494e-002	-2.6205378e-001	3.6974428e-003
3.6696510e+000	2.0745450e+000	5.9205853e-002	1.4494325e-001	5.7429607e-001	8.1919339e-003
-3.9805560e+000	2.8493370e+000	6.8753619e-002	-1.6477763e-002	-2.1108082e-001	2.9282591e-003
4.2785550e+000	2.6174610e+000	7.0445069e-002	2.2383147e-001	7.8083500e-001	1.1234374e-002
0.0000000e+000	0.0000000e+000	0.0000000e+000	5.1601940e+000	2.8911890e+000	8.1807324e-002

Figure 15. ANFIS data set structure for radial distances.

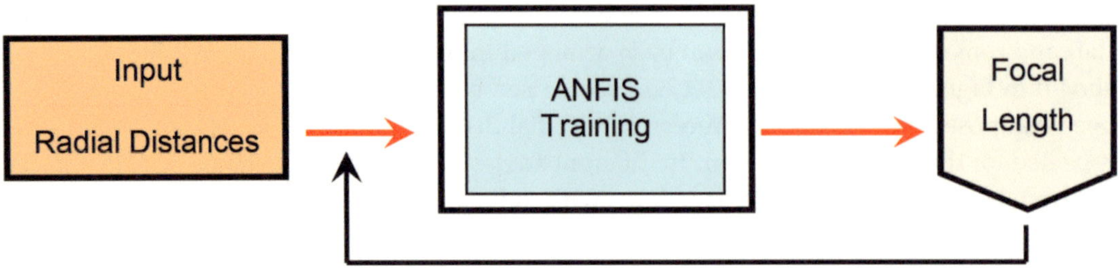

Figure 16. ANFIS algorithm.

Each data set has been subjected to process with eight different ANFIS functions (trimf, trapmf, gbellmf, gaussmf, gauss2mf, pimf, dsigmf, and psigmf) as indicated above and quadratic average error values of Artificial Neural Network and ANFIS functions have been obtained. Artificial neural network structure has been shown in **Figure 17** from MATLAB ANFIS interface.

Measured radial values and ANFIS artificial neural network structure have been indicated in **Figure 17**. Quadratic mean errors (standard deviation) of focal length differences related to radial lengths obtained for function structures used in forming artificial neural network have been generated for 10 images in total according to maximum focal length data as shown in **Table 1** and minimum focal length data as shown in **Table 2**. While there are standard deviation values here, focal length value obtained (with bundle block adjustment) in the first step has been used as exact value. At the end of the first step, inner orientation parameters obtained by conventional method will be considered as true, and they will constitute reference data to be used in comparison for the results obtained with second step in other word ANFIS.

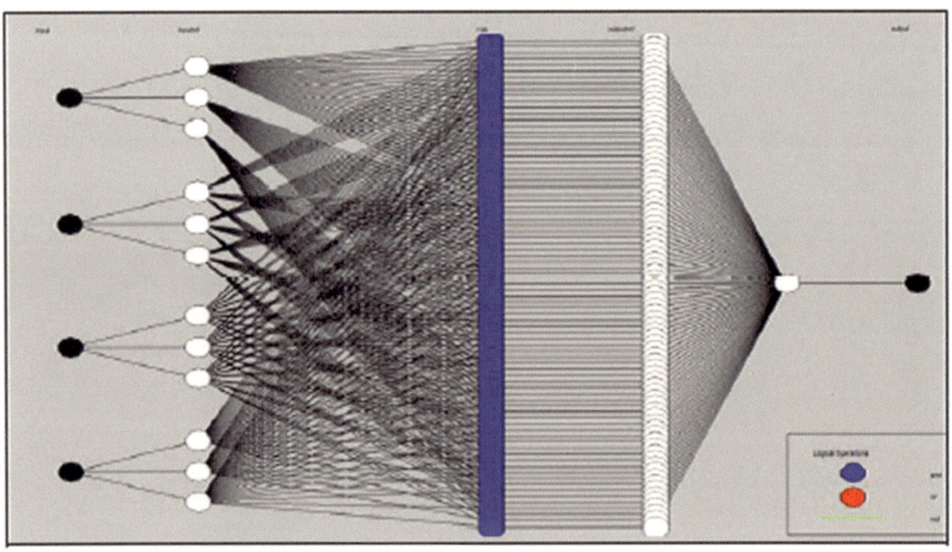

Figure 17. Artificial neural network structure generation.

Maximum focal length (71.9 mm)						
Num.	**ANFIS function**	**Image 1**	**Image 2**	**Image 3**	**Image 4**	**Image 5**
1	Trimf	0.00179	0.00173	0.00210	0.00235	0.00190
2	Trapmf	0.00373	0.00382	0.00381	0.00319	0.00441
3	Gbellmf	0.00243	0.00261	0.00283	0.00218	0.00219
4	Gaussmf	0.00234	0.00229	0.00237	0.00210	0.00210
5	Gauss2mf	0.00347	0.00530	0.00515	0.00321	0.00307
6	Pimf	0.00512	0.00344	0.00488	0.00356	0.00342
7	Dsigmf	0.00306	0.00288	0.00452	0.00327	0.00277
8	Psigmf	0.00306	0.00288	0.00452	0.00327	0.00308

Table 1. Standard deviation values of maximum focal length obtained with ANFIS functions.

The result of determining focal distance with eight ANFIS functions, which are used in the application and square mean errors which are obtained with these functions, are compared to camera calibration values which are calculated with the conventional method (bundle block adjustment), and the differences of focal distances are illustrated image by image.

The results that eight ANFIS functions have given standard deviations for first image to fifth image depending on the maximum and minimum focal lengths have been detailed in the graphs in **Figures 18** and **19**. When ANFIS method is examined in terms of reliability and inner reliability parameters between selected functions with this application realized, it has been seen that it gives the most suitable results in determination of focal length of trimf and gaussmf functions.

Minimum focal length (8.9 mm)						
Num.	**ANFIS function**	**Image 1**	**Image 2**	**Image 3**	**Image 4**	**Image 5**
1	Trimf	0.01814	0.02217	0.01626	0.02560	0.02326
2	Trapmf	0.02204	0.03220	0.02181	0.02701	0.04293
3	Gbellmf	0.01637	0.02047	0.01736	0.02276	0.03822
4	Gaussmf	0.01429	0.01982	0.01745	0.02221	0.03269
5	Gauss2mf	0.02195	0.03180	0.02524	0.02729	0.04476
6	Pimf	0.02528	0.03589	0.02648	0.02965	0.04609
7	Dsigmf	0.02323	0.03274	0.02448	0.02774	0.04494
8	Psigmf	0.02323	0.03274	0.02448	0.02774	0.04494

Table 2. Standard deviation values of minimum focal length obtained with ANFIS functions.

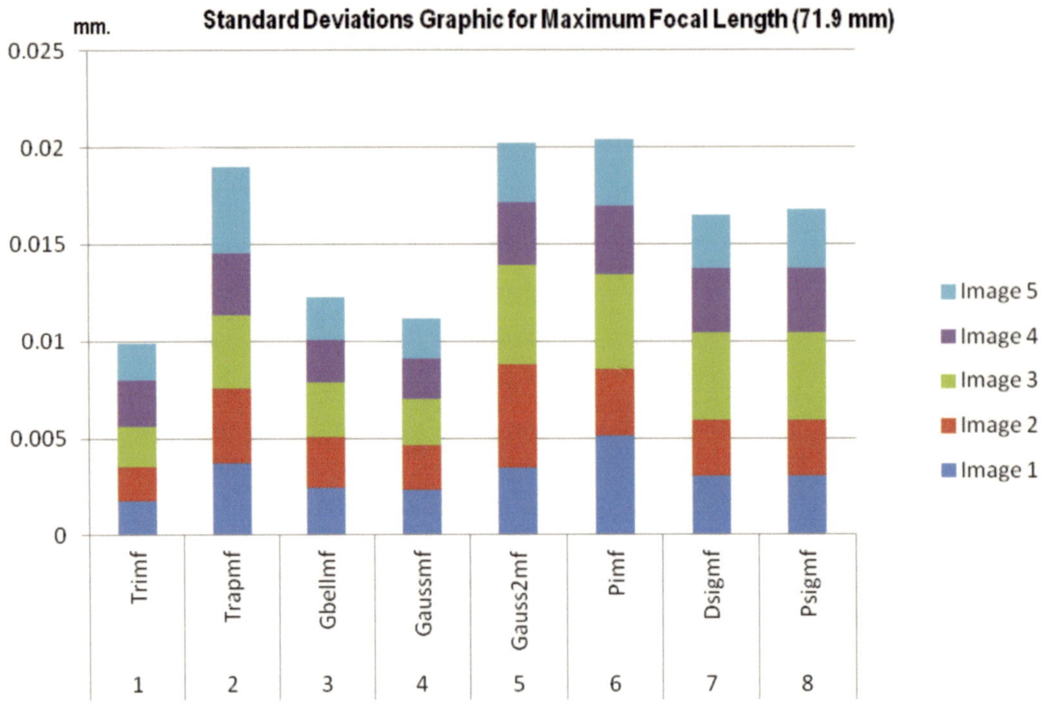

Figure 18. Standard deviations graphic for maximum focal length (71.9 mm).

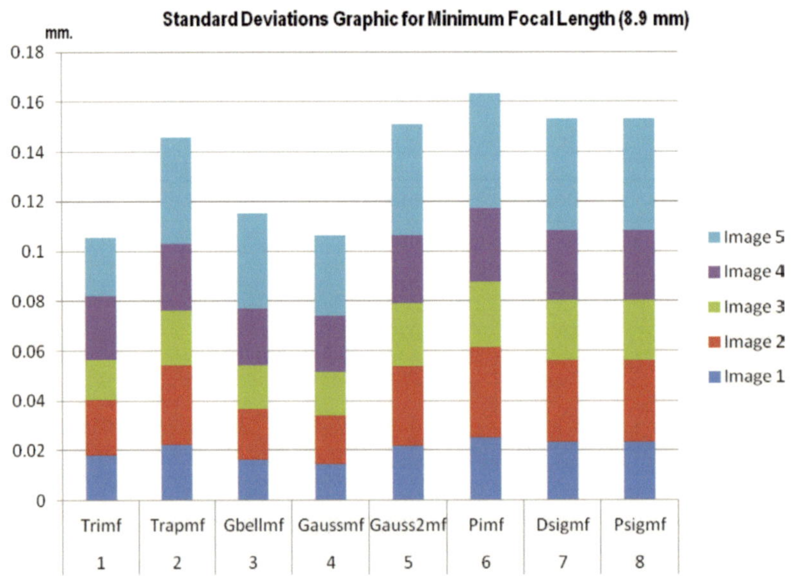

Figure 19. Standard deviations graphic for minimum focal length (8.9 mm).

4. Conclusion

When the results were obtained at the end of the application, it was seen that data with long focal lengths produce lower quadratic mean errors, and test data produced by using short focal length produce higher quadratic mean error values in the same ANFIS functions. When its mathematical structure is considered from radial distortion focal length graphics, it is seen that the shorter the focal length, the more radial distortion error happens. When this situation is considered, it is seen that radial distortion error directly influences the approach to ANFIS functions, and it has an influence on general error amount in data set. In other words, it has been seen that optical distortion law can be verified with the influence of ANFIS functions to data sets. This at the same time proves a physical fact in general table that it verifies ANFIS functions and depicts it in a very proper way.

When the graphics of the functions between each other are examined, it is seen that the most proper (with the least quadratic mean error) or the most usable ANFIS functions in a bundle block adjustment study to be modeled in ANFIS structure related to spatial resection structure are "*trimf, gbellmf and gaussmf.*"

It is seen that ANFIS functions, which depict balancing as the weakest (with the highest quadratic mean error), are "*pimf, dsigmf and psigmf*" functions with this approach. When an assessment is made concerning the general of this study to make an approach of spatial resection by using ANFIS system, advantages and disadvantages below will occur for the calibration studies.

Advantages:

- There will be no need for generating a separate unreal algorithm for spatial resection step in bundle block adjustment.
- Study will be able to be made without the need of approximate rotation and reciprocation parameters.
- It will decrease iteration number for bundle block adjustment.
- Calibration will be able to be enabled during study not only in minimum and maximum focal lengths but also in all focal lengths.

Disadvantages:

- Data sets must be prepared for ANFIS sets. This requires additional time.
- It is not possible to reach approximate rotation and reciprocation parameters.
- Since the system will directly give result, focal length necessary for calibration will be known only as approximately depending on data number.
- It is verified for more than sufficient images that focal length found with additional transaction can be used as calibrated for certain.
- All images to be determined for the project must be taken from the same focal length.

As a result of the studies conducted, it has been seen that fuzzy logic approach can be used in the study of calibration of digital cameras. It was determined that the accuracy can be increased by increasing the data number of estimation model established with ANFIS method and estimation can be made benefiting from other artificial intelligence methods. For this reason, it was concluded that it will be more beneficial to use and search fuzzy logic approach more in photogrammetric applications.

Author details

Bahadır Ergün[1]*, Cumhur Sahin[1] and Ugur Kaplan[2]

*Address all correspondence to: bergun@gtu.edu.tr

1 Department of Geodesy and Photogrammetry, Gebze Technical University, Gebze, Turkey

2 Department of Mathematics, Gebze Technical University, Gebze, Turkey

References

[1] Kajornrit J, Wong KW, Fung CC. A comparative analysis of soft computing techniques used to estimate missing precipitation records. In: Proceedings of the 19th ITS Biennial Conference; 18-21 November 2012; Bangkok, Thailand.

[2] Mamdani EH, Assilian S. An experiment in linguistic synthesis with a fuzzy logic controller. International Journal of Human-Computer Studies. 1999;**51**(2):135-147. doi:10.1006/ijhc.1973.0303.

[3] Takagi T, Sugeno M. Fuzzy identification of systems and its applications to modeling and control. IEEE Transactions on Systems, Man and Cybernetics. 1985;**15**:116-132. WOS:A1985AFX0900011.

[4] Hamam A, Georganas ND. A comparison of mamdani and sugeno fuzzy ınference systems for evaluating the quality of experience of hapto-audio-visual applications. In: Proceeding of the IEEE International Workshop on Haptic Audio Visual Environments and their Applications 18-19 October 2008; Ottawa, Canada. IEEE.

[5] Tsoukalas LH, Uhrig RE. Fuzzy and neural approaches in engineering (1st ed.). New York: John Wiley & Sons; 1996, 600 p. ISBN:0471160032.

[6] Cıtak E. Fuzzy logic approach to determine the factors affecting hand dexterity [thesis]. Unıversıty of Cukurova; 2008.

[7] Roger JS, Sun C. Functional equivalence between radial basis function networks and fuzzy inference systems. IEEE Transactions on Neural Networks. 1993;**4**:156-159. doi:10.1109/72.182710.

[8] Roger JS. ANFIS: adaptive network based fuzzy inference systems. IEEE Transactions on Systems Man and Cybernetics. 1993;**23**(3):665-685. doi:10.1109/21.256541.

[9] Perez JA, Gonzalez M, Dopico D. Adaptive neurofuzzy ANFIS modeling of laser surface treatments. Neural Computing & Applications. 2010;**19**:85-90. doi:10.1007/s00521-009-0259-x.

[10] Roger JS, Gulley N. Matlab fuzzy logic toolbox. User#x2019;s guide. Natick: The MathWorks Inc.; 2007.

[11] Jang JSR, Sun CT, Mizutani E. Neuro-fuzzy and soft computing: a computational approach to learning and machine intelligence. Englewood Cliffs, NJ: Prentice-Hall; 1997.

[12] Sabzevari R, Masoumzadeh S, Ghahroudi MR. Employing anfis for object detection in robo-pong. In: Proceeding of 5th International Symposium on Mechatronics and its Applications (ISMA08); 27-29 May 2008; Amman, Jordan, pp. 707-712.

[13] Orghidan R, Danciu M, Vlaicu A, Oltean G, Gordan M, Florea C. Fuzzy versus crisp stereo calibration: a comparative study. In: Proceeding of 7th International Symposium on Image and processing and analysis (ISPA 2011); 4-6 September 2011; Dubrovnik, Croatia, pp. 627-632.

[14] Nuno AI, Arcay B, Cotos JM, Varela J. Optimisation of fishing predictions by means of artificial neural networks, anfis, functional networks and remote sensing images. Expert Systems with Applications. 2005;**29**:356-363. doi:10.1016/j.eswa.2005.04.008.

[15] Jafari S, Jarvis R. Robotic hand-eye coordination: from observation to manipulation. In: Proceedings of the Fourth International Conference on Hybrid Intelligent Systems (HIS'04); 5-8 December 2004; Kitakyushu, Japan. IEEE. doi:10.1109/ICHIS.2004.82.

[16] Gallardo A, Taylor J, Paolini C, Lee HK, Lee GK. An ANFIS-based multi-sensor structure for a mobile robotic system. In: Proceeding of Computational Intelligence in Control and Automation; 11-15 April 2011; Paris, France. IEEE. doi:10.1109/CICA.2011.5945755.

[17] Li C, Ma H, Yang C, Fu M. Teleoperation of a virtual iCub robot under framework of parallel system via hand gesture recognition. In: Proceeding of International Conference on Fuzzy Systems (FUZZ-IEEE); 6-11 July 2014; Beijing, China, pp. 1469-1474.

[18] Orazio TD, Leo M, Mazzeo PL, Spagnolo P. Soccer player activity recognition by a multivariate features integration. In: Proceeding of Seventh IEEE International Conference on Advanced Video and Signal Based Surveillance; 29 August 2010-1 September 2010; Boston, USA, pp. 32-39.

[19] Orghidan R, Gordan M, Danciu M, Vlaicu A. A prototype for the creation and interactive visualization of 3D human face models. Advanced Engineering Forum. 2013; **8-9**:45-54. doi:10.4028/www.scientific.net/AEF.8-9.45.

[20] Navarro H, Orghidan R, Gordan M, Saavedra G, Martinez-Corral M. Fuzzy integral imaging camera calibration for real scale 3D reconstructions. Journal of Dısplay Technology. 2014;**10**(7): 601-608. doi:0.1109/JDT.2014.2312236.

[21] Ozbilge EU. An approach of matlab about calibration of digital cameras [thesis]. Gebze Institute of Technology; 2010.

[22] Sugeno M, Kang G. Structure identification of fuzzy model. Fuzzy Sets and Systems. 1988;**26**(1):15-33. doi:10.1016/0165-0114(88)90113-3.

[23] Ross TJ. Fuzzy logic with engineering applications. Chichester: McGraw-Hill; 1995, pp. 134-147. ISBN: 0-07-053917-0.

[24] Hocaoğlu FO, Kurban M. Adaptif ağ tabanlı bulanık mantık çıkarım sistemi ile Eskisehir bolgesi icin güneşlenme süreleri tahmini. Elektrik-Elektronik-Bilgisayar Mühendisliği 11. Ulusal Kongresi ve Fuarı, 22-25 Eylul 2005; Istanbul, Turkiye.

[25] Mathworks. 2017. Retrieved 25 January 2017, from https://www.mathworks.com/products/matlab.html.

Chapter 16

EMG-Controlled Prosthetic Hand with Fuzzy Logic Classification Algorithm

Beyda Taşar and Arif Gülten

Additional information is available at the end of the chapter

http://dx.doi.org/10.5772/intechopen.68242

Abstract

In recent years, researchers have conducted many studies on the design and control of prosthesis devices that take the place of a missing limb. Functional ability of prosthesis hands that mimic biological hand functions increases depending on the number of independent finger movements possible. From this perspective, in this study, six different finger movements were given to a prosthesis hand via bioelectrical signals, and the functionality of the prosthesis hand was increased. Bioelectrical signals were recorded by surface electromyography for four muscles with the help of surface electrodes. The recorded bioelectrical signals were subjected to a series of preprocessing and feature extraction processes. In order to create meaningful patterns of motion and an effective cognitive interaction network between the human and the prosthetic hand, fuzzy logic classification algorithms were developed. A five-fingered and 15-jointed prosthetic hand was designed via SolidWorks, and a prosthetic prototype was produced by a 3D printer. In addition, prosthetic hand simulator was designed in Matlab/SimMechanics. Pattern control of both the simulator and the prototype hand in real time was achieved. Position control of motors connected to each joint of the prosthetic hand was provided by a PID controller. Thus, an effective cognitive communication network established between the user, and the real-time pattern control of the prosthesis was provided by bioelectrical signals.

Keywords: EMG, fuzzy logic classification, multifunctional prosthesis hand, pattern recognition

1. Introduction

People lose limbs due to accidents and medical conditions. Robotic devices, which imitate the shape and function of a missing limb, are manufactured for use by people who lose their limb in such situations. In recent years, researchers have studied to design and control multifunctional prosthetics hand [1–7]. The complexity of the movement, that is, the number of independent movements, increases in proportion to the number of joints. There are 206 bones in the adult skeletal system. The 90 bones of the skull and face are connected to each other by non-immobilized joints, and the 33 bones of the spine are connected to each other by semi-movable joints. Movable joints are only present between the bones (except the metacarpals bones) of the arm (25) and leg (25). In light of this information, aside from the wrist joints, the human hand has 15 independent joints with three on each finger. Therefore, the biological hand movement involves the control of these joints independently. Thus, control of the hand is quite complex. Thus, of all the human parts, the hand is the most complicated in terms of kinetic analysis [8].

Two main factors enable the functional and visual prosthesis to be used like a biological hand:

- Prosthetic hand mechanical design and modeling [9, 10] and
- Perform the position and speed controls of each joint efficiently and precisely [11–19].

However, no matter how perfect the design and manufacture of the prosthetic hand may be, the utility depends on the cognitive interaction, i.e., the control algorithm, being designed properly, e.g., the type of movement and coordination between fingers. If information is not transferred to the prosthetic hand rapidly enough, then the prosthesis will not assume the desired position. Cognitive interaction is the most important factor for user to use effectively. There are many studies about cognitive interaction between human and robotic devices [20–25].

All voluntary muscle movements in humans occur as a result of bioelectrical signals transmitted from the brain through the muscle nerves. Bioelectrical electromyogram (EMG) signals transmitted to the muscles carry information about the type of movement, speed, and degree of muscle contraction or relaxation. The biological hand performs the basic tasks of holding and gripping, which involve various finger movements. The wrist movements essentially constitute the axis and assist in these gripping and holding movements. The main factor that increases the functionality of the prosthetic hand is the movement of the fingers. As the number of independent movements made by the prosthetic hand increases, it can mimic the biological hand more successfully. This study realizes the design of the bioelectrical signal control algorithm and the extension of the bioelectrical signal database with the purpose of increasing the finger motion function of bioelectrical signal-controlled prosthesis hands.

Figure 1 shows bioelectrical signals in the context of the activity of the muscle movements (e.g., flexion, relaxation force), as seen from the block flow diagram. EMG can be used to detect signals from the flexor pollicis longus, flexor carpi radialis, brachioradialis, extensor carpi radialis, extensor digiti minimi, and extensor carpi ulnaris. Bioelectrical signals were recorded with the help of four surface electrodes and subjected to a series of preprocessing and classification operations to understand the relationships between EMG signals and hand and finger movements. These signals were then applied to the prosthetic hand (space and simulator) as a reference motion signal. With the designed controllers, the position of the prosthetic hand finger joints can be controlled. Thus, a cognitive interface and communication network are established between the user and the prosthetic hand. Briefly summarized, the study creates a bioelectrical database of the activities of the hand muscles and the interaction network between the human and prosthetic hand using this database and interface to design a simulator and develop a control algorithm.

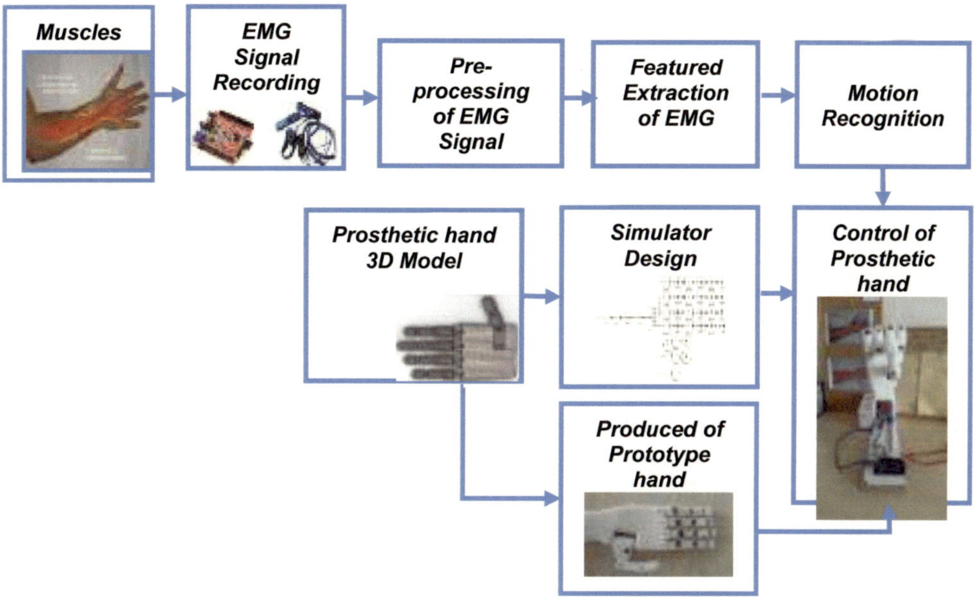

Figure 1. Control of multifunctional prosthetic hand simulator and prototype with EMG signals.

2. Recording, preprocessing, and featured extractions of EMG signal

2.1. Recording of EMG signals

EMG signals were recorded from the forearm muscles (the flexor pollicis longus, flexor carpi radialis, brachioradialis, extensor carpi radialis, extensor digiti minimi, and extensor carpi ulnaris) with the help of four surface electrodes. Electrode placements are shown in **Figure 2**. Electrode layout was chosen according to the protocol [26–28].

The signals, which support movements of the thumb, middle, ring, index, and pinkie fingers, were recorded separately for each of the respective muscles. Channels and finger relations are shown in **Table 1**.

2.2. Preprocessing of EMG signals

The recorded EMG signals also include various noise signals. It is necessary to separate the noise signals from the EMG signals, so that the characteristics of the signal can be accurately

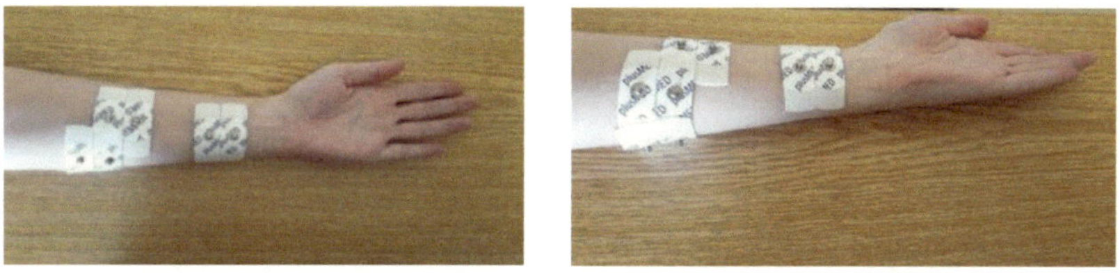

Figure 2. Placement of surface electrodes.

Channel 1	Channel 2	Channel 3	Channel 4
Pinkie finger muscle	Ring finger muscle	Middle finger muscle	İndex finger muscle

Table 1. Channel finger relations.

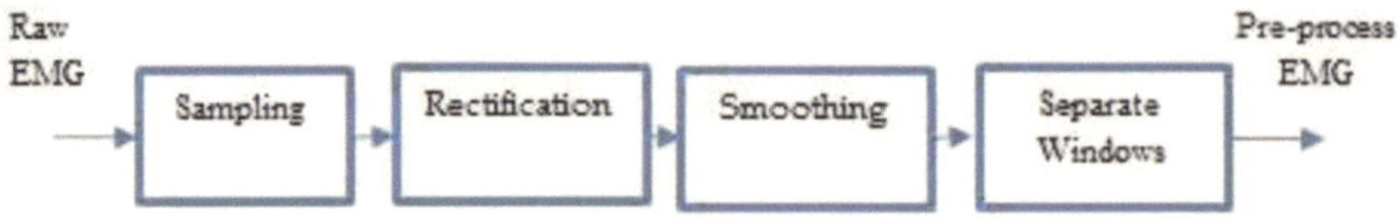

Figure 3. Preprocessing steps.

determined. For this reason, the raw EMG signal is first preprocessed. The block diagram of the preliminary preparation stage, including the separation, rectification, and sampling of the recorded EMG signals from noise, is shown in **Figure 3**.

2.2.1. Numerical sampling

EMG signals are analog voltage signals. Their amplitudes change constantly over the voltage range. Analog-to-digital conversion is the process by which the amplitude of the analog signal voltage is represented by a number sequence at specific time points [29–31]. The EMG voltage signals used in this study are converted into a number sequence by sampling with a period of 0.001 s.

2.2.2. Rectification process

Rectification is the evaluation of only the positive parts of the signal. This is done either by half-wave or full-wave rectification of the signal. A full-wave rectification method was applied to preserve the energy of the signal [25, 29–34], and the expression for the method is given in Eq. (1).

$$X\_training = |x\ (t)| \tag{1}$$

2.2.3. Smoothing of signal

A bandpass filter (50–500 Hz) was designed to soften the signal by eliminating high-frequency components.

2.2.4. Separate the signal into windows

Before the attributes of the obtained EMG signals are calculated, the frame is processed by the method adjacent to the signal. Experiments in the study of Englehart [18, 19] for framing and optimal framing values (R = 256, r = 32 ms) reached with calculations were used.

2.3. Featured extractions of EMG signal

The EMG signal is a non-stationary, time-varying signal that varies in amplitude by random negative and positive values [25, 31, 32]. Bioelectrical signals have certain characteristic values, i.e., information. Features in time domain have been widely used in medical and engineering practices and researches. Time domain features are used in signal classification due to its easy and quick implementation. Furthermore, they do not need any transformation, and the features are calculated based on raw EMG time series. Moreover, much interference that is acquired through the recording because of their calculations is based on the EMG signal amplitude. However, compared to frequency domain and time-frequency domain, time domain features have been widely used because of their performances of signal classification in low noise environments and their lower computational complexity [29]. In this study, five time domain features methods widely used in the literature have been utilized to obtain the features of the EMG signal.

2.3.1. *Signal energy*

Mathematically, the energy of the signal m (t) is calculated as in Eq. (2), where t_j and t_i denote the lower and upper bounds of the part of the signal to be integrated, respectively. The above expression represents the area below the absolute value of the signal curve at time $T = t_i – t_j$ [30–35].

$$E = \int_{t_i}^{t_j} |m(t)| dt \tag{2}$$

2.3.2. *Maximum value of signal*

The maximum value of the signal represents the largest of the sampled signal values in each packet divided by windows [29].

2.3.3. *Signal average value*

Mathematically, the average of the signal m (t) is calculated as Eq. (3) [30, 31], where t_i and t_j denote the upper and lower bounds of the part of the signal to be integrated, respectively. The above expression represents the overall average of the signal at time interval $T = t_i - t_j$.

$$AVR = \frac{1}{t_j - t_i} \int_{t_i}^{t_j} |m(t)| dt \tag{3}$$

2.3.4. *Effective value of the signal*

Effective value is a commonly used signal analysis method in the time domain, such as average rectification [29–32]. The effective value of the *m*(*t*) signal is calculated as Eq. (4).

$$RMS = \left(\frac{1}{T} \int_0^t m^2(t) dt \right)^{\frac{1}{2}} \tag{4}$$

2.3.5. *Variance of signal*

The variance value of the signal represents the amount of deviation from the mean of the sampled signal values in each packet divided by windows [30]. *p*(*t*) is the variance of the signal to represent the probability density function of *t*:

$$VAR = \left(\frac{1}{T} \int_0^t (x - ORT)^2 p(t) dt \right) \tag{5}$$

3. Pattern recognition with fuzzy logic algorithm

A classifier's function should be able to map different patterns, match them appropriately, and, in this case, select different hand grip postures. The extracted features were then fed into the fuzzy logic (FL) classifier for the developed control system. FL developed by Lofty Zadeh [35–41] provides a simple way to arrive at a definite conclusion based solely on imprecise input information. A summary of the feature extraction process from the forearm muscles is shown in **Table 2** according to motion.

In total, there are 20 features of EMG signal for four channels. In order to make relations easier, a featured function, which occurs at RMS, AVR, MAX, VAR, and E values, is defined for each channel. Finally, the number of inputs is reduced by four. The featured function is calculated as follows in Eq. (6).

$$F_i = E_i + AVR_i + MAX_i + VAR_i + RMS_i \tag{6}$$

For the FL classification analysis, the triangular shape of the membership function (MF) for the inputs (F_i) and output and the centroid method for defuzzification are used. The rules are created

	Signal	**Hand closure**	**Hand opening**	**Index-thumb touch**	**Middle-thumb touch**	**Ring-thumb touch**	**Pinky-thumb touch**
Energy	Channel 1	16,41091	9,949203	5,853087	5,405963	5,354211	**12,84222**
	Channel 2	12,48169	10,92331	7,334108	6,46115	**13,25441**	5,029002
	Channel 3	12,02946	9,254157	8,313991	**12,82708**	7,183281	4,252198
	Channel 4	14,59524	7,548085	**11,22431**	6,920272	9,376161	4,381767
Maximum value	Channel 1	2,378095	1,398911	0,822295	0,61429	0,725287	**2,255524**
	Channel 2	1,674114	1,183987	1,126519	0,961061	**1,90971**	0,609637
	Channel 3	1,606747	1,351835	1,163335	**1,60762**	1,147475	0,666139
	Channel 4	1,990469	0,844166	**1,437937**	0,906574	1,485923	0,532234
Average value	Variance	0,656436	0,397968	0,234123	0,216239	0,214168	**0,513689**
	Channel 1	0,499268	0,436932	0,293364	0,258446	**0,530176**	0,20116
	Channel 2	0,481178	0,370166	0,33256	**0,513083**	0,287331	0,170088
	Channel 3	0,58381	0,301923	**0,448973**	0,276811	0,375046	0,175271
RMS value	Channel 4	0,474695	0,273057	0,163739	0,134428	0,148438	**0,387735**
	Channel 1	0,325763	0,25909	0,207215	0,173207	**0,370618**	0,124443
	Channel 2	0,316673	0,25826	0,223731	**0,339657**	0,213173	0,122159
	Channel 3	0,383453	0,188114	**0,295885**	0,180392	0,269928	0,10675
Variance	Channel 4	0,72476	0,223357	0,08254	0,045411	0,066981	**0,508143**
	Channel 1	0,293061	0,15076	0,133987	0,086676	**0,422607**	0,038505
	Channel 2	0,281122	0,204654	0,145503	**0,326644**	0,150682	0,047588
	Channel 3	0,410777	0,089351	**0,246002**	0,089669	0,232966	0,027352

Table 2. Summary of the feature extraction process from the forearm muscles.

based on information from the states of contraction. FLC rules are shown in **Table 3**. Recorded SEMG signals have been used to initial testing. Then real time data implemented to Prosthetic hand model.

Fi Featured functions were inputs to the FL. The limits of F were set to [0, 20]. The three linguistic variables used were Small (S), Medium (M), and Big (B). The outputs of FL were Hand closure, Hand opening, Index-thumb contact, Middle-thumb contact, Ring-thumb contact, and Pinky-thumb contact. **Figure 4** shows the flow diagram of FL classification process from four SEMG signals for six hand patterns [35].

Rules	F1	F2	F3	F4	Result
1	BIG	BIG	BIG	BIG	Hand closure
2	MEDIUM	MEDIUM	MEDIUM	MEDIUM	Hand opening
3	MEDIUM	MEDIUM	MEDIUM	BIG	Index-thumb touch
4	MEDIUM	MEDIUM	BIG	MEDIUM	Middle-thumb touch
5	MEDIUM	BIG	MEDIUM	MEDIUM	Ring-thumb touch
6	BIG	MEDIUM	MEDIUM	MEDIUM	Pinky-thumb touch
7	SMALL	SMALL	SMALL	SMALL	Relax-no motion

Table 3. FL rules.

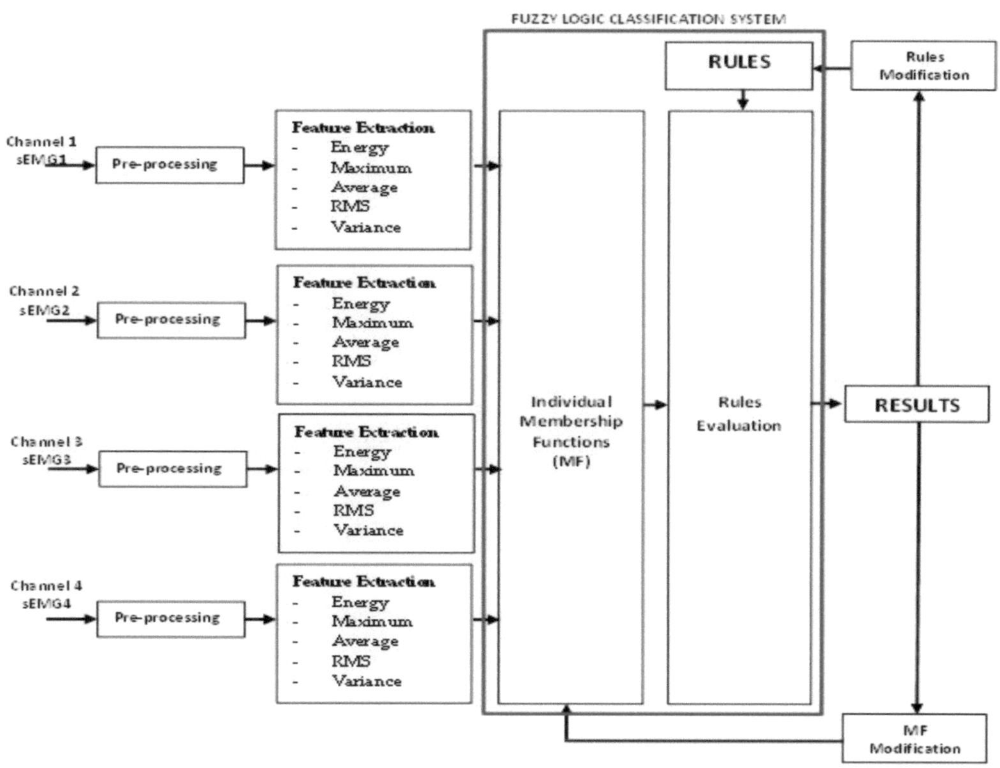

Figure 4. The flow diagram of the control system with FL classification components.

Performance of FL tested 200 hand motions. Classification performance value for the six motions is shown in **Table 4**.

In the medical decision-making process, ROC analysis method is used to determine the discrimination of the test or classification algorithm. In this study, performance of FLC algorithm for six motion class are demonstrated in **Table 5** via ROC analysis.

Performance values calculated as Eqs. (7)–(10) for each hand motion

$$Accuracy(ACC) = \Sigma True\ positive + \Sigma True\ negative / \Sigma Total\ population \tag{7}$$

$$Positive\ predictive\ value(PPV), Precision = \Sigma True\ positive / \Sigma Test\ out\ comepositive \tag{8}$$

$$True\ positive\ rate(TPR), Sensitivity = \Sigma True\ positive / \Sigma Condition\ positive \tag{9}$$

$$False\ positive\ rate(FPR) = \Sigma False\ positive / \Sigma Condition\ negative \tag{10}$$

Hand pattern	Pattern number	Tested total number of motion (A + B)	Number of true classified motion (A)	Number of wrong classified motion (B)	Average percentage of success (%)
Hand closure	MOTION 1	84	84	0	100
Hand opening	MOTION 2	84	84	0	100
Index-thumb touch	MOTION 3	84	76	8	90.476
Middle-thumb touch	MOTION 4	84	66	18	78.57
Ring-thumb touch	MOTION 5	84	72	12	85.714
Pinky-thumb touch	MOTION 6	84	76	8	90.476

Table 4. Classification achievement percentages.

ROC analysis	Motions					
Classification algorithm result	**Hand closure**	**Hand opening**	**Index-thumb touch**	**Middle-thumb touch**	**Ring-thumb touch**	**Pinky-thumb touch**
Hand closure	84	0	0	0	0	0
Hand opening	0	84	0	0	0	0
Index-thumb touch	0	0	76	6	6	4
Middle-thumb touch	0	0	1	66	2	0
Ring-thumb touch	0	0	4	10	72	3
Pinky-thumb touch	0	0	2	0	1	76
No motion	0	0	1	2	1	1

Table 5. ROC analysis.

Hand closure

TP=84	FN=0	84
FP=0	TN=420	420
84	420	504

TPR= 1.00
FPR= 0.00
PPV=1.00
ACC=1.00

Hand opening

TP=84	FN=0	84
FP=0	TN=420	420
84	420	504

TPR= 1.00
FPR= 0.00
PPV=1.00
ACC=1.00

Index -thumb touch

TP=76	FN=16	92
FP=8	TN=404	412
84	420	504

TPR= 0.826
FPR= 0.0194
PPV=0.904
ACC=0.952

Middle -thumb touch

TP=66	FN=4	70
FP=18	TN=416	434
84	420	504

TPR= 0.942
FPR= 0.041
PPV=0.785
ACC=0.956

Ring -thumb touch

TP=72	FN=17	89
FP=12	TN=403	415
84	420	504

TPR= 0.808
FPR= 0.028
PPV=0.857
ACC=0.942

Pinky -thumb touch

TP=76	FN=4	80
FP=8	TN=416	424
84	420	504

TPR= 0.95
FPR= 0.018
PPV=0.904
ACC=0.976

Table 6. Contingency matrixes.

The four outcomes can be formulated in a 2 × 2 contingency table. All contingency matrixes for each motion are shown in **Table 6**.

4. 3D modeling and manufacturing of prosthetic hand

4.1. 3D modeling of prosthetic hand via SolidWorks

In order to develop a multifunctional prosthetic hand model, the structural characteristics of the human hand must first be determined. In other words, it is necessary to determine the number of joints, the number of links, the fingers and the length and width parameters of each finger. In order to obtain a prosthetic hand the same size as a human hand, the hand characteristics of an adult male were recorded as in **Table 7** for the purposes of this study [42–44].

	First link		Second link		Third link	
	Length (mm)	**Width (mm)**	**Length (mm)**	**Width (mm)**	**Length (mm)**	**Width (mm)**
Thumb	70	30	45	30	40	30
Index	55	30	40	25	30	25
Middle	55	30	50	25	40	25
Ring	55	30	40	25	30	25
Pinky	30	30	40	25	30	25
Palm	130	120				

Table 7. Part of the hand.

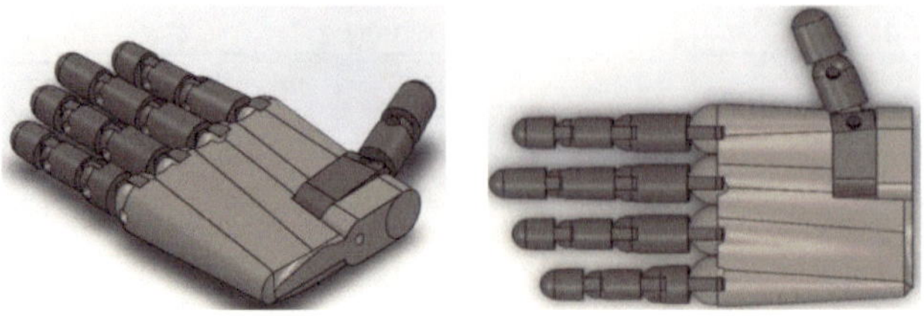

Figure 5. SolidWorks images of prosthetic hand.

Using the parameter values in **Table 5**, the prosthetic hand 3D model is designed with the help of the SolidWorks program as shown in **Figure 5**.

4.2. Manufacturing of prosthetic hand via 3D printer

The prototype of the prosthetic hand was produced with the help of the EDISON 3D printer manufactured by 3D Design Company. The necessary adjustments for the production (e.g., resolution, amount of fullness, amount of support) were made using the Simplify 3D program, which was offered by the same company as the software program. After a hand of 16 parts was produced, it was assembled as shown in **Figure 6**.

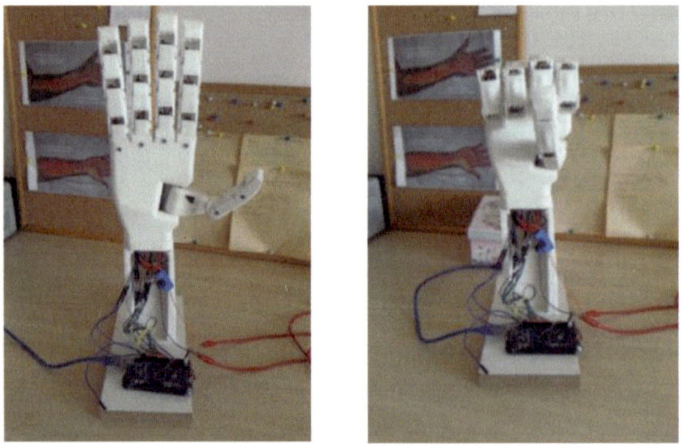

Figure 6. Prototype hand.

5. Prosthetic hand simulator design

5.1. Mechanical design of prosthetic hand simulator via SimMechanics

SimMechanics used in the realization of simulations of mechanical systems [45, 46]. By transferring the 3D CAD model of the prosthetic hand developed in the SolidWorks program to the Matlab SimMechanics program, a chain structure containing each joint and link of the prosthetic hand was obtained as shown in **Figure 7**. Five fingers connected to the palm, three rotary hinges forming each finger, and three connecting links are arranged in series to form the hand SimMechanics model.

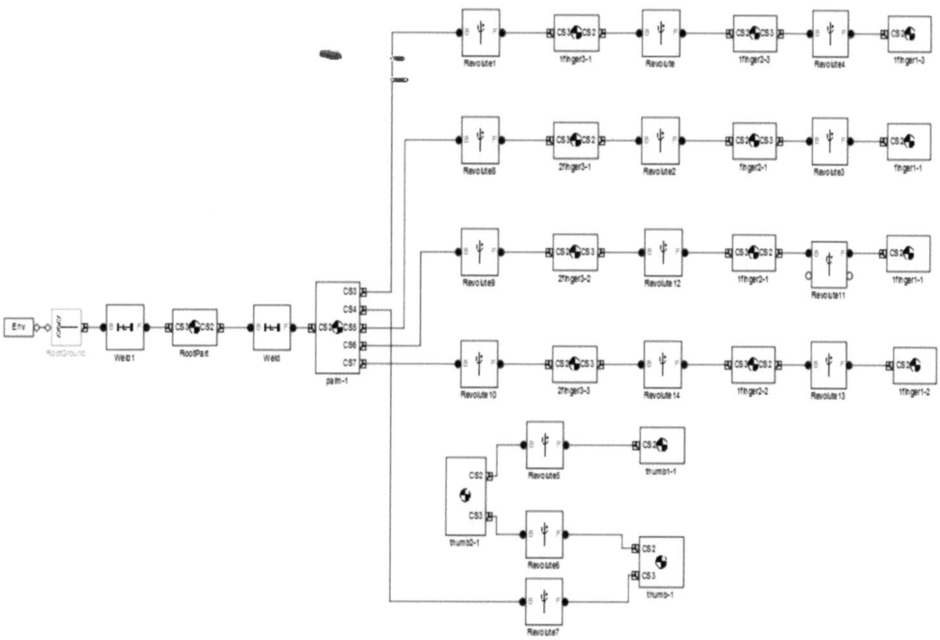

Figure 7. Prosthetic hand SimMechanics model.

As shown in **Figure 7**, when SolidWorks solid model is transferred to Matlab Program, a chain structure composed of revolute and link parts is obtained.

5.2. Modeling of the DC motor

In this study, it was decided to use a DC servo motor for movement of each joint in the prosthetic hand. The equivalent circuit of the DC servo motor is given in **Figure 8** [47–49].

Modeling equations of DC motor were expressed in terms of the Laplace variable s as Eqs. (11)–(13).

$$s(Js + B)\theta(s) = K_t I(s) \tag{11}$$

$$(Ls + R)I(s) = V(s) - K_e s\theta(s) \tag{12}$$

We arrive at the following open-loop transfer function by eliminating $I(s)$ between the two equations above, where the rotation is considered the output and the armature voltage is considered the input.

$$\frac{\theta(s)}{V(s)} = \frac{K}{s\left((Ls + R)(Js + b) + K^2\right)} \tag{13}$$

Using the mathematical model of the DC servo motor, the Matlab/Simulink model is constructed as shown in **Figure 9**.

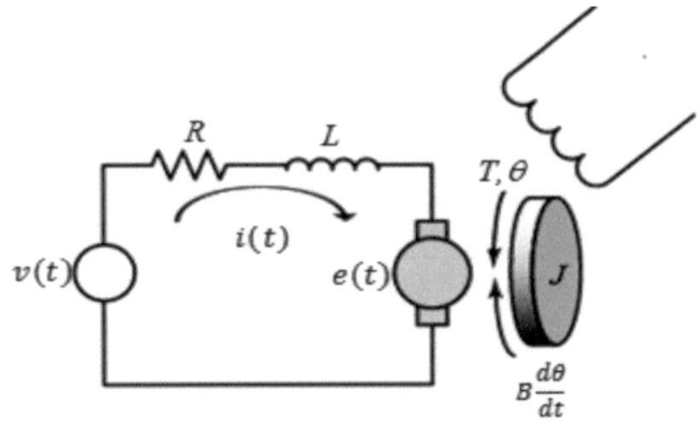

Figure 8. DC motor electrical and mechanical model.

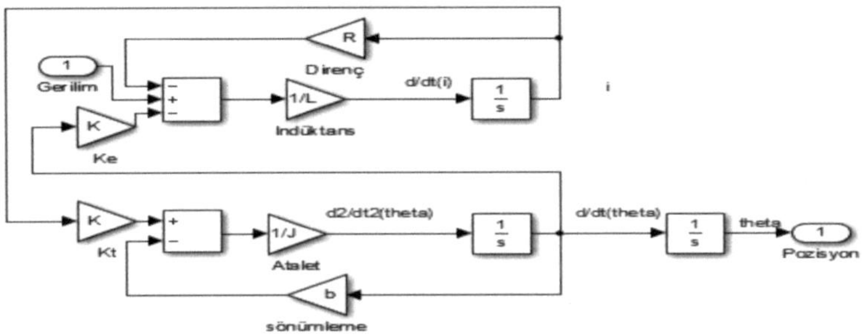

Figure 9. DC motor Matlab/Simulink model.

6. Controller design

Position of ultra-nano DC servomotors connected to joints is controlled using a PID controller. The controller's proportional gain coefficient (Kp), integral gain coefficient (Ki), and derivative gain (Kd) values are determined by Genetic Algorithm [11, 50–52] to ensure that the system quickly reaches a steady state without overshooting as shown in **Table 8**. The PID controller has an input-output relationship with input e (t) and output u (t) [53–55].

$$u(t) = K_p \cdot e(t) + K_i \cdot \int_0^t e(\tau).d\tau + Kd.\frac{de(t)}{dt} \tag{14}$$

	K_p	K_i	K_d
All DC motors connected the each finger joints	0.42176	0.75724	0.0048566

Table 8. PID parameters.

7. Graphical and numerical results

Electromyography is used to measure EMG signals, which are extracted from the forearm muscles and classified with the help of four surface electrodes. The type of motion that one wishes to perform is the perceived and designed three-dimensional prosthetic hand simulator and the five-fingered and 15-jointed hand. These movements were made in real time on the prototype. Each joint of the prosthetic hand is moved with one ultra-nano servomotor, and the position control of the motors is provided by the designed PID.

The prosthetic hand was made with hand closure, hand opening, thumb-index touch, hand opening, thumb-middle touch, hand opening, thumb-ring touch, hand opening, thumb-pinkie touch, and hand opening movements. The hand opening movement is performed after the hand closing movement and touch movement.

1. EMG signals were taken from four channels, four groups of muscles simultaneously, as shown in **Figures 10–13**, and preprocessed. First, the signal amplitude was scaled from 0 to 10 V and then filtered.

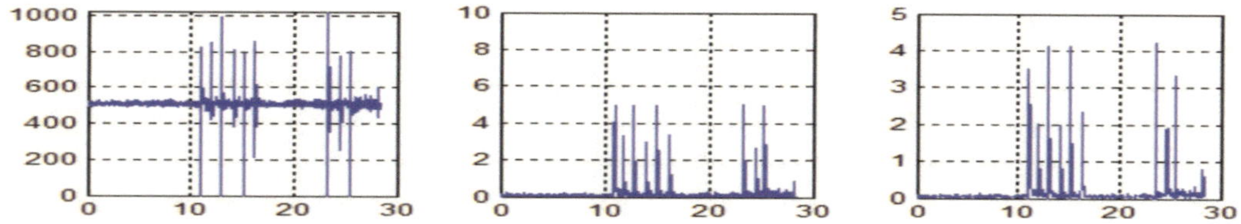

Figure 10. Preprocessing step graphics of EMG signal recorded Channel 1.

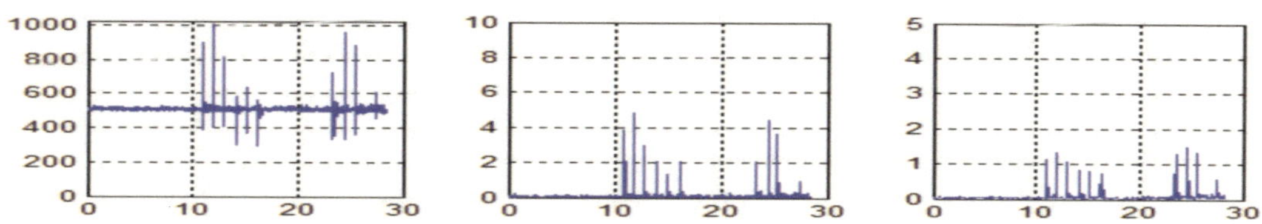

Figure 11. Preprocessing step graphics of EMG signal recorded Channel 2.

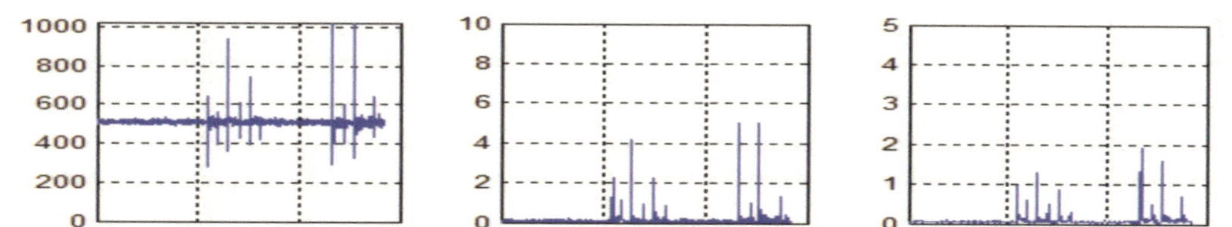

Figure 12. Preprocessing step graphics of EMG signal recorded Channel 3.

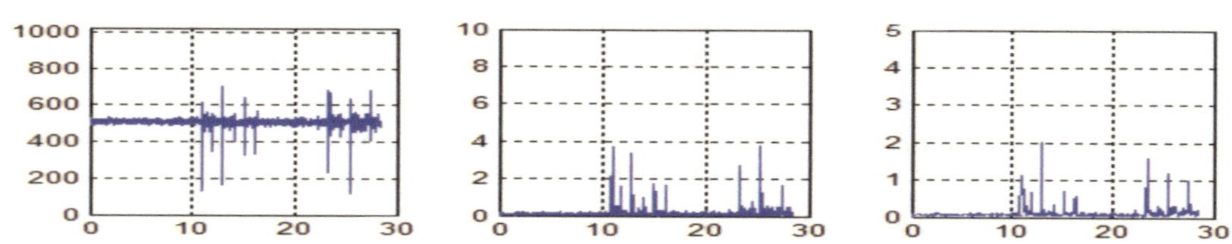

Figure 13. Preprocessing step graphics of EMG signal recorded Channel 4.

2. As shown in **Figures 14–17**, the energy, maximum, effective, mean, and variance attribute values of the respective signals were calculated.
3. Motion pattern was determined by motion classification algorithm.
4. The specified type of motion information was input to the simulator and the prototype.
5. According to the recognized hand pattern, the reference joint angles in **Table 9** were applied as the control input signal, and the closed loop position control of the DC servo-motors was performed according to feedback information from sensors connected to the simulator joints.

Position control of the finger joints for six hand patterns was provided by the PID controllers as shown in **Figures 18–23**.

For all finger joints, PID performance is shown in **Table 10**.

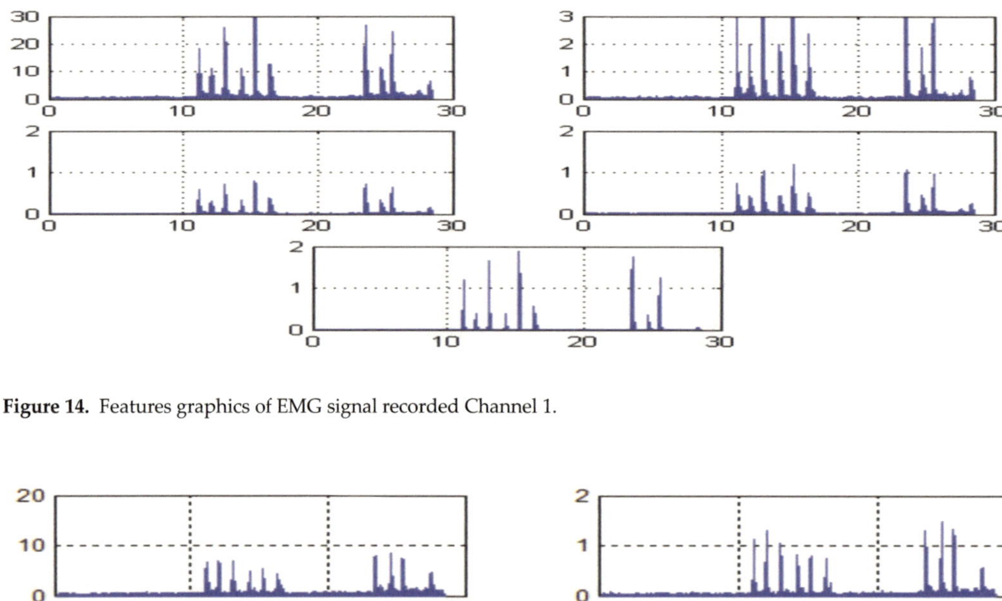

Figure 14. Features graphics of EMG signal recorded Channel 1.

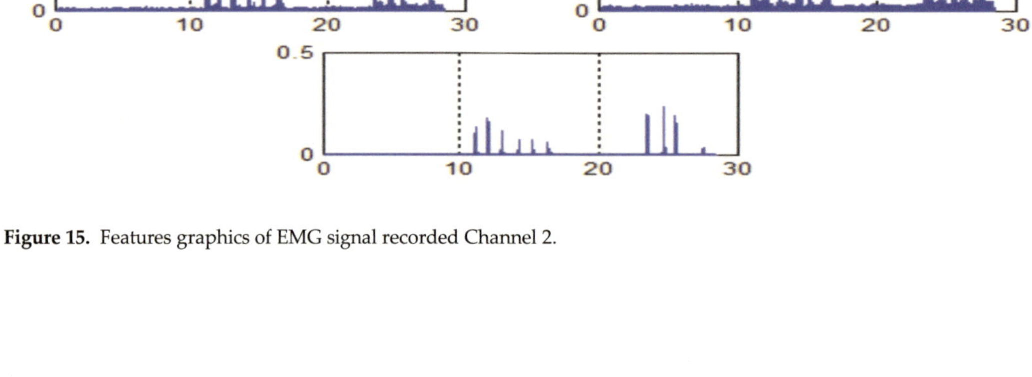

Figure 15. Features graphics of EMG signal recorded Channel 2.

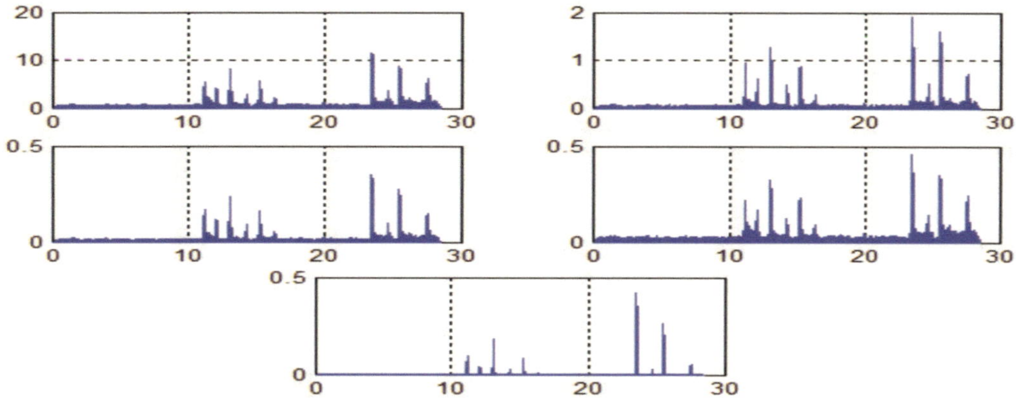

Figure 16. Features graphics of EMG signal recorded Channel 3.

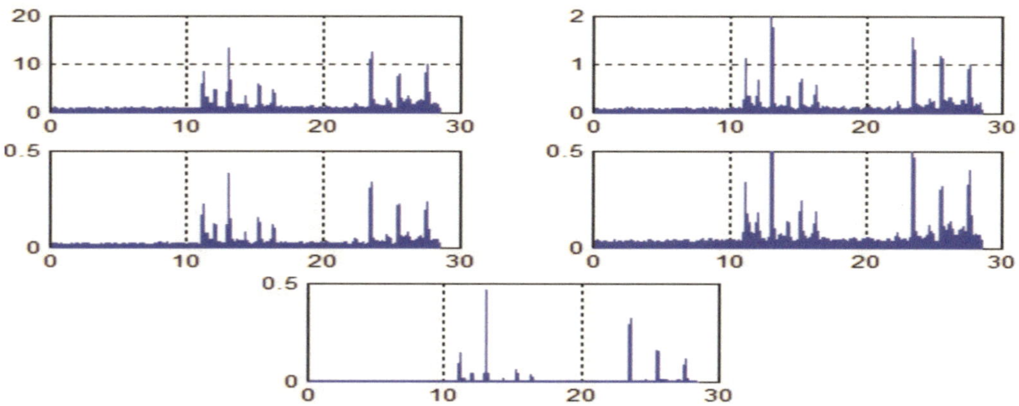

Figure 17. Features graphics of EMG signal recorded Channel 4.

		Index			Middle			Ring			Pinkie			Thumb		
		θ_1	θ_2	θ_3	θ_1	θ_2	θ_3	θ_1	θ_2	θ_3	θ_1	θ_2	θ_3	θ_1	θ_2	θ_3
1	Motion 1	90	90	90	90	90	90	90	90	90	90	90	90	90	90	90
2	Motion 2	0	0	0	0	0	0	0	0	0	0	0	0	0	0	0
3	Motion 3	90	30	30	0	0	0	0	0	0	0	0	0	70	15	5
4	Motion 4	0	0	0	90	25	25	0	0	0	0	0	0	87	5	5
5	Motion 5	0	0	0	0	0	0	90	25	10	0	0	0	105	15	5
6	Motion 6	0	0	0	0	0	0	0	0	0	90	30	5	125	15	5

Table 9. Reference value for each finger joints.

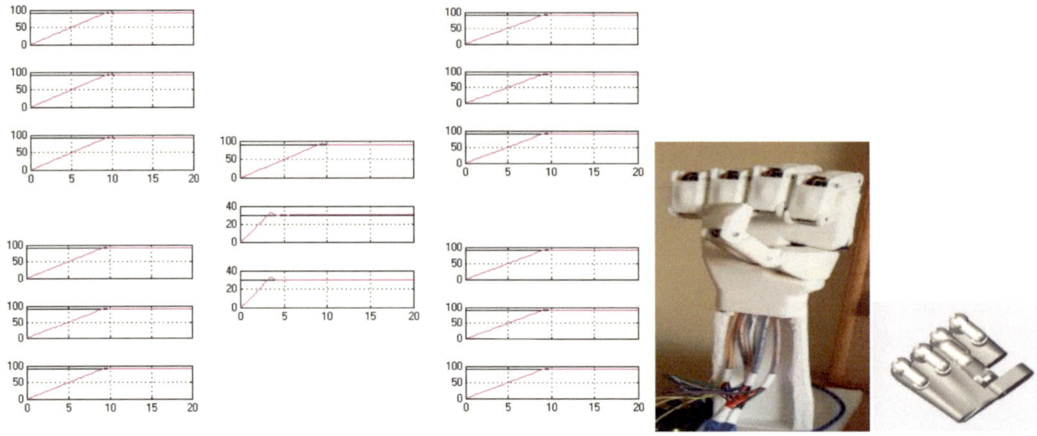

Figure 18. PID response graphics of five fingers for hand close and prosthetic hand photograph.

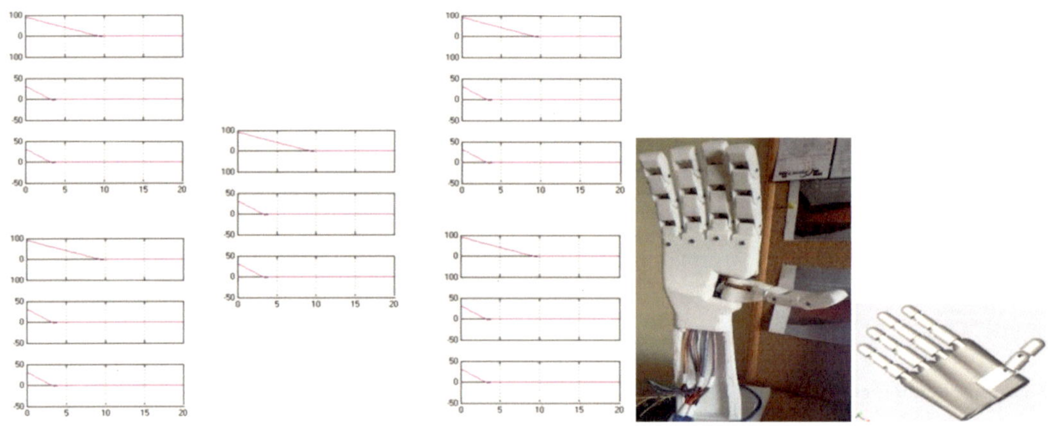

Figure 19. PID response graphics of five fingers for hand opening and prosthetic hand photograph.

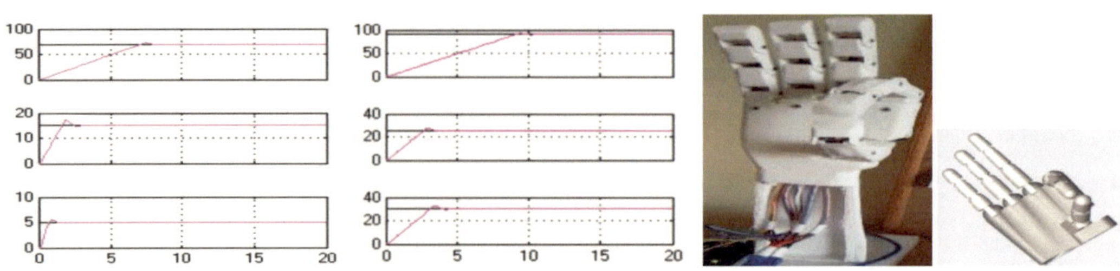

Figure 20. PID response graphics of five fingers for thumb-index touch and prosthetic hand photograph.

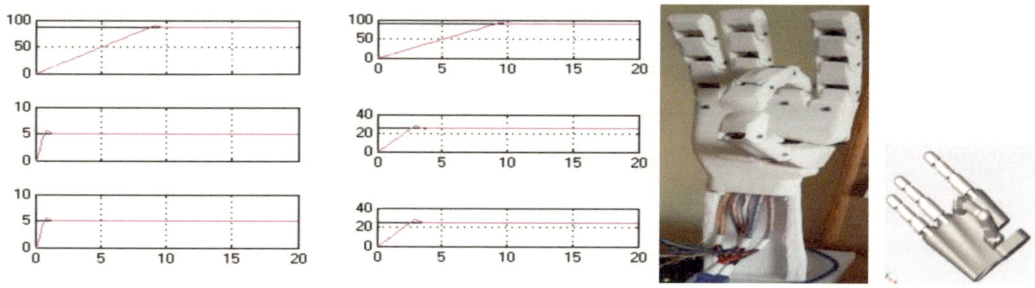

Figure 21. PID response graphics of five fingers for thumb-middle touch and prosthetic hand photograph.

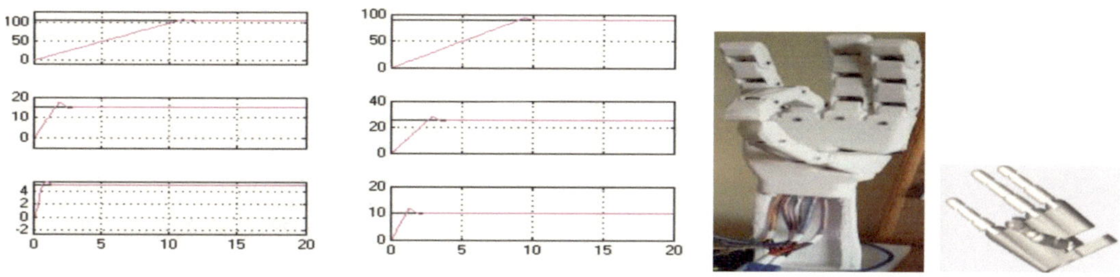

Figure 22. PID response graphics of five fingers for thumb-ring touch and prosthetic hand photograph.

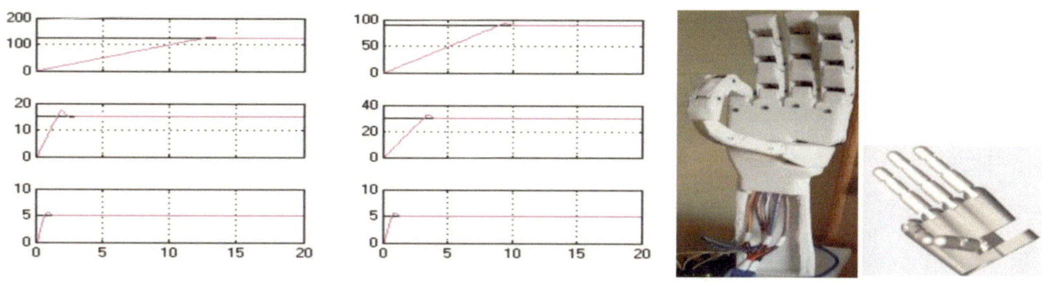

Figure 23. PID response graphics of five fingers for thumb-pinkie touch and prosthetic hand photograph.

Finger	Joint no		Motion 1	Motion 2	Motion 3	Motion 4	Motion 5	Motion 6
Thumb finger	1	Overshoot (deg.)	2.835	0.2932	2.137	2.936	3.025	3.655
		Steady state time (s)	9.8084	13.413	8.8084	9.988	10.8084	12.8084
		Steady state error (deg.)	0.046	0.041	0.037	0.027	0.021	0.024
	2	Overshoot (deg.)	2.755	0.3265	0.652	0.252	0.652	0.652
		Steady state time (s)	4.415	2.883	1.952	0.752	1.952	1.952
		Steady state error (deg.)	0.0052	2.6e-3	0.0001	0.0001	0.0001	0.0001
	3	Overshoot (deg.)	2.754	0.2696	0.357	0.357	0.357	0.357
		Steady state time (s)	4.524	1.972	0.956	0.956	0.956	0.956
		Steady state error (deg.)	0.0053	1.5e-3	1.5e-3	1.5e-3	1.5e-3	1.5e-3

Finger	Joint no		Motion 1	Motion 2	Motion 3	Motion 4	Motion 5	Motion 6
Index finger	1	Overshoot (deg.)	2.835	0.3299	2.835	0	0	0
		Steady state time (s)	9.915	9.71	9.915	0	0	0
		Steady state error (deg.)	0.045	10e-4	0.045	0	0	0
	2	Overshoot (deg.)	2.835	0.0368	2.349	0	0	0
		Steady state time (s)	9.915	4.555	7.0725	0	0	0
		Steady state error (deg.)	0.047	0.0183	0.0219	0	0	0
	3	Overshoot (deg.)	2.835	0.348	0.377	0	0	0
		Steady state time (s)	9.915	4.535	7.429	0	0	0
		Steady state error (deg.)	0.047	0.0202	0.0255	0	0	0
Middle finger	1	Overshoot (deg.)	2.8356	0.3244	0	2.8368	0	0
		Steady state time (s)	10.5022	10.279	0	10.52	0	0
		Steady state error (deg.)	0.0474	1e-3	0	0.0475	0	0
	2	Overshoot (deg.)	2.8356	0.3244	0	2.812	0	0
		Steady state time (s)	10.5022	10.279	0	3.437	0	0
		Steady state error (deg.)	0.0474	1e-3	0	0.0036	0	0
	3	Overshoot (deg.)	2.8356	0.3244	0	2.7812	0	0
		Steady state time (s)	10.5022	10.279	0	3.9265	0	0
		Steady state error (deg.)	0.0474	1e-3	0	0.0036	0	0
Ring finger	1	Overshoot (deg.)	2.8356	0.3244	0	0	2.8368	0
		Steady state time (s)	9.922	9.907	0	0	9.914	0
		Steady state error (deg.)	0.047	1e-3	0	0	0.047	0
	2	Overshoot (deg.)	2.8356	0.3244	0	0	2.781	0
		Steady state time (s)	9.915	9.9075	0	0	3.412	0
		Steady state error (deg.)	0.047	1e-3	0	0	0.0035	0
	3	Overshoot (deg.)	2.8357	0.3244	0	0	2.545	0
		Steady state time (s)	9.9055	9.906	0	0	1.884	0
		Steady state error (deg.)	0.047	1e-3	0	0	0.0005	0
Pinkie finger	1	Overshoot (deg.)	2.8356	0.3244	0	0	0	2.8368
		Steady state time (s)	9.9094	9.9122	0	0	0	9.29
		Steady state error (deg.)	0.0475	1e-3	0	0	0	0.0475
	2	Overshoot (deg.)	2.8357	0.3244	0	0	0	2.7883
		Steady state time (s)	9.9094	9.9122	0	0	0	4.8174
		Steady state error (deg.)	0.0475	1e-3	0	0	0	0.0052
	3	Overshoot (deg.)	2.8357	0.3244	0	0	0	2.636
		Steady state time (s)	9.9094	9.9122	0	0	0	1.3391
		Steady state error (deg.)	0.0475	1e-3	0	0	0	0

Table 10. PID performance value for each joint.

8. Conclusion

The main factor in increasing the functionality of the prosthetic hand to the extent of imitating biological hand functions is the movement of the fingers. The greater the number of movements the fingers can do independently of each other, the greater the ability of the prosthetic hand to move and the more successfully it can mimic the biological hand. Within the scope of this thesis, the function of the prosthetic hand is improved by six different finger movements. Bioelectrical signals of two separate users were recorded from the forearm muscles (the flexor pollicis longus, flexor carpi radialis, brachioradialis, extensor carpi radialis, extensor digiti minimi, and extensor carpi ulnaris) with the help of four surface electrode groups. Thus, a broad bioelectrical signal database was created. The recorded bioelectrical signals were subjected to a series of preprocessing and feature extraction processes to calculate the maximum, effective, mean, variance, and energy values of the EMG signals. An FL classification algorithm was developed to create an effective cognitive interaction network, and 90% classification success was obtained from these algorithms. The identified bioelectrical signals were applied to the designed three-dimensional prosthesis handheld simulator. The five-fingered and 15-jointed prosthetic hand prototypes produced with a 3D printer, and the positional control of the prosthetic finger joints was performed with the designed controllers. Each finger of the prosthetic hand was moved by an ultra-nano DC motor, and the position controls of the motors were provided by the designed PID. Thus, a cognitive interface and communication network were established between the person and the prosthetic hand with great success.

Acknowledgements

The subject of this chapter, which is Beyda TAŞAR's doctoral thesis, was supported by TÜBİTAK under the Domestic Doctoral Scholarship Program for Priority Areas in 2211 C. In addition, the study was supported by Fırat University Scientific Research Projects Management Unit within the scope of PhD Thesis Project number MF-14.25.

Author details

Beyda Taşar[1]* and Arif Gülten[2]

*Address all correspondence to: btasar@firat.edu.tr

1 Firat University, Engineering Faculty, Department of Mechatronics, Elazig, Turkey

2 Firat University, Engineering Faculty, Department of Electrical and Electronics, Elazig, Turkey

References

[1] Boostani R, Moradi MH. Evaluation of the forearm EMG signal features for the control of a prosthetic hand. Institute of Physics Publishing Physiological Measurement. Physiological Measurement. 2003;**24**:309-319. PII: S0967-3334(03)38946-4

[2] Nishikawa D, Yu W, Yokoi H, Kakazu Y. On-line supervising mechanism for learning data in surface electromyogram motion classifiers. Journal and Communication in Japan. 2002;**33**:14, pp. 2634-2643

[3] Arieta AH, Katoh R, Yokoi H, Wenwei Y. Development of multi-DOF electromyography prosthetic system using the adaptive joint mechanism. Applied Bionics and Biomechanics. 2006;**3**:101-112

[4] Zhao J, XIie Z, Jing L, Cai H, Hong L, Hırzınger G. EMG control for a five-fingered prosthetic hand based on wavelet transform and autoregressive model. International Conference on Mechatronics and Automation, China; 2006.

[5] Carrozza MC, Cappiello G, Stellin G, Zaccone F, Vecchi F, Micera S, Dario P. On the development of a novel adaptive prosthetic hand with compliant joints: Experimental platform and EMG control. In: Proceedings of the IEEE/RSJ International Conference on Intelligent Robots and Systems; April 2005. pp. 3951-3956

[6] Hocaoğlu E. Data acquisition and feature extraction for classification of prehensile SEMG signal for control of a multifunctional prosthetic hand [Master thesis]. Istanbul: Sabanci University, Mechanical Engineering; 2010.

[7] Carozza M, Cappiello G, Stellin G, Zaccone F, Vecchi F, Micera S, Dario P., On the development of a novel adaptive prosthetic hand with compliant joints: Experimental platform and EMG control, Intelligent Robots and Systems, 2005.(IROS 2005). pp. 1271–1276, 2-6 August 2005, 10.1109/IROS.2005.1545585

[8] Taylor Cl, Schwarz RJ. The anatomy and mechanics of the human hand. Artificial Limbs 2.2. 1955:**22**.

[9] Rakibul H, Vepar SH, Hujijbert H. Modeling and control of the Barrett hand for grasping. In: 15th International Conference on Computer Modelling and Simulation; 10–12 April 2013; Cambridge University.

[10] Hasan R, Rahideh A, Shaheed H. Modeling and interactional of the multifingered hand. In: Proceedings of the 19th International Conference on Automation & Computing. Brunel University, Uxbridge, UK; 2013.

[11] Michalewicz, Z. Genetic Algorithms + Data Structures = Evolution Programs. Berlin Heidelberg, New York: Springer-Verlag; 1999.

[12] Widhiada W, Douglas SS, Jenkinson ID, Gomm JB. Design and control of three fingers motion for dexterous assembly of compliant elements. International Journal of Engineering, Science and Technology. 2011;**3**(6):18–34

[13] Al-Assaf Y, Al-Nashash H. Surface myoelectric classification for prostheses control. Journal of Medical Engineering & Technology. 2005;**29**(5):203–207

[14] Parker P, Englehart K, Hudgins B. Myoelectric signal processing for control of powered limb prostheses. Journal of Electromyography and Kinesiology. 2006;**16**:541–548

[15] Englehart K, Hudgins B. A robust, real-time control scheme for multifunction myoelectric control. IEEE Transactions on Biomedical Engineering. 2003;**50**(7):848–854

[16] Hudgins B, Parker P, Scott RN. A new strategy for multifunction myoelectric control. IEEE Transactions on Biomedical Engineering. 1993;**40**(1):82–94

[17] Asghari Oskoei M, Hu H. Myoelectric control systems - a survey. Biomedical Signal Processing and Control. 2007;**4**(4):275–294

[18] Lamounier E, Soares A, Andrade A, Carrijo R. A virtual prosthesis control based on neural networks for EMG pattern classification. In: Proceedings of the Artificial Intelligence and Soft Computing; Canada; 2002

[19] Chu J, Moon I, Kim S, Mun M. Control of multifunction myoelectric hand using a real time EMG pattern recognition. In: Proceedings of the IEEE/RSJ International Conference on Intelligent Robots and Systems; China; 2005. pp. 3957–3962

[20] Mohammadreza AOI, Huosheng H. Review: Myoelectric control systems: A survey. Biomedical Signal Processing and Control. 2007;**2**:275–294

[21] Zhao Z, Chen X, Zhang X, Yang J, Tu Y, Lantz V, Wang K. Study on Online Gesture SEMG Recognition, Advanced Intelligent Computing Theories and Applications. With Aspects of Theoretical and Methodological Issues. 2007:1257–1265

[22] Mahdi K, Mehran J. A novel approach to recognize hand movements via sEMG patterns. Engineering in Medicine and Biology Society. In: 29th Annual International Conference of the IEEE; 2007.

[23] Zhizeng L, Xiaoliang R, Yutao Z. Multi-pattern recognition of the forearm movement based on SEMG, Information Acquisition. In: Proceedings of the International Conference on IEEE; 21–25 June 2004. DOI: 10.1109/ICIA.2004.1373391

[24] Roberto M. Electromyography Physiology, Engineering and Noninvasive Applications. IEEE Press, John Wiley & Sons Inc.; 2004, New Jersey, Canada

[25] De Luca CJ. Electromyography. Encyclopedia of Medical Devices and Instrumentation, John Wiley Publisher; 2006. pp. 98–109, New York City, United States, DOI: 10.1002/0471732877.emd097

[26] Hargrove L, Englehart K, Hudgins B. The effect of electrode displacements on pattern recognition based myoelectric control. In: IEEE Annual International Conference on Engineering in Medicine and Biology Society; 2006. pp. 2203–2206

[27] De Luca CJ. The use of surface electromyography in biomechanics. Journal of Applied Biomechanics. 1997;**1**(2):135–163

[28] SENIAM EMG protocol. Available from: http://www.seniam.org/(Download date: 21.03.2014)

[29] Daud WMBW, Yahya AB, Horng CS, Sulaima MF, Sudirman R. Features extraction of electromyography signals in time domain on biceps brachii muscle. International Journal of Modeling and Optimization. 2013;**3**(6)

[30] Yazıcı İ. EMG İşaretlerinin İşlenmesi Ve Sınıflandırılması [Master thesis]. Sakarya: Sakarya University, Institute of Science and Technology; 2008

[31] Taşan D. Protez Denetimi İçin Elektromiyografi (EMG)de Örüntü Tanıma, Yüksek Lisans Tezi, Ege Universty, Institute of Science and Technology, İzmir, 2008

[32] Akgün G, Demetgül M, Kaplanoğlu E. EMG Sinyallerinin Öznitelik Çıkarımı ve Geri Yayılımlı Yapay Sinir Ağı, Algoritması İle Sınıflandırılması, Otomatik Kontrol Ulusal Toplantısı, TOK2013, Malatya 26–28 Eylül 2013.

[33] Chen WT, Wang Z, Ren X. Characterization of surface EMG signals using improved approximate entropy. Zhejiang University Science B. 2006;**7**(10): 844–848

[34] Chu J, Moon I, Mun M. A real-time EMG pattern recognition system based on linear-non-linear feature projection for a multifunction myoelectric hand. IEEE Transactions on Biomedical Engineering. 2006;**53**:2232–2238

[35] Chan F, Yong-Sheng Y, Lam F, Yuan-Ting Z, Parker P. Fuzzy EMG classification for prosthesis control. IEEE Transactions on Rehabilitation Engineering. 2000;**8**(3)

[36] Yaraş B, Hüseynov R, Namazov M, ÇeliKKale İE, Şeker M. Fuzzy control and sliding mode fuzzy control of DC motor. Journal Of Engineering And Natural Sciences Mühendislik Ve Fen Bilimleri Dergisi, Sigma. 2014;**32**:97–108

[37] Ajiboye A, Weir R. A heuristic fuzzy logic approach to EMG pattern recognition for multifunction prosthesis control. IEEE Transactions on Biomedical Engineering. 2005;**52** (11): 280–291

[38] Karlik B, M.O., T., & M., A. A fuzzy clustering neural network architecture for multifunction upperlimb prosthesis. IEEE Transactions on Biomedical Engineering. 2003;**50**:1255–1261

[39] Vuskovic M, Du SJ. Classification of prehensile EMG patterns with simplified fuzzy ARTMAP networks. In: Proceedings of the International Joint Conference on Neural Networks; 2002, 3, pp. 2539–2544

[40] Khzeri M, Jahed M, Sadati N. Neuro-fuzzy surface EMG pattern recognition for multifunctional hand prosthesis control, Industrial Electronics. In: IEEE International Symposium; 2007

[41] Zheng L, He X. Classification techniques in pattern recognition. In: WSCG, Conference Proceedings; 2005, ISBN 80-903100-8-7

[42] Taşar B, Yakut O, Gülten A. Object detection and grip force control via force sensor for EMG based prosthesis hand, E205. In: International Conference on Electrical and Electronics Engineering (ICEEE); 27–28 Nisan 2015; Ankara

[43] Taşar B, Gülten A, Yakut O. Modeling controlling and simulation of 15 DOF multifunctional prosthesis hand using SimMechanics, C201. In: International Conference on Automatic Control (ICOAC); 27–28 Nisan 2015; Ankara

[44] Taşar B, Gülten A, Yakut O. Kinematic analysis of the human hand for desing prosthetic hand. In: 3rd International Congress on Natural and Engineering Sciences; September 9–13, 2015; Sarajevo, Bosnia and Herzegovina: International University of Sarajevo

[45] Dung LT, Kang HJ, Ro YS. Robot manipulator modeling in Matlab SimMechanics with PD control and online Gravity compensation. IFOST Proceedings. 2010

[46] Fedák V, Durovsky F, Üveges R. Analysis of robotic system motion in SimMechanics and MATLAB GUI Environment. MATLAB Applications for the Practical Engineer. Chapter 20. 2014

[47] Bencsik AL. Appropriate mathematical model of DC servo motors applied in SCARA robots. Acta Polytechnica Hungarica. 2004;**1**(2):2004–2099

[48] Matlan Tuturial, DC Motor Position Simulink Model. Available from: Dowload:http://ctms.engin.umich.edu/CTMS/index.php?example=MotorPosition&section=Simulink-Modeling

[49] Meier R. Modeling DC servo motors control systems, Tech Note, Milwaukee School of Engineering 1025 North Broadway, Milwaukee. https://faculty-web.msoe.edu/meier/ee3720/technotes/dcservo.pdf

[50] Haupt Randly L, Haupt Sue E. Practical Genetic Algorithms. USA: A Wiley-Interscience Publication; 1998

[51] Davis L. Handbook of Genetic Algorithms. New York, NY: Van Nostrand; 1991

[52] Wook C, Ramakr Shna RS. Elitism-based compact genetic algorithms. IEEE Transactions on Evolutionary Computation. 2003;**7**(4):367–385

[53] Astrom KJ, Hagglund T. PID Controllers: Theory, Design and Tuning. Instrument Society of America; 1995. p. 343, Research Triangle Park, North Carolina, 1995

[54] Taguchi H, Araki M. Two-degree-of-freedom PID controllers. Proceedings of the IFAC Workshop on Digital Control: Past, Present and Future of PID Control; Elsevier; pp. 91–96. 5-7 April 2000, Terrassa, Spain

[55] Bennett S. A History of Control Engineering, Peter Peregrinus; 1993. pp. 28–69, London, United Kingdom

Section 5

Applications of Fuzzy in Industrial Engineering

Fuzzy Logic Application, Control and Monitoring of Critical Machine Parameters in a Processing Company

Tawanda Mushiri

Additional information is available at the end of the chapter

http://dx.doi.org/10.5772/67762

Abstract

The processing company under study found out that the boiler was the key machine and needs artificial intelligence monitoring and control. It was simulated under Matlab software and oil level, and pressure and temperature were to be modelled and controlled using the programmable logic controller (PLC) with a fuzzy logic controller as the main brain of control. The company is for processing of fruits to produce juice.

Keywords: fuzzy logic, control, critical machinery, monitoring, processing

1. Introduction

Beverage industry denotes the industry accountable for the manufacture of drinks through the usage of highly automated systems, which are responsible for the production of beverages within a short period of time [1–3]. Beverage industry is greatly affected with profitability problems. The ever-increasing costs of new equipment and spares, perennial foreign currency shortages and the need for improved competitiveness bring about the need for more effective maintenance systems [4–6]. This results in maximum utilization of plant-installed capacity through improved reliability, uptime, quality and asset life—all achieved at optimal levels of costs versus benefits. Emphasis should be on ensuring that the correct maintenance is being done (doing the right job) rather than merely ensuring that maintenance is being done correctly (doing the job right). At the processing company, process losses have traditionally been 1.5% of total losses but increased to 10% in 2011. This has raised concern and hence the need to find new emerging maintenance management philosophies, such as improving maintenance cost effectiveness as one sure way of increasing the overall profitability [7].

1.1. Background

Currently, the plant uses run-to-failure maintenance as a strategy to take care of its machines. This approach is reported to be creating a high risk to workers, production and property. This is seen by the high rate of unscheduled events associated with breakdowns, emergency equipment, long working hours and high maintenance costs. This has lowered down the image of the company in terms of competition when compared the benchmarks with other beverage companies in developing nations. The progression towards a global economy has increased the base of competition for almost all companies. By competition, it is suggested that every organization out there is always targeting to keep a certain score. The record on a scorecard may be a ratio of more sales, increased revenue or a mounting customer base. Despite the benchmark used for the measurements, for a business to remain competitive, there is a basic company demand to improve at maintaining and taking care of machinery as well as meeting customers' requirements. The competitiveness of production corporations relies on the availability and productivity of their manufacturing equipment, which is possible if all manufacturing losses are recognized and eliminated so that the products can be sold on a marker at a minimum cost.

1.2. Fuzzy logic

The rules can process simple arithmetic and logical operations with the help of Locator Identifier Separation Protocol (LISP), but a process control algorithm that is based on fuzzy logic is called fuzzy control [8]. The oldest and most common maintenance and repair strategy are "fix it when it breaks". The appeal of this approach is that no analysis or planning is required [9]. Early detection of faults in a machinery is a key parameter in control and maintenance to avoid failure of equipment [10].

2. Critical machinery

The criticality index is regularly used to decide the extent of the nature of assessment on a given machine by checking the machine's objective [11], excess (that is on the off chance that if the machine falls flat, is there a standby machine which can assume control), repairing expenses, idle sways, well-being, security and environmental concerns and other considerable features. The criticality record puts all machines into one of three classifications as follows:

Critical equipment: This is the fundamental machinery at the processing company. With basic equipment which is the core focus of the procedure, it is understood to require adequate online state checking to ceaselessly capture much information from the mechanism as could be expected paying little heed to expenditure and is frequently indicated by the plant protection. Estimations, for example, differential development, velocity, spiral relocation and shaft pivotal, dislodging and packaging vibration, temperatures, weights and loads, are taken where conceivable. Such qualities are regularly sustained into a hardware administration programming system, which is fit for slanting the recorded information and giving the administrators data, for example, executing information and even anticipating failures and giving conclusions of breakdowns beforehand.

Vital equipment: These are units that are essential to the procedure; however, in the event that there is a breakdown, the procedure still proceeds. Excess units (if accessible) would be in this domain. Analysing and regulation of these units are additionally crucial to keep up option strategies given that critical machinery falls flat.

Balance of industrial equipment or the general purpose: These are the components that complement the whole plant and are ordinarily assessed, utilizing a handheld information gatherer, as said beforehand, to occasionally build an idea of the strength of the machine.

2.1. Types of failure caused by not maintaining machinery

Breakdowns might be characterized by their seriousness in any of these four classes and are contingent upon the characterized breakdown impacts below:

1. Insignificant breakdown: Any breakdown that does not debase the general routine and adequacy of the framework past worthy breaking points [12].
2. Key breakdown: Any breakdown that will corrupt the framework implementation and viability past worthy points of confinement, however, they can be controlled [13].
3. Critical breakdown: Any breakdown that will corrupt the framework past adequate points of confinement and could make a security danger if a prompt restorative move is not made [14].
4. Catastrophic breakdown: Any breakdown that could bring about critical framework harm, for example, to block useful achievement, and bring about death and staff wounds [15].

2.2. Types of condition monitoring techniques

In line with the research area of interest, condition monitoring will be looked at since it is a branch of Condition Based Maintenance (CBM).

2.2.1. Noise and vibration

All machines vibrate, and when they are in good condition, their frequency spectrum has a characteristic form; any departure from this form indicates that something is wrong—fatigue, or wear, or ageing of something of some component. Parameters that are useful include amplitude, frequency and phase angle [16].

2.2.2. Amplitude

This gives an indication of the stress under which a piece of rotating machinery is working, in particular, it can give a measure of the eccentricity (out-of-roundness) of a rotor [17].

2.2.3. Frequency

Through the frequency spectrum, you can detect a fault in rotating machinery. Vibrations fall into two main classes:

Synchronous: Frequencies are in multiples or sub-multiples of the frequency of rotation, that is, they are harmonics or sub-harmonics of that frequency.

Asynchronous: These are not related to the rotation frequency; they can be the natural frequencies of various parts of the system, which can be identified [17].

2.2.4. Phase angle

It locates the high point in a rotor that is not perfectly circular and thus gauges its out-of-balance characteristics. Machines must be continuously monitored to denote their state. The following machines can do these:

1. Clearance recorder: recording the actual movements of the shaft which generate the vibrations.
2. Speed recorder: mounted externally to a machine, and it gives a strong signal at medium frequencies, depending on the temperature and the general environment.
3. Accelerometer: also installed to a machine, and it gives a strong signal at high frequencies [17].

2.2.5. Temperature

Temperature recording is a relatively simple matter at the industrial level. Change of temperature in rotating machinery is often a sign of deterioration and is therefore something to which close attention should be given. This is currently practised in thermography [18].

2.2.6. Tribology

An examination of the particles suspended in oil can give very valuable information. The amount of suspended material is an indicator of the state of deterioration of the machine; the composition can identify the source of the wear and thus the component that is failing. The necessary analysis can be done in the laboratory with the electron microscope. Basically, the monitoring techniques that were listed in **Table 1** are almost the major ones currently used.

Type	Method
Visual	Eye
Temperature	Thermometer, thermocouple
Lubricant monitoring	Filtering, spectroscopy
Vibration	Signal frequency analysis
Crack	Di-penetrant analysis, radiography
Corrosion monitoring	Eye, corrosometer

Table 1. Summary of condition monitoring techniques.

http://dx.doi.org/10.5772/67762

3. Problem behind the processing company (PC)

The reactive nature of the corrective maintenance strategy at the company has resulted in loss of trivial production time. Downtime due to breakdown is greatly affecting the production targets and delivery of juice products to customers which had resulted in loss of goodwill and trust from consumers. The processing company (PC) uses a number of raw materials for the production of the fruit juice concentrates: mangoes, guavas, oranges, tomatoes, lemons, grape fruits, granadilla and other continuous inputs such as municipality and borehole water, diesel, electricity and labour. The company has got equipment in place which were mainly supplied by a company in Italy. The mission of the production department is to safely produce quality products and maintain equipment at a lowest possible cost to meet sales and marketing demands.

3.1. Failure in equipment

Using records from previous seasons, equipment with high failure rates and high downtimes is identified. For easy interpretation and identification of this equipment, the information is sorted using Microsoft Excel and graphs are drawn. From the bar graph showing the relative contributions of equipment to the total downtime, equipment that contributed the most to the total downtime is identified. Breakdown of any of these equipment results in stoppage of the production; therefore, the production loss that is represented by a breakdown of each of these machines is equivalent to the production of the diffuser line, which is 60% of total production (300 TCH of 500 TCH). The following graph in **Figure 1** shows the downtimes caused by the breakdown of each item during the last juice extraction season, and **Figure 1** shows how extraction takes place.

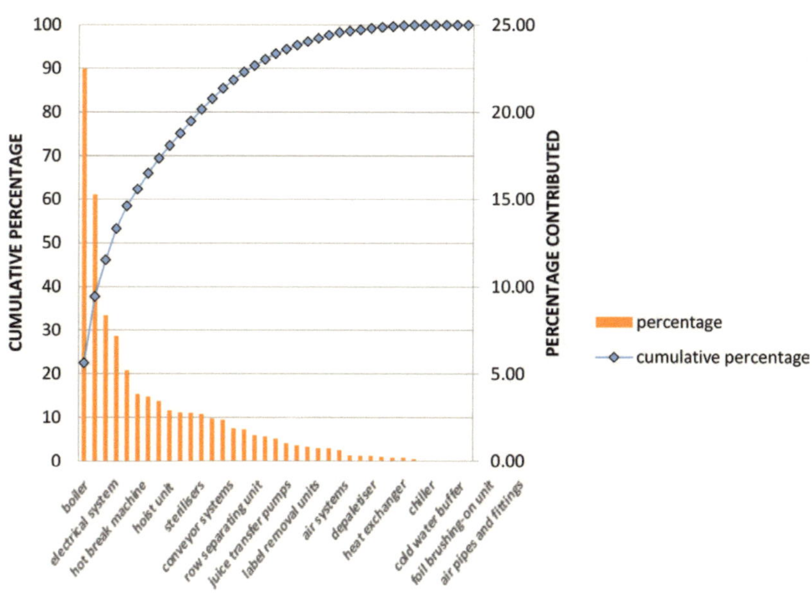

Figure 1. Unit and cumulative downtime of equipment.

3.2. Planned stoppages

The production line is stopped once every 2 weeks. The shutdown is for the whole plant, as some equipment will be due for cleaning and some equipment would be due for servicing. The shutdown starts at 0000 h and servicing starts at 1000 h. The time from 0000 to 1000 h is used to stop the processes, empty some vessels and allow some equipment to cool. This time is also used to plan some of the jobs, perform pre-task risk assessments, order and transport spares from the stores department and also to service some equipment. Full-fledged servicing kicks off at 1000 h and continues up to 1600 h. At 1600 h, the start-up process begins with steam being released into the vessels to warm them. Production is started around 1700 h when vessels are ready to receive juice, and hot imbibition water is available. The processes are then started, and the levels gradually increased. Full production levels are achieved by 1800 h. The total downtime for this planned stoppage is thus 18 h. A crushing season starts early April and ends in mid-December. On average, it has 36 weeks. The average number of hours for planned stoppages annually is thus:

$$\text{Total annual planned downtime} = \text{Number of stoppages} \times \text{Downtime per stoppage} \quad (1)$$

$$\text{Total annual planned downtime} = \frac{36}{2} \times 18 = 324\ \text{h}. \quad (2)$$

Figure 2 shows a comparison of planned stoppages and unplanned stoppages. This is a problem in this company with more stoppages. This has justified the need for my research, as shown in **Figure 2**.

Figure 2 is very critical as most of the time, the plant is always shutdown, which results in a great deficiency and loss of goodwill from the customers as the plant is failing to meet the targets.

3.3. Plant performance

From operating history, the availability performance for each item was evaluated, and the results are shown in **Figure 3**. All availability performance scores were above 90 so for clarity

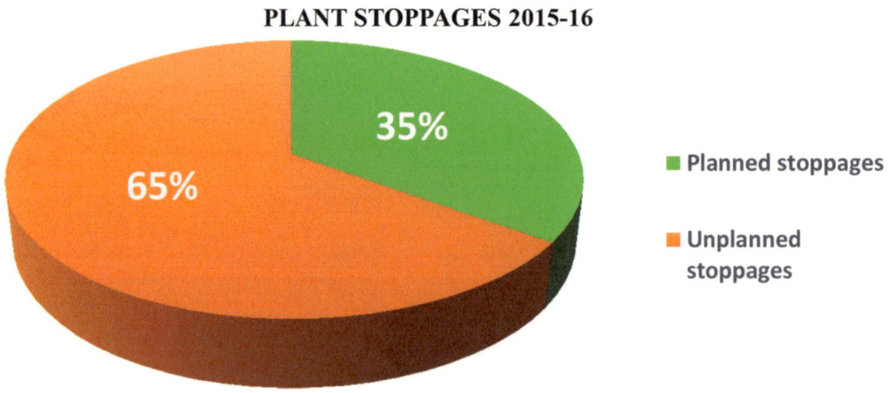

Figure 2. A pie chart for planned and unplanned stoppages.

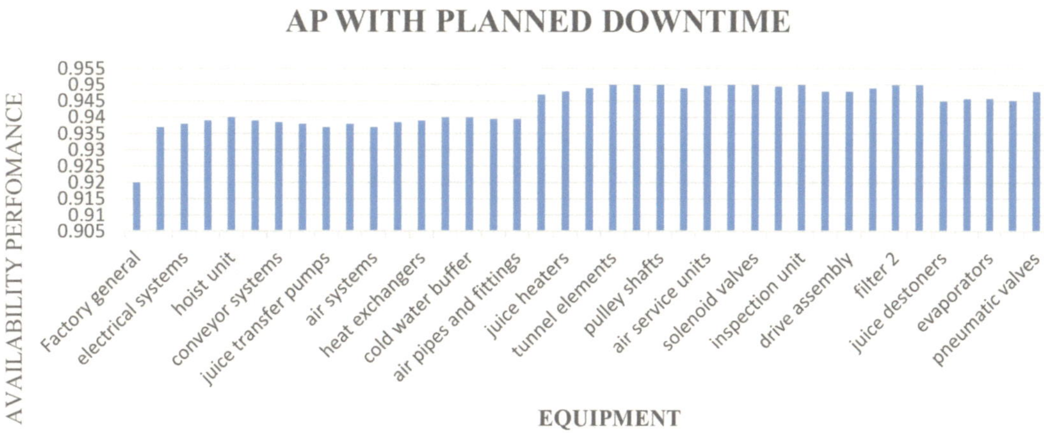

Figure 3. Equipment availability performance including planned downtime.

in comparing, the axis for availability performances was broken such that it starts at 90–100. The main contributor to downtime is planned downtime. This affects the availability of all equipment in the production line, but it is not all equipment that requires servicing every fortnight. Some of the equipment that requires servicing does not need 18 h. In view of this, maintenance schedules currently in use were analysed in order to identify equipment that really requires servicing fortnightly and has the least maintainability. **Figure 3** shows the availability performance of the plant with planned downtime.

3.4. Equipment criticality analysis

This section now seeks to identify the equipment that is critical to mainly to production, health, safety, environment and operational costs. This classification of equipment according to risk level is done so as to prioritize equipment in terms of maintenance, work orders, costs and inspection. The criticality of equipment was analysed based on two sides: the effect of failure and its consequence. Equipment history data and interviews were the methodology, which was used to find out which equipment has major effect on production. This was used to give scores to various equipment using the risk analysis matrix balanced score card by Nowlan and Heap, which is shown in Appendix.

3.4.1. Risk analysis matrix balanced score card: Appendix

As shown in Appendix, there are two categories of scores (Priority and Equipment Score), which are used to calculate the Criticality Score. The Priority Score is obtained from multiplying the results of the three factors (Factor E, Factor F and Factor G). The Equipment Score is obtained from multiplying the results of the four factors (Factor A, Factor B, Factor C and Factor D). This is summarized below.

Criticality score (The Criticality Score is obtained by multiplying the Equipment Score by the Priority Score = **ES * PS**).

Equipment score (The equipment score (ES) is obtained by multiplying the results for the four factors = **Factor A * Factor B * Factor C * Factor D**). The factors A–D are shown in Appendices 1–4.

Priority score (The priority score is obtained by multiplying the results for the three factors **(= Factor E * Factor F * Factor G)** as shown in **Table 2** and **Figure 4**.

NB: The general details on the factors used are presented in Appendices 1–4. From Table 2, it is clear that criticality equipment ranking for PC is as follows: boiler, juice transfer pumps, plant conveyor systems, sterilizers and hot break machines, and the boiler is critically affected most as shown in **Figure 5**.

The graph above shows the Key Performance Indicators (KPI) trend in the year 2010–2015. From the graph, the plant availability trend is showing an increment over the years, and the performance rate seems to be showing the same trend; thus, it seems plant availability is proportional to performance rate. Planned downtime trend has shown to be decreasing with the unplanned downtime showing an increase over the years, meaning that the maintenance effort is opposing the world-class standard as it is showing an increase in reactive domination. OEE is showing a haphazard trend as it seems to be aligning with the performance rate. OEE depends mainly on the performance rate.

Equipment	Factor A	Factor B	Factor C	Factor D	ES	Factor E	Factor F	Factor G	PS	CS
Boiler	4	4	3	4	**192**	4	4	4	64	**12,288**
Pumps and piping	4	4	3	3	**144**	4	3	4	48	**6912**
Plant conveyor Systems	4	4	3	4	**192**	4	2	4	32	**6144**
Sterilizers	4	4	3	2	**96**	2	4	3	24	**2304**
Hot break machine	4	4	3	2	**96**	2	3	3	18	**1728**
Evaporators	2	3	3	2	**36**	3	4	2	24	**864**
Heat exchangers	2	4	3	2	**48**	2	3	2	12	**576**
Super pulp creamers	4	4	3	2	**96**	2	2	1	4	**384**
Chillers	3	2	3	1	**18**	2	4	1	8	**144**
Aseptic filling machine	4	4	1	1	**16**	1	3	2	6	**96**
Sorting plant	4	3	1	2	**24**	1	1	2	2	**48**
Compressors	3	3	2	1	**18**	1	2	1	2	**36**
Washing line	2	2	2	1	**8**	1	2	1	2	**16**
Drier	1	1	1	1	**1**	1	1	1	1	**1**
CIP Plant	1	1	1	1	**1**	1	1	1	1	**1**

Table 2. Selection of critical machines.

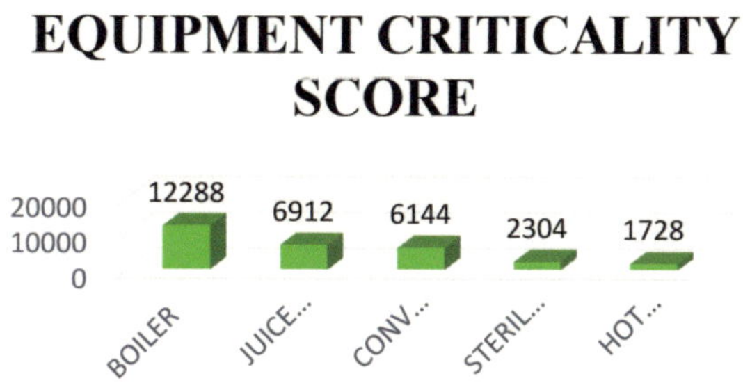

Figure 4. Critical equipment scores.

3.5. Boiler analysis

When failure occurs in any manufacturing process, it is critical to identify and analyse the root causes leading to that failure. Hybrid proactive maintenance strategy is condition-based maintenance complementing preventive maintenance (PM) strategy. PM uses mean time to failure (MTTF) as its pivot in the calculation of PM intervals. This gives calculations for MTTF on unrepairable equipment and mean time before failure (MTBF) on repairable equipment based on the assumption that the failure rate of each component is constant. Failure Modes Effects and Criticality Analysis (FMECA) analysis for the boiler is discussed in detail in a later chapter. The risk priority number chart is drawn with the highest scores corresponding to the most critical boiler component. The root cause failure analysis (RFCA) for the boiler machine is presented in the form of the Ishikawa diagram which was drawn using the Edraw max software as in **Figure 6**. The Ishikawa diagram is also called a fishbone diagram. The gearbox and feed check are reportedly failing almost every week. A Mamdani fuzzy logic controller is further presented in this chapter to monitor parameters such as moisture and dust effects on the feed check valve as well as oil level and torque control on the boiler gearbox. The fuzzy logic framework was determined among other counterfeit shrewd frameworks to be best fitting to comprehend the breakdown difficulties of the heater. A gearbox is continually sticking, and it is difficult to investigate the breakdown of fuzzy logic, and the reason is an instrument that

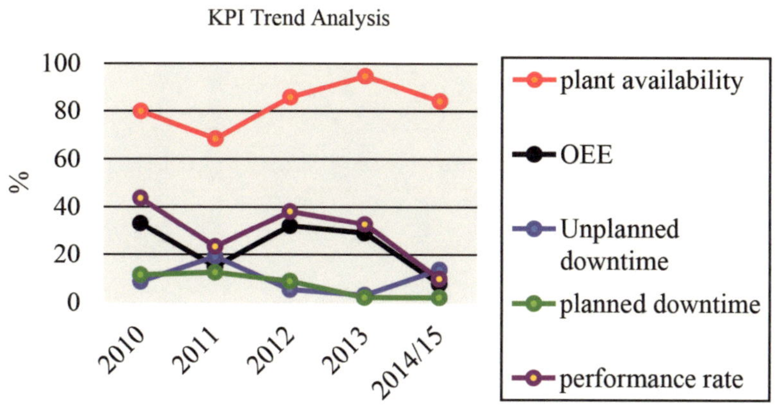

Figure 5. KPI trend analysis.

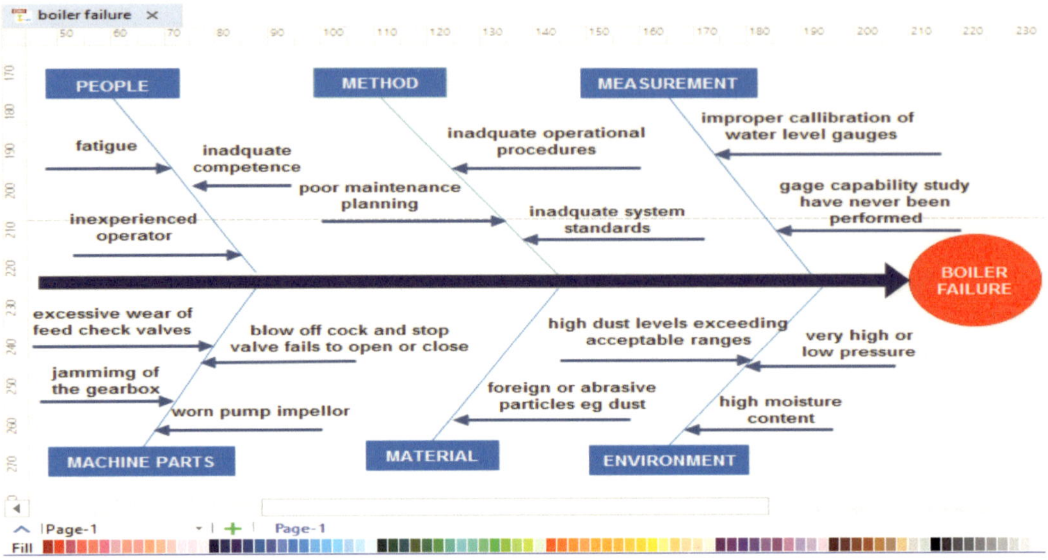

Figure 6. Ishikawa diagram for the root cause failure analysis of a steam boiler failure (Edraw Max).

was utilized for observation. On the risk priority side, it helps me knowing which segments to take and observe the most.

In general, I will do main emphasis on the raw data analysis of the overall plant performance as shown in **Figure 7**. Critical equipment has been selected using the Nowlan and Heap procedure. The critical components for the plant were found to be the boiler, juice transfer pumps, conveyor system, sterilization unit and hot break machine. This is the equipment that stands to benefit the most from a change in maintenance policy. Because of the time constraint, suitable maintenance strategies will be produced for the boiler of these critical items, which show the greatest opportunity for improvement.

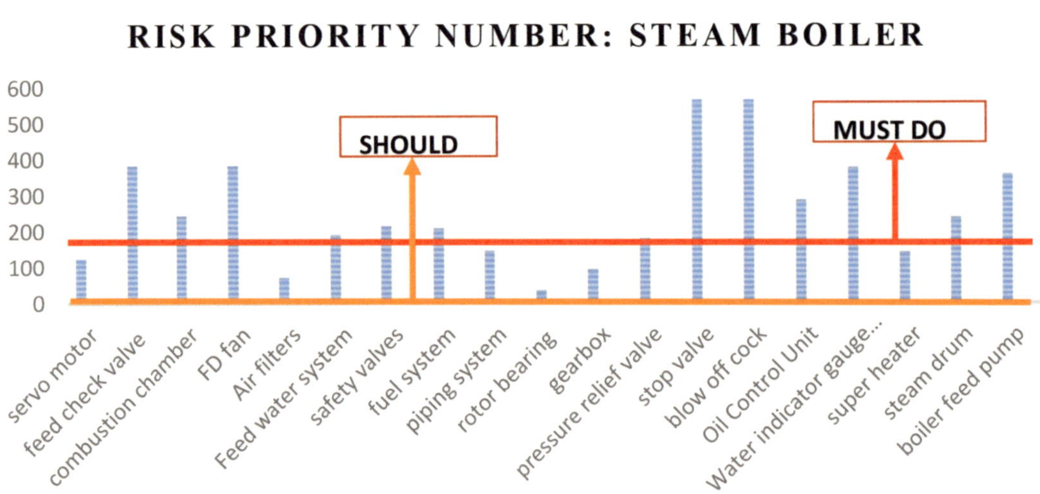

Figure 7. Risk priority number for steam boiler.

http://dx.doi.org/10.5772/67762

4. Results of the boiler for case study using fuzzy logic

Analysis of the boiler is carried out in this section, and **Figure 8** shows the risk priority number for the steam boiler.

4.1. Fuzzy logic systems (using Matlab software)

The purpose of the gearbox in a boiler setup is to change the speed ratio between the motor and kicker. The gearbox is seen to be continually sticking, and it is difficult to investigate the underlying reason for the breakdowns along these lines; fuzzy rationale framework is created to screen the torque and oil level when the boiler container is stacked. The gearbox must be halted when terrible conditions exist in the evaporator as an integrated upkeep procedure to abstain from sticking, henceforth, canny fuzzy rationale control. From the interviews with the engineers and plant artisans, as well as maintenance of history log books, it was noted down that oil level and torque are the major factors which are contributing to the jamming of the gearbox whenever the boiler is being loaded. In this chapter, the oil level is being controlled and is to be modified with specific levels: high, acceptable and low. Controlled torque is to be modified with specific levels: high, normal and low. This is highlighted beneath.

4.1.1. Oil level and torque control

4.1.1.1. Effects of oil level and torque control on gearbox jamming

Moisture inside the valves may result from the compressed air passing through the valve. The effects of moisture on the valves include:

- Moisture may cause rusting in the moving parts of the valve.

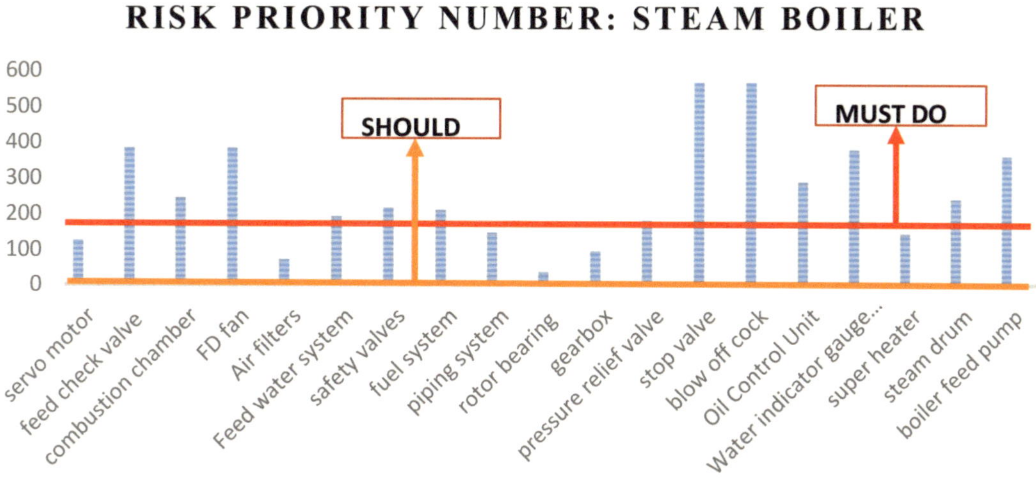

Figure 8. Risk priority number for steam boiler.

- Increased rate of wear of the valve material may also result due to moisture. Moisture washes the lubrication away, which will result in the eventual failure or malfunctioning of the valve.
- The industrial processes, which rely on the full functionality of the pneumatic control valves, may be jeopardized, and this usually results in costly breakdowns of the machine.
- Air- or gas-operated instruments may give inaccurate readings due to corrosion of the material and hence interrupting the plant processes.
- The rubber diaphragms inside the pneumatic valves can be stiffened and will eventually rupture due to the moisture flowing through them.

The following are the range of values of moisture, which are tolerated and some not tolerated inside the valve:

4.1.1.2. Oil level control

4.1.1.2.1. For a range of 0–100%

This is the aggregate scope of oil level either acknowledged or unaccepted in the gearbox.

- 0–40%

This alerts the artisans for refiling when oil is about to run out.

- 40–80%

At this level, it is acceptable, and the gearbox can continue operating.

- 80–100%

The level is fine and the gearbox has to keep running.

4.1.1.3. Torque control

4.1.1.3.1. For a range of 0–12,000 Nm

This is the total range for torque in the gearbox at any time (unaccepted or accepted).

- 0–2000 Nm

The range is very fine for starting the gearbox and is fine for running the boiler.

- 2000–8000 Nm

The torque is very fine for the gearbox to continue running.

- 8000–12,000 Nm

This range is very high, and it needs close monitoring, otherwise, stop the boiler.

4.1.1.4. *Output control*

4.1.1.4.1. *For gearbox range of 0–100%*

The gearbox has to run or stop, either of the 2, that is the meaning of 0 for stop and 100% for running that is the 0 or 1 for logic:

- 0 stop

At this range for the stop signal, the gearbox is to stop being controlled by torque and oil level

- 1 run

This involves the range 0–100%. All the conditions are being met (oil level and torque), and the gearbox can run effectively.

4.1.2. *Simulation of the effects of oil level and torque on the gearbox using fuzzy logic*

The rule base consists of a collection of expert rules, which are required to meet the control goals. These control rules can be developed from survey results, common sense, general principles and intuitive knowledge. The *IF – THEN or IF – AND – THEN* rules are mainly going to be used in designing the controller as shown in **Figure 9**. The situation for which the rules are projected is given by the IF part. The fuzzy system reaction in this state will be given by the THEN part. **Figure 9** below shows the fuzzy logic (Matlab software) screenshot for the two inputs (oil level and torque) and the output parameter (gearbox).

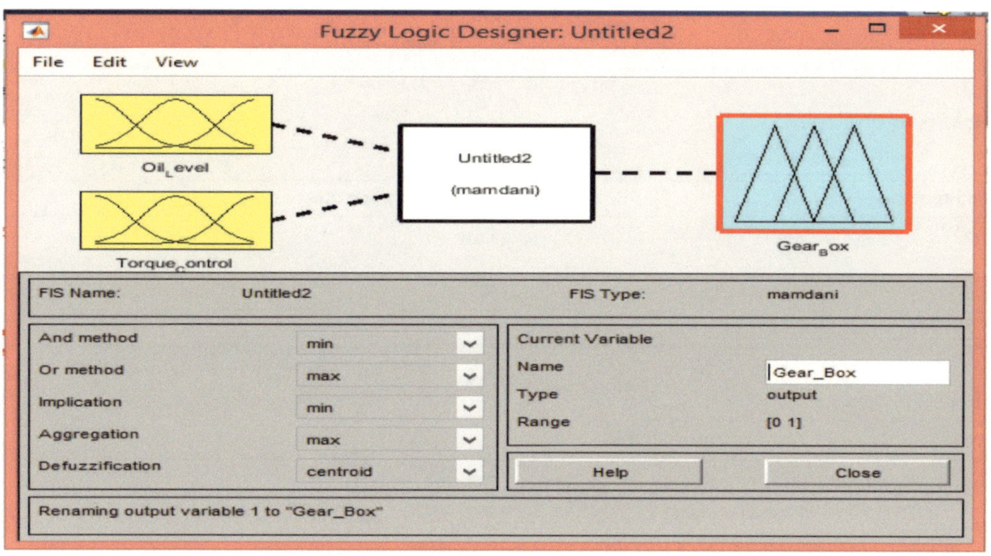

Figure 9. FIS editor for gearbox.

4.2. Control of gearbox jamming using fuzzy logic tool box

4.2.1. Membership function editor for oil level and torque monitoring

The membership function editor is the central concept of the fuzzy set, which has values ranging from 0 to 1 in the y axis. The ranges of control were done in **Tables 3–5**. Membership function editor is the stage whereby the ranges, as explained above, are inserted into the fuzzy logic Matlab software and the screenshot of the data is taken as shown in **Figure 10**. It helps to display all membership functions connected with the input and output for the entire Fuzzy Inference System (FIS). In this case, a triangular fuzzy set membership function is used for oil and torque monitoring. This is shown clearly in figures below with ranges for oil level monitoring varying from low, acceptable and high and the numerical values being added. For torque control, the corresponding numerical values range from low, normal and high. The gearbox is either run or stopped, depending on the oil level and torque conditions as in **Table 5**.

The membership functions editor for the output is shown in **Figure 11**, which is the gearbox that is plotted in the form of a triangle. Fuzzy logic is an artificial intelligence software which can store the output ranges in its memory and can learn the system to give solutions in what can be done. The Matlab software of fuzzy logic consists of the rule editor function which allows for the generation of the rules. This is done after inserting the range of values for oil level and torque monitoring and the gearbox outputs. The IF…..THEN……ELSE rules are being used for retrofitting as shown in **Figure 12**.

Input range (%)	Fuzzy variable name
0–40	Low
40–80	Acceptable
80–100	High

Table 3. Input 1: oil level.

Crisp input range (%)	Fuzzy variable name
0–2000	Low
2000–8000	Normal
8000–12,000	High

Table 4. Input 2: torque control.

Indicator	Symbol
Stop	0
Run	1

Table 5. Output 1: gearbox.

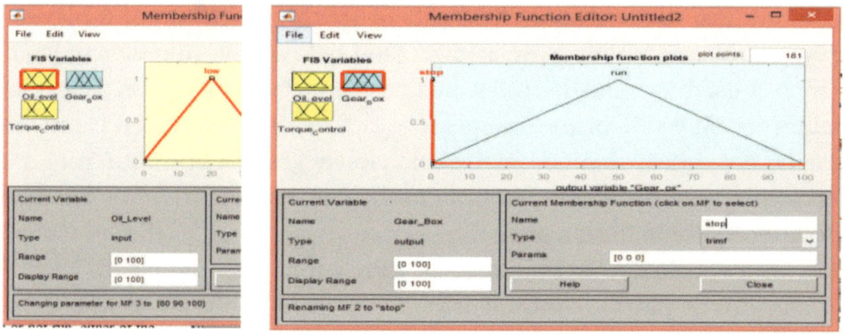

Figure 10. (a) Membership function editor for the output, gearbox and (b) membership function editor for oil level monitoring.

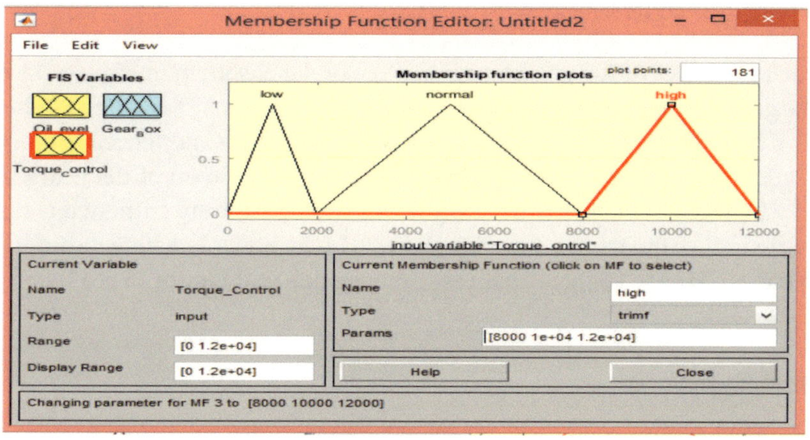

Figure 11. Membership function editor for torque control.

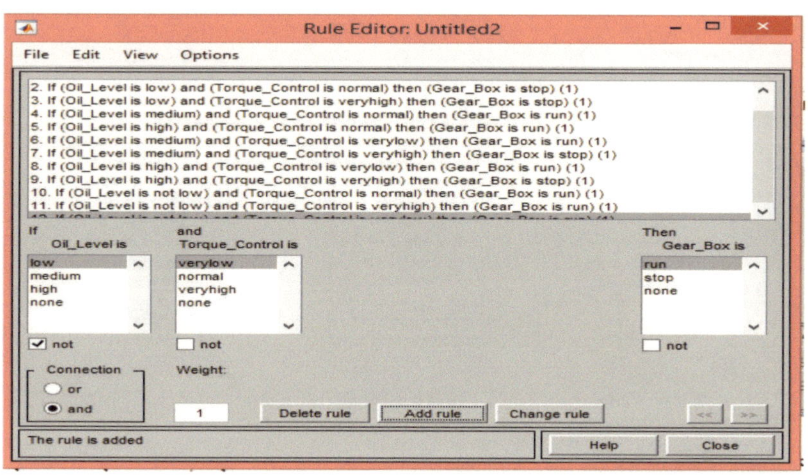

Figure 12. Rule editor for the fuzzy logic control.

After the rules are inserted, the overall results of combined effects of oil level and torque monitoring on gearbox monitoring are as discussed below. The rule viewer is an intelligent software, which is used to view the rules created using the rule editor. The rule viewer for the range values for oil level, torque and gearbox operation is discussed clearly in the figure below. However, the values obtained in the rule viewer are just optimum ranges and cannot be concluded as the best control values. Conclusions are only taken after the defuzzification process as described later in the chapter. The surface viewer shows a three-dimensional structure, which gives a conclusion to the range of values that are required to keep the gearbox functioning. This can be put in two-dimensional structure to highlight, in a simplified manner, the range of values recommended to keep the gearbox functioning as in **Figure 13**.

4.2.2. Analysis of the effects of oil level and torque on gearbox

4.2.2.1. Control of gearbox oil level

From the surface viewer of the gearbox against the oil level shown in **Figures 14** and **15**, it is seen that the gearbox is maintained and remains constant, at a 50% capacity if the oil level is 40%. If the oil level is at 40%, smooth running of the boiler is experienced. If the oil level by any means goes below 40%, stop the gearbox to avoid breakdown of the gearbox. It is seen that the fuzzy logic engine will instruct the gearbox to stop running without any human intervention so as to reduce the frequency of breakdown of the boiler machine through an effective gearbox operation. **Figure 14** shows a three-dimensional, and **Figure 15** is a two-dimensional image.

4.2.2.2. Control of gearbox torque

From the graph, **Figure 14** of gearbox operation against torque control, it is shown that the gearbox should never be operated at torque greater than 8000 Nm as shown in **Figure 16**. The gearbox is seen to run smoothly at a 50% capacity if torque ranges from 0 to 8000 Nm. From

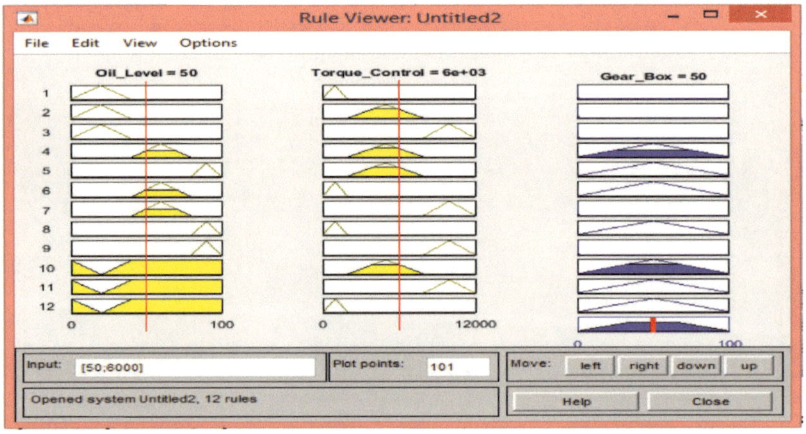

Figure 13. Rule viewer of the gearbox monitoring.

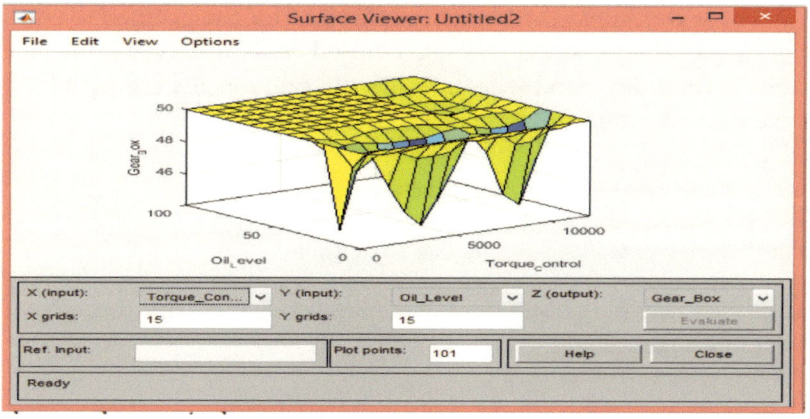

Figure 14. Surface viewer for gearbox monitoring.

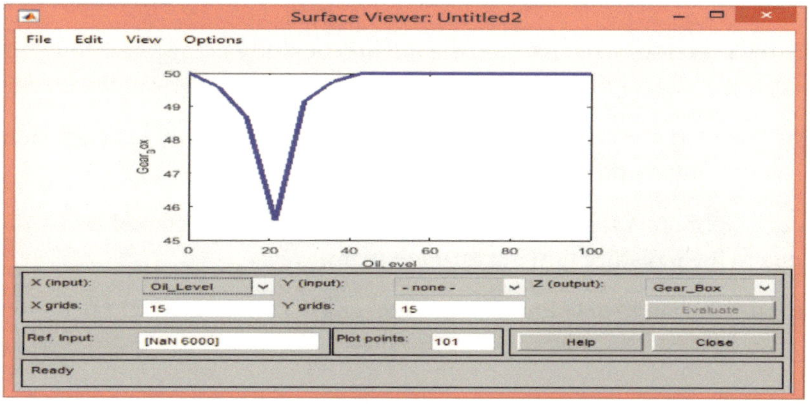

Figure 15. Surface viewer of oil level versus gearbox operation.

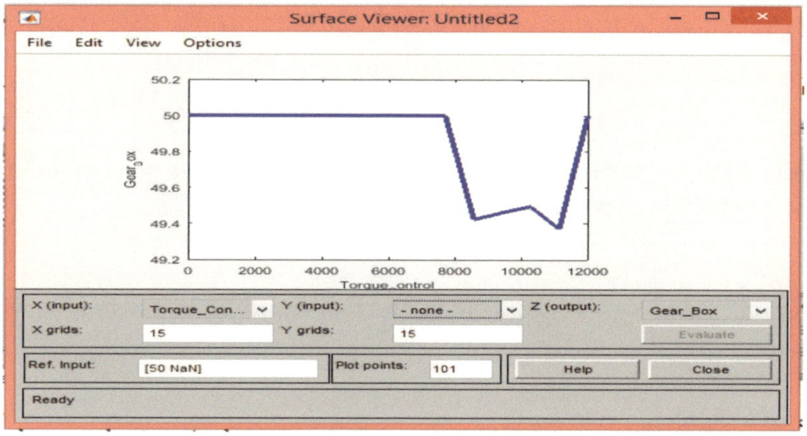

Figure 16. Surface viewer of gearbox control using torque.

this figure, it can be concluded that the required range for the gearbox to run smoothly is 0–8000 Nm of torque; otherwise, stop the gearbox and hence the boiler. Matlab software has stored this data in fuzzy logic and will automatically stop the gearbox operation and hence the boiler if the above conditions are not met.

4.2.3. Moisture and dust control on boiler feed check valves

4.2.3.1. Effects of moisture and dust levels on feed check valves

Moisture inside the valves may result from the compressed air passing through the valve. The effects of moisture on the valves include:

- Moisture may cause rusting in the moving parts of the valve.
- Increased rate of wear of the valve material may also result due to the moisture. The moisture washes the lubrication away which will result in the eventual failure or malfunctioning of the valve.
- The industrial processes which rely on the full functionality of the pneumatic control valves may be jeopardized, and this usually results in costly breakdowns of the machine.
- Air- or gas-operated instruments may give inaccurate readings due to corrosion of the material and hence, interrupting the plant processes.
- The rubber diaphragms inside the pneumatic valves can be stiffened and will eventually rupture due to the moisture flowing through them.

The following are the range of values of moisture which are tolerated and not tolerated inside the valve:

4.2.3.2. Moisture level control

4.2.3.2.1. For a range of 1–5%

This is the total range of moisture in the feed check valve at any time (either accepted or unaccepted).

- 0–1%

This defines that the amount of the moisture in the feed check valve is very minimum in such a way that it causes little or no damage to the valve.

- 1–3%

At this range, the valve works effectively but further exposure to moisture may result in a breakdown.

- 3–5%

This is the range when a maximum amount of moisture is experienced inside the valves. This range is unacceptable at all times as it leads to breakdown of the valves (rust, corrosion, wear and eventually breakdown of the boiler).

4.2.3.3. Dust level control

4.2.3.3.1. For a range of 0–1%

This is the total range of values of dust inside the feed check valve (accepted or unaccepted).

- 0–0.02%

At this range, the valves can run effectively without any damage.

- 0.02–0.5%

This amount of dust entering the valve may cause the valve to malfunction with further exposure. However, at this range, the feed check valves are seen to be operating effectively.

- 0.5–1%

These ranges of dust are not acceptable at all. The excessive exposure of dust inside the valve may result in scoring, wear and eventually total failure of the boiler machine.

4.2.3.4. Output control

The feed check valve has to either open or close, either of the two, and that is the meaning of 0 for close and 1 for open.

- 0 close

At this range, for close signal, the valve will close if the if the dust and moisture levels inside the valves are unacceptable.

- 1 open

This involves the acceptable range. All the conditions are being met (dust and moisture level), and the valve can effectively open at any angle between 0° and 90° depending on the amount of feed water to be pumped into the boiler.

4.2.4. Simulation of the effects of dust and moisture levels on feed check valves using fuzzy logic

The moisture and dust level variations in the boiler machine can be monitored using an artificial intelligence software. The inference engine of fuzzy logic can store two input parameters (moisture and dust level) and one output parameter (feed check valve operation) as shown by the print screen below from the laptop in **Figure 17**.

Control of feed check valve operation using fuzzy logic tool box is shown in **Tables 6–8**.

4.2.5. Membership function editor for moisture and dust monitoring

This is shown clearly in **Figures 18–20** with ranges for moisture level varying from low, acceptable and high and the numerical values being added. For dust control, the corresponding numerical values range from low, normal and high. The valve either closes or opens depending on moisture and dust levels. This is done to control the feed check valves so that they won#x2019;t operate in unfavourable conditions which might result in the failure of the boiler.

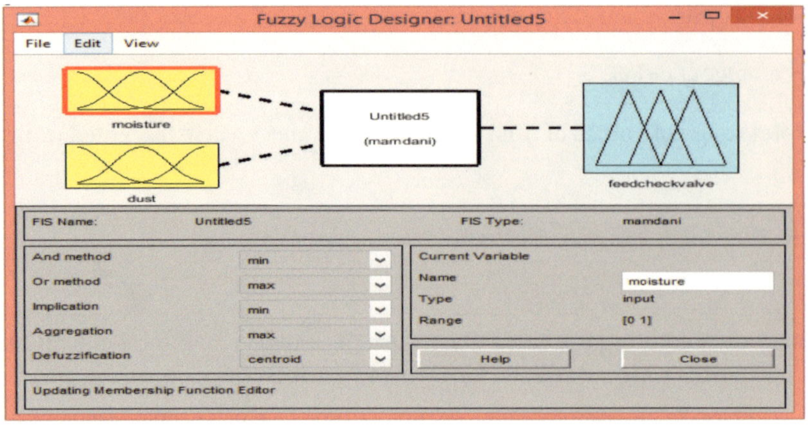

Figure 17. FIS Editor for feed check valve.

The following rules were developed in **Figure 20** and will instruct the controller what to do.

After the rules are inserted, the overall results of combined effects of moisture and dust levels on feed check valve are as discussed below. The rule viewer is an intelligent software which is used to view the rules created using the rule editor and is shown in **Figure 21**. The rule viewer for the range values for moisture level, dust and valve operation is discussed clearly

Input range (%)	Fuzzy variable name
0–1	Low
1–3	Medium
3–5	High

Table 6. Input 1: moisture level.

Input range (%)	Fuzzy variable name
0–0.02	Acceptable
0.02–0.5	Average
0.5–1	Not acceptable

Table 7. Input 2: dust level.

Output range (%)	Fuzzy variable name
0	Closed
1	Open

Table 8. Output 1: feed check valve operation.

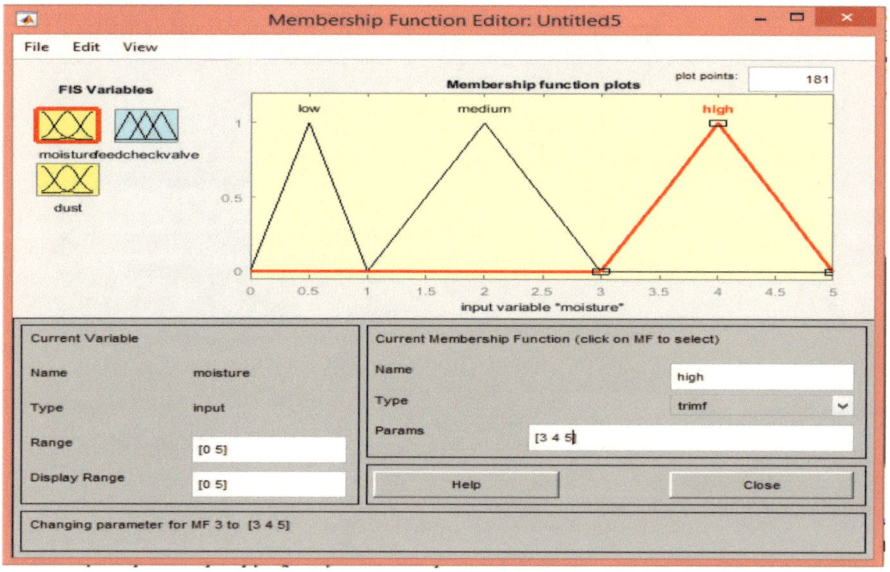

Figure 18. Membership function editor for moisture.

in **Figure 21**. The surface viewer shows a three-dimensional layout in **Figure 22**, and the blue lines indicate the values of moisture level and dust level, which may result in the failure of the valve. The required values which should be maintained are shown by the yellow surfaces. However, the values obtained from the rule viewer are just ranges and not concluded as the best control values. Conclusions are only taken after the defuzzification process.

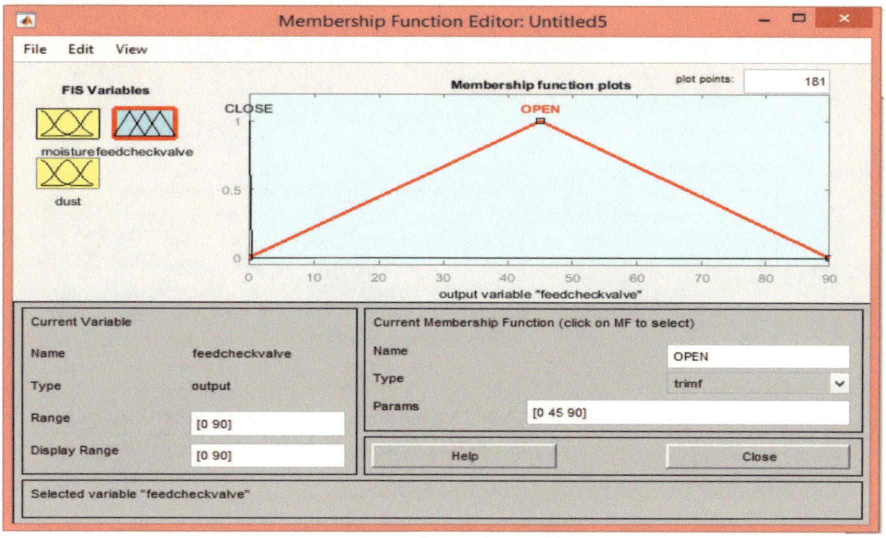

Figure 19. Membership function editor for the output and feed check valve.

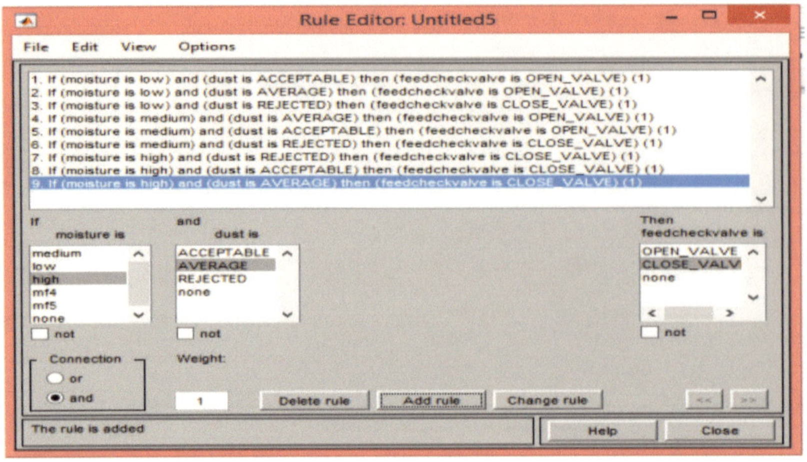

Figure 20. Rule editor for feed check valve.

4.2.6. Analysis of the effects dust and moisture on feed check valve

4.2.6.1. Control of feed check dust level

From the surface viewer of the feed check valve control against the amount of dust, it is seen that the valve remains constant at 45% corresponding to a dust range of 0–0.55%. The dust level should be maintained at this range. If the dust level is increased from 0.55 to 0.58%, there is a linear decrease in the feed check valve's opening angle from 45 to 0°. As the level of dust is further increased from 0.58 to 0.95%, the valve is kept constant at 0°. It is therefore concluded that dust level should remain between 0 and 0.55% so that the valves will run smoothly; otherwise, boiler breakdown will result. Fuzzy logic engine is an artificial intelligence system

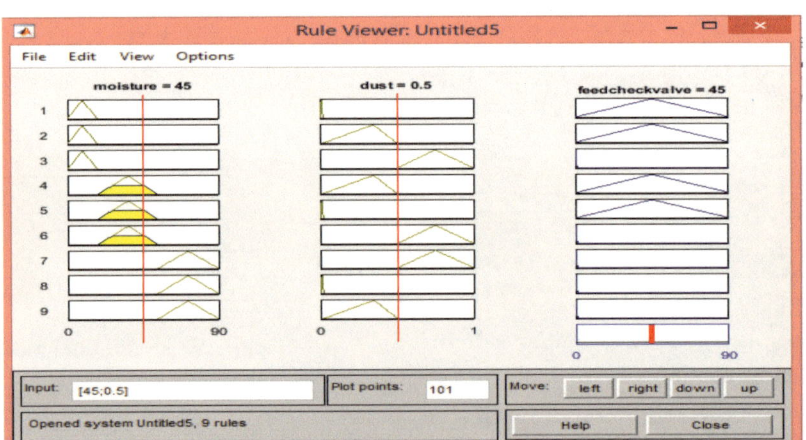

Figure 21. Rule viewer for the feed check valve monitoring.

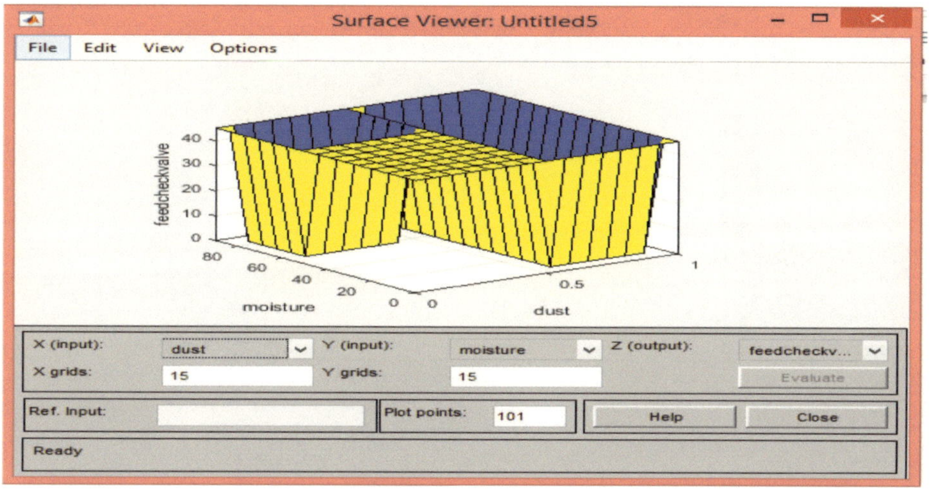

Figure 22. Surface viewer for the feed check valve monitoring.

which will instruct the valves to immediately close as soon as the dust level exceeds 0.55%, as in **Figure 23**, without any human intervention.

4.2.6.2. Control of moisture level

Figure 24 indicates that moisture level does not contribute much to the valve opening. From 0 to 85%, the feed check valves operate at a constant capacity of 45%. For smooth operation of the valves, moisture level should not exceed 85%; otherwise, fuzzy logic will set the valves to close automatically. This helps to prevent the breakdown of feed check valves.

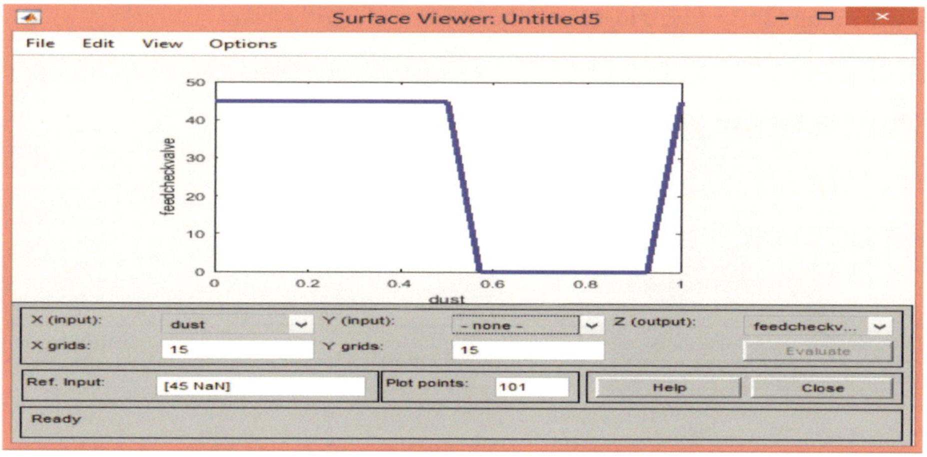

Figure 23. Surface viewer of gearbox control using dust level.

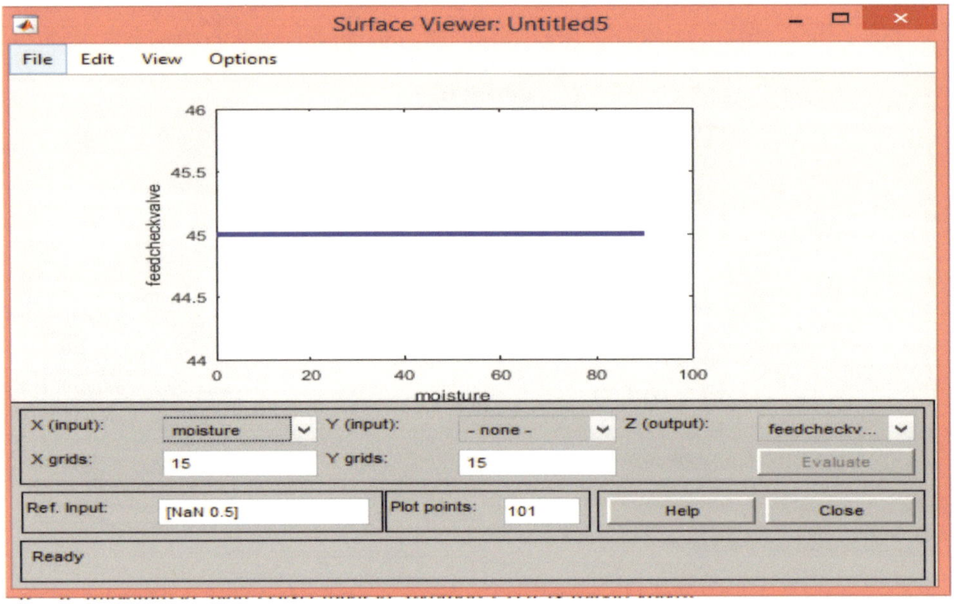

Figure 24. Surface viewer of feed check valve control using moisture.

5. Recommendations and conclusions

Improve management commitment: The success of any system depends on the commitment of the management. Design and implementation of a hybrid proactive approach require a greater initial investment for it to be successful. Since systems are people driven, it is certain to require worker participation. Management needs to look up at providing resources for in-house training for the fuzzy logic system.

Appendices for the processing company

Appendix 1: Factor A

Effect on production output (Factor A)	Factor score
No significant impact/standby equipment is available	1
Minor impact on production. Unlikely to affect other areas of the plant	2
Failure would have significant impact on output and may affect other sections	3
Major impact on the plant#x2019;s operations; failure would cause over 40% of plant production to stop	4

Appendix 2: Factor B

Utilization (Factor B)	Factor score
Equipment is used on an occasional basis	1
Equipment is required to function independently for up to 50% of available time	2
Equipment is part of a continuous process required to function for a major proportion of the planned production time.	3
Equipment is required to function for all of the major losses	4

Appendix 3: Factor C

Quality (Factor C)	Factor score
No effect on product quality	1
Minor effect on product quality	2
Critical effect on product quality and can result in major losses	3

Appendix 4: Factor D

Effect on safety or environment (Factor D)	Factor score
Little or no risk to the safety of people, equipment or the environment	1
Minor risk to people, equipment or the environment	2
Risk to people resulting in a lost-time accident and significant damage to equipment or the environment which requires notification to relevant authorities	3
Major impact on the plant#x2019;s operations; failure would cause over 40% of plant production to stop.	4

Appendix 5: Priority score

The priority score is obtained by multiplying the results for the three factors:

$$PS = \text{Factor E} * \text{Factor F} * \text{Factor G} \tag{3}$$

Factor E

Frequency of failure (Factor E)	Factor score
Failures are rare—less than once per year	1
Occasional failure between 3 and 12 months	2
Failure likely between 1 and 3 months	3
Frequent failures at least once per month	4
Frequent failures at least once per week	5

Appendix 6: Factor F

Downtime/repair time (Factor F)	Downtime	Factor score
Minor	0–30 min	1
Significant	30–120 min	2
Major	2–8 h	3
Severe	>8 h	4

Appendix 7: Factor G

Waste (Factor G)	Quantity	F
No waste is generated under normal operating conditions	0%	1
Small amounts of waste are produced by failure	2%	2
Waste is produced during production that is significant	5%	3
Quantities of waste are significant and warrant immediate attention	10%	4

Appendix 8: Fire Tube boiler's detailed FMECA analysis

Component name	Component function	Cause(s) of failure	Effect(s) of failure	Failure mode(s)	Occurrence index (O)	Severity index (S)	Detection index (D)	Risk priority number (O)*(S)*(D)
Servo motor	To generate torque to turn fan rotor	Worn bearing, lubrication failure, burnt brushes and broken fan bolts	Excessive vibration, noise. motor overheat and sparks of fires are produced	Bearing failure, coil shorts, overheating	6	7	3	126
Feed check valve	To regulate the supply of water which is pumped into the boiler by a feed pump.	Internal valve malfunction Operator error Calibration error	Overpressure in system and leakages resulting in fire	Fails to open Remains open Crack valve	6	8	2	96
Combustion chamber	To burn fuel	Leakage through the soot blower casing seal	Combustion gases entering fire room	Too much fuel being fired, and excess air	7	7	5	245

Component name	Component function	Cause(s) of failure	Effect(s) of failure	Failure mode(s)	Occurrence index (O)	Severity index (S)	Detection index (D)	Risk priority number (O)*(S)*(D)
Forced draft fan	To provide air into the combustion chamber	Dust, external shocks, vibration, and improper operation	Overheating, blocked fan, loss of performance (the fan does not start anymore)	Noisy operation, blocked fan blade	6	8	8	384
Air filters	To remove unwanted elements (dirt, dust, etc.)	Clogged air filter	Black smoke exiting the exhaust pipe and noisy operation	Dirty on surface Noisy operation	4	6	3	72
Feed water system	Used to control feed water into the steam drum and the feed tank	Pump blocked, Leakage in suction pipe, Non- return valve blocked in closed position	Pump runs but give no water	Pump trip	3	8	8	192
Safety valve	To prevent explosions due to excessive internal pressure of steam	Internal valve malfunction Operator error Calibration error	Steam burns: increased production time. Overpressure: protection compromised	Fails to open	6	9	4	216
Fuel system	Used to supply fuel (diesel or oil) to the boiler	Power supply switched on but no reset applied Relay RL6 faulty Detector not closing Open circuit on first detector or low pressure on the second	Gas leaks indicated and reset not operating	Relieve valve damage Faulty of the trip valve	5	6	7	210
Piping system	Conveys the hot combustion gasses away from the boiler to the outside	Failure to channel exhaust gases at all, gas flow restricted	Piping assembly collapses and falls to the bottom of the stack, leakages	Silencer mountings corroded away, corrosion	7	7	3	147

Component name	Component function	Cause(s) of failure	Effect(s) of failure	Failure mode(s)	Occurrence index (O)	Severity index (S)	Detection index (D)	Risk priority number (O)*(S)*(D)
rotor bearing	To absorb the axial thrust and support the rotor	Excessive vibrations	Bearing wear	Wear	3	6	2	36
Gearbox	To change the speed ratio between motor and kicker	Starting under load	Wear of gears	Impact wear	3	8	4	96
Pressure relief valve	To measure the pressure of the steam boiler	Internal valve malfunction Operator error and calibration error	Rapture of the boiler container or pipes	Jammed closed	5	9	4	180
Stop valve	To control the flow of steam from the boiler to the main steam pipe	Internal valve malfunction Operator error and calibration error	Overpressure: protection compromised	Leakages	7	9	9	567
Blow off cock	To empty the boiler when discharging mud, scale or sediments	Internal valve malfunction Operator error and calibration error	Overpressure in system resulting in fire, leakages	Fails to open Fails to close	7	6	8	336
Oil Control Unit	To circulate lubricant at specified flow rate and pressure	Blocked filters, worn impellor, high lubricant temperature	Delivery pressure lower than specified, flow rate lower than specified and cavitation	Blockage, wear and cavitation	6	6	8	288
Water indicator gauge glass	To indicate the water level inside the boiler	Salt deposition due to high water level in drum	Fails to react to water rise/ drop above/ below the preset value	Fails to react to water rise/ drop above/ below the preset value	6	9	7	378
Super heater	To increase temperature of saturated steam without raising its pressure	Blocked tube Starvation of tube Erosion of tube due to high excessive air	Cracking of carbon steel	Dirty on surface Noisy operation Impact wear Corrosion on surface	6	3	8	144

Component name	Component function	Cause(s) of failure	Effect(s) of failure	Failure mode(s)	Occurrence index (O)	Severity index (S)	Detection index (D)	Risk priority number (O)*(S)*(D)
Steam drum	To store steam before it is distributed to the plant	Excessive stresses produced by rolling tubes	Damage on the ligament areas	Wear	5	8	6	240
Boiler feed pump	To pump feed water into the boiler	Fatigue, corrosion, wear	Cavitation, pump overload, heavy vibration or imbalance	High-bearing temperature, broken axle, impellor	8	9	5	360

Author details

Tawanda Mushiri

Address all correspondence to: tawanda.mushiri@gmail.com

Faculty of Engineering and the Built Environment, University of Johannesburg, Johannesburg, South Africa

References

[1] Bond, M., Meacham, T., Bhunnoo, R. & Benton, T. (2013). Food waste within global food systems. Swindon, UK: Global Food Security Programme. A Global Food Security Report.

[2] Bose, B.K. (1994). Expert system, fuzzy logic, and neural network applications in power electronics and motion control. Proceedings of the IEEE, 82, 1303-1323.

[3] Buchanan, D., & Lo, S.-H. (1997). Reliability and integration of ultra-thin gate dielectrics for advanced CMOS. Microelectronic Engineering, 36, 13-20.

[4] Burt, D.N., & Pinkerton, R.L. (1996). A purchasing manager#x2019;s guide to strategic proactive procurement. New York: AMACOM.

[5] Christopher, M. (2016). Logistics & supply chain management. UK: Pearson.

[6] Isermann, R. (2011). Fault-diagnosis applications: model-based condition monitoring: actuators, drives, machinery, plants, sensors, and fault-tolerant systems. Springer Science & Business Media. Berlin, German.

[7] Kropp, J., Zickfeld, K., & Eisenack, K. (2002). Assessment and management of critical events: the breakdown of marine fisheries and the north Atlantic thermohaline circulation. In The science of disasters. Springer. Berlin, German.

[8] Liang, S.Y., Hecker, R.L., & Landers, R.G. (2002). Machining process monitoring and control: the state-of-the-art. In ASME 2002 International Mechanical Engineering Congress and Exposition. American Society of Mechanical Engineers, pp. 599-610. University of Missouri, United States of America.

[9] Nagilla, S.R. (2015). Non-thermal reliability considerations for. Arlington: University of Texas.

[10] Nwaobi, G.C. (2008). Energy power, digital infrastructure and e-learning platforms: African experience. Journal of Energy.

[11] Okah-Avae. (1981). Condition monitoring – a maintenance management strategy for industrial machinery. Benin: University of Benin-City.

[12] Papadopoulos, C., & Dimarogonas, A. (1992). Coupled vibration of cracked shafts. Journal of Vibration and Acoustics, 114, 461-467.

[13] Pine, B.J. (1993). Mass customization: the new frontier in business competition. Harvard Business Press. Harvard.

[14] Beverage Industry Environmental Roundtable. (2014). Beverage industry sector guidance for greenhouse gas emissions reporting. Beverage Industry Environmental Roundtable.

[15] Scheffer, C., & Girdhar, P. (2004). Practical machinery vibration analysis and predictive maintenance. Elsevier. Amsterdam.

[16] Seemann, K.W. (2000). Technacy education: towards holistic pedagogy and epistemology in general and indigenous/cross-cultural technology education. School of Tourism and Hospitality Management Papers, 9. United Kingdom.

[17] Tavner, P.J., & Penman, J. (1987). Condition monitoring of electrical machines. Research Studies Press.

[18] Zhang, Z. (2011). Energy and environmental policy in China: towards a low-carbon economy. Edward Elgar Publishing. Massachusetts, USA.

Chapter 18

Use of Fuzzy Logic for Design and Control of Nonlinear MIMO Systems

Pavol Fedor and Daniela Perduková

Additional information is available at the end of the chapter

http://dx.doi.org/10.5772/68050

Abstract

Standard analytical methods are often ineffective or even useless for design of nonlinear control systems with imprecisely known parameters. The use of fuzzy logic principles presents one possible way to control such systems which can be used both for modeling and design of the control. The advantage of using this method consists in its simplicity and easy way of developing the algorithm, which in the phase of designing the controllers and also for modeling the features of the designed structures, allows the use of computer technology. Simplicity of the proposed structure (usually with the PI controllers) and determination of their parameters without any need for complex mathematical description present another considerable advantage of the used method. This chapter presents two typical examples of designing the control of nonlinear multi-input multi-output (MIMO) systems from the field of mechatronic systems based on fuzzy logic principles.

Keywords: fuzzy PI torque controller, asynchronous motor, continuous line, black-box modeling, fuzzy model-based control

1. Introduction

There are many processes in technological practice the analytical description of which is rather complicated. This can be due to their complexity, nonlinearity, transfer lags, complicated measurement of important parameters, etc. However, information on the performance of these processes can often be obtained experimentally (by suitably chosen measurements or by monitoring their responses to the control activities of the operator). In these situations, fuzzy systems can always be considered as an alternative for system modeling and control.

Applying fuzzy logic in fuzzy controller design is very often a suitable possibility of solving problem issues in control in various fields of industry because these controllers are an effective tool for achieving high-quality properties of the controlled systems [1–9]. The disadvantage in this case is the unsystematic approach to their synthesis and a relatively demanding analysis of their stability. A fuzzy controller design is primarily based on the fuzzification of its range of inputs and the setting up of rules of its behavior within this range. The behavior of classic fuzzy controllers was designed on basis of linguistic rules obtained from experts. However, this knowledge is not always easily obtained, especially in cases of higher-order nonlinear systems [10, 11]. For this reason, special attention has been focused in recent years on the design of fuzzy control systems that are not based on the search for expert linguistic rules [12–17]. Control methods based on the controlled system fuzzy model have many modifications that depend on the particular application [6, 13–17], while the quality of the fuzzy model of the controlled system is also of significant importance.

It has been proved that fuzzy modeling can be recognized as one of the nonlinear black-box modeling techniques [11, 12, 18–20]. When designing a black-box fuzzy system, it is necessary to identify its qualitative properties only on the basis of experimentally measured data, while neither its structure nor its parameters are known. That often results in problems with inconsistency of the database, problems with covering the entire space of possible inputs, etc. [20–22], which makes the fuzzy model unusable in practical applications. In the design of a black-box fuzzy model of a dynamic system, a suitable method for the selection of qualitative properties from the collected database always needs to be applied. The functional dependencies between inputs and outputs can then be used for developing a suitable nonparametric fuzzy model of the process that can be applied in the design of their control [23–27].

Two typical examples for designing the control of nonlinear MIMO systems using fuzzy approach are presented in this chapter:

- The design of fuzzy PI torque controller of the PI type for a drive with an induction motor, whose parameters and rules are obtained by searching control input of such a vector which is optimal in terms of the selected criterion of optimality.
- The design of control for middle part of a continuous line for material processing by tension, where the continuous line presents a nonlinear MIMO system. Its control requires to ensure decoupled control of individual subsystems, because the output quality of the processed material depends directly on quality of the control. The controllers of the subsystems ensuring such decoupling usually are of complex structures and, when designing them by analytic methods, they are often unrealizable. When the continuous line is presented by a fuzzy model, it is possible to design simple controllers of the PI type ensuring high-quality dynamical properties of the controlled system.

2. Design of fuzzy torque controller for asynchronous motor drive

An asynchronous motor represents a strongly nonlinear fifth-order system, whose good quality vector torque control is solved by relatively complex mathematical transformations and leads to a complicated control structure [28, 29]. Therefore, a fuzzy system for design of torque controller for asynchronous motor drive has been used. The fuzzy controller design is based on the concept of identifying the time sequence of the input signal into the controlled system that will provide the control target in terms of the selected optimality criterion. Fuzzy controller design method is characterized by simplicity, and quality of control is appropriate to the considered drive.

API-type discrete controller is generally described by the equation:

$$u_k = u_{k-1} + q_0 e_k + q_1 e_{k-1} \tag{1}$$

where u_k and u_{k-1} are values of controller output in the relevant sampling steps, e_k and e_{k-1} are values of the control error, and q_0 and q_1 are parameters of controller [28]. From these follows, it is possible structure, shown in **Figure 1**.

A discrete fuzzy PI controller can be described, for example, by the following rules:

$$IF\ e_k\ is \ldots \wedge e_{k-1}\ is \ldots THEN\ du_k\ is \ldots \tag{2}$$

where quantities e_k, e_{k-1}, and du_k are fuzzy variables that describe the relevant workspace of the fuzzy PI controller. The fuzzy controller design procedure consists of the following three steps:

Step 1. Finding the optimal sequence of input values

Fuzzy rules and fuzzification of the fuzzy PI controller workspace can be identified by means of relations expressed by triplets [e_k e_{k-1} du_k] that have been obtained for its optimal behavior. This behavior is represented by the time sequence of input vector $d\mathbf{u}_{opt}$, at which the value of the optimality criterion is minimal. It is suitable to choose this criterion in the next form of the integral of the quadratic deviation of the system output y from the desired value w:

$$J(e) = \int \left(w(t) - y(t)\right)^2 dt \tag{3}$$

When obtaining the sequence of values of vector $d\mathbf{u}_{opt,}$ we apply to the drive input various sequences of input vector and evaluate the criterion value Eq. (3). This is a standard optimization task that can be solved, for example, by suitable geometric division of the drive workspace (which leads to rather high computing demands in the design process), or by applying the genetic algorithm method, which significantly speeds up the whole process of identification.

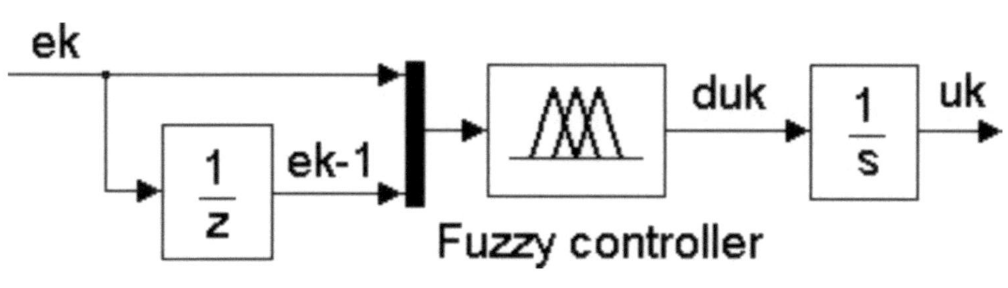

Figure 1. Block diagram of fuzzy controller.

Step 2. Finding the database of optimal data

Having found the optimal input sequence $d\mathbf{u}_{\mathrm{opt}}$ for the controlled drive, we then set up the database of triplets $[e_k\ e_{k-1}\ du_k]$, which describes relations between the inputs and the outputs of the optimal fuzzy PI controller.

Step 3. Designing the fuzzy controller from the optimal data database

From the obtained triplets $[e_k\ e_{k-1}\ du_k]$ of optimal data, it is possible to design a concrete fuzzy controller of various types using standard procedures of clustering the data into significant clusters and describing them by means of rules.

In the concrete application of the said procedure in a drive with asynchronous motor, we will use its analytical model (see Refs. [28, 29]). If we consider a rotating system which rotates with the frequency of the motor's stator field (usually marked by coordinates x, y), the mathematical model of the device is described by the following equations:

$$\begin{aligned}
u_{sx} &= R_s i_{sx} + \frac{d\psi_{sx}}{dt} - \omega_s \psi_{sy} \\
u_{sy} &= R_s i_{sy} + \frac{d\psi_{sy}}{dt} - \omega_s \psi_{sx} \\
0 &= R_r i_{rx} + \frac{d\psi_{rx}}{dt} - \omega_2 \psi_{ry} \\
0 &= R_r i_{ry} + \frac{d\psi_{ry}}{dt} + \omega_2 \psi_{rx} \\
M_e &= \frac{3}{2} p \left(\psi_{ry} i_{rx} - \psi_{rx} i_{ry} \right)
\end{aligned} \tag{4}$$

Used symbols:

i_{sx}, i_{sy} components of stator current space vector i_s

i_{rx}, i_{ry} components of rotor current space vector i_r

u_{sx}, u_{sy} components of stator voltage space vector u_s

ω_m the motor mechanical angular speed

ω_1 angular frequency of the stator voltage

p number of pole pairs ($p = 2$)

ω_2 slip angular speed $\omega_2 = \omega_1 - \omega_m$

R_s, R_r stator and rotor phase resistance

Ψ_s, Ψ_r stator and rotor magnetic flux

M_e electrical motor moment

AC drive parameters are given in the Appendix.

As a standard, asynchronous motors are supplied from static voltage frequency converters in which the stator frequency and voltage rate are U_1/ω_1.

When connected directly to the power supply network, the motor shows a large increase in torque and also in current (**Figure 2**).

Let the aim of the torque controller design be to adjust the slip ω_2 (i.e., the difference of $\omega_1 - \omega_m$) according to the desired torque value. The structure of the controlled system will be as shown in **Figure 3**.

For finding the optimal input sequence $d\mathbf{u}_{opt}$, we shall use the diagram shown in **Figure 4**.

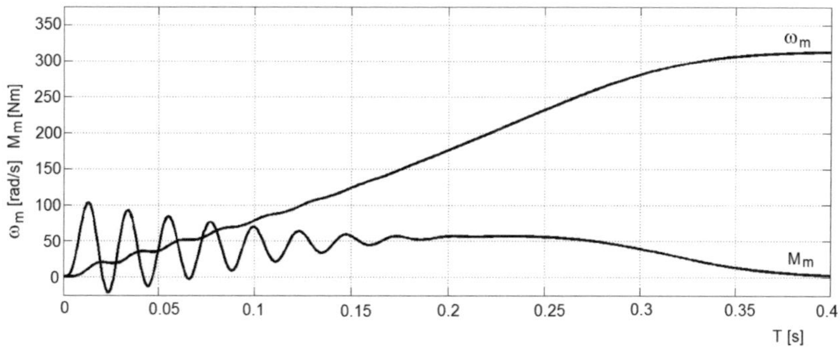

Figure 2. Direct connection of drive with AM.

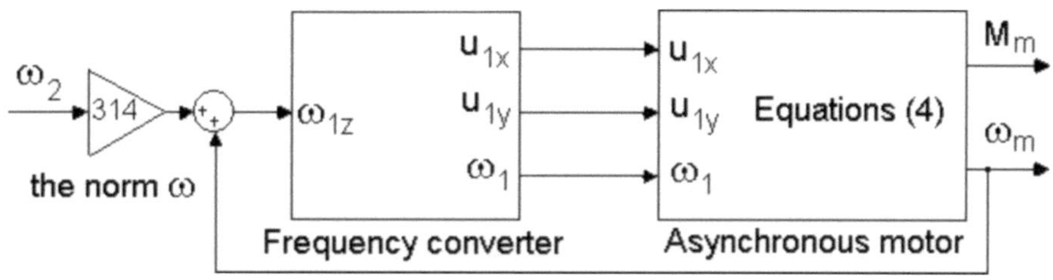

Figure 3. Structure of the controlled drive.

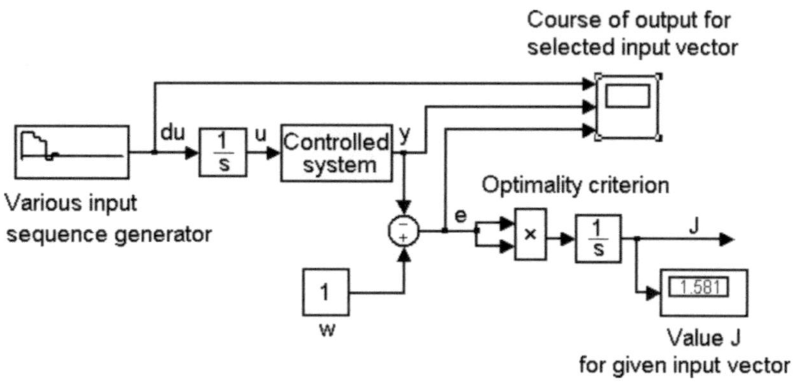

Figure 4. Block diagram for finding vector $d\mathbf{u}_{opt}$.

We search for the input signal sequence $d\mathbf{u}_{opt}$ through such changes of drive input voltage and frequency that will lead to minimal value of the criterion according to Eq. (3). This goal can be achieved, for example, by application of the genetic algorithm method, which efficiently enables finding the extreme of the selected function in a given space of mutations. Having selected input signal sampling time 50 ms and desired value of motor torque Mz = 30, we obtained the optimal input vector [0.165 0.145 0.155 0.12 0.105 0.095 0 0 0 0 0 0], as illustrated in **Figure 5**.

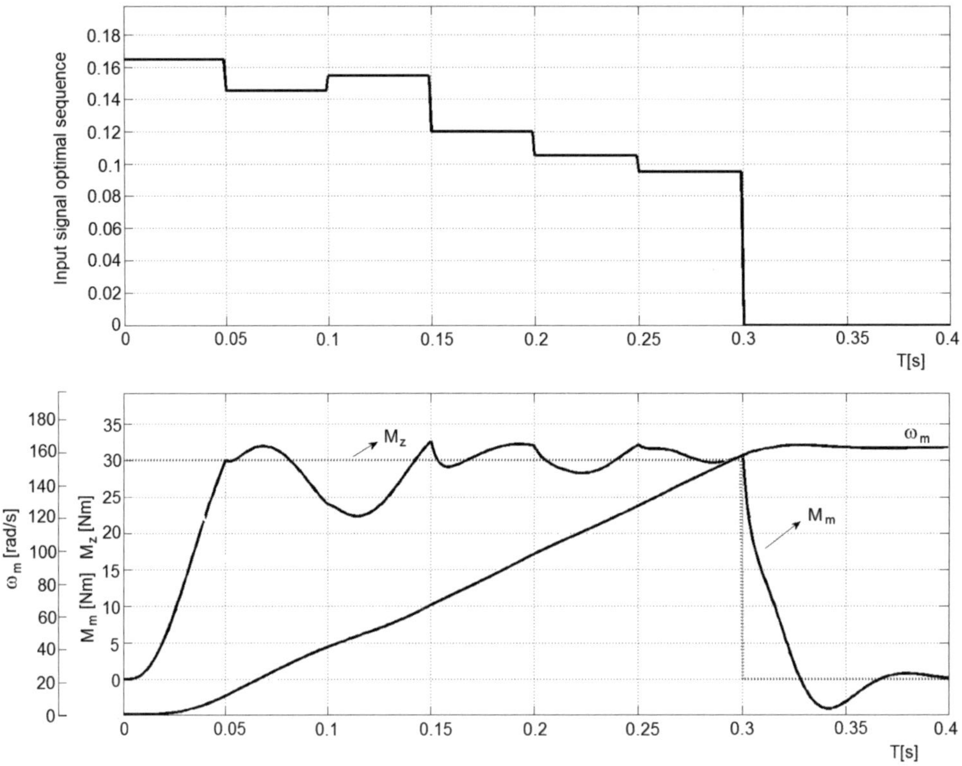

Figure 5. Optimal start-up of drive with AM.

The said optimal input signal sequence was used for generating the database of triplets $[\omega_{mk}\omega_{mk-1}\ (\omega_1-\omega_m)_k]$ for the design of the P-type fuzzy controller. Using the Anfisedit tool in the Matlab program, a static Sugeno-type fuzzy system was designed, based on the acquired database of measured data ($[\omega_{mk}\omega_{mk-1}\ (\omega_1-\omega_m)_k]$). This system has three zones for input data fuzzification, and its internal structure can be seen in **Figure 6**.

The resulting torque control structure of AM with the designed fuzzy controller is shown in **Figure 7**.

The start-up of the drive with desired torque Mz = 30 Nm using the designed fuzzy controller is shown in **Figure 8**.

The comparison of **Figures 5** and **8** shows that the application of the fuzzy controller resulted in the achievement of the desired torque responses of the AM at start-up.

2.1. Discussion

The fuzzy controller designed in this chapter is a PI controller, and it provides optimal dynamics in terms of the selected criterion Eq. (3). The design procedure consists of three steps:

In the first step, we search for such input sequence of input vector **u** of the asynchronous motor that would provide minimal value of function Eq. (3). This can be achieved by modeling

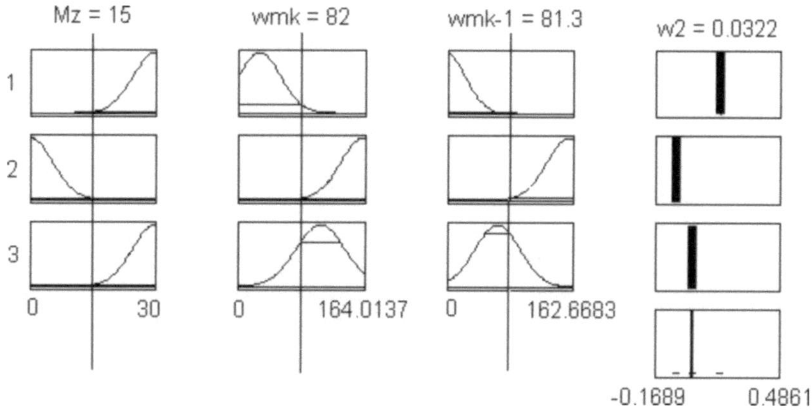

Figure 6. Structure of AM fuzzy controller.

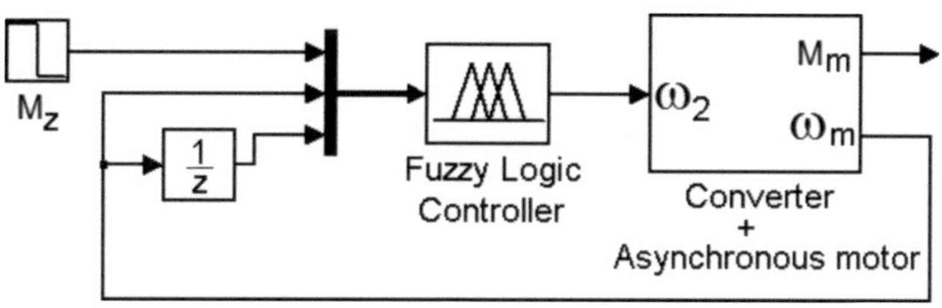

Figure 7. Structure of torque control in AM with fuzzy controller.

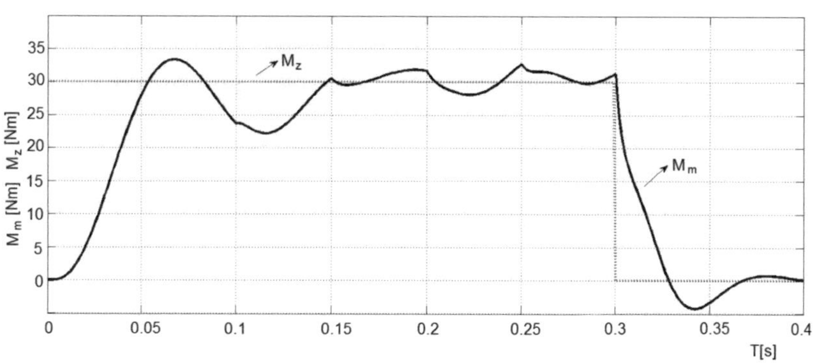

Figure 8. Optimal start-up of drive with AM with fuzzy controller.

various vector **u** input sequences in the system with asynchronous motor, and by a simulation of its behavior for a given sequence. The selection of sampling time for a change of the input vector value and by that also of the number of its samples has to be defined according to the Shannon-Kotelnikov theorem. Obviously, with an increase in the number of samples also the time required for simulation and for optimal vector $\mathbf{u}_{opt}$ identification grows. As the whole process of optimal vector identification can be fully automated and the performance of computing means nowadays is sufficient, there is no essential problem to identify the optimal control input signal for a concrete drive with asynchronous motor.

In the second step, using the database obtained by application of the input vector $\mathbf{u}_{opt}$ to the asynchronous motor, the input/output relations of the optimal PI controller for the given controlled system are defined. Various input and output signals (e.g., depending on their actual measurability …) can be chosen for setting up the suitable database. This makes it possible to adjust the controller design to a concrete drive with concrete technical capabilities.

The relations obtained in the second step of the design procedure, written down in the form of a suitable input and output signals database (mostly in a table), are described in the third step by means of fuzzy logic principles. This way a fuzzy PI controller similar to an optimal continuous PI controller is constructed. Standard computing means for working with fuzzy systems are employed in this step, such as the fuzzy toolbox in Matlab.

The whole procedure of fuzzy PI controller design was verified by simulation of its properties in the concrete control of a drive with asynchronous motor. The results of simulation experiments show that the controller, in spite of its simplicity and the uncomplicated computer oriented design procedure applied, enables considerable improvement in the control circuit dynamic properties also in case of strongly nonlinear higher order controlled systems.

3. Multi-motor drive optimal control using a fuzzy model-based approach

Typical representative of multi-motor drive is the middle part of the continuous line, where the individual working machines are coupled with each other through the material. It can be lines for processing continuous flows of material (e.g., sheet metal strips, tubes, processing lines in paper mills, and printing works) by material traction in the field of elastic or plastic deformation, which influences the material's mechanical properties. It means that the multiple motor drives are complex and coupled MIMO nonlinear systems. Therefore, due to the complexity of their mathematical models, which parameters are difficult to identify, the development of effective control systems is quite complicated task. This chapter presents the design of optimal control of continuous production line using a fuzzy model-based approach.

The structure of the middle part of the continuous line (further referred to as CL) is shown in **Figure 9**. The structure includes DC motors powered through static transistor converters TC. The working machines of the line are driven by the motors through gearbox j; v_1 and v_2 are machine rolls circumferential velocities, and F_{12} is the tension in the web of material between the two machines. The main line disturbances are tensions before and after the middle part of the considered line which are affecting the first and second drive (F_{01} and F_{23}). K_v is circumferential velocity sensors, K_F is tension sensor, u_{v1} and u_{v2} are outputs from velocity sensors, and u_{F12} is output of tension sensor. The controlling voltages u_1, u_2 of converters present the input variables of the system. The tension in the web of material F_{12} and the web of material velocity v_2 is the output variables (let us consider $y_1 = u_{F12}$ and $y_2 = u_{v2}$).

The described system with the mechanical coupling of two machines presents a third-order nonlinear MIMO system with two inputs and two outputs (**Figure 10**), the parameters of the

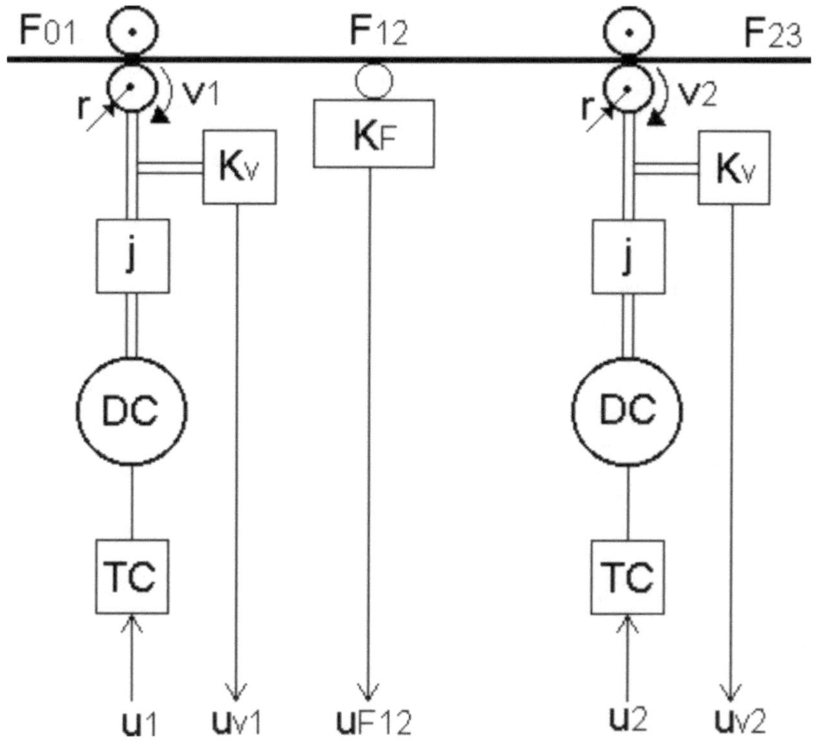

Figure 9. Structure of middle section of continuous line.

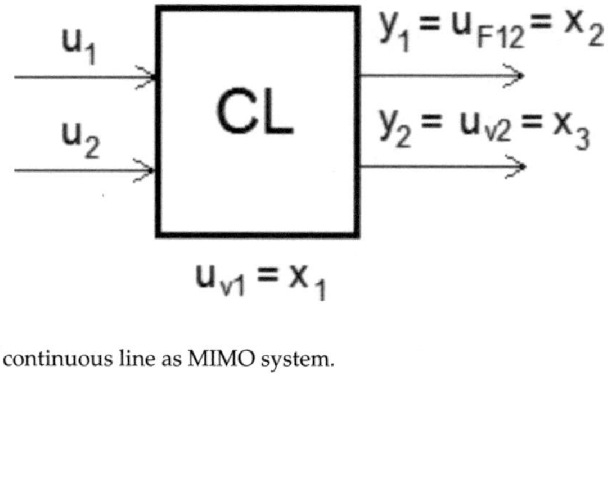

Figure 10. Middle section of continuous line as MIMO system.

system change depending on the mechanical properties of the material and on the speed of its motion. Defining precise parameters of this nonlinear system analytically presents a rather demanding task, and therefore, it is suitable to use for its description a fuzzy system (model) built only on basis of its measured input/output data.

Various fuzzy system structures consisting of static fuzzy subsystems and their dynamic parts can be found in the literature. In setting up the structure of the fuzzy model of a continuous line, we used its state description, where the given state of the system and the given input allow us to define the subsequent state, which can be expressed mathematically by the following equation:

$$\begin{aligned} \boldsymbol{x}_{k+1} &= \boldsymbol{x}_k + \Delta\, \boldsymbol{x}_k \\ \Delta\, \boldsymbol{x}_k &= f\left(\boldsymbol{u}_{k-1}, y_{k-1}\right) \end{aligned} \tag{5}$$

where **u** is the model's input quantities vector, **x** is the state quantities vector, *f* is the searched for static vector function of the controlled system, and *k* is representing the sampling step.

Construction of the CL fuzzy model consists in determining the fuzzy approximation of this function on basis of the obtained CL inputs and outputs database. Considering the choice of CL input, state and output quantities presented in **Figure 10**, the structure of the proposed CL fuzzy model is shown in **Figure 11**.

The whole design of CL optimal control consists of two steps:

Step 1. The design of the fuzzy model for the middle section of the continuous line.

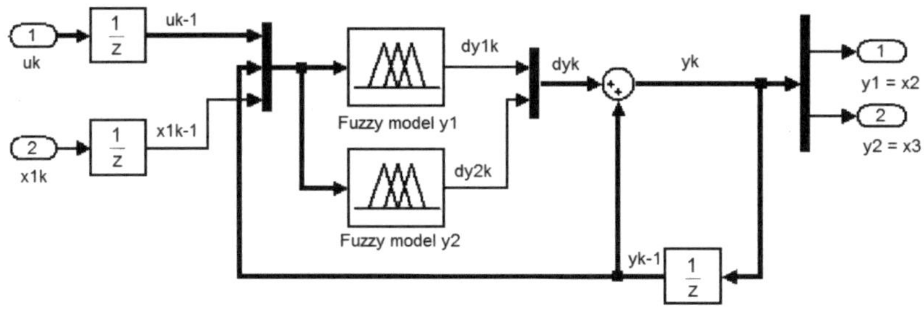

Figure 11. Structure of the discrete CL fuzzy model.

The first step in the design of the fuzzy model for the middle section of the continuous line is the establishment of a consistent database from measured inputs and their corresponding outputs, which covers its entire assumed work space and describes the behavior of the modeled system. For establishing a consistent database, we can use, for example, the method of dividing the input range into *n*-levels and generating $n(n-1)$ transient trajectories between them [12, 31, 32], or the method of exciting the system by a random input signal [10, 30] in case it is not possible (e.g., for operational reasons) to apply a pre-defined input signal at the system's input. Knowledge of the structure or of the parameters of the modeled system is not required in either of these methods. To define suitable sampling time *T* (according to the Shannon-Kotelnikov theorem) and approximate times for transitions for database measurement, we performed identification measurements on the physical model of the CL with input signals u_1 and u_2. Their value and performance are shown in **Figure 12**. Responses of the CL physical model output quantities to the input signals are illustrated in **Figure 13**.

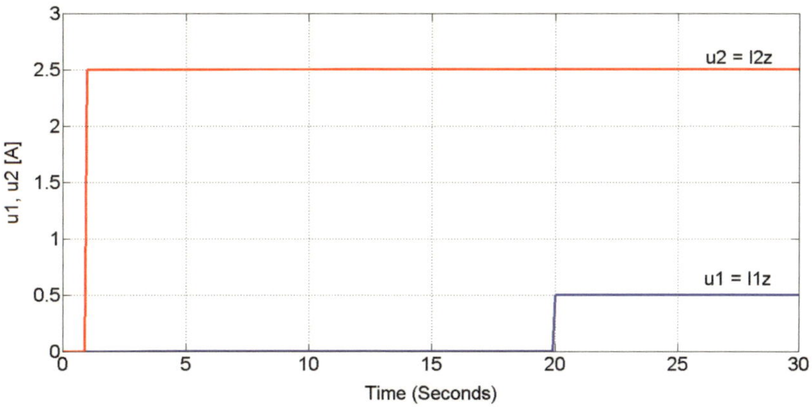

Figure 12. Input signals u_1 and u_2 for identification measurements.

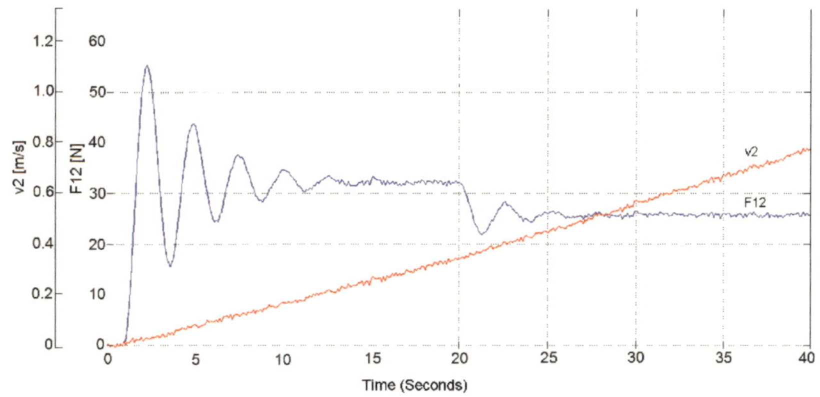

Figure 13. Identification response of CL middle section outputs.

The responses of the CL to inputs u_1 and u_2 (**Figure 13**) show that this system includes a fast tension subsystem and a slow speed subsystem. The range of input values for input u_1 (web tension) is assumed within the interval [−1, 1] and the range of input values for input u_2 (line speed) within the interval [−4, 4]. The database for CL fuzzy model set up will be generated so that line speed (input u_2) will increase in steps each 12 s, and each one second the faster (oscillating) part of the system will be excited by input u_1. The plot of input signals for generating the CL fuzzy model database is shown in **Figure 14**; **Figure 15** shows the output quantities corresponding to these inputs.

The database for CL fuzzy model was generated as demonstrated in **Figure 16**. With sampling time $T = 0.1$ s, we obtained a database with 1000 samples.

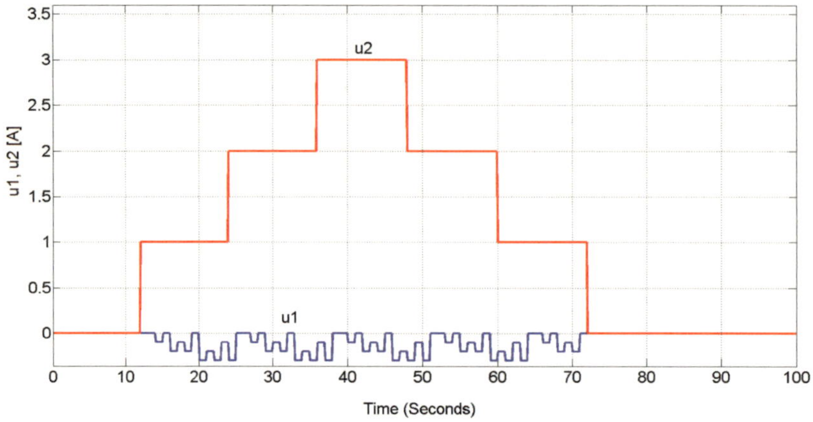

Figure 14. Identification of transitions for CL fuzzy model database generation.

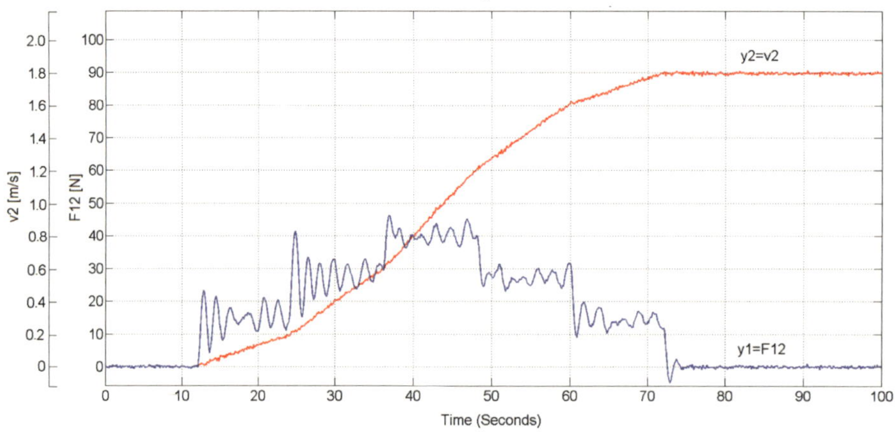

Figure 15. CL output variables corresponding identification transitions of inputs.

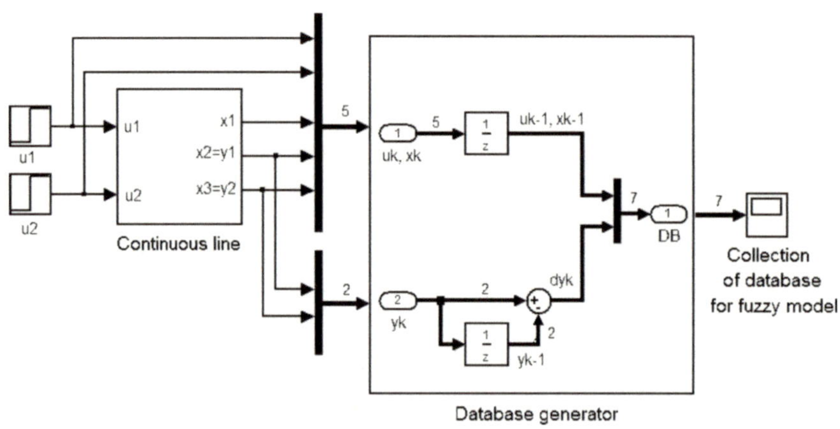

Figure 16. Generating database for CL fuzzy model.

http://dx.doi.org/10.5772/68050

This measured database can be used to search for two FIS structures of the given nonlinear system which best describe the measured relations between $[u_{1k-1}, u_{2k-1}, x_{1k-1}, x_{2k-1}, x_{3k-1}] \rightarrow \mathrm{d}y_{1k}$, and $[u_{1k-1}, u_{2k-1}, x_{1k-1}, x_{2k-1}, x_{3k-1}] \rightarrow \mathrm{d}y_{2k}$.

Using the measured database, the particular fuzzy model can be designed by standardly known procedures of cluster analysis and adaptive approaches to improve the quality of modeling and reduce development time. The fundamental features of cluster analysis are reduction of the number of fuzzy rules and provision of good initial rule parameters. For our purpose from the large number of methods for adaptive fuzzy networks development [33–36], we chose the adaptive neuro-fuzzy inference system (ANFIS) with subtractive clustering [14], which is a fast and robust data analysis method, having the following parameters: range of influence = 0.4, squash factor = 1.25, accept ratio = 0.4, reject ratio = 0.01. Subtractive clustering determines the optimal clusters [34] in a multi-dimensional input/output space that accurately represent the data [34, 37] and CL behavior. The ANFIS approach uses Gaussian functions for fuzzy sets, linear functions for the rule outputs, and Sugeno's inference mechanism [15]. The results were two static Sugeno-type fuzzy systems with two rules for each output quantity as is shown in **Figure 17**.

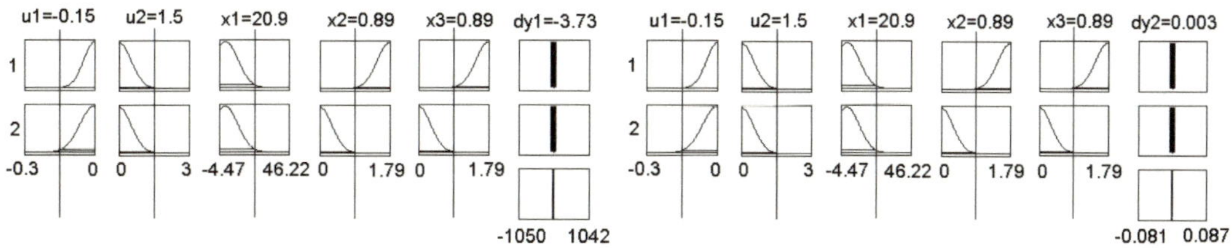

Figure 17. CL fuzzy model—SUGENO type with 2 rules.

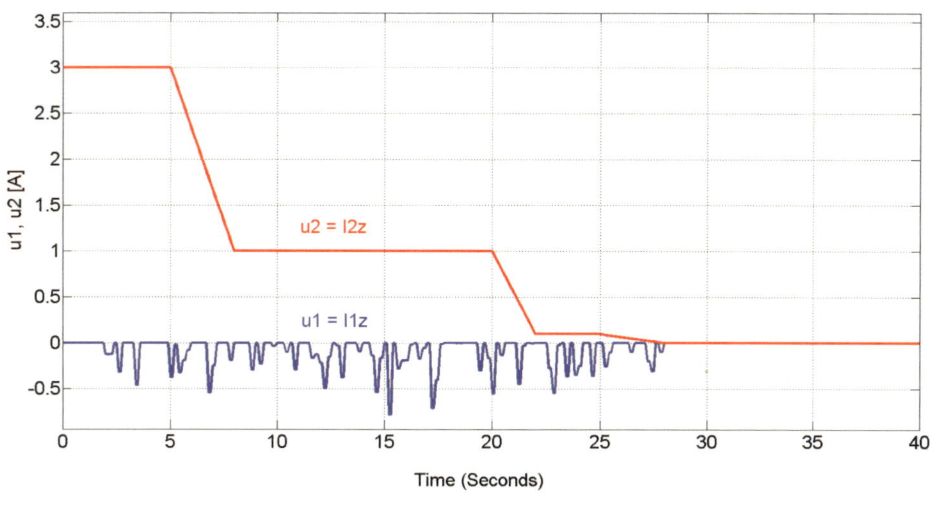

Figure 18. Performance of randomly generated signals u_1 and u_2.

The thus obtained fuzzy systems were implemented into the final continuous line fuzzy model structure, as illustrated in **Figure 11**.

To verify the correctness of the CL fuzzy model, randomly generated signals u_1 and u_2 were applied to its input, as demonstrated in **Figure 18**.

The comparison of the fuzzy model outputs and CL physical model outputs for these inputs is shown in **Figure 19**.

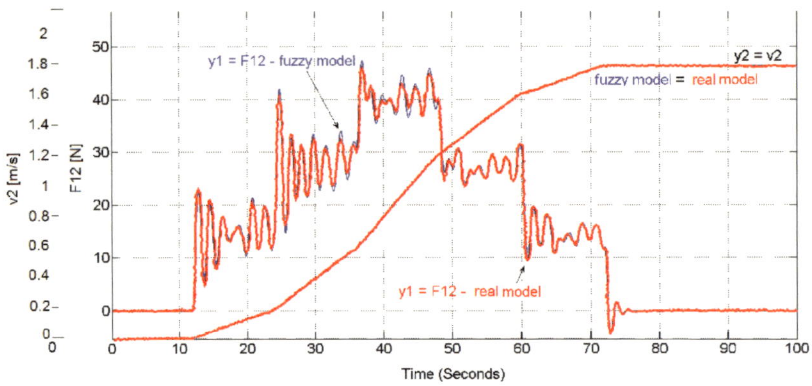

Figure 19. Performance of CL fuzzy model outputs y_1 and y_2 for randomly generated inputs u_1 and u_2.

The obtained results confirm that the designed fuzzy model very well approximates the performance of the continuous line also for randomly generated inputs and can be further used for the design of CL control.

Step 2. Design of optimal controller for middle section of continuous line.

The principal aim of CL control consists in achieving good dynamic control of tension in the material, with the speed of material movement being in accord with the pre-set CL speed. As it has been said above, this is in fact a nonlinear MIMO system an important feature of which is mutual influencing of the individual input and state quantities that can result in bad quality or even in the destruction of the material being processed. This fact makes the controller design methods and their subsequent resulting structures often very complex and presents an obstacle to their wider practical application in industry. Therefore, our aim was to design a simple CL controller that would ensure the desired dynamics in terms of the selected criterion for systems that are only described by input/output relationships, that is, on basis of their fuzzy model.

For control of middle section of CL (for which fuzzy model was designed), we chose the simplest control structure consisting of two standard PI controllers (one for tension control F_{12} and one for output velocity control v_2), as illustrated in **Figure 20**.

Processing of material in a CL is usually carried out in operation cycles during which a required amount of prepared material is processed (e.g., a roll of paper, a sheet metal coil.)

An operation cycle includes three stages—line start-up, line running at constant processing speed, and line delayed shut-off.

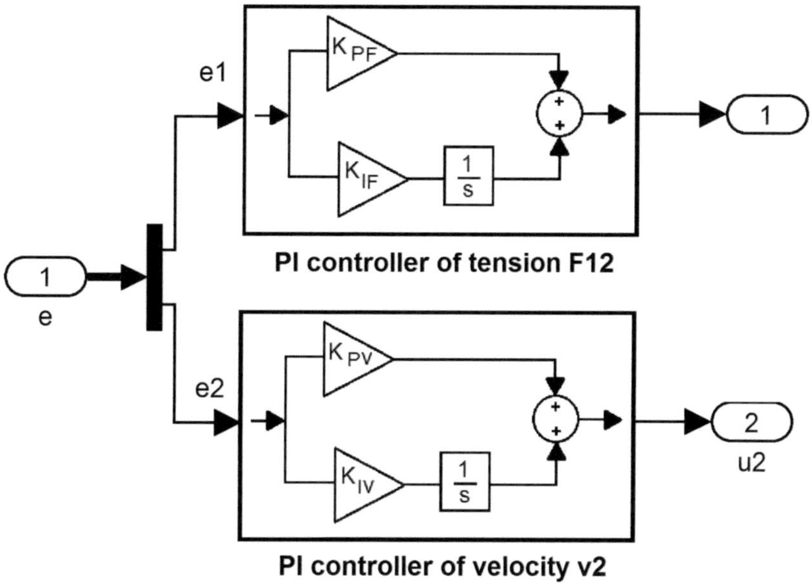

Figure 20. Continuous line PI controller structure.

The objective of the optimization is to find such vector **K**[K_{PP}, K_{IP}, K_{PV}, K_{IV}] of the CL controller parameters for which the selected optimization criterion for a given CL operation cycle would be minimum. Most often this criterion is selected in quadratic form according to following equation

$$J(K) = \int \left(C_1 * e_1{}^2 + C_2 * e_2{}^2\right) dt \tag{6}$$

where e_1 is the deviation between the desired and real tensile force of CL, e_2 is the deviation between the desired and real output velocity of the CL. Coefficients C_1 and C_2 determine the importance placed on the control errors of the particular outputs. In this case, we chose $C_1 = 5$ and $C_2 = 1$, which in terms of physics can be interpreted as larger emphasis put on the quality of regulation of error e_1 (tension in the strip of material which primarily determines its final quality). The optimal value of parameters of vector **K** is identified in the space of real values of gain of proportional and integrating elements of the particular controllers.

Let us note that what we are looking for is the extreme of the function of various variables, where the value of the criterial function for the individual vector **K** is determined by simulation on basis of the experimentally constructed CL fuzzy model according to **Figure 21**.

Several procedures can be applied for the purpose of optimization (e.g., genetic algorithm methods, and network charts). Thanks to today's availability of high-performance computing means, we chose the method of even geometrical division of the parameter space into equal intervals and of systematic searching within the whole range of the space. The advantage of this approach is that we can always identify the global minimum of the function Eq. (6), the disadvantage may be the time and computing demands in case vector **K** has several parameters, and the division of the space is more dense.

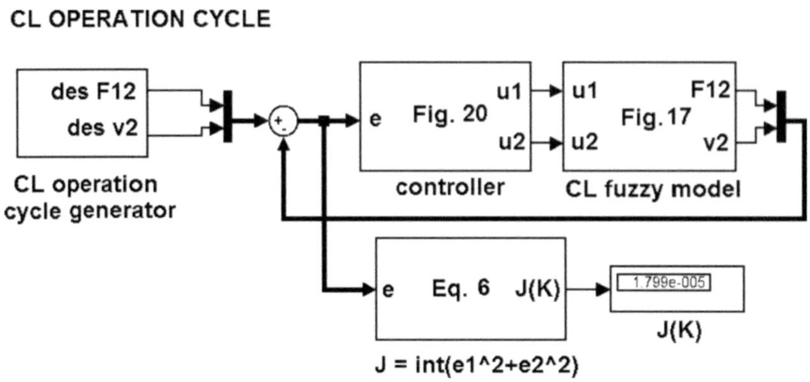

Figure 21. Simulation diagram for computing the value of criterial function $J(\mathbf{K})$.

At the start of the optimization process, we determined the initial values of controller parameters $K_{PF0} = 1$, $K_{IF0} = 1$, $K_{PV0} = 1$, $K_{IV0} = 10$, and we divided the parameter space of individual gains of controllers by increments $\Delta K_{PF} = 1$, $\Delta K_{IF} = 5$, $\Delta K_{PV} = 1$, $\Delta K_{IV} = 5$. The maximum values of individual parameters were defined as $K_{PFmax} = 10$, $K_{IFmax} = 100$, $K_{PVmax} = 10$, $K_{IVmax} = 100$—this was based on physically standard values of proportional and integrating elements of PI controllers.

The value of the criterial function for initial values of vector $\mathbf{K} = \mathbf{K}_0$ and for the chosen CL operation cycle is equal to $J(\mathbf{K}_0) = 37.19$. The time responses of CL physical model output quantities corresponding to this initial PI controller parameter setting are illustrated in **Figure 22**.

For finding optimal controller parameters, an m-file was created in Matlab program environment. At the end of the search process for optimal vector of CL controller parameters, the value of the optimization criterion was $J(\mathbf{K}_{opt}) = 0.3906$, which corresponds with optimal CL controller parameter values $\mathbf{K}_{opt}$ [9, 20, 7, 80]. Time responses of CL physical model output quantities for optimally set parameters of its controller are shown in **Figure 23**.

3.1. Discussion

The proposed controller has been verified by experimental measurements on a real system which presented the physical model of the continuous line (the parameters of the CL physical model are specified in the Appendix). **Figures 22** and **23** illustrate experimental results of the control of the continuous line middle section for selected operational cycle. In industrial practice, the required tension in the strip of material is set the first and then the line starts up to the desired operational speed.

Figure 22 shows the selected CL operation cycle in which first the desired value of tension in the strip of material is set to 0.8 N and at time 4 s the line starts up to reach the operational

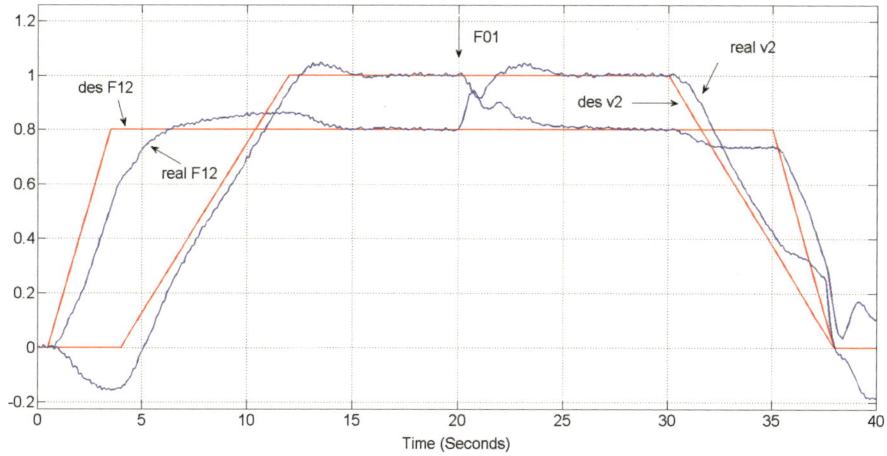

Figure 22. Time responses of CL outputs during operation cycle for $\mathbf{K}_0$.

speed. At time 20 s, failure $F_{01} = 0.7$*FN occurs at entry of the line middle section (caused e.g., by a change in material thickness—material weld). The Figure shows that at initial values of parameters of CL speed and tension PI controllers, the deviation in tension from the desired value is up to 10%, the speed deviation is up to 25%, and at certain moments, the line also runs in the opposite direction. Autonomy and invariance in terms of failures are poor in dynamic states. On the contrary, when we set optimal values of parameters of CL speed and tension PI controllers, we can see—as illustrated in **Figure 23**—that tension in the line is maintained also in dynamic states within the range of 2% (which ensures high material processing quality during the whole operation cycle), and line speed only briefly falls outside the desired value by approx. 8% at a moment of influence by an external step disturbance. Optimal setting of the CL PI controller parameters therefore ensures good quality dynamics, autonomy and invariance of the controlled system against failures.

For the control of continuous line tension and velocity, a very simple control structure with two PI controllers was designed. We looked for four optimal parameters in the structure, such that would best satisfy the chosen quadratic optimality criterion for the given operation cycle of the line.

The quality of the designed controllers depends on the quality of the constructed fuzzy model which very well approximates the performance of the modeled system and can be further employed in the design of various CL control structures and also in the identification of non-measurable additive disturbances influencing the system, principally in real time.

The quality of the proposed controller depends on a large extent on a good quality of the nonlinear system fuzzy model which is constructed in the first step of the design procedure The model is constructed only on basis of suitably measured relations between the system's inputs and outputs, without the necessity of preliminary knowledge of its internal structure and parameters. The fuzzy model design is based on the basic idea of dynamic system description in state space.

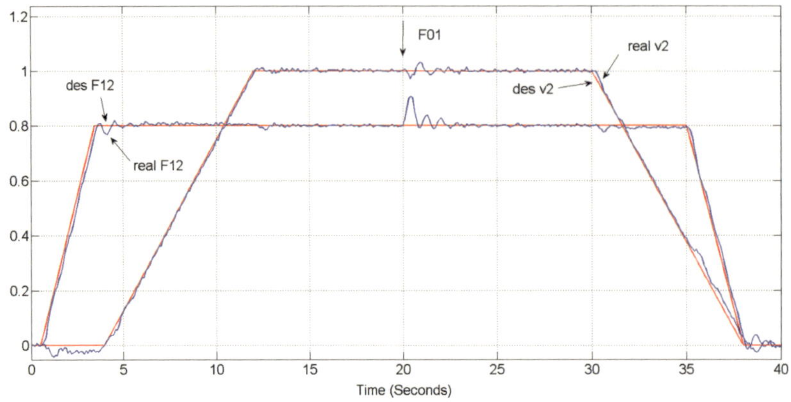

Figure 23. Time responses of CL outputs during operation cycle for $\mathbf{K}_{opt.}$

The quality of the proposed controller depends on a large extent on a good quality of the nonlinear system fuzzy model which is constructed in the first step of the design procedure (see Section 4). The model is constructed only on basis of suitably measured relations between the system's inputs and outputs, without the necessity of preliminary knowledge of its internal structure and parameters. The fuzzy model design is based on the basic idea of dynamic system description in state space.

With this method, no principal limitations for the investigated system's nonlinearities are defined, and therefore, there is good reason to assume that the presented method will find wide use in multi-motor drives in steel industry, paper-making, printing and textile industries, in the production of synthetic fibers and foils in the chemical industry and in other industries.

Acknowledgements

The authors wish to thank the project VEGA 1/0464/15 for its support.

Appendix

AC drive parameters:

$P_N = 15$ kW	$U_{1N} = 220$ V	$I_{1N} = 29.5$ A	$n_N = 1450$ rev/min
$M_N = 98.8$ Nm	$J = 0.11$ kgm^2	$M = 0{,}064$ H	$R_1 = 0.178\ \Omega$
$R_2 = 0.36\ \Omega$	$K_{11} = 277.08$ H^{-1}	$K_{12} = -269$ H^{-1}	$K_{12} = -269$ H^{-1}

Stator phase resistance: $r_1 = 0.267\ \Omega$, rotor phase resistance: $r_2 = 0.54\ \Omega$
Main inductance: $L_h = 96$ mH, leakage inductance: $L_{S1} = L_{S2} = 2.75$ mH
Slip angular speed: $\omega_2 = \omega_1 - \omega_g$, $\omega_0 = K_{11}\ (R_1 + (M^2/L_2) \cdot \omega_g) = 143.33$ s^{-1}, $\omega_g = 5.46$ s^{-1}
Mechanical angular speed of the motor: ω_m, angular frequency of the stator voltage: ω_1
Number of pole pairs: $n_p = 2$
Parameters of the CL physical model:
DC motors:

$U_N = 24$ V	$n_N = 3650$ rpm-s	$R_a = 0.7\ \Omega$	$I_N = 8.5$ A	$P_N = 140$ W	$F_N = 250$ N
$L_a = 0.1$ Mh	$M_N = 0.39$ Nm	$J = 0.002$ kgm^2	$j = 24$	$c\phi = 0.043$ Vs	$I_{max} = 20$ A

Converters: $T_{TM} = 0.1$ ms
Current sensor: $K_I = 2$V/A, velocity sensor: $K_v = 6.6$ V/m s^{-1}, tension sensor: $K_F = 0.022$ V/N
Working rolls: $r = 0.04$ m, $v_{max} = 1.5$ m s^{-1}.

Author details

Pavol Fedor and Daniela Perduková*

*Address all correspondence to: daniela.perdukova@tuke.sk

Department of Electrical Engineering and Mechatronics, Technical University of Kosice, Kosice, Slovak Republic

References

[1] Bai Y., Zhuang Q., Roth Z.S. Fuzzy logic control to suppress noises and coupling effects in a laser tracking. IEEE Transactions on Fuzzy Systems. 2005;**13**(1):113-121.

[2] Baturone I., Moreno-Velo F.J., Sanchez-Solano S., Ollero A. Automatic design of fuzzy controllers for car-like autonomous robots. IEEE Transactions on Fuzzy Systems. 2004;**12**(4):447-465.

[3] Chang C.Y., Chen B.S. Intelligent robust tracking controls for holonic and nonholonomic mechanical systems using only position measurements. IEEE Transactions on Fuzzy Systems. 2005;**13**(4):491-507.

[4] Kim E., Lee S. Output feedback tracking control of MIMO systems using a fuzzy disturbance observer and its application to the speed control of a PM synchronous motor. IEEE Transactions on Fuzzy Systems. 2005;**13**(6):725-741.

[5] Mousavi S.A., Hashemipour M., Sadeghi M., Petrofsky J.S., Prowse M.A. A fuzzy logic control system for the rotary dental instruments. Iranian Journal of Science & Technology, Transaction B: Engineering. 2010;**34**(No. B5):539-551.

[6] Amaral da Silva J., Luiz de Oliveira Serra G. A new methodology for fuzzy model based robust control for nonlinear systems. In: IEEE International Conference on Fuzzy Systems (FUZZ-IEEE); 2012. pp. 1-8. doi:10.1109/FUZZ-IEEE.2012.6251329

[7] Beirami H., Zerafat M.M. Self-tuning of an interval type-2 fuzzy PID controller for a heat exchanger system. Iranian Journal of Science & Technology Transactions of Mechanical Engineering. 2015;**39**(No. M1):113-129.

[8] Fedor P., Perduková D., Ferková Ž. Optimal input vector based fuzzy controller rules design. Advances in Intelligent Systems and Computing. 2013;**189**:371-380.

[9] Fedor P., Perduková D. A simple fuzzy controller structure. Acta Electrotechnica et Informatica. 2005;**5**(4):53-56.

[10] Babuška R. Fuzzy modeling for control. Boston: MA: Kluwer; 1998.

[11] Fedor P., Perduková D. Fuzzy model of a nonlinear mechatronic system. Procedia Engineering. 2014;**96**:91-100.

[12] Feng G. A survey on analysis and design of model-based fuzzy control systems. IEEE Transactions on Fuzzy Systems. 2006;**14**(5):676-697.

[13] Qi P., Liu Ch., Ataka A., Lam H.K., Althoefer K. Kinematic control of continuum manipulators using a fuzzy-model-based approach. IEEE Transactions on Industrial Electronics. 2016;**63**(9):5022-5035.

[14] Kiriakidis K. Fuzzy model-based control of complex systems. IEEE Transactions on Fuzzy Systems. 1998;**6**(4):517-529.

[15] Wu Ch., Wang J., Li H., Liang H. Fuzzy-model-based control for nonlinear networked systems with random packet losses. In: 5th International Conference on Information Science and Technology (ICIST); 2015. pp. 455-460. doi:10.1109/ICIST.2015.7289015

[16] Fedor P., Perduková D. Model based fuzzy control applied to a real nonlinear mechanical system. Iranian Journal of Science and Technology, Transactions of Mechanical Engineering. 2016;**40**(2):113-124.

[17] Li B., Fan X., Zhang D., Jiang G. Modeling and fuzzy sliding decoupling control of looper multivariable system. In: Chinese Control and Decision Conference (CCDC); 2016. pp. 2467-2472. doi:10.1109/CCDC.2016.7531400

[18] Sjoberg J., Zhang Q., Ljung L., Benveniste A., Delyon B., Glorennec P., Hjalmarsson H., Juditsky A. Nonlinear black-box modeling in system identification: a unified overview. Automatica. 1995;**31**:1691-1724.

[19] Tečec Z., Petrović I., Matuško J. A Takagi-Sugeno fuzzy model of synchronous generator unit for power system stability application. AUTOMATIKA: Journal for Control, Measurement, Electronics. Computing and Communications. 2010;**51**(2):12-137.

[20] Fedor P., Perduková D. Fuzzy model for middle section of continuous line. International Journal of Engineering Research in Africa. 2015;**18**:75-84.

[21] Liu X., Xioung Z., Chen L., Zhu Z. A new Takagi-Sugeno fuzzy approach of process modeling and fault detection. In: 35th Chinese Control Conference – CCC; 2016. pp. 7126-7130. doi:10.1109/ChiCC.2016.7554483

[22] Johansen T.A. Fuzzy model based control: Stability robustness and performance issues. IEEE Transactions on Fuzzy Systems. 1994;**2**(1):221-233.

[23] Girovský P. Fuzzy control of synchronous motor with permanent magnet. Acta Electrotechnica et Informatica. 2016;**16**(4):17-20.

[24] Tang K.S., Man K.F., Chen G., Kwong S. An optimal fuzzy PID controller. IEEE Transactions Industrial Electronics. 2001;**48**(4):757-765.

[25] Tang W., Rui Z., Jiang H., Gao H. Coagulation control fuzzy modeling based on cluster analysis and optimization technique. In: 35th Chinese Control Conference (CCC); 2016. pp. 2154-2159. doi:10.1109/ChiCC.2016.7553687

[26] Tao C.W., Taur J.S. Robust fuzzy control for a plant with fuzzy linear model. IEEE Transactions on Fuzzy Systems. 2005;**13**(1):30-41.

[27] Zeng K., Zhang N.Y., Xu W.L. A comparative study on sufficient conditions for Tagaki-Sugeno fuzzy sytems as universal approximators. IEEE Transactions on Fuzzy Systems. 2000;**8**(6):773-780.

[28] Vo H.H., Brandstetter P., Dong C.S.T. MRAS observers for speed estimation of induction motor with direct torque and flux control. Lecture Notes in Electrical Engineering. 2016;**371**:325-335.

[29] Vittek J., Butko P., Ftorek B., Makyš P., Gorel L. Energy near-optimal control strategies for industrial and traction drives with a.c. motors. Mathematical Problems in Engineering. 2017;**2017**:22 pages/ID 1857186. doi:10.1155/2017/1857186

[30] Tseng Ch.S., Chen B.S., Uang H.J. Fuzzy tracking control design for nonlinear dynamic systems via TS fuzzy model. IEEE Transactions on Fuzzy Systems. 2001;**9**(3):381-392.

[31] Zadeh L.A. Outline of a New Approach to the Analysis of Complex Systems and Decision Processes. IEEE Transactions on Systems, Man, and Cybernetics. 1973;**1**:28-44.

[32] Brandstetter P., Dobrovsky M., Kuchar M. Implementation of Genetic Algorithm in Control Structure of Induction Motor AC Drive. Advances in Electrical and Computer Engineering. 2014;**14**(4):15-20.

[33] Jang J.S.R. ANFIS: Adaptive-Network-Based Fuzzy Inference System. IEEE Trans. Syst., Man, and Cyber.. 1993;**23**(3):665-685.

[34] Chiu S.C. Extracting Fuzzy Rules from Data for Function Approximation and Pattern Classification, Fuzzy Information Engineering: a guided tour of applications. New York: John Wiley & sons; 1997.

[35] Takagi T., Sugeno M. Fuzzy identification of systems and its application to modeling and control. IEEE Trans. Syst., Man and Cyber. 1985;**15**:116-132

[36] Yager R.R. Essentials of Fuzzy Modeling and Control . New York: Wiely; 1994.

[37] Hüseyin A., Faruk O. ANFIS & PID controller design and comparison for overhead cranes. Indian Journal of Engineering & Materials Sciences. 2011;**18**:191-203.

Chapter 19

Vibration Suppression Controller of Multi-Mass Resonance System Using Fuzzy Controller

Hidehiro Ikeda

Additional information is available at the end of the chapter

http://dx.doi.org/10.5772/intechopen.68319

Abstract

Vibration suppression control of the mechanical system is a very important technology for realizing high precision, high speed response and energy saving. In general, the mechanical system is modeled with a multi-mass resonance system, and vibration suppression control is applied. This chapter presents a novel controller design method for the speed control system to suppress the resonance vibration of two-mass resonance system and three-mass resonance system. The target systems are constructed by a motor, finite rigid shafts, and loads. The control system consists of a speed fuzzy controller and a proportional-integral (PI) current controller to realize precise speed and torque response. In order to implement the experimental system, the system is treated as the digital control. This chapter also utilizes a differential evolution (DE) to determine five optimal controller parameters (three scaling factors of the fuzzy controller and two controller gains of PI current controller. Finally, this chapter verified the effectiveness to suppress the resonance vibrations and the robustness of the proposed method by the computer simulations and the experiments by using the test experimental setup.

Keywords: multi-mass resonance system, vibration suppression control, fuzzy controller, differential evolution

1. Introduction

Recently, motor drive system, which consists of several motors, shafts, gears, and loads, is widely utilized in industrial fields. These mechanical systems are made a request the high-speed response, weight reduction, miniaturization, and high precision requirements for various industrial applications.

Hence, in industrial field, the system is treated as a multi-mass resonance system, which consists of several inertial moments, torsional shafts, and gear coupling. The first-order approximation model of multi-mass resonances model is two-mass resonance model. For instance, several control methods, which are PID control (Proportional plus Integral plus Derivative Control) with a resonance ratio control using the disturbance observer, coefficient diagram method (CDM), full state feedback control with the state observer, the pole placement method, fractional order PID_K control, and H_∞ control method, are effective to control for two-mass resonance system [1–3]. Ikeda et al. [4] have explained the effectiveness of the controller design technique using the pole placement method for the two-mass position control system.

However, the resonance system is required more high precision and high response speed control in recent years. Therefore, it is necessary to deal with a higher order model of the resonance system. For instance, the drive train of the electric vehicle is constructed the four-mass system. Likewise, the ball screw drive stage is typically four-mass system. The thermal power generation system composed of multiple turbines and generators is modeled as twelve-mass resonance system. Thus, several vibration suppression control methods on three-mass resonance system or more have been proposed [5, 6]. Here, modified-IPD speed controller using Taguchi Method has been proposed in Refs. [7, 8].

Meanwhile, the state equations of the controlled object and its parameters are required to design the control systems. Refs. [9, 10] previously proposed a controller gain tuning method for a vibration suppression-type speed controller using fictitious reference iterative tuning (FRIT) for single-input multi-variable control objects without knowledge of the system state equations and the parameters.

In contrast, a fuzzy control system can be assumed as one method for solving these problems. A fuzzy control system using a fuzzy inference is the embodiment of non-mathematical control algorithm, which is constructed by experience and intuition. Several applications brought in the fuzzy control system to motor drive system [11–14].

This chapter proposes a vibration suppression controller by using a fuzzy inference. The control system consists of a speed fuzzy controller and a proportional-integral (PI) current controller to realize precise speed and torque response on two or three inertial resonance system. In the control system, only motor side state variables are utilized for controlling the resonance system. Additionally, this chapter treats with the proposed control system as the digital control system. Here, the proposed control system is new system that I improved to apply the control system which I already proposed for simulation model in Refs. [13, 14] to experimental actual equipment.

The fuzzy controller has three scaling factors, and the PI current controller has two controller gains. In this chapter, a differential evolution algorithm (DE) is utilized the determination of these five controller parameters [13–18]. DE, which was proposed by Price and Storn, is one of the evolutionary optimization strategies. By using DE, it is easy and fast to determine the proper controller parameters.

Lastly, the validity of the controller design, the robustness, and the control effectiveness of the proposed method was verified using the simulations and the experiments by using the test experimental set up.

2. Multi-mass vibration suppression control system

2.1. 2-mass model

Figure 1 shows the two-mass resonance model. The model is configured of two rigid inertial masses with a torsional shaft, where ω_M, T_{dis}, ω_L, T_{in}, J_M, J_L, K_s, and T_L denote the motor angular speed, the torsional torque, the load angular speed, the input torque, the inertia of motor, the inertia of load, the shaft torsional stiffness, and the load torque, respectively.

If all the state variables can be observed by several sensors and all the system parameters are known or identified, it is easy to construct the optimal control system. However, in general, it is difficult to measure the state variables of the load side due to constraints on scarce measurement environment and sensor installation location. Therefore, in this chapter, we use only the motor side variables. Furthermore, we contemplate for the current minor control in order to compensate torque response. Eq. (1) shows the continuous state equation of two-mass resonance model, where the viscous friction is not considered.

$$\frac{d}{dt}\begin{pmatrix}\omega_M\\ \omega_L\\ T_{dis}\end{pmatrix}=\begin{pmatrix}0 & 0 & -\frac{1}{J_M}\\ 0 & 0 & \frac{1}{J_L}\\ K_s & -K_s & 0\end{pmatrix}\begin{pmatrix}\omega_M\\ \omega_L\\ T_{dis}\end{pmatrix}+\begin{pmatrix}\frac{1}{J_M}\\ 0\\ 0\end{pmatrix}T_{in}+\begin{pmatrix}0\\ -\frac{1}{J_L}\\ 0\end{pmatrix}T_L \tag{1}$$

Eq. (2) shows the transfer function of two-mass model, which input signal is T_{in} and output signal is ω_M.

$$\frac{\omega_M}{T_{in}}=\frac{s^2+\omega_a^2}{J_M s(s^2+\omega_r^2)} \tag{2}$$

where ω_r is a resonance frequency and ω_a is an anti-resonance frequency. Here, we use the DC servo motor as the driving motor. Eq. (3) is the voltage equation of dc servo motor, where R_a is the armature resistance, L_a is the armature inductance, K_e is the back-emf constant, and K_0 is the converter gains of the DC power supply. Input torque is calculated by $T_{in} = K_t i_a$, where K_t is the torque constant.

$$L_a\frac{di_a}{dt}+R_a i_a=K_0 u_c-\omega_M \tag{3}$$

Figure 2 is indicative of the block diagram of the two-mass resonance system.

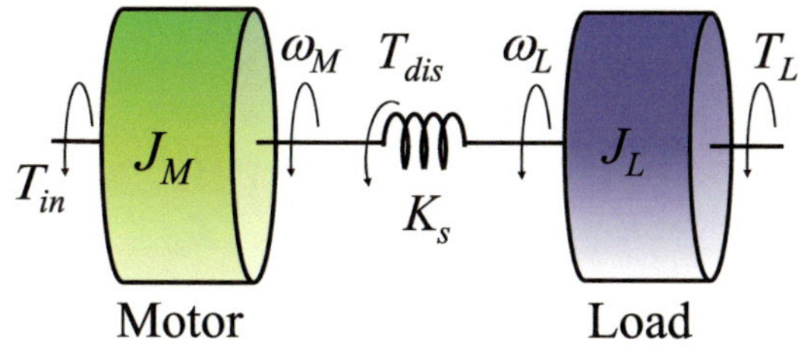

Figure 1. 2-mass model.

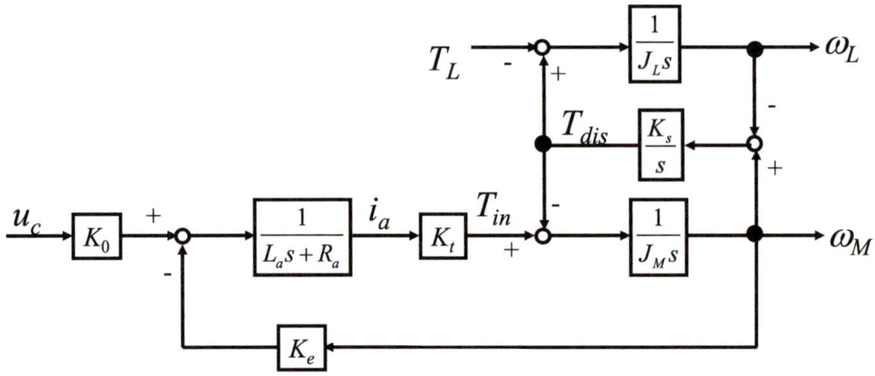

Figure 2. Block diagram of two-mass resonance model.

The inertia ratio R of two-mass model is given by Eq. (4), where J_{Mn} and J_{Ln} represent the nominal values of the motor and load inertias, respectively.

$$R = \frac{J_{Ln}}{J_{Mn}} \tag{4}$$

2.2. Three-mass model

Similarly to two-mass resonance model, **Figure 3** reveals the three-mass model. The model consists of three rigid inertias and two shafts. Here, J_c and J_L are the load 1 inertia moment and the load 2 inertia moment, respectively. Furthermore, ω_c, ω_L, T_{dis1}, T_{dis2}, K_{s1}, and K_{s2} denote load 1 angular speed, load 2 angular speed, shaft 1 torsional torque, shaft 2 torsional torque, the shaft 1 stiffness, and the shaft 2 stiffness, respectively.

$$\frac{d}{dt}\begin{pmatrix} \omega_M \\ \omega_M \\ \omega_M \\ T_{dis} \\ T_{dis} \end{pmatrix}\begin{pmatrix} \omega_M \\ \omega_L \\ T_{dis} \end{pmatrix} = \begin{pmatrix} 0 & 0 & 0 & -\frac{1}{J_M} & 0 \\ 0 & 0 & 0 & \frac{1}{J_c} & -\frac{1}{J_c} \\ 0 & 0 & 0 & 0 & \frac{1}{J_L} \\ K_{s1} & -K_{s1} & 0 & 0 & 0 \\ 0 & K_{s2} & -K_{s2} & 0 & 0 \end{pmatrix}\begin{pmatrix} \omega_M \\ \omega_M \\ \omega_M \\ T_{dis} \\ T_{dis} \end{pmatrix} + \begin{pmatrix} \frac{1}{J_M} \\ 0 \\ 0 \\ 0 \\ 0 \end{pmatrix} T_{in} + \begin{pmatrix} 0 \\ 0 \\ -\frac{1}{J_L} \\ 0 \\ 0 \end{pmatrix} T_L \tag{5}$$

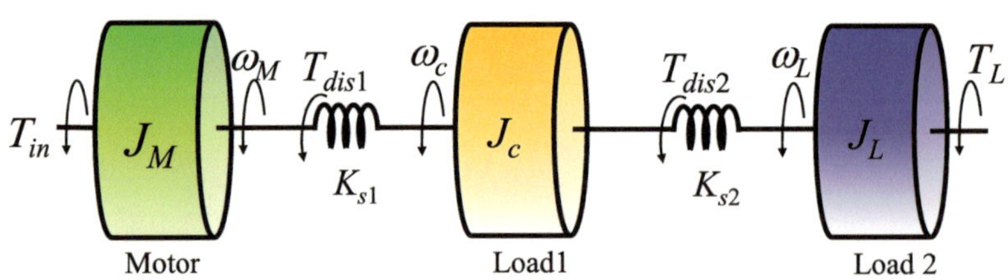

Figure 3. Three-mass model.

http://dx.doi.org/10.5772/intechopen.68319

The state equation of three-mass resonance model is shown in Eq. (5). Then, Eq. (6) shows the continuous transfer function of three-mass resonance model, which input signal is T_{in} and output signal is ω_M.

$$\frac{\omega_M}{T_{in}} = \frac{(s^2 + \omega_{a1}^2)(s^2 + \omega_{a2}^2)}{J_M s(s^2 + \omega_{r1}^2)(s^2 + \omega_{r2}^2)} \tag{6}$$

In this equation, ω indicates the angular frequency, where ω_{r1}, ω_{r2}, ω_{a1}, and ω_{a2} are the resonance frequencies, and anti-resonance frequency, respectively. Then, the block diagram realized by using above equations is shown in **Figure 4**.

2.3. Experimental set up

This chapter confirms the effectiveness and performance of the proposed method by experiments using the experimental equipment.

Figure 5 is the appearance of the experimental system constructed in this research. The two-mass resonance system is simulated by utilizing the dc servo motor and the dc generator with a finite rigid coupling. The controller is realized on a digital signal processor, which calculates the PWM signal to a four-quadrant dc chopper.

The DSP board (PE-PRO/F28335 Starter Kit, Myway Plus Corp.) consists of the DSP (TMS320F28335PGFA), a digital input/output (I/O), ABZ counters for encoder signals, analog-to-digital (AD) converters and digital-to-analog (DA) converters [19]. The motor and load angles and angular speeds are detected using 5000 pulses-per-revolution encoders. The current of dc servo motor is measured by the current sensor and AD converter.

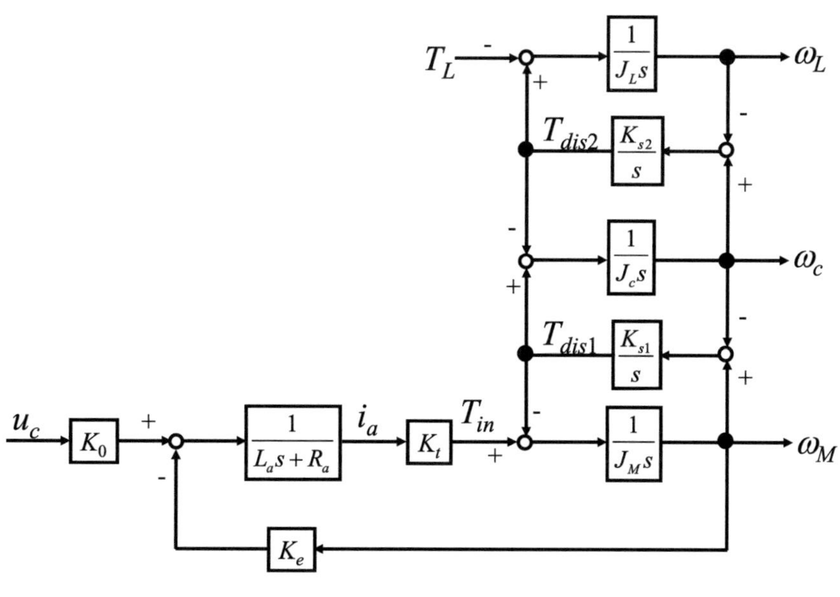

Figure 4. Block diagram of three-mass model.

Figure 5. Experimental apparatus.

The control frequency and the detection frequency of the encoder are both 1 ms, and the detection period for the current is 10 μsec. The design language used was C. Then, while considering the application of the system to specific apparatus, we constructed a digital control system that contains a discrete controller. In addition, we used MATLAB/Simulink software for the proposed off-line tuning process based on simulation and constructed the fuzzy control system as a continuous system [20]. The disturbance is added to the dc generator as the torque by using the electric load device on constant current mode. **Figures 6** and **7** show the apparatus of the two-mass model and three-mass model used in the experimental set up, respectively. **Figure 8** shows the experimental system configuration. For reference, the nominal parameters

Figure 6. Photograph of two-mass resonance model.

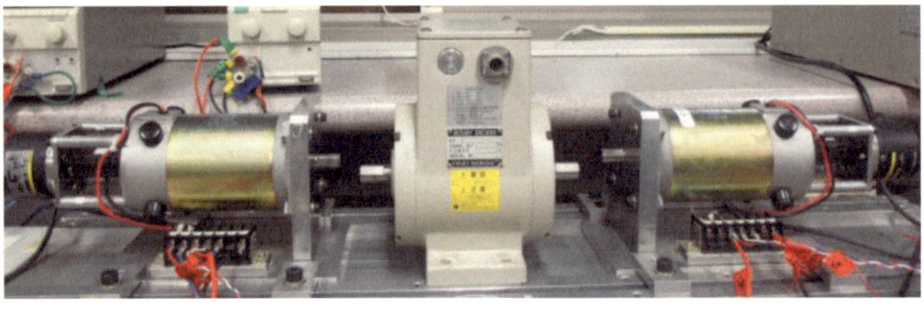

Figure 7. Photograph of three-mass resonance model.

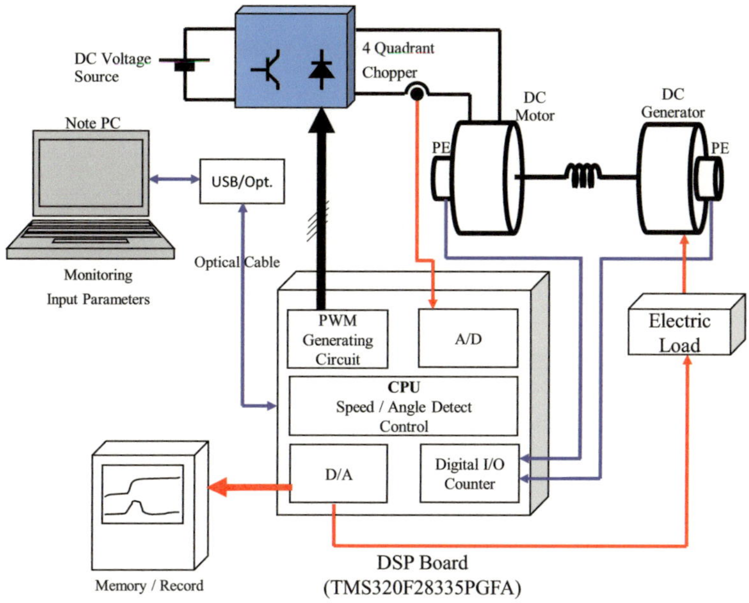

Figure 8. Configuration of experimental system (two-mass resonance model).

of the experimental two-mass model and three-mass model are given in **Tables 1** and **2**, respectively.

Figure 9 shows an example of experimental result using two-mass model. These step waves are the motor and load angular speeds with direct current voltage input. Similarly, **Figure 10** shows an example of experimental result using three-mass model, which are the motor and

Symbol	Parameter	Value
J_{Mn}	Motor inertia	2.744×10^{-4} (kgm^2)
J_{Ln}	Load inertia	2.940×10^{-4} (kgm^2)
K_{sn}	Shaft stiffness	18.5 (Nm/rad)

Table 1. Nominal parameters of two-mass experimental model.

Symbol	Parameter	Value
J_{Mn}	Motor inertia	2.744×10^{-4} (kgm^2)
J_{cn}	Load 1 inertia	1.112×10^{-4} (kgm^2)
J_{Ln}	Load 2 inertia	2.940×10^{-4} (kgm^2)
K_{s1n}	Shaft stiffness 1	18.5 (Nm/rad)
K_{s2n}	Shaft stiffness 2	18.5 (Nm/rad)

Table 2. Nominal parameters of three-mass experimental model.

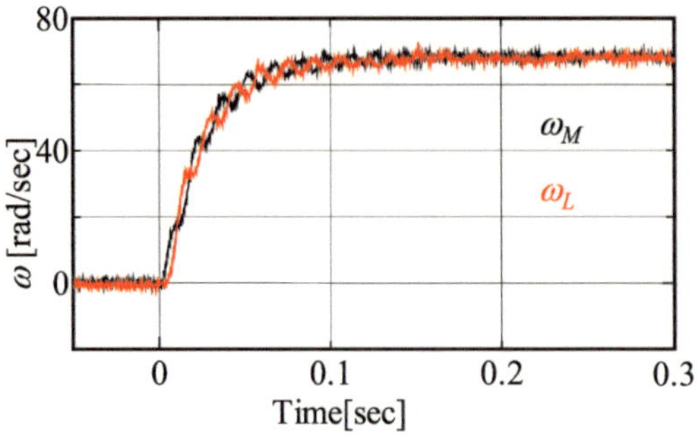

Figure 9. Angular speeds (ω_M and ω_L) of the step responses to a DC voltage input (two-mass model).

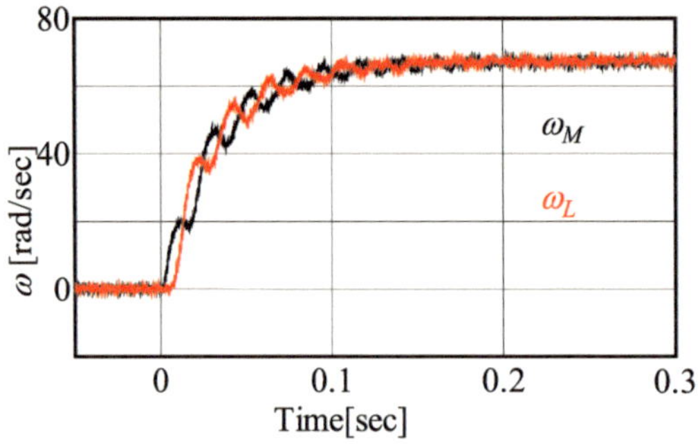

Figure 10. Angular speeds (ω_M and ω_L) of the step responses to a DC voltage input (three-mass model).

load angular speeds with same above condition. In these figures, the resonance vibrations can be observed. The purpose of this research is to suppress these resonance vibrations.

3. Proposed fuzzy control system

3.1. Fuzzy speed controller

Fuzzy controller, which is executed by the fuzzy set and the fuzzy inference, can control for nonlinear systems or uncertain model. **Figure 11** indicates the proposed fuzzy speed controller in this chapter. The speed controller is based on fuzzy control. The current controller is typical PI controller. Furthermore, the load side state variables are not utilized for control, where S_1, S_2, and S_3 are the parameters to determine the scale of the membership function, which are called scaling factors or scaling coefficient. K_{pc} and K_{ic} are the current PI controller gains. Eq. (7) shows the transfer function of current PI controller. Additionally, this chapter uses the discrete control system.

$$u_c(k) = \left(K_{pc} + \frac{1}{s}K_{ic}\right)e(k) \tag{7}$$

Figure 12 is indicative of the membership function for the premise variables. This membership function is a shape of triangle with a dense center. **Figure 13** indicates the membership

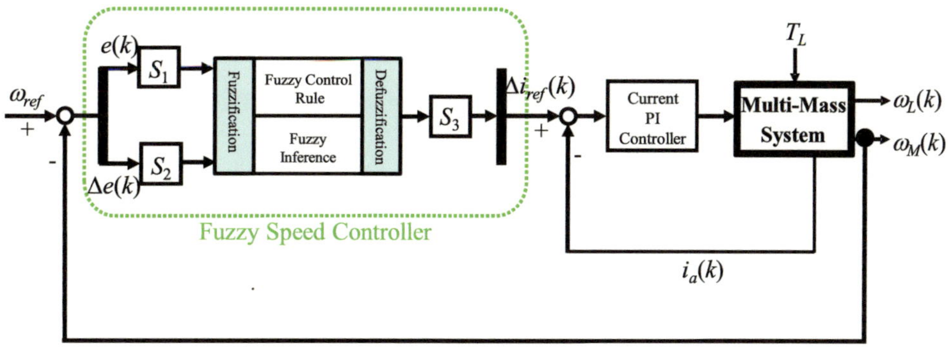

Figure 11. Block diagram of the proposed control system.

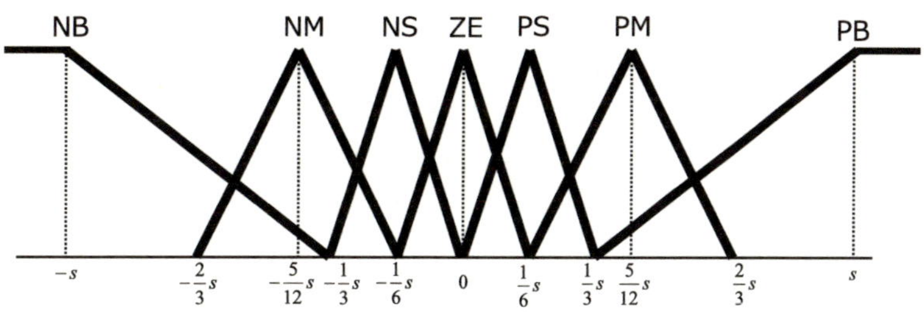

Figure 12. Membership functions of the antecedence.

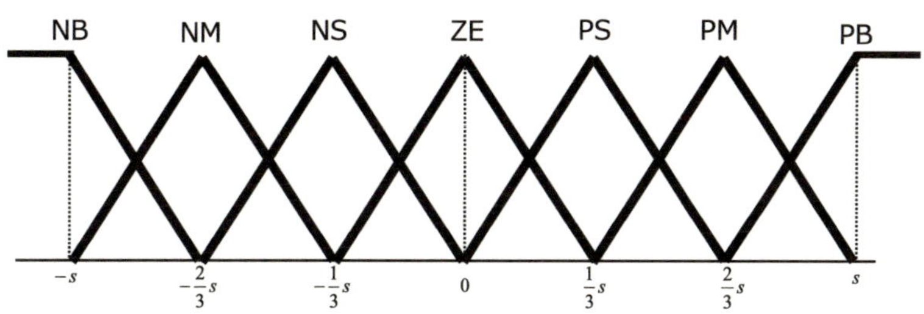

Figure 13. Membership functions of the consequence.

function, which is formed uniformly triangle for the consequent variable. Here, the s denotes the scaling factor. PB, PM, PS, ZE, NS, NM, and NB are the linguistic variables of the fuzzy control where, PB indicates positive big, PM indicates positive medium, PS indicates positive small, ZE indicates zero, NS indicates negative small, NM indicates negative medium, and NB indicates negative big, respectively. The premise variables are $e_{\omega M}(k)$ and $\Delta e_{\omega M}(k)$.

$$e_{\omega M}(k) = \omega_{ref} - \omega_M(k) \tag{8}$$

$$\Delta e_{\omega M}(k) = e_{\omega M}(k) - e_{\omega M}(k-1) \tag{9}$$

Then, the consequence variable is the variation width of the current input $\Delta i_{ref}(k)$. Therefore, the proposed fuzzy controller is nearly same as the proportional-derivative (PD) type controller.

Figure 14 is indicative of the fuzzy rule table. The rule is included the rising correction of the angular speed response.

e \ Δe	NB	NM	NS	ZE	PS	PM	PB
NB				NB	NM		
NM				NM			
NS				NS	ZE		PM
ZE	NB	NM	NS	ZE	PS	PM	PB
PS	NM		ZE	PS			
PM				PM			
PB			PM	PB			

Figure 14. Control rule table.

3.2. Design method of controller parameters by differential evolution

In this chapter, five parameters (S_1, S_2, S_3, K_{pc} and K_{ic}) of the proposed controller have to be designed. However, it is difficult to determine them by trial and error or some. Therefore, this chapter proposes the differential evolution (DE) to search the optimal controller parameters. Here, DE is one of evolutionary optimized solution search methods. DE is the optimization method-based multi-point search method. In particular, basic GA expresses parameter by binary coding, whereas DE uses the parameters by real variable vector. The DE design is conducted by the initial population, the mutation, the crossover, and the selection. The design flow of DE is shown in **Figure 15**. In this chapter, DE/rand/1/bin design strategy is used for the determination of five controller parameters.

where D is the number of design parameter vectors, NP is the number of members in each population. Each parameter vector is represented by the parameter vector (target vector) $\boldsymbol{x}_{i,G}$, where G denotes one generation. The mutation vector $\boldsymbol{v}_{i,G}$ is calculated by Eq. (10). From this equation, F indicates the step width (scaling factor) of DE design, and CR indicates of the crossover rate, where r_1, r_2, and r_3 are different values.

$$\boldsymbol{v}_{i,G+1} = \boldsymbol{x}_{r1,G} + F(\boldsymbol{x}_{r1,G} - \boldsymbol{x}_{r3,G}), \quad r_1 \neq r_2 \neq r_3 \neq i \tag{10}$$

$$\boldsymbol{u}_{j,G+1} = \begin{cases} \boldsymbol{v}_{j,G+1} & \text{rand} \leq CR \text{ or } j = \text{ST} \\ \boldsymbol{x}_{j,G} & \text{rand} > CR \text{ or } j \neq \text{ST} \end{cases} \tag{11}$$

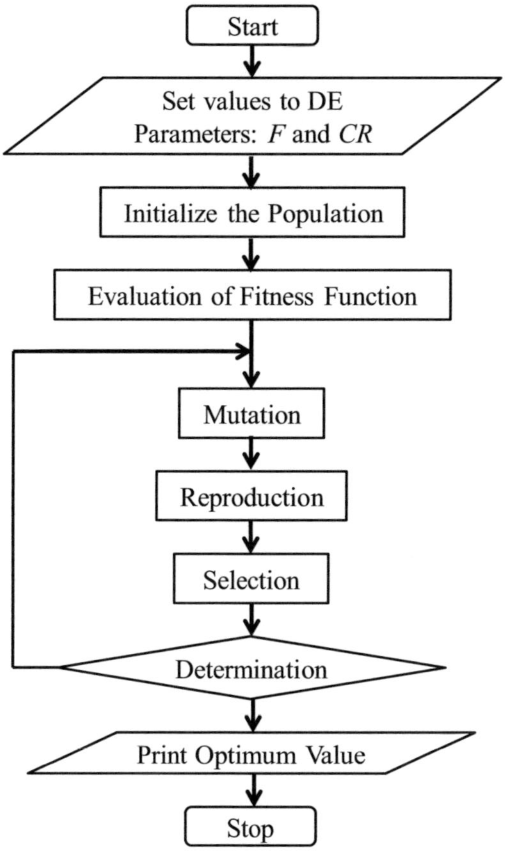

Figure 15. Flow of DE algorithm.

In Eq. (11), $\boldsymbol{u}_{j,G+1}$ is the vector of trial parameter, the rand is random value, and ST indicates the start point. The selection is utilized next algorithm,

$$\boldsymbol{x}_{i,G+1} = \begin{cases} \boldsymbol{u}_{i,G+1} & \text{if } y(\boldsymbol{u}_{i,G+1}) > y(\boldsymbol{x}_{i,G}) \text{ for maximaization problems} \\ \boldsymbol{x}_{i,G} & \text{otherwise} \end{cases} \tag{12}$$

As previously described, the proposed method uses five control parameters (S_1, S_2, S_3, K_{pc}, and K_{ic}). The population size is 2000, the order of each vector is 20, and the coefficient of membership function F is 0.5. Moreover, the rate of crossover CR is 0.9. Then, the performance index function is shown in Eq. (13). Meanwhile, this chapter utilizes the inverse of y as a fitness function.

$$y = \int_0^\infty t\sqrt{(\omega_{ref} - \omega_L)^2}dt \tag{13}$$

4. Simulation and experimental results

4.1. Verification results of computer simulation

Next, the simulation results of the proposed method are demonstrated by computer simulation.

Table 3 shows the results of design parameter using the proposed method for two-mass model. **Figure 16** is indicative of the transition of the maximum fitness function. In this simulation design, the step response and the disturbance response have been evaluated. Furthermore, the inertia ratio R is 1.07, and the stiffness of shaft K_{sn} has been set to 18.5 Nm/rad in the simulation design.

S_1	S_2	S_3	K_{pc}	K_{ic}
8.486	0.4802	0.4001	4.678	1.0×10^{-6}

Table 3. Results of design parameter calculated by DE.

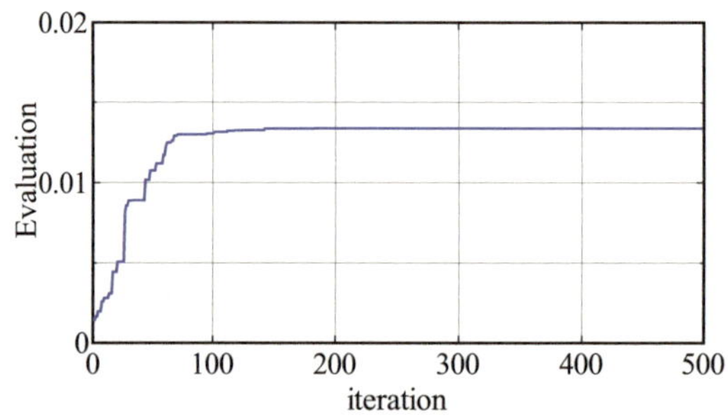

Figure 16. Convergence of index function y.

Figures 17 and **18** show the step responses that were obtained for the motor and load angular speeds, and armature current when using the proposed method. In this chapter, ω_{ref} is 30 rad/s, the DC voltage input is 25 V, and the disturbance input T_L is changed from 0 to 20% at t = 0.3 s. As shown by these figures, good waves are observed for the reference-following, vibration

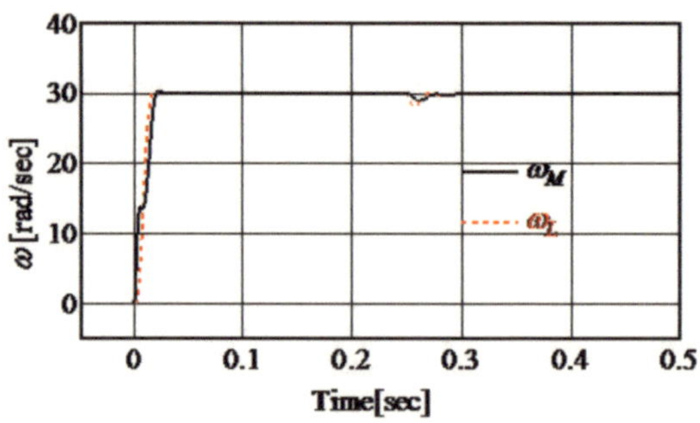

Figure 17. Simulation results ω_M and ω_L (two-mass, R = 1.07, K_{sn} = 18.5 Nm/rad).

http://dx.doi.org/10.5772/intechopen.68319

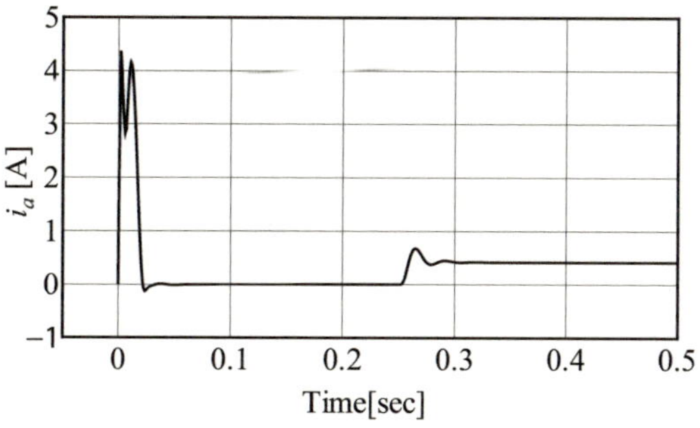

Figure 18. Simulation results i_a (two-mass, $R = 1.07$, $K_{sn} = 18.5$ Nm/rad).

suppression, and the disturbance performance. **Figure 19** is indicative of the search process of the S_1 vector. Similarly, **Figures 20–23** show the transition of the S_2 vector, S_3 vector, K_{pc} vector and K_{ic} vector, respectively. In particular, from **Figure 23** and **Table 3**, K_{ic} is 1.0×10^{-6} of the design limitation value. Therefore, integral gain of the current PI controller can be omitted for this control object.

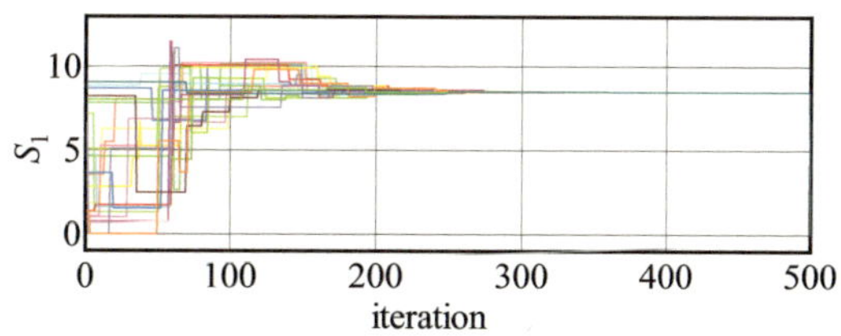

Figure 19. Transition of scaling factor S_1.

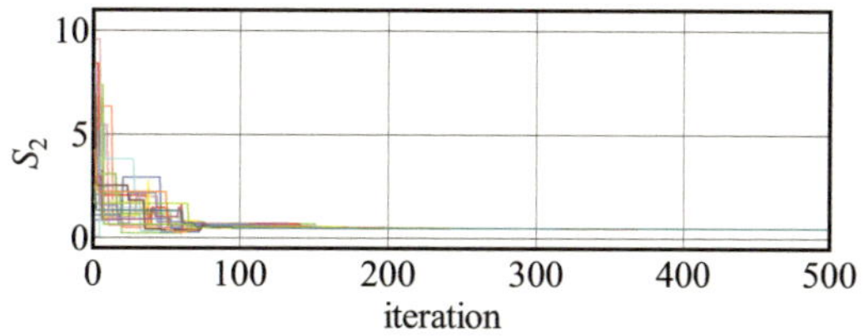

Figure 20. Transition of scaling factor S_2.

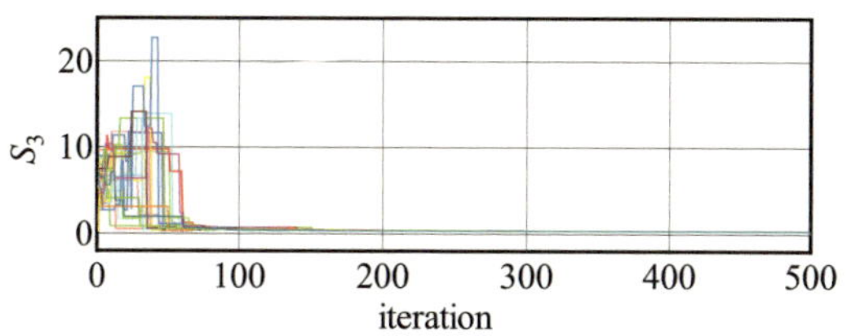

Figure 21. Transition of scaling factor S_3.

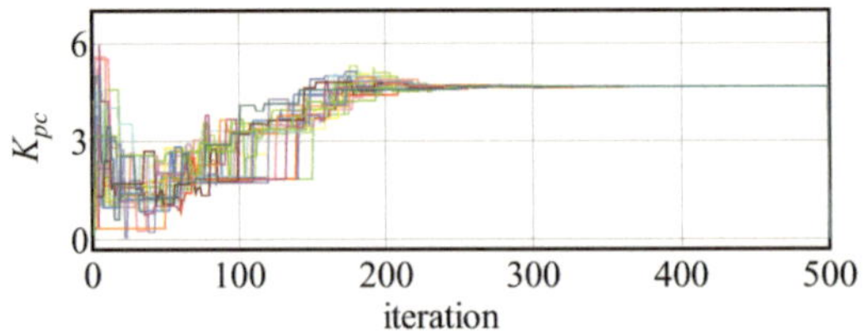

Figure 22. Transition of current proportional gain K_{pc}.

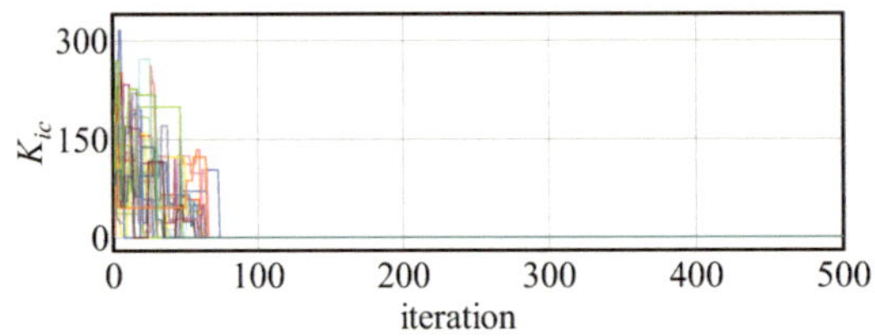

Figure 23. Transition of current integral gain K_{ic}.

4.2. Experimental results

4.2.1. 2-mass model

Next, the experimental results by using the proposed method are illustrated in this section. **Figures 24** and **25** show the experimental results of two-mass model using the proposed method, where the condition (R = 1.07, K_{sn} = 18.5 Nm/rad) is same as the above simulation results shown in **Figures 17** and **18**. From these figures, it is observed that the resonance vibrations between the motor and the load angular speed (ω_M and ω_L) have been suppressed very well. Furthermore, after inputting disturbance, it can be seen that the angular speeds immediately have followed the reference speed ω_{ref} without resonance vibrations. Hence, the validity of the control system, which consists of the proposed method, can be confirmed.

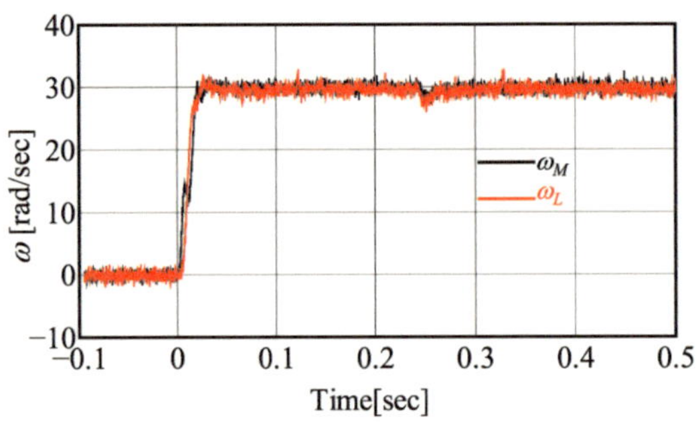

Figure 24. Experimental results for ω_M and ω_L obtained using the proposed method (two-mass, R = 1.07, K_{sn} = 18.5 Nm/rad).

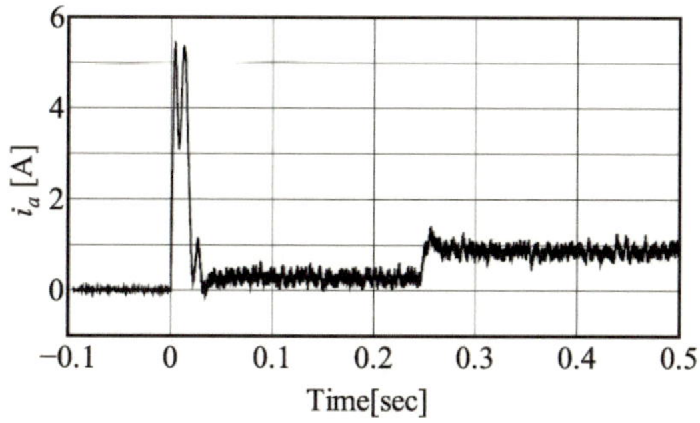

Figure 25. Experimental results for i_a obtained using the proposed method (two-mass, $R = 1.07$, $K_s = 18.5$ Nm/rad).

5. Effects of parameter variation

Next, it is described the effectiveness of robustness by the proposed design method. This section evaluates the robustness to variations in the ratio of inertia and the stiffness of the rigid shaft based on a nominal value.

Figures 26 and **27** show the experimental results of the motor and load angular speeds obtained for the inertia ratio variation when using the same controller gains that were designed using the

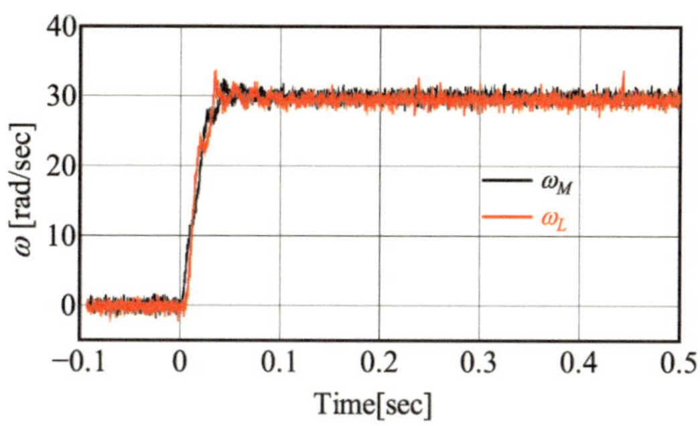

Figure 26. Robustness verification results (two-mass, $R = 0.42$, $K_{sn} = 18.5$ Nm/rad).

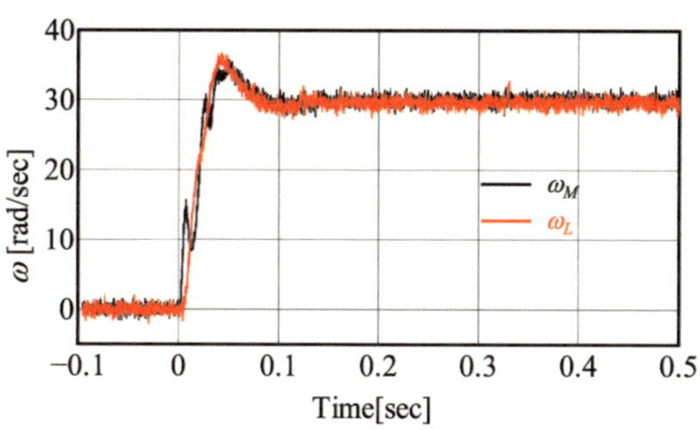

Figure 27. Robustness verification results (two-mass, $R = 2.67$, $K_{sn} = 18.5$ Nm/rad).

proposed method, when $R = [0.42, 2.65]$, where the disturbance torque input was skipped. From these figure, although it can be observed some overshoot and resonance vibration, the good results can be confirmed that were obtained for the design condition.

Figure 28 shows the experimental results of the motor and load angular speeds obtained for the stiffness of shaft variation using the same controller gains, when $K_{sn} = 70.7$. From this figure, it can be seen some resonance vibrations. However, the vibrations rapidly have been suppressed well.

Similarly, **Figure 29** shows the experimental results when $K_{sn} = 3.1$. As can be seen, the motor and load angular speeds oscillated and overshot. Therefore, if the stiffness of shaft of the experimental model is less than the design value, the settling time to suppress the resonance vibration becomes longer, although the proposed control system is not unstable. In addition, **Figure 30** shows the experimental result when the control parameter redesigned with the stiffness of shaft K_{sn} as the nominal value of experimental model. Good responses can be observed in this figure.

Furthermore, the proposed fuzzy control system is applied to a three-mass resonance model. **Figure 31** shows the experimental results of the motor and load angular speeds when using the same controller gains designed for two-mass model (R = 1.07, K_{sn} = 18.5 Nm/rad, where the nominal parameters of the three-mass experimental setup are $J_{Mn} = 2.774 \times 10^{-4}$ kgm^2, $J_{Ln} = 2.940 \times 10^{-4}$ kgm^2, K_{s1n} = 18.5 Nm/rad, K_{s2n} = 18.5 Nm/rad. From this figure, the effectiveness of the proposed method can be confirmed in a similar manner to the two-mass model case.

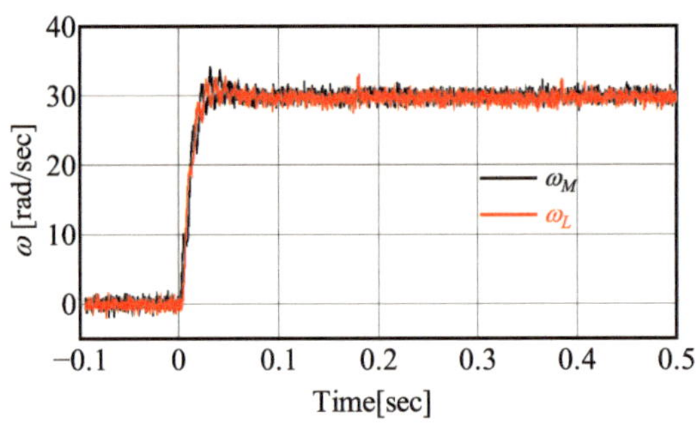

Figure 28. Robustness verification results (two-mass, R = 1.07, K_{sn} = 70.7 Nm/rad).

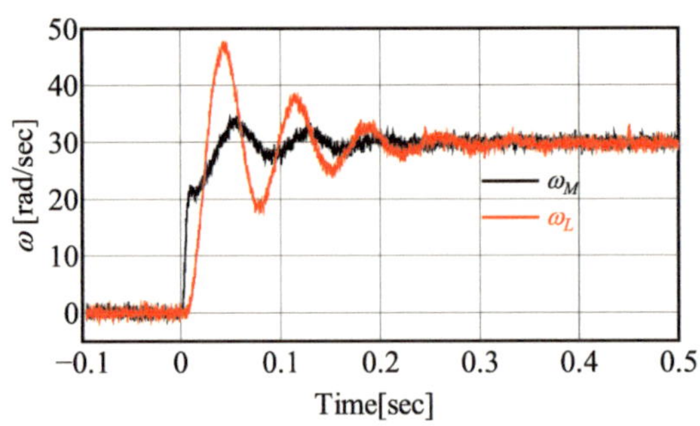

Figure 29. Robustness verification results (two-mass, R = 1.07, K_{sn} = 3.1 Nm/rad).

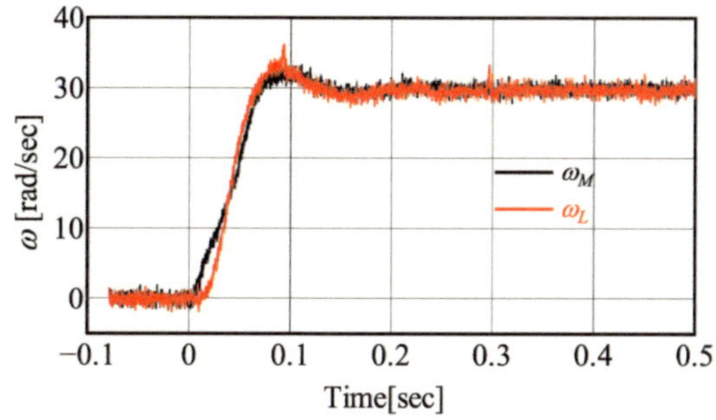

Figure 30. Experimental results for ω_M and ω_L redesigned using the proposed method (two-mass, $R = 1.07$, $K_{sn} = 3.1$ Nm/rad).

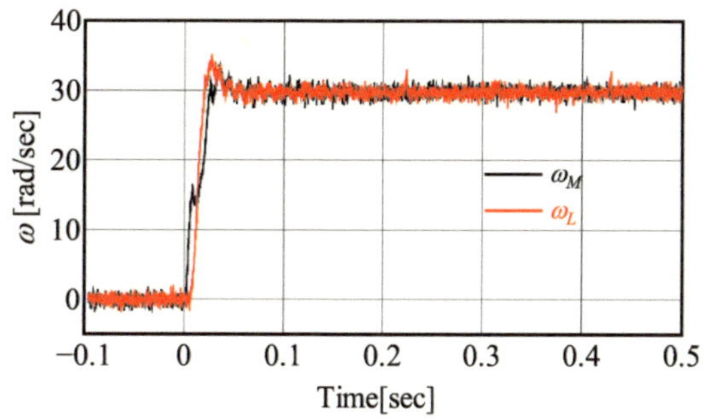

Figure 31. Robustness verification results (three-mass, $J_{Mn} = 2.774 \times 10^{-4}$ kgm^2, $J_{cn} = 1.112 \times 10^{-4}$ kgm^2, $J_{Ln} = 2.940 \times 10^{-4}$ kgm^2, $K_{s1n} = 18.5$ Nm/rad, $K_{s2n} = 18.5$ Nm/rad).

6. Conclusions

This chapter proposed the speed control system to suppress the resonance vibration of multi-inertial model, especially two-mass system and three-mass system. The controller has been constructed with the digital fuzzy controller for speed control and the digital PI controller for current control. In the control system, only motor side state variables have been used for controlling the resonance system. Additionally, this chapter utilized the DE to determine these five controller parameters. Finally, the validity of the controller design, the robustness, and the control effectiveness of the proposed method has been verified using the simulations and the experiments by using the test experimental set up.

Author details

Hidehiro Ikeda

Address all correspondence to: ikeda@nishitech.ac.jp

Nishinippon Institute of Technology, Kitakyushu, Japan

References

[1] Hori Y, Sawada H, Chun Y. Slow resonance ratio control for vibration suppression and disturbance rejection in torsional system. IEEE Transactions on Industrial Electronics. 1999;**46**(1):162–168

[2] Kwang-Ho Yoon, Jong-Kwang Lee, Ki-Ho Kim, Byung-Suk Park, Ji-Sup Yoon. Hybrid robust controller design for a two mass system with disturbance compensation. In: Proc. ICCAS 2008, pp. 1367–1372

[3] Szabat K, Kowalska TO. Vibration suppression in a two-mass drives system using PI speed controller and additional feedbacks – comparative study. IEEE Transactions on Industrial Electronics, 2007;**54**(2):1193–1206

[4] Ikeda H, Hanamoto T, Tsuji T, Tanaka Y. Position control of 2-mass systems with speed minor loop designed by pole placement method. IEEJ Transactions on Industry Applications. 1999;**119-D**(4):544–545

[5] Zhang G, Furusho J. Control of three-inertia system by PI/PID control. Transactions of IEE Japan. 1999;**119-D**(11):1386–1392

[6] Eker I, Vural M. Experimental online identification of a three-mass mechanical system. Proceedings of 2003 IEEE Conference. 2003;**1**:60–65

[7] Ikeda H, Hanamoto T, Tsuji T. Design of multi-inertia digital speed control system using Taguchi method. In: Proceedings of ICEM 2008, Paper ID 1167, PB.3.9; 2008. pp. 1–6

[8] Ikeda H, Hanamoto T, Tsuji T, Tomizuka M. Design of vibration suppression controller for 3-inertia systems using Taguchi method. In: Proceedings of SPEEDAM 2006, Mechatronic Systems, S10-19 to S10-25; 2006.

[9] Ikeda H, Hanamoto T. Design of vibration suppression controller for 2-inertia system by fictitious reference iterative tuning. In: Proceedings of ICEE 2015, ICEE15A-123; 2015. p. 6

[10] Ikeda H, Ajishi H, Hanamoto T. Application of fictitious reference iterative tuning to vibration suppression controller for 2-inertia resonance system. In: Proceedings of IECON 2015, TS-48, YF-008451; 2015. pp. 1825–1830

[11] Malhotra R, Kaur T, Deol GS. DC motor control using fuzzy logic controller. International Journal of AEST. 2011;**8**(2):291–296

[12] Chakravorty J, Sharma R. Fuzzy logic based method of speed control of DC motor. International Journal of ETAE. 2013;**3**(4):64–66

[13] Ikeda H, Hanamoto T. Fuzzy controller of multi-inertia resonance system designed by differential evolution. In: Proceedings of ICEMS 2013, MC-1883; 2013. pp. 2291–2295

[14] Ikeda H, Hanamoto T. Fuzzy controller of three-inertia resonance system designed by differential evolution. Journal of International Conference on Electrical Machines and Systems. 2014;**3**(2);184–189

[15] Yadav JS, Patidar NP, Singhai J. Controller design of discrete system by order reduction technique employing differential evolution optimization algorithm. International Journal of IMS. 2010;**6**(1):43–49

[16] Yamaguchi S. An automatic control parameters tuning method for differential evolution. Transactions of IEE Japan. 2008;**128-C**(11):1696–1703

[17] Brest J, et al. Self-adapting control parameters in differential evolution: A comparative study on numerical benchmark problems. IEEE Transactions on Evolutionary Computation. 2006;**10**(6):646–657

[18] Ikeda H, Hanamoto T. Design of m-IPD controller of multi-inertia system using differential evolution. In: Proceedings of IPEC-Hiroshima 2014 – ECCE ASIA, 21J1-2; 2014. pp. 2476–2482

[19] Myway Plus Corporation, https://www.mayway.co.jp/, Yokohama, Japan.

[20] The Mathworks, https://www.mathworks.com, Massachusetts, U.S.A.

Chapter 20

Design and Stability Analysis of Fuzzy-Based Adaptive Controller for Wastewater Treatment Plant

Mao Li

Additional information is available at the end of the chapter

http://dx.doi.org/10.5772/intechopen.68411

Abstract

In this chapter, design and stability analyses of direct model reference control system based on wastewater treatment plant are addressed. The purpose of controller design includes input saturation control and two-level control system with fuzzy supervisor control. The wastewater treatment plant is a highly uncertain non-linear system and the plant parameter are unknown, therefore controller design are under those condition.

Keywords: fuzzy control, fuzzy supervisor, wastewater treatment plant, adaptive control, model reference adaptive control, Lyapunov function

1. Introduction

The problem to be solved for this chapter is the dissolved oxygen reference trajectory tracking in an aerobic reactor for nutrient removal using direct model reference adaptive controller at the activated sludge wastewater treatment plant (WWTP). The reference trajectory is provided on-line by upper control layer of the overall control system. The controller design utilizes a different time scale in the internal dissolved oxygen dynamic and in disturbance inputs. In this chapter, we introduce two kinds of adaptive control, one is Direct Model Reference Adaptive Control (DMRAC) and another one is fuzzy logic based on DMRAC with two-level control.

2. Adaptive control

The basic concept of adaptive control is that it comprises of two main types. The first is called model reference adaptive control (MRAC) mode whereas the second is called self-tuning mode. An adaptive control characteristic is that the control parameter are variable, and those parameters are updated online with the signal in the system.

2.1. Model reference adaptive control

A model reference adaptive control can be divided into four parts, such as plant, reference model, control law and controller. A plant includes unknown parameters. The reference model is described as control system output. The closed-loop control law is adjusting mechanism for adjustable control parameters. The controller updates the adjustable parameter with time-varying control system.

The plant is supposed to exist with known system structure, but plant parameters are unknown in the real. The structure of the dynamic equation is known with some unknown parameters in the nonlinear plants. The number of poles/zero are supposed to be known with unknown location poles/zero.

The reference model is used in order to obtain assignation ideal response of adaptive control system to control output. For the adaptive control system, that mean it is supply the ideal response by adjusting ideal plant parameters in the adaptation mechanism. To design the adaptive control system, first step is choice of the reference model. It is needed to meet two following clauses.

- The reference model should satisfy the performance of adaptive control system such as rise time, overshoot and settling time.
- The ideal plant parameter should be implemented by the adaptive control system.

The controllers are composition of several adjustable parameters. This implies that the controllers are distribution signal to each adjustable parameter with online update. The controller needs to have good tracking performance which means it can achieve tracking convergence behaviour. To design controller, two conditions need to be considered.

- If the plant parameters are known, then the plant out should track model reference trajectory by relevant controller parameters.
- If the plant parameters are unknown, then the plant out should track model reference trajectory by adjusting the controller parameters.

The linearly parameterized that mean is the control law with linear term of the adjustable parameters. To guarantee stability and tracking performance, adaptive control design is used by linear parameterization of the controller.

Adjusting parameters in adaptive control law is call adaptation mechanism. In the MRAC systems, the adaptation law is used in order to search the plant parameters, therefore the plant out can track set-point (model reference) with good performance by adaptive controller. The difference between ideal adaptive control parameter and real plant parameter is call tracking error. The tracking error converge to zero that implies that adaptive control system is stable.

2.2. Self-tuning model reference adaptive control

The control parameters estimate plant unknown parameters in control system. If a plant parameter is unknown, then a parameter estimator provides estimation values to those plant unknown parameters. If a plant parameter is known, then control parameters would transmit plant parameters by on-line update on model reference. The estimator provides estimation control parameters with on-line update from model reference, it is called self-tuning controller. The self-tuning controller is estimation unknown parameter in the plant at the same time.

The self-turning MRAC manipulate processes:

- The estimator transfer estimated plant parameters to controller; therefore, it can compute the plant corresponding unknown parameters at the same time. The plant estimation parameters depend on the past plant input and output.
- Computes a control input and rely on control parameters and measured signal, and this control input rely on new plant output.
- The close-loop parameters and plant input are updated on-line with time-varying adaptive control system.

The estimation parameter can be taken from an ideal parameters and real parameters by plant input/plant output data that are updated on-line with time-varying adaptive control system. The error dynamic is described as the difference between ideal plant parameters and real plant parameters; this implies that if tracking errors converge to zero by adjusting parameters adaptation then plant output complete tracking reference model. It is purpose of self-turning adaptive control design.

The self-turning control includes two types of adaptive controllers, one is called Indirect Model Reference Adaptive Control (IMRAC) and the another one is called Direct Model Reference Adaptive Control. The plant unknown parameters are provided by adaptive controller estimation of those plant parameters. If the estimation plant parameters need transfer into controller parameters,furthermore control law parameters can influent plant unknown parameters. This implies that the control parameters can adjust plant unknown parameters with standard estimation approach. It is called IMRAC. On the other hand, if it does not need transfer process, this method is called DMRAC.

2.3. Direct model reference control design

The property of adaptive control is used for plant with unknown parameters; therefore, choosing the adaptive control law is more implicated in controller design. Since we mention before, adaptive control law produce controller parameters. Also the stability analysis for control system need to be considered in controller design. In this chapter, we used Lyapunov theory to analyse control system whether stable or unstable. The process of adaptive control design includes three steps. The first step is choosing control law (include plant variable parameters). The second step is choosing adaptation law. The final step is stability analyses to guarantee convergence of control system.

3. DMRAC with input saturation apply on WWTP

3.1. Introduction

An activated sludge wastewater treatment plant (WWTP) is a complex nonlinear system due to multiple time scale and unmeasurable state variables. In addition, it has time-varying input disturbances and saturation during the WWTP operation; hence, the hierarchical structures which were considered in Refs. [1, 2]. The two-level controller of tracking prescribed a concentration of the dissolved oxygen (*DO*) trajectory, while the reference of concentration dissolved oxygen (DO^{ref}) was developed in Refs. [3, 4]. The activated sludge plant contained two main components, such as bioreactor and settler as illustrated in **Figure 1**.

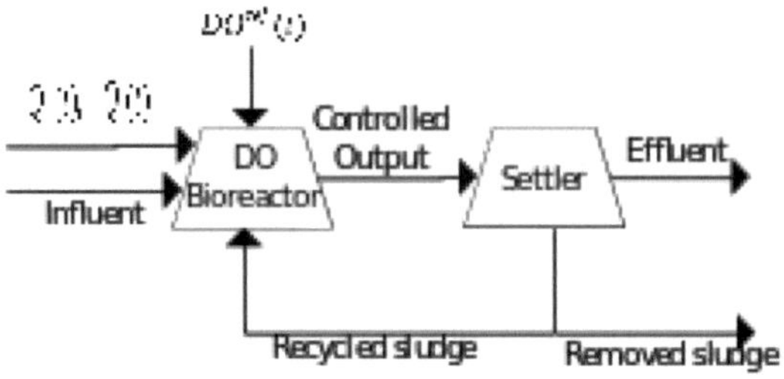

Figure 1. Structure of WWTP for nutrient removal.

The microorganism produced the biomass to nutrient removal in the bioreactor. The concentration of dissolved oxygen control is an important state parameter that feeds the microorganisms. The concentration of *DO* control was considered in Ref. [5]. The upper level controller produced airflow $Q_{air}^{ref}(t)$ into the aerobic biological reactor zone. The lower level controller produced the concentration of *DO* to track the $Q_{air}^{ref}(t)$ set-point trajectory. The airflow is the control input, and the concentration of *DO* is the control output.

The dissolved oxygen reference trajectory $DO^{ref}(t)$ set-point was optimized by the upper control layer which was the medium control layer in overall WWTP [1]. The clean water came out from the settler after being separated from the biomass and sludge. The concentration of substrate and biomass were unmeasurable state variables; hence, they were not able to be online updated. The upper layer control with input saturation was presented in Ref. [6]. The saturation function was assumed considering that aeration system controller was ideal and thus the airflow was equal to airflow reference. However, the physical modelling of the wastewater treatment plant was used to design the controller. In this chapter, we consider the upper level controller with input saturation by designing the new direct model reference adaptive control (DMRAC).

3.2. Problem statement

A mathematical model of the WWTP is based on the mass balance equations, which are illustrated in **Figure 1**. Hence, they represent the plant variables to produce the model in the state-space format [8]:

$$\frac{dX}{dt} = \mu(t)X(t) - D(t)(1+r)X + rD(t)X_r(t) \tag{1}$$

$$\frac{dS}{dt} = -\frac{\mu(t)X(t)}{Y} - D(t)(1+r)S(t) + D(t)S_{in}(t) \tag{2}$$

$$\frac{dDO}{dt} = -\frac{K_0\mu(t)X(t)}{Y} - D(t)(1+r)DO(t) + k_{La}(Q_{air}(t))(DO_{\max} - DO(t)) + D(t)DO_{in}(t) \quad (3)$$

$$\frac{dX_r}{dt} = D(t)(1+r)X(t) - D(t)(\beta + r)X_r(t) \quad (4)$$

$$\mu(t) = \mu_{\max}\frac{S(t)}{K_s + S(t)} \times \frac{DO(t)}{K_{DO} + DO(t)} \quad (5)$$

where

$$D(t) = \frac{Q_{in}}{V_a}; \quad r = \frac{Q_r}{Q_{in}}; \quad \beta = \frac{Q_w}{Q_{in}} \quad (6)$$

$X(t)$, $S(t)$, $DO_{\max}$, $X_r(t)$, $D(t)$, $S_{in}(t)$, $DO_{in}(t)$, Y, $\mu(t)$, $\mu_{\max}$, K_S, K_{DO}, $Q_{air}(t)$, K_0, r, β, $Q_{in}(t)$, $Q_r(t)$, $Q_w(t)$, V_a are biomass concentration, substrate concentration, maximum dissolved concentration, recycled biomass concentration, dilution rate, substrate concentration in the influent, dissolved oxygen concentration in the influent, biomass yield factor, biomass growth rate, maximum specific growth rate, affinity constant, saturation constant, aeration rate, model constant, recycled sludge rate, removed sludge rate, influent flow rate, effluent flow rate, recycled flow rate, waste flow rate and aerator volume, respectively.

The function $k_{La}(Q_{air}(t))$ is the oxygen transfer, which depends on the aeration actuating system and sludge conditions [4]. In this chapter, it is assumed that

$$k_{La}(t) = \alpha Q_{air}(t) + \delta \quad (7)$$

where α and δ are two known constant values relating to oxygen transfer.

As only the DO output is considered in this chapter, the model is sufficiently accurate. Otherwise, more detailed model, for example, the ASM3, than it should be utilized as it has been done in Ref. [7]. In **Figure 2**, the detailed structure of the activated sludge WWTP for nutrient removal with the airflow actuator is illustrated. The actuator dynamics are described by a complex hybrid model. The output of aeration control system airflow output $Q_{air}(t)$ needs to

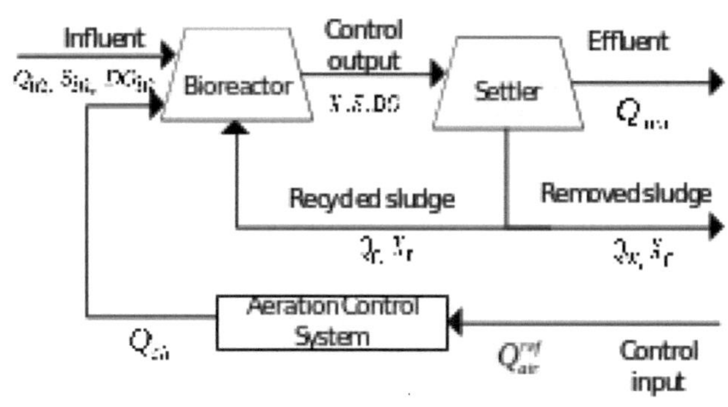

Figure 2. Structure of wastewater treatment plant.

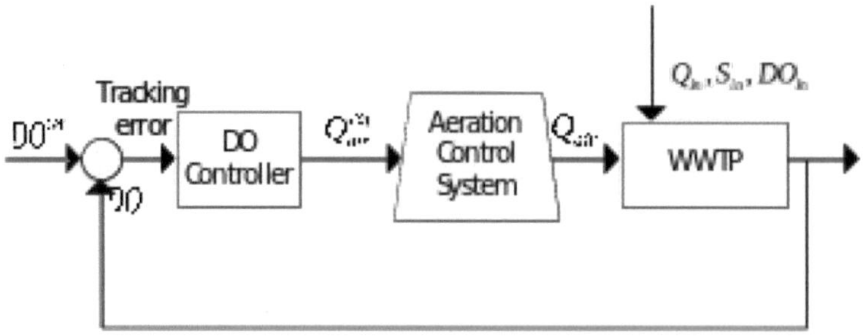

Figure 3. Structure of DO control system with input saturation.

follow the reference airflow input $Q_{air}^{ref}(t)$, which was described in Ref. [8]. The plant input with time-varying disturbances are dissolved oxygen concentration in the fluent $DO_{in}(t)$, the substrate concentration in the influent $S_{in}(t)$ and influent flow rate $Q_{in}(t)$. The controller needs to have high performance to enables the airflow output to track the reference airflow input.

The structure of *DO* control system for nutrient removal with input saturation at activated sludge WWTP is illustrated in **Figure 3**.

The aeration controller designed in this chapter considers the aeration control system as the input $Q_{air}(t)$ with input saturation to achieve $Q_{air}^{ref}(t) = Q_{air}(t)$. This is the main difference in comparison with Ref. [8]. If the plant dynamics have several serially coupled reactors, the decentralized controller needs to consider the input saturation [6]. Previous papers [1–3] considered the two-level controller to remove the nutrient in the activated sludge WWTP. The upper layer controller generated the expected $Q_{air}^{ref}(t)$ into each of the aerobic biological reactor zones. The input of the lower layer controller was $Q_{air}(t)$, which needs to track prescribed upper layer output $Q_{air}^{ref}(t)$ for each of the reactor zones. If the upper layer controller had an input saturation condition, it influenced global control system stability and performance. As the plant dynamics have very high order and nonlinear dynamics as in Eqs. (1)–(5). The fixed parameter linear controller could not continue to keep the expected performance under full range of operating conditions. This was verified in Ref. [4] by using fixed parameter PI controller in low layer control. The upper layer controller used a fuzzy supervised controller. It obtained the expected performance. In practice, if the disturbance of the input becomes large, fast varying and with saturation input, the PI controller becomes very complex. The DMRAC with input saturation in upper layer control is considered in this chapter, which is not based on previous papers [7]. The $DO(t)$ of the DMRAC input-output model rearranges the state-space model from Eqs. (1)–(4). As the state variable are not measureable, the unknown quantities in this input-output model will integrate into one term known as respiration. The parameter adaptation laws of the adaptive controller enable the respiration to be estimated indirectly and automatically.

3.3. DMRAC design

The direct state-space model of WWTP is represented in Eqs. (1)–(5). The dynamics are uncertain and nonlinear. The state variables $X_r(t)$, $X(t)$, $S(t)$ are unmeasurable. The state variables $Q_{\mathrm{air}}(t), Q_{\mathrm{in}}(t), Q_{\mathrm{w}}(t), DO(t)$ are measured by on-line updates. To design the direct model reference adaptive controller, we shall derive dissolved oxygen dynamics model in input-output format of first-order and with input disturbance and input saturation. We shall rewrite Eq. (3) by substituting Eqs. (5) and (7). The term $D(t)DO_{\mathrm{in}}(t)$ can be neglected in state-space model, since it is very small in comparison with other state variables. The dissolved oxygen input-output model (DOIOM) is derived as follows:

$$\frac{dDO}{dt} = -a_p(t)DO(t) - c_p(t)f(DO(t)) + b_p(t)Q_{\mathrm{air}}(t) + d_p \tag{8}$$

where $a_p(t), c_p(t), b_p(t), d_p$ are DOIOM parameters and

$$\begin{aligned} a_p(t) &= \frac{Q_{\mathrm{in}}(1+r)}{V_a} + \delta \\ c_p(t) &= \frac{K_0 X(t)}{Y} \frac{\mu_{\max} S(t)}{(K_S + S(t))} \\ f(DO(t)) &= \frac{DO(t)}{(K_{DO} + DO(t))} \\ b_p(t) &= \alpha(DO_{\max} - DO(t)) \\ d_p &= \delta DO_{\max} \end{aligned} \tag{9}$$

The parameters $a_p(t)$ and $c_p(t)$ are slowly varying and unknown. The parameters $a_p(t)$ is dependent on upper control layer which operates in the time scale and is slower in comparison with $DO(t)$ control time scale (6) and (9). The parameter $c_p(t)$ is dependent on X and S, and is slower in comparison with $DO(t)$ (1) and (15). The parameter $b_p(t)$ is dependent on the fast internal dynamics of $DO(t)$ time scale (9). Hence, $DO(t)$ is fast varying and known. The parameter d_p is slowly varying and known. The model reference dynamics (MRD) generate achieved $DO(t)$ dynamics.

The $DO(t)$ tracks the prescribed dissolved oxygen trajectory $DO^{\mathrm{ref}}(t)$ by the controller. The MRD equation is as follows:

$$\frac{dDO_{m,\mathrm{ref}}}{dt} = -a_m DO_{m.\mathrm{ref}}(t) + b_m DO^{\mathrm{ref}}(t) \tag{10}$$

where $DO_{m.ref}(t)$ is the reference dynamics output and the parameters a_m and b_m are constant.

The $DO(t)$ dynamics SISO input-output model with input saturation yields is as follows:

$$\frac{dDO}{dt} = -a_p(t)DO(t) - c_p(t)f(DO(t)) + b_p(t)W(t) + d_p \tag{11}$$

where $W(t)$ is assumed saturation control input with constraint (SCIC) $W(t) = \mathrm{saturation}(Q_{\mathrm{air}}(t))$,

$$W(t) = \begin{cases} Q^{\mathrm{L}}_{\mathrm{air}}(t), & \text{if} \quad Q_{\mathrm{air}}(t) > Q^{\mathrm{L}}_{\mathrm{air}}(t) \\ Q_{\mathrm{air}}(t), & \text{if} \quad Q^{\mathrm{L}}_{\mathrm{air}}(t) \le Q_{\mathrm{air}}(t) \le Q^{\mathrm{U}}_{\mathrm{air}}(t) \\ Q^{\mathrm{U}}_{\mathrm{air}}(t), & \text{if} \quad Q_{\mathrm{air}}(t) < Q^{\mathrm{U}}_{\mathrm{air}}(t) \end{cases} \tag{12}$$

where $Q_{\mathrm{air}}^{\mathrm{L}}(t)$ and $Q_{\mathrm{air}}^{\mathrm{U}}(t)$ are actuator lower and upper constant bounds. If under the input saturation condition, the filter tracking error $n(t)$ is increasing, then global stability will be unstable for the control system. The model reference adaptive control law without input saturation was proposed in Ref. [5]. This motivates us to develop a new control law in comparison with Ref. [8] by explicitly considering the influence of the actuator input saturation nonlinearity.

The filter tracking error is applied as follows:

$$n(t) = e(t) - \lambda(t) \tag{13}$$

where $e(t)$ represents the difference between $DO(t)$ and $DO_{\mathrm{m,\,ref}}(t)$ with on-line update.

$$e(t) = DO(t) - DO_{\mathrm{m,\,ref}}(t) \tag{14}$$

We define the auxiliary signal as follows:

$$\frac{d\lambda}{dt} = b_p \Delta Q_{\mathrm{air}}(t) - \Phi\lambda(t) \tag{15}$$

where Φ is small position constant parameter. The parameter $\Delta Q_{\mathrm{air}}(t)$ is the difference between SCIC $W(t)$ and control input $Q_{\mathrm{air}}(t)$.

The affine MRAC law is applied as follows:

$$\begin{aligned} W(t) = {} & \frac{1}{b_p} a_{\mathrm{DO}}(t)DO(t) + \frac{1}{b_p} a_f(t) f(DO(t)) \\ & + \frac{1}{b_p} a_{\mathrm{DO^{ref}}}(t) DO^{\mathrm{ref}}(t) - \frac{1}{b_p} d_p \\ & - \frac{1}{b_p} \Phi(t)(DO(t) - DO_{\mathrm{m,\,ref}}(t)) \end{aligned} \tag{16}$$

The MRD in Eq. (10) has linear dynamics. The terms $\frac{1}{b_p} a_f(t) f(DO(t))$ and $-\frac{1}{b_p} d_p$ in Eq. (16) can be cancelled by closed-loop with an impact of the nonlinear and additive terms in Eq. (8). The control input saturation is described by the last term in Eq. (16) which is retained in DOIOM. The fifth term in Eq. (16) is updated on-line. The parameters $a_{\mathrm{DO}}(t)$, $a_{\mathrm{f}}(t)$ and $a_{\mathrm{DO^{ref}}}(t)$ are updated by adaptive control law. The MRD is achieved in the closed-loop for ideal parameter. Closing the loop by Eq. (16) yields:

$$\begin{aligned} \frac{dDO}{dt} = {} & -(a_p(t) - a_{\mathrm{DO}}(t))DO(t) \\ & - (c_p(t) - a_f(t)) f(DO(t)) \\ & + a_{\mathrm{DO^{ref}}}(t) DO^{\mathrm{ref}}(t) \\ & - \Phi(DO(t) - DO_{\mathrm{m}}(t)) \end{aligned} \tag{17}$$

and

$$-(a_{\mathrm{p}}(t) - \hat{a}_{\mathrm{DO}}(t)) = -a_{\mathrm{m}} \tag{18}$$

$$-(c_{\mathrm{p}}(t) - \hat{a}_{\mathrm{f}}(t)) = 0 \tag{19}$$

$$\hat{a}_{\mathrm{DO^{ref}}}(t) = b_{\mathrm{m}} \tag{20}$$

$$-\Phi(DO(t) - DO_{\mathrm{m}}(t)) = -\Phi e(t) \tag{21}$$

where $\hat{a}_{\mathrm{DO}}(t)$, $\hat{a}_f(t)$ and $\hat{a}_{\mathrm{DO^{ref}}}(t)$ are the ideal parameters, which can now be obtained as follows:

$$\hat{a}_{\mathrm{DO}}(t) = -a_m + a_p(t) \tag{22}$$

$$\hat{a}_{\mathrm{f}}(t) = c_{\mathrm{p}}(t) \tag{23}$$

$$\hat{a}_{\mathrm{DO^{ref}}}(t) = b_{\mathrm{m}} \tag{24}$$

The parameter adaption laws which can achieve stability for a DMRAC system with SISO-controlled plant were derived in Ref. [6]. It was a first-order dynamic system composed of the mixed linear uncertainty in constant but not time-varying parameters and additive structured nonlinear. Applying these laws to Eq. (8) yields:

$$\frac{da_{\mathrm{DO}}}{dt} = -\gamma_1 e(t) DO(t) \tag{25}$$

$$\frac{da_f}{dt} = -\gamma_2 e(t) f(DO(t)) \tag{26}$$

$$\frac{da_{\mathrm{DO^{ref}}}}{dt} = -\gamma_3 e(t) DO^{\mathrm{ref}}(t) \tag{27}$$

γ_1, γ_2 and γ_3 are small enough positive constants representing the parameter adaptation gains which are used to control the parameter adaptation rates. In order to guarantee the stability of the closed-loop system, these rates shall be harmonized with the process variable rates. The DMRAC structure is presented in **Figure 4**.

3.4. Stability analysis

The estimated parameters $a_{\mathrm{DO}}(t)$, $a_{\mathrm{f}}(t)$ and $a_{\mathrm{DO^{ref}}}(t)$ are updated on-line by the adaptation laws (25)–(27). The error between estimated parameter and ideal parameters are denoted as $\Delta a_{\mathrm{DO}}(t)$, $\Delta a_{\mathrm{f}}(t)$ and $\Delta a_{\mathrm{DO^{ref}}}(t)$:

$$\Delta a_{\mathrm{DO}}(t) = a_{\mathrm{DO}}(t) - \hat{a}_{\mathrm{DO}}(t) \tag{28}$$

$$\Delta a_{\mathrm{f}}(t) = a_{\mathrm{f}}(t) - \hat{a}_{\mathrm{f}}(t) \tag{29}$$

$$\Delta a_{\mathrm{DO^{ref}}}(t) = a_{\mathrm{DO^{ref}}}(t) - \hat{a}_{\mathrm{DO^{ref}}}(t) \tag{30}$$

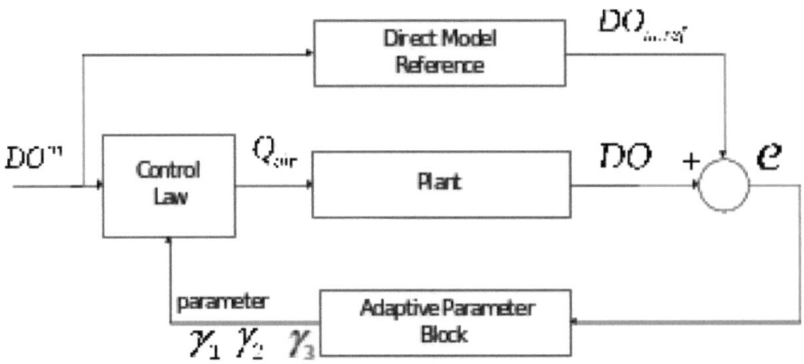

Figure 4. DMRAC structure.

Considering the following Lyapunov function:

$$V(t) = \frac{1}{2}n^2(t) + \frac{1}{2}\Delta a_{\mathrm{DO}}{}^2(t) + \frac{1}{2}\Delta a_{\mathrm{f}}{}^2(t) + \frac{1}{2}\Delta a_{\mathrm{DO^{ref}}}{}^2(t) \tag{31}$$

Hence,

$$\begin{aligned}\frac{dV(t)}{dt} &= n(t)n(t) + (a_{\mathrm{DO}}(t) - \hat{a}_{\mathrm{DO}}(t))(a_{\mathrm{DO}}(t)) - a_{\mathrm{DO}}(t)) = \frac{1}{\gamma 1} \\ &+ (a_{\mathrm{f}}(t) - a_{\mathrm{f}}(t))(\hat{a}_{\mathrm{f}}(t) - a_{\mathrm{f}}(t))\frac{1}{\gamma_2} \\ &+ (a_{\mathrm{DO^{ref}}}(t)) - \hat{a}_{\mathrm{DO^{ref}}}(t))(a_{\mathrm{DO^{ref}}}(t) - a_{\mathrm{DO^{ref}}}(t))\frac{1}{\gamma_3}\end{aligned} \tag{32}$$

It follows from Eqs. (13), (15), (10) and (17) that

$$\begin{aligned}n(t) &= (a_{\mathrm{DO}}(t) - \hat{a}_{\mathrm{DO}}(t)DO(t)) \\ &+ (a_{\mathrm{f}}(t) - a_{\mathrm{f}}(t))f(DO(t)) \\ &+ (a_{\mathrm{DO^{ref}}}(t)) - \hat{a}_{\mathrm{DO^{ref}}}(t))DO_{\mathrm{m,ref}}(t) \\ &- \Phi n(t)\end{aligned} \tag{33}$$

Applying Eqs. (33), (25) (26) and (27) into (32), yields:

$$\frac{dV(t)}{dt} = -\Phi n^2(t) \tag{34}$$

Summarizing the result of the Lyapunov function (RLF) with input saturation closed-loop $DO(t)$ dynamic system, it can be seen that RLF progressively approaches zero. If the RLF approaches zero, then filter tracking error approaches zero, when time approaches infinity, Φ is small position constant and $n^2(t)$ is positive variable. To find the bounded saturation control input by limiting error between control output and dissolved oxygen trajectory reference with auxiliary signal, yields:

$$\text{limit}\{e(t) - \lambda(t)\} \leq 0 \tag{35}$$

If time approaches infinity, the $e(t)$ approaches zero. To confirm whether the auxiliary signal is negative or positive when time goes to infinity by considering the following Lyapunov function, yields:

$$V_\lambda(t) = \frac{1}{2}\lambda(t) \tag{36}$$

Hence

$$\frac{dV_\lambda(t)}{dt} = \lambda(t)\lambda(t) \tag{37}$$

Applying Eq. (12) into Eq. (37), yields:

$$\frac{dV_\lambda(t)}{dt} = \lambda(t)b_{\text{p}}(t)\Delta Q_{\text{air}}(t) - \lambda(t)\lambda(t)\Phi \tag{38}$$

$$\frac{dV_\lambda(t)}{dt} = \lambda(t)$$

Assume term

$$b_{\text{p}}(t)\Delta Q_{\text{air}}(t) = \Delta Q_{\text{air}}^{\text{plant}}(t) \tag{39}$$

It follows Eqs. (38) and (39) so that

$$\frac{dV_\lambda(t)}{dt} = -\lambda^2(t)\Phi + \frac{1}{2}\lambda^2(t) + \frac{1}{2}\Delta Q_{\text{air}}^{\text{plant}^2}(t) \tag{40}$$

Now we assume

$$\Phi = \frac{1}{2}a_0, \quad a_0 < 0, \tag{41}$$

where a_0 is small positive constant value. Applying Eq. (41) into Eq. (40) yields:

$$\frac{dV_\lambda(t)}{dt} = -2V_\lambda(t)a_0 + \frac{1}{2}\Delta Q_{\text{air}}^{\text{plant}^2}(t) \tag{42}$$

The second term $\frac{1}{2}\Delta Q_{\text{air}}^{\text{plant}^2}(t)$ is bounded. To find bound of first term by integral, yields:

$$V_\lambda(t) = \frac{\Delta Q_{\text{air}}^{\text{plant}^2}(t)}{4a_0} + (V_{\lambda.0}(t) + \frac{\Delta Q_{\text{air}}^{\text{plant}^2}}{4a_0})e(t)^{-2a_0} \tag{43}$$

where the $V_{\lambda.0}(t)$ is the initial value of the $V_\lambda(t)$. As time approaches infinity and a_0 is large enough for a positive value, then second term is equal to zero in (37).

$$V_{\lambda}(t) = \frac{\Delta Q_{\text{air}}^{\text{plant}^2}(t)}{4a_0} \tag{44}$$

It follows from Eqs. (36) and (44) and the limitation $V_{\lambda}(t)$ as negative or zero by squared.

$$V_{\lambda}{}^2(t) = \frac{\Delta Q_{\text{air}}^{\text{plant}^2}(t)}{4a_0}$$

$$\frac{1}{2}\lambda^2(t) = \frac{\Delta Q_{\text{air}}^{\text{plant}^2}(t)}{4a_0} \tag{45}$$

$$\lambda(t) = \sqrt{\frac{\Delta Q_{\text{air}}^{\text{plant}^2}(t)}{2a_0}}$$

It follows from Eqs. (35) and (41) that

$$e(t) \leq \sqrt{\frac{\Delta Q_{\text{air}}^{\text{plant}^2}(t)}{2a_0}} \tag{46}$$

If the value of a_0 is large enough, then the tracking error $e(t)$ is closer to zero. The control system will be more stable. Finally, the standard application of the Barbalat's lemma allows concluding the adaptive control system that achieves the asymptotic tracking of $DO^{\text{ref}}(t)$ under-bounded $a_{DO}(t)$, $\hat{a}_{\hat{D}\hat{O}}(t)$, $a_f(t)$, $a_f(t)$, $a_{DO^{\text{ref}}}(t)$, $a_{DO^{\text{ref}}}(t)$ if (a) the parameter in adaptive control law are close enough to the set-point in initial condition; (b) the parameter adaptation rates are positive small enough and (c) the saturation input is small enough, the control parameters bounded are stabilized

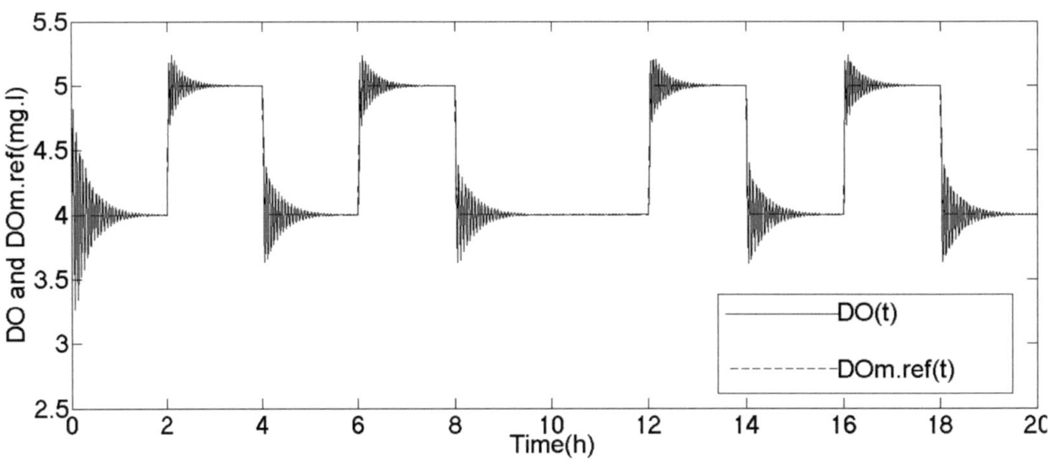

Figure 5. DO and DO^{ref} with input saturation.

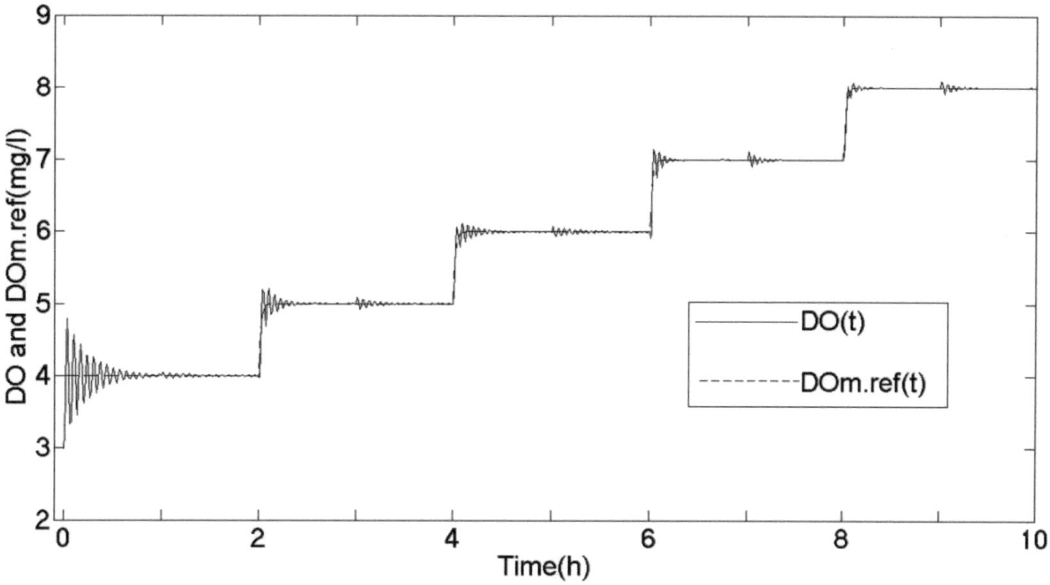

Figure 6. DO and DO^{ref} with input saturation and disturbances.

3.5. Simulation results

The simulation data are based on the real record. We assumed WWTP without disturbance, the very good $DO(t)$ tracking model reference $DO^{ref}(t)$ performance has been shown in **Figure 5**. The real plant contained some disturbances such as effluent flow rate, recycled flow rate and waste flow rate. The controller that we have designed still indicated perfect tracking performance in **Figure 6**.

4. Fuzzy supervisor based on multiple DMRAC

4.1. Introduction

In this section, we consider that fuzzy control are based on multiple DMRAC. The fuzzy control represents upper level control and DMRAC represents lower level control. More detail information are described in the next section.

4.2. Problem statement

In Ref. [4] is descried two-level controller tracking previously set-point of DO trajectory in several serially coupled reactors for the nutrient removal served by one actuator system with several air blower at WWTP. The upper level control delivers airflow into each bioreactors to be bioreactor set-point trajectory close to ideal trajectory. The lower level control is used for the concentration of DO trajectory flowing the set-point. The structure of WWTP with coupled reactors is illustrated in **Figure 7**. The structure of WWTP is different from Section 5 that

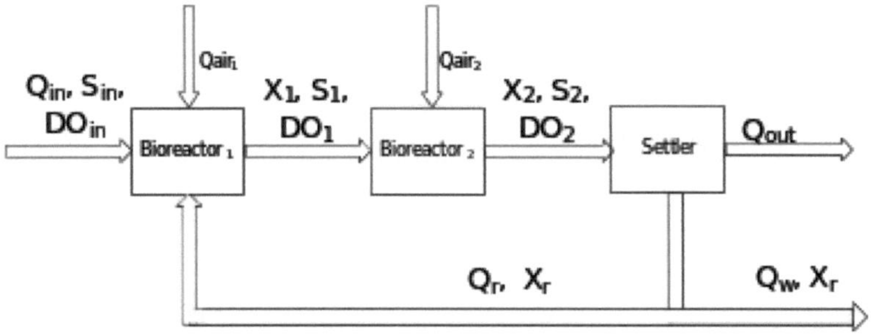

Figure 7. The structure of wastewater treatment plant for nutrient removal with coupled reactors.

contains two bioreactors. The capacity limit is that total airflow $Q_{air}^{max}(t)$ should be small or equal to sum of all the adaptive control signals $Q_{air.k}(t)$.

$$\sum_{k=1}^{q} Q_{air.k}(t) \leq Q_{air}^{max}(t) \tag{47}$$

where k is the number of bioreactor.

4.3. Controller design

4.3.1. Lower DMRAC design

As mentioned in Section 3, the process of DMRAC design is explained in detail; therefore, in this section we provide essential equations. The state-space format is same with last section for single reactor as in Eqs. (1)–(6).

- The dissolved oxygen input-output model (DOIOM) with coupled bioreactors is derived as follows:

$$\frac{dDO_i}{dt} = -a_{p.i}(t)DO_i(t) - c_{p.i}(t)f(DO_i(t)) + b_{p.i}(t)Q_{air.i}(t) + d_{p.i} \tag{48}$$

where $a_{p.i}(t), c_{p.i}(t), b_{p.i}(t), d_{p.i}$ are DOIOM parameters and

$$\begin{aligned} a_{p.i} &= \frac{Q_{in}(1+r_1)}{V_{a.i}} + \delta_i \\ c_{p.i} &= \frac{K_{0.i}X_i(t)}{Y_i}\frac{\mu_{max.i}S_i(t)}{K_{s.i}+S_i(t)} \\ b_{p.i} &= \alpha_i(DO_{max.i} - DO_i(t)) \\ d_{p.i} &= \delta_i DO_{max.i} \end{aligned} \tag{49}$$

where $i = 1, 2$

- The plant parameters status are exactly same with single reactor. The model reference dynamics equation is set as:

$$\frac{dDO_{m.\text{ref}.j}}{dt} = -a_{m.j}DO_{m.\text{ref}.j}(t) + b_{m.j}DO_j^{\text{ref}}(t) \tag{50}$$

where $j = 1, 2$

- The affine model reference adaptive control law is applied as follows:

$$Q_{\text{air}.k}(t) = a_{DO.k}(t)DO_k(t) + a_{f.k}(t)f(DO_k(t)) + a_{DO_k^{\text{ref}}}(t)DO_k^{\text{ref}}(t) - \frac{\delta_k DO_{\max.k}}{b_{p.k}(t)} \tag{51}$$

where $k = 1, 2$

- Model reference adaptive control law is used as:

$$\frac{da_{DO_n}}{dt} = -\gamma_{\text{zone}.z.1}e_l(t)DO_i(t) \tag{52}$$

$$\frac{da_{fn}}{dt} = -\gamma_{\text{zone}.z.2}e_l(t)f(DO_i(t)) \tag{53}$$

where $n = 1, 2; z = 1, 2; l = 1, 2; i = 1, 2.$

4.3.2. *Fuzzy supervisor design*

The purpose of fuzzy supervisor is to divide total airflow $Q_{air}^{\max}(t)$ into two lower control signal, but those should satisfy capacity limit (47). Each of the bioreactors airflow restrict lower control output by MRAC. This implies that if fuzzy supervisor delivers airflow big enough then bioreactor output is more close to set-point trajectories (model reference). The error dynamic described each bioreactors output approaching uniform level. The fuzzy supervisor is designed as following:

4.3.2.1. *Step 1: Fuzzification*

Linguistic variable is at lower level for each DMRAC error dynamics. Those error dynamics are divided into three types such as small, medium and big by percentage of lower level error dynamics (54). Membership function used in this chapter are Sigmoidal condition (55) and Gauss condition (56).

$$V(t) = e_i(t)^{-1}\text{g}\sum_{i=2}^{q} e_i(t) \tag{54}$$

where $V(t)$ is percentage of lower level error dynamics for each airflow.

$e_i(t)$ is DMRAC error dynamics for each bioreactors.

$$\mu(v(t)) = \frac{1}{1 + \exp\left(-a(v(t)) - c\right)} \tag{55}$$

where a, c are membership function shape parameters.

$$\mu(v(t)) = \begin{cases} \exp\left(-(v(t)) - c_1\right)^2/\sigma_1^2) \\ 1; \\ \exp\left(-(v(t)) - c_2\right)^2/\sigma_2^2) \end{cases} \tag{56}$$

where a, σ are membership function shape parameters.

4.3.2.2. Step 2: Fuzzy rule

4.3.2.2.1. First rule

If error dynamics is small and sum of level airflow is greater than total airflow,

Then bioreactor receives corresponding percentage of total airflow.

If $V(t)$ is small and $\sum_{i=1}^{q} Q_{air.i}(t) \geq Q_{air}^{max}(t)$

Then

$$Q_{air.i.2}^{supervisor}(t) = \frac{Q_{air}^{max}(t)}{\sum_{i=1}^{q} Q_{air.i}(t)} \times Q_{air.i}\, g10\% \tag{57}$$

4.3.2.2.2. Second rule

If error dynamic medium and sum of level airflow is greater than total airflow,

Then bioreactor receives corresponding percentage of total airflow.

If $V(t)$ is medium and $\sum_{i=1}^{q} Q_{air.i}(t) \geq Q_{air}^{max}(t)$

Then

$$Q_{air.1.2}^{supervisor}(t) = \frac{Q_{air}^{max}(t)}{\sum_{i=1}^{q} Q_{air.i}(t)} \times Q_{air.i}\, g30\% \tag{58}$$

4.3.2.2.3. Third rule

If error dynamic is big and sum of level airflow is greater than total airflow,

Then bioreactor receive corresponding percentage of total airflow.

If $V(t)$ is big and $\sum_{i=1}^{q} Q_{air.i}(t) \geq Q_{air}^{max}(t)$

Then

$$Q_{\text{air}}^{\text{supervisor}}(t) = \frac{Q_{\text{air}}^{\max}(t)}{\sum_{i=1}^{q} Q_{\text{air}.i}(t)} \times Q_{\text{air}.i}\, g60\% \tag{59}$$

4.3.2.3. Step 3: Defuzzification

Each of the bioreactors obtain airflow by a fuzzy value.

5. Summary

In this chapter, we considered two different adaptive control. The first adaptive control is applied on WWTP with control input saturation. The second adaptive control descried that how upper level fuzzy control working is based on lower level DMRC applied on the coupling bioreactors of WWTP.

Author details

Mao Li

Address all correspondence to: limaomxl554@gmail.com

The University of Birmingham, Birmingham, UK

References

[1] Brdys MA, Grochowski M, Gminski T, Konarczak K, Drewa M. Hierarchical predictive control of integrated wastewater treatment system. Control Engineering Practice. 2008;**16** (6):751–767.

[2] Piotrowski R, Brdys MA, Konarczak K, Duzinkiewicz K, Chotkowski W. Hierarchical dissolved oxygen control for activated sludge processes. Control Engineering Practice. 2008;**16**:114–131.

[3] Brdys MA, Diaz-Maiquez. Application of fuzzy model predictive control to the dissolved oxygen concentration tracking in an activated sludge process. In: Proceedings of the 15th IFAC World Congress; 21–26 July; Barcelona.

[4] Brdys MA, Chotkowski W, Duzinkiewicz K, Konarczak K, Piotrowski R. Two-level dissolved oxygen control for activated sludge processes. In: Proceedings of the 15th IFAC World Congress; 21–26 July 2002; Barcelona.

[5] Olsson G, Newell R. Wastewater Treatment System. Modelling, Diagnosis and Control. London, UK: IWA Publishing; 1999.

[6] Zubowicz T, Brdys MA, Piotrowski R. Intelligent PI controller and its application to dissolved oxygen tracking problem. Journal of Automation Mobile Robotics & Intelligent Systems. 2010;**4**(3):16–24.

[7] Li M, Brdys MA. Direct model reference adaptive control of nutrient removal at activated sludge wastewater treatment plant. In: The 20th International Conference on Methods and Models in Automation and Robotics; 24–27 August 2015; Miedzyzdroje, Poland.

[8] Duzinkiewicz K, Brdys MA, Kurek W, Piotrowski R. Genetic hybrid predictive controller for optimized dissolved-oxygen tracking at lower control level. IEEE Transaction on Control System Technology. 2009;**17**:1183–1192.

Chapter 21

A Model for Evaluating Soil Vulnerability to Erosion Using Remote Sensing Data and A Fuzzy Logic System

Ignacio Meléndez-Pastor, Jose Navarro Pedreño,
Ignacio Gómez Lucas and Antonis A. Zorpas

Additional information is available at the end of the chapter

http://dx.doi.org/10.5772/67989

Abstract

Soil vulnerability is the capacity of one or more of the ecological functions of the soil system to be harmed. It is a complex concept which requires the identification of multiple environmental factors and land management at different temporal and space scales. The employment of geospatial information with good update capabilities could be a satisfactory tool to assess potential soil vulnerability changes in large areas. This chapter presents the application of two land degradation case studies which is simple, synoptic, and suitable for continuous monitoring model based on the fuzzy logic. The model combines topography and vegetation status information to assess soil vulnerability to land degradation. Topographic parameters were obtained from digital elevation models (DEM), and vegetation status information was derived from the computation of the normalized difference vegetation index (NDVI) satellite images. This spectral index provides relevance and is updated for each scene, evidences about the biomass and soil productivity, and vegetation density cover or vegetation stress (e.g., forest fires, droughts). Modeled output maps are suitable for temporal change analysis, which allows the identification of the effect of land management practices, soil and vegetation regeneration, or climate effects.

Keywords: soil vulnerability, fuzzy logic, remote sensing, soil erosion, soil degradation

1. Introduction

Soil is considered a nonrenewable resource, and it is essential for food security and for our sustainable future and is defined as the top layer of the earth's crust formed by mineral particles, organic matter, water, air, and living organism [1]. Desertification associated to soil erosion processes is the main topic to protect and maintain soils. Moreover, desertification and land degradation are terms that are used in order to indicate ecosystem productivity and are associated with loses of vegetation covered [2]. The United Nations program against desertification is crucial as it focuses on fighting hunger and poverty, foresting stability, and building resilience to climate change in some of the world's most vulnerable areas [3]. Soil erosion, water-holding capacity, salinity, sodicity, losses of nutrients, etc. are common indicators used to categorize land degradation [1]. The genesis of soil is a long process, and the formation of a layer of 30 cm depth takes from 1000 to 10,000 years [4]. This scenery created between soil degradation and the long process of formation gives us few alternatives about soil conservation, which is essential for humans.

Soil vulnerability is the capacity of one or more of the ecological functions of the soil system to be harmed, i.e., biomass production, filtering, buffering and transformation medium, gene reserve, and protection of flora and fauna [5]. In other words, it is the sensitivity of soil against degrading processes.

The most important degrading process worldwide related to desertification, among others, is soil erosion as a result of climate change. Climate change is one of the major drivers of an ecosystem shift in a decertified state [6]. The vegetation plays a key role protecting soil. Moreover, soil erosion is associated to biomass production which has an important role in the future energy situation [7]. There are several factors affecting soil erosion. However, vegetation and topography play a crucial role in the vulnerability of soil. The first one is the most important source of organic matter to soil, protecting it against rain and wind (water and wind erosion) [8]. The second one determines and facilitates the losses of soil due to erosive processes and transport.

Soils with low vegetation cover and high slopes are more vulnerable against degrading processes. For this reason, it is very important to know these factors to establish the vulnerability of a soil in order to protect it, prevent erosion, and keep resources for a sustainable future use.

Many countries like Cyprus, Greek Islands, Spain, and Italy are affected by soil erosion, and there are many problems to access or obtain soil data because of the orography, accessibility, or other environmental factors. It is estimated that more than 115 million hectares which represents 12% of European total land are subject to erosion [9]. Moreover, it is estimated that, in the Mediterranean area (consists of a vulnerable area due to climate change effects), water erosion could affect the loss of 20/40 ton/ha of soil after a single cloudburst, and in extreme cases, the erosion could be even more than 100 ton/ha [10, 11]. For this reason, it is important to have techniques that facilitate the creation and population of spatial information. These techniques are englobed in digital soil mapping (DSM) which are systems composed by numerical models inferring the spatial and temporal variations of soil types and soil properties from soil observation and knowledge and from related environmental variables [12]. It is important to have an overview of the state and properties of soils for decision-makers and to facilitate the population and access to soil data. We cannot forget that land and water resources are central to agriculture and rural development and are intrinsically linked to global challenges of food insecurity and poverty, climate change adaptation and mitigation, as well as degradation and depletion of natural resources that affect the livelihoods of millions of rural people across the world. [13].

The use of easy models that can be applied by administration for decision-making plays a key role for resource conservation and people. In this sense, the models based on remote sensing data and mathematical models as those based on fuzzy logic systems can be helpful.

The models based on fuzzy logic systems are related to fuzzy set theory which plays a primary role in fuzzy logic. Zadeh developed the theoretical basis of this theory and defined the concept of fuzzy sets as "A fuzzy set A in X is characterized by a membership function fA(x) which associates with each point in X a real number in the interval [0, 1], with the value of fA(x) at x representing the 'grade of membership' of x in A. The nearer the value of fA(x) to unity, the higher the grade of membership of x in A" [14]. The more interesting property is that fuzzy sets work well with data uncertainties. A strict categorization (with discrete classes) of environmental/soil factors falls within subjectivity and carelessness of uncertainties (e.g., error measurements, selection of decimals, missed data, etc.). Fuzzy sets are used for classifications where the classes do not have sharply defined boundaries [15]. This approach is easily applied for environmental phenomena study, such as soil nutrient losses, atmospheric pollutant dispersion, forest productivity, etc., due to the great spatial variability and possible discontinuity of those phenomena.

Fuzzy theory provides a rich mathematical basis for understanding decision problems and for constructing decision rules in multi-criteria evaluation and combination [16]. A wide variety of fuzzy logic approaches have been developed in order to expand the concept of fuzzy sets. Fuzzy measure refers to any set of function which is monotonic with respect to set membership [17]. The variety of functions that can be applied is great. In fact, a fuzzy function is a generalization of the concept of a classical function. A classical function f is a mapping (correspondence) from the domain D of definition of the function into a space S; $f(D) \subseteq S$ is called the range of f. Different features of the classical concept of a function can be considered fuzzy [18]. Fuzzy measures include Bayesian probabilities, beliefs and plausibilities of Dempster-Shafer theory, and membership grades of fuzzy sets, providing a framework for the methodologies of uncertainty studies [19].

In this process, fuzzy sets and soil erosion are combined in order to facilitate decision-making based on easy criteria that can be applied by administration. Desertification, climate change, and soil productivity have to be considered. The need of global understanding of local processes and the possibilities of the computers and remote sensing techniques are the bases of this methodology. For these environmental, social, and economic reasons, it is necessary to have a tool for taking decisions. On the one hand, decision theory is concerned with the logic by which one arrives at a choice between alternatives [20]. On the other hand, the procedure by which one selects among different alternatives is enabled by a decision support system (DSS). The study of decision support systems is an applied discipline that uses knowledge and especially theory from other disciplines [21]. In this case, the knowledge of soil science and mathematical modeling helps us build an effective DSS.

A type of decision support system widely used is the multi-criteria evaluation (MCE) based on multi-criteria analysis (MCA) of analytical hierarchy process (AHP) [22, 23]. The AHP analysis is based on three basic principles as indicated by Zorpas and Saranti [22], and those include the preferences configuration, the interruption of the problem into subproblems, and finally the pair-wise comparison of criteria/subcriteria with the proposed alternative scenarios. The MCA [22, 23] focuses on four main stages starting with recognition of the problem (if possible to split in subproblems) and the formation of a hierarchical structure, followed by the pair-wise comparison of decision elements used to derive normalized absolute scales of numbers whose elements are then used as priorities, and continued with the control of the priorities set (to solve the problem) and finally the assessment and the evaluation of the alternative scenarios to solve the specific problem. This methodology can be used from decision-makers, engineers, consultants, researchers, and governmental and local authorities in order to evaluate or propose specific solution in a specific problem [21].

The MCE explores how to combine the information from several criteria to form a single index of evaluation using discrete or continuous factors [21–23]. For this purpose in environmental sciences, thematic information aggregation procedures are used in the process of criteria combination. Two traditions of aggregation procedures have been extensively used for MEC [24]: (1) Boolean overlay where all criteria are assessed by thresholds of suitability to produce Boolean maps using logical operators such union, intersection, or negation and (2) weighted linear combination (WLC) where continuous criteria are standardized (generally by a simple linear transformation) to a common numeric range and then combined by weighted averaging. Both aggregation criteria are rather inflexible as consequence of their inherent logic aggregation (type of operators and properties).

Aggregation operations on fuzzy sets are operations by which several fuzzy sets are combined in a desirable way (assuming some rules and operators) to produce a single fuzzy set [25]. Fuzzy measures provide a theoretical base to explore an expand understanding of MCE processes and the design of new aggregation operators [24]. Two traditions of logic operators have been extensively used since decades [26]: (1) MIN and MAX operators for Boolean overlay and fuzzy membership aggregation and (2) averaging operator for weighted linear combination. Jiang and Eastman [24] suggested the use of weighted linear combinations as a fuzzy set membership operator together with the MIN and MAX operators, in the framework of fuzzy measures. This is a very flexible approach for fuzzy set aggregations. In this chapter, the proposed soil vulnerability model related to soil erosion is based on this approach for fuzzy set aggregation and easy functions for the analysis.

2. A fuzzy logic model to evaluate soil vulnerability

Soil erosion may be considered the most important degradation processes in this approach, and the vulnerability of soils is based on the most important criteria affecting this process. Two types of information used by this model are (1) topographic parameters and (2) vegetation. Remote sensing techniques allow us to have data from any part of the Earth. For instance, digital elevation models (DEMs) which can describe the topography of any region can be derived from data obtained by the Shuttle Radar Topography Mission of NASA [18] and many other platforms. Remote sensing techniques can also provide vegetation status data for the analysis. For instance, data obtained from Landsat missions [27] or other programs facilitate the calculus or vegetation index like normalized difference vegetation index (NDVI).

http://dx.doi.org/10.5772/67989

The analytical capabilities derived from the use of DEM are enormous, ranging from basic topographic feature estimation, to flood simulations, and others. Moreover, remote sensing has a great potential to obtain imagery from optical, thermal, and microwave spectral regions across wide regions of the planet with a great temporal frequency, enough accuracy, and open source. Image processing methods of remotely-sensed data could extract valuable and specialized information of selected targets (e.g., soils, vegetation, waters, etc.). An added value of image processing techniques is the capability to analyze temporal series of data which can be useful to study the vulnerability and changes occurred in soils along time.

The proposed model is based on a set of three initial inputs, selected considering the previous indications (**Figure 1**):

- Slope: This topographic parameter could be derived from a DEM, and it is defined as the variation of altitude between two points in relation with the distance that separate them. Slope parameter reports information about the roughness of the territory and plays a key role in soil erosion because it could increase or reduce the effect of water [28]. The universal soil loss equation (USLE), and derived equations from this, considers slope parameter as a key factor for superficial soil water erosion [29].
- Aspect: This topographic parameter could be derived from a DEM. It reports information about the geographic orientation of the slopes. As a general rule (in the north hemisphere), a slope closer to a north aspect presents lower temperature and higher moisture levels

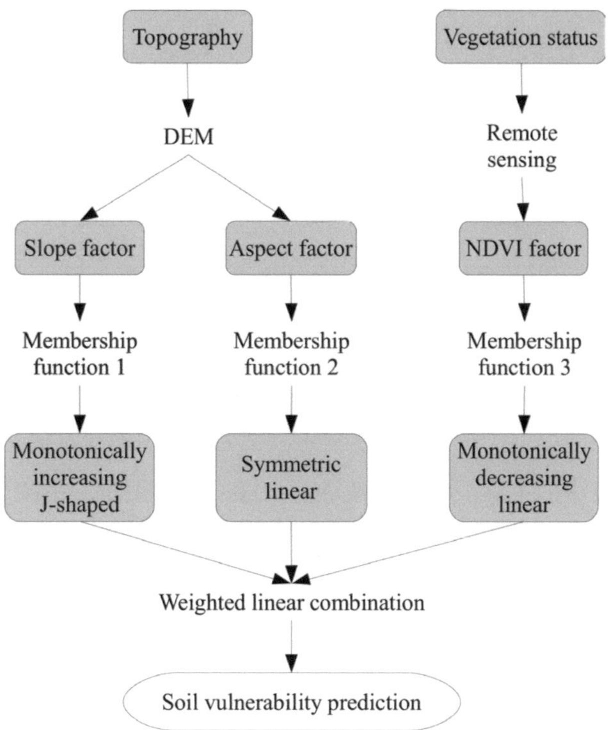

Figure 1. Flowchart of the fuzzy logic model to soil vulnerability evaluation.

(lower incident radiation levels). This parameter could suggest information about soil moisture, water availability by plants, etc. This general rule could be dismissed in special local situations (e.g., for places where wet maritime winds impact against slopes closed to a south aspect). This factor is of great importance in plant distribution and growth and soil development and properties [30].

- Normalized difference vegetation index (NDVI) is computed as a normalized ratio between near infrared (NIR) and red spectral bands of remotely-sensed imagery. NDVI values range between −1 and +1. This vegetation index is strongly related with several vegetation parameters such as changes in biomass and chlorophyll content [31, 32]. NDVI is related with other vegetation parameters too [33] such as leaf water content, CO_2 net flux, absorbed photosynthetically active radiation (APAR), and leaf area index (LAI), among others. A high NDVI value (usually over 0.3–0.5) indicates a vigorous and dense vegetal coverage.

Slope, aspect, and NDVI are the parameters computed in our model and could be extracted for wide regions from remote sensing data. Those parameters are computed following the flowchart as **Figure 1** presents, with an easy function that can be implemented. The simplicity of the model may be the limiting factor because only three inputs are taken into account.

However, this tool is good enough to have an overview of the soil erosion and could be a good tool for regional soil vulnerability assessment and long-term monitoring with a low economic cost. A near real-time monitoring of soil vulnerability, depending on the availability of data from satellite, could be performed and used by land managers and scientist. The main advantage in this sense is that field campaign to check soil, vegetation status, slope, and other filed parameters is initially not necessary. Moreover, the centered field work in determined areas reduces the cost of large field campaigns of wide territories.

The proposed model considers input parameters as fuzzy sets (factors) to be standardized by the definition of individual membership functions. A great variety of fuzzy set membership functions have been developed.

Based on Eastman [16], fuzzy set membership functions can be defined by three parameters:

- Shape of the function that provides three possible options: (1) monotonically increasing which begins at 0 and then rise and stay at 1, (2) monotonically decreasing which is opposite to the previous one, and (3) symmetric which firstly rise and then fall again
- Control points that govern the shape of the curve
- Type of function with three possible options: (1) sigmoidal produced by a cosine function, (2) J-shaped produced by a sine function, and (3) linear produced by a linear function

The definition of these parameters provides a great number of possibilities to define the effect of individual factors in the observed phenomena. The shape and type of membership functions should be defined according to the knowledge of environmental phenomena. Weighted linear combination is used for the standardized aggregation factors. This method allows the possibility to assign equal or differential weights to factors in function of experts' knowledge about their relative importance for the studied phenomena. Finally, it is important to mention that this simple model approach could be completed with other factors to be considered in the aggregation process or in other stages of the model. Equilibrium must be found between model simplicity and valuable obtained data to soil conservation.

3. Soil vulnerability under high slope changes: a case study

A test of soil vulnerability evaluation of this model after a fire event was done, considering that the area to test the model has many changes in slope and is close to the sea, affected from marine breeze and moisture. The selected area is located on the south-east coast of Spain (Alicante province) in the area of "La Granadella" (38.73°N, 0.19°E). This test site supported a fire event from 26 to 30 on August 2000. The topography is characterized by a highly rough relief (cliffs above the sea of more than 150 m) and a wet Mediterranean climate (700–1000 mm/year of rainfall). This is an interesting test area which combines a wet climate and a complex and precipitous relief and exhibits a great recurrence of fires.

The hypothesis to verify with the fuzzy logic model is that vegetation regeneration has a primary role in soil vulnerability reduction. The potential advantage for the use of fuzzy sets with respect to other methods of factor combination must be owed to their great flexibility derived from the defined membership functions.

Vegetation status information was obtained from two satellite scenes acquired by the multispectral advanced spaceborne thermal emission and reflection radiometer (ASTER): (http://asterweb.jpl.nasa.gov/) sensor. The first scene was acquired 4th October in 2000, some days after the forest fire, while the second scene was acquired 7th June in 2003, being a reasonable period for a substantial evolution in the landscape.

ASTER sensor is composed of three subsystems in function of their spectral and spatial resolution characteristics: visible near-infrared radiometer (VNIR) with 15 m of spatial resolution and stereoscopic capability, short-wave infrared radiometer (SWIR) with 30 m of spatial resolution, and thermal infrared radiometer (TIR) with 90 m of spatial resolution [34]. Only VNIR system bands were employed for our analyses. Both scenes were in origin preprocessed to an ASTER high-level product denominated AST_07 which contain data of surface reflectance [35].

Topographic information was obtained from a high-resolution digital elevation model (**Figure 2**) that was computed using the previous vector cartography (scale 1:10,000). Triangulated irregular network (TIN) polygons were computed with the nodes of the vector cartography and used to develop the raster DEM. The spatial resolution of the DEM was adjusted to ASTER-VNIR imagery (15 m). Satellite images were geometrically corrected using the high-resolution DEM and additional cartography in order to minimize the positional errors among the different sources of information. The root-mean-square error (RMSE) of the geometric correction was less than half a pixel for both ASTER scenes.

Topographic parameters (slope and aspect) were computed using the DEM. Both parameters were quantized as degrees. Normalized difference vegetation index is derived from a spectral band transformation between near-infrared and red bands. The original NDVI formulation has been attributed to Rouse et al. [36] and their research with early Landsat images. The original NDVI formulation has been adapted to subsequent sensors whose spectral characteristics are

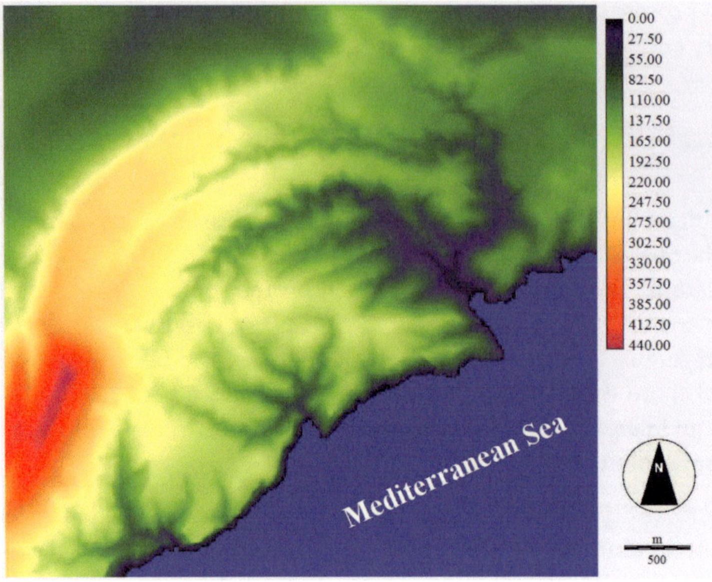

Figure 2. Digital elevation model (DEM) of "Sierra de Escalona" test site.

different among them. The following equation is the adaptation of NDVI for ASTER spectral bands (Eq. (1)):

$$\mathrm{NDVI} = \frac{\rho_{\mathrm{VNIR3}} - \rho_{\mathrm{VNIR2}}}{\rho_{\mathrm{VNIR3}} + \rho_{\mathrm{VNIR2}}} \tag{1}$$

where ρ_{VNIR2} is the spectral reflectance for the second band of the ASTER-VNIR subsystem and ρ_{VNIR3} is the spectral reflectance for the third band of the same subsystem. The computation of NDVI is more suitable with surface reflectance data (like AST_07 high-level ASTER product) in order to minimize differential wavelength-dependent atmospheric disturbances. NDVI was computed in the same way for the 2000 and 2003 scenes (**Figure 3 (a)** and **(b)**).

Soil vulnerability estimations were computed for both 2000 and 2003 scenarios. Individual fuzzy set membership functions are characterized by their shape and type and were defined for each of the considered soil vulnerability factors (i.e., slope, aspect, and NDVI). In this sense, **Table 1** provides a synthesis of model parameters used with the fuzzy logic model. The proposed model employed the following membership function for input variables: (1) NDVI, a monotonically decreasing linear function; (2) slope, a monotonically increasing J-shaped function; and (3) aspect, a symmetric linear function. Membership functions required the definition of several control points (CP) for a potential model generalization.

Fuzzy membership functions were combined weighting the relative importance of each one using a linear system. Jiang and Eastman [24] suggested the use of weighted linear combinations as a fuzzy set membership operator together with the MIN and MAX operators, as a very flexible approach for fuzzy set aggregations.

This weighting process considers the relative importance of each variable in the modeling process. Vegetation and slope were considered the most important factors for the soil vulnerability model.

http://dx.doi.org/10.5772/67989

Further details of the definition of membership functions and model calibration are available in Melendez-Pastor et al. [37]. Finally, a temporal change analysis of soil vulnerability was computed using the percentage of change procedure employing the following formulation (Eq. (2)):

$$\text{Temporal change (\%)} = \frac{(t_2 - t_1)}{t_1} \cdot 100 \tag{2}$$

where t_1 and t_2 are the soil vulnerability estimations for 2000 and 2003, respectively.

The application of the fuzzy logic model for this study area covers two environmental facts, an (almost) invariant one associated to the slope and orientation (which are mainly determined by the geomorphology of the landscape) and a highly changing parameter as the vegetation through NDVI. Landscape topography greatly affects soil profile formation [30, 38], while vegetation status and dynamics are highly related with the quality of the soils [39]. The DEM analysis revealed that the study area has a mean slope value of 16 degree, with a maximum slope value of 78 degrees. Relief is mainly configured on a NE-SW direction with preferential slope orientation to the SE direction. NDVI mean values varied from 0.31 (standard deviation of 0.12) on 2000 to 0.38 (standard deviation of 0.09) on 2003.

Soil vulnerability simulations (**Figure 3 (c)** and **(d)**) revealed the severe impact of the forest fire on vegetation. The model marks the importance of vegetation dynamics, the relation with the presence of soil although both are limited to the position as the model indicates (slope and

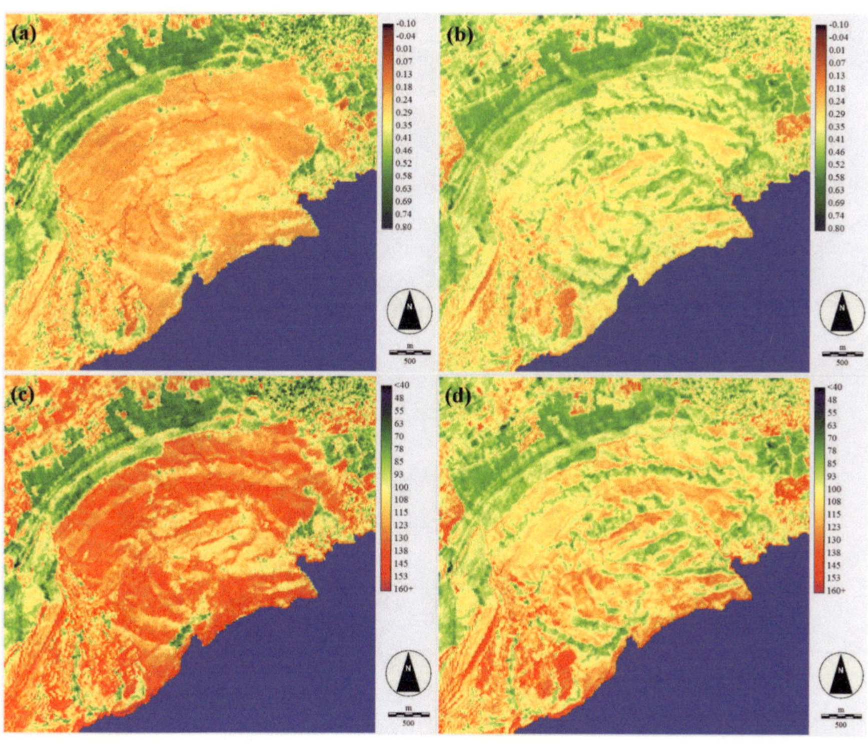

Figure 3. NDVI estimations for 2000 (a) and 2003 (b). Soil vulnerability estimations for 2000 (c) and 2003 (d). Soil vulnerability results area in 8 bits of quantization.

Factors		Slope	Aspect	NDVI
Membership functions	Shape	Monotonically increasing	Symmetric	Monotonically decreasing
	CP 1	0	0	0
	CP 2	90	180	1
	CP 3	No	180	No
	CP 4	No	360	No
	Type	J-shaped	Linear	Linear
Weights		0.35	0.05	0.6

Table 1. Model parameters for the soil vulnerability factors slope, aspect, and NDVI.

aspect). A characteristic pattern of soil vulnerability mitigation by intense vegetation regeneration could be advertised along the valleys where NDVI has increased faster since 2000. The temporal change estimation of soil vulnerability simulations (**Figure 4**) remarks the local high postfire ecosystem regeneration. Change rates are up to a 30% of less soil vulnerability within burned area. Soil vulnerability change map also highlighted areas where soil vulnerability has increased. Those changes correspond to land use conversions to urban (change rates up to a 30% of more soil vulnerability) and vegetation status variations.

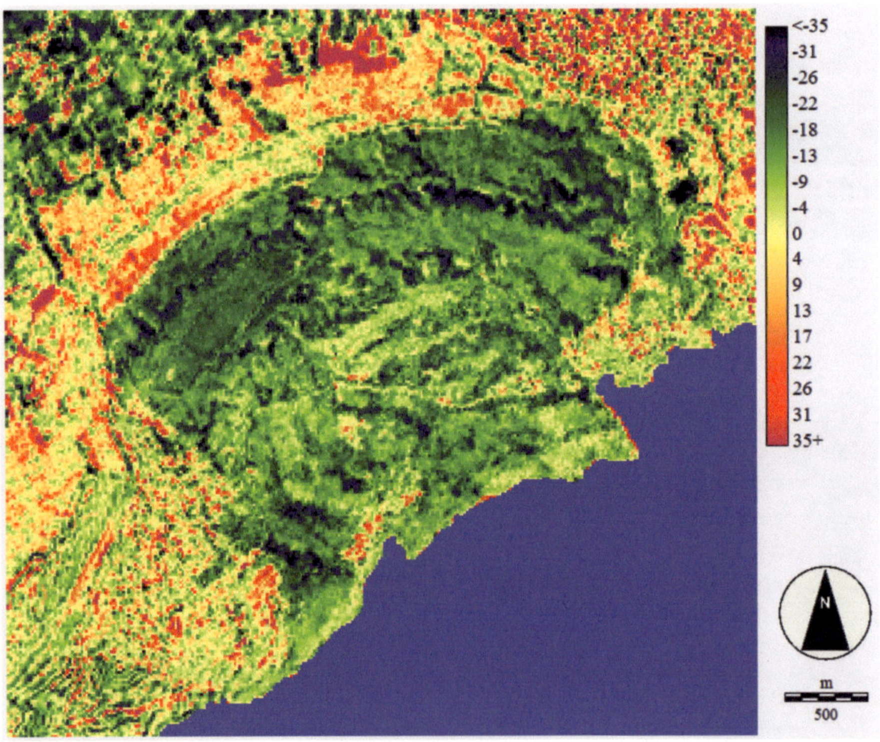

Figure 4. Temporal change (%) of soil vulnerability estimation. Green pixels correspond to less vulnerable areas on 2003.

http://dx.doi.org/10.5772/67989

4. Soil vulnerability and vegetation as a limiting factor

The proposed model was also engaged in a research of soil vulnerability evaluation in a semiarid area in the south of Alicante province (Spain). This study area is located in a portion of the "Sierra de Escalona" (37.97°N, 0.86°W), an area with altitudes ranging from 80 to 340 m.a.s.l., a semiarid Mediterranean climate with hot summers and scarce precipitations (less than 250–300 mm), and fragile soils severely affected by erosion and land degradation processes. Dominant land cover classes are intensive agriculture (citrus and almond trees) at low to medium slopes, xerophytic shrubs in abandoned fields and marginal areas, and sparse *Pinus halepensis* stands at the highest slopes. A large reservoir fed by a transbasin diversion is located in the north of the study area. This is an interesting test area which combines a semiarid climate, intensive exploitation of soil resources by agriculture, and high erosion rates resulting transport of sediments and nutrients to the reservoir.

The hypothesis to verify with this fuzzy logic model application is that soil vulnerability is enhanced during drought periods when vegetation status is less protective against land degradation drivers. The engagement of satellite images allowed the estimation of soil vulnerability changes within a drought period and between hydrologic years with different precipitation patterns (drought vs. normal year).

A comparison of the changes regarding soil vulnerability between late spring/early summer and late summer for different hydrologic years was included in this research. The selections of the dates were based on meteorological information and satellite image availability. Meteorological information for the nearby Pilar de la Horadada town meteorological stations was obtained from the Spanish Agroclimatic information System for Irrigation (Ministry of Agriculture and Fisheries, Food and Environment). Precipitation data indicated that the 2000 water year was characterized by a severe drought (169.8 mm), while 2003 was a more regular hydrologic year (253.0 mm). This information was the starting point for the compilation of vegetation and topography datasets.

Four satellite scenes acquired by the multispectral ASTER sensor were employed to obtain vegetation information. Two scenes were acquired for the 2000 hydrologic year (June 30 and August 1) and two other images for the 2003 water year (May 22 and August 10). The first images correspond to the end of spring and early summer, while the second scenes correspond with the end of summer. Only VNIR system bands (15 m of spatial resolution) were employed for our analyses. Both scenes were in origin preprocessed to the ASTER high-level product of surface reflectance (AST_07).

Topographic information was obtained from a high-resolution digital elevation model (**Figure 5**) that was computed using the previous vector cartography (scale 1:10,000). Triangulated irregular network (TIN) polygons were computed with the nodes of the vector cartography and used to develop the raster DEM in the same way as the previous case study. The spatial resolution of the DEM was adjusted to ASTER-VNIR imagery (15 m). Satellite images were geometrically corrected using the high-resolution DEM and additional cartography in order to minimize the positional errors among the different sources of information. The root-mean-square error (RMSE) of the geometric correction was less than half a pixel for both ASTER scenes.

Figure 5. Digital elevation model (DEM) of "La Granadella" test site.

Slope and aspect were computed using the DEM and quantized as degrees. Normalized difference vegetation indices were computed with the surface reflectance data spectral band transformation between near-infrared and red bands according to Eq. (1). Soil vulnerability estimations were computed for the four dates of the ASTER scenes (**Table 2**). The proposed model employed the following membership function and weights for input variables: (1) NDVI, a monotonically decreasing linear function with a weight value of 0.4, (2) slope, a monotonically increasing J-shaped function with a weight value of 0.4, and (3) aspect, a symmetric linear function with a weight value of 0.2. Membership function control points were the same of the previous case study. Further details of the definition of membership functions and model calibration are available in Melendez-Pastor et al. [40]. Finally, a temporal change analysis of soil vulnerability was computed using the percentage of change procedure employing the formulation of Eq. (2). Temporal change analyses were done between different seasons for the same year estimations (E01 vs. E02 and E31 vs. E32) and between the same seasons for different years (E01 vs. E31 and E02 vs. E32).

NDVI estimations highlighted the critical importance of hydrologic year accumulated precipitation for the maintenance of vegetation status for nonirrigated areas. In **Figure 6 (a)** and **(b)**,

Estimation	Year	Season	ASTER scene	Accumulated precipitation
E01	2000	Spring summer	06-30-2000	167.2
E02	2000	Late summer	08-01-2000	167.8
E31	2003	Spring summer	05-22-2003	246.4
E32	2003	Late summer	08-10-2003	250.0

Table 2. Soil vulnerability estimations and employed ASTER scenes. Hydrologic year accumulated precipitation for Pilar de la Horadada meteorological station is shown.

the largest portion of the study area corresponds with nonirrigated crops, sclerophyllous vegetation, and some coniferous forest areas. They had very low NDVI values, while irrigated areas and some wet ravines exhibited quite high NDVI values (0.4–0.6). These NDVI images correspond with an intense drought water year with less than 168 mm of accumulated precipitation.

Climatic conditions for the hydrologic year 2003 were very different with almost 50% more precipitation. Greener areas in **Figure 6 (c)** and **(d)** correspond with the irrigated crops and also with some coniferous forest and dense sclerophyllous vegetation areas. More intense NDVI changes for irrigated crops were associated with the harvest of seasonal crops.

High values of soil vulnerability estimations were obtained for almost the whole study area in all the scenes (**Figure 7**), due to the abrupt orography and the lack of vigorous vegetation cover. Southern slopes with high slope values are shown as the most vulnerable areas since its vegetation cover, and edaphic development is very scarce. The lowest soil vulnerability estimations were obtained for some crop fields with permanent crops in flat areas.

Temporal change analyses between different seasons for the same year estimations were done. **Figure 8 (a)** corresponds with the change in 2000 (E01 vs. E02), and **Figure 8 (b)** is for the change in 2003 (E31 vs. E32). The most remarkable changes were associated with the irrigated crops. Temporal crops had been collected within the period of time of the change analysis, and the soil vulnerability increased by the elimination of the vegetation cover. On the other hand, temporal change between the same seasons but for different years (2000 vs. 2003) was analyzed. **Figure 8 (c)** corresponds with the change analysis for late spring-early summer (E01 vs. E31), while **Figure 8 (d)** corresponds with the change analysis for late summer (E02 vs. E32).

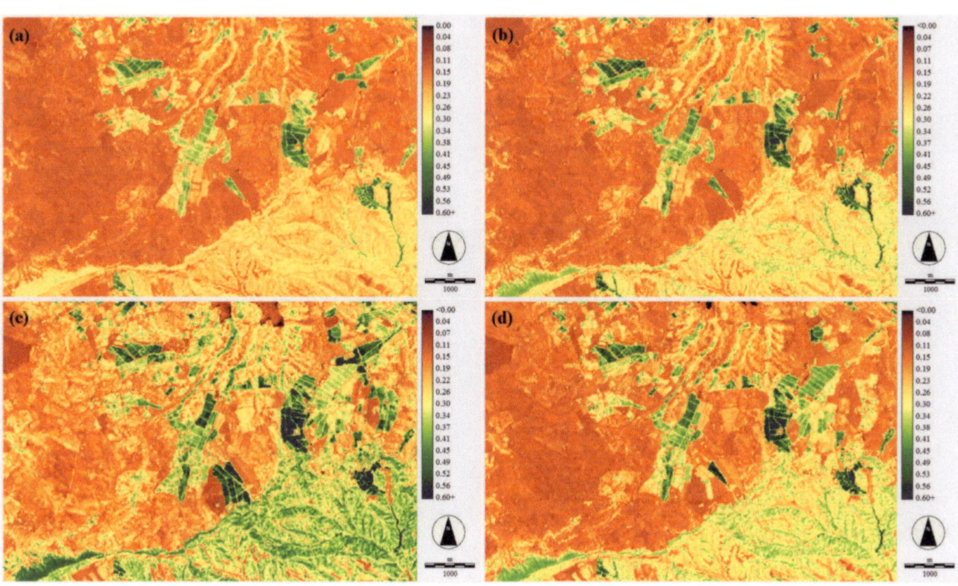

Figure 6. NDVI images for the estimations E01 (a), E02 (b), E31 (c), and E32 (d).

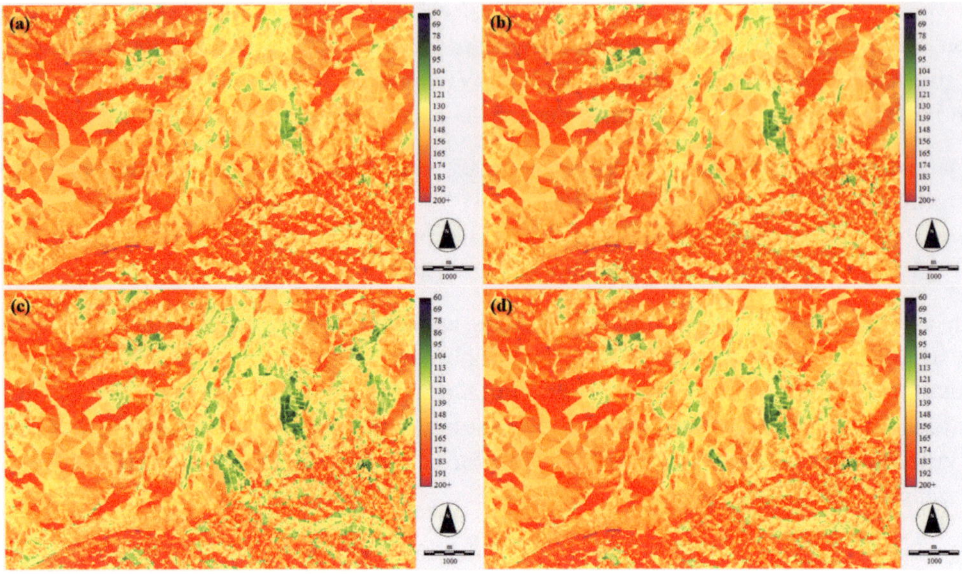

Figure 7. Soil vulnerability estimations E01 (a), E02 (b), E31 (c), and E32 (d).

The intense effect of drought mitigation in soil vulnerability values was estimated. This effect was more intense for the change analysis of late spring-early summer when very few areas exhibited an increase of soil vulnerability estimations. **Figure 8 (c)** indicated a large increase of soil vulnerability in areas of the center and east of the image, which are produced by crop

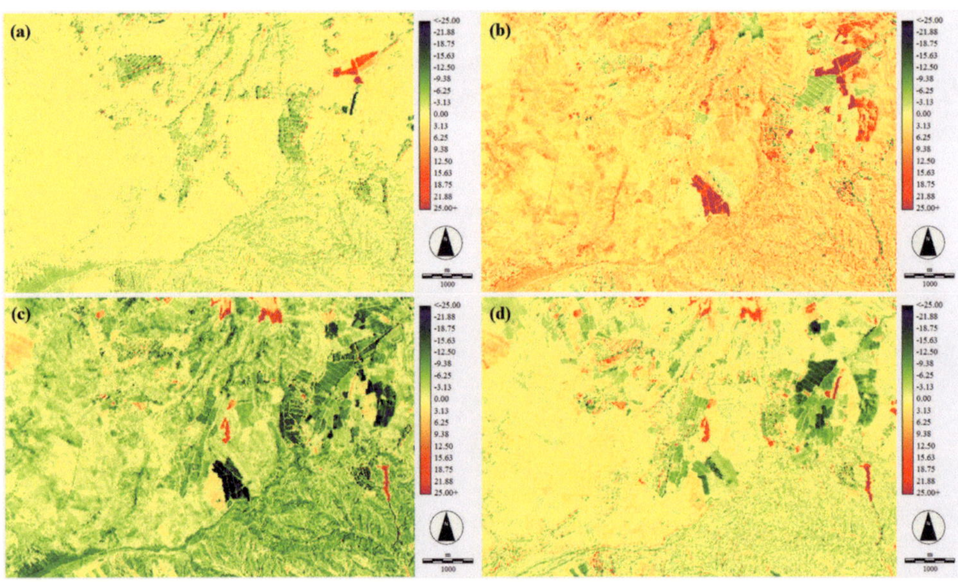

Figure 8. Temporal change (%) of soil vulnerability estimation E01 vs. E02 (a), E31 vs. E32 (b), E01 vs. E31 (c), and E02 vs. E32 (d).

replacements. Two higher soil vulnerability areas in the north of the image are part of the reservoir basin without a soil profile. Late summer temporal changes were less evident for the nonirrigated areas by the absence of precipitations in 2000 and 2003. Soil vulnerability reductions were associated to irrigated areas less affected by the typical summer drought of the study area.

5. Conclusions

The main advantage of the proposed model is the fact that it is associated to the easy data collection and computation, which can be used to evaluate soil vulnerability and erosion. The tests applied as an example remarks the great potential of the proposed approach and its great sensitivity to evaluate the actual state and detect temporal changes of soil vulnerability as a dynamic key parameter which plays significant role for soil conservation.

Our studies verified the utility of this simple and easy tool to update with satellite images and fuzzy logic model approach. The applicability of this approach for large and sparsely populated areas with limited field information could be useful in order to promote better land management strategies and more in-depth analysis of soil degradation processes. Decision support systems for local and regional authorities and land management can be tools that help decision-makers arrive to an adequate response for environment, society, and sustainable future.

Author details

Ignacio Meléndez-Pastor[1], Jose Navarro Pedreño[1]*, Ignacio Gómez Lucas[1] and Antonis A. Zorpas[2]

*Address all correspondence to: jonavar@umh.es

1 Department of Agrochemistry and the Environment, Miguel Hernández University of Elche, Elche, Spain

2 Faculty of Pure and Applied Sciences, Environmental Conservation and Management, Laboratory of Chemical Engineering and Engineering Sustainability, Cyprus Open University, Nicosia, Cyprus

References

[1] Doula MK, Sarris A. Soil Environment. In: Poulopoulos SG, Inglezakis JV editors. Environment and Development: Basic Principles, Human Activities and Environmental Implications. Netherlands: Elsevier; 2016. pp. 213-216.

[2] D'Odorico P, Ravi S. Land degradation and environmental change. In: Sivanpillai R, Shroder JF Jr editors. Biological and Environmental Hazards, Risks, and Disasters. Netherlands: Elsevier. 2016. pp. 219-227.

[3] FAO. Soil Is A Non-Renewable Resource. FAO fact sheet. Job Number I4373E/1/2015; 2015. 4 p. Available from: http://www.fao.org/documents/card/en/c/ec28fc04-3d38-4e35-8d9b-e4427e20a4f7/ [Accessed: 2017-04-03]

[4] Häberli R, Luscher C, Praplan Chastonay B, Wyss C. L'affaire sol. Pour une politique raisonnée de l'utilisation du sol. Rapport final du programme national de recherche "Utilisation du sol en Suisse" (PNR 22), Switzerland: Georg Editeur SA; 1991. 192 p.

[5] Blum WEH. The challenge of soil protection in Europe. Environmental Conservation. 1990;**17**:72-74.

[6] D'Odorico P, Bhattachan A, Davis KF, Ravi S, Runyan CW. Global desertification: Drivers and feedbacks. Advances in Water Resources. 2013;**51**:326-344.

[7] Kort J, Collins M, Ditsch D. A review of soil erosion potential associated with biomass crops. Biomass and Bioenergy. 1998;**14**(4):351-359.

[8] Department of Agriculture RSA. Soil erosion. Resource Centre, Directorate Agricultural Information Services. South Africa: Department of Agriculture; 2008. 7 p.

[9] COM/2006/0231. Communication from the Commission to the Council, the European Parliament, the European Economic and Social Committee and the Committee of the Regions—Thematic Strategy for Soil Protection [SEC(2006)620] [SEC(2006)1165]. Available from: http://eur-lex.europa.eu/legal-content/EN/TXT/?uri=CELEX:52006DC0231 [Accessed: 2017-04-03]

[10] Franchis L, Ibañez F, Fox T. Plan Blue Papers-2, Threats to Soil in Mediterranean Countries-Document Review. France: UNEP; 2003. 78 p.

[11] COM 2006/0232. Proposal for a Directive of the European Parliament and of the Council establishing a framework for the protection of soil and amending Directive 2004/35/EC. Available from: http://eur-lex.europa.eu/legal-content/EN/TXT/?uri=CELEX%3A52006PC0232 [Accessed: 2017-04-03]

[12] Lagachierie P. Digital soil mapping: A state of the art. In: Hartemink AE, McBratney A, Mendonça-Santos ML editors. Digital Soil Mapping with Limited Data. Netherlands: Springer; 2008. pp. 3-14.

[13] FAO. The State of the World's Land and Water Resources for Food and Agriculture (SOLAW)—Managing Systems at Risk. Italy and United Kingdom: Food and Agriculture Organization of the United Nations and Earthscan; 2011. 308 p.

[14] Zadeh LA. Fuzzy sets. Information and Control. 1965;**8**:338-353.

[15] Urbanski JA. The use of fuzzy sets in the evaluation of the environment of coastal waters. International Journal of Geographical Information Sciences. 1999;**13**(7):723-730.

[16] Eastman JR. IDRISI Kilimanjaro. Guide to GIS and Image Processing. USA: Clark University; 2003. 328 p.

[17] Dubois D, Prade H. A class of fuzzy measures based on triangular norms. International Journal of General Systems. 1982;**8**:43-61.

[18] Zimmermann HJ. Fuzzy set theory. WIREs Computational Statistics. 2010;**2**:317-332.

[19] Bonissone PP, Decker K. Selecting uncertainty calculi and granularity: An experiment in trading-off precision and complexity. In: Kanal LN, Lemmer JF editors. Uncertainty in Artificial Intelligence. Netherlands: Elsevier Science; 1986. pp. 217-247.

[20] Power DJ. Decision support systems: A historical overview. In: Burstein F, Holsapple CW editors. Handbook on Decision Support Systems 1. USA: Springer; 2008. pp. 121-140.

[21] Department for Communities and Local Government. Multi-Criteria Analysis: A Manual. UK: Communities and Local Government Publications; 2009. 168 p.

[22] Zorpas AA, Saranti A. Multi-criteria analysis of sustainable environmental clean technologies for the treatment of Winery's wastewater. International Journal of Global Environmental. 2016;**15**:151-168.

[23] Zorpas AA, Pociovălisteanu DM, Georgiadou L, Voukkali I. Environmental and technical evaluation of the use of alternative fuels through multi criteria analysis model. Progress of Industrial Ecology. 2016;**10**(1):3-15.

[24] Jiang H, Eastman JR. Application of fuzzy measures in multi-criteria evaluation in GIS. International Journal of Geographical Information Sciences. 2000;**14**(2):173-184.

[25] Sasikala KR, Petrou M. Generalized fuzzy aggregation in estimating the risk of desertification of a burned forest. Fuzzy Sets and Systems. 2001;**118**:121-137.

[26] Buckley JJ. The multiple judge, multiple criteria ranking problem: A fuzzy set approach. Fuzzy Sets and Systems. 1984;**13**:25-37.

[27] Jet Propulsion Laboratory. Shuttle Radar Topography Mission. Available from: http://www2.jpl.nasa.gov/srtm/[Accessed: 2017-01-25].

[28] USGS. Landsat Missions. Available from: https://landsat.usgs.gov/[Accesed: 2017-01-25].

[29] Antolin C (coord.). El suelo como recurso natural en la Comunidad Valenciana. Spain: Conselleria d'Obres Públiques, Urbanismo i Transports de Valencia; 1998. 187 p.

[30] Wischmeier WH, Smith DD. Predicting Rainfall Erosion Losses from Cropland East of the Rocky Mountains. US: USDA, U.S. Government Printing Office, Agric. Handbook 282; 1965. 47 p.

[31] Carter BJ, Ciolkosz EJ. Slope gradient and aspect effects on soils developed from sandstone in Pennsylvania. Geoderma. 1991;**49**(3-4):199-213.

[32] Jensen JR. Remote Sensing of the Environment: An Earth Resources Perspective. (2nd edition). USA: Pearson; 2007. 592 p.

[33] Chuvieco E. Teledetección Ambiental. Madrid, Spain: Ariel; 2010. 528 p.

[34] ERSDAC ASTER Level 1 Data Products Specifications (GDS Version). Version 1.3. Japan: Earth Remote Sensing Data Analysis Center (ERSDAC); 2001. 126 p.

[35] Abrams M, Hook S, Ramachandran B. ASTER User Handbook. Version 2; USA: Jet Propulsion Laboratory-NASA/California Institute of Technology; 2002. 135 p.

[36] Rouse JW, Hass RH, Schell JA, Deering DW. Monitoring vegetation systems in the great plains with ERTS. In Proceedings, Third Earth Resources Technology Satellite-1 Symposium. NASA SP-351. USA: NASA; 1974; pp. 3010-3017.

[37] Melendez-Pastor I, Navarro-Pedreño J, Koch M, Gómez I, Hernández EI. Evaluation of land degradation after forest fire with a fuzzy logic model. Environmental Engineering and Management Journal. 2013;**12**:2087-2096.

[38] Jenny H. Factors of Soil Formation. USA: McGraw-Hill; 1941.

[39] Morgan RPC. Erosión y conservación del suelo. Spain: Mundi-Prensa; 1997. 348 p.

[40] Melendez-Pastor I, Córdoba Sola P, Navarro-Pedreño J, Gómez I. Evaluación de la vulnerabilidad a la degradación por erosión en suelos mediante un modelo de lógica borrosa. Revista de Ciências Agrarias. 2010;**33**:171-181.